7th International Conference on Mechanical Engineering, Materials Science and Civil Engineering (ICMEMSCE2019)

IOP Conference Series: Materials Science and Engineering
Volume 758

Sanya, China
17 - 18 December 2019

ISBN: 978-1-7138-4703-8
ISSN: 1757-8981

Printed from e-media with permission by:

Curran Associates, Inc.
57 Morehouse Lane
Red Hook, NY 12571

Some format issues inherent in the e-media version may also appear in this print version.

Institute of Physics
Dirac House, Temple Back
Bristol BS1 6BE UK

Phone: 44 1 17 929 7481
Fax: 44 1 17 920 0979

techtracking@iop.org

Additional copies of this publication are available from:

Curran Associates, Inc.
57 Morehouse Lane
Red Hook, NY 12571 USA
Phone: 845-758-0400
Fax: 845-758-2633
Email: curran@proceedings.com
Web: www.proceedings.com

TABLE OF CONTENTS

IOP Conf. Series: Materials Science and Engineering **758** (2020) 011001 doi:10.1088/1757-899X/758/1/011001

Preface

Dear Distinguished Authors and Guests,
The Organizing Committee warmly welcomes you to 2019 the 7th International Conference on Mechanical Engineering, Materials Science and Civil Engineering (ICMEMSCE2019), held on 17th-18th, December, 2019, Sanya, China.

The aim of ICMEMSCE2019 is to present the latest research and results of scientists (professors, students, PhD Students, engineers, and post-doc scientist) related to Mechanical Engineering and Materials topics. This conference provides opportunities for the different areas delegates to exchange new ideas and application experiences face to face, to establish business or research relations and to find global partners for future collaboration.

After the peer-review process, the submitted papers were selected on the basis of originality, significance, and clarity for the purpose of the conference. The selected papers and additional late-breaking contributions to be presented as lectures will make an exciting technical program. The conference program is extremely rich, featuring high-impact presentation. We hope that the conference results constituted significant contribution to the knowledge in these up to date scientific field.

The proceeding records the fully refereed papers presented at the conference. The main conference themes and tracks are Mechanical Engineering and Composite materials, nano-materials, bio-materials, Material processing, molding technology, Advanced Materials Science, Materials Processing Technology, Electrical Engineering, Vehicle and Motor, Metal Alloys and Simulation. Hopefully, all participants and other interested readers benefit scientifically from the proceedings and also find it stimulating in the process.

On behalf of the organizing committee, I would like to especially thank all the editors from IOP Conference Series for their great support to ICMEMSCE2019. Without their excellent editorial work, ICMEMSCE2019 will not be published so timely and successfully.

Finally we wish all the authors and attendees of ICMEMSCE2019 a unique, rewarding and enjoyable memory at ICMEMSCE2019 in Sanya, China. We look forward to your participation in the 8thICMEMSCE in 2020.
With our warmest regards,
JingWei Zhao
Conference Organizing Chair
December 30th 2019

IOP Conf. Series: Materials Science and Engineering **758** (2020) 011002 doi:10.1088/1757-899X/758/1/011002

ICMEMSCE2019 Committee

Conference Chairs

Prof. JingWei Zhao, University of wollongong, Australia

Technical Program Committee Chairs

Hyun-Do YUN, Chungnam National University, Korea

THANH-LAM NGUYEN, Lac Hong University, Vietnam

Mou Chen, Nanjing University of Aeronautics and Astronautics, China

Keishi Matsuda, Matsuyama University, Japan

International Program Committee Chairs（America）

Jeremy (Zheng) Li, School of Engineering, University of Bridgeport, America (Guest editor for 3rd ICMEMSCE2014)

International Program Committee Chairs（Japan）

Ikuo IHARA, Nagaoka University of Technology, Japan (Guest editor for 2nd ICMEMSCE2013)

International Program Committee Chairs（Malaysia）

Norhaliza Abdul Wahab, University Teknologi Malaysia, Malaysia

Abdul L Mohd Tobi, University Tun Hussein Onn Malaysia, Malaysia

Aishah Najiah Dahnel, International Islamic University Malaysia, Malaysia

Ahmad Baharuddin Abdullah, University Sains Malaysia, Malaysia

Toh Tien Choon, University Tunku Abdul Rahman, Malaysia

Goh KaiChen, University Tun Hussein Onn Malaysia, Malaysia

Yusri Yusof, University Tun Hussein Onn Malaysia, Malaysia

Mohammad Jawaid, University Putra Malaysia, Malaysia

Associate Prof. Sivakumar Dhar Malingam, Universiti Teknikal Malaysia Melaka, Malaysia

International Program Committee Chairs（Srilanka）

K.M. Muditha Dassanayake, University of Peradeniya, Srilanka

International Program Committee Chairs（Slovak Republic）

Michal Kelemen, Technical University of Kosice, Slovak Republic

International Program Committee Chairs （South Korea）

Sangyoon Lee, Konkuk University, South Korea

International Program Committee Chairs（Hungarian）

László Horváth, óbuda University, Hungarian

International Program Committee Chairs（Thailand）

Achanai Buasri, Silpakorn University, Thailand

International Program Committee Chairs（Taiwan）

Ming-Hung Shu, National Kaosiung University of Applied Sciences, Taiwan

Chih-Ta Tsai, National Cheng Kung Universit, Taiwan

International Program Committee Chairs（Iran ）

Seyed Hashem Eslami, University of Tehran, Iran

International Program Committee Chairs（Brazil）

Walter Jesus Paucar Casas, Universidade Federal do Rio Grande do Sul, Brazil

International Program Committee Chairs（India）

Sandip Patil, Bharat Forge Ltd, Pune, India

Prof.Vijayan Gurumurthy Iyer, Department of Civil Engineering, Koneru Lakshmaiah Education Foundation (KLEF) Deemed To Be University

Dr. Subhasis Roy,Department of Chemical Engineering University of Calcutta

Dr. Padmakumar Muthuswamy, Research & Development Engineer in Kennametal Inc, India

International Program Committee Chairs (Turkey)

Osman Adiguzel, Firat University, Turkey

PINAR TERZIOGLU, Mugla Sitki Kocman University University,Turkey

Prof. Dr. Osman Adiguzel, Firat University, Turkey

International Program Committee Chairs (Poland)

Magorzata asiulewicz – Kaczmarek, Poznan University of Technology

IOP Conf. Series: Materials Science and Engineering **758** (2020) 011002 doi:10.1088/1757-899X/758/1/011002

Miroslaw Kwiatkowski, AGH University of Science and Technology

Krzysztof Kuś, University of Warmia and Mazury in Olsztyn

Grzegorz Woroniak, Bialystok University of Technology

International Program Committee Chair (Romanian)

Stefan Talu, University of Cluj-Napoca, Romanian

International Program Committee Chair (UK)

Yingang Du, Anglia Ruskin University, Chelmsford

International Program Committee Chair (Portugal)

F.Pacheco Torgal, University of Minho

International Program Committee Chair (Czech Republic)

Fatma Yalcinkaya, Technical University of Liberec

International Program Committee Chair (Paris)

GHORBEL Elhem, University of Cergy-Pontoise

Internation Program Committee Chair (Algeria)

Dr. Beddiaf ZAIDI, University of Batna 1, Algeria
Associate Prof. Mamoun FELLAH, Khenchela University, Algeria

First name	Surname	Institution	Email address
Kaijun	Zhou	Nantong Vocational University	765801038@qq.com
Fei	Shuai	Special Equipment Safety Supervision Inspection Institute	2028346158@qq.com
Jing	Wu	Naval University of Engineering	prettygirl.wj@163.com
Jiajun	Ma	National University of Defence Technology	miles001VIP@126.com
Adil	Saleem	Shandong University	adilsaleemuaf@gmail.com
Chengfei	Fan	Academy of Military Science PLA China	tjufcf@163.com
Guobo	Luo	Institute of CSCEC	745630455@qq.com
Yuhan	Huang	Institute of CSCEC	386389880@qq.com
Camila Ysabel	Pichardo Neyra	Universidad Peruana de Ciencias Aplicadas	u201412410@upc.edu.pe
Xianzhang	Li	Guilin University of Electronic Technology	y-lianfa@163.com
Angela Tatiana	Espinoza Trujillo	Peruvian University of Applied Sciences	u201319891@upc.edu.pe
Chengfei	Fan	Research Institute for National Defense Engineering	tjufcf@163.com
Shiling	Zhang	Chongqing Electric Power Research	526793305@qq.com
Meier	Belizario	Universidad Peruana de Ciencias Aplicadas	u201415921@upc.edu.pe
Vidal	Zamora	Universidad Peruana de Ciencias Aplicadas	u201412090@upc.edu.pe
Pablo Olivas	Maldonado	Universidad Peruana de Ciencias Aplicadas	u201412012@upc.edu.pe
Kelong	Liu	Company	liuklong@126.com
Yongjing	Peng	Reduction for Power Grid Transmission and Distribution	352739788@qq.com
Nyuk Yoong	Voo	Universiti Brunei Darussalam	nyukyoong.voo@ubd.edu.bn
Carlos	Raymundo	Universidad Peruana de Ciencias Aplicadas	carlos.raymundo@upc.edu.pe
Qiaodan	Ye	Tongji University	yqd134167@163.com
Lijun	Peng	GRIMAT Engineering Institute Co., Ltd	sklcopper1967@163.com
Huifeng	Zhang	Shandong Xiandai University	zhf.cla.y@163.com
Huifeng	Zhang	Shandong Xiandai University	zhf.cla.y@163.com
Huifeng	Zhang	Shandong Xiandai University	zhf.cla.y@163.com
Kene	Li	Guangxi University of Science and Technology	likene@qq.com
Zhihua	Zhu	Guangdong Mechanical & Electrical Polytechnic	184823174@qq.com
Lei	Xu	Northeastern University	xul@smm.neu.edu.cn
Do Duc	Trung	Ha Noi university of Industry	doductrung@haui.edu.vn
Xin	Qiu	Zhejiang Normal University	xqiu@zjnu.cn
Xin	Qiu	Zhejiang Normal University	xqiu@zjnu.cn
Bo	Yin	TECHNOLOGY	zhaozy@sziit.edu.cn
Jie	Lai	Xi'an Research Institute of High-Tech	513516059@qq.com
Seung Gyu	Kim	Konkuk University	sgdady@konkuk.ac.kr
Fang	Xie	Huizhou University	xiefang4498@126.com
Qiyu	Zou	Guangdong University of Technology	ygn@gdut.edu.cn
Zhanbin	Chen	Guilin University of Electronic Technology	y-lianfa@163.com
Zhenyu	Zhao	Shenzhen Institute of Information Technology	yxpzzy01@163.com
Julio Villalta	Vergara	Universidad Peruana de Ciencias Aplicadas	gary.duran@upc.edu.pe
HOSVICK JEFFER	RAMIREZ MEJIA	Universidad Peruana de Ciencias Aplicadas	u201417593@upc.edu.pe
Milagros Castro	Gonzales	Universidad Peruana de Ciencias Aplicadas	gary.duran@upc.edu.pe
José Saavedra	Salazar	Universidad Peruana de Ciencias Aplicadas	gary.duran@upc.edu.pe
Sandra	Rodriguez	Universidad Peruana de Ciencias Aplicadas	gary.duran@upc.edu.pe
Hongjun	Li	Zhejiang Sci-Tech University	lihongjun@zstu.edu.cn
Claudia Salguero	Sandoval	Universidad Peruana de Ciencias Aplicadas	u201613500@upc.edu.pe
Liangliang	Bao	China University of Petroleum East China	baoliangfighting@163.com
Zhihua	Zhu	Guangdong Mechanical & Electrical Polytechnic	184823174@qq.com
Gabriel Rivera	Saavedra	Universidad Peruana de Ciencias Aplicadas	aldo.bravo@upc.pe
Shih Tsung	Hsu	Chaoyang University of Technology	sthsu@cyut.edu.tw
Victor Elias	Torres Diaz	Peruvian University of Applied Sciences	victor.torres@upc.pe
Dikshit Babu	Nepal	Shanghai JiaoTong University	nepaldikshit@gmail.com
Kedar	Baral	Shanghai Jiao Tong University	kedar@sjtu.edu.cn
Hsuhui	Cheng	School of Environmental and Chemical Engineering, Zhao	hhcheng1126@gmail.com
Tiancheng	Fang	Northeast Petroleum University	fangtc211@163.com
Artemov Arsenii	Olegovich	Perm National Research Polytechnic University	artemoff87@mail.ru
Naumov Stanislav	Valentinovich	Perm National Research Polytechnic University	NaumovStanislav@yandex.ru
Igor	Maslenikov	Technological Institute for Superhard and Novel Carbon	i.i.maslenikov@gmail.com
Gary Duran	Ramirez	Universidad Peruana de Ciencias Aplicadas	gary.duran@upc.edu.pe
Gary Duran	Ramirez	Universidad Peruana de Ciencias Aplicadas	gary.duran@upc.edu.pe
Victor Elias	Torres Diaz	Peruvian University of Applied Sciences	victor.torres@upc.pe
Carlos Alberto	Huerta Campos	Peruvian University of Applied Sciences	carlos.huerta@upc.edu.pe
Haibing	Xiao	Shenzhen Institute of Information Technology	xiaohb@sziit.edu.cn
Shengrong	Shu	East China Jiaotong University	shushengrong201@163.com
Hongru	Ji	Chang'an University	zgfeng@chd.edu.cn

Salomatova Ekaterina	Sergeevna	Perm National Research Polytechnic University	weld-katy@mail.ru
Edwin Jean Marco	Galarza Mateo	Universidad Peruana de Ciencias Aplicadas - UPC	u201322444@upc.edu.pe
Xiaoli	Liu	Zhengzhou University of Light Industry	liuxiaolizzuli2018@163.com
Dong Won	Jung	Jeju National University	jungdw77@naver.com
Fengjie	Cai	Chang' an University	zgfeng@chd.edu.cn
Dong Won	Jung	Jeju National University	jungdw77@naver.com
Ming	Zhang	Yangzhou University	zhangmingjob@sina.com
Ming	Zhang	Yangzhou University	zhangmingjob@sina.com
Ming	Zhang	Yangzhou University	zhangmingjob@sina.com
Mohamed	Moustafa	HEU	moustafa@hrbeu.edu.cn
Xuefeng	Bai	Institute of Petrochemistry,Heilongjiang Academy of Scie	tommybai@126.com
Gary Duran	Ramirez	Universidad Peruana de Ciencias Aplicadas	gary.duran@upc.edu.pe
Baojin	Wang	Northeast Petroleum University	bjwangbaojin@126.com
Tiefeng	Zhang	Nanjing University of Science and Technology	809758956@qq.com
Jianpeng	Huang	Nanjing University of Science and Technology	2176609317@qq.com
Kunyan	Zhang	Guilin University of Electronic Technology	y-lianfa@163.com
Wei	Wu	School of Chemistry and Material Sciences, Heilongjiang	marywu64@126.com
Manish	Prasad	Shanghai Jiao Tong University	xbsong@sjtu.edu.cn
Di	Zhang	Dalian University of Technology	di.zhang@dlut.edu.cn
Lidia Pacheco	Miranda	Universidad Peruana De Ciencias Aplicadas	pccilpac@upc.edu.pe
Zenagebriel	Gebremedhn	Harbin Institute of Technology	zencheboyy@gmail.com
Mamoun Idris	Ali Sagiroun	Harbin Engineering University	sagiroun005@hotmail.com
Jian	Sun	Technical Institute of Physics and Chemistry CAS	sunjian16@mails.ucas.ac.cn
Huanyou	Wang	Xiangnan University	whycs@163.com
Yuanqi	Tong	Harbin University of Science and Technology	3066995232@qq.com
Zhaojie	Tong	Shenzhen Municipal Design & Research Institute Co., Ltd	zongshi2006@163.com
Santiago Vizcarra	Arapa	Universidad Peruana de Ciencias Aplicadas	u201523064@upc.edu.pe
Junior Genaro	Cruz Diaz	Universidad Peruana de Ciencias Aplicadas	u201213193@upc.edu.pe
Shah	Bikram	SJTU	bikram025@sjtu.edu.cn
Mayela Danixa	Castillo Gonzales	Universidad Peruana de Ciencias Aplicadas	u201410388@upc.edu.pe
Mengya	Wang	Tibet University	wuqi_zangda@163.com
XueZhi	Wei	Huzhou University	tangps@zjhu.edu.cn
WeiChao	Du	Xi'an Shiyou University	duweichao@xsyu.edu.cn
BaoJun	Di	National University of Defense Technology	zq_shen@outlook.com
WeiChao	Du	Xi'an Shiyou University	duweichao@xsyu.edu.cn
Quang	Pham	Hanoi University of Science and Technology	quang.pham@hust.edu.vn

IOP Conf. Series: Materials Science and Engineering **758** (2020) 011004 doi:10.1088/1757-899X/758/1/011004

Peer review statement

All papers published in this volume of *IOP Conference Series: Materials Science and Engineering* have been peer reviewed through processes administered by the proceedings Editors. Reviews were conducted by expert referees to the professional and scientific standards expected of a proceedings journal published by IOP Publishing.

IOP Conf. Series: Materials Science and Engineering **758** (2020) 012001 doi:10.1088/1757-899X/758/1/012001

Design and Analysis of Rotor Line Type for Three-lobe Roots Blower

Gu Jian-feng[1], Zhou kai-jun[2]

1 School of Electromechanical and Traffic Engineering, Nantong Vocational College of Science & Technology, Jiangsu Nantong 226007, China

2 School of mechanical Engineering, Nantong Vocational University, Jiangsu Nantong 226007, China

E-mail: 765801038@qq.com

Abstract. In order to improve and solve the capability of small and medium enterprises to design new products, the principle of operation of the three-lobe Roots blower rotor was firstly analyzed, and the characteristics of meshing of the rotor profile were analyzed. The meshing design formula was given. Based on this, the 3L41WD involute Roots blower design was taken as an example to calculate its full line type and to analyze and determine the meshing tolerance. Finally, the field test proved the rationality of the design, and provided guidance for SMEs to rapidly design the involute Roots blower rotor.

1. Introduction

The Roots blower is a type of rotary blower, which was invented in 1854. The volumetric fan has the advantages of simple structure, strong pressure bearing capability, good hard air performance, low energy consumption and easy-to-manufacture, and thus becomes the most widely used blower [1]. The rotor of the Roots blower is the core component, and the two rotors combined with the pump body form a closed section, which transports gas from the inlet to the outlet, so the rotor's linear shape is essential for operating efficiency, overall performance and safety. Although the researchers have done a lot of research on the profile of the rotor [2-4], there is still a lack of a complete process from the working principle → meshing situation → profile design → field test. Small and medium-sized roots blower companies have certain development difficulties due to the limitation of technical strength.

In view of the weak design ability of small and medium-sized enterprises, this paper focuses on the optimization design of 3L4 series Roots blower of a certain fan company. Firstly, based on the working principle of the three-leaf Roots blower rotor, the characteristics of the rotor-type wire meshing are analyzed, and the meshing design formula is given. In this case, the 3L41WD involute Roots fan design is taken as an example, and all the line types are given. Finally, the field test proved the rationality of the design, optimized the existing products for SMEs, and provided guidance and technical support for the rapid development of new energy-efficient products.

2. Analysis of meshing condition of Roots blower

2.1 working principle

Figure 1. Working principle of Roots blower

As shown in figure 1, the two rotors of the Roots blower move synchronously in opposite directions, and the rotor and the fan casing form a closed space to surround a certain amount of gas, and the gas is transported from the inlet to the outlet as the rotor rotates (The shaded portion of figure 1 shows the process of a rotor conveying air volume.). The amount of air delivered is proportional to the number of revolutions. At present, the common line types of Roots blower rotors are: involute, arc, cycloid and the combination of the above-mentioned line types, wherein the arc-shaped rotor type wire tip seal performance is poor, and the cycloid type area utilization coefficient is low. Involute type is easy to cause linear interference due to improper design. Therefore, the most widely used combination of various line types is the involute.

Figure 2 shows the operation of the multi-line combined Roots blower with the involute as the main body in the 0°-60° meshing section. The rotor meshing involute follows the following rules during the meshing process:

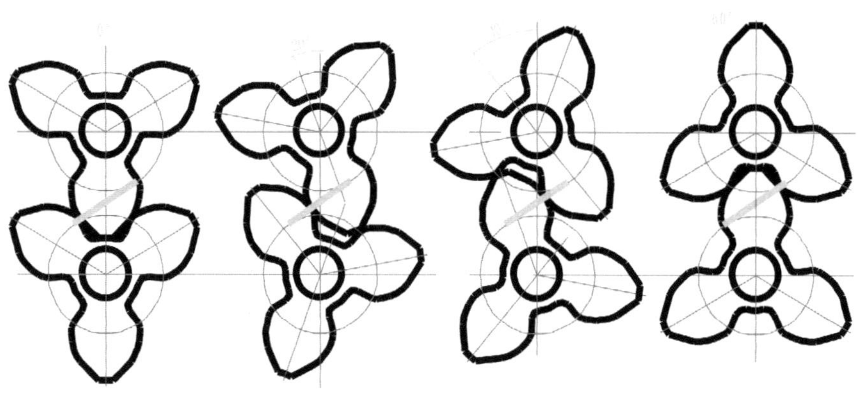

Figure 2. Schematic diagram of the involute type of leaf rotor

1) The meshing points on the two involute rotor blades always slide on the inner tangent line of the two involute base circles, and the meshing points do not exceed the range of the inner tangent line and the two base circle tangent points;

2) Each of the rotor blades of the three-leaf Roots blower is responsible for meshing 120° during the rotation, and one side of the involute profile of each of the rotor blades is responsible for the

engagement angle of 60°, thereby completing the 360° engagement process for one week. The operation of the fan is the cyclical engagement of the synchronous rotor blade and the involute profile of the rotor.

3) The radius of the base circle of the two involute lines determines the length of the inscribed line of the two base circles [5], and the length of the inscribed line affects the length of the involute. The smaller the radius of the involute base circle, the farther the initial meshing point of the two rotors is from the pitch circle position, so the wider the rotor bottom (the area formed on the base circle) and the sharper the top of the rotor; Conversely.

2.2 Meshing analysis

As shown in figure 3, the inner tangent of the two rotor base circles has a base circle radius of R_0 ;

$2a$ is the center distance of the two rotors, The meshing point of the two rotor involutes at 0° is K_1, and the meshing point at 60° is K_2。 When the rotor is in operation, the involute line is switched from one side to the other side every 60°, and the meshing relationship satisfies the requirement of equation (1).

$$\begin{cases} \dfrac{PP'+K_1K_2}{2R_0} = \alpha'+\alpha \\ \cos\alpha = R_0 / a \\ PP' = 2\sqrt{a^2 - R_0{}^2} \\ K_1K_2 = \pi R_0 / Z \end{cases} \tag{1}$$

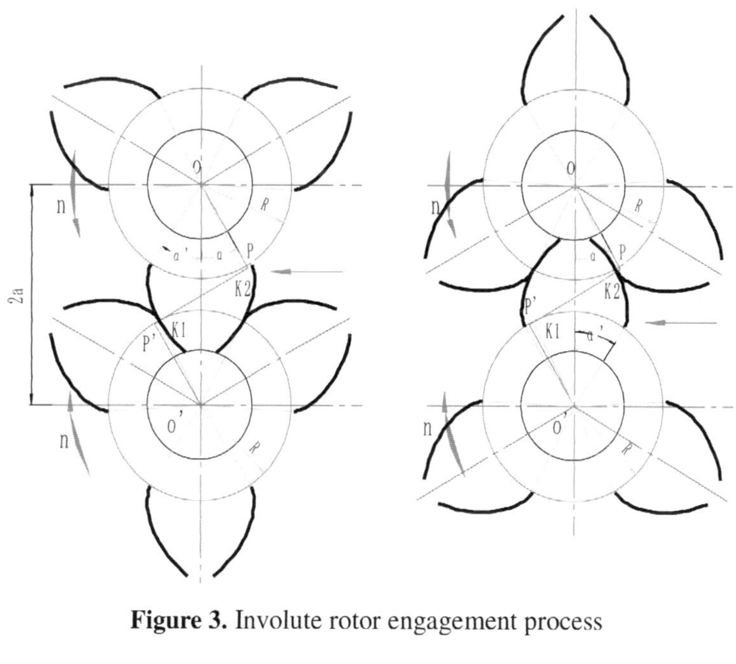

Figure 3. Involute rotor engagement process

According to the document [7]: $R_0 = \frac{2Z}{\pi}(R-a)\cdot\xi$, ξ should be between 0.7878~0.9277. In order to reduce the amount of machining of the rotor and increase the area utilization factor of the rotor, the blade shape with a larger radius of the base circle should be selected as much as possible.

3. Typical rotor design analysis

Taking a 3L41WD rotor of wind turbine company as an example, the number of rotor blades is 3, the center distance is 2a=160, and the outer diameter is 2R=240. The radius of the base circle that satisfies the meshing requirement can be selected in [60.1858, 70.8726]. According to the analysis of the second part, the larger value R0=70. Figure 4 shows the blade pattern of the rotor in the range of 30o~150o. The control points in the 30°~90° range are A, B, C, D, E, F, G. The specific coordinates are as follows: A(0, 40), B (5.4168, 39.6315), C (25.9049, 50.2885), D (32.0557, 62.2289), E (71.2138, 76.6541), F (100.6405, 63.3567), G (103.9230, 60).

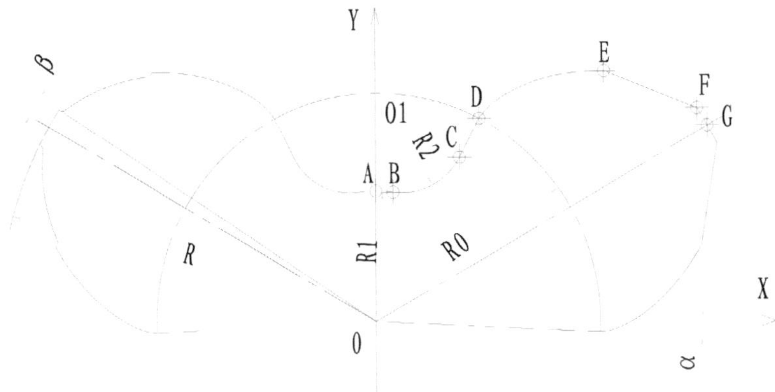

Figure 4. 3L4 series involute Roots fan rotor line type and control point (30°~90°)

3.1 Rotor blade equation

1) The AB segment is an arc of radius R_1 and the equation is as shown in equation (2):

$$x^2 + y^2 = R_1^{\ 2} \tag{2}$$

among them: $\begin{cases} 0 \le x \le 5.4168 \\ 39.6315 \le y \le 40 \end{cases}$, R_1=40。

2) The BC segment is an arc of radius R2 and the equation is as shown in equation (3):

$$(x - x_{o1})^2 + (y - y_{o1})^2 = R_2^{\ 2} \tag{3}$$

among them: $\begin{cases} 5.4168 \le x \le 25.9049 \\ 39.6315 \le y \le 50.2885 \end{cases}$; $\begin{cases} x_{o1} = 8.1253 \\ y_{o1} = 59.4473 \end{cases}$; R_2=20。

3) The CD segment is a straight line and the equation is as shown in equation (4):

$$\frac{y - y_c}{x - x_c} = \frac{y_d - y_c}{x_d - y_d} \tag{4}$$

among them: $\begin{cases} 25.9049 \le x \le 32.0557 \\ 50.2885 \le y \le 62.2289 \end{cases}$; $\begin{cases} x_C = 25.9049 \\ y_C = 50.2885 \end{cases}$; $\begin{cases} x_D = 32.0557 \\ y_D = 62.2289 \end{cases}$

4) The DE segment is the involute of the base circle R0, and it's equation consists of equations (5)(6)(7):

$$\begin{cases} x_1 = R_0(\cos\theta + \theta\sin\theta) \\ y_1 = R_0(\sin\theta - \theta\cos\theta) \end{cases} \tag{5}$$

among them: $0 \le \theta \le \dfrac{l}{\pi R_0} \times 180°, l = 77.7641, R_0 = 70$

$$\begin{cases} x_2 = x_1\cos\alpha + y_1\sin\alpha \\ y_2 = y_1\cos\alpha - x_1\sin\alpha \end{cases} \tag{6}$$

among them: $\alpha = -2.7732°$

$$\begin{cases} x = x_2 - L\sin 30° \\ y = y_2 + L\cos 30° \end{cases} \tag{7}$$

among them: $L = \sqrt{(x - x_2)^2 + (y - y_2)^2}$

5) The EF segment is a straight line and its equation is as shown in equation (8):

$$\frac{y - y_E}{x - x_E} = \frac{y_F - y_E}{x_F - x_E} \tag{8}$$

among them: $\begin{cases} 71.2138 \le x \le 100.6405 \\ 65.3567 \le y \le 76.6541 \end{cases}$; $\begin{cases} x_E = 71.2138 \\ y_E = 76.6541 \end{cases}$; $\begin{cases} x_F = 100.6405 \\ y_F = 63.3567 \end{cases}$。

6) The FG segment is an arc of radius R, and its equation is as shown in equation (9):

$$x^2 + y^2 = R^2 \tag{9}$$

among them: $\begin{cases} 100.6405 \le x \le 103.9230 \\ 60 \le y \le 65.3567 \end{cases}$; R=120。

3.2 Rotor blade engagement and tolerance control

Figure 5 shows the complete meshing of a rotor of the linear Roots blower rotor within 90 degrees. The rotor meshing clearance varies continuously from 0.036-0.0121-0.0105-0.0129-0.037. The problem is that the two ends are small (0°, 60°), the middle gap is larger and more uniform. The reason for the formation is to ensure the full meshing closure, the length of the involute has a certain amount of redundancy, at 0°, 60°, the meshing between the involute with a large radius of curvature and the involute with a small radius of curvature, The meshing gap is uneven, which can be compensated by the control of the meshing gap or the error compensation during processing. Under normal circumstances, the side clearance of the rotor should be between 0.20-0.30. Considering the assembly error of the rotor and the influence of the medium during the working process, the nominal width of the maximum width of the rotor index circle is set to 80mm. The machining error is controlled between (-0.075, -0.150).

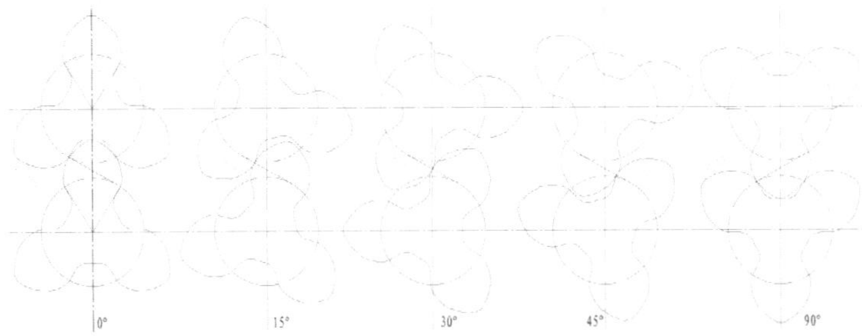

Figure 5. 3L4 series involute roots fan rotor 0°~90° meshing diagram

4. Experiment analysis

In order to test the actual operation of the fan, according to JB/T 8941.2-1999 "General Purpose Roots Blower Part 2 Performance Test Method" and JB/T 8941.1-1999 "General Purpose Roots Blower Part 1 Technical Conditions" standard, the 3L41WD fan was tested on-site. The atmospheric pressure is 101.3 kPa, the intake air temperature is 20C°, and the air density is 1.2kg/m3. The inspection site is shown in figure 6.

Figure 6. 3L41WD type fan inspection site map

The overall inspection of the 3L41WD Roots blower is shown in table 1. The overall operation of the fan is good, and all the indicators are within the technical requirements. Fan flow, volumetric efficiency and boosting changes are shown in figure 7. When the boost is 10 kPa, the fan flow is 11 mm3/min, and the volumetric efficiency can reach about 83%. When the boost reaches 60 kPa, the fan flow rate is 8.4 mm3/min, and the volumetric efficiency is about 64%. Compared with the original line type, the temperature rise, vibration, noise and other aspects have different degrees of decline, the overall performance of the fan has been further improved, so the parameter control of this design is relatively successful.

Table 1. Overall inspection of 3L41WD Roots blower

Serial number	Test items	Unit	skills	test result	Remarks

IOP Conf. Series: Materials Science and Engineering **758** (2020) 012001 doi:10.1088/1757-899X/758/1/012001

			requirement		
1	Boost	kPa	9.8-58.8	9.8-58.8	-
2	Flow	m3/min	11.0-8.0	11-8.4	-
3	Flow deviation	%	±5.0	0-0.5	-
4	Bearing temperature	C°	≤95.0	18.0-55.0	< 56
5	Axle box temperature	C°	≤65.0	12.0-40.0	< 48
6	Maximum vibration speed	mm/s	≤11.2	3.3	< 3.5
7	Noise	dB(A)	≤100	81	< 86

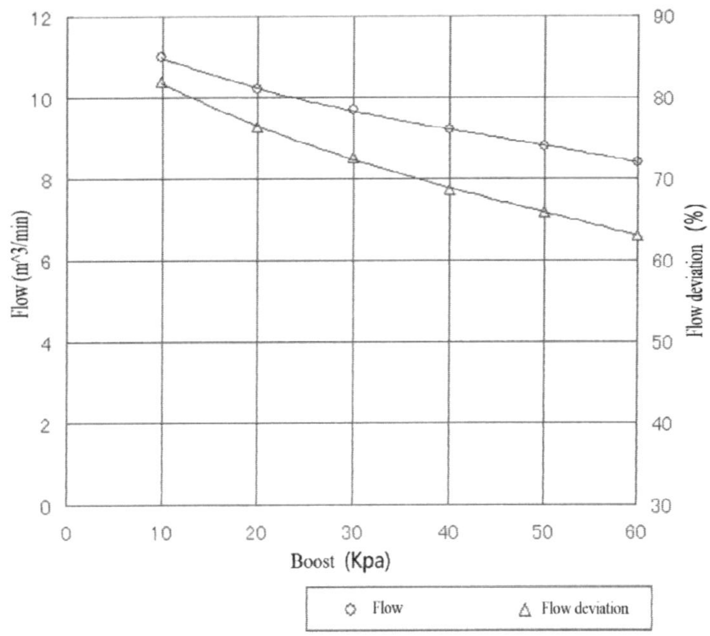

Figure 7. 3L41WD fan boost, flow and volumetric efficiency graph

5. Conclusion

The design of the rotor blade of the three-leaf involute Roots blower involves a large amount of calculation and verification. The workload is large and the cycle is long. It is difficult for SMEs to fully grasp the reasons for their own design strength. This paper combines the optimization design of a 3L4 series Roots blower of a wind turbine company, and introduces the process of meshing analysis, line type calculation and on-site inspection in detail, and provides technical guidance for the rapid development of new product development for SMEs.

References

[1] Sun S K, Jia X H, Xing L F, et al. Numerical study and experimental validation of a Roots blower with backflow design[J]. Engineering Applications of Computational Fluid Mechanics, 2018, 12(1): 282-292.

[2] Eybergen W N, Walker J G. Air supply system with two-stage roots blower: U.S. Patent 9,074,524[P]. 2015-7-7.

[3] Tien T N, Thai N H. A NOVEL DESIGN OF THE ROOTS BLOWER[J]. Vietnam Journal of Science and Technology, 2019, 57(2): 249.

[4] Sun S H, Kovacevic A, Bruecker C, et al. Numerical and Experimental Analysis of Transient Flow in Roots Blower[C]//IOP Conference Series: Materials Science and Engineering. IOP Publishing, 2018, 425(1): 012024.

[5] Zhou Kaijun,Tong Yifei,Li Yenong, et al. Numerical Analysis and Optimization for Three-Lobe Involute Roots Blower[J]. Journal of Graphics, 2014, 35(2): 214-220.

IOP Conf. Series: Materials Science and Engineering **758** (2020) 012002 doi:10.1088/1757-899X/758/1/012002

Characteristics analyses of energy consumption for bridge crane based on the energy flow

Shuai Fei[1], Ge Zhihao[2], Tong Yifei[2] and Li Xiangdong[1*]

[1] Special Equipment Safety Supervision Inspection Institute of Jiangsu Province, Nanjing, China
[2]School of Mechanical Engineering, Nanjing University of Science and Technology, Nanjing, China
E-mail: 2028346158@qq.com
*Corresponding author

Abstract. Aiming at the problems of energy consumption disorder and energy waste of bridge crane, based on the characteristics analysis of working condition and energy consumption, the power balance equation is established combining with the energy transmission process of bridge crane, which provides an important theoretical basis for the evaluation of crane energy consumption.

1. Introduction

Bridge cranes are indispensable special equipments for national economic construction and are widely used in industrial and mining enterprises, port terminals, logistics turnover and other fields. In the face of increasingly harsh environmental problems and the inevitable trend of energy conservation and emission reduction, exploring the energy-saving measures and detection and evaluation technologies of bridge cranes has become a hot research topic [1-3].

In recent years, some scholars have conducted some research on the energy consumption characteristics of cranes. Li used the bridge crane as the evaluation object. Based on the principle of energy conservation during the use of the bridge crane, the energy consumption characteristics of the hoisting mechanism, the trolley running mechanism and the running mechanism of the cart were studied respectively [4]. Fu taken the variable frequency gantry crane as the research object. From the perspective of energy flow, the energy consumption characteristics of the main working links of the running mechanism and the hoisting mechanism are studied respectively, and the frequency conversion gantry crane mechanism system is established [5]. The start and brake of the bridge crane is the most important component of energy consumption. Ge focused on the energy consumption of these two processes [6].

In this paper, the bridge crane is taken as the research object, and the working characteristics and energy consumption components of the bridge crane are analysed. On the basis of this, combined with the characteristics of energy transmission of the bridge crane, the power balance equation of the bridge crane is established and the characteristics analysis of energy consumption is completed.

2. Overview of bridge crane

2.1. Composition of bridge crane

Bridge crane is a kind of mechanical equipment that moves materials in three-dimensional space by vertical lifting and horizontal movement of the lifting device such as hooks. It consists of working mechanism, metal mechanism, power and electrical control system, and safety protection system [7]. The working mechanism is composed of a lifting and running mechanism, and the lifting mechanism is used to realize the vertical lifting movement of the material; the running mechanism refers to an actuator that realizes the horizontal movement of the material (or the whole machine) by the movement of the crane trolley (or the large vehicle). The metal structure serves as a support platform and skeleton for forming a working space of the whole machine, and is used to support the working mechanism and bear the load. The power and electrical control systems include power sources and drives, as well as lighting, steering, and control tasks. The safety protection system usually consists of safety devices, guards and safety indicating devices. It is mainly used for the comprehensive safety system of the other three parts.

2.2. Working characteristics of bridge crane

The operation of the overhead crane refers to the cycle of taking, transporting, unloading and resetting. From the perspective of mechanical movement characteristics: in a working cycle of the bridge crane, the working mechanism is in the forward and reverse alternate operation, such as the vertical lifting movement of the lifting mechanism, the horizontal reciprocating motion of the operating mechanism. From the running time characteristics: a work cycle is relatively short, usually tens of seconds, a few minutes, and each organization has a short pause at different times, and there is a sharp difference from the mechanical equipment that works continuously. From the perspective of load and usage: crane working load is not always full load, but also light load and half load; the working environment also changes with the change of work requirements and location; frequent forward and reverse operation makes power system and working structure bear certain impact and vibration; the overall performance is intermittent, random, and dynamic shock.

Therefore, the operation requirements and working process of the bridge crane determine its unique working characteristics: intermittent work, variable load, frequent positive and negative, short-duration, cycle.

2.3. Energy composition of bridge crane

From the whole life cycle of bridge crane, its energy consumption mainly consists of energy consumption in design stage, manufacturing stage, use stage and maintenance stage. This paper focuses on the analysis of the largest proportion of energy consumption in the use stage, which can be divided into standby energy consumption and work energy consumption according to their different working conditions. Among them, standby energy consumption components include electrical control system, lighting and safety protection system monitoring system; work energy consumption can be divided into lifting energy consumption, reducing energy consumption and running energy consumption according to the working characteristics of the working mechanism.

3. Energy consumption analysis of working mechanism

The working mechanism of bridge crane mainly includes two parts: motor and transmission system, which are important energy consumption units. The internal consumption of working mechanism mainly includes the loss of motor and transmission system.

3.1. Energy consumption of motor

During the operation of the overhead crane, the electric motor converts electrical energy into mechanical energy. The basic working principle is that a current generates a rotating magnetic field in the air gap through the stator winding, and there is relative motion between the magnetic field and the rotor. The rotor winding cuts the magnetic field to generate an electromotive force, and generates a current in the rotor winding, and the rotor charged conductor will be in a changing magnetic field. The electromagnetic force generates electromagnetic torque, which drives the rotor to rotate. The structure and working principle of the three-phase asynchronous motor determine that the speed control method of the bridge crane can be divided into three basic types, namely, variable pole speed regulation, variable slip speed regulation and frequency conversion speed regulation [8]. The variable pole speed regulation is generally only used on the electric hoist motor with a small capacity. The variable speed regulation is widely used on bridges and gantry cranes because of its simplicity and reliability. The common form is the control method of the rotor string resistance. The frequency conversion speed regulation is characterized by changing the frequency of the motor power supply and making it have a large speed regulation range, hard mechanical characteristics and smooth speed regulation. The use of frequency conversion technology is beneficial to save energy and the service life of the whole machine. At present, crane manufacturers are also actively promoting the frequency conversion transformation of cranes. The equivalent schematic diagram and energy flow diagram of the three-phase asynchronous motor are shown in figure 1.

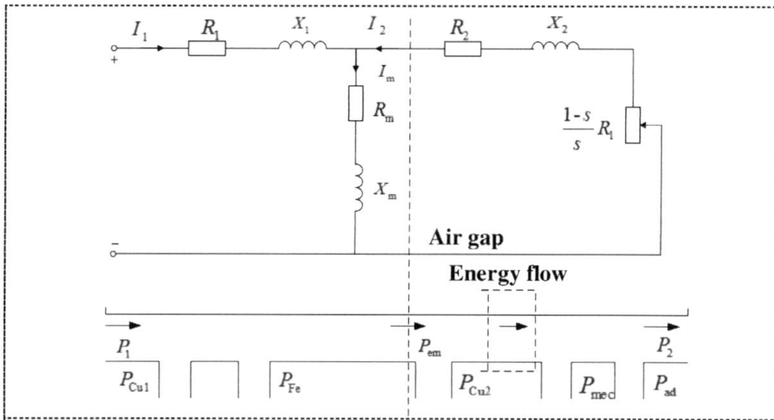

Figure 1. Equivalent circuit diagram and energy flow chart of three-phase asynchronous motor.

When the motor is working, the current passes through the stator winding, which will cause the winding to generate heat and lose some power. This is called the stator copper consumption P_{Cu1}. At the same time, due to the existence of eddy current phenomena and hysteresis, a certain eddy current loss and hysteresis loss will be collectively referred to as core loss P_{Fe}. The copper loss, core loss, and energy stored in the electromagnetic field stored in the stator winding E_m are removed, and the rest will be transferred to the rotor portion through the air gap. In normal operation, since the difference between the rotor speed and the magnetic field speed is small, the rotor iron loss is small and negligible. Therefore, the power transmitted to the rotor is converted into mechanical energy by the motor, except that part of the power (copper consumption P_{Cu2}) is consumed by the rotor winding heat, which is called the total mechanical power P_M of the motor. The total mechanical power mainly includes the mechanical power P_2 transmitted to the transmission system, and the mechanical loss P_{mec} and the additional loss (stray loss) P_{ad} caused by the friction torques such as bearings and wind resistance. Based on the above analysis, the energy loss of the three-phase asynchronous motor can be divided into two parts: the power loss P_{Ld} and the mechanical energy loss P_M, namely:

$$P_1 = P_{Ld} + P_M \tag{1}$$

In which, power loss includes stator winding loss, rotor winding loss, stator core loss and electromagnetic field energy storage change are expressed by P_{Ld}:

$$P_{Ld} = \frac{dE_m}{dt} + P_{Cu1} + P_{Fe} + P_{Cu2} \tag{2}$$

The mechanical energy loss includes the mechanical loss of the rotor, the stray loss of the mechanical system of the motor, the change of the kinetic energy of the mechanical system and the power transmitted to the transmission system. Therefore, the mechanical energy loss P_M is as follows:

$$P_M = P_{mec} + P_{ad} + \frac{dE_{km}}{dt} + P_2 \tag{3}$$

3.2. Energy consumption of driving system

The transmission mechanism of the working mechanism is the most important part of the energy transmission of the mechanism. In the energy transmission process of the transmission system, there are various kinds of energy loss, which can be mainly divided into two parts, one part is the Coulomb friction power loss P_k which is approximately proportional to the angular velocity, and part of which is proportional to the square of the angular velocity. Viscous friction power loss P_c. The Coulomb friction power loss can be divided into the load power loss P_a related to the transmission system output power (load power) and the load-independent non-load Coulomb friction power loss P_{ua}. The sum of the non-load Coulomb friction power loss and the viscous friction is called the non-load power loss. According to the above analysis, the power composition of the transmission system energy flow process is shown in figure 2.

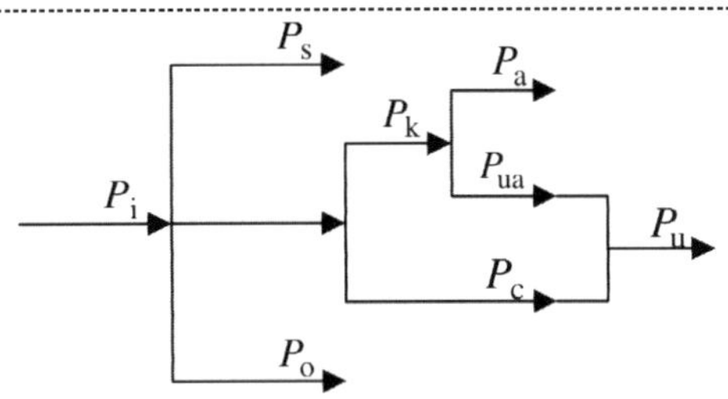

Figure 2. The composition of transmission system power

Further, the energy flow of the transmission mechanism of the bridge crane working mechanism is shown in Gigure 3. In the figure, P_{ii} is the input power of the *i*-th transmission link, P_{mi} is the mechanical loss of the *i*-th transmission link, and P_{si} is stored for each component of the *i*-th transmission link.

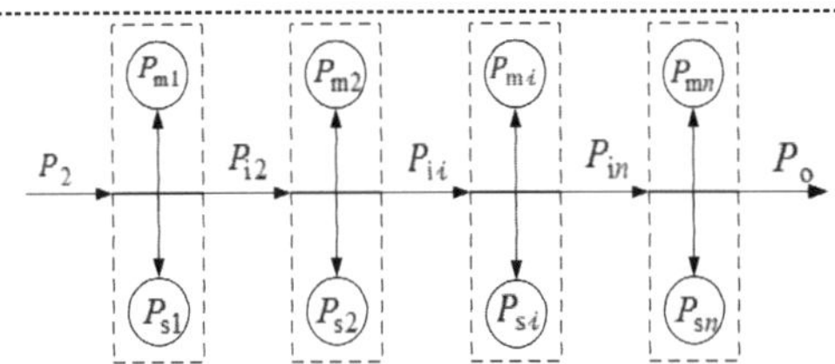

Figure 3. Dynamic energy flow diagram of transmission system

3.3. Power balance equation of working mechanism

The above not only analyses the internal loss of the working mechanism, but also establishes the power equation of the electric motor and the transmission system. The power balance equation for the entire working mechanism is established below. Based on the above analysis, the energy loss of the working mechanism is divided into two parts according to the transmission form. One is that the power loss part includes the iron loss and copper loss of the motor, and the other is the mechanical energy loss part, including the mechanical and stray losses of the motor and the energy loss of the transmission system.

Considering the stray loss of the motor P_{ad} will generate a braking torque of the braking property. Therefore, it is studied to divide the stray loss of the motor into the energy loss P_{ad0} related to the motor output power P_2 and the energy loss P_{ad1} independent of P2. Among them, P_{ad1} is incorporated into the mechanical power loss P_{mec} of the rotor, which is counted as P_{m0}, and is processed according to the method of analysing the mechanical loss of the transmission system, and there are equations (4) and (5). The power distribution of the mechanical drive of the asynchronous motor is shown in figure 4.

$$P_{\mathrm{M}} = P_2(1+b_{\mathrm{m}}) + M_{\mathrm{m}}\omega + B_{\mathrm{m}}\omega_{\mathrm{m}}^2 + J_{\mathrm{m}}\omega_{\mathrm{m}}\frac{d\omega_{\mathrm{m}}}{dt} \tag{4}$$

$$b_{\mathrm{m}} = \frac{P_{\mathrm{ad0}}}{P_2} \tag{5}$$

In which, b_{m} is the motor load factor; M_{m} is the motor non-load Coulomb friction torque; B_{m} is the motor damping coefficient; J_{m} is the moment of inertia on the motor shaft.

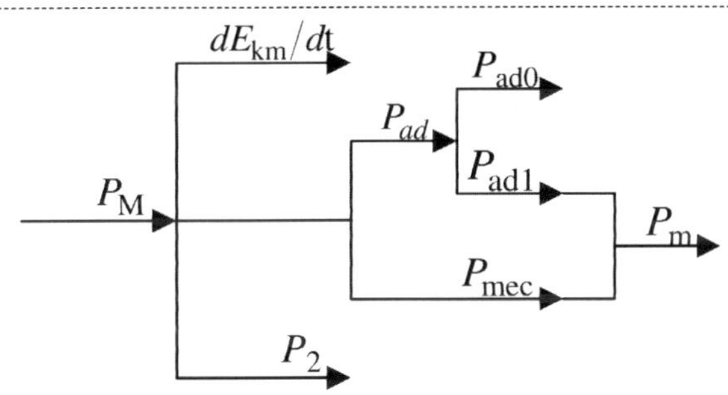

Figure 4. The composition of the total mechanical power of motor

Substituting equation (4) into equation (1), the power balance equation of the working mechanism is obtained:

$$P_1 = P_{\mathrm{Ld}} + \alpha_{\mathrm{t}}P_{\mathrm{o}} + \beta_{\mathrm{t}}\omega_{\mathrm{m}} + \gamma_{\mathrm{t}}\omega_{\mathrm{m}}^2 + \lambda_{\mathrm{t}}\omega_{\mathrm{m}}\frac{d\omega_m}{dt} \tag{6}$$

$$\alpha_{\mathrm{t}} = (1+b_{\mathrm{m}})\alpha_{\mathrm{m}} \tag{7}$$

$$\beta_{\mathrm{t}} = M_{\mathrm{m}} + (1+b_{\mathrm{m}})\beta_{\mathrm{m}} \tag{8}$$

$$\gamma_{\mathrm{t}} = B_{\mathrm{m}} + (1+b_{\mathrm{m}})\gamma_{\mathrm{m}} \tag{9}$$

$$\lambda_{\mathrm{t}} = J_{\mathrm{m}} + (1+b_{\mathrm{m}})\lambda_{\mathrm{m}} \tag{10}$$

4. Conclusion

In this paper, the energy consumption of bridge crane in the whole life cycle is discussed, and the energy consumption in the use stage is mainly analysed. According to the operation status of bridge crane, the energy consumption is divided into standby energy consumption and working energy consumption. And, based on the working characteristics of the working mechanism, the working energy consumption is divided into lifting energy consumption, reducing energy consumption and running energy consumption, and the causes of energy consumption are analysed. At the same time, starting from the energy consumption analysis of motor and transmission system, the internal losses of the working mechanism are mainly discussed, and the power balance equation of the working mechanism is established.

According to the power balance equation of the working mechanism, the higher the mechanism transmission efficiency, the lower the energy consumption. Moreover, the power balance equation of the working mechanism consists of a plurality of transmission links, so under the same transmission efficiency, the less the transmission link, the lower the overall energy consumption. In summary, the efficiency of transmission link and the design scheme of transmission system are important factors of crane energy consumption, as well as an important basis for analysing energy consumption factors, establishing energy consumption index system and realizing energy consumption evaluation of bridge crane.

Acknowledgments

This work was financially supported by the National Key Research and Development Program of China (2017YFF0209804), and the Program of Science Foundation of General Administration of Quality Supervision and Inspection of Jiangsu Province (KJ175940). The supports are gratefully acknowledged. The authors would also acknowledge the Editor.

References

[1] Sun N Fang Y 2012 New energy analytical results for the regulation of underactuated overhead cranes: An end-effector motion-based approach *IEEE Transactions on Industrial Electronics* (vol 59) pp 4723-4734

[2] Corral-Vega P J Fernández-Ramírez L M García-Triviño P 2019 Hybrid powertrain, energy management system and techno-economic assessment of rubber tyre gantry crane powered by diesel-electric generator and supercapacitor energy storage system *Journal of Power Sources* (vol 412) pp 311-320

[3] Iris Ç Lam J S L 2019 A review of energy efficiency in ports: Operational strategies, technologies and energy management systems *Renewable and Sustainable Energy Reviews* (vol 112) pp 170-182

[4] Li X D Ye W Yuan X C 2013 Energy-saving evaluation of bridge crane based on energy consumption factors of whole life cycle *Hoisting and Transport Machinery* (vol 12) pp 60-62

[5] Fu W G Cheng W M Ruan D J 2010 Research on energy-saving evaluation system based on energy consumption model of variable frequency gantry crane *30 years of technological innovation development of logistics engineering*

[6] GE X CHEN Y F 2010 Measurement and Analysis of Energy Consumption Status of Cranes *Hoisting and Transport Machinery* (vol 3) pp 6-9

[7] Liu H 2018 Introduction of Heavy Bridge Crane and Design of Main Girder *Internal Combustion Engines and Parts* (vol 3) pp 90-91

[8] Krause P Wasynczuk O Sudhoff S 2013 Analysis of Electric Machinery and Drive Systems

IOP Conf. Series: Materials Science and Engineering **758** (2020) 012003 doi:10.1088/1757-899X/758/1/012003

A Mechanical Model of Carbon Fiber Composite Materials based on Interfacial Effect

Jing Wu 1, a, Xiangming Zhang1,b

1College of Science, Naval University of Engineering, Wuhan, 430033, China
aprettygirl.wj@163.com,bzhxm312@sina.com

Keywords: Carbon fiber; cement concrete; strain; compression sensitivity; interface

Abstract: The carbon fiber cement has compression sensibility. In this paper, the compression sensitivity of carbon fiber concrete is analyzed based on shear-lag theory and single fiber pull-out testing on the basis of predecessors' researches. The results show that the change in interface resistance plays an important role on the change in resistance. The resistance increases with interfacial strain increasing. In the meanwhile, it is shown that the present research provides a comprehensive understanding on compression sensitivity of carbon fiber concrete.

Introduction

The chopped carbon fiber to be added uniformity in the cement structure material not only makes the material more strong, but also can improve its conductivity. Research at home and abroad has shown that certain ratio and process conditions of carbon fiber concrete specimen with strain-resistance effect. Many kinds of model is put forward by predecessors about the pressure sensitive mechanism of carbon fiber concrete, including the insert and pull out model of carbon fiber, conductive channel model, the tunneling effect model and series model. The resistance of the carbon fiber concrete is composed of three parts, which respectively are the resistance of carbon fiber, matrix and interface. Experiment and calculation show that when the specimen change deformation, the change of the resistance of carbon fiber and cement matrix is very small. The resistance change of sample is mainly caused by the interfacial interaction. On the basis of experimental analysis this paper compare the results of numerical simulation and set up a carbon fiber concrete mechanics model[1-2].

Experimental

A.Specimen Preparations

The sample making: the carbon fiber bundle (main technology parameter table1) is oxidized [3-4]. After drying out, draw out independent one monofil under the microscope from long carbon fiber bundle. Cement matrix coated directly on the side of rectangular strip of cardboard middle opening. The carbon fiber end is embedded in the cement matrix. Use the four-electrode resistance measurement in order to eliminate the contact resistance of the test results, shown in Figure 1.

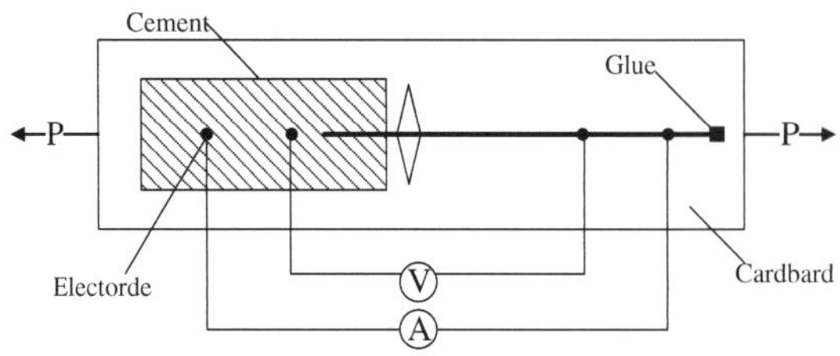

Fig1. Sample sketch

B. Carbon fiber monofilament puling out experiments

Under tensile load, Resistance rate formula the carbon fiber which is not embedded in concrete was derived and ohm's law is obeyed: $\Delta R / R = (1+\varepsilon)(1-2\upsilon_a\varepsilon)-1$ (υ_a is the Poisson's ratio of carbon fiber, $\upsilon_a = 0.27$). To compare calculation results with the experimental results, the impact of the change of carbon fiber resistance is caused by the change of the outside deformation is shown in table 1. Obviously, the change of resistance caused by the outside deformation of carbon fiber is very small, and the cement matrix changes of resistance can be ignored. Thus, in the process of the carbon fiber pulling out test, the change of resistance is mainly caused by the interface of carbon fiber cement matrix factors.

Table 1. the Influence coefficient

No.	P (mN)	Resistance rate of change $\varDelta_1$ (10^{-3})	External Resistance rate of change $\varDelta_2$ (10^{-3})	Influence coefficient $\varDelta_2/\varDelta_1$ (%)
a	24.87	129	4.82	3.74
b	28.83	144	5.59	3.88
c	31.58	154	6.12	3.97
d	39.92	185	7.74	4.18

The mechanical model of CFRC

A.The micro unit mechanical model of CFRC

(1): The basic assumptions

Simplify the sample of carbon fiber pull out test for the axisymmetric concentric cylindrical model. Where L is the deepness of carbon fiber be buried, r_0 is the radius of carbon fiber radius, P is the tensiont. It is shown in figure 2.According to characteristics of its boundary, do the assumptions. (1) the carbon fiber and concrete substrate are the isotropic linear elastic material; (2) the interface between the carbon fiber and concrete substrate is ideal interface, which is only effected by the shear stress. On the interface the stress and displacement are all continuous; (3) the surface of the concrete substrate is free, regardless of the physical strength; (4) the load play on the top of the fiber and is parallel to the fiber without sliding friction at the bottom of the fibe; (5) because the carbon fiber is very thin and the stiffness is large, the radial displacement can be ignored. which w is the direction of displacement.

IOP Conf. Series: Materials Science and Engineering **758** (2020) 012003 doi:10.1088/1757-899X/758/1/012003

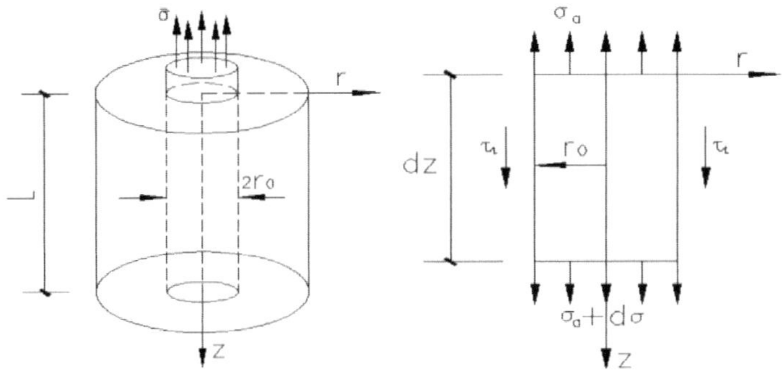

Fig2. the micro carbon fiber concrete fiber monofilament drawing model

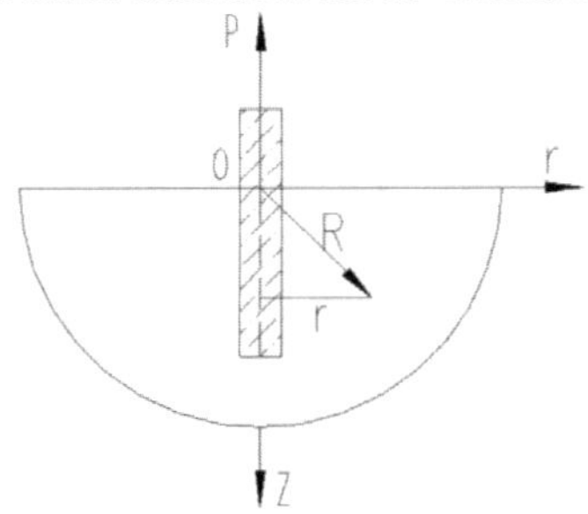

Fig3.the half space body with the concentrated tension

(2): The axial strain formula of interface eduction

According to this model, the stress and strain in the carbon fiber and matrix is all symmetry, using the cylindrical coordinates system (r, ϕ, z). All amount in the axisymmetric problem has nothing to do with coordinates ϕ. So the shear stress r_{zr} and the displacement of toroidal v are all zero. Remove the micro sectiong dz to analysis. Then remove the concrete and carbon fiber micro unit cell in the micro sectiong, which the radius is r, as shown in figure 3.

Consider the axial displacement of concrete matrix to $\tau_i = G_b \frac{\partial w}{\partial r}\Big|_{r=r_0}$

(1)

Where G_b is the shear modulus of concrete, τ_i is the shear stress of interface.

According to the hypothesis(2), there is $w\big|_{r=r_0} = w_i = w_0$.

Whih carbon fiber as the reasearch object, anslysis can get it, which is $\sigma_a = E_a \frac{dw_a}{dz} = E_a \frac{dw_i}{dz}$ (2)

According to the hypothesis(5), the axial stress σ_z of cement matrix near can be expressed, which is

$$\sigma_z = 2G_b(\frac{1-\upsilon_b}{1-2\upsilon_b})\varepsilon_z = 2G_b(\frac{1-\upsilon_b}{1-2\upsilon_b})\frac{dw}{dz} \tag{3}$$

According to the shear-lag theory

$$\begin{cases} \frac{d^2 w_i}{dz^2} + \frac{2G_b}{r_0 E_a}\frac{\partial w}{\partial r}\Big|_{r=r_0} = 0 \\ \frac{d^2 w_i}{dz^2} + \frac{1-2\upsilon_b}{2(1-\upsilon_b)}\frac{1}{2}\frac{\partial}{\partial r}(r\frac{\partial w}{\partial r})\Big|_{r=r_0} = 0 \end{cases} \tag{4}$$

According to the formula (4), We have the formula which is $w = A_2 ch\left[\sqrt{A_1}(L-z)\right]e^{-\frac{A_1}{K_1}r}$
(5)

Where they are $A_1 = \frac{K_1}{r_0} - K_1^{\ 2}K_2$, $K_1 = \frac{2G_b}{r_0 E_a}$, $K_2 = \frac{1-\upsilon_b}{1-2\upsilon_b}$.

Makeing $z = z$, $\sigma_a\big|_{r=r_0} = \overline{\sigma}$ and puting the faomula (5) into the formaula (3),

$$A_2 = -\frac{\overline{\sigma}}{E_a\sqrt{A_1}\,sh\left[\sqrt{A_1}(L-z_0)\right]}e^{\frac{A_1}{K_1}r_0} \tag{6}$$

In combination with the formula (5) and (6), we have the formula which is

$$w = -\frac{\overline{\sigma}}{E_a\sqrt{A_1}}\frac{ch\left[\sqrt{A_1}(L-z)\right]}{sh\left[\sqrt{A_1}(L-z_0)\right]}e^{\frac{A_1}{K_1}(r_0-r)} \tag{7}$$

Then puting the faomula (7) into the formaula (2), we have the calculationg fromulas of the interface shear stress and the axial tensile stress, which are

$$\tau_i = \overline{\sigma}\frac{-G_b\sqrt{A_1}}{K_1E_a}\frac{ch\left[\sqrt{A_1}(L-z)\right]}{sh\left[\sqrt{A_1}(L-z_0)\right]},\sigma_a = \overline{\sigma}\frac{-K_2G_b}{E_a}\frac{sh\left[\sqrt{A_1}(L-z)\right]}{sh\left[\sqrt{A_1}(L-z_0)\right]}e^{\frac{A_1}{K_1}(r_0-r)} \tag{8}$$

According to the boundary condition of $z=0$, the formaula does not meet this conditons of $\tau_{rz}\big|_{z=0} = 0$ and $\sigma_z\big|_{z=0} = 0$. Using the analysis method of the half space body with the concentrated tension (shown in figure 3) which can be obtained is

$$\tau_{rz}^{'} = \frac{Bz^3}{(r^2+z^2)^{5/2}}, \sigma_z^{'} = \frac{Drz^2}{(r^2+z^2)^{5/2}} \tag{9}$$

Assuming that when it is $z < z_0$ or $z > z_0$, the faomula (9) or (8)is expressed the stress of the interface. When it is $z = z_0$, satisfy the following conditiongs.

$$\begin{cases}\tau_{rz}^{'} = \tau_i \\ \sigma_z^{'} = \sigma_b\end{cases} \tag{10}$$

we have the formula of z_0 which is $z_0 = \frac{r_0K_1K_2}{\sqrt{A_1}}th\left[\sqrt{A_1}(L-z_0)\right]$ (11)

Then puting the faomula (11) into the formaula (7) and combined with $\varepsilon_z = \frac{\partial w}{\partial z}$,We have the formula which is $\varepsilon_z = \overline{\sigma}sh\left[\sqrt{A_1}(L-z)\right]e^{\frac{A_1}{K_1}(r_0-r)}\Big/E_a sh\left[\sqrt{A_1}(L-z_0)\right]$
(12)

The formula of the axial strain in interface ($r = r_0$)is

$$\varepsilon_i = \overline{\sigma}sh\left[\sqrt{A_1}(L-z)\right]\big/E_a sh\left[\sqrt{A_1}(L-z_0)\right] \tag{13}$$

B. The theoretical model calculation contrast analysis

Based on results of calculating the interface strain, under the action of the force the interface shear strain gradually decreases along the axis direction. The significant change happen on where is Z=0.05mm~0.3mm, and ε_z sharply reduced by orders of magnitude. The figure 4 shows the interface strain distribution of the monofilament tensilet specimens under the action of 62.51 mN. The figure 5 shows the trend of strain curve under the condition of four kinds of tension(a,b,c,d). Obviously, that the trend curve is upward, which proved the strain of each point gradually increases as the tension increases.

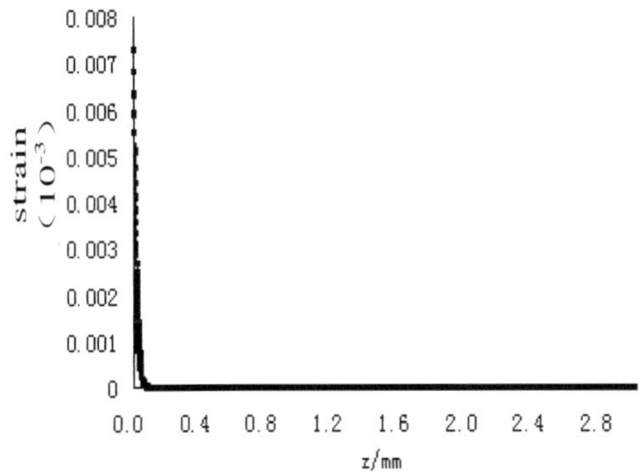

Fig4. the interface strain distribution

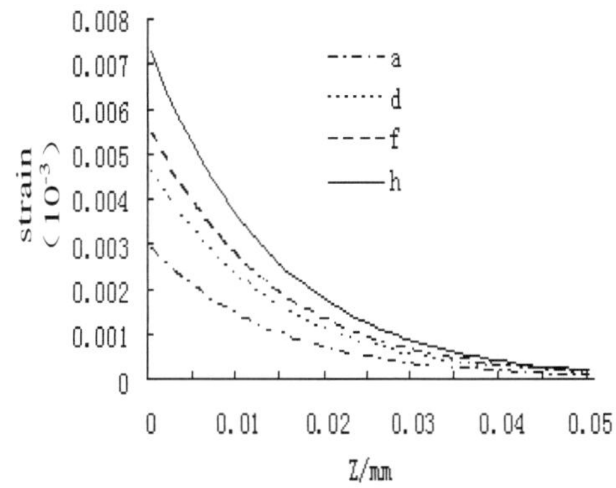

Fig5. interface strain distribution under different tensions

Conclusions

In monofilament drawing experiment, the overall resistance variation is made up of changes the resistance of the carbon fiber, cement matrix and interfacial resistance. The former occurs because carbon fiber and cement matrix geometric size change under external force. Under the action of external force of 50 mN, the change of the whole specimen geometry are extremely small. So the resistance change of carbon fiber and cement matrix can be neglected. The total resistance change of the carbon fiber reinforced concrete specimen under the effect of external force is mainly interface resistance changes, which have the direct relationship with the interface deformation. The total resistance change gradually increase along with the increase of interface strain.

Acknowledgements

The work was supported by the National Natural Science Foundation of College of Science, Naval University of Engineering (No. HGDJCB17ZK009) .

References

[1] Pu-Wei Chen, D.D.L.Chung. Carbon fiber reinforced concrete as an intrinsically smart concrete for damage assessment during dynamic loading. J.American Ceramic Society. 1995, 78(3) : 816-818

[2] Jing Wu, Aiqin Xu. Micro-experimentation research on Compression Sensitivity of Carbon Fiber Concrete. Applied Mechanics and Materials Vols. 2014, 137-140.

[3] Xu Y S, Chung D.D.L. Cement-based materials improved by surface treated admixtures. Aci Mater. J.2000, 97(3):333-342

[4] Agarwal B D. elastic-platic finite element analysis of short fiber composites. Fiber Science and Technology, 1974,7,45-58.

IOP Conf. Series: Materials Science and Engineering **758** (2020) 012004 doi:10.1088/1757-899X/758/1/012004

Penetration resistance of concrete reinforced by hybrid fibre subjected to multiple projectile impacts

Jiajun Ma[1],Luhui Yan[2] and Mohan Xu[3]

[1,3]College of Aeronautics and Astronautics, National University of Defence Technology, Changsha 410072, Hunan, China
[2] College of Military Education and Training， National University of Defence Technology, Changsha 410072, Hunan, China
Email: miles001VIP@126.com
Abstract. Penetration resistance of steel-ultrahigh molecular weight polyethylene (S-UHMWPE) hybrid-fibre-reinforced concrete under multiple projectile impacts are investigated. In this paper， 16 sets of hybrids with various hybridizations were designed and prepared considering 0%-1.5% steel fibre volume and 0%-0.3% UHMWPE fibre volume. Each hybrid-fibre concrete target was stroke triple times sequentially under 12.7mm armor-piercing projectile. The projectile impact locations are among the middle of the target in equilateral triangle shape. After each penetration, depth of penetration (DOP) and crater area were investigated. Results show that hybrid-fibre concrete performs much better than steel or UHMWPE fibre reinforced concrete with less crater area and lower DOP for hybrid-fibre concrete combines the reinforcement and toughening effect of UHMWPE fibre on concrete`s tensile strength and strengthening effect of steel fibre on concrete`s compressive strength. Besides, the crater area caused by each impact decreased with the number of strikes increases.

Key words: steel fibre; hybrid-fibre-reinforced concrete; multiple impacts; penetration performance

1. Introduction

With the development of high-tech weapons, direct hits and multiple impacts have become the main ways to strike important targets, posing a serious threat to the safety of military engineering and equipment.

In order to improve the anti-penetration ability of engineering project, many scholars have focused on high-strength and high-performance concrete, carrying out researches on the anti-penetration problem of various fibre-reinforced concrete. For instance, steel fibre`s effects on concrete had been researched by Ji C[1]、Wu H[2], indicating that the penetration resistance of concrete increased as the volume of steel fibre rose and were affected by fibre`s length and shape obviously. Zhang Y W[3,4] conducted experimental research on the mechanical properties and anti-penetration performance of ultrahigh molecular weight polyethylene (UHMWPE) fibre-reinforced concrete. Experimental results indicate that UHMWPE fibres as flexible fibres, though not readily apparent to the reinforcing effect of concrete`s compressive strength, can significantly improve the tensile strength and toughness of concrete because of their high elastic strength. Besides, they can suppress the cracking damage of fibre concrete under impact penetration well thus can effectively improve the anti-penetration ability of concrete. Based on the study of single fibre -reinforced concrete, considering one fibre can exploit the potential mechanical properties of the other fibre during hybrid-fibre reinforced concrete, many scholars studied the effect of hybrid fibre on penetration resistance of concrete. Investigations focused on concrete with steel fibres of different length and configurations[5];concrete incorporated with flexible and rigid fibres such as steel-polypropylene or polyvinyl alcohol fibres[6], steel-polypropylene or Kevlar fibres[7]. It was proved that hybrid fibre could exploit the advantages of different fibre`s properties and characteristic parameters, which resulted in better anti-penetration ability than single fibre.

Despite various studies focusing on hybrid-fibre concrete anti single projectile impact, there are some studies investigated on concrete subjected to multiple impacts due to the needs to improve engineering project penetration resistance under multiple strikes. There are mainly two patterns of multiple impacts under studied. The first pattern demonstrates multiple projectile striking at the same place on target, like the multiple impacts experiments conducted by Gomeza[8],who hit semi-infinite reinforced concrete target with 6.4mm projectile and Lai J Z [9],who hit concrete target reinforced by hybrid fibres and corundum aggregates with 14.5mm projectile. The second pattern describes projectile impact locations distributed on target, like multiple impacts on UHMWPE fibre-reinforced concrete conducted by Xu M H [10] and experiments performed by Jun F[11] to study penetration resistance of multiple strikes on 12 sets hybridizations of hybrid-fibre-reinforced concrete.

Experimental results show that the second pattern will cause larger crater area than the first one. Besides, depth of penetration will increase as the latter projectile strike at the damage area of the previous projectile, however, it will not increase multiply.

This paper concentrated on multiple-impact performances of steel-UHMWPE hybrid-fibre-reinforced concrete based on the researches about multiple-strike penetration tests of concrete reinforced by UHMWPE fibre[10]16 sets of hybrids with various hybridizations were designed and prepared. Each target were stroke sequentially with projectile impact locations distributed in equilateral triangle like the experiments[10, 11]show, whereas target`s damage appearance, crater area and DOP were under estimated and studied. Based on these results, effects of the number of impacts on the penetration damage and the influence of fibre ratio on penetration performances were studied. The results showed important theoretical guidance and reference significance to the development of high-strength fibre concrete and was beneficial to the application in the field of engineering project.

2. Materials and concrete production

The hybrid-fibre-reinforced concrete was designed according to C70 grade high strength concrete. The raw materials are as follows: Portland cement (P.I 42.5). Coarse aggregate, which contains continuous grading pebbles with particle size between 5 and 10mm. Fine aggregate, which is medium-sized river sand with mud content of 0.8%, it`s fineness modulus is 2.5, the apparent density is 2.68g/cm^3.Admixture is micro silica fume with SiO_2 content no less than 95%, specific surface area is 25000m^2/kg, bulk density is 200-250 kg/m^3, and particle diameter is 0.1-0.3 μm. For the sake of better performance, early strength and high efficiency naphthalene superplasticizer is added, which may contribute to the reduction of water up to 30%. The details of mix proportions are listed in Table 1.

Table1 Mixture proportions of plain concrete（kg/m^3）

Water	Cement	Coarse aggregate	Sand	Fly ash	Superplasticizer
185	547	936	615	57.5	12

Researches[5] indicate when rigid and flexible fibres with two different properties are mixed in large content, flexible fibres are prone to agglomerate, which may lead to performance degradation. In order to ensure the uniform dispersion of fibres incorporated into concrete, the volume of UHMWPE fibre is limited within 0.3%[3], including 4 gradients of 0%, 0.1%, 0.2%, 0.3%; steel fibre is limited within 1.5%, including 4 gradients of 0%, 0.1%, 0.2%, 0.3%, the total hybridizations are 16 sets. UHMWPE fibre is twisted by ZTX99-400D fibre and steel fibre is short curved steel fibre, the geometric and mechanical properties information of these two fibres are listed in Table 2.

The production procedures of hybrid-fibre-reinforced concrete need to be controlled carefully to ensure good workability and particle distribution. The exact procedures are as follows. ①mix the dry cementitious materials like cement, flash ash, coarse aggregate, sand, steel fibre evenly ②add water and superplasticizer along with UHMWPE fibre into the mixture carefully and mixed them for 5 min to ensure fibres are randomly distributed ③ add the remaining water, stir evenly, unload the mold, and tamper with a vibrator until the surface is flooded ④ cure the specimens at the temperature of 20°C±2°C with humidity over 95%.

Table2 Fibre information

Fibre type	Diameter(um)	Length(mm)	Density (g/cm^3)	Elastic modulus (GPa)	Tensile strength (MPa)	Fracture elongation (%)
Steel(Q195)	500	30	7.8	200	1300	2
UHMWPE	60	30	0.97	100	3000	2.8

In this paper, we use S_xP_y to represent each concrete target, where S represents steel fibre and P represents UHMWPE fibre and the subscript relates to the content of each fibre. For example, $S_{0.5}P_{0.2}$ represents hybrid-fibre-reinforced concrete with 0.5% steel fibre and 0.2% UHMWPE fibre.

3. Penetration tests of concrete targets

Penetration tests were completed at the Impact Penetration Laboratory of National University of Defense Technology. 12.7mm projectiles were fired to penetrate concrete targets, the mass of the projectile is 47.3-49g. The size of concrete target is Φ400mm×250mmin, molded by 2mm steel plate where fibre concrete targets were cast. During penetration tests, concrete targets were constrained into the mold. Concrete targets are avoided from breaking apart and spreading out with the constraint of steel plate. The incident velocity of the projectile was measured by a dual-light curtain target and a six-channel electronic time-meter. The test velocity is 781^{+9}_{-11} m/s. Diagram of the test device is shown in Figure 1.

This paper use triple projectile strike distributed on the target to conduct multi-impact penetration experiments. According to the dispersion radius of projectile described in article[12] and the crater area in single penetration of hybrid-fibre-reinforced concrete, we set each projectile`s locations` distance is 4D, where D represents the diameter of projectile. The specific processes of positioning are as follows. Firstly, select the first hit point on the target surface. Secondly, take the first shot location as the center of circle with R=4D and position the second hit location on the edge of circle. Afterwards, take the second shot location as the center of circle with R=4D and position the third hit location on the point of intersection of the above two circles. The position of the three strike locations is shown in Figure 2.

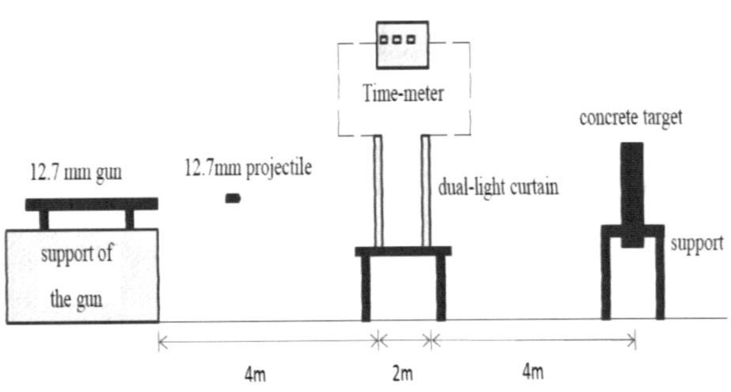

Figure1 Diagram of the test device

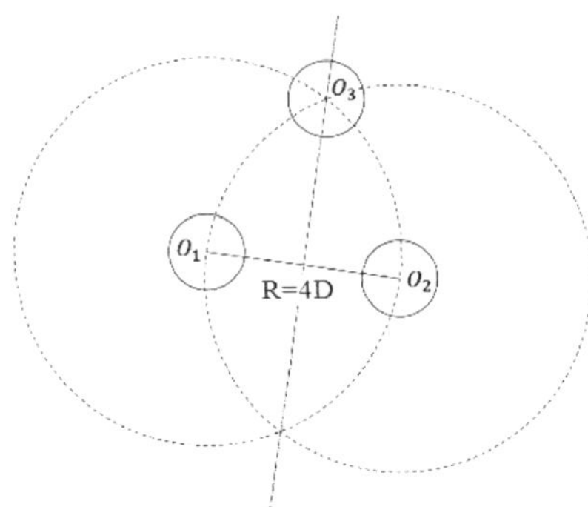

Figure2 Diagram of multiple-impact locations

During the experiments, the three strikes were carried out one by one. After each strike, the target damage and the size of damage crater area were recorded in time. After experiments, according to each penetration`s trajectory, the ballistic profile was cut and the depth of each penetration was measured one by one.

4. Results and Analysis

4.1 Pattern of target damage

The diagrams of the pattern of 16 different hybridizations of concrete targets after 3 projectile impacts are shown in Figure 3. From the diagrams, we can find that:

(1) After three strikes, those concrete targets reinforced by single fibre of steel fibre or UHMWPE fibre, outperformed 6-8 large cracks that adjoin to the outer edge of the concrete targets. Steel fibre can constrain the development of these large cracks as there are only small crackles left when the content of steel fibre reaches 1.5%. Unlike steel fibre, UHMWPE fibre shows much less constraint to the development of large cracks. Nevertheless, UHMWPE fibre show a preventive effect on the germination of the initial crack. Cracks and crackles of concrete targets reinforced by two mixed fibres are less than that of concrete targets reinforced by single fibre. Besides, the number of cracks or crackles decreased with the increase of fibre`s content.

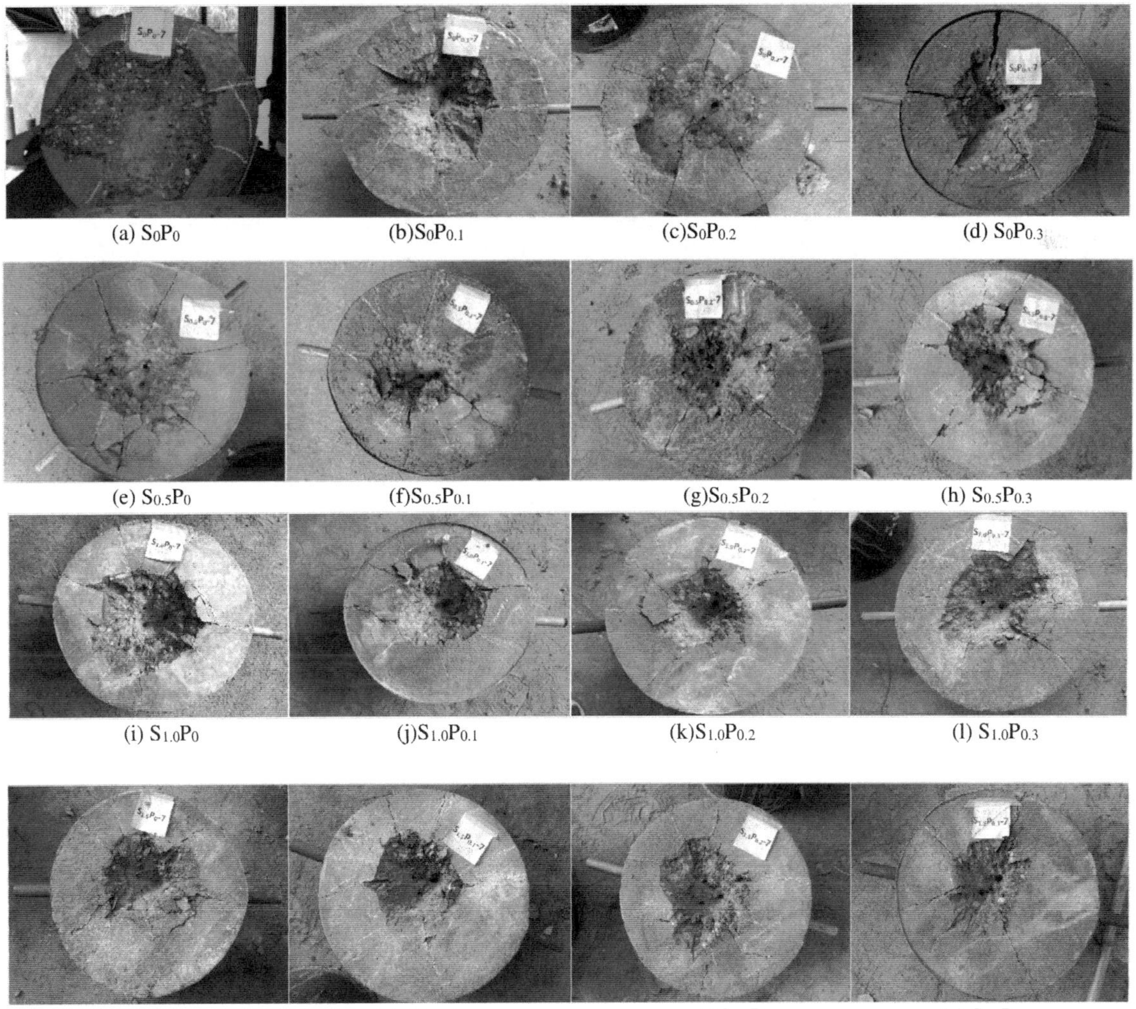

(a) S_0P_0 (b)$S_0P_{0.1}$ (c)$S_0P_{0.2}$ (d) $S_0P_{0.3}$

(e) $S_{0.5}P_0$ (f)$S_{0.5}P_{0.1}$ (g)$S_{0.5}P_{0.2}$ (h) $S_{0.5}P_{0.3}$

(i) $S_{1.0}P_0$ (j)$S_{1.0}P_{0.1}$ (k)$S_{1.0}P_{0.2}$ (l) $S_{1.0}P_{0.3}$

(m) $S_{1.5}P_0$ (n)$S_{1.5}P_{0.1}$ (o)$S_{1.5}P_{0.2}$ (p) $S_{1.5}P_{0.3}$

Figure3 Diagrams of the pattern of 16 concrete targets after triple projectile impacts

(2) There are funnel crater areas on the front of each concrete target. Compared to the crater area covering up almost the whole front face of plain concrete target without fibre（S_0P_0）, crater area on concrete target reinforced by single steel fibre or UHMWPE fibre or hybrid fibre becomes much smaller. Nevertheless, in the case of steel fibre or UHMWPE fibre, fibre mixing and increased dosage will further reduce the crater size.

4.2 Analysis of target damage

To analysis the effect of hybrid fibre in concrete target under multiple-impact, this paper assumes the crater area to be approximately oval. After each penetration, we measured the longest and shortest length of the crater area on the front face and set them as oval`s long and short axis length d_{i1} and d_{i2}. The subscripts "1" and "2" indicate the long axis and the short axis respectively, i indicates the i-th impact. The crater area ratio r_i is defined to represent the ratio of the crater area to the whole target area after i-th impact:

$$r_i = \frac{\pi d_{i1} d_{i2} / 4}{\pi \overline{D}^2 / 4} = \frac{d_{i1} d_{i2}}{\overline{D}^2} \tag{1}$$

Where $\overline{D}$ represents the diameter of target. Obviously, we can use r_i to judge the degree of local damage to the target.

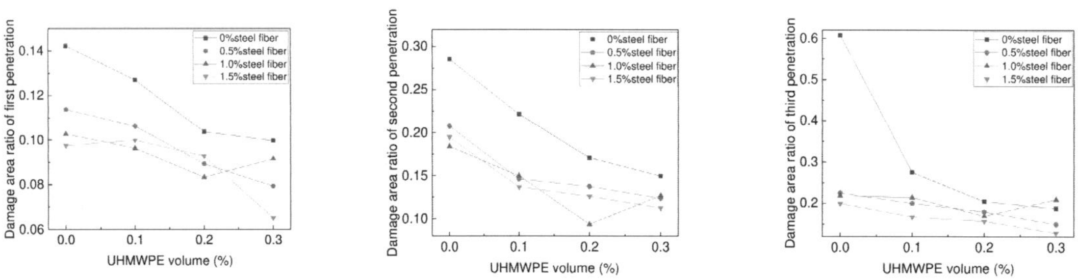

(a) first penetration (b) second penetration (c) third penetration

Figure 4 Diagrams of the relationship between each target`s damage crater area ratio and UHMWPE volume

Each target`s damage crater area ratio varies from fibre content after each penetration, the relationship of them is demonstrated in Figure 4. The following phenomenon can be observed from these diagrams:

(1) After single impact (first penetration), the crater area decreased with the increasement of fibre`s content, which illustrated slighter damage crater of target. When adding single fibre of UHMWPE fibre, the damage area ratio decreased linearly with the increase of the dosage. Although the hybrid-fibre concrete had a little fluctuation due to the interaction of the two fibres, in general, the larger the content, the smaller the damage of the crater. In this experiment, the minimum crater area of hybrid-fibre concrete is $S_{1.5}P_{0.3}$, as small as 6.51%, equal to 45.8% of plain concrete`s, 66.8% of concrete reinforced by single steel fibre, 65.3% of concrete reinforced by single UHMWPE fibre.

(2) After multiple impacts, all hybridizations of concrete target`s crater area increased to some extent. After second shot, the damage area ratio increased an average of 0.59 times by the first shot, that is, the additional damage of the second shot was about 59% of the first shot. On the basis of the second cumulative damage, the third issue has again increased damage by 38%. Notably the additional damage will gradually decrease with the increase of the number of strikes. The range of change of S_0P_0 exceeds the average value, mainly due to its large first strike crater, so that the subsequent strike locations are all located in the crater.

The relationship between penetration depth of hybrid-fibre concrete under the impact of three-shot projectiles and the fibre content is shown in Figure 5.

We can observe from these diagrams that:

(1) After first penetration, the depth of penetration (DOP) decreased with the increasement of fibre`s content. DOP decreased linearly with the increase of fibre content in concrete reinforced by single steel or UHMWPE fibre. The main influencing factor of the depth of penetration in hybrid-fibre concrete is the steel fibre content. In general, the depth of penetration decreases as the steel fibre content increases. In this experiment, the minimum depth of penetration of hybrid-fibre concrete is $S_{1.5}P_{0.1}$, as small as 118mm, equal to plain concrete`s 75.6%, 93.6% of concrete reinforced by single steel fibre, 81.9% of concrete reinforced by single UHMWPE fibre. The depth of

penetration of $S_{1.5}P_{0.2}$ was abnormal. It may be caused by the concrete stirred uniformly, and the fibre agglomeration effect is obvious, resulting in an abnormal increase in concrete DOP.

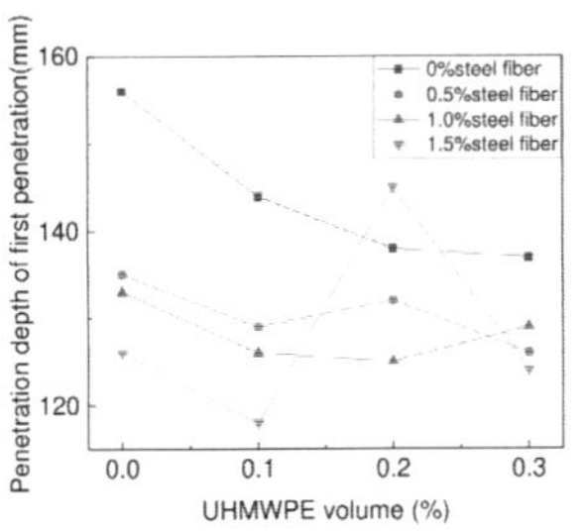

(a) DOP of first penetration

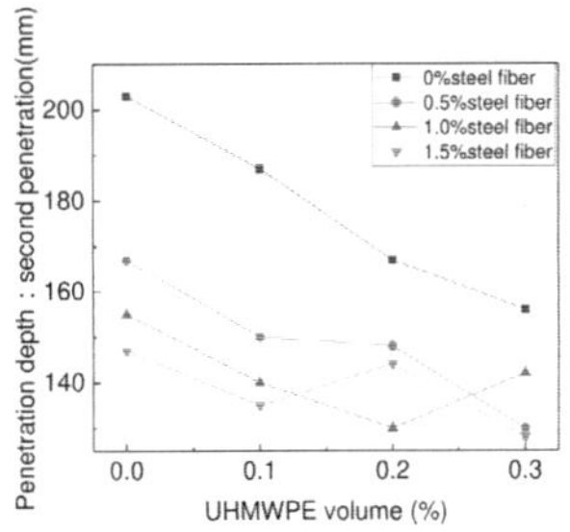

(b) DOP of second penetration

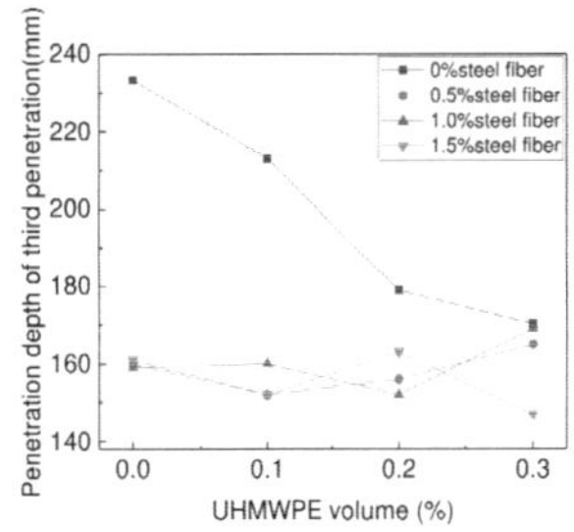

(c) DOP of third penetration

Figure 5 Diagrams of the relationship between each target`s DOP and UHMWPE volume

(2) After multiple shots, all concrete target`s depth of penetration increased to some extent. Depth of penetration decreased more regularly on multiple-impact penetration compared to individual anomaly test data of single penetration. The depth of penetration of the second impact increases 16.7% averagely compared to the first impact, and the third increased 8.7% compared to the second. After the second and third penetration experiments, the minimum depth of concrete is $S_{1.5}P_{0.3}$, equal to 63.9% of plain concrete averagely, 90.5% of single steel fibre reinforced concrete and 85.5% of single UHMWPE fibre reinforced concrete.

5. Conclusions

From the triple-impact penetration experiments conducted on hybrid-fibre-reinforced concrete with steel fibre and UHMWPE fibre, pattern of target damage is compared. DOP and crater areas were examined after each penetration. The main conclusions are drawn as follows:

(1) Hybrid-fibre concrete exploits the advantage of UHMWPE fibre in preventing initial crack germination and the effect of steel fibre on crack opening and expansion. Compared to plain concrete, concrete reinforced by single steel fibre or UHMWPE fibre had better penetration resistance with slighter crater damage and crack damage.

(2) Under the multiple strike mode demonstrated in this paper, with the increase of the number of strikes, the crater damage area is gradually reduced. The DOP increases compared to the single strike, but the increasement mainly originates from the previous projectile impact.

References

[1] Ji C, Long Y, Wan W Q and Shao L Z 2008 *Explosion and Shock Waves* **28** 178-185
[2] Wu H, Fang Q, Gong Z M and Peng Y 2015 *Int. J. Impact Eng.* **84** 88-95
[3] Zhang Y W, Yan L H, Li L F and Zhu L 2014 *Protective Eng.* **36** 10-14
[4] Zhang Y W, Yan L H and Li L F 2017 *Journal of Vibration and Shock* **36** 92-96
[5] YU R, Spiesz P and Brouwers H J H 2016 *Cement and Concrete Composites* **68** 109-122
[6] Feng J, Gao X D, Li J Z, Dong H L, He Q, Liang J Q and Sun W W 2019 *Composites Part B: Eng.* **202** 487-496
[7] Almusallam T H, Abadel A A, Alsalloum Y A, Siddiqui N A and Abbas H 2015 *Int. J. Impact Eng.* **81** 61-73
[8] Jason-T Gomez, Arun Shukla 2001 *Int. J. Impact Eng.* **10** 965-979
[9] Lai J Z, Yang H R, Wang H F, Zheng X B and Wang Q 2019 *J. Mater. Civil Eng.* **30** 4018256
[10] Xu M H, Yan L H, Liang Q H and Wang S S 2017 *International Conference on Material Science, Energy and Environmental Engineering*
[11] Feng J, Gao X D, Li J Z, Dong H L, He Q, Liang J Q and Sun W W 2019 *Constr. Build. Mater.* **202** 341-352

Effect of SiC on the dielectric and microwave absorption performance of F-doped Si_3N_4 ceramics in X-band

Adil Saleem[1,2], Yujun Zhang[1,2,4], Hongyu Gong[1,2,4], Muhammad K Majeed[3], Xiao Lin[1,2], M Zeeshan Ashfaq[1,2], and Zhang Xinfeng[1,2]

[1]Key Laboratory for liquid-solid Structural Evolution & Processing of Materials of Ministry of Education School, Shandong University, Jinan 250061, PR China
[2]Key Laboratory of special Functional Aggregated Materials, Ministry of Education, Shandong University, Jinan 250061, PR China
[3]Key Laboratory of Colloid and Interface Chemistry, School of Chemistry and Chemical Engineering, Ministry of Education, Shandong University, Jinan 250100, PR China
Email: adilsaleemuaf@gmail.com, yujunzhangcn@sdu.edu.cn, gong_hongyu@163.com, drmkashifmajeed@mail.sdu.edu.cn, linxiao050994@sina.com, zee4194202@gmail.com, 17862703596@163.com

Abstract: In this study, silicon carbide reinforced silicon nitride (SiC/Si_3N_4) ceramics were fabricated by hot press (HP) sintering with the addition of fluorides (F= AlF_3, MgF_2). In order to investigate the dielectric and microwave (MW) absorption properties of SiC/Si_3N_4 ceramics in X-band, various concentrations of silicon carbide (SiC) have been used. The results demonstrate a complete phase transformation and the intergranular behavior of the SiC/Si_3N_4 ceramics with large and irregular grains of β-Si_3N_4. The relative complex permittivity and dielectric loss tangent increase significantly with the increasing fraction of SiC. The Cole-Cole semicircle analysis for SiC/Si_3N_4 ceramics shows a ternary dielectric behavior with the addition of SiC. Meanwhile, the reflection loss (R_L) decreases with the increase of SiC content and the lowest value of R_L was achieved -8.2 dB at 12.4 GHz, which reveal a favorable prospect as MW absorbers.

Keywords: SiC/Si_3N_4; ceramic composites; fluoride additives; dielectrics; MW absorption properties

1. Introduction

Si_3N_4 ceramic is one of the most favorable multi-functional structural material for numerous industrial applications because of its excellent mechanical, thermal, and electrical properties, with the additional characteristics of low permittivity and good MW absorption. [1, 2]. In recent years, MW absorbers have achieved huge attention due to the development of electronic and communication devices. Generally, the MW absorbers are comprised of absorption agent and matrix. In order to fabricate an excellent MW absorber, the dielectric constant should be ~1 and the dielectric loss should be greater, therefore to meet these requirements the MW absorbing agent is usually reinforced with a non-conductive matrix to prepare MW absorber [2]. The Si-based ceramics (such as SiC and SiCN) are usually preferred to fabricate the MW absorbers. A lot of investigations have been dedicated to develop the materials with exceptional MW absorption properties, such as carbon-based, ferromagnetic and SiC-based materials. The SiC possesses excellent MW absorption properties due to its wide band gape semi-conduction, good electrical and dielectric properties [3]. Furthermore, Li et al. [4] and Zheng et al. [5] synthesized the Si_3N_4-SiC ceramics and improved the dielectric and MW absorption performance of the ceramics. Therefore, SiC/Si_3N_4 ceramics can be a good MW absorber, which can be used in an open environment [6]. In this paper, the dielectric and MW absorption properties of F-doped SiC/Si_3N_4 ceramics have been investigated at various concentrations of SiC.

[4]Corresponding author. Tel / Fax: +86-531-88399760;

E-mail: yujunzhangcn@sdu.edu.cn, gong_hongyu@163.com

2. Experimental

The starting materials were α-Si_3N_4 (0.5 μm, >93%, Beijing Ziguang Co.), SiC (Shanghai Haiyi Technology) and sintering additives (AlF_3 and MgF_2, Shanghai Macklin Biochemical technology). The detail of different compositions of raw materials is illustrated in **Table 1**. The mixed raw powders were wet milled with ethyl alcohol for 30 minutes, Resulted powders were dried in an oven at 70°C, ground and sieved with a mesh size of 154 μm. Afterward, the as-prepared powders were densified in HP sintering furnace at 1700°C for 1h in a high purity N_2 atmosphere at a pressure of 30 MPa (heating rate 10°C min^{-1}). The synthesized materials were characterized by X-ray diffraction (XRD, D/MAX, Ultima-IV, Rigaku, Japan), scanning electron microscope (FE-SEM, JSM-7800F) and Raman analysis was performed at room temperature using Raman spectroscopy (RMS, Renishaw, UK). Furthermore, the bulk density of ceramics was measured by Archimedes' principle in distilled water. The dielectric and MW absorption parameters were analyzed by the transmission line and free space method with a vector analyzer (Agilent Technologies, N5222A) in the frequency range 8.2-12.4 GHz (X-band). The specimens for MW dielectric parameters test were prepared in a rectangular shape with the dimension of $22.88 \times 10.16 \times 3$ mm^3.

Table 1. Compositions of the composites, density and relative density.

Composites name	Compositions				Density (g/cm^3)	Relative density (%)
	Si_3N_4 wt%	SiC wt%	AlF_3 wt%	MgF_2 wt%		
S1	95	0	2.5	2.5	3.10	97.3
S2	90	5	2.5	2.5	3.08	96.7
S3	85	10	2.5	2.5	3.04	95.3
S4	80	15	2.5	2.5	3.05	95.4

3. Results and discussion

XRD patterns of hot pressed F-doped SiC/Si_3N_4 ceramics can be observed in **Fig. 1a**. A complete phase transformation from α to β-Si_3N_4 have been detected with few extra peaks of $MgSiN_2$ and SiC. The presence of $MgSiN_2$ relates to the reaction of doping materials with Si, present on the surface of Si_3N_4 due to the liquid phase sintering phenomenon, which is beneficial for the interfacial interaction between grain boundaries. The Raman spectra for all the composites can be seen in **Fig. 1b**. The Raman spectra for S2, S3 and S4 show two peaks of SiC (at ~787 and 972 cm^{-1}) [7]. Furthermore, the S2, S3, and S4 also showed D (~1400 cm^{-1}) and G (~1519 cm^{-1}) bands with a low intensity suggesting a very less content of free carbon [8]. The intensity of the G band increases slightly with the increase of SiC content attributing to the graphitization degree. SEM images (**Fig. 1c-d**) shows the intergranular behavior of the SiC/Si_3N_4 ceramics with large and uneven grains of β-Si_3N_4 and SiC. Furthermore, the presence of SiC contents within the grain boundaries shows a strong interfacial interaction, as well as promoted the β-Si_3N_4 grains and helps to reduce its grain size. The bulk and relative densities (**Table 1**) of the ceramics decreases with the increasing contents of SiC, attributing to the more significant interaction of grains rearrangements and volumetric retraction of ceramic composites. Previously reported that the presence of fluoride additives decrease the oxygen contents and help to improve the density of a material, therefore the overall improvement in the composites can also be attributed to the presence of fluoride additives [9]. The dielectric properties of SiC/Si_3N_4 ceramics with various SiC content were examined at the frequency range of 8.2-12.4 GHz. The relative complex permittivity ($\varepsilon_r = \acute{\varepsilon} - j\ddot{\varepsilon}$) and magnetic permeability ($\mu_r = \mu' - j\mu''$) are the key parameters to characterize the dielectric and MW absorption performance. The $\acute{\varepsilon}$ and $\ddot{\varepsilon}$ are related to polarization and capability of dielectric loss, also correspond to the ability to store and dissipate electromagnetic (EM) wave energy, respectively. Although, the high $\ddot{\varepsilon}$ implies good MW absorption properties, whereas the too high value of $\ddot{\varepsilon}$ can affect the impedance match and absorption capacity

[10]. As exhibited in **Fig. 2a** and **b**, both $\acute{\varepsilon}$ and $\ddot{\varepsilon}$ decreased with the increase of frequency, which

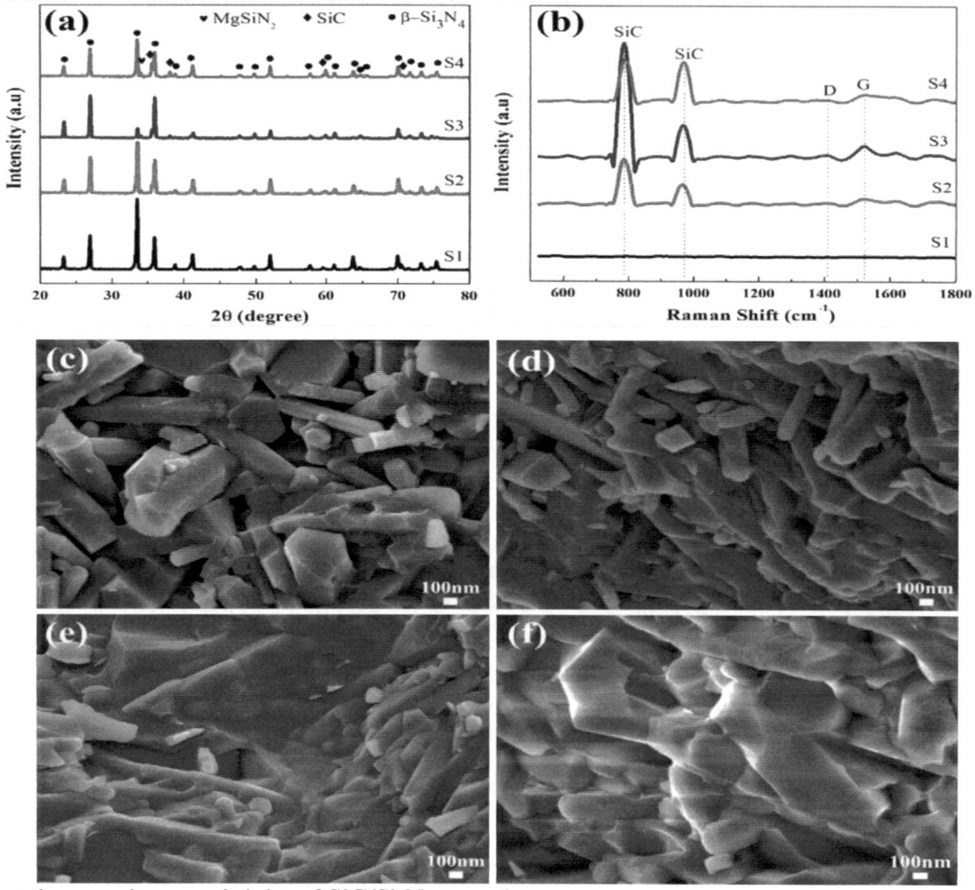

affect the complex permittivity of SiC/Si_3N_4 ceramics.

Fig. 1: XRD patterns (a) and Raman spectra (b), SEM images of S1 (c), S2 (d), S3 (e) and S4 (f) of SiC/Si_3N_4 ceramics.

The performance of the real part (μ') and imaginary part (μ'') of the permeability of SiC/Si_3N_4 ceramics are shown in **Fig. 2c** and **d**. The resonance peaks of both μ' and μ'' appeared for S3 and S4 composites at high frequency, however other composites showed no intense change. Generally, the MW absorption performance of ceramics can be improved with the reduction of complex permittivity. In this case, it can be observed that the complex permittivity and tangent loss decreases as the SiC content decreased to the lowest value and increased rapidly with the increase of SiC content.

The dielectric loss ($\tan\delta_\varepsilon = \frac{\varepsilon''}{\varepsilon'}$) tangent and magnetic loss ($\tan\delta_\mu \frac{\mu''}{\mu'}$) tangent, represent to the MW attenuation capability of materials. The greater values of $\tan\delta_\varepsilon$ and $\tan\delta_\mu$ means good MW attenuation capability. The behaviour of $\tan\delta_\varepsilon$ and $\tan\delta_\mu$ can be observed in **Fig. 3a** and **b**, the MW attenuation capability of SiC/Si_3N_4 ceramics increase with the increase of SiC. The resonance peaks of $\tan\delta_\varepsilon$ were shifts towards high frequency with a maximum value of 0.36 at 12.4 GHz and the average value of $\tan\delta_\mu$ for composites, S1, S2, and S3 was ~ 0, whereas composite S4 shows a small resonance peak with a maximum value of 1.19 at 10.24 GHz. The overall behavior of $\tan\delta_\mu$ was closely related to μ'' at high frequency. The dielectric loss phenomenon is related to the Debye relaxation ($\acute{\varepsilon} - \ddot{\varepsilon}$), generally examined by the Cole-Cole semicircle equation ($(\varepsilon' - \varepsilon_\infty)^2 + (\varepsilon'')^2 = (\varepsilon_s - \varepsilon_\infty)^2$, where ε_s and ε_∞ are the stationary and optical dielectric constants) in which each semicircle relates to a Debye relaxation [11]. The Cole-Cole semicircles for SiC/Si_3N_4 ceramics can be observed in **Fig. 3c**, in which all composites with SiC show three semicircles for each, attributing to the ternary dielectric behavior, which could lead to a favorable perception for a MW absorber.

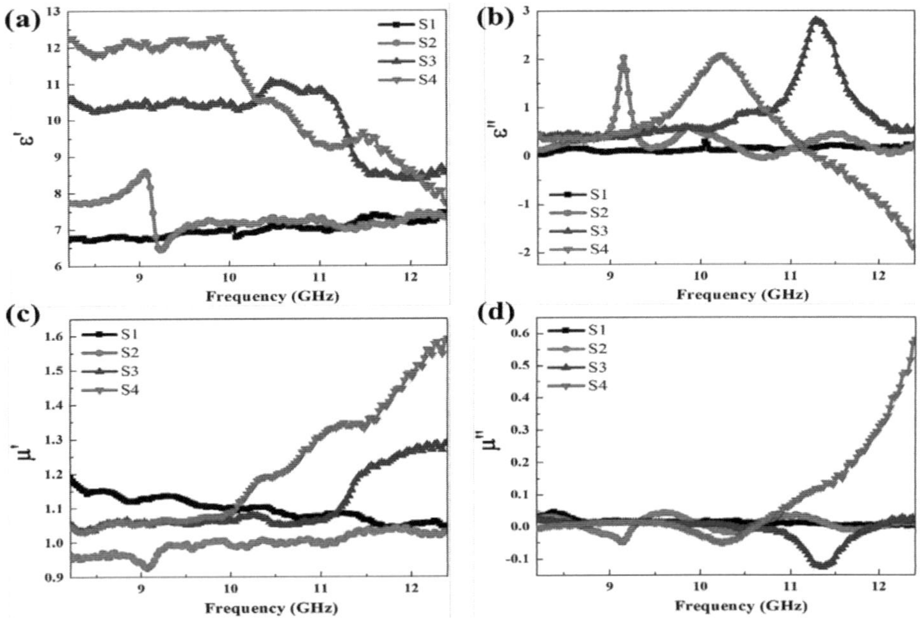

Fig. 2: Frequency dependence on the real part of permittivity (a), imaginary part of permittivity (b), real part of permeability (c) and imaginary part of permeability (d).

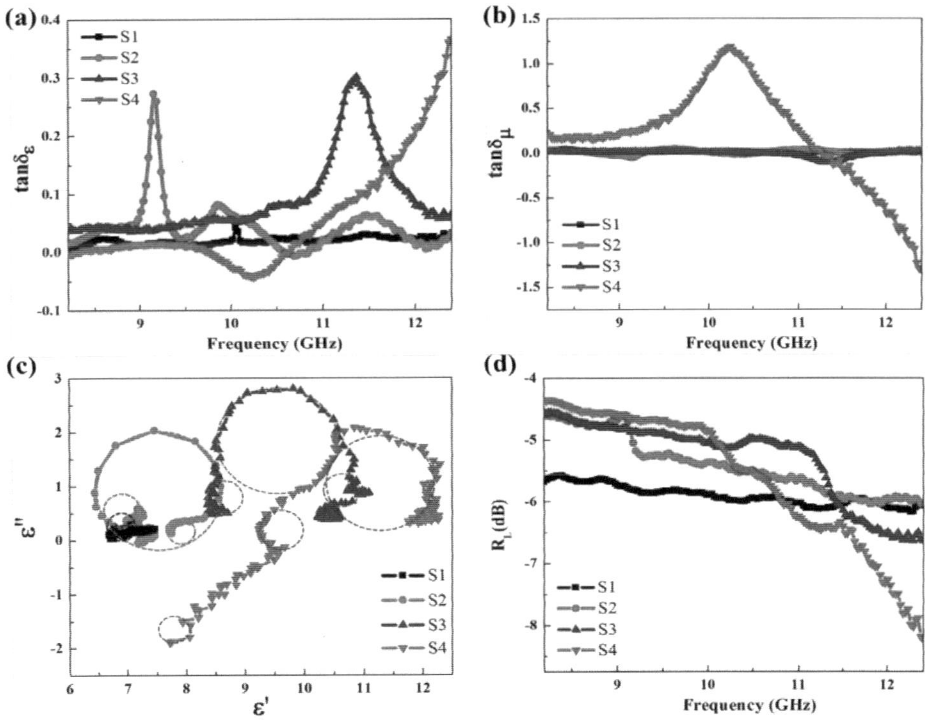

Fig. 3: Frequency dependence on the dielectric loss tangent (a) and magnetic loss tangent (b). The Cole-Cole analysis (c) and frequency dependence R_L (d).

The MW absorption properties were calculated according to the transmission line theory, using R_L, which was calculated by using these equations;

$$R_L = 20\log_{10}|(Z_{in} - 1)/(Z_{in} + 1)|$$

$$Z_{in} = \sqrt{\mu_r/\varepsilon_r}\tanh\left[j(2\pi fd/c)\sqrt{\mu_r\varepsilon_r}\right]$$

Where Z_{in} is input impedance, f is the frequency, d is the thickness of composite and c is the speed of light. Generally the values of μ_r affect the R_L, as the increase in μ_r shifts the reflection loss peak towards low frequency. The lower value of R_L means good MW absorption, μ_r for SiC is considered to be 1, because magnetic properties of SiC are poor. **Fig. 3d** shows the R_L for all the composites with a thickness of 3mm, where it can be seen that the R_L first increase with the increasing content of SiC and then start to decrease at high frequency. The lowest R_L peak was achieved -8.2 dB at 12.4 GHz (~ 90 % absorption) for S4 composite, which shows a good MW absorbance.

4. Conclusions

The F-doped SiC/Si_3N_4 ceramics were well prepared by HP sintering with different SiC content. The microstructural results confirmed a complete phase transformation from α to β-Si_3N_4 and revealed that the increasing content of SiC improved the dielectric properties of the SiC/Si_3N_4 ceramics. The complex permittivity and tangent loss was decreased at the low content of SiC and increased rapidly with the increase of SiC content. Furthermore, to investigate the MW absorption properties the lowest R_L was recorded -8.2 dB at 12.4 GHz (~ 90 % absorption). Hence, the as-synthesized F-doped SiC/Si_3N_4 ceramics can be a good candidate with the excellent characteristics of low permittivity and good MW absorption.

Acknowledgment

The authors would like to extend their sincere appreciation to the financial supports from Inorganic Non-metal Institute of Shandong University, Shandong Provincial Science and Technology Department Project (No.31370004011501) and Shandong Province Key research and Development Program (No.2016ZDJS05A05) for its funding.

References

[1] Riley F L 2000 *J. Am. Ceram. Soc.* **83**, pp 245-265.
[2] Li Q, Yin X, Duan W, Hao B, Kong L, and Liu X 2014 *J. Eur. Ceram. Soc.* **34** pp 589-598.
[3] Duan S, Zhu D, Dong J, Lu X, Ding D, Zhou W, *et al* 2019 *J. Alloy. Comp.* **790** pp 58-69.
[4] Li X, Zhang L, Yin X, Feng L, and Li Q 2010 *Script. Mater.* **63** pp 657-660.
[5] Zheng G, Yin X, Wang J, Guo M, and Wang X 2012 *J. Mater. Sci. Tech.* **28** pp 745-750.
[6] Li Q, Yin X, and Feng L 2012 *Ceram. Int.* **38** pp 6015-6020.
[7] Qi X, Zhai G, Liang J, Ma S, Liu X, and Xu B 2014 *CrystEngComm* **16** pp 9697-9703.
[8] Cho W S, Oh Y S, Kim C S, Osada M, Kakihana M, Lim D S, *et al* 1999 *J. Alloy. Comp.* **285** pp 255-259.
[9] Tatlı Z, Çalışkan F, Butler J, Crowley C, and Hampshire S 2014 *Ceram. Int.* **40** pp 1399-1404.
[10] Guo X, Feng Y, Lin X, Liu Y, Gong H, and Zhang Y 2018 *J. Eur. Ceram. Soc.* **38** pp 1327-1333.
[11] Feng Y, Guo X, Gong H, Zhang Y, Liu Y, Lin X, *et al* 2018 *Ceram. Int.* **44** pp 15686-15689.

IOP Conf. Series: Materials Science and Engineering **758** (2020) 012006 doi:10.1088/1757-899X/758/1/012006

Research on Influence Law of Explosion Load on New Subway Cross-structure

Fan Chengfei, Zhou Bukui, Tao Xigui, Hao Lubo, Zhang Xiaochen, Li Chaofeng

Research Institute for National Defense Engineering of Academy of Military Science PLA China, Beijing 100850, China

E-mail: tjufcf@163.com

Abstract: In order to reveal the influence law of explosion load when new subway under-crossing existing subway, the research with the method of combining theoretical calculation and numerical simulation was developed. With theoretical calculation method, the reasonable structural parameters and material parameters were chose, then the equivalent static load of under-crossing structure was obtained at the condition of explosion wave. By using the ANASYS/LS-DYNA software, the propagation law of explosion wave and the load change law of structure were simulated in four different models. Through comparison analysis, it showed the theoretical calculation method and numerical simulation method were rational, and the influence that existing subway on new under-crossing subway was only about the explosion wave propagation velocity except the explosion wave form and amplitude value, and the influence that new under-crossing subway on existing subway about explosion wave was very small. It provided a theoretical basis for design and construction for new under-crossing subway.

1 Introduction

Due to the needs of underground space planning and development, there are more and more crossovers in underground civil air defence projects[1]. The new subway could pass through the existing subway, thus changing the basic status of the existing subway. Improper handling would reduce the resistance level of the existing subway and destroy the stability of the existing subway. The existing subway would also have an impact on the resistance of the new subway [2,3].

Liu Jie[4] et al. relied on the Nanjing Ninglang Railway to penetrate the deep foundation pit of the Nanjing South Railway Station. The combination of theoretical analysis and on-site monitoring was used to analyse the deformation of the deep foundation pit of the subway tunnel. The internal force, displacement and axial force of the steel support in the retaining pile were numerically simulated. The influence of shield tunnel construction on the settlement of surrounding buildings, and the stresses and strains of the shields in different directions in the shield tunnelling process were studied by Xu Zemin of Tianjin University[5]. Yan Guodong[6] took a subway construction tunnel in a city as an example. The finite numerical software FLAC3D was used to carry out three-dimensional numerical simulation of construction and mining, and analyze the deformation of surrounding rock. It provided a theoretical basis for similar construction in China for great promotion value.The static and dynamic characteristics of the three-dimensional cross railway tunnel and its engineering application were researched by Yu Heran[7]. The dynamics research mainly focused on the response analysis of different structural forms to the dynamic load of high-speed trains. However, there were relatively few

About the author: Fan Chengfei (1986-), male, engineer, Ph.D., main research direction: blasting and protection, E-mail: tjufcf@163.com.
Address: No. 24, Taiping Road, Haidian District, Beijing;
Zip code: 100850; Contact: 13716171007

studies on the load variation law of subway tunnels under the blast loading. Especially for subway tunnels belonging to civil air defense engineering, it is especially important to analyze the response of subway tunnels under blast loading. .

In this paper, for the subway station of intersection of subway lines, the combination of numerical simulation and theoretical calculation were used to study the variation law of the explosion loading when the new subway underpassed the existing subway. It would ensure that the cross structure meets the specified resistance requirements and provide a theoretical basis for solving the design and construction problems for structural crossover.

The new A-line section line is from west to east. After passing through the existing subway station of B-Line, the tunnel depth was 17~21m. The section was constructed by shield method. The A-line crossed the existing subway stations and other underground pipelines. The resistance level of the two subway lines should be not reduced by the crossover structure. In this section of the project, the plane relationship of the A subway line crossing the B subway line is shown in Figure 1. The relationship between the section of the tunnel of Line A below the line B is shown in Figure 2.

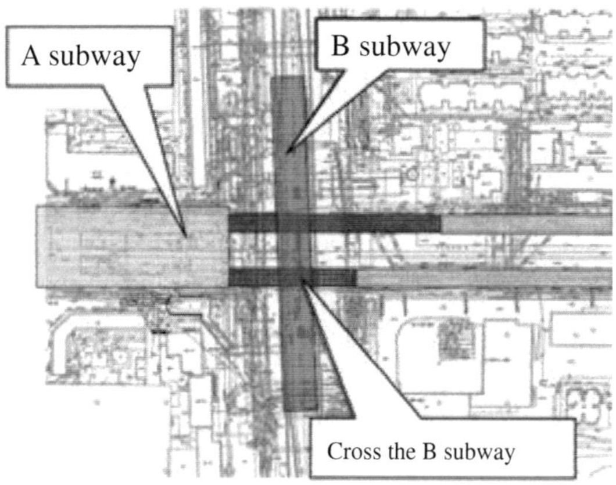

Fig.1 The plane relation between running tunnel of A subway and the subway station of B subway

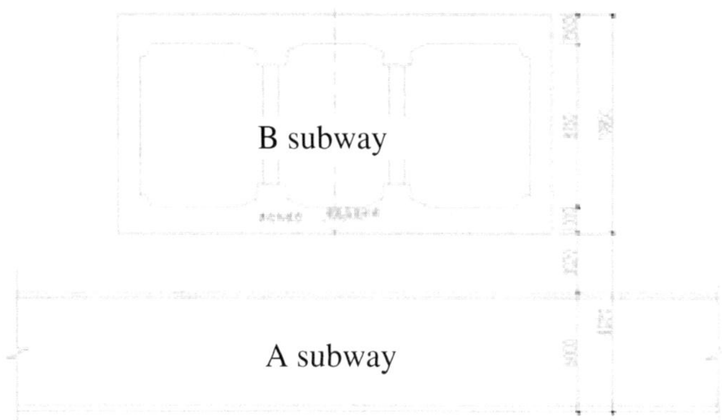

Fig.2 The profile relation between running tunnel of A subway and the subway station of B subway

2 Theoretical calculation

Assumed that the explosion shock wave overpressure of the B-line was p, and the equal-pressure action time was t.

2. 1 Structure self-vibration frequency

The natural frequency of the circular structure is calculated as follows:

- $$\omega = \frac{2.68}{R^2}\sqrt{\frac{\psi EJ}{\overline{m}}} \tag{1}$$

in which,

R——calculated radius of the ring (*cm*);

ψ ——the stiffness reduction factor generally taken as 0.85;

E——material elastic modulus (kg/cm^2);

J——section moment of inertia (cm^4); $J = \dfrac{bd^3}{12}$

$\overline{m}$ ——mass per unit length ($kg\ sec^2/cm^2$); $\overline{m} = \dfrac{\gamma F}{g}$

γ ——material bulk density (kg / cm^3);

g——gravity acceleration g=981cm/sec^2;

b——longitudinal unit length (*cm*);

d——section thickness (*cm*).

2. 2 Calculation of compression wave in soil

The compression waveform in the soil (rock) is simplified into a triangle. The maximum pressure and boosting time is calculated according to the following formula:

- $$P_h = \left[1 - \frac{h}{c_1 t_2}(1-\delta)\right]\Delta P_m \quad (2)$$
- $$t_{0h} = (\gamma_c - 1)\frac{h}{c_0} \quad (3)$$
- $$\gamma_c = \frac{c_0}{c_1} \quad (4)$$

in which,

P_h ——the peak pressure of the compression wave (*MPa*);

ΔP_m —— ground air shock wave overpressure peak (*MPa*);

h —— the calculated depth of the soil (rock) (*m*); when calculating the roof, the thickness of the roof should be took; when calculating the outer wall, the depth from the midpoint of each layer to the surface;

c_0 —— the initial pressure wave velocity of the soil (*m/s*);

c_1 —— the peak pressure wave velocity of the soil (*m/s*);

γ_c —— wave speed ratio;

δ —— the strain recovery ratio of soil;

t_{0h} —— compression wave boost time (*s*);

t_2 ——buck time (s), the equivalent positive pressure action time can be simplified according to the equal impulse.

2. 3 Structural dynamic coefficient

When the waveform of explosion loading was reduced to a triangle, the power factor should be calculated according to the following formula:

- $$K_d = \frac{[\beta] + \sqrt{[\beta]^2 - (2[\beta]-1)(1-\varepsilon^2)}}{2[\beta]-1} \quad (5)$$

2. 4 Structurally unfavorable soil thickness

Unfavorable soil thickness was calculated by the following formula:

- $$h_m = \frac{2\pi}{\omega_1} \cdot \frac{\alpha c_0}{\gamma_c - 1}$$

2. 5 Structural integrated reflection coefficient

The integrated reflection coefficient of structural top cover was related to the unfavorable soil thickness. When the thickness of the top cover soil is equal to or greater than the structural unfavorable soil thickness, the comprehensive reflection coefficient of the arch structure can be calculated according to the following formula:

- $$k_r = 2 - 0.4\left(1 - \frac{1}{\gamma_c^2}\right) \quad (6)$$
- $$\alpha_w = 1 - \frac{\rho c_1}{\overline{m}\overline{c}}(1 - n\alpha_k)\left[1 - \frac{1}{\overline{c}t_{oh}}\left(1 - e^{-\overline{c}t_{oh}}\right)\right] \quad (7)$$
- $$K = 0.5(1 + k_r\alpha_w) \quad (8)$$

k_r——fixed rigid body reflection coefficient;

α_w—— structural displacement influence coefficient, when the value of k_r α_w is less than 1.0, it is 1.0;

2. 6 Equivalent Static Load Calculation

When the underground engineering structure is designed by equivalent static load, the load should be designed according to the simultaneous action. The standard value could be calculated according to the following formula:

- $$q_1 = K_{d1}KP_h \quad (9)$$
- $$q_2 = K_{d2}\xi P_{\overline{h}} \quad (10)$$
- $$q_3 = K_{d3}\eta\frac{q_1}{K_{d1}} \quad (11)$$

in which,

q_1、q_2、q_3—— the equivalent static load standard values of the structural top cover, outer wall and bottom plate respectively (N/mm^2);

K_{d1}、K_{d2}、K_{d3}——the dynamic coefficients of the structural roof, outer wall and floor;

P_h——the peak pressure of the compression wave at the depth of the overburden soil (MPa);

$P_{\bar{h}}$——the peak pressure of the compression wave at the midpoint of each layer of the outer wall (MPa);

K——the comprehensive reflection coefficient of the top cover;

ξ——the lateral pressure coefficient of the soil;

η——the base pressure coefficient.

Taking the relevant structural parameters, material parameters [8-10] into above formulas, the equivalent static loads q_1, q_2, q_3 of the structural dynamic load are calculated.

3 Numerical simulation

The numerical simulation calculation software ANSYS/LS-DYNA is a general-purpose nonlinear dynamic analysis finite element software with explicit and implicit supplements. It is especially suitable for solving high-speed collisions of various 2D and 3D nonlinear structures. Nonlinear dynamic shock problems such as explosion, metal forming, heat transfer, fluid and fluid solid coupling problems also would be solved[11,12].

This paper intended to use the combination method of numerical simulation and comparative to carry out related research. It was proposed to carry out numerical simulations of the following modes.

A. The propagation of an explosion shock wave in a soil-covered medium;

B. Regardless of the existing subway structure, the propagation of the explosion shock wave in the overburden and underlying structures;

C. Regardless of the new undercut structure, the propagation of the explosion shock wave in the overburden medium and the existing subway structure;

D. Propagation of blast shock waves in earth-covering media and intersecting structures.

The four calculation models and boundary definitions of numerical simulation are shown in Figure 3. The explosion shock wave propagation process are shown in Figure 4. Figure 5 shows the key measurement points and their pressure time history curves selected in Model A.

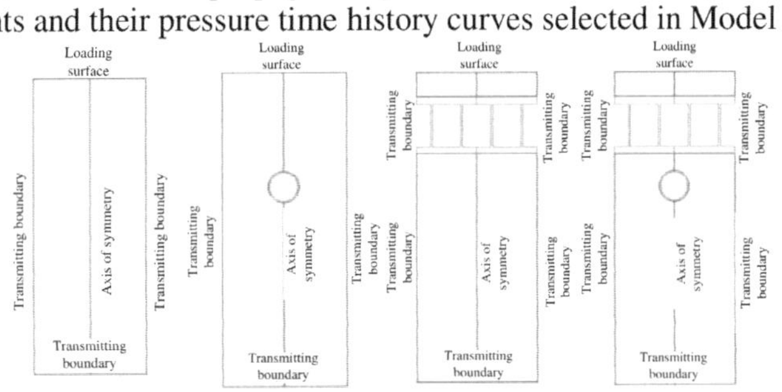

Model A Model B Model C Model D

Fig.3 Four calculation models and relevant border

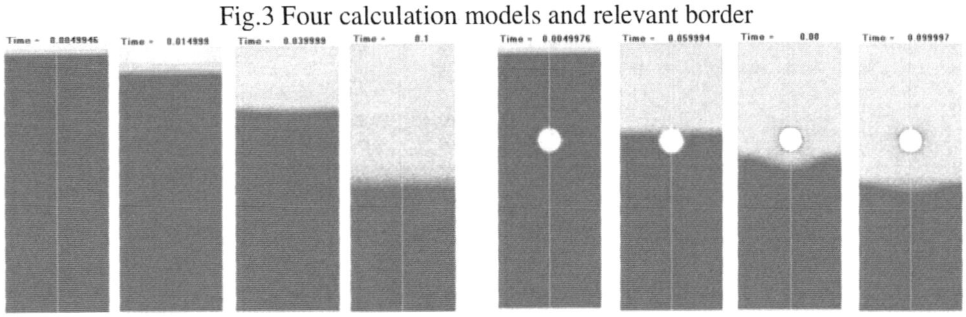

Model A Model B

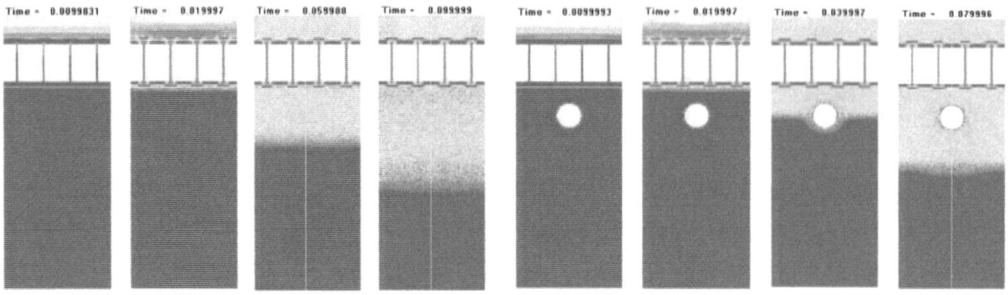

Model C Model D

Fig.4 The explosion shock wave propagation process

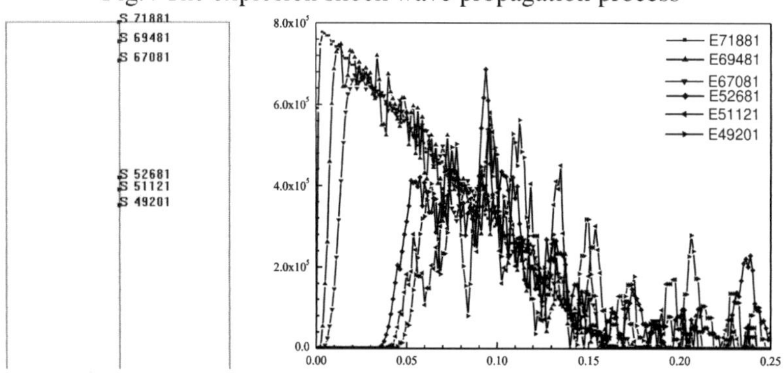

Fig.5 The place and the pressure curve of measuring point in model A

4 Results and analysis

4. 1 Comparison of theoretical calculation results and numerical simulation results

The peak value of the explosion shock wave after the attenuation of 18.9m was 0.47MPa, which was very close to the theoretical calculation result of 0.45MPa. The correctness of the numerical simulation and theoretical calculation method was verified, and the relevant parameters could be reasonable.

4. 2 Influence of station structure on pressure wave propagation

The new subway A-line passed through the existing subway station of B-line. The compression wave of explosion load propagated downward as a plane wave. When the wave met the station, the reflection waves would be occurred. Under different model conditions of different ground depths, the pressure curves of different measuring points were shown in Fig. 6. It could be seen from the figure that the peak value of the overburden pressure in the upper part and the lower part of the station was larger than that in the no station. It was mainly caused by the superposition of incident waves and reflected waves. Compared with the overburden, the station structure could be regarded as a rigid body, and the pressure acting on the top of the station would be transmitted directly to the bottom of the station through the column. Therefore, the compression wave at the bottom of the station would arrive firstly, when there were no station. It could be seen from Fig. 6 that there were little effect that the impact of the new A-line on the peak of the compression of the top and bottom of the station.

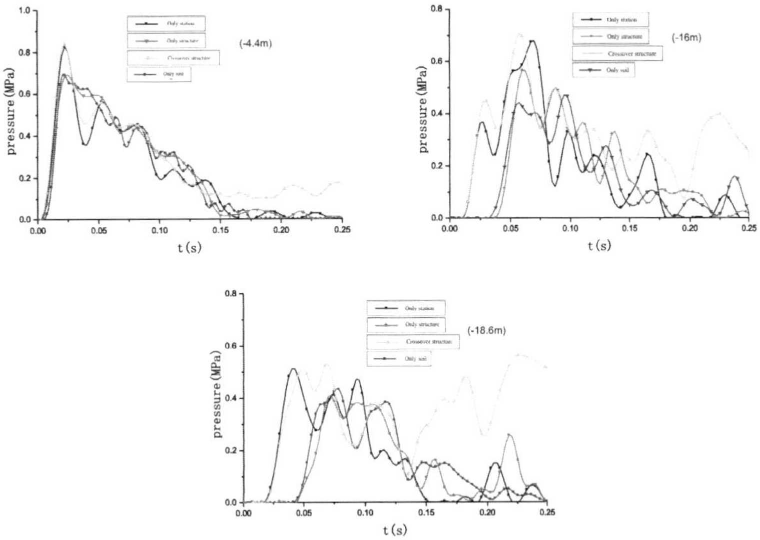

Fig.6 The earth pressure curve in different models

4. 3 Influence of the existing station built on the force of the new underpass subway line structure

E1 to E6 in Fig. 7 were the key stress measurement points of the structure. The pressure and effective stress curves were shown in Fig. 8 and Fig. 9 about the six measurement points in the structure with and without the station. From the pressure and stress distribution of the new subway line structure, it was easy to see that the pressure and effective stress curves were very close in the case of station and no station under the explosion load. It indicated that there was little effect that the existing station on the dynamic response of the structure. It could be ignored. Therefore, in the calculation of the structural dynamics of the newly built subway under the built subway, the influence of the station could be disregarded, and the upper part of the structure could be considered as pure covering soil.

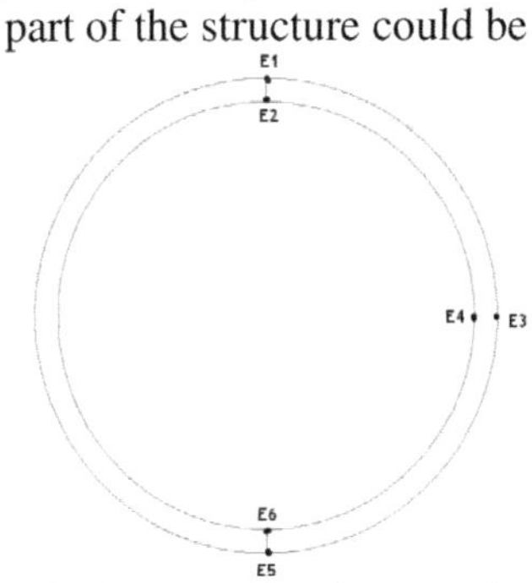

Fig.7 The key structural measuring point

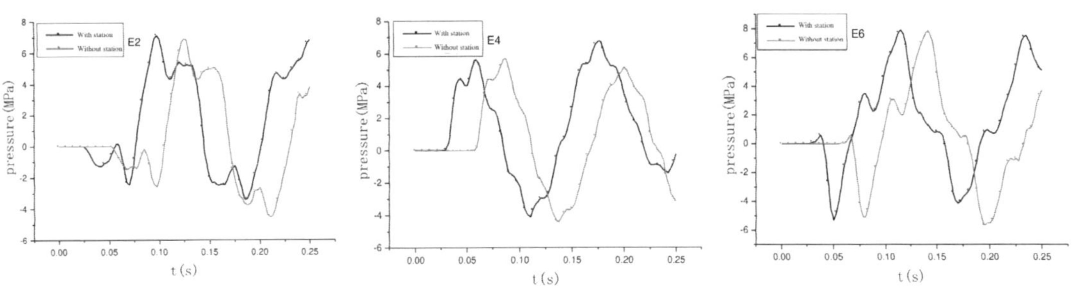

Fig.8 The pressure curve of key structural measuring points

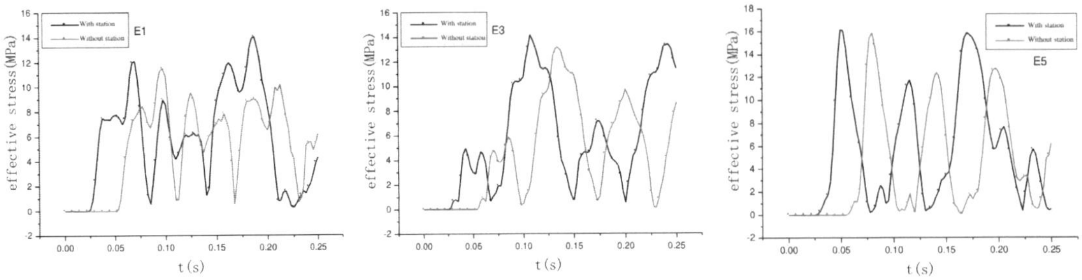

Fig.9 The effective stress curve of key structural measuring points

5 Conclusion

In this paper, the reasonableness and correctness of theoretical calculation methods and numerical simulation methods were verified by the combination of theoretical calculation and numerical simulation. By analysing and comparing the propagation modes of four different explosion shock waves, it was revealed that the propagation law of the built-up station structure on the explosion shock wave and showed the influence law of the built station structure on the new underpass subway structure. The conclusions of this paper could be used to solve the problems of new underpass subway structures. The theoretical basis was also benefit for future design and construction practices.

Acknowledgements

This work was financially supported by National Natural Science Foundation of China(No.51608524).

References

[1] JI Cheng. The Beijing Yizhuang subway station transfer and integration of design[J]. Shanxi Architecture, 2013, 17(39), 5-8. (in Chinese)

[2] LI Runjun. Study on the key technology of shallow mining in Beijing metro line.7 Da Guan Ying station and Wan Da section[D]. Beijing: China University of Mining Technology, 2014. (in Chinese)

[3]—KUI Jianbo. Study on the stability of enclosure structure for an open-cut metro station[D]. Chengdu: Southwest Jiaotong University, 2010. (in Chinese)

[4] LIU Jie. The research on deep foundation pit excavation optimization and the lower subway tunnel's deformation control[D]. Hefei: Heifei University of Technology, 2012. (in Chinese)

[5] XU Zemin. Study on effects of metro tunnel under historical building[D]. Tianjin: Tianjin University, 2013. (in Chinese)

[6] YAN Guodong. Study on settlement control of existing building induced by horsehead excavation from transverse gallery to main tunnels[J]. Journal of Shijiazhuang Tiedao University, 2013(1): 31-36. (in Chinese)

[7] YU Heran. Study on static and dynamic mechanical properties of railway over cross tunnels and it's application[D]. Chengdu: Southwest Jiaotong University, 2008. (in Chinese)

[8] LI Feng . Study on personal injury and structural damage due to urban traffic underground explosion[D]. Xian: Changan University, 2014. (in Chinese)

[9] JIANG Xinliang, JIA Yong, ZHAO Baojian, et al. Analysis of influence of metro tunnel construction on adjacent buildings[J]. Rock and Soil Mechanics, 2008, 29(11): 3047-3053. (in Chinese)

[10]-LI Qiang, WANG Mingnian, LI Decai, et al. Effects of subsurface construction of metro station tunnel on existing pile foundation[J]. Chinese Journal of Rock Mechanics and Engineering, 2006, 25(1): 184-190. (in Chinese)

[11]WANG Xinmin . Numerical analysis of ANSYS engineering structure[M] . Beijing: People's Communication Publishing House, 2007. (in Chinese)

[12]LI Ya . Research on the seismic performance of subway station structure in Tianjin soft soil area[D]. Tianjin: Tianjin Institute of Urban Construction, 2012. (in Chinese)

IOP Conf. Series: Materials Science and Engineering **758** (2020) 012007 doi:10.1088/1757-899X/758/1/012007

Reinforcement research of a built project due to functional transformation

Guo Bo Luo[1] ,Zheng Bin PAN[2] And Xue Rui Yang[1]

[1]Technician,Guizhou Construction Science Research and Design Institute of CSCEC, Guiyang,550000,China
[2] Graduate Student, Guizhou University, Guiyang,550000,China.
E-mail:745630455@qq.com

Abstract. According to the functional requirements, the construction unit intends to dismantle the 7/B and 8/B columns of the first to second floors of a Renovation of an existing project. The force of the structure has changed after the demolition, and the structure needs to be strengthened.In this paper, 7/AC and 8/AC beams of Negative third floors are reinforced with an enlarged section. After the section is enlarged, the negative moment reinforcement of adjacent beams 7/C-D and 8/C-D do not meet the requirements. The two beams has being enlarged the section at the upper part. Due to the large cross-sectional area of the beam after reinforcement, in order to form a strong column weak beam, the beam lower column 7/A, 8/A, 7/C, 8/C is strengthened by the enlarged section method, and the section is enlarged to make the column line stiffness Greater than the beam line stiffness.The carbon fiber sheet is pasted for reinforced with the beam-slab members in the range of 6-7/A-C and 8-9/A-C. According to the experience calculation, the reinforcement effect is better and meets the requirements of the force.

1. Introduction

With the continuous advancement of society, the construction industry continues to develop, and the demand for building functions varies.Although some of the existing projects have been built, the demand for their functions will change due to various factors. At this time, the structure needs to be functionally modified and the structure will change.The increase or decrease of the components will affect the stress of the overall structure, and some parts will also be insufficiently loaded due to the transformation. At this time, the modified structure needs to be reinforced.Now has an project,Due to functional needs, the construction unit plans to dismantle the 7/B and 8/B columns of first to second floors.The first floor is the exhibition hall, and the rest of the floors are offices. The structure needs to be strengthened after the transformation. Research.

2. Project Overview

The total construction area of the project is 7080 square meters, and a total of 5 exhibition halls will be built. The structure is a frame structure, and the seismic fortification intensity is 6 degrees.The construction unit is now planning to change the function of the building.The 1-6 axis of the building is 5 layers, the 6-7 axis is two layers, and the right side of the 8 axis is symmetrical with the left side.Due to functional needs, the construction unit plans to dismantle the 1st to 2nd floor 7/B and 8/B columns, the first floor is the exhibition hall, and the remaining floors are offices.After the transformation, the load of the third floor 7/AC and 8/AC beams is 3.5kN/m, and a large screen is added to the beam which weight is 5t. the beam length is 16.2m, and the conversion load is 50/16.2=3.083.5kN/m; the floor load is 1.5 kN/m^2 (excluding the board weight), the live load is $2.0kN/m^2$ for the office; and the constant load for the roof is 4.0 kN/m^2 (excluding the board weight),the live load is $2.0kN/m^2$.

3. Take reinforcement measures

The reinforcement measures taken for this project are as follows:

（1）After the dismantling, the longitudinal steel bars at the bottom of the 7/AC and 8/AC are discontinuous and the span becomes larger. The beam is reinforced with an enlarged section. After the section is enlarged, the adjacent beams 7/C-D and 8/C-D negative moment reinforcement do not meet the requirements. The two beams are thickened in the upper part for reinforcement.

（2）Because the cross-sectional area of the beam is large, in order to form a strong column weak beam, the 7/A, 8/A, 7/C, 8/C column wich lower beam is reinforced by the enlarged section method to make the column line stiffness greater than the beam line stiffness.

（3）Bonding carbon fiber sheets to the beam-slab members in the range of 6-7/A-C and 8-9/A-C.

4. Structural reinforcement check

4.1. Basic review and component reinforcement calculation

Check 7/A, 7/C pile bearing capacity, 7/ A and 7/C piles are elliptical piles, pile reinforcement is 21 B12, concrete strength is C30, and pile cross-sectional area is $1.18m^2$.The compressive bearing capacity of the stressed pile section is：

$$N \le \varphi_c f_c A_{ps} + 0.9 f_y' A_s' = 0.8 \times 14.3 \times 1.18 \times 10^6 + 0.9 \times 300 \times 2375$$
$$= 14140.45\text{KN}$$

In the formula：

N-Design value of axial pressure of pile top under basic combination of load effects;

φ_c -Foundation pile forming process factor;

f_c -Concrete axial compressive strength design value;

f_y' -Longitudinal main rib compressive strength design value

A_S' -Longitudinal main rib sectional area;

The standard axial force of the 7/C column bottom is 950.5kN, and the standard value of the axial force of the two columns on the two sides of the elliptical pile is 950.5×2=1901 kN, then the axial force design value N=1901×1.35=2566kN <14140.45 kN, the pile bearing capacity meets the requirements.The 7/C column vertical force is greater than 7/A, that is, the 7/A pile bearing capacity also meets the requirements.

4.2. Beam bearing capacity review

The 7/A-C beam is used for bearing capacity review. The section size of the beam is 500×1300mm, The reinforcement between the ends of the top of beam is 8Φ25,and the bottom is 10Φ25, the beam concrete strength is C30, and the beam maximum bending moment is 1077kN. m, the maximum negative bending moment is 1072 kN·m, and the maximum shear force is 424 kN.

Bending capacity calculation(positive bending moment):

$$\alpha_1 f_{c0} bx = f_y A_{s0} + f_y a_s A_s - f'_{y0} A'_{s0}$$

Due to the large size of the reinforcement, the original longitudinal reinforcement is not considered.

$$M \le afA(h_0 - x/2) + f_{y0} A_{s0}(h_{01} - x/2) + f'_{y0} A'_{s0}(x/2 - a')$$

$$x = (f_y A_{s0} + f_y a_s A_s - f'_{y0} A'_{s0}) / a_1 f_{c0} b = 44.49mm$$

Take $a_s = a'_s = 50mm$

Then $x \le (2a_s = 100mm)$

Taking the moment of pressed steel bar:

$$M = 0.9 f_y A_s h_0 = 1988\, KN \cdot m > 1077\, KN \cdot m$$

In the formula：

M–Design value of bending moment after component reinforcement(KN·m);

α_s–New steel bar strength utilization factor,take α_s=0;

f_y–New cross-sectional area of tensile reinforcement(N/mm^2);

A_s–New cross-sectional area of tensile reinforcement(mm^2);

h_0、h_{01}–Effective height of the section after reinforcement of the component and before reinforcement;

x-Height of concrete compression zone(mm);

f_{y0}、f'_{y0} –Design value of tensile and compressive strength of original steel bars(N/mm^2);

A_{s0}、A'_{s0} –Cross-sectional area of original tensile reinforcement and compression steel(mm^2);

a' –The distance from the longitudinal compression joint force point to the edge of the concrete compression zone(mm);

α_1 –The ratio of the stress value of the concrete rectangular stress diagram of the compression zone to the design value of the concrete axial compressive strength.When the concrete strength grade does not exceed C50, take $\alpha_1 = 1.0$ .When the concrete strength grade is C80, take $\alpha_1 = 0.94$ .Linear interpolation between them to take;

f_{c0}-Original structural concrete axial compressive strength design value(N/mm^2);
b-Rectangular section width(mm);

ξ_b-The relative limit of the nip area after the member is enlarged by the section.

That is, the bending bearing capacity meets the requirements.
Shear capacity calculation:

$$h_w / b = 1300/500 = 2.6 \leq 4$$

$$v = 424KN \leq (0.25\beta_c f_c bh_0 = 0.25 \times 0.80 \times 14.3 \times 500 \times 1250 = 1787KN)$$

The section size meets the requirements.

$$v = 424KN \leq \left(0.7 f_t bh_{01} + 0.7 a_c f_t A_c + a_s f_{yv} \frac{A_{sv}}{s} h_0 + f_{yv} \frac{A_{sv0}}{s_0} h_{01} = 793KN \right)$$

In the formula：

V-Shear design value after component reinforcement(KN);

β_c -Concrete strength influence factor;

b-The width of the rectangular section or the web width of the T-shaped, I-shaped section(mm);

h_w -The height of the web of the section; for the rectangular section, take the effective height; for the T-section, take the effective height minus the height of the flange; for the I-section, take the net height of the web.

The shear capacity is satisfactory.

4.3. Column bearing capacity review

The bearing capacity of the 7/C column composite column is taken. The section size of the column after reinforcement is 700×900mm, the axial force of the column is 1242.1kN, the moment of x-axis and y-axis is Mx=63.9 kN.m, My=15.1kN.m, and the bearing capacity of the Mx-oriented column is taken.

$$e_1 = e_0 + e_a = 12 + 30 = 42mm$$

$$e = e_i + h/2 - a = 442mm$$

$$N \leq a_1 f_{CC} bx + 0.9 f_y' A_s' - 0.9 f_y A_s \text{ (Reinforcement reduction calculation).}$$

The result is:
x=125mm

$$Ne \leq a_1 f_{cc} bx(h_0 - x/2) + 0.9 f_y' A_S' (h_0 - a_s) - 0.9 f_y A_s (h_0 - a_s)$$

$$549 KN \cdot m \leq 985 KN \cdot m$$

After the reinforcement, the bearing capacity of the column meets the requirements.

5. Conclusion and Prospect

（1）The 7/C pile foundation bearing capacity meets the requirements, and the 7/C column vertical force is greater than 7/A, that is, the 7/A pile bearing capacity also meets the requirements.

（2）The 7/A-C beam section size meets the requirements, and the shear capacity meets the requirements.

（3）The column 7/C is taken for calculation. It can be seen from the calculation that the column bearing capacity of the column after reinforcement is satisfactory.

References

[1] Bosco C , Carpinteri A .*Fracture behavior of beam cracked across reinforcement[J]*.Theoretical and Applied Fracture Mechanics, 1992, 17(1):61-68.

[2] Lee T K , Pan A D E . *Estimating the relationship between tension reinforcement and ductility of reinforced concrete beam sections[J]*. Engineering Structures, 2003, 25(8):1057-1067.

[3] Zhang B , Guan H , Loo Y C . *UST failure criterion for punching shear analysis of reinforcement concrete slab-column connections[J]*. Computational Mechanics–new Frontiers for the New Millennium, 2001:299-304.

[4] Lee H J , Aschheim M , Enrique Hernández-Montes, et al. *Optimum RC column reinforcement considering multiple load combinations[J]*. Structural and Multidisciplinary Optimization, 2009, 39(2):153-170.

IOP Conf. Series: Materials Science and Engineering **758** (2020) 012008 doi:10.1088/1757-899X/758/1/012008

Research on anti-floating reinforcement of an existing project

Yu Han Huang[1], Zheng Bin PAN[2],Guo Bo Luo[1] And Long Chen[1]

[1]Guizhou Construction Science Research and Design Institute of CSCEC, Guiyang,550000,China
[2] Graduate Student, Guizhou University, Guiyang,550000,China.
E-mail:386389880@qq.com

Abstract. Due to the lack of structural anti-floating, a certain part of the main structure has cracks, and the structure needs to be reinforced by buoyancy to make the project meet the requirements.After research, the Negative second floor of the original structure is reinforced with add the anti-floating anchors, apply pressure grouting for cracks in the plate, Paste carbon fiber cloth for the bottom of the plate and the damaged shear wall, externally-paste profile steel for the damaged column, and paste Carbon fiber sheet attached to damaged beam.the calculated of shear bearing and flexural capacity show that the board meets the requirements.

1. Introduction

With the continuous development of the construction industry, more and more buildings are standing in the crowd, and problems are emerging.Especially in Guizhou Province, the geological environment and hydrological conditions are complicated, and it is difficult to obtain accurate geological surveys. Therefore, some structural bearing capacity is likely to occur during the design and construction process, and reinforcement measures are required for buildings with insufficient structural capacity.An existing project in Guizhou has found that cracks have occurred in structural components due to insufficient structural anti-floating, and the structure needs to be anti-floating. For the reinforcement measures, this paper has made corresponding research.

2. Project Overview

The construction area of the project is 25,808.73m^2, and the building height is 8.55m. The basic form is pile foundation plus waterproof board, independent column base plus waterproof board and raft foundation.The foundation bearing layer is a medium-grade limestone, and the main structure is a two-layer underground frame structure. Because of insufficient anti-floating , Some frame columns, frame beams and bottom plates in the range of (1)~(10)/(A)~(M) axis of the negative one floor and Negative second floor of the project are deformed and cracked. Therefore, the damage status of

IOP Conf. Series: Materials Science and Engineering **758** (2020) 012008 doi:10.1088/1757-899X/758/1/012008

concrete members in (1) ~(10)/(A)~(M) axis of the negativ one floor and Negative second floor of the project is detected.According to the test, the damage range of the structure is (1)~(10)/(A)~(M) axis; some column has cracking phenomenon, and there are mutation and dislocations in the cross section of individual frame columns; the negative second floor shear wall (1)~(7)/(A),and negative one floor of shear wall (1)/(A)~(B) has cracking phenomenon; a few beams have cracking phenomenon; negative second floor (1)~(2)/ (A)~(C), (3)~(4)/(J)~(K), (6)~(7)/(E)~(G) Plate has being crack.

3. Take reinforcement measures

In view of the engineering problems, the following reinforcement measures are proposed. pressure grouting is applied to the bottom and basis of the plate before reinforcement. anti-floating anchors are added to the negative second floor, and the head height of the water for the anti-floating is 4.3 m.Pressure grouting treatment for the board of crack, and the carbon fiber cloth is reinforced on the bottom of the board.the damaged column is reinforcement by externally pasted steel, externally attached carbon fiber sheet, and the enlarged section method. Before the reinforcement of damaged column, the pressing should be removed.The carbon fiber cloth method is used for reinforcement for the damaged beam. Before the reinforcement of damaged beam, the crack on the beam should be pressure grouted. the damaged shear wall should be reinforced with carbon fiber cloth.

The original structure is reinforced and the reinforcement measures are shown in the figure below.

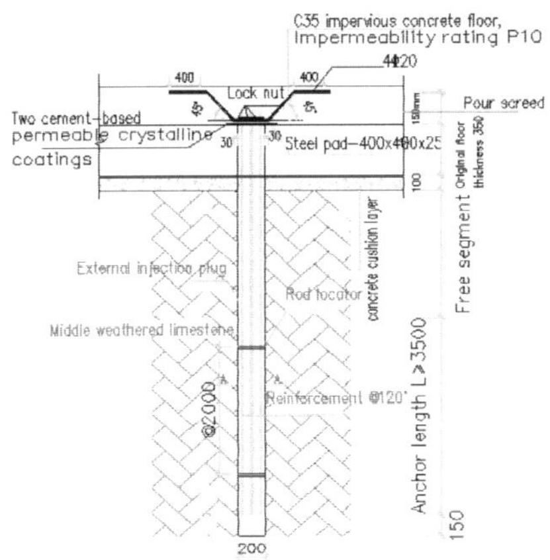

Figure 1. Uncasting concrete screed

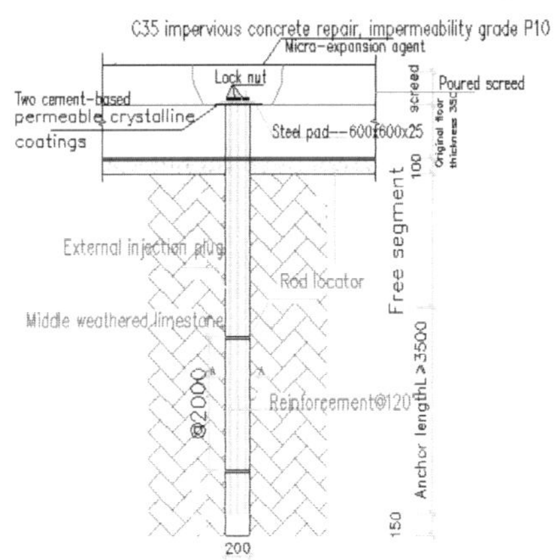

Figure 2. Pour screed concrete anchor practice

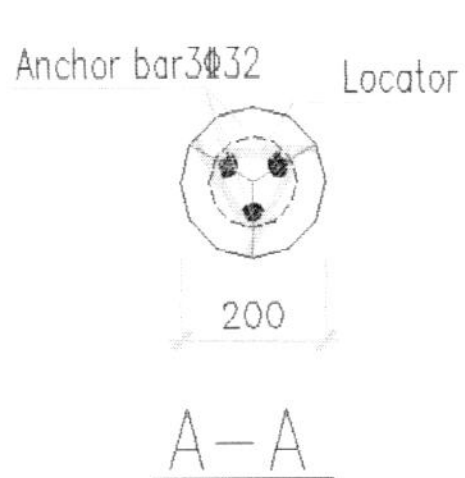

Figure 3. Anchor section

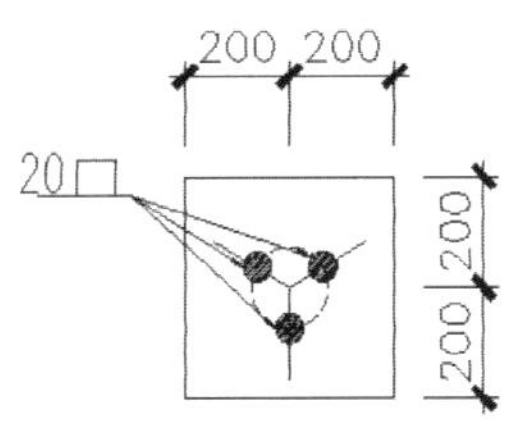

Figure 4. Uncasted water stop plate steel plate

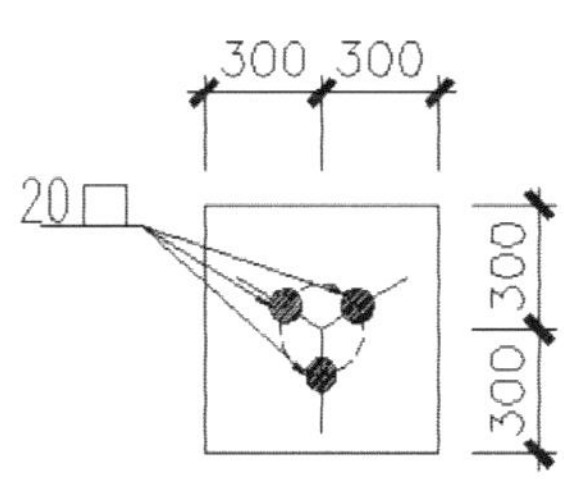

Figure5. Poured water-stop

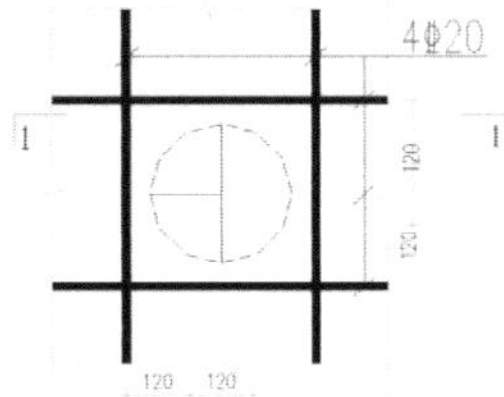

Figure 6. Anti-shearing steel bar arrangement in the board

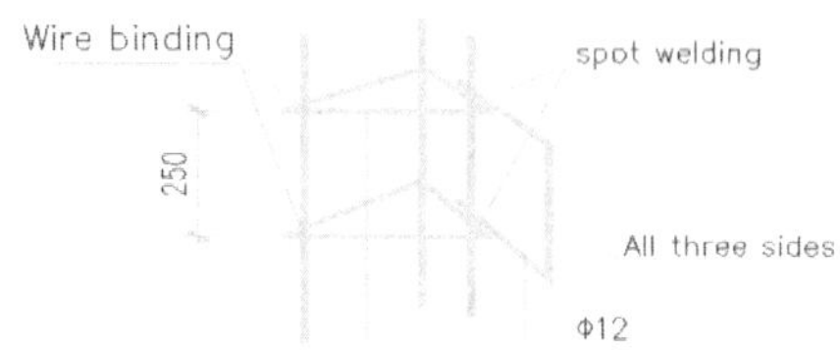

Figure 7. Locator

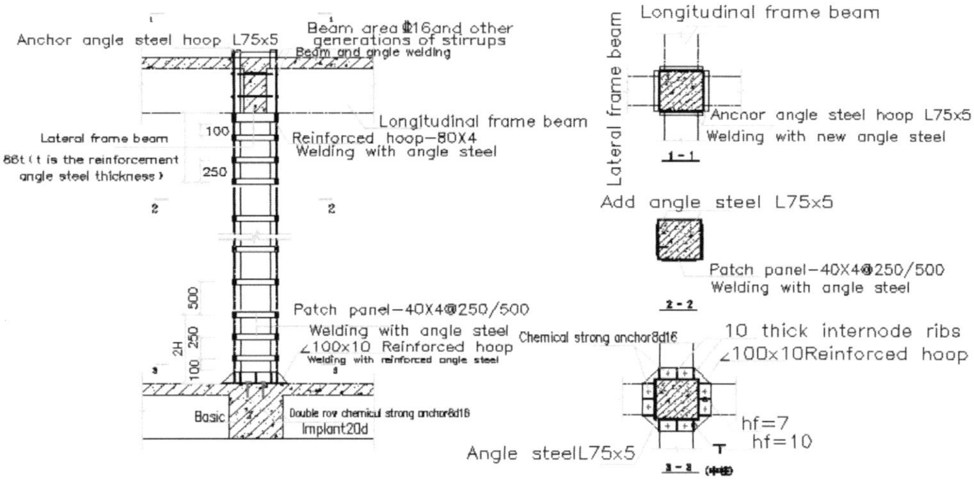

Figure 8. Bonded steel reinforcement column practice

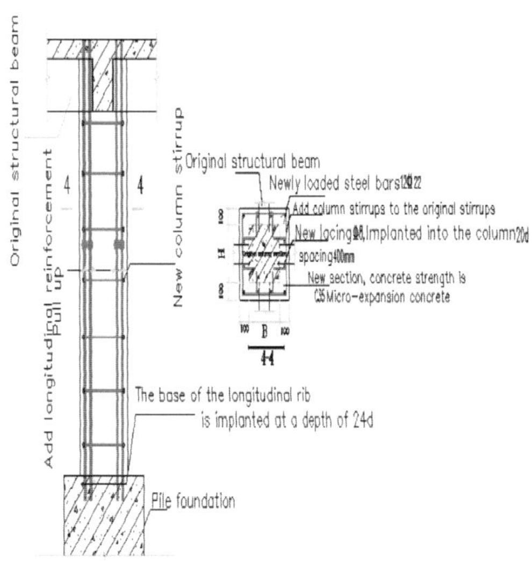

Figure 9. Increase section reinforcement

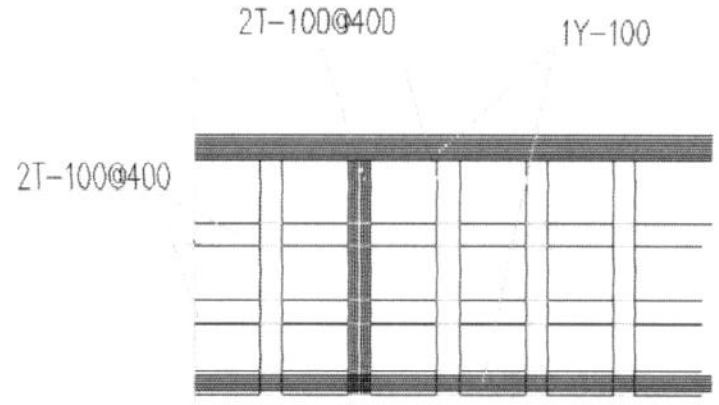

Figure 10. Carbon fiber cloth reinforcement Hear wall practice

IOP Conf. Series: Materials Science and Engineering **758** (2020) 012008 doi:10.1088/1757-899X/758/1/012008

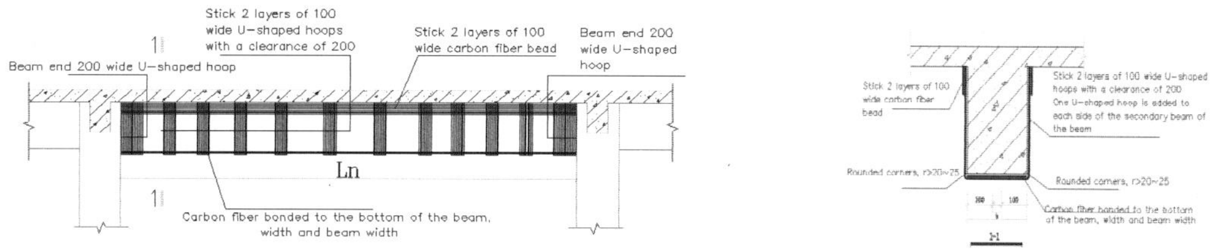

(a) (b)

Figure 11. Paste fiber cloth

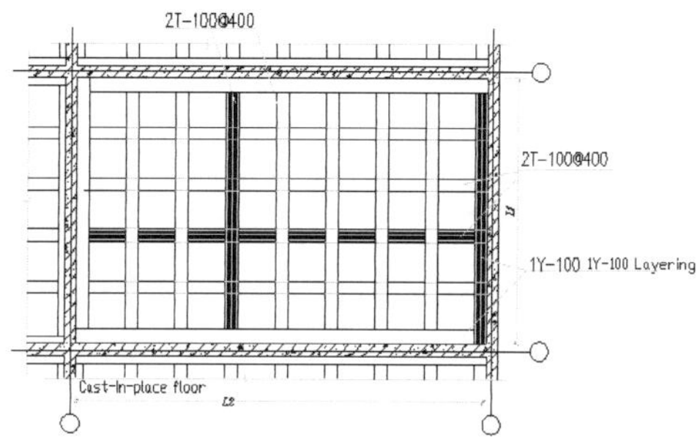

Figure 12. Bottom plate carbon fiber layout method

4. Structural reinforcement check

4.1. Anti-floating design water level calculation

According to the geological survey report, the anti-floating level is H=1469.4m, and The basement floor bottom level of the two-layer part of 1-32 is 1465.1m (the lane is also considered to be within this part), and the head height is 4.3m. The 32-42 axis basement floor has a bottom elevation of 1467.6m and a head height of 1.8m.

Upper load statistics:

(1)1-32 two-layer part

This load calculation is calculated by taking the 8.1×8.1 plate as an example.

Covering soil: $18\times0.7=12.6$ kN/m^2.

Reinforced concrete slab weight: $0.71\times25=17.75$ kN/m^2.

The thickness of the bottom plate is 350mm, the thickness of the negative two-layer top plate is 200, and the thickness of the negative one is 160mm.

Concrete plain screed, 7~15cm on site, calculated according to 5cm: $0.05\times20=1$ kN/m^2.

Beam and column calculation:

$(0.3\times(0.6-0.2)\times4\times2+0.4\times(0.9-0.2)\times2\times2+0.6\times0.6\times2\times8.1)\times25/(8.1\times8.1)=3.0$ kN/m^2.

Dead load calculation (favorable load): $12.6+17.75+1+3.0=34.35$ kN/m^2.

So 34.35 kN/m$^2<43.0\times1.05=45.15$ kN/m^2

The area's anti-floating does not meet the requirements.

(2)32-42of One layer part

This load calculation is calculated by taking the 8.1×8.1 plate as an example.

Covering soil: $18\times0.7=12.6$ kN/m^2.

Reinforced concrete slab weight: $0.05\times20=1\ kN/m^2$

The thickness of the bottom plate is 350mm, and the thickness of the negative layer is 160mm.

Concrete plain screed, 7~15cm on site, calculated according to 5cm: $0.05\times20=1\ kN/m^2$

Beam and column calculation:

$$(0.3\times(0.6-0.2)\times4+0.4\times(0.9-0.2)\times2+0.6\times0.6\times2\times4.0)\times25/(8.1\times8.1)=1.5\ kN/m^2.$$

Dead load calculation (favorable load): $12.6+12.75+1+1.5=27.85\ kN/m^2$.

So $27.85\ kN/m^2 > 18\times1.05=18.9kN/m^2$

The area is resistant to floats.

4.2. Anti-floating anchor calculation

(1) Buoyancy calculation of anchor

The original structure has a dead weight of $34.35\ kN/m^2$.The head height of groundwater buoyancy of 4.3m is $4.3\times10=43\ kN/m^2$.

(The anti-buoyancy (kN) of the anchor rod) $=1.05\times$(the groundwater buoyancy-the original structure is dead load)$=10.8kN/m^2$.

(2)Calculation of tensile bearing capacity of single anchor

Using a bolt with a diameter of 200mm, the circumference of the bolt is Ur(m)=0.628m.The effective anchorage length hr(m)=2.6 m in the anchorage section embedded in the rock stratum.Empirical coefficient ξ = 0.8.The bond strength characteristic value of cement mortar and concrete to rock is f(MPa)=0.4 (less than the value of the geological survey report).

Anchor rod pullout bearing characteristic value Rt(kN)=Ur hrξf X1000= 522.496 kN

Required bolt root n=the anti-buoyancy/anchor pull-out bearing capacity characteristic value of the anchor= $10.8 \times 8.1 \times 8.1 / 522.496 = 1.36$.

The actual 8.1×8.1 unit takes 2 anchors (pile base parts), and the 1-11 axis independent base part spans 4 anchors.

The bearing capacity of the overall anti-floating single anchor is: $10.8\times8.1\times8.1/2=354.294$ kN.

Review the anti-buoyancy of a single anchor: the head area of the anchorage of the 8.1m×8.1m plate is $8.1\times8.1-2\times2.7\times2.7=51.03m^2$, and the standard of anti-floating force of each anchor is: $(43-0.35\times0.25-0.05\times20)\times51.03/4=424$ kN.Then the actual bearing capacity of each anchor is 424 kN.

(3)Calculation of cross-sectional area of anchor bar

Anchor rod body tensile safety factor K b=1.8.Anchorage tensile strength design value fy=360 N/mm^2.

The total area of the anchor steel bar section:

$$As\geq K\ b\times NaK/\ fy(mm^2)=1000\times1.8\times424/360=2120\ mm^2$$

Actual matching steel area As (332) = $2413mm^2$.

(4)Calculation of the anchor length of the anchor anchor and the formation

Anchor diameter D = 200 mm.Standard value of ultimate bond strength of rock and soil layer and anchor frb (kPa) = 1800kPa.Bolt anchor solid pullout safety factor K=2.2.

Bolt anchor and length of anchorage section of the formation

$$La1\geq K\ Nak/(\pi Dfrb)=2.2\times397.3/(3.14\times0.2\times1800)=0.77m.$$

Calculation of anchorage length between anchor rod body and anchor mortar

Anchor bar diameter d=32mm.Anchor rod number n=3 (root).Design value of bond strength between steel bar and anchor mortar fb=2.1 (KPa).

Anchorage length between anchor rod body and anchoring mortar

$$La2 \geq K\ Nak/(n\pi d\ fb) = 2.2\times424/(0.032\times3.14\times3\times2.4\times1000) = 1.28m.$$

The bolt insertion length is 3.5m.

4.3. Base plate punching calculation

Under the action of concentrated reaction, the punching capacity of the plate without the stirrups or the bent steel bars shall comply with the following provisions:

$$F\ L \leq (0.7\beta hft + 0.25\ бpc,\ m)\ \eta umh0$$

Section height influence coefficient βh=1.0.The weighted average of the effective preloading stress of the critical section concrete by length бpc, m=1.0N/mm^2.Concrete axial tensile strength design value ft=1.43N/mm^2.Section effective height h0=300-50=250(mm).

Critical section perimeter um=3.14×2 (100+250/2)=1413mm.

η=min{η1,η2}

η1=0.4+1.2/βs βs=2

η1=0.4+1.2/βs=0.4+1.2/2=1.0

η2=0.5+αsh0/4um αs=40

η2=0.5+αsh0/4um = 0.5+40×250/(4×1413)=2.269

η=min{η1,η2}=1.0

(0.7βhft+0.25бpc， m)ηumh0

=(0.7×1.0×1.43+0.25×1.0) ×1.0×1413×250

=441.916kN

Concentrated reaction force design value = actual single anchor bolt tensile capacity = 397.3 <441.916KN.Therefore, the punching resistance meets the requirements, and it is not necessary to configure the stirrups or bend the reinforcing bars.

4.4. Floor bearing capacity calculation

Take the 8.1m×8.1m meter span for calculation. After the anti-floating anchor is added to the bottom plate, the plate span is calculated according to 4.05m, and the one-meter wide strip is taken for calculation.

The height of the water head is 4.3m, the weight of the plate is 0.350×25×1=8.75 kN/m, and the water buoyancy is 43×1=43kN/m. The force of the plate is calculated according to the basic combination:

$$1.4\times43-8.75=51.45$$

The maximum shear force of a 1 meter wide strip is: ql/2=51.45×4.05/2=104.2 kN.

The maximum bending moment of a 1 m wide strip is: ql2/12=51.45×4.052/12=70.32 kN·m.

The 1 m wide strip is subjected to the maximum bending capacity (calculated by the ⏀14@200 board with smaller reinforcement):

$$M \leq f_y A_s (h_0 - a_s^{'}) = 77.6kN \cdot m$$

The shear capacity is: $$a_{cv} f_t bh_0 = 315kN$$

It can be seen from the calculation that the shear bearing capacity of the plate is satisfactory.

5. Conclusion and Prospect

the Negative second floor of the original structure is reinforced with add the anti-floating anchors, apply pressure grouting for cracks in the plate, Paste carbon fiber cloth for the bottom of the plate and the damaged shear wall, externally-paste profile steel for the damaged column, and paste Carbon fiber sheet attached to damaged beam.After calculation, the following conclusions are drawn:

(1)1-32 two-layer partial area anti-floating does not meet the requirements, 32-42 layer area anti-floating meets the requirements.

(2)The actual bearing capacity of each anchor is 424 kN, and the bolt insertion length is 3.5 m.

(3)The punching of the bottom plate meets the requirements, and it is not necessary to configure the stirrups or bend the steel bars.

(4)The shear bearing capacity of the plate is satisfactory.

References

[1] Yu Y , Gu S , Wu Z , et al. *Experimental study of load transfer law of prestressed cables under loess stratum[J].* Chinese Journal of Rock Mechanics & Engineering, 2010, 29(12):2573-2580.

[2] Xing J Y , Yong Z X , Lin C P . *Safety Analysis of Shield Tunnel Segment Lining Based on Field Test[J].* IOP Conference Series Earth and Environmental Science, 2017, 94(1):012200.

[3] Sun S R , Wu J M , Wei J H . *Ground settlement research of excavation for doubled arch tunnels in expressway[J].* Rock and Soil Mechanics, 2006.

[4] Qingfeng D . *Anti-floating Reinforcement Design of Underground Engineering[J].* Chinese and Overseas Architecture, 2010.

[5] Wang F X , Yuan Z R , Cai H X . *The Method Comparison of Anti-Floating of Reinforced Concrete Slab at Underground Business Street[J].* Key Engineering Materials, 2017, 730:429-434.

IOP Conf. Series: Materials Science and Engineering **758** (2020) 012009 doi:10.1088/1757-899X/758/1/012009

Evaluation of the curvature ductility ratio of a circular cross-section of concrete reinforced with GFRP bars

C Pichardo[1], W Tovar[1] and V I Fernandez-Davila[2]

1 Research Student. Faculty of Civil Engineering, UPC, Lima, Peru
2 Professor. Faculty of Civil Engineering, UPC, Lima, Peru
E-mail: pccivfer@upc.edu.pe

Abstract. The present study deals with the use of fiberglass reinforced polymer bars (GFRP) as a replacement for the common steel of a reinforced concrete circular pile, in order to avoid the corrosion of durability of reinforcing bars and thus improve them. The comparative analysis was carried out between a pile reinforced with GFRP and another with steel, where the ductility was evaluated by obtaining moment-curvature diagram. As a result, said idealized moment-curvature diagrams and ductility indices are presented, concluding the ductility of the section reinforced with GFRP in 20% more than that of steel.

1. Introduction

Fiberglass composite bars have emerged in recent years as a possible and efficient alternative to replace steel in reinforced concrete elements due to the problem corrosion [1]. Some of the advantages of this material are the high resistance-to-weight ratio, high corrosion resistance and low maintenance cost [2]. These bars are made up of fiberglass, which in combination with a polymer improves stress resistance and can be used as structural reinforcement [3].

A methodology for the construction of the interaction diagram for a composite section with GFRP bars subjected to eccentric axial force was proposed [4]. An experimental research on moment-resisting frame reinforced with GFRP bars and steel bars subjected to lateral cyclic loads and axial loads was conducted [5]; it was observed that the GFRP-reinforced section developed greater elastic deformations than steel creep deformations. Finally, a series of tests on columns reinforced with fiberglass bars and carbon bars subjected to axial load was carried out and it was developed an analytical method for determining the moment-curvature diagram and the lateral displacement [6].

In this investigation, a comparative analysis of the moment-curvature diagrams of a circular cross-section of concrete reinforced with steel bars and GFRP bars subjected to axial load and lateral force was carried out. In addition, the diagrams were idealized into trilinear curves in order to assess the index of ductility curvature for both types of reinforcement through a program code made in MATLAB [7].

2. Method

2.1. Procedure Diagram Moment-Curvature

The fiber method was used for the analysis of a circular section of reinforced concrete. This analysis was carried out computationally as follows:

i. Define the mechanical properties of materials, dimensions, and distribution of rebar in the section. Divide the section in 1 mm thick fibers.
ii. Determine the design resistance compression of the section ϕPn. It will depend of the reinforcement material (steel or GFRP). The section is subjected to a constant axial load P equal to 15, 30, 45 and 60% of ϕPn.

iii. Assume an initial value of strain to compression of the top fiber (ε_{c1}) and a value of the distance to the neutral axis of that fiber (c_1).

iv. In each fiber, calculate the core area (A_{core}) and the area of the cover (A_{cover}) and the deformation of the fiber (ε_{ci}) with the $c_1$assumed. Calculate the stress of the concrete (f_c) unconfined for the cover according to [8] and for confined core according to [9].

v. Calculate the resulting force as the product of the stress by the fiber area, both in the core and cover area. Then, sum of all the resulting forces in each fiber to determine the total force of the concrete (Fc); also, multiplicate the resulting forces by the distance from the centroid and sum the moments in each fiber to determine the resulting moment relative to the neutral axis (Mc).

vi. Calculate the contribution on each axis of rebar by calculating the reinforcement area (A_{sj}), the deformation (ε_{sj}), stress (f_{sj}) and force (F_{sj}) from its respective elasticity module ($E_s\ o\ E_f$), Finally, sum the forces and moments relative to the neutral axis to determine the resulting force (Fs) and the resulting moment relative to the neutral axis (Ms).

vii. P1 is obtained from the sum of the resultant forces Fc and Fs. In turn, the sum of the resultant moments Mc and Ms is M1.

viii. If the value of P1 is verified to be equal to P, then the assumed value c_1 is correct, otherwise iterations of step 5 to 8 are performed with different values of "c" until you find the right one.

ix. Once the value of c is set, the curvature Ψ is determined as $\frac{\varepsilon_{ci}}{c}$ and with its respective M1, the ordered pair is formed (Ψ, M1) and a point of the curve is obtained.

x. This process is repeated for a range of ε_{c1}values. In this case, values from 0.0005 to 0.008 were chosen with increments of 0.00025 in order to build the diagram until the higher fiber reaches the last deformation of the unconfined concrete.

It is important to consider the degradation of the concrete, that is, if the upper fiber has a deformation equal to or greater than the maximum deformation of concrete ε_{cu} equal to 0.003, its contribution is disregarded. In addition, the contribution of GFRP bars in compression is not considered.

2.2. *Idealization del diagram moment-curvature*

According to [10], the curvature moment diagram can be idealized as a trilinear diagram (figure 1).

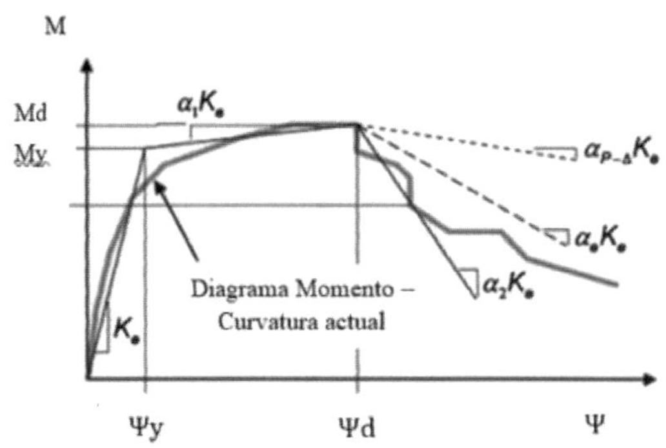

Figure 1. Idealized curve.[10]

The first line of the idealized curve [0;Ψy] has a slope equal to effective lateral stiffness and represents the area where the section has elastic behavior. The second line [Ψy, Ψd] represents the post-yield of positive slope and part of the yield point (Ψy;My) until the start of the degradation zone (Ψd;Md) and represents the plastic behavior zone of the section. Finally, in the third line of negative slope, the value of which is estimated between the starting point of the degradation to the point where it is intercepted with the horizontal line at 60% of the yield moment [10]. It is important that the areas below the idealized curve and the actual curve approximately equal in order to have about the same energy dissipation.

3. Results

The analysis of a circular section reinforced with steel bars and the same with GFRP bars was performed. This section has 40 cm diameter, spirals and rebar are 3/8" and 5/8" in diameter, respectively. It is considered a 7.5 cm of cover. The compression resistance of the concrete is 350 kg/cm^2 and the yield strength of the steel and the ultimate strength of the GFRP are 4200 kg/cm^2 and 10193 kg/cm2, respectively. The elasticity modulus of steel and GFRP are 2000000 kg/cm^2 and 713634 kg/cm^2, respectively; and the yield strain of the steel and ultimate strain of the GFRP is 0.0021 and 0.015. It is considered an environmental reduction factor CN equal to 0.8 set out in the [3].

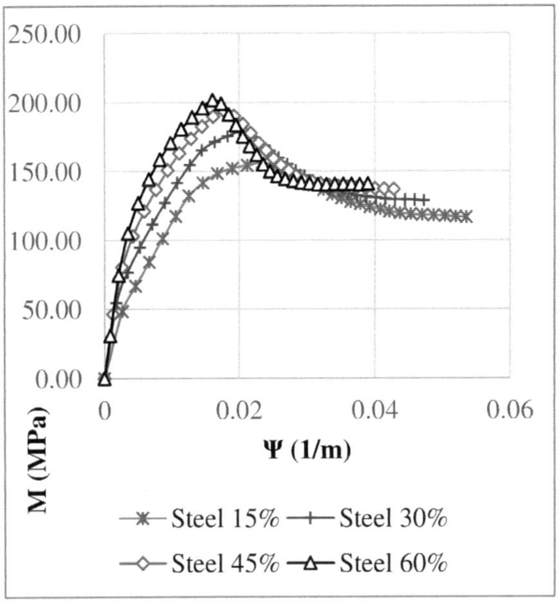

Figure 2. Moment-Curvature diagram for circular section with steel

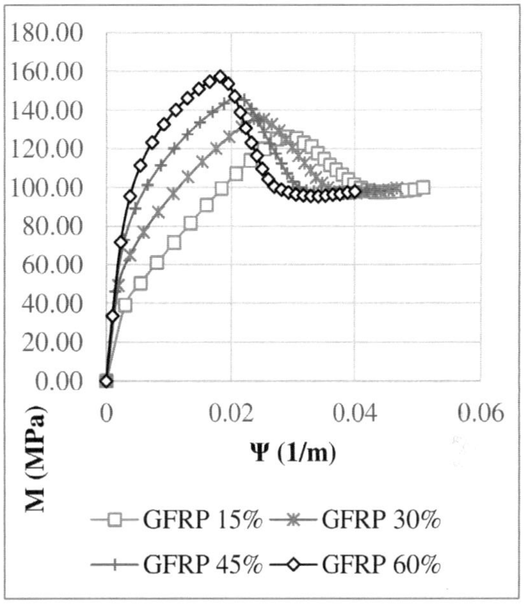

Figure 3. Moment-Curvature Diagram for circular section with GFRP

In figure 2 and 3, it is observed that, for both types of reinforcements, the axial load is higher, the maximum moment of the section increases, while, its respective curvature decreases for both types of reinforcements.

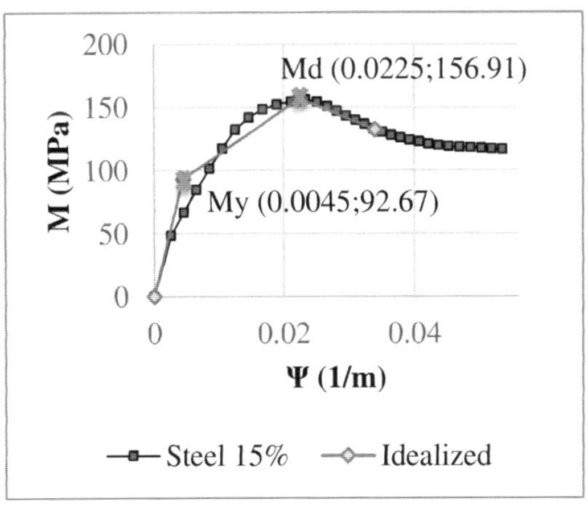

Figure 4. Idealized diagram Moment-Curvature for steel reinforcement subjected to an axial load of 15% of ϕPn.

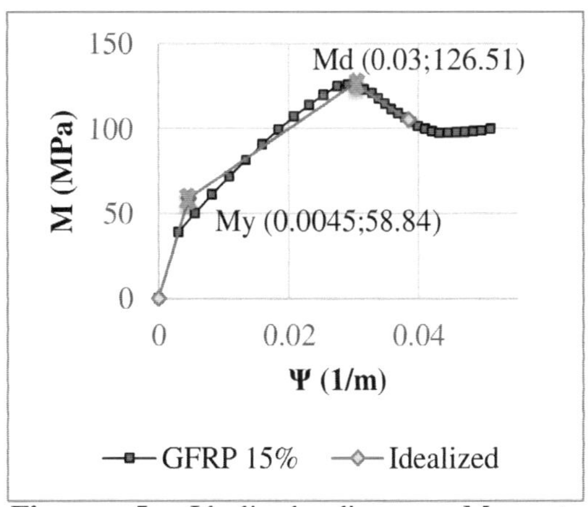

Figure 5. Idealized diagram Moment-Curvature for GFRP reinforcement subjected to an axial load of 15% of ϕPn.

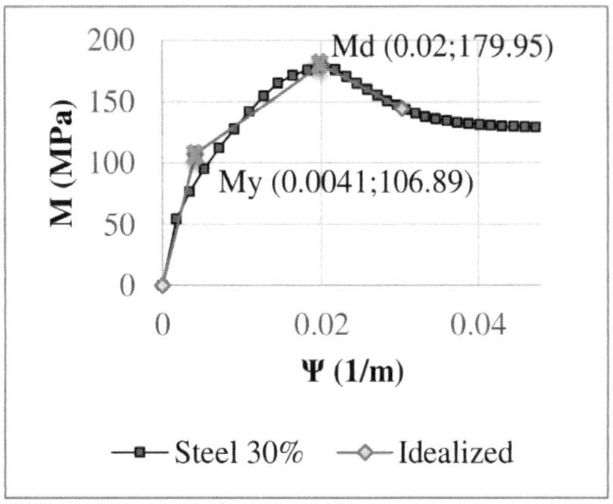

Figure 6. Idealized diagram Moment-Curvature for steel reinforcement subjected to an axial load of 30% of the ϕPn.

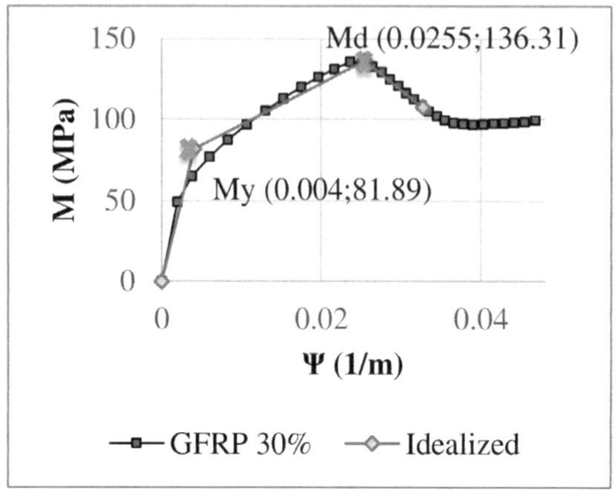

Figure 7. Idealized diagram Moment-Curvature for GFRP reinforcement subjected to an axial load of 30% of ϕPn.

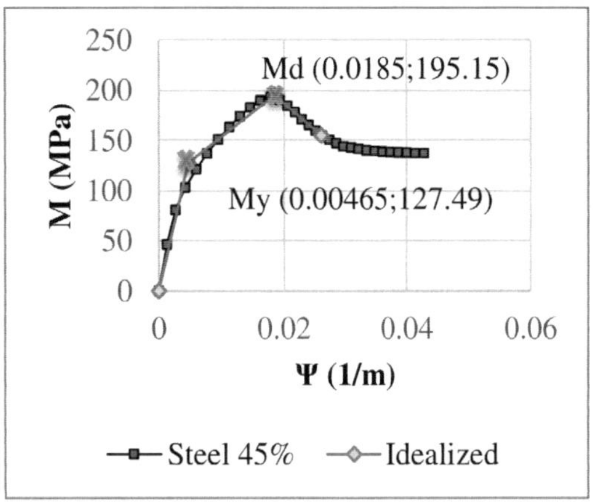

Figure 8. Idealized diagram Moment-Curvature for steel reinforcement subjected to an axial load of 45% of ϕPn.

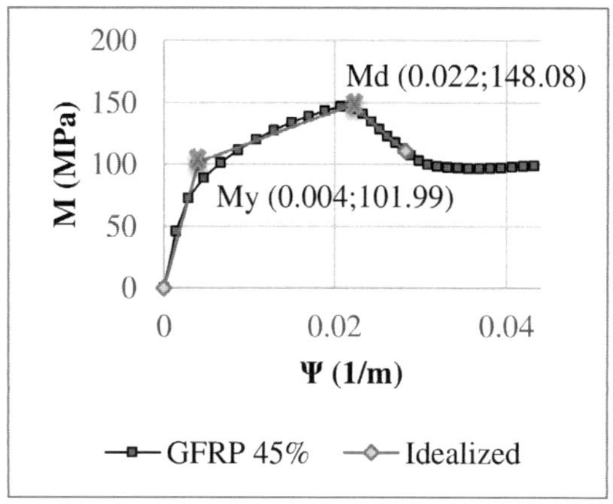

Figure 9. Idealized diagram Moment-Curvature for GFRP reinforcement subjected to an axial load of 45% of ϕPn.

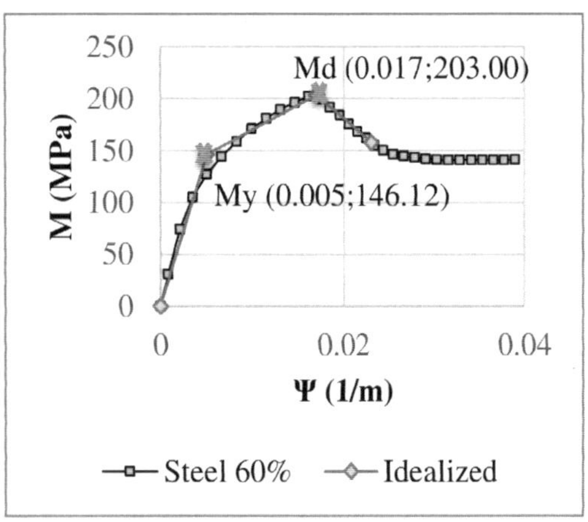

Figure 10. Idealized diagram Moment-Curvature for steel reinforcement subjected to an axial load of 60% of ϕPn.

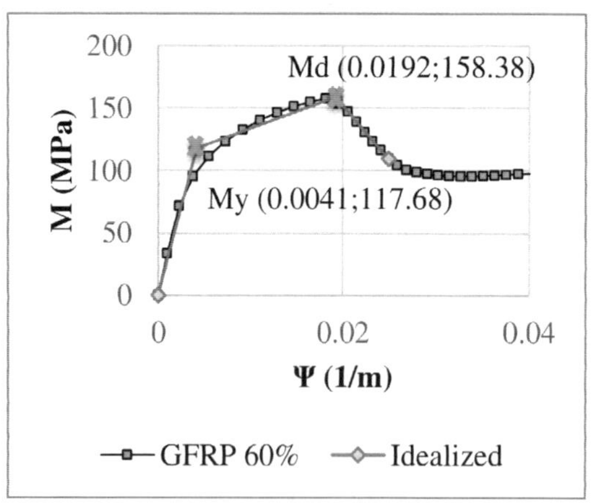

Figure 11. Idealized diagram Moment-Curvature for GFRP reinforcement subjected to an axial load of 60% of ϕPn.

The Moment-Curvature diagrams were idealized and a ductility curvature of 5.00 for steel (figure 4) and 6.67 for GFRP (figure 5) was obtained for 15% ϕPn. Meanwhile, for 30% ϕPn, a ductility curvature of 4.88 for steel (figure 6) and 6.38 for GFRP (figure 7) was obtained. In addition, a ductility curvature is obtained of 3.98 for steel (figure 8) and of 5.50 (figure 9) for 45% ϕPn. Finally, for the section loaded with 60%Pn, a ductility curvature of 3.40 for steel (figure 10) and 4.68 for GFRP (figure 11).
In addition, the slope of the second line of the idealized Moment-Curvature diagram for the steel reinforced section represents the bending stiffness in the plastic range, the which is on average 17.2% the slope of the first line ,that is, the effective bending stiffness in the elastic range of the section in each case.On the other hand, pfor the GFRP-reinforced section, the slope in the plastic range of the second line of the idealized Moment-Curvature diagram is 20.3%, 17.3%, 10.0% and 9.4% the slope of the first one, respectively.

4. Conclusions

The analysis of a circular cross-section of concrete reinforced with steel bars and GFRP bars was carried out and the following conclusions were drawn [11]:

- The maximum resistance moment increases, while, their respective curvature decreases in both types of rebar.
- Although GFRP-bars has greater curvature compared to steel-bars, the maximum resistance moments associated with such curvatures are less than 22% on average.
- The curvature ductility ratio obtained for the GFR-rebar were 26% higher on average compared to the steel-rebar.
- The slope of the second branch, which represents the bending stiffness in the plastic range, is on average 15.5% of the slope of the first branch, which represents the effective bending stiffness for cases studied with steel rebar.
- The bending stiffness in the plastic range for GFRP-reinforced section for 15, 30, 45 and 60% of maximum design compression resistance is 20.3%,17.3%,10.0% and 9 .4% of the elastic effective bending stiffness, respectively. Therefore, it is concluded that for lower axial load, this section can develop greater tension input of GFRP bars and the stiffness to effective bending increases.

5. References

[1] Hadhood A, Mohamed H and Benmokrane B 2018 *J. ACI Structural* **115** S27
[2] Nanni A, De Luca A and Jawaheri H 2014 *Reinforced concrete with FRP bars* (London: CRC Press) p 24
[3] Zogui M 2014 *The international handbook of FRP composites in civil engineering* (London: CRC Press) p 16
[4] ACI Committe 440 2015 *Guide for the design and construction of structural concrete reinforced with FRP bars* p 18-21
[5] Jawaheri H and Nanni A 2013 *J. Composites for construction* **17** 294-304
[6] Mohammed H, Farghaly A and Benmokrane B 2018 *J. ACI Structural* **115** 337–349
[7] MATLAB and Statistics Toolbox Release 2012b, The MathWorks, Inc., Natick, Massachusetts, United States.
[8] Popovics S 1973 *J. Cement and Concrete Research* **3** pp. 583-599
[9] Mander J, Priestley J and Park R 1988 *J. Structural Engineering*, ASCE **114** 1804-1826
[10] ASCE (41-13) 2013 *Seismic Evaluation and Retrofit of Existing Buildings* (EE.UU) p 147
[11] Pichardo C and Tovar W 2019 *Structural comparative study of a circular section of concrete armed with fiberglass bars (GFRP) instead of steel bars exposed to chloride corrosion on the Peruvian coast* (Undergraduate thesis, Peruvian University of Applied Sciences).

Acknowledgments
Thanks to Prof. César Morales, Ph.D. for his help in the drafting of this paper.

IOP Conf. Series: Materials Science and Engineering **758** (2020) 012010 doi:10.1088/1757-899X/758/1/012010

Determination of expansion pressure of heat exchange based on three material models

Xianzhang Li[1, a], Lianfa Yang[2, b*], Jingyu Jiang[3, c], Jinjie Huang[4, d]

[1,2,3,4]School of Mechanical and Electrical Engineering, Guilin University of Electronic Technology, Guilin, Guangxi, 541004

E-mail: [a]229198641@qq.com,[b]y-lianfa@163.com, [c]j_jingyu@163.com, [d]JinjieHuang231@126.com

Abstract. The hydraulic expansion of the heat exchange tube-to-tubesheet is to apply the liquid pressure to expand the inner wall of the tube uniformly through the device. It make the tube wall deformed and contact the tube sheet. The expansion pressure which affects the quality of the expansion of the heat exchanger is one of the important factors for the hydraulic expansion of the heat exchanger. Therefore, in order to meet the expansion requirements, it is necessary to determine the expansion pressure of the heat exchange. The determination of hydraulic expansion pressure by some scholars are introduced in this paper.

1. Introduction

In hydraulic expansion[1], a uniform liquid pressure is applied on the inner surface of heat exchange tube Initially. It make the tube start to happen plastic deformation, until the outer wall of the heat exchange tube contacts the inner wall of the tube sheet. However, there is no deformation to tubesheet. This stage is called the deformation stage of the heat exchange tube. As the hydraulic expansion is increased, the tubesheet partly or all happen plastic deformation, this stage is called the loading to the tube sheet. Finally, liquid pressure is unloaded when the expansion pressure reaches a predetermined value. Because the tubesheet which happen elastic deformation or partly happen plastic deformation occur a elastic rebound and the tube which happen full plastic deformation can't occur rebound, The outer wall of heat exchange tube and the inner wall of tubesheet are tightly attached together. Hydraulic expansion has the advantages of simple operation and high production efficiency. What's more, There is not cold hardening which is produced by mechanical extrusion and less residual stress and corrosion for the hydraulic expansion. At the same time, the expansion pressure can be set precisely, the pull-out force can be predicted, and the quality of expansion can be guaranteed.

Since born to the present, the theory of expansion joint has been continuously developed and improved. Next, the determination of the expansion pressure of hydraulic expansion joints by some scholars are introduced and the three model which these scholars apply to determine the hydraulic expansion are introduced simply. These three models are elastic-pefectly plastic model, hardening model and bilinear model.

2. determination expansion pressure using elastic-pefectly plastic model

Some scholars applied the elastic-pefectly plastic model for determining the expansion pressure of the of heat exchanger. This model neglects the affect of the hardening of material in the expansion of the heat exchange. The assumption of the plane stress is used to determine the expansion pressure during the expansion process of the heat exchange tube. Fig.1 is the geometrical character of the heat exchange expansion model.Where the r_i, r_o are respectively indicate the inner and outer tube radius, R_i, R_o are respectively indicate the inner and outer tubesheet radius, c_0 is indicate the clearance of the tube and tubesheet.

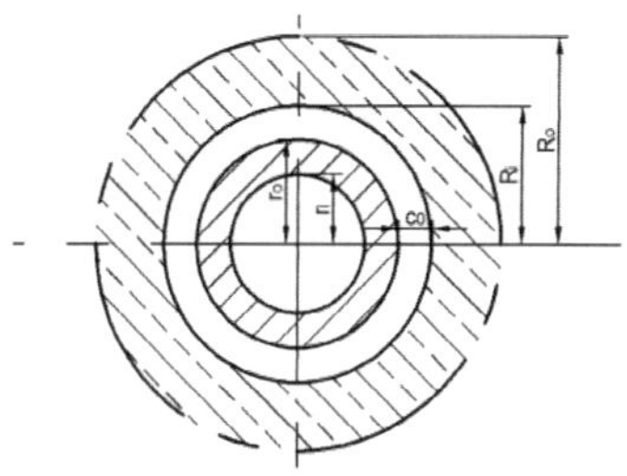

Fig 1. geometrical character of the expansion

In 1976, Krips and Podhorsky [2] applied the Mises yield criteria and Lamé equation to determine the expansion pressure during the expansion process of the heat exchange tube. And the minimum expansion pressure p_{imin} and maximum expansion pressure p_{imax} of the heat exchange tube are given:

$$p_{i\min}=\frac{2\sigma_{st}}{\sqrt{3K_t^4+1}}\bullet\frac{K_s^2-1}{K_s^2(1+\mu)+1-\mu}+\frac{\sigma_{st}(K_t^2-1)}{2} \tag{1}$$

$$p_{i\max}=(\frac{\sigma_{st}}{\sigma_{ss}}\bullet\frac{K_t^2+1}{K_t^2-1}+\frac{K_s^2-1}{2})\sigma_{ss} \tag{2}$$

Where K_t, K_s are respectively indicate the ratio of radius, $K_t= r_o/r_i$, $K_s=R_o/R_i$, σ_{st}, σ_{ss} are respectively the yield stress of tube and tubesheet, μ is the Poisson's ratio of the material.

In 2001, Based on the formula proposed by Krips and Podhorsky, Cheng gang et al.[3] Introduced a coefficient β when he considered the length of the expansion joint is generally smaller than the thickness of the tube sheet, and the influence of the material becoming rigid, etc. It make the results closer to reality. The coefficient β is generally selected between 0.8 and 0.9. After rewriting, the minimum expansion pressure and ultimate expansion pressure of the heat exchange tube are:

$$p_{i\min}=\frac{1}{\beta}\bullet\frac{2\sigma_{st}}{\sqrt{3K_t^4+1}}\bullet\frac{K_s^2-1}{K_s^2(1+\mu)+1-\mu}+\frac{\sigma_{st}(K_t^2-1)}{2} \tag{3}$$

$$p_{i\max}=\frac{\sigma_{st}}{\beta}\bullet\frac{K_t^2+1}{K_t^2-1}+\frac{\sigma_{ss}(K_t^2-1)}{2} \tag{4}$$

In 1996, Yan Huigeng et al.[4] used the single tube model which is two concentric circles with a initial clearance to determine the expansion pressure. An assumption which The material is the elastic-pefectly plastic material and the ratio of the inner to outer diameters of the heat exchange tube is still K_t after deformation is made.Axial stress of tubesheet and heat exchange tube during expansion is neglected and Von mises yield criteria are applied to determine expansion pressure at each deformation stage:

$$p_{i\min} = \frac{2}{\sqrt{3}}\sigma_{st} \ln K_t \Big/ (1-2c) \tag{5}$$

$$p_{imanx} = \frac{2}{\sqrt{3}}\sigma_{st} \ln K_t + \frac{2}{\sqrt{3}}\sigma_{ss} \ln K_s \tag{6}$$

Where c is a material constant associated with elastic modulus of the tube and tubesheet, K_t, K_s and the poisson's ratio of the tube and tubesheet.

In 2009, Laghzale and Bouzid [5] used the Tresca yield criteria to study the expansion pressure of the hydraulic expansion process of the heat exchange tubes. and the expansion pressure at each stage are given:

$$p_{isy} = \sigma_{st} \ln K_t + \frac{\sigma_{ss}}{2}\frac{(K_s^2-1)}{K_s^2} \tag{7}$$

$$p_{i\max} = \sigma_{st} \ln K_t + \sigma_{ss} \ln K_s \tag{8}$$

3. determination expansion pressure using strain hardening model

Metal materials generally have power strain-hardening properties.The material properties have a significant effect on the expansion process, which leads to a larger error in the determination of the expansion pressure using the elastic-pefectly plastic model .Some scholars have considered the material hardening properties and use the material hardening model to determine the expansion pressure during the expansion process.

In 2007, Wang Haifeng et al.[6] comprehensively considered the material strain-hardening properties, heat exchange tube and tubesheet geometry and other factors. power strain-hardening properties are used to describe the mechanical behavior of the heat exchange tubes and tube sheets during the expansion process. Constitutive relationship of the materials power strain-hardening is given:

$$\sigma_i = \begin{cases} E\varepsilon_i & \sigma_i \le \sigma_s \\ A\varepsilon_i^m & \sigma_i > \sigma_s \end{cases} \tag{9}$$

Where E is the elastic modulus of material, m is material exponential, A is the material constant, ε_i is the material strain, σ_i, σ_s are respectively indicate the material stress and yield stress.

Moreover, Von mises yield criteria and Hencky deformation theory are applied to determine the expansion pressure of each expansion joint in the expansion process of the heat exchange tube :

$$p_{i\min} = k_2(c_0 r_o)^{m_t}(r_o^{-2m_t} - r_i^{-2m_t}) \tag{10}$$

$$p_{i\max} = \frac{\sigma_{ss}}{\sqrt{3}m_s}(K_s^{2m_s} - 1) + k_2[r_o(c_0 + \frac{\sqrt{3}\sigma_{ss}R_o^2}{2E_s R_i})]^{m_t}(r_o^{-2m_t} - r_i^{-2m_t}) \tag{11}$$

Where c_0 is the initial clearance, m_t, m_s is material exponential of the tube and tubesheet, k_2 is the constant.

In 2010, Xie T et al.[7] applied a general strain-stress model which is proposed by XP Huang et al.[8] in 2005 to describe the strain hardening during the expansion of the heat exchange tube.Constitutive relationship of the materials general strain-stress model is given:

$$\sigma_i = \begin{cases} E\varepsilon_i & \sigma_i \le \sigma_s \\ A_1 + A_2\varepsilon_i^m & \sigma_i \ge \sigma_s \end{cases} \tag{12}$$

Where A_1 and A_2 are the material constant, A_{t1}, A_{t2} are respectively indicate the tube material constant,

A_{s1}, A_{s2} are respectively indicate the tubesheet material constant.

Xie applied Von mises yield criteria and Hencky deformation theory to determine the expansion pressure of heat exchange and the expansion pressure of each expansion joint in the expansion process of the heat exchange tube are given:

$$p_{i\min} = \frac{2}{\sqrt{3}} A_{t1} \ln K_t + (\frac{2}{\sqrt{3}})^{m_t+1} A_{t2} (c_0 r_o)^{m_t} (\frac{1}{-2B_1})(r_o^{-2m_t} - r_i^{-2m_t}) \quad (13)$$

$$p_{i\max} = \frac{2}{\sqrt{3}} A_{s1} \ln K_s + \frac{2}{\sqrt{3}} A_{t1} \ln K_t - \frac{1}{-2B_2}(\frac{2}{\sqrt{3}})^{m_s+1} A_{t2}\left[r_o(c_o + \frac{\sqrt{3}\sigma_{ss} R_o^2}{2E_s R_i})\right]^{m_s} (R_o^{-2m_s} - R_i^{2m_s}) - \frac{1}{2B_1}(\frac{2}{\sqrt{3}})^{m_t+1} A_{t2}\left[r_o(c_0 + \frac{\sqrt{3}\sigma_{ss} R_o^2}{2E_s R_i})\right]^{m_t} (r_o^{-2m_t} - r_i^{-2m_t}) \quad (14)$$

In 2016, based on the general strain-stress model which is proposed by XP Huang, Bouzid and Mourad[9] determined the expansion pressure of each expansion joint in the expansion process of the heat exchange tube by using Von mises yield criteria, Hencky deformation theory and Tresca yield criteria:

$$p_{i\min} = \frac{2A_{t1}}{\sqrt{3}} \ln K_t + (\frac{2}{\sqrt{3}})^{m_t+1} \frac{A_{t1}}{-2m_t} (c_0 r_o)^{m_t} (r_o^{-2m_t} - r_i^{-2m_t}) \quad (15)$$

$$p_{i\max} = \frac{2A_{s1}}{\sqrt{3}} \ln K_s - \frac{A_{s2}}{\sqrt{3}B_2} (\frac{\sigma_{ss}}{E_s})^{m_s} (1 - K_s^{2m_s}) + \frac{2A_{t1}}{\sqrt{3}} \ln K_t - \frac{A_{t2} r_o^{m_t}}{\sqrt{3}m_t} (\frac{2c}{\sqrt{3}} + \frac{\sigma_{ss} R_c^2}{E_s R_i})^{m_t} (r_o^{-2m_t} - r_i^{-2m_t}) \quad (16)$$

4. determination expansion pressure using bilinear model

Strain-hardening model which consider the strain hardening of the material and the effect of the initial clearance is close to the real performance of the expanding material. The complexity of the model ,the cumbersome calculation and the relevant parameters which are difficult to determine in the formula make the strain-hardening model hard to apply to engineering practice. Based on this situation, it is necessary to simplify the strain-hardening model.

In order to simplify the material stress-strain hardening curve, a bilinear simplify model is proposed by Hong ying et al.[10] in 2018. The bilinear simplified model consists of two straight lines with different slopes. The slope of the first line is the elastic modulus of the material.The slope of the second line reflects the strength of the stress and strain hardening ability of the material. The stress-strain model of the model material is characterized by bilinearity as:

$$\sigma_i = \begin{cases} E\varepsilon_i(\sigma_i \le \sigma_s) \\ A_1 + A_2\varepsilon_i(\sigma_i \ge \sigma_s) \end{cases} \quad (17)$$

After simplification, the expansion pressure of each expansion joint during the expansion process of the heat exchange tube is:

$$p_{i\min} = \frac{2}{\sqrt{3}} A_{t1} \ln K_t - \frac{2}{3} A_{t2} c_0 r_o (r_o^{-2} - r_i^{-2}) \quad (18)$$

$$p_{i\max} = \frac{2}{\sqrt{3}} A_{s1} \ln K_s - \frac{2}{3} A_{s2}[r_o(c_0 + \frac{\sqrt{3}\sigma_{ss} R_c^2}{2E_s R_i})](R_c^{-1} - R_i^{-2}) + \frac{2}{\sqrt{3}} A_{t1} \ln K_t - \frac{2}{3} A_{t2}[r_o(c_0 + \frac{\sqrt{3}\sigma_{ss} R_c^2}{2E_s R_i})](r_o^{-2} - r_i^{-2}) \quad (19)$$

5. Conclusions

Elastic-pefectly plastic model is simple, but the calculation results of the expansion pressure based on the elastic-pefectly plastic model has a large error. The strain hardening model is more in line with the real properties of materials, the calculation results of the expansion pressure has a high accuracy, but this model is complex and has a cumbersome calculation to determine the expansion pressure. The bilinear model had simplified the strain-hardening model. After simplifying the strain hardening model, the determination of the expansion pressure which is close to the determination based on the strain hardening model become simple. So, the determination of the expansion pressure based on the strain hardening model is convenient for practical application. The determination of the expansion pressure which is mentioned in the paper are based on the single tube model. In the practical production , it need to expand multi tubes at the same time. So, It's necessary to determine the expansion pressure with multi tubes further. In the hydraulic expansion, the problem how to clean and dry the heat exchanger tube need to be solved.

Acknowledgments
This work was financially supported by the National Natural Science Foundation of China (51564007), Natural Science Foundation of Guangxi Province (2017GXNSFAA198133), and Guangxi Key Laboratory of Manufacturing System & Advanced Manufacturing Technology (14-045-15-005Z).

References

[1] Laghzale N, Bouzid A. Theoretical analysis of hydraulically expanded tube-to-tubesheet joints with linear strain hardening material behavior. ASME J PressVessel Technol 2009;131(6):061202-8.

[2] Krips H, Podhorsky M. Hydraulic Expansion—A New Procedure for Fastening Tubes[J]. VGB Kraftswerkstech, 1976, 56(7): 456-464.

[3] CHEN Gang, LI Wei, WANGXiu- li. Study and application of hydraulic expansion tube for the joint of tubeplate[J]. Chlor-Alkali Industry, 2001,(03):40-45.

[4] Huigeng Yan, Bingsheng Zhang, Letong Ge and Peiling Li. Research on Hydraulically Expanding of Heat Exchangers (l) Determination of the Expanding Pressure[J].China Pressure Vessel Technology. 1996(2):126-130.

[5] Laghzale N and Bouzid A. Analytical Modelling of Hydraulically Expanded Tube-To-Tubesheet joint[J]. ASME J. Pressure Vessel Technol. 131,p. 011208

[6] Haifeng Wang, Zhifu Sang. Theory solution of Hydraulic expansion tube-to-tubesheet joint residual contact pressure for power-hardening material[J]. China petroleum machinery 2007(11):24-28.

[7] Tian X , Huang X P , Fu Z Y . Determination of Expanding Pressure of Hydraulically Expanded Tube-to-Tubesheet Joint[J]. Advanced Materials Research, 2010, 97-101:2898-2902.

[8] Huang X P. A General Autofrettage Model of a Thick-Walled Cylinder Based on Tensile-Compressive Stress-Strain Curve of a Material[J]. Journal of Strain Analysis for Engineering Design, 2005, 40(6):599-607.

[9] Bouzid A H , Mourad A H I , El Domiaty A . Influence of Bauschinger effect on the residual contact pressure of hydraulically expanded tube-to-tubesheet joints[J]. International Journal of Pressure Vessels and Piping, 2016:S0308016115300132.

[10] HONG Ying, WANG Xuesheng, CHEN Qinzhu, WANG Jianfu. Bilinear Simplification of Material Model in Theoretical Calculation of Hydraulic Expansion[J]. Machine Design and Research,2018,34(01):199-202.

Mechanical Properties of an *Eco-friendly Concrete* with partial replacement of POC and Rubber

A Espinoza[1,3], B Jiménez[1,4], J Rodríguez[2] and C Eyzaguirre[2]

1 Bachelor, Civil Engineering Program, Universidad Peruana de Ciencias Aplicadas, Lima, Perú
2 Full Professor, Civil Engineering Program, Universidad Peruana de Ciencias Aplicadas, Lima, Perú
3 Email: u201319891@upc.edu.pe
4 Email: u201214486@upc.edu.pe

Abstract. Concrete plants consume 10 billion tons of natural aggregates annually from quarries and gravel plants for produce concrete, this demand requires exploiting natural resources from mountains and rivers producing an ecological imbalance. One solution is to use Palm Oil Clinker (POC), which is eliminated in large quantities in the dumps and rivers without taking advantage of its puzolanic, binding and resistance properties as an aggregate in the concrete; another alternative is to apply rubber from abandoned and discarded tires as waste in landfills or burned, without taking advantage of its performance of improvement in concrete, increasing its resistance to impact and fatigue. Unable to find joint POC and rubber information, this research studies its influence replacing 2.5% rubber (grained and crushed) with 10%, 12.5% and 15% POC in the fine aggregate on traditional concrete; results indicate that with 12.5% of POC as the ideal percentage, the compressive strength, tensile strength and flexural strength rise between 2.16 - 9.54%, so the concrete obtained has a cost of less than 4.09% and has 3.65% less CO_2 emission.

1. Introduction

The construction industry is the largest consumer of natural resources on the planet; where the most commonly used building material is the concrete [1]. Deforestation of green hills to obtain limestones to satisfy the demand for coarse aggregates leads to an ecological imbalance [2]. Counteracting the fast depletion of natural aggregates requires partially replacing sand with recycled aggregates or by-products from industries. The oil palm industry produces several by product and waste, one of them is POC, which is obtained after the oil extraction process, after the pressed layer and fibers are burned in the incinerator [3]. POC is used as an aggregate in the production of CT [4]. [3] with an addition of 12.5% POC finds a greater compressive strength of 3.5%, compared to CT. Tensile strength was studied with different additions of 12.5%, 25%, 37.5% and 50% obtaining values greater than CT [5]. Rubber in concrete provides greater durability and elasticity when is replaced as a fine aggregate. [6] studies ground rubber in 10% replacement of the fine aggregate obtaining a compressive strength of 53.13 MPa at 60 days, which represents 4% more than CT. With this same percentage of replacement in fine aggregate according to [7], compressive strength at 28 days is 1% lower than CT. From a wide bibliographic review, we show that we have not found joint work on POC and rubber in the elaboration of the concrete, what exists are separate works of this waste. This is why the combination of POC and rubber (granulated and screening) is used as a replacement for the natural fine aggregate, using 10%, 12.5% and 15% POC and 3%, 5% and 7% rubber to evaluate compressive strength, tensile strength, flexural strength , the cost per m_3 of concrete and the amount of CO_2 removal.

2. Materials and Method

2.1 Materials

The POC used has a fineness module of 3.79, the rubbers used are screening (0.15-19 mm) and granules (0.75-4.75 mm). The cement used is Portland type I, the natural aggregates are coarse sand and ¾" stone. The water was drinkable and a plasticizer additive was used with a dosage of 250 ml per cement bag.

2.2 Méthod

The mixing design was made according to the [8] for a concrete of f'_c 210 kg/cm_2 (called Traditional Concrete, CT). Table 1 shows the doses performed and table 2 shows the tests and number of specimens used. The specimens were cured in the laboratory up to 28 days according to [9], then demolded and tested.

Table 1. Dosage of concrete mixtures for a f'_c 210 kg/cm_2

	Code	POC (kg/m3)	Rubber (kg/m3)	Fine agregate (kg/m3)	Water (lt/m3)	SP (lt./m3)	Cement Port. Tipo I (kg/m3)	Coarse aggregate (kg/m3)	Air (%)
CT	0%	-	-	799.78	213.56	2.16	366.73	972.75	2
P-1	10%	79.98	-	719.80					
P-2	12.50%	99.97	-	619.83					
P-3	15%	119.97	-	499.86					
C-1	3%	-	2.40	797.38					
C-2	5%	-	39.99	759.79					
C-3	7%	-	55.98	743.80					
PC-1	10% - 2.5%	79.98	19.99	699.81					
PC.2	12.5% - 2.5%	99.97	19.99	679.81					
PC-3	15%-- 2.5%	119.97	19.99	659.82					

Table 2. Table of tests and specifications

Estate	Test type	Specimens number	Time/ Age	Method
Hardened Concrete	Compressive Strength	30	28 d	[10], [13]
	Tensile Strength	30	28 d	[11], [14]
	Flexural Strength	20	28 d	[12], [15]

3. Results y Analysis

3.1 Compressive Strength

Figure 3 shows the incidence of the percentage of PC on compressive strength. The results indicate that the resistance increases when the PC percentage is increased, with the maximum value being 356.75 Kg/cm_2 for PC2 blending with 12.5% and a variation on CT of 9.54%; from this value the resistance decreases. [3] replaces 12.5% POC and gets the resistance of 80 MPa at 28 days, being 3% more than CT. [16] replaces 10% of PC and obtains a resistance of 55 MPa at 28 days, which represents 17% more than CT. This increase is related to the expansión of the POC as a fine aggregate since it allows the the concrete to achieve máximum compaction by filling its pores, as well as becoming active as a pozzolanic material and improving compressive strength. On the other hand, [17] performs the replacement of 5% rubber as a fine aggregate obtaining a resistance of 27 Mpa at 28 days, which is 20% less than CT. [18] with a 5% replacement of granulated and screening rubber, obtains a resistance of 48.8 MPa at 28 days, which is equal to CT. Also, [19] replaces 10% rubber as a fine aggregate and achieves a resistance of 22.4 MPa at 28 days, which is 7.91% less than CT. This behavior is explained by the SEM (Scanning Electron Microscope) analysis that characterizes the morphology and porosity of the interface between the rubber particle and the cement matrix, observing the presence of a particle space and the matrix of cement and the rubber content leading to the reduction of resistance [19].

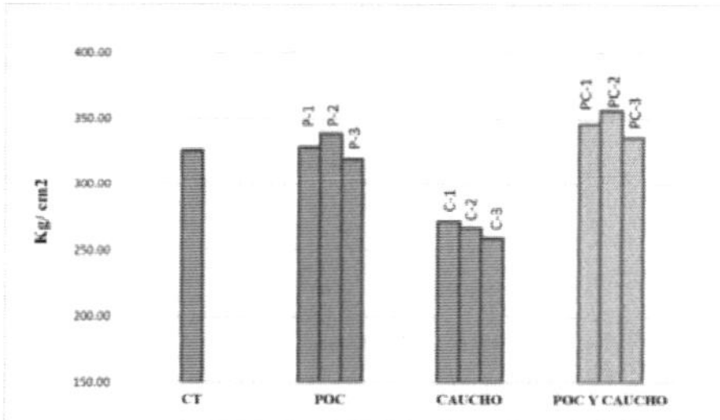

Figure 1. PC Compressive strength respect to CT

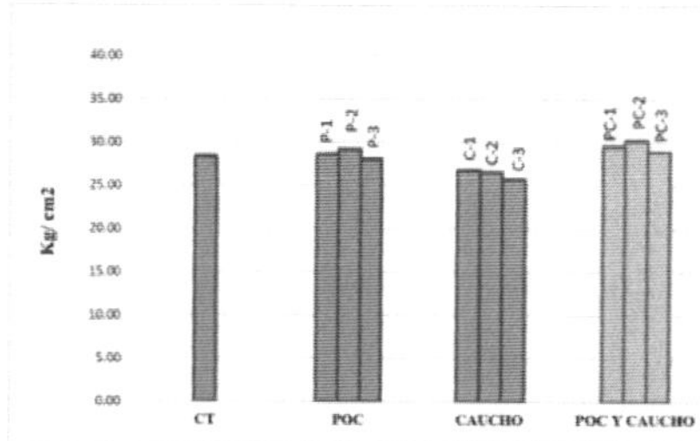

Figure 2. PC Tensile strength respect to CT

3.2 *Tensile Strength*

Figure 4 shows the influence of the PC percentage on tensile strength. The results show that the resistance is increased with the increase in the percentage of PC, reaching the maximum value of 30.28 Kg/cm2 for PC2 mixture with 12.5% and a variation with respect to CT of 6.6%; from this value the resistance decreases. [7] replaces 50% POC and obtains a resistance of 2.17 MPa at 28 days, reaching a resistance of 23% less than CT. This behavior attributes it to the presence of porous characteristics in the matrix produce microcracks in time that decrease resistance with increased POC, reason also reported by [4]. [5] replaces 10% granulated rubber and achieves a resistance of 3.6 MPa at 28 days, which represents 34.60% less than CT. So is [17] with a 5% rubber replacement achieves a resistance of 2.5 MPa at 28 days, which represents 17% less than CT. This decrease is related to the low density and softness of rubber particles, which form a non-uniform mixture that leads to lower resistance of the rubber parts of the concrete; this reduction in strength increases with the increase in the percentage of rubber. [17].

3.3 *Flexural Strength*

Figure 5 shows the influence of the PC percentage on flexural strength. It sees that the strength increases when the PC percentage is increased, being the maximum value of 71.09 Kg/cm2 for PC2 mixture with 12.5% and a variation from CT of 5.21%; from this value the resistance decreases. [3] replaces 12.5% POC for a resistance of 13.5 MPa at 28 days, reaching a resistance of 4% more than CT. This increase is related to the existence of a good interlock rate between the POC and the cement matrix. That is, the irregular shape of the POC with its scatable structure increases the interaction or joining of the interface, producing a higher load support capacity per bending [3]. [5] with a replacement of 10% granulated rubber obtains a resistance of 2.5 MPa at 28 days, which represents 20% less than CT; this behaviour is due to the performance of small rubber particles in the isolation of the aggregates with each other and also to the binder paste that leads to weak adhesion between the particles in the mixture [5]. There are other studies [11] that with a replacement of 5% rubber as a fine aggregate obtains a resistance of 5.97 MPa at 28 days, which is 25% more than CT.

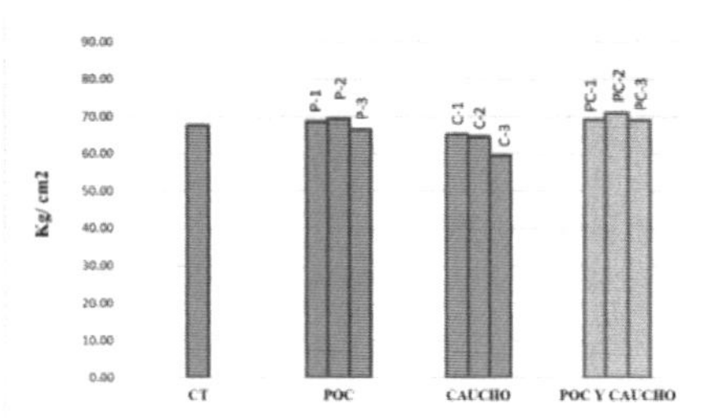

Figure 3. PC flexural strength respect to CT

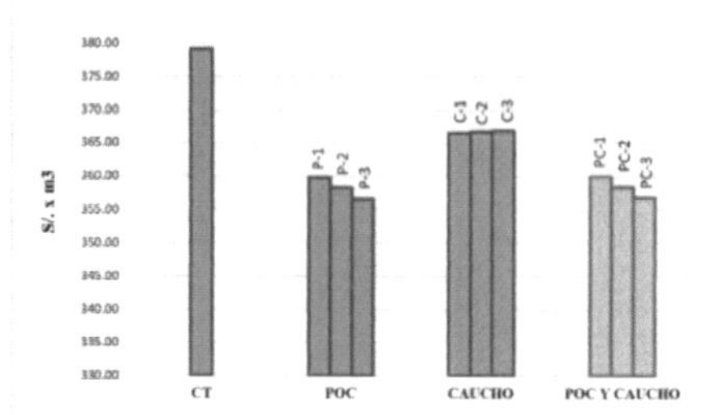

Figure 4. PC cost respect to CT

3.4 Cost and Sustainability

3.4.1 Cost evaluation

Figure 6 shows the influence of the percentage of PC related to the specific cost/m3. It sees that the cost decreases with the increase in the percentage of PC, reaching the value of S/. 356.76 for 15% PC and representing a 3.9% decrease over CT. [3] develops concrete with 12.5% POC replacement, resulting in a 3% decrease from CT. This is because the sand being replaced in a certain percentage by a much lower or non-existent price material, the cost of mixing decreases relative to conventional aggregates. [3]

3.4.2 CO_2 emission

Figure 7 shows the influence of the percentage of PCs related to CO2 emission. The results indicate that CO_2 emission decreases by increasing the percentage of PCs, obtaining the value of 547.95 for 15% of PC and representing a decrease of 3.8% compared to CT. [3] with a 12.5% POC replacement reduces concrete CO_2 emissions by 4% over CT. From an environmental point of view, the use of waste materials generally reduce the carbon footprint, in the case of POC its processing stages do not involve the production of significant carbon emissions. [3]. [19] with a 5% rubber replacement reduces concrete CO_2 emissions by 1% over CT. The same author, with a replacement of 5% rubber as a fine aggregate gets 6,626 kg of concrete CO_2/m3, which is 0.73% higher than CT. This CO_2 footprint increase behavior is due to the need for increased energy to crush rubber compared to natural aggregate sieving [19].

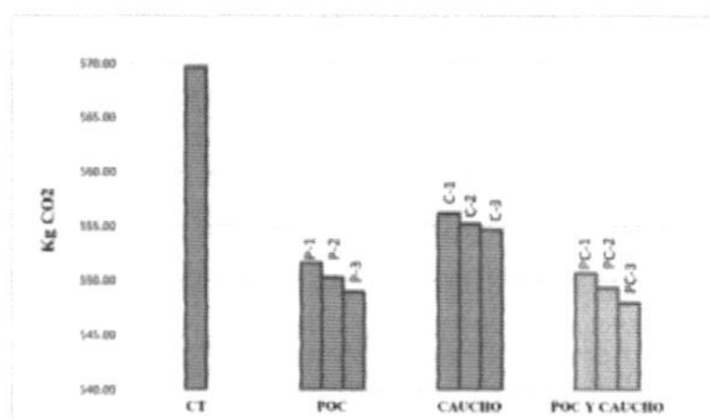

Figure 5. CO_2 emission respect to CT

4. Conclusion

- The main contribution of using POC + rubber in concrete is to reduce the CO_2/m3 emission of concrete by 4% and obtain an improved concrete by 6% in its strength.
- The mix design that shows a better mechanical performance compared to the CT is achieved with 2.5% rubber and 12.5% POC as a replacement for fine aggregate in concrete.
- The use of POC + rubber represents a lower cost per m3 of concrete compared to a CT, being an interesting alternative for the recycling and reuse of these highly polluting waste and a great opportunity for use in sustainable housing construction.

5. References

[1] S.M. Alamgir Kabir, U. Johnson Alengaram, Mohd Zamin Jumaat, Sumiani Yusoff, Afia. Sharmin, Iftekhair Ibnul Bashar 2017 J. Cleaner Production 161 477-492

[2] H. Mahmud, M.Z. Jumaat, U.J. Alengaram 2009 J. Applied Sciences 9 1764-1769

[3] Kanadasan J, Razak HA, Subramaniam V 2017 J. Cleaner Production 170 1244-1259

[4] Karakoç, M.B. Türkmen, I. Maras, M.M. Kantarci, F. Demirbogia, 2016. Ceram. Int. 42 1254-1260.

[5] Aly Muhammed Aly 2019 Construction and Building Materials 207 136-144

[6] D. Flores-Medina, N.F. Medina, F. Hernández 2014 Mater. Struct 47 (7) 1185-1193

[7] Rasel Ahmmad, U. Johnson Alengaram 2017 Const. and Building Materials 135 94–103

[8] ACI Committee 211.1-91 *Standard Practice for Selecting Proportions for Normal, Heavyweight, and Mass Concrete.* (Reapproved 2009).

[9] NTP 339.033 CONCRETO. *Práctica normalizada para la elaboración y curado de especímenes de concreto en campo*. 2015.

[10] NTP 330.034 CONCRETO. *Método de ensayo normalizado parala determinación de la resistencia a la compresión del concreto en muestras cilíndricas*. 2015.

[11] NTP 339.084 CONCRETO. *Método de ensayo normalizado para la determinación de la resistencia a la tracción simple del concreto, por compresión diametral de una probeta cilindrica.* 2012.

[12] NTP 339.079 CONCRETO. *Método de ensayo para determinar la resistencia a la flexión del concreto en vigas simplemente apoyadas con cargas en el centro del tramo*. 2012.

[13] ASTM Designation: C39/ C39M *Standard Test Method for Compressive Strenght of Cylindrical Concrete Specimens. 2018.*

[14] ASTM Designation: C496/ C496M *Standard Test Method for Splitting Tensile Strenght of Cylindrical Concrete Specimens. 2017.*

[15] ASTM Designation: C78/ C78M *Standard Test Method for Flexural Strenght of Concrete. 2018.*

[16] R. Ahmmad , U. Alengaram, M. Jumaat, N.H. Sulong, M. O. Yusuf, M. Rehman 2017 Construction and Building Materials 135 94-103

[17] F. Jokar, M. Khorram 2019 Construction and Building Materials 208 651–658

[18] C.W. Chan, T. Yu, S.S. Zhang 2019 Const. and Building Materials 211 416–426

[19] K. Rashid, A. Yazdanbakhsh, M. Rehman 2019 J. Cleaner Production 224 396-410

[20] ASTM Designation: C1064/ C1064M *Standard Test Method for Temperature of Freshly Mixed Portland Cement Concrete. 2001.*

[21] ASTM Designation: C 143 *Standard Test Method for Slump of Portland Cement Concrete. 1978.*

[22] ASTM Designation: C 231/ C231M *Standard Test Method for Air Content of Freshly Mixed Concrete by the Pressure Method. 2017.*

[23] BS EN 12350-1 *Sampling and common apparatus.* Testing fresh concrete. *2019*

[24] BS EN 12350-2 *Slump-test.* Testing fresh concrete. 2009

[25] BS EN 12350-7 *Air content. Pressure methods.* Testing fresh concrete. 2009

IOP Conf. Series: Materials Science and Engineering **758** (2020) 012012 doi:10.1088/1757-899X/758/1/012012

Research on Attenuating Mechanism of Wave-eliminating Interceptor on Explosion shock wave in Tunnel Engineering

Fan Chengfei, Yang Kezhi, Mao Wei, Huang Jinghua, Liu Xinwei
Research Institute for National Defense Engineering, Beijing 100850, China
E-mail: tjufcf@163.com

Abstract: The research on the explosion shock wave propagation and attenuation technology in tunnel engineering had been a hot issue at home and abroad. The installation of a series of wave-eliminating interceptor in tunnel engineering was beneficial to attenuate explosion shock wave in tunnel engineering. It was rapidly achieved to attenuate the explosion shock wave. In this paper, based on the clarification of the propagation law explosion shock wave in air, explosion load in tunnel engineering was analysed. And the waveform characteristics and propagation law of explosion shock wave were revealed. Then the rectification mechanism was got. The interaction between the explosion shock wave and the interceptor was analysed. Then the reflection mechanism and the disturbing mechanism were obtained. With the numerical simulation method, the propagation law of explosion shock wave under different angles of the interceptor was studied. The attenuation law of explosion shock wave was revealed. The optimal angle of interceptor was achieved.

1 Introduction

With the improvement of precision and power of precision guided weapons, more powerful weapons would be exploded in the mouth of tunnel engineering, and the high-strength explosion shock waves would be entered the interior of tunnel engineering. It was a great threat to the facilities and personnel in the engineering. Because of the constraint of the wall surface, the attenuation and duration of the explosion shock wave were slower and longer. And the killing and damaging effect of the explosion shock wave in tunnel engineering is much larger than the free space. In order to quickly attenuate the explosion shock wave and mitigate the explosion damage effect, propagation and attenuation technology of the explosion shock wave were studied by more scholars at home and abroad.

Extending the tunnel length, excavating the T-shaped tunnel, increasing the foam aluminium buffer layer, arranging the water curtain or water spray and other explosion shock wave attenuation measures in tunnel engineering were propose, so that to attenuating the explosion shock wave. However, disadvantages and shortcomings were existed, such as high cost, high requirements for mountain conditions, unresponsiveness. In order to improve the attenuation efficiency of explosion shock wave in tunnel engineering, a new type of interceptor was developed. It was installed on the both walls of the tunnel engineering. And the interceptor was made of light high-strength material and processed into a standard component. It was usually put up and attached to the wall surface, and the normal use of the tunnel engineering could be not affected. If needed, the interceptor would be opened, and a series of prominent flats would be formed in the tunnel engineering. The explosion shock wave flow field would be reflected, disturbed and obstructed by interceptors, the explosion shock wave power would be consumed, the propagation of the explosion shock wave would be affected.

2 Explosion shock wave load in free space

When explosives exploded, huge amounts of energy in a very short time and space would be released. The substances in the explosives would be converted into high-temperature and high-pressure gases. It would be accumulated, and the surrounding air would be compressed, then explosion shock waves

About the author: Fan Chengfei (1986-), male, engineer, Ph.D., main research direction: blasting and protection, E-mail: tjufcf@163.com.
Address: No. 24, Taiping Road, Haidian District, Beijing;
Zip code: 100850; Contact: 13716171007

would be generated. The pressure, density, temperature and speed of air particles would be sharply increased. The typical explosion shock wave pressure-time curve in free space was shown in Figure 1.

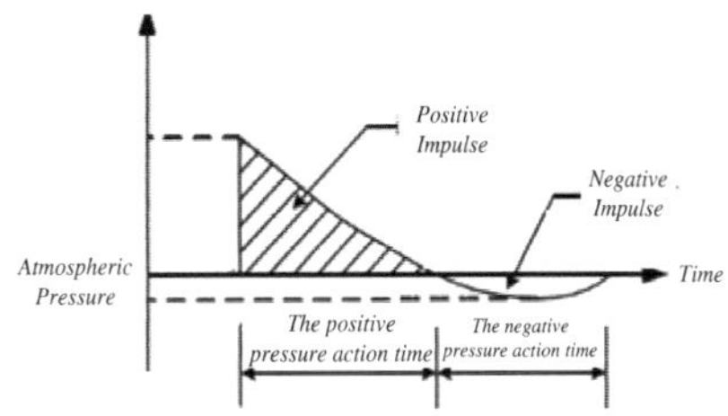

Fig.1 The typical explosion shock wave pressure-time curve

When the explosion shock wave front was reached the peak value P_{so} at time t_A, the incident wave was attenuated to the ambient atmospheric in positive action time t_0. When the reverse flow was happened, the negative pressure peak was P_{so-}, the negative action time was t_{0-}. The damage of the explosion shock wave in free space could be measured by three parameters: (1)The pressure peak of the explosion shock wave P_{so}. (2) The positive pressure action time t_0. (3) The specific impulse i_s, it could be the area enclosed by pressure-time curve in the positive pressure section.

The empirical calculation formulas were obtained by lots of scholars such as Brode, M.A.Sadovskyi, Josef Henrgeh, Baker, Chengqing&Hong Hao. In order to compare the differences of different empirical calculation formulas, the calculated results were placed in the same coordinate system, and the comparison results were shown in the Figure 2. It could be known that the deviation of the results from different formula was decreased by increasing of proportional distance R. The value of the Mills formula was higher than other formulas, and the value of the Henrych formula was lower. When the proportional distance R was greater than $1m/kg^{1/3}$, the predicted results from different formula were relatively close.

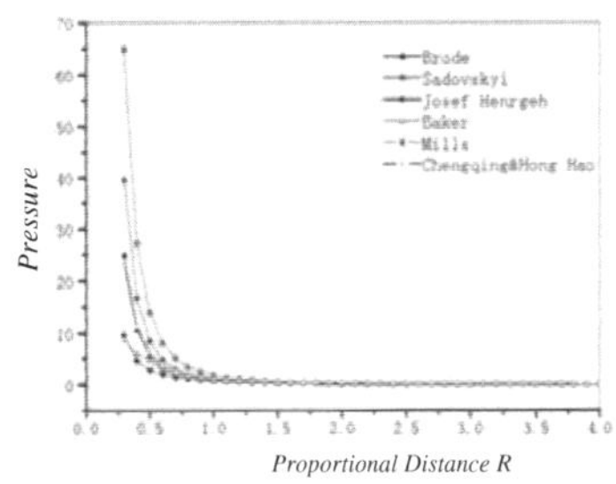

Fig.2 the relationship curve between pressure peak and proportional distance

3 Attenuating mechanism

3. 1 Integration mechanism of the tunnel engineering on explosion shock wave

The explosion effect in tunnel engineering was very different from in the free space. In the infinite space, the explosion shock wave was expanded outward in a spherical shape. The area of the wave front was added by the increasing of propagation distance. Therefore, the energy distributed per unit area of the wave front must be reduced. However, the attenuation of explosion shock wave in the tunnel engineering were mainly due to the continuous widening of the positive pressure region of the explosion shock wave with the wave propagation, and the irreversible energy loss due to the adiabatic compression of the air during its propagation.

When the explosives explode in the tunnel engineering, the pressure, impulse and heat generated by the explosion could not be spread out in time, because of the limitation of the wall surface of the tunnel engineering. The pressure action time became longer by the multiple reflection of the explosion shock wave on the tunnel wall. The personnel and equipment in the tunnel would be damaged seriously. When the same dose and the same blasting distance, the peak value of the explosion shock wave pressure, the action time and the impulse ratio in the tunnel were significantly increased, and the peak attenuation was much slower, so the damage effect was significantly enhanced.

IOP Conf. Series: Materials Science and Engineering **758** (2020) 012012 doi:10.1088/1757-899X/758/1/012012

In the blasting center, due to the combined action of detonation products and explosion shock waves and the reflection and superposition of explosion shock waves on various surfaces in the tunnel, the distribution of the explosion shock wave flow field was complicated, the waveform was relatively disordered, and the peak value of pressure at different points on the same section of the tunnel was different. Through multiple reflections on the wall of the tunnel and the upper and lower sides, a relatively regular plane explosion shock wave waveform was generated in the distance of the blasting center (generally, when the burst distance L ≥ 6D, the blast was in the distance, and D was the equivalent diameter of the tunnel engineering). At this time, the difference about the peak value of the explosion shock wave pressure and the arrival time were gradually reduced, and the explosion shock wave gradually propagated in the form of stable fluctuation. Its pressure peak attenuation was extremely slow, and its waveform could be approximated as a regular plane explosion shock wave waveform.

3. 2 Reflection mechanism of the interceptor on the explosion shock wave

When the explosion shock wave was propagated in the tunnel engineering, the air particle behind the explosion shock wave front was moved with the wave front at a high speed. The explosion shock wave propagating toward the interceptor or the protective door installed in the tunnel engineering was the incident explosion shock wave. When the incident explosion shock wave meets the interceptor or the protective door, the movement of the air particle was blocked, and the air near the interceptor or the protective door were squeezed to increasing the pressure. The pressure and the particle velocity were changed into reverse. That was the reflected explosion shock wave.

The comparison curves of reflected explosion shock wave and incident explosion shock wave were shown in Figure 3. It could be known that the peak pressure of the reflected explosion shock wave was much larger than the peak pressure of incident explosion shock wave. And the negative pressure peak of the reflected explosion shock wave was also correspondingly larger than the incident pressure in the negative pressure section.

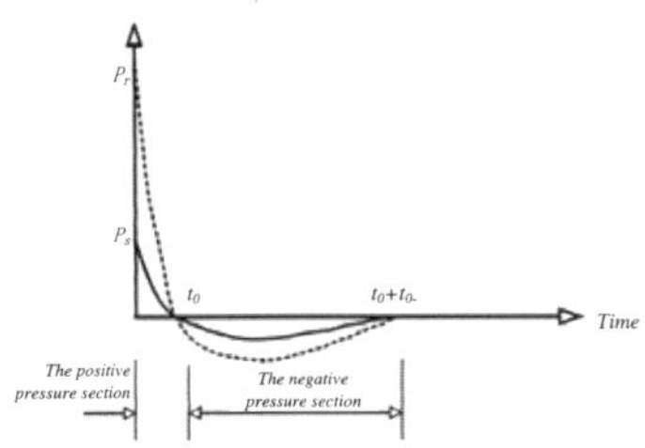

Fig.3 Comparison curves of reflected explosion shock wave and incident explosion shock wave

For the regular reflection, the ideal gas equation could be solved jointly to obtain pressure calculation formula of the reflection explosion shock wave:

$$P_r = 2P_s + \frac{6P_S^2}{P_S + 7P_0} \tag{1}$$

In which, the reflection pressure is P_r, the incident pressure is P_s, and the atmospheric pressure of the air is P_0. It could be known that the reflection pressure was nearly twice to the incident pressure when P_r was smaller; the reflection pressure was nearly eight times to the incident pressure when P_r was lager. Even in the actual explosion process, the gas was subjected to the high temperature and high pressure of the explosion shock wave, and its state equation was no longer an ideal gas state equation, and the reflected pressure could reach several times of the incident pressure.

For regular oblique reflection, pressure calculation formula of the reflection explosion shock wave was that:

$$P_r = \left(1+\cos\theta\right)P_s + \frac{6P_S^2}{P_S + 7P_0}\cos^2\theta \tag{2}$$

In which, the incidence angle isθ.

3. 3 Disturbing mechanism of the interceptor on the explosion shock wave

When the explosion shock wave in the tunnel engineering was acted on the interceptor, the disturbing effect would be happened. It was shown in Figure 4.

When the incident explosion shock wave was encountered by the interceptor, oblique reflection would be occurred on the front surface of the interceptor, and the pressure of the explosion shock wave would be increased, a high-pressure region reflecting in the front surface of the interceptor would be formed.

The unobstructed incident wave was propagated forward, the pressure was not changed. And it became a low-pressure zone comparing to the front surface of the interceptor.

And a pressure difference was formed, by which the generation of air flow and sparse waves were caused. The phenomenon that the explosion wave current flows around the interceptor was the disturbing phenomenon. When the explosion shock wave was passed through the upper surface and the rear surface, an air vortex was generated and suction was generated, which would affect the pressure distribution of the upper surface and the rear surface.

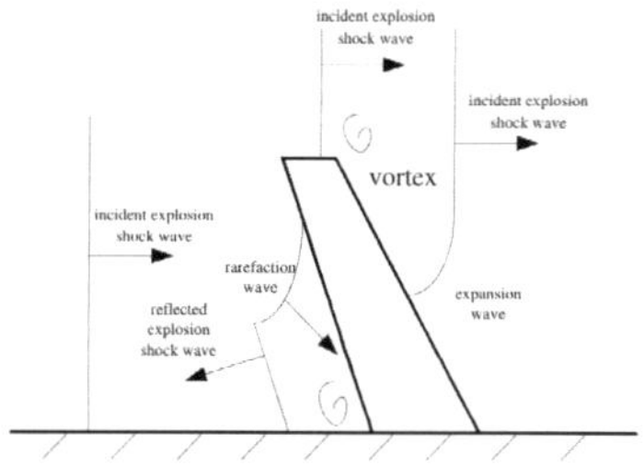

Figure 4 Disturbing effect of the interceptor on the explosion shock wave

4 Numerical simulation of attenuating explosion shock wave

The numerical simulation method was used to analyse the attenuation law of the explosion shock wave with the same explosive load in different working conditions. The peak pressure and impulse of the explosion shock wave were compared. The inclination angle θ between the interceptor and the direction of the explosion shock wave was shown in Figure 5. The distribution of the explosion shock wave flow field in the tunnel engineering was shown in Figure 6. The peak pressure and impulse of the explosion shock wave were shown in Figure 7 and Figure 8 at the different inclination angles.

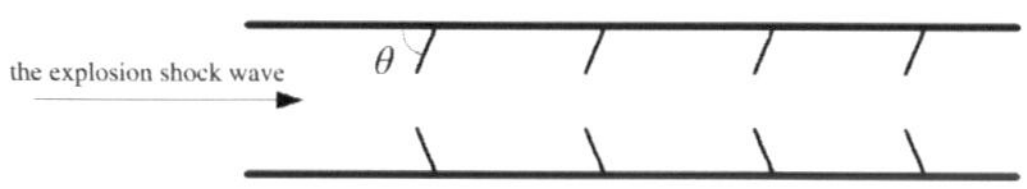

Fig. 5 The explosion shock wave propagated in the tunnel engineering

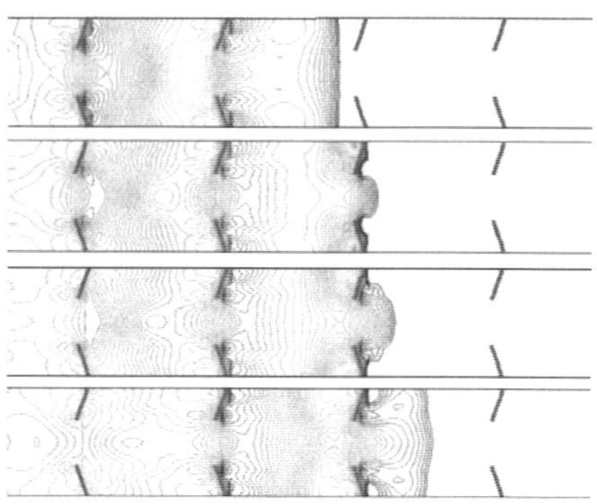

Fig. 6 The distribution of the explosion shock wave flow field

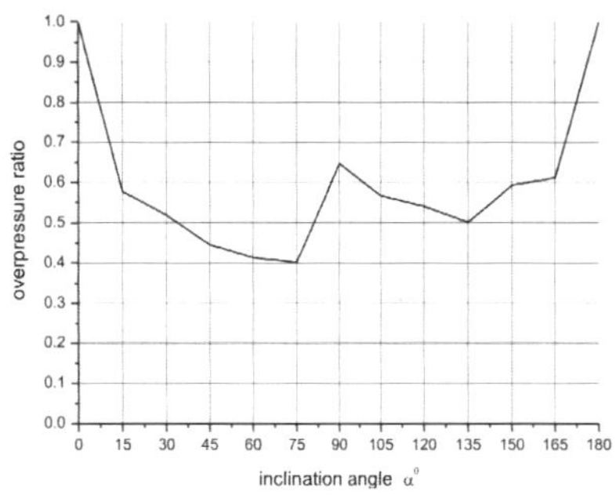

Fig. 7 The curve between inclination angle and pressure ratio

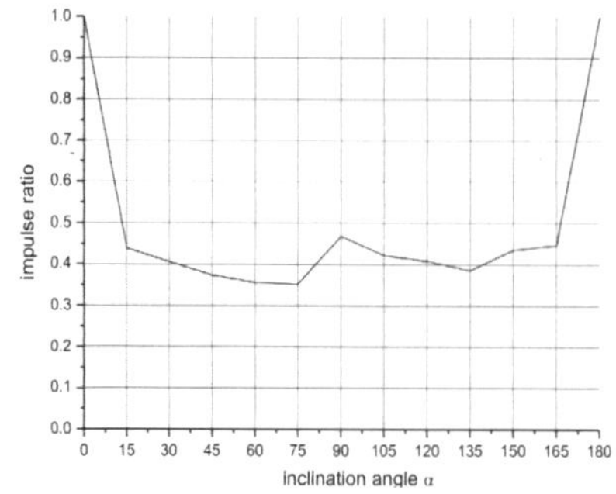

Fig. 8The curve between inclination angle and impulse ratio

The schematic diagram of the numerical simulation results of the interceptor attenuating the explosion shock wave were shown in Figure 7 and Figure 8. It was shown that the explosion shock wave flow field in the tunnel engineering would be interfered by the interceptor, the propagation of the explosion shock wave would be delayed, and the pressure peak and impulse of the explosion shock wave would be effectively attenuated. According to the curves in Figure 7 and Figure 8, the optimal inclination angle of the interceptor attenuating the explosion shock wave was about 75 degrees.

5 Conclusion

Based on the clarification of the propagation law explosion shock wave in air, explosion load in tunnel engineering was analysed. The integration mechanism, reflection mechanism and disturbing mechanism were revealed. A complete mechanism of attenuating explosion shock wave by the interceptor in the tunnel engineering was obtained. With the numerical simulation method, the propagation law of explosion shock wave under different angles of the interceptor was studied. The attenuation law of explosion shock wave was revealed. The optimal angle of interceptor was achieved. There were lots of advantages, such as high wave elimination efficiency, simple structure, low cost and convenient installation. It would be significant economic benefits and wide application value.

Acknowledgements

This work was financially supported by National Natural Science Foundation of China(No.51608524).

References

[1] Li Xiudi. Numerical simulation for blast propagation and attenuation inside T-shaped tunnel from HE-charges detonation[J]. Journal of Logistical Engineering university, 2011, 27(4): 8-13. (in Chinese)
[2] Qu Kangkang, Yan Yunju, Liu Yang. Comparative study on the propagation rules of explosive shock-wave in long straight and complex tunnel [J]. Chinese Journal of Applied Mechanics, 2011, 28(4): 434-440. (in Chinese)
[3] Zhang Yao, Nian Xinzhe. Mitigation effects of explosion-proof water walls and explosion-proof concrete walls on blase shock wave[J]. Journal of Vibration and Shock, 2014, 33(18): 214-221. (in Chinese)
[4] Yang Kezhi, Zhou Bukui, Yang Haijie. Water curtain speedily attenuation blast wave in tunnel[J]. Protective Engineering, 2016, 38(2): 29-35. (in Chinese)
[5] Yan Qiushi. Flow field distribution and dynamic response analysis of typical subway structure subjected to internal explosion [D]. Beijing: Tsinghua University, 2011. (in Chinese)
[6] Zhang Yadong, A Shuailei, Zou Bin. Shock wave loads on the blast door in straight tunnel [J]. Explosion and Shock Waves, 2017, 37(6): 1057-1065. (in Chinese)
[7] Zhen Zhengang, Li Xiudi, Miao Chaoyang. Propagation of blast wave of thermobaric explosive inside a tunnel [J]. Journal of Vibration and Shock, 2017, 36(5): 23-30. (in Chinese)
[8] Yu Wenhua, Zhang Yadong. Study on propagation laws of explosion shock wave in tunnels [J]. Explosive Materials, 2013, 42(3): 1-4. (in Chinese)
[9] Xu Lina, Yan Shunning, Wang Fengdan. Study of blast wave overpressure propagation inside straight tunnel [J]. Journal of Test and Measurement Technology, 2014, 28(2): 114-119. (in Chinese)
[10] Wu Shiyong, Zhang Wei, Lu Fangyun. Study on the propagation laws of blast in tunnels [J]. Chinese Journal of Underground Space and Engineering, 2010, 6(2): 396-401. (in Chinese)

IOP Conf. Series: Materials Science and Engineering **758** (2020) 012013 doi:10.1088/1757-899X/758/1/012013

Analysis of Insulation Structure Design for High Altitude UHV Reactor Bushing

Zhang Shiling

1 State Grid Chongqing Electric Power Company Chongqing Electric Power Research Institute, Chongqing, 401123
E-mail: 526793305@qq.com

Abstract. The ultra high voltage reactor bushing is the important equipment to realize the electrical connection between the reactor and the valve tower. The high altitude application environment puts forward higher requirement for electrical structure design of reactor bushing. The paper mainly discusses the design of the internal and external insulation structure. The design and analysis of the tail of the bushing, the corona ring and the inner insulation of bushing have been done. The three dimensional electric field simulation model has been set up to calculate and analyze the insulation performance of the bushing. The results show that the elevation correction of the external ring and the flash-over voltage of the porcelain under high altitude should be done, and the calculation shows that the correction coefficient are 2.07 and 1.36 respectively, and the end of bushing should install with double type ring. The insulation distance of the hollow composite insulator is designed as 8500mm. The insulation distance of the bushing tail is designed as 1955mm. The maximum field strength of the corona ring is 1.251kV·mm^{-1}, and maximum field strength on the surface of tail corona ring is 16.96kV·mm^{-1}. The axial field strength of the bushing core near center conductor is higher than the flange side, and the maximum field strength is 0.55kV·mm^{-1}. The results of the paper can provide the theoretical basis for the design of the insulation of the reactor transformer bushing used in the UHV transmission project.

1. Introduction

The structure of the high-voltage capacitive bushing is complex, involving many insulating media and withstanding high voltage in narrow and long insulating space. Therefore, it is difficult to design the internal and external insulation structure of high-voltage capacitive bushing. For reactor bushing, the structure of the outer ceramic sleeve with the oil-immersed paper capacitor core is generally adopted at present. The insulation coordination between outer insulation of ceramic sleeve and inner insulation of core is the key link in the design[1,2]. On the other hand, the research object of this paper is high altitude and UHV reactor bushing. High altitude and UHV indicate that the operation environment of this bushing is special[3,4]. In high altitude condition, the structural parameters of outer insulation of bushing need to be revised to ensure that no flash-over accident occurs. In UHV condition, the inner insulation of bushing needs to be designed compactly to ensure that no partial discharge and internal breakdown occurs in the inner insulation of core under UHV application environment.

In view of this, the electrical structure parameters of UHV reactor bushing are firstly analyzed, and the elevation correction factor K is introduced to modify the ceramic bushing structure of reactor bushing. At the same time, the top equalizing cover structure of reactor bushing is designed according to Peek formula. Considering that the overall immersion of bushing tail structure in transformer oil is less affected by elevation, and its electrical structure parameters are designed according to the traditional

calculation method[5-7]. Based on the reasonable design of the inner and outer insulation structure of the reactor bushing, the internal insulation structure is optimized by using the equal margin method, and the reactor bushing pressure equalizing ring with reasonable structure is configured. The three-dimensional electric field simulation model of high altitude reactor bushing is established, and the electric field distribution at the key position of bushing is checked and calculated. The insulating performance of outer ceramic sleeve of high altitude reactor bushing under lightning impulse voltage is discussed, and the corona condition of the voltage equalizing cover and ceramic sleeve at the end of bushing under actual operation conditions is also discussed.
This paper focuses on the design method of internal and external insulation structure of the reactor bushing under high altitude and UHV conditions, discusses the design scheme of altitude correction coefficient and internal insulation margin under UHV conditions, and uses three-dimensional finite element method to check and calculate. The research results and data in this paper can provide reference for the design of internal and external insulation structure of reactor bushing under high altitude and UHV conditions, and also provide theoretical and practical basis for the operation and maintenance of oil-immersed paper bushing.

2. Electrical structure parameters of UHV reactor bushing

2.1. Design of dry arc distance for outer porcelain sheath of reactor bushing
The outer insulation level of reactor bushing in the high altitude substation is 960kV, the peak value of lightning impulse test voltage is 2100kV, the peak value of operating impulse test voltage is 1550kV, and the main arc distance of bushing is more than 7400mm. The structural parameters of the outer contour of the UHV reactor bushing are shown in Fig. 1.

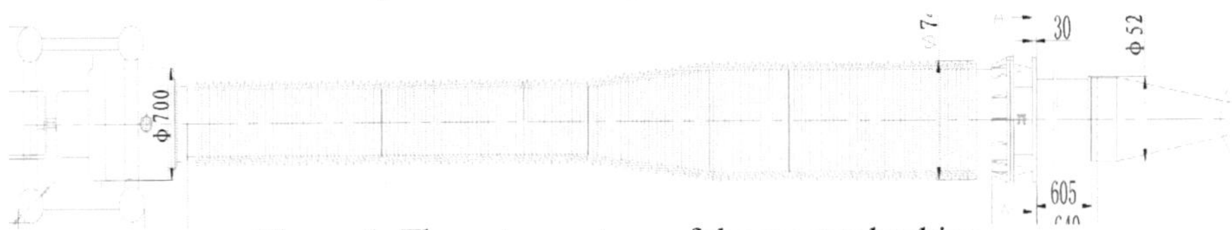

Figure.1 The outer contour of the reactor bushing

According to the requirements of GB/T4109 "insulating bushing with AC voltage higher than 1000V", altitude correction should be carried out for the external insulation level above 1000m. The altitude correction factor K is calculated according to equation (1)[8-10]:

$$K = e^{q\frac{(H-1000)}{8150}} \qquad (1)$$

For the lightning impulse voltage, withstand voltage coefficient q=1, the short-term power frequency withstand voltage q = 1 for air gap and bushing hollow composite insulator. The elevation of high-altitude substation is H = 3500m, and K = 1.36 is calculated in the entrance formula (1). The valid value of power frequency test voltage is 1305kV, the peak value of lightning impulse test voltage is 2855kV, and the peak value of operation impulse test voltage is 1950kV.
Under the power frequency and impulse voltage, the electrical performance of bushing requires that flash-over discharges can not occur on the outer surface of hollow ceramic sleeve in air or on the outer surface of the bushing core in oil. The key to the design of external insulation of the bushing is the coordination of the internal and external insulation of bushing and the regulation of main insulation of core. The dry flash-over voltage of hollow composite insulator is close to the air gap breakdown voltage between the upper flange and the middle flange, which mainly depends on the dry flash-over distance L_d. The relationship between the flash-over voltage and the dry flash-over distance is shown in Fig.2. The above figure shows that the flash-over voltage of the bushing hollow composite insulator increases with the increase of insulation distance, and there is the saturation effect. Existing formula (2) quantitatively fits the curve of Figure 2[11-13], and the fitting results are shown in Table 1.

$$U_{f50} = AL_d^B \qquad (2)$$

In the formula: U_{f50} is the flash-over voltage value (discharge probability 50%)/kV; L_d is insulator dry flash-over distance/m; A and B are undetermined coefficients.

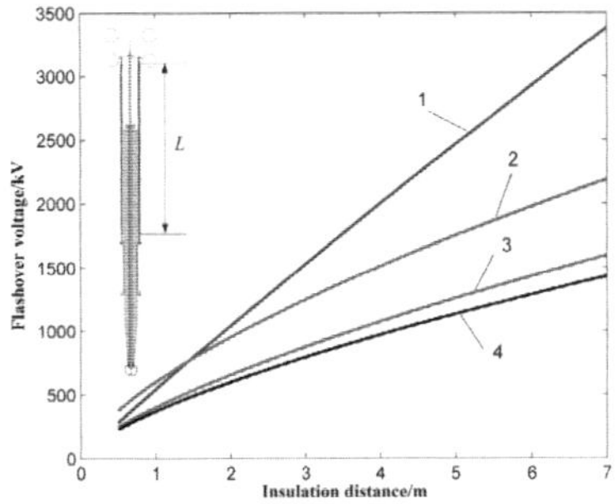

Figure.2 The relationship between flash-over voltage and insulation distance of hollow composite insulator

Table.1 Quantitative relationship between flash-over voltage and insulation distance of composite insulator

Test Voltage Type	*A*	*B*
Lightning Full Wave Shock Dry/Wet Flash Voltage	545.5	0.9377
Positive Operating Dry/Wet Flash Voltage	603.4	0.6624
Power Frequency Dry Flash-over Voltage (RMS)	404.6	0.7053
Power Frequency Wet Flash-over Voltage (RMS)	372.3	0.6921

Combined formula (1) and formula (2) are used to calculate the dry arc distance of outer insulation of bushing and porcelain sleeve (3):

$$L_d = (Te^{q\frac{H-1000}{8150}})^{\frac{1}{B}} \qquad (3)$$

Coefficient among them $T = (MU_{design})/A$, M is the insulation design margin, U_{design} is the voltage value under the different voltage types. Formula (3) shows that the altitude, test voltage and insulation design margin are positively correlated with the distance of the bushing dry arc. Setting the insulation margin as M=1.2, combined with the test data in Table 1 and (3) formula, the dry arc distance of the outer insulation of UHV reactor bushing under lightning impulse test voltage 2100kV is 7093 mm. Similarly, the dry arc distance of the outer insulation of bushing porcelain bushing under the power frequency wet flash-over voltage 960kV and the positive operation wave dry/wet flash-over voltage 1550kV can be calculated. The arc distances are 7967 mm and 8494 mm, respectively. Considering the manufacturing specifications, the dry flash-over distances of the ceramic sleeve of the reactor sleeve are designed to be 8500 mm. The outline of the ceramic sleeve is shown in Figure 3.

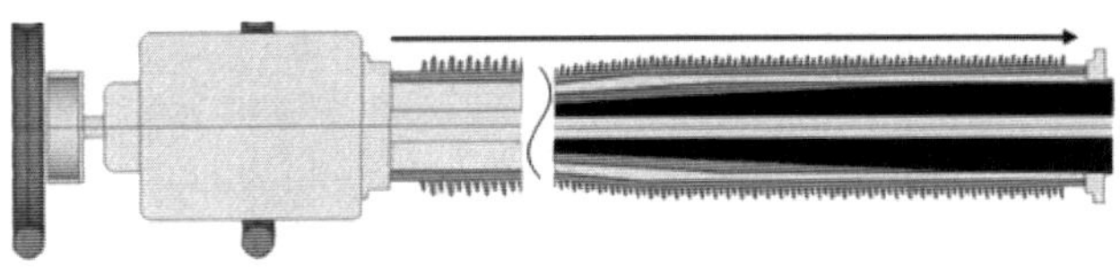

Figure.3 The outline of the outer porcelain of reactor bushing

2.2. Design of Reactor Bushing Tail Distance

Reactor bushing tail is immersed in transformer oil, so its flash-over voltage is not limited by bushing operation in the high altitude area. In insulation structure design, it can be considered according to conventional bushing size[14]. In order to ensure that no flash-over occurs at the tail of capacitor core under the RMV of power frequency test voltage 960kV (without altitude correction), the axial field strength is set to 0.6kVmm^{-1} in the main insulation design of bushing, so the total length of lower step is 960/0.6 = 1600 mm. Considering the actual winding conditions of bushing core, the insulation distance at the tail of bushing is determined. The design is 1955 mm, and the tail design is shown in Fig. 4. The bushing tail equalizing cover is used to shield the connection area between the central guide rod and the transformer winding, and the insulation medium at the bushing tail is complex, including bushing oil, porcelain bushing, oil-immersed insulation paper, etc. [15]. The electric field distribution of the bushing tail equalizing cover needs to be further checked and calculated.

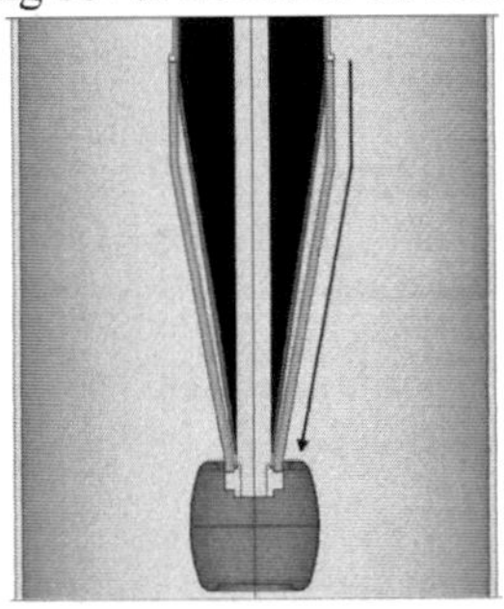

Figure.4 The condenser tail of the reactor bushing

2.3. Design of corona equalizing cover for bushing

In order to improve the flash-over voltage of the outer ceramic sleeve of the reactor bushing, it is necessary to configure the reasonable end pressure equalizing cover to uniformly distribute the voltage along the ceramic sleeve. In addition, there are some key structural components at the end of the sleeve, such as the junction plate, confluence fittings, bus lead-out line, etc. After installing pressure equalizing ring at the end of the sleeve, the sharp corners and protrusions on the surface of the above irregular conductors can be effectively shielded, and the corona discharge and the resulting external insulation flash-over can be suppressed .

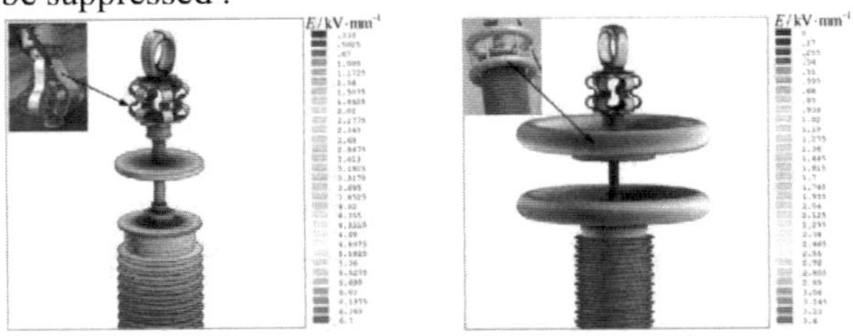

(a) No corona equalizing (b) Double-ring corona equalizer

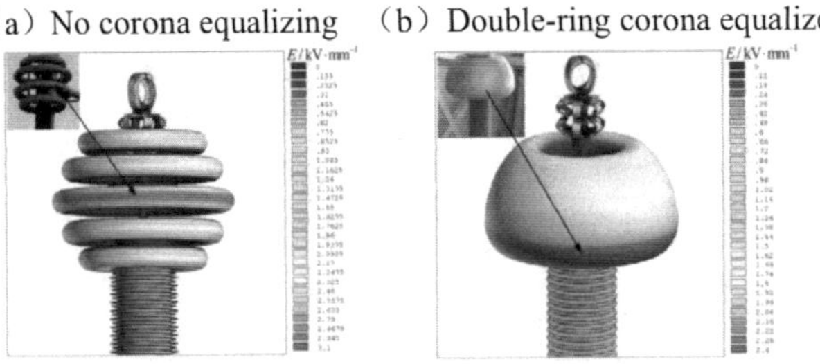

(c) Multi-ring corona equalizer (d) Apple type corona equalizer

Figure.5 The structure of high voltage bushing corona ring

As shown in Fig. 5 (a), in absence of the pressure equalizer, the high field strength area is concentrated at the edge of the confluence metal, the maximum field strength is 6.7kV mm^{-1}, which is higher than the breakdown field strength of air 3kVmm^{-1}. Therefore, corona discharges will occur at the edge of the metal. After installing the double-ring pressure equalizer, the confluence metal is located at the low field strength inside pressure equalizer. The maximum field intensity is reduced to 3.4kVmm^{-1}, which appears on surface of the equalizing cover, as shown in Fig. 5 (b). In order to further improve the electric field distribution at the end of the bushing, the structure of multi-ring pressure equalizer can be adopted. As shown in Fig. 5 (c), it can be seen that the position where the maximum electric field intensity occurs shifts to the middle part of the pressure equalizer, and the maximum electric field intensity is 3.1kVmm^{-1}. The double-ring and the multi-ring pressure equalizing covers have practical application in high-voltage substation. Apple-shaped pressure equalizing covers are generally used at the end of power equipment in valve hall of high-voltage converter station, as shown in Fig. 5 (d). It can be seen that the apple-type pressure equalizer can completely cover the confluence metal in the low field strength area, and the electric field distribution on the surface of the pressure equalizer is uniform. The maximum electric field strength is 2.4kVmm^{-1}, which is lower than the breakdown field strength of the air 3kVmm^{-1}. However, it is difficult to manufacture the apple-type pressure equalizer. Considering the manufacturing cost and installation difficulty, the shroud of transformer substation reactor bushing end equalizer generally adopts double-ring structure, but the structure size needs to be further enlarged.

2.4. Insulation structure design of reactor bushing

On the basis of the insulation structure design of the outer porcelain sleeve and the tail of the core of the reactor bushing, the internal insulation structure design of the reactor bushing is further developed. Because the running environment of the bushing is at high altitude, the coordination between the inner insulation structure and the outer insulation of the reactor bushing should be considered in particular. In the design, the dry arc insulation length of the outer ceramic bushing is 8500 mm, the length of the inner plate is 6645 mm, and the length of the outer plate is 1665 mm. The inner plate shields the outer ceramic bushing. The percentage of distance is (6645-1955)/8500=55.2%. It has better shielding effect in theory to lengthen the insulating distance of dry arc, but the longer zero-layer plate makes it more difficult to roll capacitor core. After detailed design and calculation, the design parameters of the insulation structure in the bushing of high-altitude reactor are shown in Table 2.

Table.2 Design parameters of insulation structure of high altitude reactor bushing

Plate sequence number	Plate length	Plate diameter	layer thickness
0	6645	115.0	/
10	6293	133.0	1.0
30	5504	180.0	1.4
50	4602	243.6	1.8
70	3625	319.2	1.9
90	2671	392.2	1.7
110	1843	446.2	1.1

Table 2 shows that the number of design layers of high altitude reactor bushing reaches more than 100 layers. This design can maximize the potential distribution on the surface of the bushing porcelain jacket by using the voltage equalization effect of capacitance structure, and raise the critical flash-over voltage of the bushing porcelain jacket. At the same time, Table 2 shows that the thickness of the inner insulation structure increases first and then decreases with the inverted "U" distribution. The actual design calculation results are consistent with the theoretical calculation, which proves to some extent the rationality of the design of the inner insulation structure of the reactor bushing.

3. Design Checking Calculation Based on Finite Element Electric Field Simulation

3.1. Establishment of calculation model for reactor bushing

According to the design dimensions of the outer ceramic sleeve dry arc distance, the tail distance of the reactor sleeve, the pressure equalizing cover of the sleeve and the internal insulation structure, the simulation calculation model of the high altitude reactor sleeve is established under the finite element simulation environment, as shown in Figure. 6.

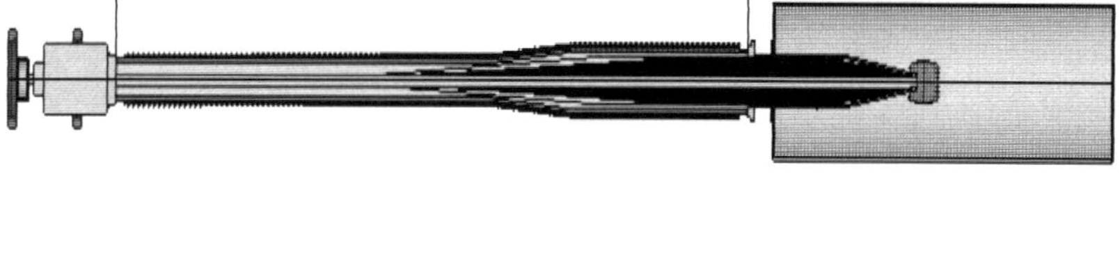

IOP Conf. Series: Materials Science and Engineering **758** (2020) 012013 doi:10.1088/1757-899X/758/1/012013

Figure.6 The FEM calculation model of the reactor bushing

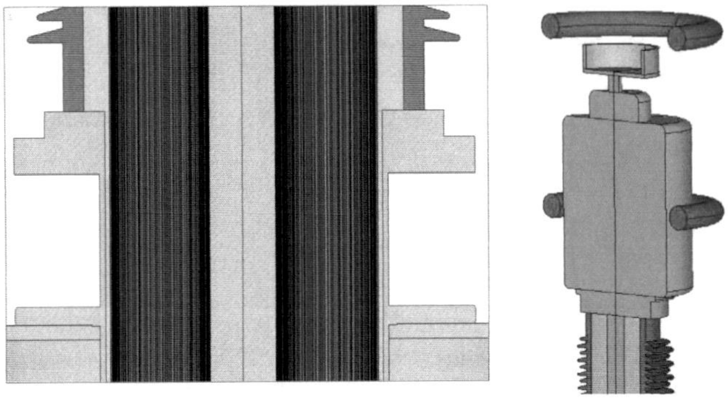

Figure.7 The FEM model of main insulation of reactor bushing

In simulation calculation model, the insulation medium of reactor bushing, the top voltage equalizing cover and the tail voltage equalizing cover of bushing are also considered, especially the modeling details of the insulating pole plate in bushing are shown in Fig. 7. The model considers more than 100 layers of plates into the checking calculation, which makes the simulation results closer to the actual operating environment of UHV reactor bushing.

3.2. Calibration of Insulation Electric Field inside and outside Reactor Bushing

Lightning impulse test voltage is applied to high-potential conductor of high-altitude reactor bushing to calculate its overall potential distribution. The distribution of the overall equipotential line of the porcelain bushing outside the bushing and the distribution of the equipotential line at the tail of the bushing are shown in Figure 8. It shows that the equipotential line distribution is more uniform, and the inner pole plate of the main insulation of the bushing has better modulation effect. The equipotential line is evenly dispersed from the edge of the pole plate to gather at the top and tail of the bushing equalizing cover, and the equipotential line in the inner pole plate of the bushing is forced to be parallel to the axis of the bushing core, which ensures that the electric field distribution in the bushing core is basically uniform.

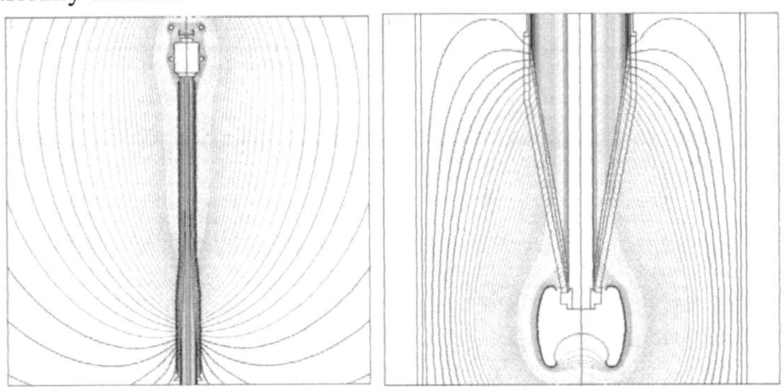

Figure.8 The distribution of equipotential line of porcelain and bushing tail

Focus on the distribution of electric field in the top and tail of reactor bushing to ensure that no visible corona occurs on the surface of the shroud. The results of E-field verification are shown in Figure 9.

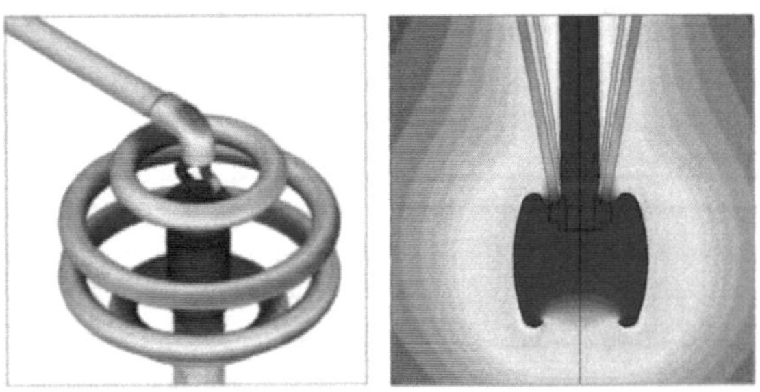

(a) Top equalizing ring (b) Tail equalizer

Figure.9 The E-field distribution of the busing corona ring

A large number of experiments conducted by Peek show that the initial field strength of corona is related to tool size, atmospheric state and surface state. Summarizing the experimental results, the corona initial field strength (peak value) can be obtained as follows:

$$E_c = 30.3m\delta(1+\frac{0.298}{\sqrt{\delta r}}) \qquad (4)$$

According to formula (4), it can be seen that the atmospheric state has a great influence on the initial field strength of corona. The reactor bushing runs at an altitude of about 3500M. When Peek formula is used, the design margin can be taken as 0.72, the altitude factor can be taken as 0.7416, the radius of equalizing ring R is 10cm, and the corona onset field strength E_c is about 1.81 kVmm^{-1}. According to the calculation and statistics, the maximum electric field intensity on the surface of the equalizing ring, which can be controlled by 0.8, can be calculated to be 1.45 kVmm^{-1} at the peak value. The corona phenomenon can hardly be seen on the top of the equalizing ring at 3500m altitude. Figure 9 shows that the electric field distribution on the surface of the reactor bushing pressure equalizing cover is uniform. The maximum electric field intensity on the surface of the top pressure equalizing ring is 1.251 kVmm^{-1} and the maximum electric field intensity on the surface of the tail pressure equalizing ring is 16.962 kVmm^{-1}, which are lower than the control electric field intensity on the surface of the metal fittings in the air and the transformer oil, respectively.

3.3. Calibration of Insulation Electric Field in Reactor Bushing

The distribution curves of potential and electric field are intercepted along the surface of the ceramic sleeve of the reactor bushing and the surface of the ceramic sleeve in the oil at the tail of the bushing respectively, as shown in Figure. 10 and 11.

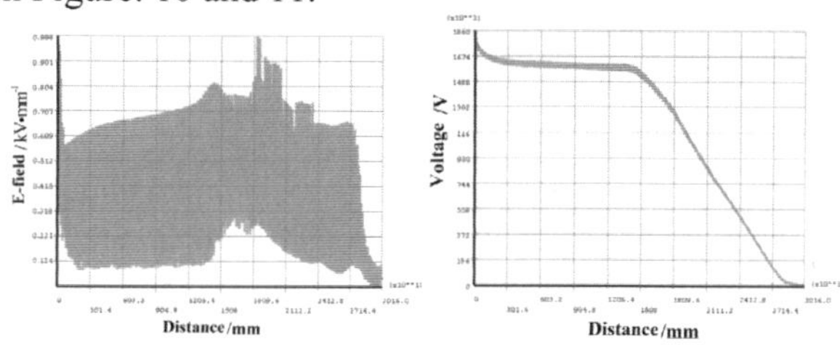

(a) E-field distribution (b) Voltage distribution

Figure.10 The E-field distribution of the bushing porcelain

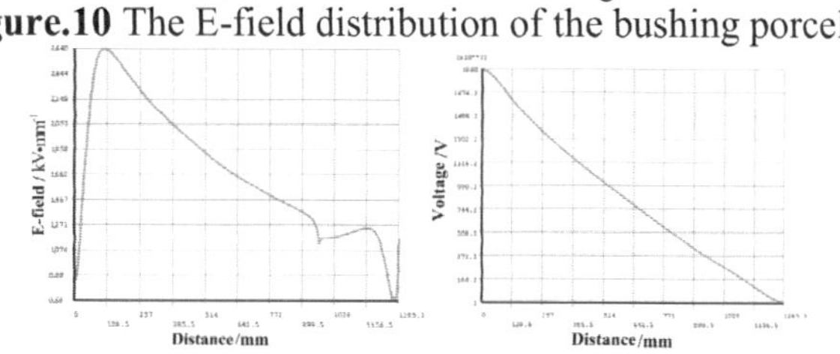

(a) E-field distribution (b) Voltage distribution

Figure.11 The E-field distribution of the bushing tail

The figure shows that the potential distribution on the outer and tail of the reactor bushing presents the linear distribution characteristics due to modulation of inner plate, while the electric field distribution increases gradually and then decreases. However, the electric field distribution on the surface of the ceramic bushing has large jump, which is mainly caused by the umbrella ridge structure on the surface of the ceramic bushing. On the other hand, the radial field intensity distribution in the capacitor core is shown in Figure 12.

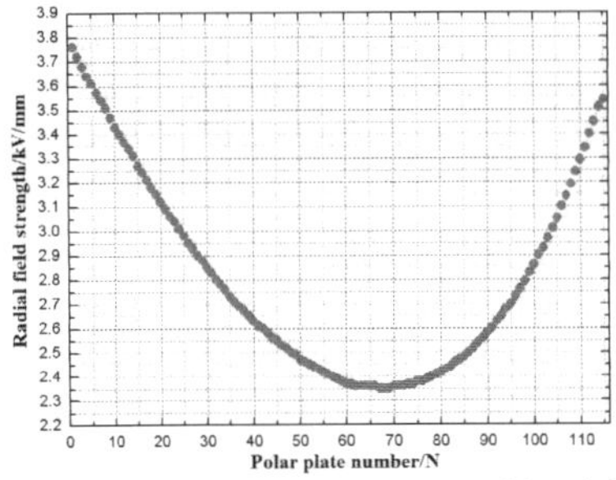

Figure.12 The radial field distribution of bushing condenser

Figure 12 shows that radial field intensity of reactor bushing core presents the typical "U" distribution characteristics of high end and low middle, and the field intensity at the interface between core and bushing core is higher. Therefore, the insulation strength at the interface between core and core conductor should be strengthened. At the same time, the distribution of field intensity at the edge of plate and the distribution of partial discharge margin between plates should be strengthened. The situation is shown in Figure 13.

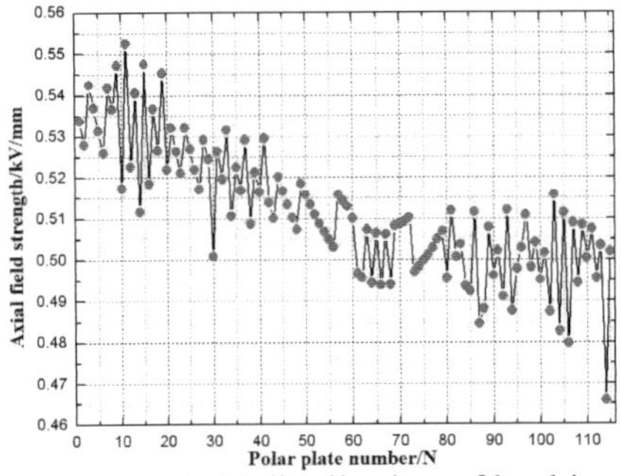

Figure.13 The axial field distribution of bushing condenser

Figure 13 shows that the axial field strength of the reactor bushing core basically presents uniform distribution, and the axial field strength near the central guide rod side is slightly higher than that at flange side, and the maximum field strength is 0.55kVmm^{-1}, which is lower than the field strength control value. And the partial discharge margin between the plates is basically uniform distribution, and its value is stable around 2.0, which has the high safety margin. The maximum field strength at the key position of the reactor bushing at high altitude under various operating conditions is summarized in Table 3. The field strength at the key position under other voltage levels can be converted according to the above data.

Tab.3 The maximum field strength at each key position of the reactor bushing/V/mm

Equalizing ring in air	Ball in oil	Porcelain Covers in Air	Surface of Porcelain Sheath in Oil	Maximum radial field strength	Maximum Axial Field Strength
1.251	16.962	0.998	2.64	3.75	0.55

4. Conclusion

1) Under the condition of high altitude, the flash-over voltage of metal fittings and porcelain jackets outside the reactor bushing should be corrected at altitude. The calculation shows that the correction coefficients of metal fittings and flash-over voltage are 2.07 and 1.36, respectively.

2) Through checking calculation, it is determined that the dry arc distance of outer porcelain sleeve of high-altitude reactor is 8500mm, the insulation distance of casing tail is 1955mm, and the double-ring pressure equalizing cover is adopted for the pressure equalizing cover.

3) Finite element calculation shows that the maximum field strength of the top equalizing ring surface is 1.251kVmm^{-1}, the maximum field strength of the tail equalizing ring surface is 16.962kVmm^{-1}, which is lower than the control field strength of the metal surface in air and transformer oil; the axial field strength near the central guide rod side is slightly higher than that on the flange side, and the

maximum field strength is 0.55kVmm^{-1}, which is lower than that in the air and transformer oil. The high-altitude reactor bushing successfully passed the lightning impulse test of 2855kV, and the corona phenomenon could not be seen at the end of the high-altitude reactor bushing under actual operation conditions, which proved that the structure design of the reactor bushing was reasonable and the electrical performance was stable under the high-altitude conditions.

Reference

[1] Zhang Xuecheng, Tan Jinhua, Niu Wanyu. Bushings design of converter transformer's valve side of UHVDC transmission project [J]. High Voltage Engineering, 2012, 38(2): 393-399.
[2] Chen Zhong, Wu Heng, Huang Heyan, et al. Cause analysis and improvement measure of ± 800kV DC wall bushing occurring flashover during overvoltage withstand test[J]. High Voltage Engineering, 2011, 37(9): 2133-2139.
[3] Toshiaki rokunohe, Tatsuro kato, Makoto hirose, et al. Development of insulation technology in compact SF6 gas-filled bushings: development of compact 800kV SF6 gas-filled bushings [J]. Electrical Engineering in Japan, 2010, 171(1): 19-27.
[4] Hosokawa M, Okumura K, Yamagiwa T, et al. Dielectric performance of improved gas insulated bushing for UHV GIS[J]. IEEE Transaction on Power Delivery, 1987, 2(2): 359-366.
[5] Dexin Nie, Hailong Zhang, Zhong Chen, et al . Optimization Design of Grading Ring and Electrical Field Analysis of 800kV UHVDC Wall Bushing[J]. IEEE Transaction on Dielectrics and Electrical Insulation, 2013, 20(4): 1361-1368.
[6] Wei Xiaoguang, Zong Wenzhi, Wang Gaoyong, et al. Experimental studies on switching impulse flashover characteristics of external insulation and altitude correction methods for ±1100kV UHVDC converter valves[J]. Proceedings of the CSEE, 2014，34(12)：1996-2003.
[7] Wang Yongqiang, Ouyang Baolong, Lian Shasha, et al. Study on electric field distribution along the surface of snowed composite insulator in high altitude area[J]. High Voltage Apparatus, 2016，52(6)：115-123.
[8] Wang Jianyi, Li Jinzhong, Li Jun et al. Modification of condition assessment for HVDC equipment used in high altitude area[J]. Proceedings of the CSEE, 2016，36(17)：4786-4792.
[9] Xiao Gan, Chang Minghui, Wu Xueen et al. Analysis of insulation length change of bushing used in high altitude area[J]. Transformer, 2017，54(10)：10-13.
[10] Luo Xiaoqing, Hu Wei, Xu Zuoming, et al. Air gap flashover characteristics and selection of gap distances for ±1100kV DC U-shaped wall bushing[J].High Voltage Engineering, 2017，43(3)：946-952.
[11] Xu Jianyuan, Ren Chunwei, Si Binge, Lin Xin. Three dimensional electric field analysis of 40.5kV SF6 cubicle gas-insulated metal-enclosed switchgear [J]. Proceedings of the CSEE, 2008, 28(15): 136-140.
[12] Zhang Shiling, Peng Zongren, Wu Hao. Analysis on power loss of valve-side RIP bushing for ± 800kV converter transformer [J]. Power System Technology, 2014, 38(7): 1758-1764.
[13] Lin Xin, Li Xintao, Xu Jianyuan, et al . Research on numerical computation of SF6 breakdown voltages and spectral experiment in uniform electric fields[J]. Proceedings of the CSEE, 2016，36(1)：301-309.
[14] Wen Miao, Lin Xin, Zhong Simeng. Numerical simulation and experimental study of heat transfer characteristics in transformer bushing [J].High Voltage Engineering, 2016，42(9)：2956-2961.
[15] Zhang Jia, Lin Xin, Su An, et al. Experiment on breakdown characteristics of SF6/CF4 gas mixture under two typical electrode arrangements[J].High Voltage Apparatus, 2016，52(12)：93-98.

IOP Conf. Series: Materials Science and Engineering **758** (2020) 012014 doi:10.1088/1757-899X/758/1/012014

Empirical and Numerical Finite-Element-Based Model to Improve Narrow Vein Mine Design in Peruvian Mining

M Belizario-Calsin[1], R Condori-Cardenas[1], H Pehovaz-Alvarez[1], C Raymundo-Ibañez[2], Moises Perez[3]

[1] Ingeniería de Gestión Minera, Universidad Peruana de Ciencias Aplicadas (UPC), Lima 15023, Perú.
[2] Dirección de Investigación, Universidad Peruana de Ciencias Aplicadas (UPC), Lima 15023, Perú.
[3] Escuela Superior de Ingeniería Informática, Universidad Rey Juan Carlos, Mostoles, Madrid, España.
E-mail: Carlos.raymundo@upc.edu.pe

Abstract. This paper proposes a numerical finite-element-based model aimed at optimizing narrow-vein stope stability. This model combines empirical and numerical methods to develop a sequence, which may determine an acceptable stope safety factor. A stope stability analysis was conducted through the Mathews stability graph method, which requires two factors: the hydraulic radius (HR) and stability number (N'). The Mathews stability graph method is used to assess the stability of an underground design. Variations in stope dimensions are estimated by changing the HR and Factor A within the N', which is determined through numerical methods. The results of the numerical simulation indicate that the HR increases with an increase in stope dimensions, while Factor A maintains an inverse relationship with the maximum stress induced on the excavation walls. This document demonstrates the potential of combining empirical and numerical methods in stope design optimization, especially when developed in small narrow vein mines.

1. Introduction

In recent decades, the depletion of near-surface mineral ore resources has prompted mining companies to invest more heavily in underground exploitation methods. In this context, underground work instability is inherent to the exploitation methods used and worsens based on mine depth [1]. In fact, mining companies have identified instability as one of the main issues that threaten mining operations. Some examples of instability issues are the 2010 collapse of the San José mine in Chile, where 33 miners were trapped for 70 days; the collapse in the Villa de la Paz mine in Mexico in 2015; and the collapse in the San José de Tacopilla mine in Chile in 2019, where three people died.

In general, most small and medium-scale mining companies design mining pits based on empirical and analytical criteria without support from scientific studies, which may lack accuracy, especially since the execution of empirical designs is based on the subjectivity of geomechanical engineers [2]. The instability issues are mostly due to various factors, such as mine depth; operations near blasting areas; water; geological discontinuities; ground stress; creep zones; relaxation zones; the physical and mechanical properties of the rock; rock mass qualities; earthquakes; regional, local, and structural faults; and inadequate geometric pit parameters [3][4].

This study seeks to combine empirical and numerical methods to propose better geometric stope design parameters and increase their stability.

2. State of the art

2.1. Empirical Stope Design Methods

Some empirical methods such as the Mathews stability graph allow for stope design, while others such as the Q system provide recommendations as support [5]. The science of empirical design follows six steps: problem identification, end user identification, construction of a rock mechanics conceptual model, identification of key model parameters and values for each key parameter, and the development of new measures [6]. To verify design dimensions as well as the support systems implemented, the SF, which must be greater than 1.5 to guarantee work stability, is determined through numerical methods. In addition, empirical relationships between classification systems and the properties of intact rock are also used to determine rock mass properties [7][8].

2.2. Empirical and Numerical Methods in Stope Design

Sepehri, Apel, and Liu [5] validate this in their research studies, where they assess stope stability and the effect of the "k" ratio, which is the relationship between horizontal and vertical stresses, both in the creep and relaxation zones around the stopes. To address the issue, the authors first used the Mathews stability graph method to assess stope stability based on the N' and the HR. The authors conclude that the numerical model provides a better understanding of effort distribution and that the relaxation areas are easily identified. Other authors such as Iglesias Comesaña et al. [1]; Kozyrev et al. [4]; and Basarir, Su and Li [9] argue that design results from empirical methods must be validated through numerical results because both methods supplement each other.

3. Contribution

3.1. Proposed Model

The geomechanical characterization of the rock mass incurs in the use of classification systems such as Rock Mass Rating (RMR), Tunneling Quality Index (Q) System and Geological Strength Index (GSI) to determine rock mass quality. Stability analysis consists of using the Mathews stability graph method to determine whether the current design is located in a stable area or not. The empirical design is performed once the current design has been deemed unstable through the manipulation of the independent variables of the HR. The purpose of determining rock mass properties is to provide the input parameters required for the numerical model to perform simulations, which is achieved through the empirical relationships between the classification systems and the properties of the intact rock. The numerical model seeks to design dimensions and use them to provide the SF of the structure, which is assessed either with or without support. (Fig. 1)

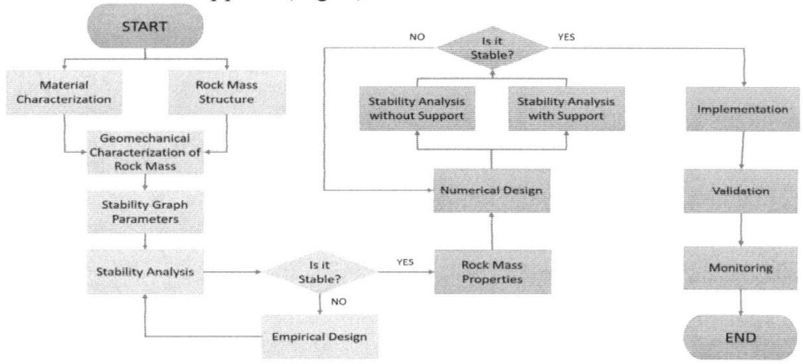

Figure 1. Proposed model

3.2. Proposed Method

Figure 2 depicts the proposed method.

IOP Conf. Series: Materials Science and Engineering **758** (2020) 012014 doi:10.1088/1757-899X/758/1/012014

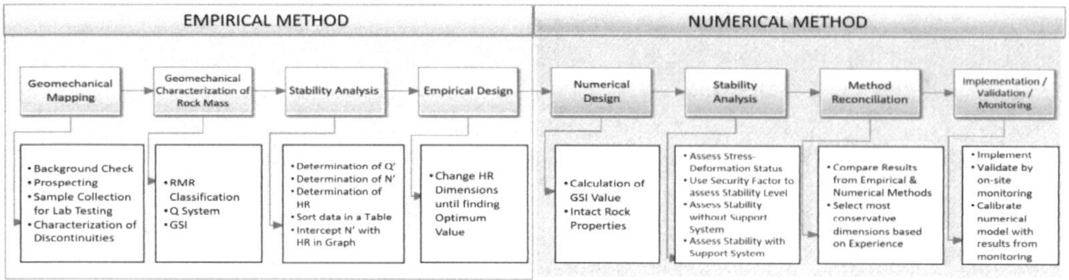

Figure 2. Proposed method

3.3. Indicators

3.3.1. Hydraulic radius

The HR allows modification of the pit's geometric dimensions so that this HR ratio may be entered into the Mathews stability graph to assess how stable the design is.

$$HR = \frac{Area}{Perimeter} = \frac{w \times h}{2w + 2h} \tag{1}$$

Interpretation: The higher the HR value, the greater the geometric dimension of the pit, but HR growth proportionality depends on the geometric shape of the design. For example, the HR ratio for a square shape is greater than that for a rectangular shape.

3.3.2. Stability number

The N' indicates the geological and geomechanical conditions of the land, and this value is used in the Mathews stability graph to assess the stability level of the geometric design.

$$N' = Q' \times A \times B \times C \tag{2}$$

Interpretation: a higher N' means that the ground conditions are more apt to support larger pit designs.

3.3.3. Security factor

The SF assesses the acting and resistant forces present in the excavation, which means that it is a pit design stability level indicator.

$$FS = \frac{Strength\ (MPa)}{Induced\ Stress\ (Mpa)} \tag{3}$$

Interpretation: If SF > 1, the stope is stable; if SF = 1, the stope is at its stability threshold; and if SF < 1, the stope is unstable. For security purposes, an SF ⩾ 1.5 is recommended.

4. Validation

4.1. Case Study

The model proposed seeks to assess, through numerical and empirical methods, stope instability at a narrow vein mine within the Eliana mining unit, Compañía Minera Escorpión S.A., located in the department of Ayacucho, province of Parinacochas, district of Pullo. As the mining unit could not provide any previous geomechanical studies, it was necessary to perform geomechanical mapping, laboratory tests and numerical modeling.

4.2. Initial Assessment

During the site walkthrough, it was found that stopes developed in the mining unit were 1.75 meters wide and 1.85 meters high, which means that their Hydraulic Radius (HR) was 0.45. (Fig. 3)
Regarding the calculation of the N′, first, as part of the empirical methods, rock mass classification was performed at the ceiling and floor cavities of the geomechanical stations.
Next, Factor A was calculated as per Equation 4. Thus, numerical methods are used to calculate the maximum effort induced on the stope.

$$A = 0.1125 \times \left(\frac{UCS}{\sigma_{max}}\right) - 0.125 = 0.1125 \times \left(\frac{45}{5.62}\right) - 0.125 = 0.78 \tag{4}$$

Then, Factor B is calculated based on the orientation of the most unfavorable discontinuity (72°) on the stope surface. This figure depicts the interpolation of Factor B calculation: $\alpha = 90° \rightarrow B = 1.0$ and $\alpha = 60° \rightarrow B = 0.8$. Hence, B is found for $\alpha = 72°$, then $B=0.88$.
Factor C is given by the following equation:

$$C = 8 - 7 \times \cos(\theta) = 8 - 6 \times \cos(90°) = 8 \tag{5}$$

Once all the factors have been obtained, the N′ is 31.57 in the floor cavity and 60.73 in the ceiling cavity.
Finally, to calculate the SF through numerical methods, laboratory tests had to be performed.
For these purposes, representative samples were taken from the ceiling and floor cavities, which were prepared in the Rock Mechanics Laboratory at the Peruvian University of Applied Sciences and were tested for uniaxial compression, indirect traction, triaxial compression, and physical property determination. (Fig. 4). The numerical model analysis was performed through the Examine 8.0 and UnWedge 7.0 software using the parameters depicted in Table 1.

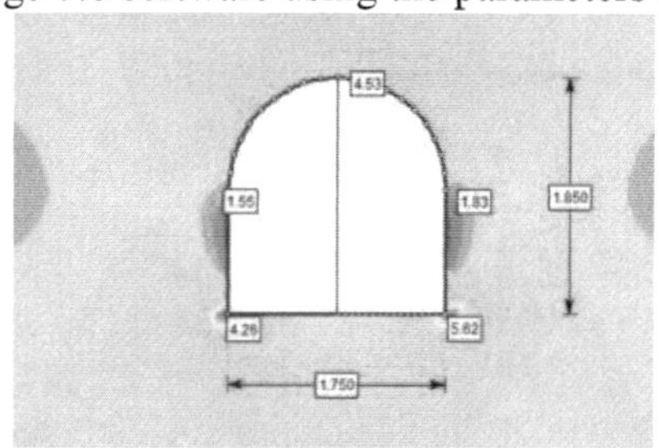

Figure 3. Initial Numerical Analysis of the Stope

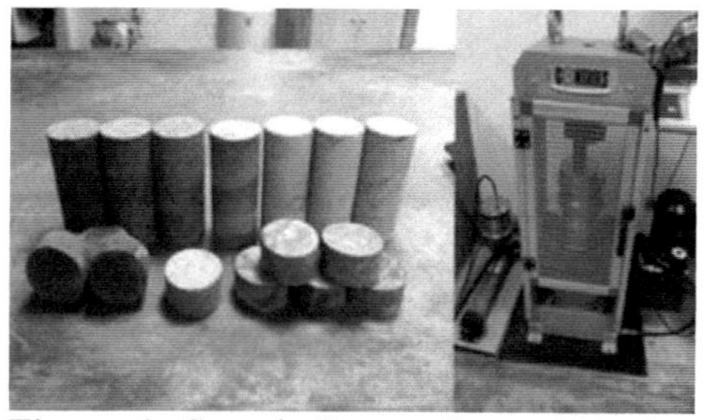

Figure 4. Sample Preparation and Testing

4.3. Numerical Model Application

For application of the numerical model, the only values that will be modified will be the dimensions of the stope, which now is 1.85-meter-wide and 2.4 meters high, with a HR of 0.52. Other values that were modified are Factor A of the N′ and the SF. From (4), $UCS=45$, $\sigma_{max} = 6.23$, then $A=0.69$
Regarding the new dimensions, the numerical analysis was performed again using the Examine 8.0 software to obtain the new value for the maximum stress induced.
Once the values of the HR and N' were obtained, the stability graph method proposed by Mathews (1981) was applied. (Fig. 6).
Finally, the SF was calculated again through the Examine 8.0 and UnWedge 7.0 software.

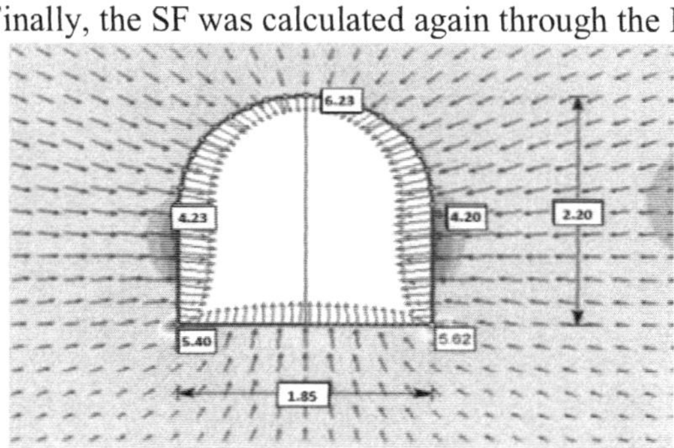

Figure 5. Final Numerical Analysis

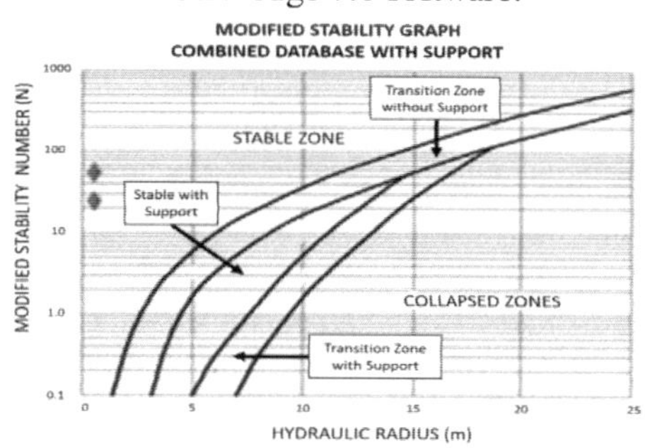

Figure 6. Mathews Stability Graph

The final indicator values are shown below (Table 2).

Table 1. Initial Indicator Value

Hydraulic Radius (HR)	Stability Number (N′)	Safety Factor (SF)

Table 2. Final Indicator Value

Hydraulic Radius (HR)	Stability Number (N′)	Safety Factor (SF)

Footwall	0.45	31.57	1.22
Hangingwall	0.45	60.73	1.35

Footwall	0.52	27.83	2.27
Hangingwall	0.52	53.53	2.34

5. Conclusions

The objective of the present study was complementary to the empirical method with the numerical methods in order to obtain better geometric parameters of the design of the cuttings that increase its stability. The results showed that both the motivation and the objective of the research were achieved.
The proposed model was based on the assumption that the empirical method can analyze in narrow veins, since it was only used in large-scale exploitation methods.
The use of numerical methods for the structural control in mines of narrow veins guarantees a support system and size of cut appropriate to use appropriately the resistance of the rock mass.
The TJ 55 N mineralized block of the Eliana mining unit evidences its safety factor went from 1.35 to 2.34, representing a 73% increase in the design.

6. References

[1] C. Iglesias Comesaña, J. Taboada Castro, J. Arzúa Touriño, E. Giráldez Pérez, and J. E. Martín Suárez, "Geotechnical Design in Underground Slate Mines," Boletín Geológico Y Min., vol. 128, no. 2, pp. 405–420, 2017.
[2] S. Heidarzadeh, A. Saeidi, and A. Rouleau, "Evaluation of the effect of geometrical parameters on stope probability of failure in the open stoping method using numerical modeling," Int. J. Min. Sci. Technol., 2018.
[3] Z. Li, G. Xu, P. Huang, X. Zhao, and Y. Fu, "Experimental Study on Anisotropic Properties of Silurian Silty Slates," Geotech. Geol. Eng., 2017.
[4] A. A. Kozyrev, I. E. Semenova, and A. V Zemtsovskiy, "Investigation of Geomechanical Features of the Rock Mass in Mining of Two Contiguous Deposits under Tectonic Stresses," Procedia Eng., vol. 191, pp. 324–331, 2017.
[5] M. Sepehri, D. Apel, and W. Liu, "Stope Stability Assessment and Effect of Horizontal to Vertical Stress Ratio on the Yielding and Relaxation Zones Around Underground Open Stopes Using Empirical and Finite Element Methods," Arch. Min. Sci., vol. 62, no. 3, pp. 653–669, 2017.
[6] C. Mark, "Science of empirical design in mining ground control," Int. J. Min. Sci. Technol., vol. 26, no. 3, pp. 461–470, 2016.
[7] B. Vásárhelyi and D. Kovács, "Empirical methods of calculating the mechanical parameters of the rock mass," Period. Polytech. Civ. Eng., 2016.
[8] R. Bertuzzi, "Revisiting rock classification to estimate rock mass properties," J. Rock Mech. Geotech. Eng., no. xxxx, 2018.
[9] H. Basarir, Y. Sun, and G. Li, "Gateway stability analysis by global-local modeling approach," Int. J. Rock Mech. Min. Sci., vol. 113, no. November 2018, pp. 31–40, 2019.

IOP Conf. Series: Materials Science and Engineering **758** (2020) 012015 doi:10.1088/1757-899X/758/1/012015

Optimal mesh design methodology considering geometric parameters for rock fragmentation in open-pit mining in the Southern Andes of Peru

V Zamora-Paredes[1], L Arauzo-Gallardo[1], C Raymundo-Ibañez[2], Moises Perez[3]

[1] Ingeniería de Gestión Minera, Universidad Peruana de Ciencias Aplicadas (UPC), Lima 15023, Perú.
[2] Dirección de Investigación, Universidad Peruana de Ciencias Aplicadas (UPC), Lima 15023, Perú.
[3] Escuela Superior de Ingeniería Informática, Universidad Rey Juan Carlos, Mostoles, Madrid, España.
Email: carlos.raymundo@upc.edu.pe

Abstract. Blasting is one of the most important stages in the productive process of a mine due to its direct impact on rock fragmentation, which determines the degree of productivity of operations and the extraction costs generated. In this scenario, an optimized methodology is presented for designing blasting meshes by using mathematical models that help calculate the geometric parameters of a blasting mesh, such as burden, considering the variables of the rock mass and the type of explosive to measure its impact on rock fragmentation and loading productivity (tons/hour). The main advantage of this method is the reliability of the design, which takes into account a greater number of variables that influence fragmentation and uses the principle of distribution and amount of energy in an optimal way. The results obtained in the case of application show that a change in design (2.7 x 2.7 square mesh to 2.2 x 2.5 triangular mesh) reduces P80 by 65%, from 17 to 6 inches, approximately. Additionally, the results show that greater operational efficiency was achieved by increasing excavator productivity by approximately 15.6%.

1. Introduction

Currently, most mining companies seek significant unit cost savings in order to maintain a sustainable market position [1]. Because of this, continuous improvement processes are sought within mining operations under a holistic approach [2]. Fragmentation of blasting rocks has been studied because it can add great value by reducing costs, as it has a high impact on the productive and economic performance of loading, carrying (50–60% mine costs), crushing, and milling [3]. Several studies have been conducted to measure the relationship of these parameters with P80 (an indicator of the degree of fragmentation). However, an integrated approach is necessary in order to obtain more reliable results, since each mine has different characteristics in geological, structural, and operational areas [4], [5].
An optimal methodology that uses the Pearse mathematical model is proposed for the design of blasting meshes. The methodology is based on the principle of distribution and exploitation of the explosive energy within the drill, for which a triangular mesh model is proposed, depending on burden and spacing. To analyze the results, the Wipfrag software.

2. State of the art

2.1. Blasting mesh designs
Studies carried out to find solutions optimize rock fragmentation [6] determined the direct relationship between the resistance to the uniaxial compression of the intact rock and the power factor (amount of explosives used/tons of material in the mined bank), and the authors concluded that a greater power factor improves optimization, based on the theory of amount of energy required. Moving a step further, [7] analyzed rock mass, including intact rock and discontinuities, which affects energy distribution and the entire detonation process. The authors conclude, through fragmentation image analysis, that indicators such as the Bienowski rock mass rating and RQD must be present in the calculation of blast mesh designs in order to achieve the expected results.

2.2. Blasting mesh designs based on geometric parameters
Currently, all research allows current mines to use models such as Pearse (1955) and Langefors (1963) to determine geometric parameters such as burden, which serves as input for the calculation of spacing, block, and overdrilling. Considering the explosive factor, [8] proposed blasting mesh designs that take into account the implementation of higher energy explosives, which also use new geometric parameters, according to the areas of exploitation within the mine under study. The conclusion stated that a greater amount of energy and its better distribution are necessary to obtain optimal fragmentation in order not to have excessive costs in the drilling and blasting processes. Moreover, in [9] with the use of integral mesh designs, achieved more uniform fragmentation, contributing to an increase in loading, hauling, and milling performance.

3. Contribution
This research uses a methodology to perform reliable blasting mesh designs based on geometric design parameters, including loading, spacing, block length, overdrilling, and retardation times, taking into account the physical and mechanical properties of the rock mass, such as lithology, compressive strength, traction, and Deere RQD, in order to optimize rock fragmentation. This methodology is based on the principle of better distribution and greater amount of chemical energy used to fragment and displace the rock mass. For this, an equilateral triangular perforation is executed, as shown in Figure 1, which permits better distribution of the energy inside the blasting bank, a product of the existing geometric disposition. In (1), the geometrical relationship between burden and spacing is shown, and we can see how they impact energy distribution, affecting the coverage area.
Therefore, the area of each drill is represented in (2). The influence area of the explosive per drill is represented in (3). The coverage percentage of the explosive per drill is *I1*=90.69% (4).

$\frac{S}{B} = \frac{2R}{\sqrt{3}R} = 1.15; \quad R = \frac{B\sqrt{3}}{3}$ (1)	$A_E = \pi * R^2$ (3)
$A_H = 2R\left(\sqrt{3}R\right) = 2\sqrt{3}R^2$ (2)	$\%I = \frac{\pi * R^2}{2\sqrt{3}R^2} * 100\% = 90.69\ \%$ (4)

Area F_1, where there is no interaction between the explosives of the drills on the triangular mesh, is

$$F_1 = A_T - 3A_S = \frac{2R * R\sqrt{3}}{2} - \frac{3\pi * R^2 * 60}{360} = \sqrt{3}R^2 - 1.57R^2 = 0.1621R^2 = 0.054\,B^2 \quad (5)$$

where B is the burden, S is spacing, R is the radius of influence of the drill, and F1 is the area where there is no interaction of drill holes in triangular mesh during blasting.
On the other hand, Figure 2 shows a square mesh design, which is used in medium open-pit mining operations and quarries, because it allows a reduction in drilling and blasting costs. However, the geometric influence of burden and spacing negatively impacts fragmentation, because the percentage of explosive coverage per drill is less than that of a triangular mesh.
Area F2, where there is no interaction between the explosives of the drill the triangular mesh, is equal to where *%I2* = 78.54%.

$$F_2 = A_C - 4\,A_S = (2R)^2 - 4 * \frac{\pi * R^2 * 90}{360} = 4R^2 - 3.14R^2 = 0.8585 * \left(\frac{B}{2}\right)^2 = 0.2146\,B^2 \quad (6)$$

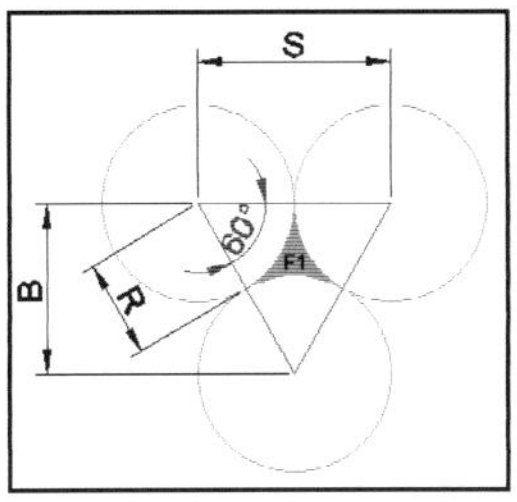

Figure 1. Triangular mesh design

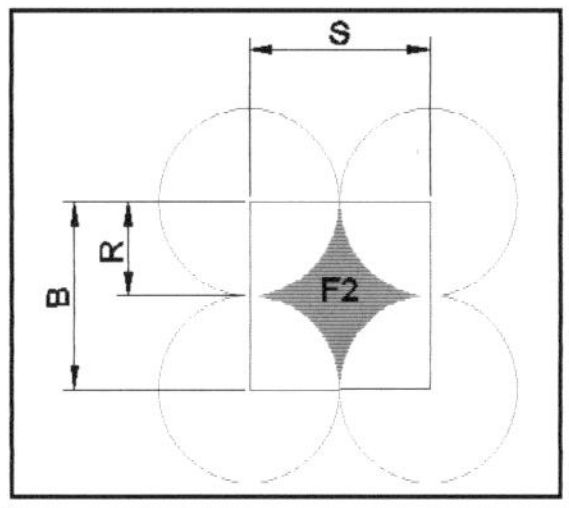

Figure 2. Square mesh design

where F2 is the area where there is no interaction of square mesh holes during blasting. It is determined that the percentage of explosive coverage per drill in a triangular mesh is 12% higher than in a square mesh and that F1 > F2 is, therefore, more likely to result in finer rock fragments.
When using the equilateral triangular mesh, the best use of chemical energy is achieved with a spacing/burden ratio equal to 1.15.
The research methodology is shown in Figure 3 and begins with the comparison of mesh designs (square and triangular). From this, it is concluded that the triangular pattern generates a better energy distribution, which favors the optimization of rock fragmentation.

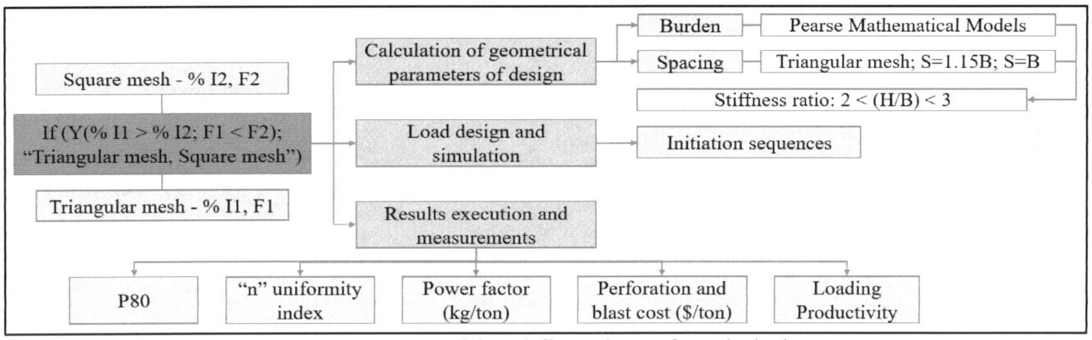

Figure 3. Conditional flow chart of mesh design

Among the indicators to be evaluated is P80, as a measure of rock fragmentation, based on the granulometric curves of Kuz Ram. This varies between 0.6 and 2.2; a value close to the lower limit means that there is a variable percentage between fine and coarse fragments. Conversely, a value close to the upper limit indicates that the stack of material has fine or coarse fragments. Furthermore, the power factor (kg/tons) is evaluated in order to obtain the relationship between the quantity of explosive used and the fragmentation of rocks in P80 and to evaluate the quantity of energy used.

4. Validation

4.1. Case study

The investigation was conducted in a medium-sized mining operation (approximately 800 tpd), which exploits copper ore using the open-pit method in the Southern Andes of Peru. The deposit is located in the Tacaza volcanic formation. Table 1 presents the physical and mechanical properties of the rock mass. In order to achieve the objectives proposed in this paper, a field study was carried out on 10 blasting banks in an open-pit mine in the Andes of Peru.
In Table 2, a comparative table is presented with the mesh design parameters using a square mesh pattern with an empirical model (baseline) and the equilateral triangular pattern with a mathematical model (proposed methodology) in order to obtain an optimal and reliable result involving the largest number of variables for the design.

Table 1. Characteristics of the rock mass in the area under study

Domain	Lithology	t/m^3	Rc (MPa)	Rt (MPa)	RQD %
III-B	Tuff breccia and andesite augitic	2.1	90.2	9.02	50

IV-A	Feldspatic Andesite	2.5	107.2	10.72	42
I-B	Reworked tuff	1.8	9.61	0.98	15

Table 2. Design parameters of the conventional mesh and the proposed mesh

Parameters	Units	Square mesh -empirical model	Triangle mesh–mathematical model
Drill diameter	mm	101.6	101.6
Bank height	m	5	5
Burden	m	2.7	2.2
Spacing	m	2.7	2.5
Length of block	m	1.0	1.5
Over perforation	m	0.3	0.5
Stiffness ratio		1.85	2.27
Type of initiation		Pyrotechnic detonators	Pyrotechnic detonators
Power factor	Kg/t	0.26	0.36
Retardation times between rows	ms	25 t 42	17 to 25
Retardation times between drills	ms	17	17

5. Results

Below is a summary of the results for each of the lithologies, n estimate of unit costs in the drilling and blasting process, and measurement of the variations with the proposed new mesh design. Figure 4, shows that an optimal fragmentation was obtained, according to those compared to the baseline, where the fragments exceeded 11 inches, which was the limit standard for the concentrator plant.

Table 3 compares both the mesh designs in terms of $PR_{80}R$, uniformity index, power factor, unit cost of drilling, and blasting. Comparisons are also made at the loading stage to measure the performance of excavators during the mining cycle, as shown in Table 4.

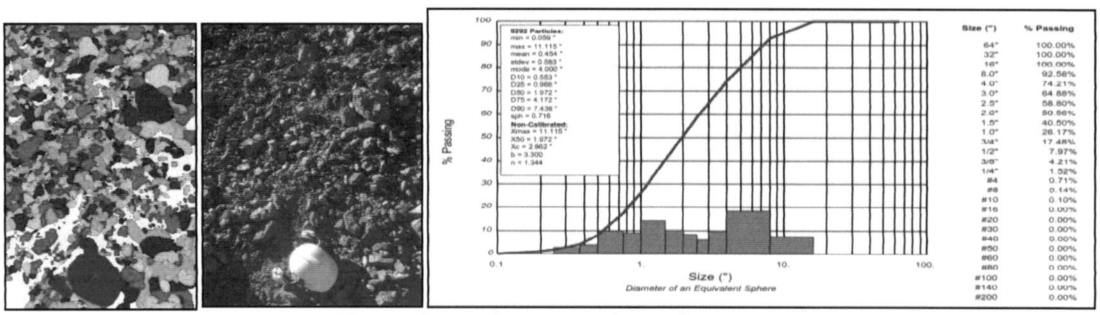

Figure 4. Kuz Ram granulometric curve

Table 3. Comparative indices of fragmentation, and drilling, and blasting costs

Model	Unit	Baseline	Proposed design
Type		CAT 336D2L	
Capacity	m^3	2.6	2.6
Filling factor	%	80	90
Density	ton/m^3	2.3	2.3
Sponging		1.2	1.2
Cycle	seg	40	35
Productivity	ton/hour	358.8	403.7

Table 4. Comparative productivity between the designs presented

Description	Unit	Square mesh	Triangle mesh
Average volume per blast	m^3	4200	4200
Tons	ton	8880	8880
No. of drills	tal	108.00	133.00
Unit cost per perforation	$/ton	0.10	0.13
Unit cost per blast	$/ton	0.32	0.36
Power factor	kg/ton	0.24	0.36
Average P80	inch	17.0	6.0
Uniformity index		1.0	1.4

6. Data analysis

Table 5 shows the average results of the 10 blasting operations in ore banks, where the average mesh design is 2.0 x 2.50 m, resulting in P80 of 5.9 inches and uniformity index "n" of 1.5, which indicates a large percentage of material with similar size of fine fragments. Additionally, higher productivity compared with blasting with square mesh designs are obtained, as the fragmentation obtained is finer, which allows the loading of ore in less time and with a higher filling factor in the excavator buckets.

Table 5. Comparative productivity between the designs presented

Statistics	Burden (m)	Espacing (m)	P_{80} (inches)	n	Power factor (kg/t)	T/hour - excavator	Perforation cost ($/t)	Blasting cost ($/t)
No. of samples	10	10	10	10	10	10	10	10
Mínimum	2.1	2.4	4.0	1.3	0.35	403.0	0.11	0.35
Maximum	2.3	2.55	7.0	1.8	0.38	435.0	0.15	0.39
Medium	2.2	2.5	5.9	1.5	0.37	416.2	0.13	0.37
Standard deviation	0.07	0.06	0.94	0.13	0.01	9.73	0.01	0.01

7. Conclusions

The research concludes that triangular mesh designs and mathematical models are favorable for rock fragmentation due to better energy distribution. In addition, this impacts the unit costs of drilling and blasting but allows an increase in the productivity of the loading equipment, which, in turn, decreases hauling cycles (fixed time), impacting the reduction of loading and hauling costs. This generates greater profitability in the mining operation.

- Triangular mesh designs generate a better use of energy in the drill compared with square meshes (+15.68%). The area of non-interaction is reduced by 75 %, which favors rock fragmentation.
- The power factor increases by 50% due to greater number of holes drilled in a triangular mesh and decrease in burden and spacing.
- Average P80 is reduced by approximately 65%, from 17 to 5.9 inches.
- Drilling and blasting costs increased due to higher steel consumption and power factor (+30% and +12% respectively).
- Loading productivity increased by 15.61% (ton/hour).

8. References

[1] D. Whittle, "CHAPTER 10.2: Open-Pit Planning and Design."
[2] D. Murr, Workman Lyall, J. Eloranta, and P. Katsabanis, "Blasting Influence on comminution," in SAG Conference, 2015, pp. 1–21.
[3] E. Rybinski et al., "Optimisation and Continuous Improvement of Antamina Comminution Circuit," in SAG Conference, 2011, p. 19.
[4] P. K. Singh, M. P. Roy, R. K. Paswan, M. Sarim, S. Kumar, and R. Ranjan Jha, "Rock fragmentation control in opencast blasting," J. Rock Mech. Geotech. Eng., vol. 8, no. 2, pp. 225–237, Apr. 2016.
[5] G. L. Gomes-Sebastiao and W. W. De Graaf, "An investigation into the fragmentation of blasted rock at Gomes Sand," J. South. African Inst. Min. Metall., 2017.
[6] F. Mohamed, A. Hafsaoui, K. Talhi, and K. Menacer, "Study of the Powder Factor in Surface Bench Blasting," Procedia Earth Planet. Sci., 2015.
[7] K. Nur Lyana, Z. Hareyani, A. Kamar Shah, and M. H. Mohd. Hazizan, "Effect of Geological Condition on Degree of Fragmentation in a Simpang Pulai Marble Quarry," Procedia Chem., vol. 19, pp. 694–701, Jan. 2016.
[8] A. Aguirre, "Optimización de parámetros de tronadura en función de explosivos de alta energía en Sociedad Contractual Minera El Abra," Universidad de Chile, 2016.
[9] P. Zuñiga, "Diseno de tronaduras y su impacto en el costo mina de Compañía Minera Cerro Negro," Universidad de Chile, 2015.

IOP Conf. Series: Materials Science and Engineering **758** (2020) 012016 doi:10.1088/1757-899X/758/1/012016

Optimized Method of Estimating Mineral Reserves Using the O'Hara Mathematical Model for Underground Mining in Peru

V Torres-Sanchez[1], P Olivas-Maldonado[1], G Diaz-Huaina[1], C Raymundo-Ibañez[2], Moises Perez[3]

[1] Ingeniería Industrial, Universidad Peruana de Ciencias Aplicadas (UPC), Lima 15023, Perú.
[2] Dirección de Investigación, Universidad Peruana de Ciencias Aplicadas (UPC), Lima 15023, Perú.
[3] Escuela Superior de Ingeniería Informática, Universidad Rey Juan Carlos, Mostoles, Madrid, España.
Email: carlos.raymundo@upc.edu.pe

Abstract. This research project is aimed at applying the proposed methodology to optimize the calculation of mining reserves using an empirical relationship for dilution, as proposed by O'Hara. This methodology uses three key parameters: dilution, mineral value, and total costs. In general, costs include mining costs, processing within plants, and administrative expenses related to producing a mineralized block. Similarly, to calculate the mineral value, studies on changes in mineral prices over the last 30 years were included to provide more representative values. The research was applied to mineralized block number 1000, which yielded positive results since dilution could be easily calculated using three key parameters: operating method, dip, and vein width. The dilution value of 12.25% was valid, as it was within the range of 10% to 23%. This range was considered based on other studies that apply this operating method. Then, the methodology was analyzed in three more settings, each represented by a mineralized block. The blocks assessed were 1001, 1002, and 1003, and the results indicated that the dilution found was acceptable at values of 18%, 19%, and 22%, respectively.

1. Introduction

Estimating mineral reserves in marginal resources is of utmost importance for a mining company because materials deemed below the cutoff grade could become mineral reserves in the future. According to [1], in the international context where prices increase, not only would proven reserves start producing higher production volumes but also work in probable reserve areas would be justified.
This research proposes using a methodology that can validate the change from a mineralized resource to a mineral reserve. For these purposes, the three key parameters used are cutoff grade, international price of the mineral ore, and dilution. However, the dilution parameter may vary depending on the method used for its calculation. For example, [2] developed an empirical method where dilution directly depended on the mining method, stope width, dip angle, and rock's competence. [3] proposes an empirical method based on gathering information from different mines, [4] studies a variation of the Mathews stability graph method, and concludes that the higher the stability, the lower is the dilution.
Within this context, this research study seeks to propose a method that simplifies dilution calculation in order to estimate mining reserves, considering the main parameters of vein thickness, dip, operating

method, and wall competence. Obtaining dilution can be associated with the NSR cutoff grade and the mineral ore's economic value to verify if the block is considered a mineral resource or reserve.

2. State of the art

2.1. Estimating Reserves

Mineral reserves are the economically exploitable part of a measured or indicated mineral resource. According to [5], mineral ore's value is an essential assessment factor in the short, medium, and long term, because it is the only factor that companies selling the mineral cannot directly control.
According to [6], the level of influence of the reserves calculated based on geological models improves the management of mining cycle processes with precision of the extracted mineral ore's uniform grade.

2.2. Calculating Classic Dilution and Using Mining Software

When assessing mining reserves, it is necessary to understand some geological parameters specific to each operating method. The dilution present during blasting and transporting minerals is one of them. As stated by [7], the processing costs of the diluted material can range from USD 30 to 40 per ton.
The method proposed by O'Hara can be used graphically and mathematically by applying the formula that was developed, which points to a more exact value.

2.3. Cutoff Grade and Operating Costs

The secondary grade, also known as operational grade, is below the cutoff grade but related to sterile material. The secondary material is not taken to the dump site but stored in specially designed places for future treatment. Estimating costs is a very important factor taken into account by [8], [9], and [10] when it comes to creating value, because this parameter is compared to the mineral ore's value and helps determine if the business is profitable or not. As noted by [11], the costs that should be assessed during an economic analysis concern preparation, operating, mining, processing, energy, and administrative.

3. Contribution

3.1. Proposed Method

The reserve calculation method proposed begins by calculating the dilution, continues by determining the mineral ore's value, and develops the NSR cutoff grade.
As mentioned earlier, the dilution method we are putting into practice is from O'Hara. This empirical method can be applied graphically (Figure 1) or mathematically (Equation 1). In order to apply the graph method, the three variables studied by O'Hara are assessed.

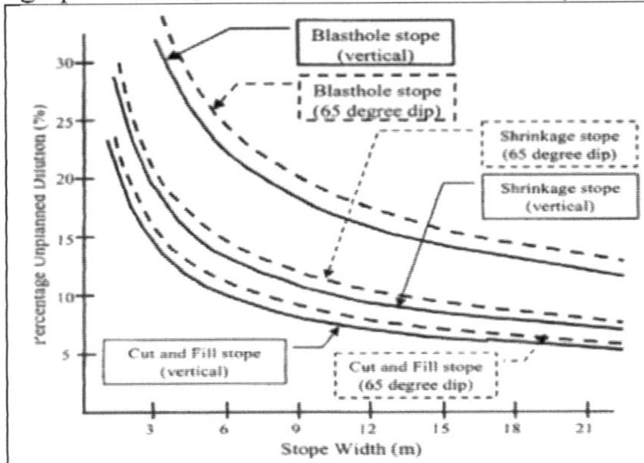

Figure 1. Estimating dilution caused by detachment of the stope walls

$$(\%)Dil = \frac{K}{W^{0.5}\sin A^{\circ}} \quad (1)$$

Where:

- W is mineralized body dip,
- A is stope width,
- D is % dilution,
- K is constant for each operating method.

The first variable, and the most general, is the operating method that the mining unit uses. In Fig. 1, the operating method is represented using different curves, which include four operating methods: cut and fill, shrinkage, room and pillar, and blasthole. The second variable is the stope width. In Fig. 1, the horizontal axis is the stope width variable, which is assessed in feet. The third variable is the dip angle of the mineralized body. In Fig. 1, the dip angle varies depending on the operating method, and there are two curves: one for a vertical angle or 90-degree dip and another for a minimum of 65 degrees.
When the mathematical formula is applied (1) (Empirical formula to find the O'Hara dilution), a similar reasoning is followed, since the formula is a mathematical adjustment of the graph. Its three variables are as follows: dip angle of the body, operating method, and stope width. The only difference is that in the formula, the operating method is connected using the K parameter, which is a factor included in the formula according to each method: cut and fill=45, shrinkage=60, room and pillar=70, blasthole=100.
Once the dilution is obtained, the mineral ore's value is assessed using (2) (Calculating mineral value in polymetallic mines). Two parameters are necessary for this calculation: diluted grade and mineral factor. The diluted grade can be found quickly using (3) (Diluted grade equation), where the grade is multiplied by geological tonnage and divided by geological tonnage plus tons of waste.

$$VM = diluted\ grade\ (Cu)\ x\ FM\ (Cu) + diluted\ grade\ (Pb)\ x\ FM\ (PB + diluted\ grade\ (Ag)\ x\ FM\ (Ag) + diluted\ grade\ (Zn)\ x\ FM\ (Zn) \quad (2)$$

$$Dil.grade = \frac{Grade\ x\ geological\ tonnage}{Geological\ tonnage\ +\ waste\ tonnage} \quad (3)$$

As the last step, the NSR cutoff grade is assessed (4) (Cost equation applied for underground mining), which is understood as the sum of all costs from the different operations during the mining and processing cycle.

$$Cost\ =\ mining\ cost\ +\ cost\ \frac{profits}{processing} + G.A \quad (4)$$

In the practical sense, the methodology below is the following: if the mineral value is less than the NSR cutoff grade, the block is considered not to have economic value; if the mineral value is greater than 1.5 times the cutoff grade, it is considered economically valuable; and if it is greater than the NSR cutoff grade but less than 1.5 times, it is considered marginal material.
Finally, the methodology to be followed is summarized in Figure 2.

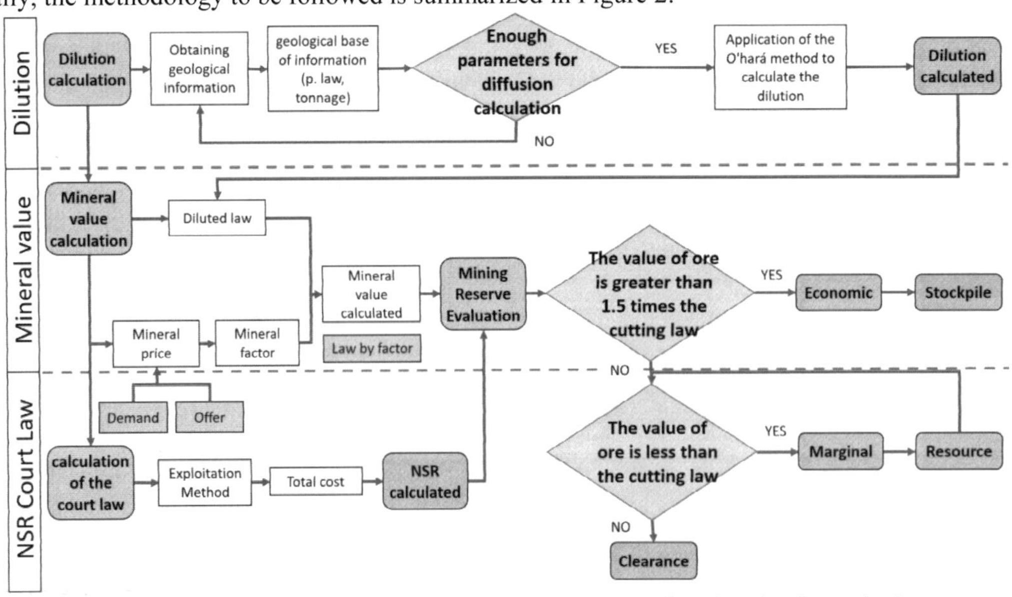

Figure 2. Process for estimating mineral reserves using the classic method

3.2. Results Indicators

The indicators to measure the results are based on the three stages of the proposed method: dilution, mineral value, and NSR cutoff grade. Thus, they are quantitative results of the mineralized reserve estimates. The three indicators to be measured and their corresponding units of measurement are presented below: The dilution percentage found in each mineral block , The mineral ore's economic value (USD/MT), NSR cutoff grade (USD/MT).

4. Validation

4.1. Study Area

The case study where the research was conducted is the mining company Raura, has an existing capacity of 2,000 tons per day. The Raura mining deposit is located between the Departments of Huánuco, Lima, and Pasco. The medium-sized underground mining company specializes in mining copper, lead, silver, and zinc minerals (polymetallic), and obtaining copper, lead, and zinc concentrates.

4.2. Initial Assessment

Using the initial calculation, block 1000 had values similar to those obtained by the proposed method; however, the two indicators that changed the most were the dilution percentage and the mineral ore's economic value shown in dollars per ton. In Table 1, the parameters mentioned can be identified.

4.3. Results from the Applied Method

First, after applying the O'Hara formula, a dilution of 12.25% was found for block 1000 from the Veronica Vein. This value was found using the methodology proposed by O'Hara. The values of the parameters: K = 24, 75-degree dip and vein thickness = 4.26 feet.
Second, the mineral value for block 1000 was found. The diluted grades are as follows: 2.72 oz/MT silver, 0.33% copper, 2.76% lead, and 8.31% zinc. Diluted grades are the product of multiplying geological grades by the dilution level obtained, in this case by 12.25%.
The resulting NSR cutoff grade was USD 77 per ton. The area with the highest costs is the area related to mining. The main costs are maintenance, operating, transportation, loading and hauling, and preparation, with almost 78% of the total (USD 35.84 per ton). The second highest area is general mining services. Its main costs are maintenance and energy, environment, and geology, which is 71% of the total (USD 14.35 per ton).
Finally, observing the final results, the dilution percentage is 12.25%, the cutoff grade is still USD 77 per MT, and the mineral value amounts to USD 197.24 per MT. Another value that varies is the dollar value of the mineralized block, since it changes from USD 1,717,068 to USD 2,011,806 due to the variation in dilution (Table 2).

Table 1. Initial research indicators

Structure	Lode Veronica
Method	CRBR
Block	1000
%Dilution	15.59%
Cutting law (US$/TMS)	77
Mineral value (US$/TMS)	168.34
Block value (US$)	S/. 1,717,068.00

Table 2. Final research indicators

Structure	Lode Veronica
Method	CRBR
Block	1000
%Dilution	12.25%
Cutting law (US$/TMS)	77
Mineral value (US$/TMS)	197.24
Block value (US$)	S/. 2,011,806.17

The proposed method achieved its objective of optimizing the calculation of mineral reserves; however, more tests must be completed in different mineralized blocks to verify that different case studies will work, since this method was adapted to an underground mine in Peru with specific geological and geomechanically conditions. Statistical parameters such as the mean, median, and geometric mean were assessed as representative measurements of the blocks (Table 3).
As mentioned above, the blocks vary from 10% to 23%, which is considered an acceptable dilution according to the operating method. The three central tendency values of mean, median, and geometric

mean can be analyzed, which summarize the dilution values found in the four mineral blocks. The geometric average is the value that best represents the four blocks and has a value of 17.75%.

Table 3. Settings and final research indicators

Scenarios	Structure	Tons	%Dilution	Cutting law (US$/TMS)	Mineral value (US$/TMS)	Block value (US$)
Block 1000	Lode Veronica	8,200	12.25%	76.77	208.98	1,713,629.49
Block 1001	Lode Veronica	10,200	18.94%	76.77	197.24	2,011,806.17
Block 1002	Lode Veronica	10,200	19.22%	76.77	196.78	2,007,106.72
Block 1003	Lode Veronica	20,900	22.24%	76.77	169.61	3,544,911.81
Mean		12,375	18.16%		193.15	2,319,363.55
Median		10,200	19.08%		197.01	2,009,456.45
Geometric mean		11,556	17.75%		192.59	2,225,459.74

5. Conclusions

The reserve estimate results confirm the validity of the proposed methodology. Using the O'Hara method generated an acceptable dilution of 10% to 23%, the average value being 18%. Applying the method estimated very high mineral values. Among them, block 1003 obtained a value of USD 3,544,911.81 due to its high tonnage and considerably high diluted grades. Compared with average block values of USD 2,319,363.55, it is clear that the block has an extremely high economic value. It would have been optimal to have costs for each block; however, it was used as a constant for this study because the four blocks were mined using the long drill (breasting) method. The cost in each case was USD 76.77 per MT.

The method has only been tested on a mineral body that has the same operating method in all its blocks, in addition to similar geological characteristics. However, the dilution formula's K parameter should be adjusted if the method is applied to another body or another section of mineral operations because the empirical values assigned in the literature are specific to the geology studied in each research study. It is not feasible to expect to use that parameter because Peru's geology differs significantly, so it would be appropriate to research adapting the K constant for underground polymetallic mines in Peru.

6. References

[1] M. R. Ahmadi and R. S. Shahabi, "Cutoff grade optimization in open pit mines using genetic algorithm," Resour. Policy, vol. 55, pp. 184–191, Mar. 2018.
[2] T. O. Alan, H. Manager, D. Group Hudson Bay Mining, and S. Co, "Quick guides to the evaluation of orebodies."
[3] T. Pakalnis, "EMPIRICAL STOPE DESIGN AT THE RUTTAN MINE," vol. 1986, p. 614, 1986.
[4] Y. Potvin, "Empirical open stope design in Canada," Univ. Br. Coloubmi.,Nov, p. 350, 1988.
[5] Y. R. M. IDROGO, "ESTIMACIÓN DE RESERVAS MINERALES DE ORO Y PLATA EN LA VETA KARINA - LOS PIRCOS, SANTA CRUZ – CAJAMARCA," 2017.
[6] A. A. R. Baylón, "Aplicación del modelo geológico en la gestión por procesos para la extracción de oro de sulfuros, en las Minas del Perú," Ef. un Model. Interv. Educ. para Fam. pacientes oncológicos sometidos a quimoterapia en el Hosp. Daniel Alcidez Carrion, p. 142, 2014.
[7] L. M. Clark, "Minimizing Dilution in Open Stope Mining," p. 336, 1998.
[8] P. J. Le Roux and T. R. Slacey, "Value creation in a mine operating with open sloping mining methods," J. South. African Inst. Min. Metall., vol. 117, no. 2, pp. 133–142, 2017.
[9] K. Bannister, "Estimation of Open Cut Mining Recovery and Mining Dilution," pp. 1–6.
[10] M. Thompson and D. Barr, "Cut-off grade: A real options analysis," Resour. Policy, vol. 42, pp. 83–92, 2014.
[11] J. G. Vásquez Rivas, "'Elección Y Aplicación Del Método Tajeo Por Subniveles Con Taladros Largos Para Mejorar La Producción En La Veta Gina Socorro Tajo 6675 - 2 De La U.E.a. Uchucchacua De La Compañía De Minas Buenaventura S.a.a.,'" p. 126, 2015.

Research on key technology of jacking cable tunnel construction in mudstone and gravel composite stratum

Kelong Liu[1,2,4], Zhiming Lu[1,2], Chunguang He[1,2], Xiaoxing Zhong[1,2], Yang Jiao[1,2] and Liangzhong Fan[3]

[1]State Grid Ningbo Zhenhai District Power Supply Company, Ningbo, China.
[2]Ningbo Sanming Power Development Co. Ltd., Ningbo, China.
[3]Ningbo Institute of Technology, Zhejiang University, Ningbo, China.
[4]Corresponding Email: 3867018@qq.com.

Abstract. In this paper, the key technology of jacking cable tunnel construction in mudstone and gravel composite stratum is studied. According to the conditions of the jacking cable tunnel crossing the mudstone gravel composite stratum, the selection of the jacking method, the design of the relay and the construction technology of the mud jacket were analyzed. This has important practical significance for promoting the construction of ecological underground engineering and reducing the impact of underground crossing construction on the ground environment and underground environment. It also has great social and economic benefits.

1. Introduction

With the rapid development of the national economy, in order to solve the contradiction between power grid construction and land supply and demand, and meet the requirements of environmental landscape, high-voltage cable lines have been widely used in urban power grid construction. Commonly used laying methods for high-voltage cables include direct burial laying, protective pipe laying, cable trench laying, tunnel laying, and bridge laying and shaft laying under special engineering conditions. Among them, the tunnel laying method has the following advantages: (1) The number of cables to be accommodated is large, (2) the scalability is good, (3) the reliability is high, and (4) the operation and maintenance are convenient. With the vigorous development of the national grid, the construction of cable tunnels in special composite formations will become more and more common, and related engineering problems must be paid enough attention and attention.

According to the construction method, tunnels can be divided into open tunnels and undercut tunnels. Pipe jacking technology is one of the methods of underground excavation. Its development has been more than 100 years old. Compared with the open-buried pipelines constructed by open excavation and tunneling, the pipe jacking construction has the advantages of uninterrupted traffic and no danger to the foundation of the houses on both sides of the street. Therefore, the application of pipe jacking technology at home and abroad is increasing year by year.

Mudstone is a complex formation between soft soil and rock, belonging to soft rock. Soft rock is widely distributed in China, and its engineering properties are obviously different from hard rock and soft soil. Many scholars at home and abroad have carried out experimental research on local soft rock [1-5], and obtained the physical and mechanical properties of soft rock. Soft rock has obvious elastoplastic deformation characteristics, and as the strain increases, strain softening and dilatancy appear. Soft rock also has strong creep, which is an important cause of instability and deformation of surrounding rock [6]. However, the compactness of the round gravel is not uniform and the pebble is partially clamped. In this stratum, the implementation of the jacking pipe has high permeability coefficient, large top force, difficulty in axis control, and less redundancy of correction, and less domestic related applications and research. The related technical problems need to be solved urgently. In the project, the pipe diameter is large, the pipeline is long, and the pipe cost is high. It is necessary to carry out research on key technologies of pipe jacking cable tunnel construction to bring better economic and social benefits.

It is becoming more and more common to construct cable tunnels in special composite formations. Many scholars have conducted research on related methods and techniques for the construction of a specific cable tunnel. In [7], the authors introduce the design and construction method of the 400kV cable system for the Severn Tunnel. In [8], Anchor spacing design of pre-stressed tunnel concrete lining with un-bonded annular anchors for Songhua River water supply project is studied. In [9], the soil pressure balance shield tunneling parameters of the water-rich gravel stratum and the gravel mudstone composite stratum are compared and analyzed, and the calculation method of the bucket soil pressure value range is discussed. In the literature [10], the author takes a certain gravel and mudstone composite stratum in Nanning as an example to study the construction technology of the mud tunnel shield of the subway tunnel under such stratum conditions. Although there have been many studies on the construction of cable tunnels in special composite strata, from the current research status analysis, there are few comprehensive studies on theoretical analysis, experimental techniques, design methods, construction techniques, etc. for the construction of composite strata such as mudstone gravel. There are many imperfections in theory and practice. The application of domestic engineering practice has just begun. With the current design and construction technology level, there is still a large technical difficulty in the underground crossing and micro-disturbing construction without stopping.

Therefore, in combination with the underground underground crossing project that is constantly emerging in the construction of underground cable tunnels in China, it is very necessary and urgent to study the construction technology of reasonable composite structures such as mudstone and gravel. The research results are to promote the construction of ecological underground engineering and reduce underground crossing. The impact of engineering construction on the ground environment and the underground environment has important practical significance, and it also has great social and economic benefits.

This paper focuses on the conditions of pipe jacking cable tunnel crossing mudstone gravel composite stratum, and analyzes the choice of pipe jacking method, the design of relay room and the construction technology of mud bushing.

2. Research on Design Method of Pipe Jacking Cable Tunnel

This section studies the design method of cable tunnels and lays the foundation for the subsequent construction techniques. The design of the cable tunnel includes the design of the main line, the design of the pipe joint, and the structural design of the tunnel.

2.1. Total line plane and profile design

The line inside the tunnel should be designed as a straight line. When a curve is required, the curve radius is generally not less than 400m. The maximum slope of the cable tunnel constructed by the pipe jacking method should not exceed 30‰. In the section of the cable tunnel located in the high drop section, the passage should not be stepped, and the longitudinal slope should not be greater than 15°. The minimum slope in the cable tunnel should not be less than 5‰. In the difficult area, under the condition of ensuring drainage, the slope of less than 5‰ can be used, but should not be less than 3‰. The buried depth of the cable tunnel structure should reduce the adverse impact on the environment during and after construction. Consider the effect of the surrounding environment on the structure caused by urban planning. The roof of the tunnel under the main road of the city should not be less than 3m, and the top cover soil under the urban secondary road should not be less than 2m. For special locations, the thickness of the covering soil should be adjusted accordingly according to the opinions of the planning department.

2.2. Tube design

The most important two points in the design of the pipe jacking section are the calculation of the load and top thrust of the vertical segment. For the theoretical analysis of the pipe jacking load model and its pipe jacking force calculation, please refer to the literature [11,12].

Through the analysis of the load bearing capacity of the vertical pipe jacking, the precautions for the calculation of the force of the jacking pipe and the selection principle of the pipe wall thickness are as follows:

(i) The calculation of earth pressure should adopt the principle of water and soil division, and the water and soil calculation should not be used. The main reason is that the effect of water pressure on the pipeline is uniform, and the vertical and horizontal effects of earth pressure are different, so the water and soil is not applicable. Calculated for such pipelines.

(ii) When calculating the strength of the steel pipe, if the expansion joint is provided, it is not necessary to consider the influence of the temperature difference on the longitudinal stress.

(iii) When calculating the steel pipe, if it is the working condition, the corrosion margin of the wall thickness should be deducted, generally 2mm, but when the construction stage is jacked, the allowance can be deducted.

2.3. Structural design of the tunnel

According to different functional requirements and construction methods, general cable tunnels are divided into two types: circular tunnels and rectangular tunnels. The circular tunnel section has a diameter of 3m. Mainly set up lamps, cables, drains and corresponding fire-fighting facilities, the cable tunnel channel is about 1m wide. In this paper, a circular tunnel is adopted for the jacking tunnel of the mudstone gravel composite stratum. The circular tunnel has strong adaptability to the stratum, and the construction of a tunnel with a large depth in the area with high groundwater level has high technical and economic superiority. However, when encountering soft and hard stratum, high-strength rock stratum, pebble layer, lone stone, knife-cutting mud cake, etc., it will affect the tunneling speed. Therefore, the selection of the shield machine and the configuration of the cutterhead are relatively high.

According to the previous engineering experience, considering the feasibility of construction and the cost of engineering measures, when the tunnel adopts a small diameter (3m~4m diameter) circular section, when the section length is greater than 1.2km, the shield method is more economical and the interval length is smaller than The jacking method is more economical at 1.2km.

3. Research on Key Technology of Pipe Jacking Construction

In order to adapt to the complex geological environment, it is necessary to study a set of efficient and reliable cable tunnel construction methods. This section is mainly for the conditions of the jacking cable tunnel crossing the mudstone gravel composite stratum, and the selection of the jacking method, the design of the relay and the construction technology of the mud jacket are analyzed. The key technologies for the construction of jacking cable tunnels under the combined geological conditions of mudstone and gravel are given below.

3.1. Selection of pipe jacking method

According to the different methods of unearthing, the pipe jacking construction methods are mainly divided into two categories: muddy water balance pipe jacking method and earth pressure balance pipe jacking method. The muddy water balance pipe jacking machine and the earth pressure balance pipe jacking machine have different applicable conditions respectively. If the pipe jacking construction method and the pipe jacking machine suitable for the site conditions are not selected, it will not only be difficult to jack up, but also cause the jacking of the pipe piece to be damaged.

At the same time, the cutterhead setting is also a very important aspect. The choice of the cutter head should not only consider the compressive strength of the rock formation, but also the rock quality and composition of the rock formation. In general, the choice of commonly used cutter heads is divided into the following cases: 1) soft rock selection cutters with a compression strength of $20MN/m^2$ or less, 2) soft rock or medium hard rock selection hobs with a compression strength exceeding 20MN/m^2, 3) A complex type of cutterhead is selected for complex rock formations containing the above two rock formations.

For the mudstone and gravel composite stratum, the stratum strength is relatively large, and the gravel is from fine gravel with a diameter of 2mm to coarse gravel, and even contains boulders with a particle size of 30cm or more. Therefore, it is suitable to select a soil pressure balance pipe jacking machine with a large opening. Figure 1 below is a schematic diagram of the earth pressure balance top pipe method. The earth sand cut by the cutter head is filled in the soil bin at the front end of the tunnel, and has a certain fluidity, which balances the water and earth pressure at the front end of the top pipe, and discharges the spoil through the earthmoving machine connected thereto. The earth pressure type can effectively control the cost because it does not require mud water treatment equipment.

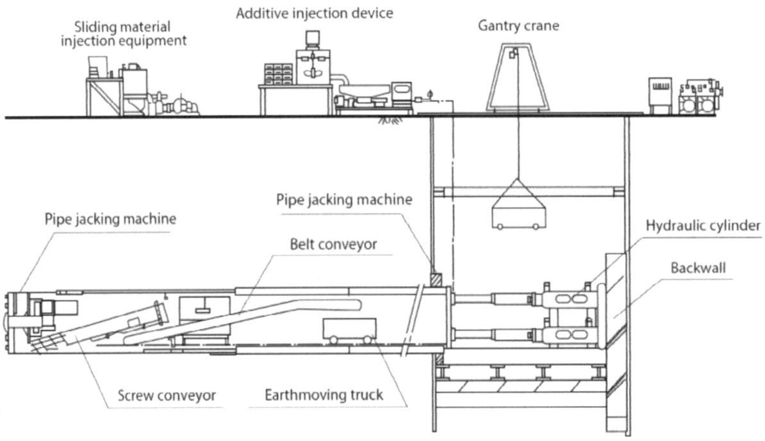

Figure 1. Earth pressure balance top pipe method.

3.2. Design of cutter opening rate

The cutterhead opening rate is an important parameter to characterize the geological adaptability of the shield machine. The cutterhead opening rate has a significant influence on the silo pressure and the cutterhead torque. If the opening position is not set properly, the bauxite cannot enter the earthen space, which will accelerate the wear of the cutter tool. The shield machine tool is installed on the shield machine cutter disc. The cutter disc form can be divided into panel type and spoke type according to engineering geological conditions and construction control requirements. Generally, the panel type cutterhead has an opening rate of about 30%. The spoke type cutterhead has a large opening rate of about 60% to 95%.

The reasonable opening rate of the cutterhead is closely related to the geological conditions of the passage of the tunnel. The sensitivity of the different strata to the cutter disc structure is quite different. The clay and fine sand stratum have low sensitivity to the cutter disc structure, and almost all kinds of cutterheads can be used. The structural form, while the gravel-pebble formation is highly sensitive to the structure of the cutterhead, and the poor design of the cutterhead structure will have a serious impact on the excavation and even cause engineering accidents. Table 1 summarizes the cutterhead opening rate and geological comparison of the construction of the nose at home and abroad.

Considering the fact that the actual project is in the composite stratum of gravel and mudstone, and referring to the existing engineering cases at home and abroad, the spoke-type cutterhead with moderate opening ratio is adopted, and the cutterhead is used as an auxiliary measure for the stability of the excavation face. The cutterhead opening rate is set in the range of 25%~35%.

Table 1. Comparison of cutterhead opening rate and geological conditions.

Soil layer	Geological condition	Opening rate(%)
Flood rock	Clay soil	25~30
	Sandy soil	15~20
	Gravel soil	20~25
	Soft and hard interlayer	30~35
Impact rock	Clay soil	20~25
	Sandy soil	10~15
	Gravel soil	15~20
	Soft and hard interlayer	20~25

3.3. Design of relay

For the special formation of mudstone gravel, special processing is used for the relay between the steel relay rooms. The relay top adopts a two-segment and one-shoulder retractable sleeve socket structure, and the deflection angle $\alpha=20^{o}$. The end structure is the same as the selected pipe joint, and the outer shape is basically the same as the pipe joint. Two sealing devices for radially adjusting the sealing gap are arranged at the hinge to ensure no leakage when jacking, and a nozzle capable of injecting lithium grease No.1 at the socket is provided to reduce the sealing ring when jacking abrasion. Four grouting holes are provided at the hinge of the relay top, and simultaneous grouting can be performed when jacking, reducing the jacking resistance.

In normal jacking, only the first seal is used and the second is used as a reserve. When the first seal ring is worn, it is found that there is a slurry leakage point, and the radial adjustment device can be used to adjust the seal gap so that the slurry leakage phenomenon can be stopped in time. When the first seal fails, the second seal can be activated to ensure continuous advancement. Due to the long penetration distance, the sealing ring wears quite badly. To prevent the first sealing device, the first sealing device is designed to be detachable, and it is easy to replace the sealing ring, thus achieving foolproof. The schematic diagram (unit: mm) of the relay is shown in the figure 2.

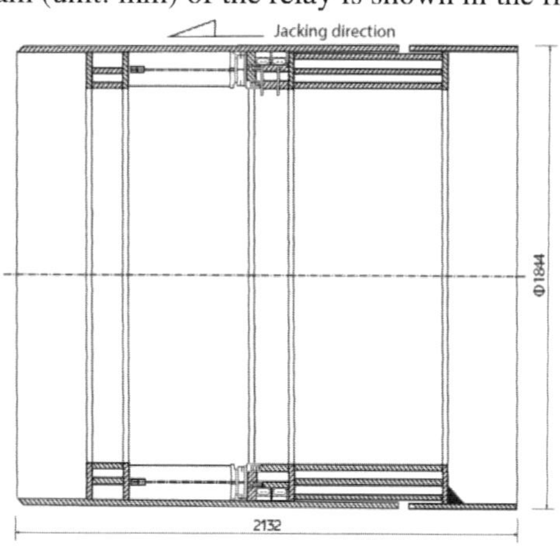

Figure 2. Relay diagram with double-layer inflatable seal expansion ring structure.

According to the "Technical Regulations for Water Supply and Drainage Pipe Jacking Engineering CECS246:2008" and the literature [13,14] for the number of calculation formula for relays:

$$n=\frac{\pi D f_k (L+50)}{0.7 f_0}-1,$$

where D indicates the outer diameter of the pipe (m); L indicates the jacking length (m); f_k indicates the average frictional resistance between the outer wall of the pipe and the soil (KN/m^2); f_0 indicates the allowing force between relays (KN).

3.4. Design of grouting system

The earthwork cut by the pipe jacking machine is conveyed into the dumping tank through the screw machine. After being stirred, it is transported to the ground sedimenting tank through the conveying pipeline by muddy water. The muddy water is discharged after sedimentation, and the clean water is transported to the dumping tank through the returning pipeline for reuse. The principle of jacking system is shown in the figure 3.

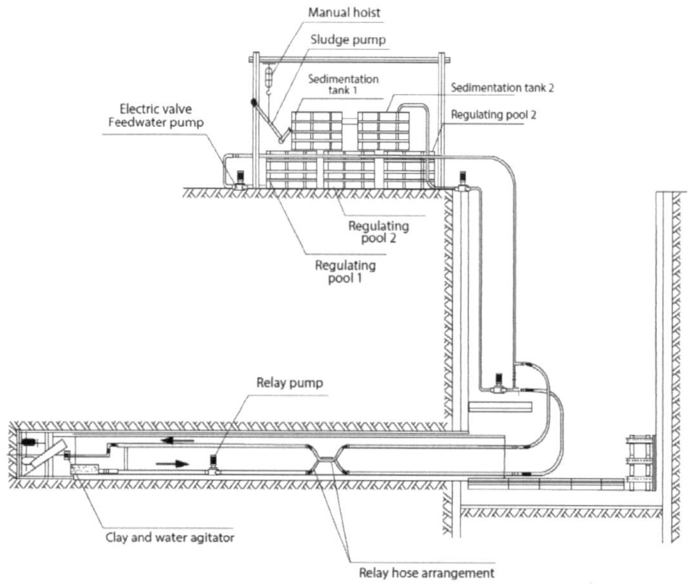

Figure 3. Schematic diagram of the jacking system.

In jacking construction, the use of drag reduction mud is the main measure to reduce the jacking resistance. When jacking, through the grouting hole on the pipe joint, a quantitative drag reduction mud is injected into the outer wall of the pipe to form a mud ring around the pipe. Reduce the friction between the outer wall of the pipe joint and the soil layer, thereby reducing the top force when jacking. The effect of thixotropic mud drag reduction is related to the quality of the mud itself (the amount of mud blending and the thickness of the mud jacket), as well as the grouting pressure, grouting procedure, grouting amount, and placement of grouting holes.

To determine the grouting pressure, you must first know the water and earth pressure at the top of the pipe, that is, the water pressure at the top of the mud jacket:

$$P_A = \gamma_w H_1 + \gamma H \tan^2(45^\circ - \frac{\phi}{2}) \cdot$$

When there is an unloading arch,

$$P_A = \gamma_w H_1 + \gamma h_0,$$

where

$$h_0 = \frac{D[1 + \tan(45^\circ - \frac{\phi}{2})]}{2\tan\phi} \cdot$$

In the above equations, P_A indicates the water pressure and main power (kPa) at the top of the mud jacket; γ_w indicates the gravity density of water (kN/m^3); H_1 indicates the height of the water column above the earth surface or the unloading arch; H indicates that the top of the pipe covers the soil height; h_0 indicates the height of the unloading arch; D indicates the outer diameter of the pipe; γ indicates the gravity of the soil; ϕ indicates the internal friction angle of the soil.

If the pressure of the mud sleeve $P < P_A$, the soil will collapse to the mud, so it is required that $P \geq P_A$. But it can't be too big, too much mud loss is increased. In case of thin cover layer, it is also possible to arch the soil surface, and at the same time, the mud sleeve is enlarged, and the amount of pulp is greatly increased. Therefore, the grouting pressure P is required to remain in the following ranges:

$$P_A \leq P \leq P_A + 30(kPa).$$

4. Conclusions

In this paper, we studied the key technology of jacking cable tunnel construction in mudstone and gravel composite stratum. The following aspects were analyzed for the conditions of the pipe jacking cable tunnel crossing the mudstone gravel composite stratum.

(1) Selection of pipe jacking method. For mudstone gravel composite stratum, the stratum strength is relatively large. It is suitable for selecting earth pressure balance pipe jacking machine with large opening.

(2) Design of the inter-relay. For the special formation of mudstone gravel, the special processing of the relay room uses steel relay room. The relay top adopts a two-stage one-hinge retractable sleeve socket structure. Two sealing devices for radially adjusting the sealing gap are arranged at the hinge to ensure no leakage when jacking, and a nozzle capable of injecting lithium grease No. 1 at the socket is provided to reduce the sealing ring when jacking abrasion. Four grouting holes are provided at the hinge of the relay top, and simultaneous grouting can be performed when jacking, reducing the jacking resistance.

(3) Design of the grouting process. When the jacking is passed through the grouting hole on the pipe joint, a quantitative drag reducing mud is injected into the outer wall of the pipe, and a mud ring sleeve is formed on the outer periphery of the pipe to reduce the gap between the outer wall of the pipe joint and the soil layer. Friction, thereby reducing the top force when jacking. The grouting pressure is required to remain in the following ranges: $P_A \leq P \leq P_A + 30(kPa)$.

Acknowledgments

This work was supported by the management consulting project (project number: (F)YYKJ-JY-Z-201906-0013.) of state grid Ningbo power supply Co. Ltd., entitled by “pipe jacking technology of shale sandstone address and its impact on adjacent structures”.

References

[1] Shun’Ichi F, Yoshida N and Sahara M 2009 *J. Jpn. Assoc. Earth. Eng.* **9** 46-64
[2] Walkden M and Hall J 2011 *J. Coast. Res.* **27** 529-43
[3] Tomanovic Z 2006 *Mec. Time-Depen. Mater.* **10** 135-54
[4] Lu Z, Chen Z and Cao J 2001 *Rock Soil Mech.* **22** 339-42
[5] He M, Leng X and Heng C 2003 *J. Rock Mech. Eng.* **22** 1151-55
[6] Zhang X, Li Y and Zhang S 2004 *J. Rock Mech. Eng.* **23** 1635-39
[7] Arkell, Hance G and Blake W 1977 *Proceed. Inst. Elect. Eng.* **124** 303-16
[8] Pi J, Zhao Yand Cao R. 2017 *Key Eng. Mater.* **737** 505-10
[9] Yang Y, Tan Z and Peng B 2017 *J. Civil Eng.* **50** 94-98
[10] Yu X 2016 *Urban Rapid Rail Transit* **29** 57-60
[11] Reed M 1986 *J. Appl. Mathe.* **36** 223-45
[12] Ying J 2018 *Tunnel Rail Transit*, **4** 26-29
[13] Gunduz D, Yener A and Goldsmith A 2013 *IEEE Trans. on Infor. Theory* **59** 51-63
[14] He L 2004 *Water Supply Drain.* **6** 93-96

Study on ZnO Varistors Doped with Y_2O_3 by Microwave Two-step Sintering

Yongjing Peng, Zhen Fang, Bowen Wang, Jianping Hu, Pengkang Xie

State Key Laboratory of Disaster Prevention & Reduction for Power Grid Transmission and Distribution Equipment, State Grid Hunan Electric Power Company Disaster Institute Institute Institute Institute Institute Institute Institute
Project Supported by Science and Technology Department of SGCC (5216A01700UD)
Email: danielpeng88@qq.com

Abstract. The zinc oxide varistor prepared in this paper is mainly composed of submicron zinc oxide powder and dopants, and is prepared by two-step sintering method, which is in contrast with traditional sintering method and ordinary microwave sintering method. The effects of microwave sintering temperature and holding time on the microstructure and electrical properties of ZnO varistors were investigated. The results show that the potential gradient of microwave sintering is higher than that of traditional sintering, but the leakage current also increases. Compared with ordinary microwave sintering, the two-step sintering can further effectively reduce the grain size, enhance densification, not only improve the potential gradient, but also reduce leakage current. In two-step sintering, increasing sintering temperature can effectively reduce leakage current and non-linear coefficient, but the potential gradient is also greatly reduced. When the holding time is more than 1 hour, there is little difference in electrical performance. When the holding time is too small, its electrical performance is weak overall.

Keywords: Two-step sintering; ZnO varistors; Dielectric relaxation; Electrical performance; Microstructure.

1. Introduction

Sintering, as the last step in the preparation of ceramic materials, is the most important and critical step. In traditional sintering, heat diffuses from the surface to the inside through radiation, while microwave energy absorbed by the material during microwave sintering is converted into kinetic energy and potential energy of molecules inside the material, heating has integrity and uniformity, temperature gradient inside the material is very small, and heating and sintering speed are very fast. Under the action of microwave electromagnetic energy, the kinetic energy of molecules or ions in the material increases and the diffusion coefficient increases, so that low-temperature rapid sintering can be realized, sintering can be completed before crystal grains can grow up, and the performance of ceramic materials is obviously improved [1-4].

Using microwave sintering and nano-sized zinc oxide powder to produce varistor has very high density and high potential gradient, but leakage current is also very large [5-7]. The two-step sintering process was first proposed by Chen et al. in 2000 and applied to the densification of Y2O3 ceramics. The densification of ceramics and the inhibition of grain growth can be achieved at a lower sintering temperature by exploring the difference in activation energy required for grain boundary migration or grain boundary migration [8]. At present, there is little research on the performance of ZnO varistors produced by two- step sintering of micron-sized zinc oxide powder, and there is little discussion on the influence of two-step sintering temperature and holding time.

2. Test Sample Preparation and Sintering Methods

Sub-micro varistor powder is prepared with analytically pure materials in the following molar ratio by normal solid phase reaction technology:

95.09mol%ZnO+0.8mol%Bi2O3+0.8mol%Sb2O3+1.0mol%Co2O3+0.8mol%MnO2+0.6mol%NiO+0.2mol%SiO2+0.4mol%Y2O3+0.3mol%Cr2O3+0.01mol%Al （NO3） 3·9H2O.

Grinding in a high-speed ball mill for 3 hours and carrying out centrifugal spray granulation to obtain varistor powder, and pressing into a disc under 200 MPa. The organic binder (polyvinyl alcohol) added to the mixture was removed at 600°C. After sintering, the upper and lower polished surfaces of varistor samples were covered with silver paste, and then the samples were heated in air at 600°C for 10 minutes to form electrodes. The final disc has a diameter of 30mm and a thickness of 4mm.

The schematic diagram of the two-step sintering is shown as follows: firstly, the sintering temperature *t*1 is increased at *v*1 heating rate for a duration T1, and then the sintering temperature *t*2 is decreased at *v*2 cooling rate for a duration T2.

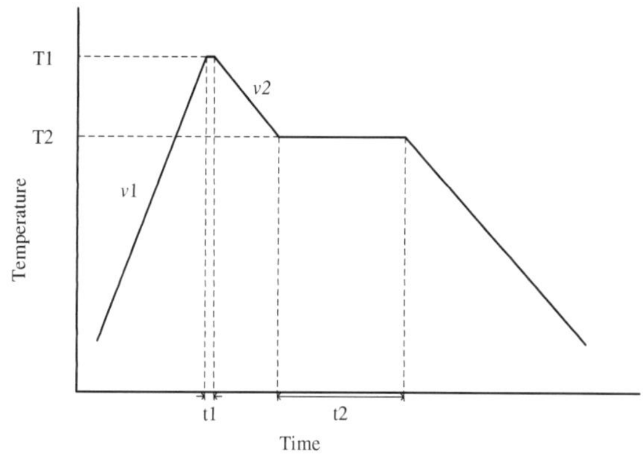

Figure1． Schematic Diagram of Microwave Two-step Sintering

3. Result and Analysis

3.1 Influence of Sintering Temperature on Varistor Performance

T1 is set to 850, 950, 1050°C, *v*1 is 10°C/min, *t*1 is 5 min; T2 is 800°C, *v*2 is 5°C/min, *t*2 is 240min. The measured electrical performance parameters are shown in the following table, and the microstructure is shown in the following figure:

Table 1 Comparison of Non-linearity Coefficient, Leakage Current and Voltage-sensitive Voltage at Different Sintering Temperatures

T1/°C	α	J_L/（μA·cm^{-2}）	V_{1mA}/（V·mm^{-1}）
850	16.8	18.3	733

IOP Conf. Series: Materials Science and Engineering **758** (2020) 012018 doi:10.1088/1757-899X/758/1/012018

950	31.0	13.0	435
1050	48.0	2.11	303

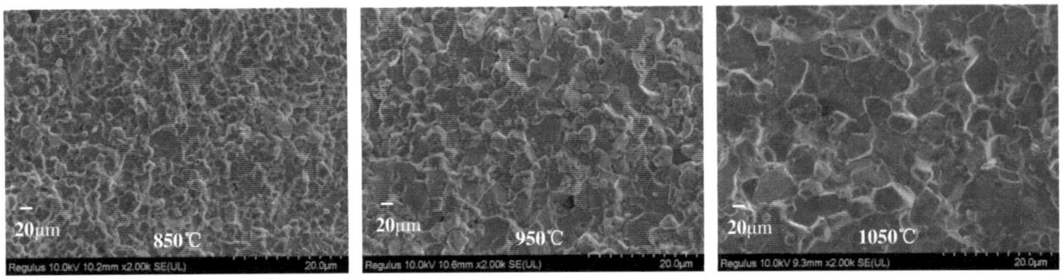

Figure2. SEM photos of samples at different sintering temperatures

From the above figure, it can be seen that the crystal grain grows continuously with the increase of T1, the porosity decreases, the porosity decreases, the density increases, the nonlinear performance improves, and the leakage current decreases. At the same time, since the voltage-sensitive potential depends on the number of grains per unit length, the voltage-sensitive voltage decreases as T1 increases.

3.2 Influence of Holding Time on Varistor Performance

T1 is 1025°C, *v*1 is 10°C/min, *t*1 is 5 min; T2 is 800°C, *v*2 is 5°C/min, *t*2 is set to 60, 120 and 240min respectively. The measured electrical performance parameters are as follows:

Table 2 Comparison of Non-linearity Coefficient, Leakage Current and Voltage-sensitive Voltage under Different Insulation Time

t2/min	α	J_L/（$\mu A\cdot cm^{-2}$）	V_{1mA}/（$V\cdot mm^{-1}$）
30	41.9	3.71	286
60	53.8	1.69	342
120	46.8	1.62	361
240	47.5	1.60	345

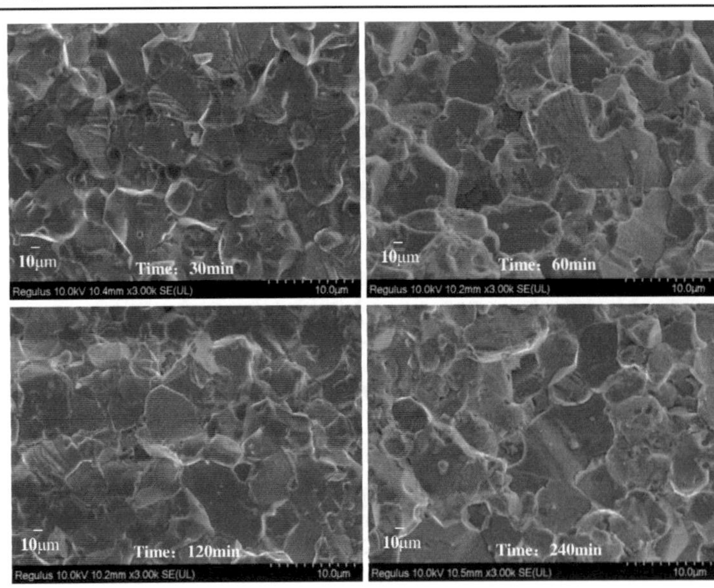

Figure3. SEM photos of samples at different holding times

As can be seen from the above figure, when the holding time is increased from 30min to 60min, the grain size slightly increases and the densification further increases. When the holding time is further prolonged, the porosity tends to be saturated and the properties tend to be stable.

3.3 Influence of Sintering Methods on Varistor Performance

Comparison of two-step sintering with traditional sintering and microwave One-step sintering is as follows:

Table 3 Comparison of Sex Coefficient, Leakage Current and Voltage-sensitive Voltage under Three Sintering Non-linear Conditions

Sintered Powder	T/℃	*t*/min	α	J_L/（μA·cm^{-2}）	V_{1mA}/（V·mm^{-1}）
One-step Sintering	1050	240	42.2	5.05	203
Two-step Sintering	1050	240	48.0	2.11	303
Ordinary Sintering	1050	240	40.7	1.71	266

As can be seen from the above table, compared with traditional sintering, the potential gradient of wave sintering is increased, but the leakage current also increases. Compared with ordinary microwave sintering, the two-step sintering can further effectively reduce the grain size, enhance densification, not only improve the potential gradient, but also reduce leakage current.

4. Conclusion

In this paper, submicron zinc oxide powder varistor was prepared by two-step sintering. The effects of microwave sintering temperature and holding time on the microstructure and electrical properties of zinc oxide varistor were investigated and compared with traditional sintering method and ordinary microwave sintering. The conclusions are as follows:

1) Compared with traditional sintering, microwave sintering has higher potential gradient, but leakage current also increases;

2) Compared with ordinary microwave sintering, the two-step sintering can further effectively reduce the grain size, enhance densification, not only improve the potential gradient, but also reduce leakage current;

3) In two-step sintering, increasing the sintering temperature can effectively reduce leakage current and nonlinear coefficient, but the potential gradient is also greatly reduced. When the holding time is more than 1 hour, there is little difference in electrical performance. When the holding time is too small, its electrical performance is generally weak.

5. References

[1] Rodolfo F.K. Gunnewiek, Ruth H.G.A. Kiminami. Two-step microwave sintering of nanostructured ZnO-based varistors

[2] Yulong Zhang, Ye Yang, Xianpeng Zhang, et al. Two-Step Sintering of Pristine and Aluminum-Doped Zinc Oxide Ceramics [J]. International Journal of Applied Ceramic Technology,

2012,9 (5)： 960-967.

[3] Wei Chen, X.-H. Wang. Sintering dense nanocrystalline ceramics without final-stage grain growth [J]. Nature, 404 (9)： 168-171.

[4] Two-step flash sintering of ZnO: Fast densification with suppressed grain growth

[5] Faical Kharchouche, Etienne Savar, Anthony Thuault. Effects of microwave sintering on intrinsic defects concentrations in ZnO-based varistors

[6] Pedro Durán, Francisco Capel, Jesús Tartaj. A Strategic Two-Stage Low-TemperatureThermal Processing Leading to Fully Dense and Fine-Grained Doped-ZnO Varistors [J]. Advanced Materials, 2002，14(2)： 1376-141.

[7] Rodolfo F.K., Gunnewiek, Ruth H.G.A. Kiminami. Two-step microwave sintering of nanostructured ZnO-based varistors [J]. Ceramics International，2017 (43)： 847-853.

[8] Chen I W, Wang X H. Sintering Dense Nanocrystalline Ceramics without Final-stage Grain Growth. Nature. 2000,404:168~171.

IOP Conf. Series: Materials Science and Engineering **758** (2020) 012019 doi:10.1088/1757-899X/758/1/012019

Growth and infrared switching properties of deposited VO_2 films at various sputtering power with a VO_2 target by RF magnetron sputtering

N Muslim[1], Y W Soon[1] and N Y Voo[1]

1 Physical and Geological Sciences, Faculty of Science, Universiti Brunei Darussalam, Jalan Tungku Link, Gadong BE 1410, Brunei Darussalam
E-mail: noormariah.muslim@ubd.edu.bn, noormariahmuslim219@gmail.com

Abstract. In this work, structural and morphological properties as well as phase transition temperature and hysteresis width of VO_2 thin films grown with a VO_2 target by RF magnetron sputtering were studied under the influence of relatively low sputtering power of 80 – 140 W. It was observed that as sputtering power increased, the crystallinity of the films improved with the presence of obvious diffraction peaks of VO_2, and a slight increase in the average crystallite and grain sizes of the films. The deposited films revealed a slight change in infrared transmittance during heating and cooling cycles. Films deposited at sputtering power of 140 W exhibited the lowest Tt of 60.5 °C with a hysteresis width of 41.0 °C.

1. Introduction

Vanadium dioxide or Vanadia (VO_2), due to intelligent thermochromic properties, has received a great attention over the last few decades. The properties make it very beneficial for the electrochemical energy storage, smart windows as well as aerospace and microelectronics industries [1-2]. However, studies on optical properties such as infrared transmittance, and surface properties such as morphology, topography and crystal structure, of VO_2 thin films are crucial for those applications [3-4].
Many investigations on preparations and developments of VO_2 thin films have been reported [5]. For example, studies on surface and optical properties of direct-current (DC) [6-7] and radio-frequency (RF) [8-9] magnetron sputtered VO_2 thin films have been investigated. VO_2 films have also been prepared using RF magnetron sputtering to study its structural, morphological, optical and thermochromic properties for the application of solar control coatings [10]. Meanwhile, VO_2 thin films deposited with high sputtering power of 330 – 400 W by RF magnetron sputtering have also been studied [11]. Control of sputtering power was highly significant to achieve high purity monoclinic VO_2 films. In addition, at such high sputtering power of 400W, target poisoning happened due to a chemisorption of oxygen on the surface of the target causing a reduction in deposition rate and film thickness with large grain size, and an increase in formation of defects. Thus, this leads to films with porous structure [12]. Hence, in order to prevent porous films, this present work will study depositions of VO_2 films at low sputtering power to ensure pores filled during the film depositions.
This work will investigate the influence of low sputtering power of 80 – 140 W on structural, morphological and infrared switching properties of VO_2 films grown with a VO_2 target by RF magnetron sputtering, which has not been considered yet.

2. Details of experiments

2.1. Preparation of films

The method for preparing VO_2 films was reported as in our previous papers [6,10]. VO_2 films were prepared on Corning 2947 glass substrates by RF magnetron sputtering using a high purity VO_2 target (99% purity) of 76 mm diameter and 5 mm thickness. The substrates were ultrasonically cleaned with ethanol for about 15 min and rinsed with distilled water. The substrates were then blow-dried using high purity nitrogen gas (99.9% purity), and screwed onto a substrate holder in the sputtering chamber. Commercial high purity argon, Ar (99.9% purity) was supplied into the chamber as the sputtering gas. The sputtering chamber was evacuated to 10^{-4} Pa by a turbomolecular pump that was backed by a mechanical pump. The Ar gas flow was kept constant at a gas flow rate of 40 sccm by a mass flow controller. Before depositing VO_2 onto the substrates, Pre-sputtering of the VO_2 target in Ar atmosphere was performed for 10 min to eliminate any oxide layers and impurities. The substrates were rotated at 10 rpm throughout the film deposition in order to get films with uniform thickness. The film thickness during deposition was maintained at 50 nm and monitored using a quartz crystal thickness meter with 1 Angstrom (Å) resolution reliably supplied by Inficon (Model SQM-160). During sputtering, the chamber pressure and substrate temperature were 10^{-1} Pa and 500 °C, respectively. The films were grown on the substrates at various sputtering power of 80 – 140 W.

2.2. Characterization of films

A surface profilometer (Alpha-Step IQ) was employed to ensure film thickness of the deposited VO_2 films by applying a stylus force of 0.03 N with measurement accuracy of ± 1.5 nm. Crystal structure of the films was analyzed by grazing incidence X-ray diffraction (GIXRD) on a Shimadzu XRD-7000 using monochromatic high-intensity Cu Kα (λ = 1.54056Å) radiation operated at 40 kV and 30 mA. GIXRD patterns of the films were detected in the range of 2θ = 20 – 40° and an incident angle of 1.0 ° at a scanning rate of 2 °/min. Surface morphology was studied by field-emission scanning electron microscopy (FE-SEM) using a JEOL JSM-7610F field-emission scanning electron microscope. The infrared switching properties, i.e. phase transition temperature and hysteresis width in infrared region, were recorded using a CARY 5000 UV-Vis-NIR spectrophotometer (Agilent Technologies) assembled with a SPECAC 4000 Series™ high stability temperature controller. The temperature-dependent transmittance spectra were taken at a wavelength of 2.5 μm with temperature varied from room temperature of 25 °C to 100 °C.

3. Results and discussion

3.1. Crystal structure

The GIXRD patterns of the VO_2 thin films deposited at sputtering power of 80 – 140 W are detected as shown in figure 1. It could be seen that all films have an obvious diffraction peak at 2θ = 27.9 ° of monoclinic VO_2(M) phase. Thus, it can be deduced that the deposited films have a prominent peak of VO_2(M) corresponding to (011) orientation plane. The peak enhanced with increasing sputtering power. As the sputtering power stepped-up from 100 to 140 W, another peak appeared at 2θ = 29.2 ° of metastable VO_2(B) phase. When the sputtering power was 80 W, the deposition rate was 0.72 nm/min. The films were found to have low film crystallinity as seen from the low peak intensity. This was also reported by Ahmadipour et al. [13], in which the low sputtering power leads to very low formation energy which was insufficient to cause production of a high crystalline thin film. As the sputtering power rose up to 140 W, the deposition rate seemed to increase up to 1.32 nm/min. This might be due to more energetic particles bombarded with high driving energy and settled on suitable lattice sites resulting in high film crystallinity [14]. In addition, the average VO_2 crystallite size can be estimated using the (011) crystallite orientation and Scherrer formula. The average crystallite size of the films at sputtering power of 80, 100, 120 and 140 W was 14.9, 25.6, 39.0 and 43.1 nm, respectively. It has also been reported by several studies that enhancement of crystallite size was obtained by increasing the RF sputtering power [13,15-16].

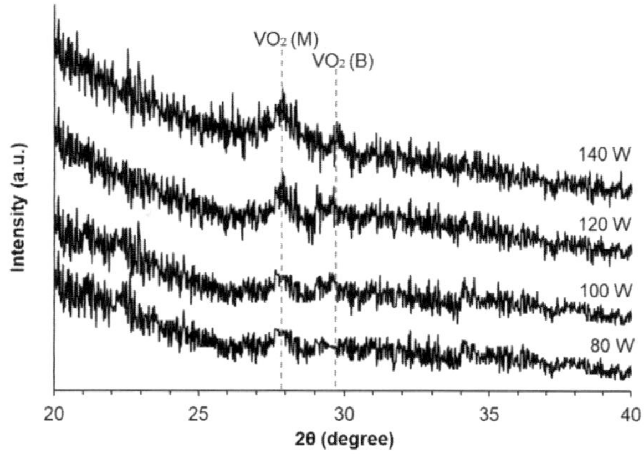

Figure 1. GIXRD patterns of monoclinic VO_2(M) and metastable VO_2(B) thin films at sputtering power of 80 – 140 W.

3.2. Surface morphology

Figure 2 represents the FESEM images of the films at sputtering power of 80 – 140 W. Spherical-like grains in VO_2(M) phase are grown on the film surface with a slight increment in the grain size of the VO_2 as the sputtering power increased. Irregular and dark areas, which indicate deep and wide trenches, are seen on the surface of the films grown at sputtering power of 80 W. As the sputtering power stepped-up, from 100 to 140 W, uniform grain distributions are observed on the film surface. It has been reported that, at high sputtering power, more ionized Ar gas particles bombarded on the VO_2 target, and hence more energetic target atoms are dominant causing the formation of nuclei on the film surface [14,16]. Thus, this indicates more sites for growth of the grains. In other words, when the energetic atoms implanted into the surface of the films, this induced a thermal effect in the substrate [11]. This will cause the grain size to increase and the defect density to decrease with elevating sputtering power.

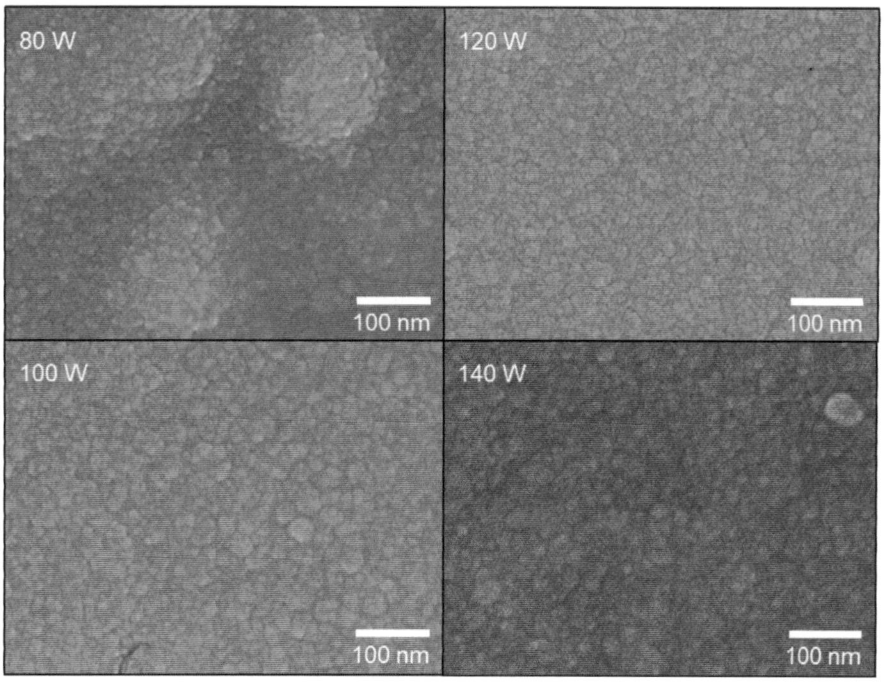

Figure 2. FESEM images of VO_2 thin films grown at sputtering power of 80 – 140 W.

3.3. Temperature-dependent transmittance spectra

Figure 3 exhibits transmittance spectra measured in heating and cooling cycles at a wavelength of 2.5 μm for the films prepared at various sputtering power. Hysteresis loops could also be obviously observed. According to Luo et al. [11], grain size could influence the hysteresis width. When the films have bigger grain size, the hysteresis width will be narrower. Particularly, the transition temperature of the VO_2 films was reported to correlate with microstructures such as defect density and grain boundary of the films [11]. By increasing sputtering power, the defect density could be reduce and lower transition temperature could be achieved. In addition, Table 1 shows the infrared switching properties i.e. phase transition temperature and hysteresis width, of the prepared films in infrared region with respect to the figure 3. Infrared transmittance of the films varied with sputtering power at the measured wavelength. The film deposited at 80 W showed a slightly high transmittance than the other films. The Tt of the films prepared at 80 W is 65.5 °C with a hysteresis width of 57.0 °C; while that of the films deposited at 100 W is 69.5 °C with a hysteresis width of 43.0 °C; and the films deposited at 120 W is 65.0 °C with a hysteresis width of 44.0 °C. The lowest Tt of 60.5 °C with a narrowest hysteresis width of 41.0 °C was observed for the films prepared at sputtering power of 140 W. This might due to high crystallinity with larger average crystallite size of the VO_2 films sputtered at sputtering power of 140 W. Furthermore, the nonmonotonic variation of transition temperature against sputtering power obtained in this study was also reported in a study whereby VO_2 films prepared at sputtering power of 330 – 400 W [11].

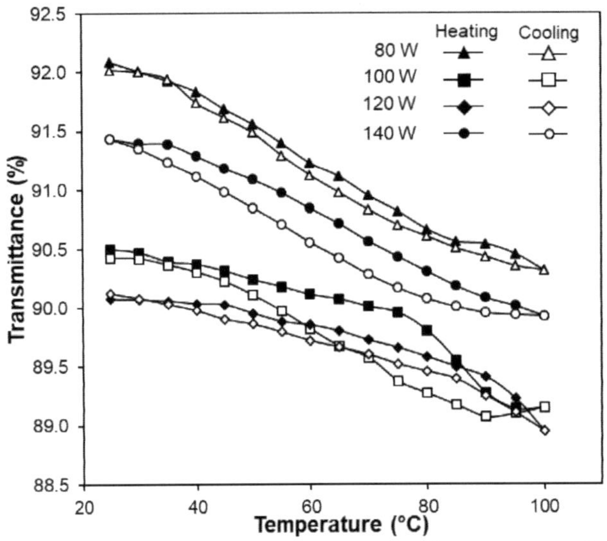

Figure 3. Temperature-dependent transmittance of VO_2 thin films at sputtering power of 80 – 140 W. Colored and uncolored triangles represent heating and cooling cycles, respectively.

Table 1. Phase transition temperature and hysteresis width of VO_2 thin films deposited at various sputtering power, measured at a wavelength of 2.5 μm.

Sputtering power (W)	Phase transition temperature (°C)	Hysteresis width (°C)
80	65.5	57.0
100	69.5	43.0
120	65.0	44.0
140	60.5	41.0

4. Summary
The influence of relatively low sputtering power on several properties of VO_2 thin films prepared with a VO_2 target by RF magnetron sputtering were analysed. This study also showed the ability of the technique employed to control sputtering power in order to obtain desired VO_2 thin films. As sputtering power increased, the film crystallinity enhanced revealing diffraction peaks of VO_2, and a slight increase in the average crystallite sizes. A slight change in infrared transmittance of the deposited films during heating and cooling cycles was observed. By increasing sputtering power, the average crystallite and grain sizes of the films rose which contributes to narrow hysteresis width. In addition, the enhancement in sputtering power resulted in lower transition temperature due to decrease in defect density. Among the prepared films, films prepared at sputtering power of 140 W have the lowest Tt of 60.5 °C with a hysteresis width of 41.0 °C.

5. References
[1] Prasadam V P, Bahlawane N, Mattelaer F, Rampelberg G, Detavernier C, Fang L, Jiang Y, Martens K, Parkin I P and Papakonstantinou I 2019 *Materials Today Chemistry* **12** 396.
[2] Liu M, Su B, Tang Y, Jiang X and Yu A 2017 *Advanced Energy Materials* **7** 1700885.
[3] Strelcov E, Davydov A V, Lanke U, Watts C and Kolmakov A 2011 *ACS Nano* **5** 3373.
[4] Ho H C, Lai Y C, Chen K, Dao T D, Hsueh C H and Nagao T 2019 *Applied Surface Science* **495** 143436.
[5] Muslim N, Soon Y W, Lim C M and Voo N Y 2018 *IOP Conf. Series: Materials Science and Engineering* **409** 012016.
[6] Muslim N, Idris M N S M, Soon Y W, Chou Chau Y F, Lim C M and Voo N Y 2018 *IOP Conf. Series: Materials Science and Engineering* **409** 012025.
[7] Fang B, Li Y, Tong G, Wang X, Yan M, Liang Q, Wang F, Qin Y, Ding J, Chen S, Chen J, Zheng H and Yuan W 2015 *Optical Materials* **47** 225.
[8] Reddy I N, Sreedhar A, Shim J and Gwag J S 2019 *Journal of Electroanalytical Chemistry* **835** 40.
[9] Sullivan M, Son T V, Beaudoin N and Haché A 2015 *Optics Communications* **356** 395.
[10] Muslim N, Soon Y W and Voo N Y 2019 *IOP Conf. Series: Materials Science and Engineering* **538** 012022.
[11] Luo Y Y, Pan S S, Xu S C, Zhong L, Wang H and Li G H 2016 *Journal of Alloys and Compounds* **664** 626.
[12] Zhang X, Ren W, Kong W, Zhou Q, Wang L, Bian L, Xu J, Chang A and Jiang C 2018 *Applied Surface Science* **435** 815.
[13] Ahmadipour M, Ayub S N, Ain M F and Ahmad Z A 2017 *Materials Science in Semiconductor Processing* **66** 157.
[14] Li S, Jiao S, Wang D, Gao S and Wang J 2018 *Journal of Alloys and Compounds* **753** 186.
[15] Srinivas K, Manivel Raja M, Sridhara Rao D V and Kamat S V 2014 *Thin Solid Films* **558** 349.
[16] Lu Y M, Hwang W S, Liu W Y and Yang J S 2001 *Materials Chemistry and Physics* **72** 269.

Acknowledgments
This research study was financed by the Brunei Research Council (S&T 17), and Universiti Brunei Darussalam Block Grant (UBD/RSCH/1.4/FICBF(b)/2019/020).

IOP Conf. Series: Materials Science and Engineering **758** (2020) 012020 doi:10.1088/1757-899X/758/1/012020

Clamp bending machine and annealed wire cutter for reinforced concrete columns

J Marron[1], G Quispe[2], C Raymundo-Ibañez[3,*], Moises Perez[4]

[1] Facultad de Ingeniería, Universidad Continental, Lima 15046, Perú.
[2] Ingeniería Industrial, Universidad Nacional Autónoma Altoandina de Tarma, Tarma 12651, Perú.
[3] Dirección de Investigación, Universidad Peruana de Ciencias Aplicadas (UPC), Lima 15023, Perú.
[4] Escuela Superior de Ingeniería Informática, Universidad Rey Juan Carlos, Mostoles, Madrid, España.

*carlos.raymundo@upc.edu.pe

Abstract. This study developed a reinforced steel rod bending machine for rods with diameters of up to 8 mm and annealed wire cutter for up to 5 kg for replacing manual intervention required to bend rods in reinforced concrete columns. This study aims to reduce the physical effort that could lead to occupational diseases, such as tenosynovitis, bursitis, muscle disorders. Clamp manufacturing possesses great risk for workers, who are exposed to injuries while using different cutting devices, such as grinders and electric saws. They also face potential problems such as muscular fatigue due to the nonergonomic and repetitive work positions. The proposed machine features a mechanical dragging and bending systems and manual shears. Additionally, the proposed machine has been designed theoretically and its effectiveness has been assessed through simulations conducted using the SolidWorks CAD software. A bending machine prototype for producing clamps is developed and its machine productivity is measured. Using this machine, approximately 300 clamps can be bent per hour without possessing any risk to the worker.

1. Introduction

Steel bars that are used as reinforcements in building and construction structures to prevent natural phenomena, such as earthquakes, are frequently used in Peru. Owing to the geographical location of Peru, such practice has become mandatory in all types construction works. Civil engineers, who are generally in charge of these projects, build structural plans, listing steel bar measurements and designs according to the requirements and needs of a particular construction project. When the construction is ready to commence, the best method for performing this task quickly, safely, and efficiently must be selected (Figure 1) [1].

This study presents one of the three alternatives that currently exist for the preparation of reinforcement steel bars for reinforced concrete as an alternative solution for the traditional model used today. This way, a more reliable, technical, and safe process may be implemented. This process would also allow for the production of shapes with diameters larger than standard, which cannot be created regularly due to the individual limitations of each worker [2] [3].

This project is expected to contribute to the improvement of construction techniques and creation of different shapes based on the proposed design of a machine that performs this work. This machine is described below based on data, plans, and guidelines [4].

2. Description

2.1. Reinforcing Steel

Reinforced concrete is defined as the material resulting from the union of cast concrete, i.e., a proportional mixture of Portland cement, or of any other hydraulic cement, with sand, gravel, and clean water with or without additives, which when setting and hardening becomes more resistant and armoring or reinforcing steel bars, combined in such a way that they create a solid, monolithic and unique element in terms of their physical characteristics, thus leveraging the individual features of both materials [5][6].

2.2. Annealed wire no. 16

Generally, wire is a filament or thin rod of a flexible metal with a uniform section. The metals usually used for wire production are copper, aluminum, steel, brass, iron, gold, silver, and platinum. Specifically, wire No. 16 is made of low carbon steel (Figure 2). The most modern methods for wire manufacturing consist of stretching a metal rod, passing it through increasingly thin, tapered holes, made of tungsten carbide dies, until the required diameter is obtained (wire drawing)[7][8][9].

Figure 1. Manual bending of rods

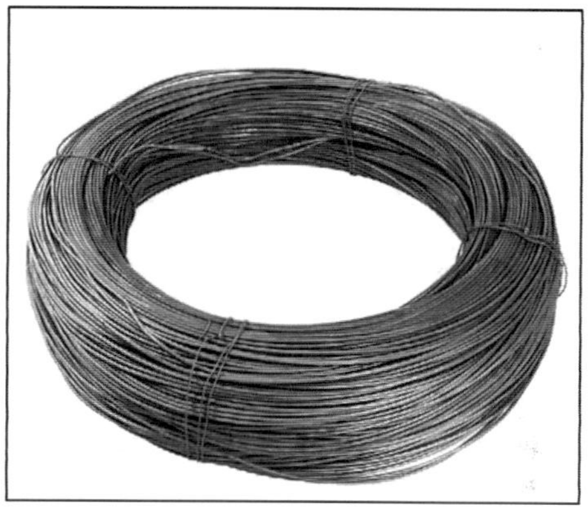

Figure 2. Annealed wire no. 16

2.3. Design and Construction

The design of the machine begins with the calculation of the bending force of the corrugated rods.

$$F = \frac{4 \,.\, K \,.\, \sigma máx. \, I}{L \,.\, c} \tag{1}$$

To bend the rods, the formula of maximum flexion or effort is applied in relation to the plastic moment as follows:

$$\sigma máx = \frac{M \,.\, c}{I} \tag{2}$$

The necessary force required to bend the corrugated rod is calculated. For this calculation, the flexion formula was used, which was combined with the plastic factor.

To determine the bending force, the following indicators will be needed: maximum yield stress, inertia, supports length, distance from the neutral axis, and K factor. With this result, only the starting bending

IOP Conf. Series: Materials Science and Engineering **758** (2020) 012020 doi:10.1088/1757-899X/758/1/012020

force is calculated but the required result is not achieved yet. To achieve the expected result, weights must be applied to a mechanical lever system to determine the bending factor (fd = 1.32) and, in turn, the total force.
The necessary force required to bend the corrugated rod is calculated as follows:

$$K = \frac{Mp}{P} \tag{3}$$

The cut separates the punch and screen that is embedded in a section of the material along an area defined by the periphery of both elements. The cut is a mechanical action that includes splitting or separating a longitudinal element or material into several parts without causing damages or producing shavings. Additionally, the sheet metal cutting formula is used.

$$V = T\,.\,A \tag{4}$$

After bending and cutting calculations, the machine mechanism is designed by determining the most important forces, i.e., the driving and cutting forces. The cutting force meets the need and driving force is proportional to the torque provided by the reduction motor.
The following procedure and formula are used to calculate the required power:

$$M_T = \frac{63025{,}35 * P}{N} \tag{5}$$

where
MT = Torsion moment [kg.cm],
P = System power in [HP], and
N = Number of revolutions [rpm].

To calculate the required axis diameter, first, the existing forces of the axis and its flexing momentum are analyzed.

When assessing the position and direction of the forces, it is observed that all calculations are made using the XZ plane as a referential point. The force distribution is shown in Figure 3.

With this data, the axis diameter can now be calculated but not before selecting a safety factor of three due to design issues. Replacing this value in the equation, we have:

$$n_y = \frac{S_y}{\sigma_{máx}} \tag{6}$$

where
Sy = axis yield strength, as per AISI 1018 standards.

2.4. Operations and Parts
The machine will be designed to bend clamps and cut annealed wire number 16 using a 4 HP engine. The torque produced by a rotary movement is converted to a linear movement using a crankshaft-rod device, as shown in Figure 4.

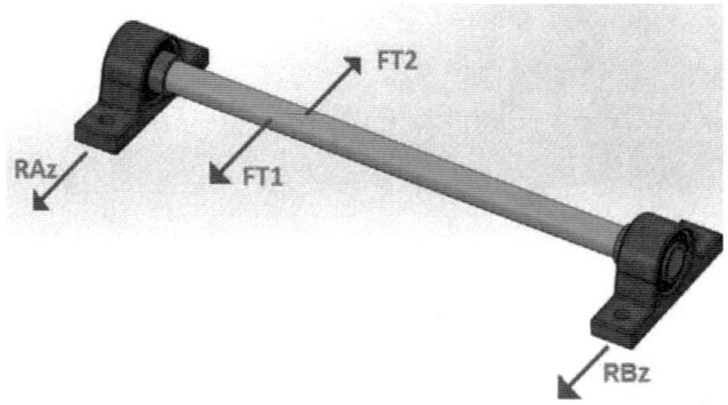

Figure 3. Distribution of Forces

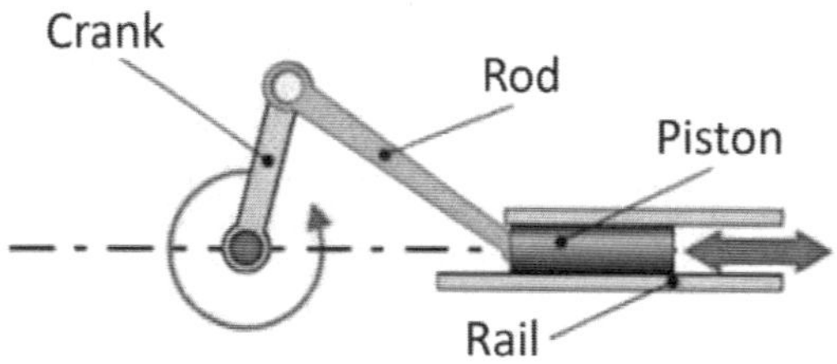

Figure 4. Crankshaft-Rod device

The force is transmitted to the bending arm, generating movement in the rear links, which move and create reactions necessary to cut the corrugated rod at the rear end of the machine.

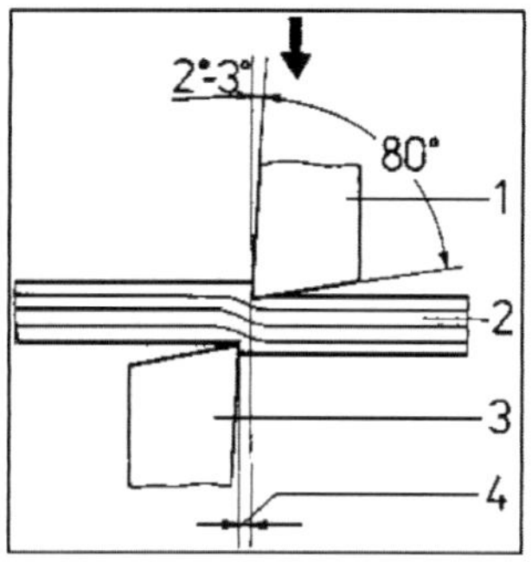

Figure 5. Shear cut

The main parts of the machine are cutting and bending systems, which are adjustable to the required limited distances.

3. Results

The machine will be designed to bend several clamps of up to 8 mm in diameter and cut annealed wires of up to 5 kg in different sizes.

In functional tests, 300 bent rods must be obtained in 1 h, which would increase production and improve the finishing and shape (Figure 6). In addition, these results must exceed the results obtained manually, implying that the worker is free to work in another activity for about 5 h, thus preventing fatigue and reducing operating costs.

Figure 6. Clamp bending machine

IOP Conf. Series: Materials Science and Engineering **758** (2020) 012020 doi:10.1088/1757-899X/758/1/012020

4. Conclusions

Based on the functional-technical criteria for the final design of the rod bender, the shape and dimensions of the machine were determined, along with the component strength required. Through the manufacturing criteria, steel profiles available in the market and used by local workshops were selected. Additionally, the economic factor has been important in the design, since the cost of profiles, screws, bearings, and other elements, as well as the manufacturing, must be affordable for the company.

One of the options that meets profile production needs is bending through semiautomatic machines since it bends reinforced steel bars, from smaller diameters to standard commonly used sizes.

The design of the clamp bending machine seeks to increase productivity, since an operator was able to bend 500 units manually per day, whereas the semiautomatic machine might bend 300 clamps in an hour. This will help prevent the physical exhaustion of workers.

5. References

[1] CONFER S.A. ALAMBRE RECOSIDO COMPER. CONFER S.A. [En línea] [Citado el: 20 de 04 de 2019.] http://www.comfer.com.pe/procesos/alambre-recocido-comfer/16/.
[2] Curbelio, Basilio J. 2015. CONCRETO ESTRUCTURAL. Quindio: Armenia, 2015.
[3] DETERMINACIÓN DE LA RESISTENCIA A LA FLUENCIA DE UN ACERO AISI 1040. M. Torres, V. Di Graci y G. González. 2000. Caracas: RLMM, 2000.
[4] Diseño de una máquina para cortar y doblar estribos en seriee para columnas de hormigon armado. Pérez Silva, José Olger y Trujillo Zurita, Oscar Paul. 2016. 16, Quito: Zaimella, 2016. 1390-650X.
[5] DUCTILIDAD Y TENACIDAD DE UN BORDE CORTADO POR PLASMA EN UNA CHAPA DE ACERO S460M. J Aldazabar, y otros. 2014.
[6] G. Budinas, Richard y J. Keith, Nisbett. 2008. DISEÑO EN INGENIERÍA MECÁNICA DE SHIGLEY. MEXICO: MCGRAWHILL, 2008.
[7] G. Budynas, Richard y J. Keith, Nisbett. 2008. DISEÑO EN INGENIERÍA MECÁNICA. MEXICO: MC GRAW-HILL INTERAMERICANA, 2008.
[8] Máquina dobladora para fabricar estribos cuadrados de varilla trefilada de acero de 6 mm de diámetro. Guerrero, Diego. 2017. Ibarra : FICA,, 2017.
[9] Trujillo Zurita, Oscar Paul. 2014. DISEÑO Y CONSTRUCCIÓN DE UNA MÁQUINA PARA CORTAR Y DOBLAR ESTRIBOS EN SERIE PARA COLUMNAS DE HORMIGON ARMADO. QUITO: s.n., 2014.

Application and Development of the Super-large Diameter Column Foundation of Bridge

Ye Qiaodan [1*], Wang Junjie [1] and Wang Changjiang [2]

1 Department of Bridge Engineering, Tongji University, Shanghai, 200092, China

2 Zhejiang Provincial Institute of Communications Planning, Design&Research, Hangzhou, Zhejiang, 310000, China

E-mail: yqd134167@163.com

Abstract. With the increase of span, the structural form of bridge foundation is seeking for development. The super-large diameter column foundation has become one of the best alternatives for bridge substructures. With the development of bridge foundation, the application status of the super-large diameter column foundation is combed. Its construction methods are briefly summarized, such as artificial digging, embedding prestressed pipe and super-large diameter bored pile, which promote the application of single column foundation. In the future, engineers can also get inspiration from the offshore wind turbine and seek new development ideas. Combined with the application of super-large diameter column foundation in the cross-sea bridge at home and abroad, the unique advantages and the problems needed to be solved in the practical application of the super-large diameter column foundation of bridge are pointed out, and its development is prospected.

1.Introduction

Column foundation of bridge includes pile-column foundation and pipe-column foundation, which are deep foundation. Pile-column foundation mainly refers to solid-pile foundation, which is usually constructed by digging and drilling. Pile-column foundation has the characteristics of high bearing capacity and wide application range. Pipe-column foundation mainly refers to hollow circular foundation, which contains prestressed concrete pipe-column and steel pipe-column. Pipe-column foundation has the characteristics of reducing masonry volume, improving construction quality and realizing prefabrication of factories, which is suitable for bridge site conditions such as deep water and undulating rock surface. In the bridge engineering, the foundation with diameter larger than 2.5m is generally defined as the large diameter foundation. But in order to realize the single-column foundation without cap for deep-water cross-sea bridges, the diameter usually reaches 4m or more, which is called "super-large diameter foundation". This paper introduces the development of super-large diameter column bridge foundation and briefly summarizes its construction methods. Combining with the application of super-large diameter column foundation in overseas and domestic bridges, especially the unique advantages of single-column foundation in practical application and the problems that need to be solved, the development prospects of it are prospected.

2.Development of super-large diameter column foundation

2.1. Pile-column foundation

The diameter of the bored-pile foundation of the main bridge of the New Bayi Bridge in Nanchang, Jiangxi Province, which opened to traffic in 1997, reaches 4m[1]. The layout of main bridge foundation caps of New Bayi Bridge in Nanchang is shown in figure 1[2]. Since then, bored and cast-in-place pile foundation has been widely used in bridges in China. The main tower foundation of Poyang Lake Bridge built in 1999 is super-large diameter bored piles with rock-socketed depth of 7m[3].

Because of the difficulty, the high risk and the high cost of underwater construction of cap, engineers in China have developed the technology of composite section pier without cap, which has been first developed and applied on land and cross-river bridges. In 1986, a new concept of variable cross-section pile-column foundation without cap was put forward for Jiujiang Bridge in Guangdong Province. After many engineering practices, in 1991, Xiangtan Second Bridge realized the use of single row piles with super-large diameter column without cap, which diameter reached 5.0m, as shown in figure 2[4, 5].

2.2. Pipe-column foundation

During the construction of Wuhan Yangtze River Bridge in 1953, Chinese pioneered the construction method of pipe-column foundation. In 1958, the pipe-column foundation was applied to the new bridge of Zhengzhou Yellow River Bridge, and its diameter increased to 3.6m. In 1993, the Yabadu Bridge in Hunan Province adopted a variable cross-section of single pipe-column foundation structure with diameter of 3.8m/3m. In 1995, the Taoyuan Yuanshui Bridge adopted two separated non-capped pipe-column foundation with a maximum diameter of 7.5m, as shown in figure 3[6].

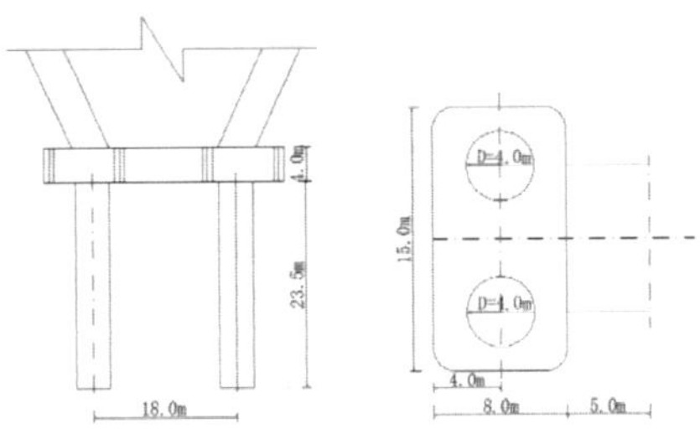

Figure 1. Layout of main bridge foundation cap of New Bayi Bridge

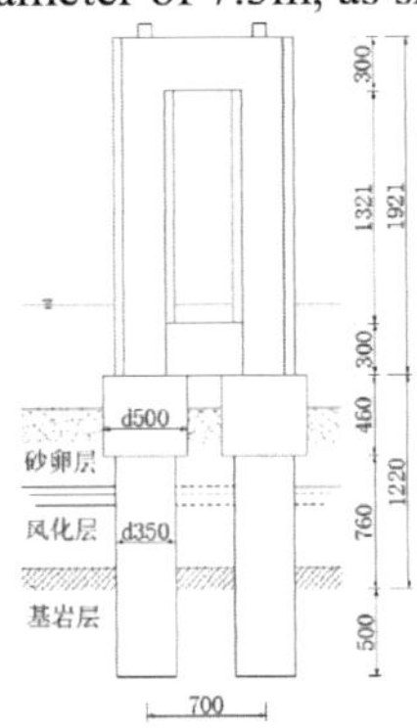

Figure 2. Pile-column foundation of Xiangtan Second Bridge

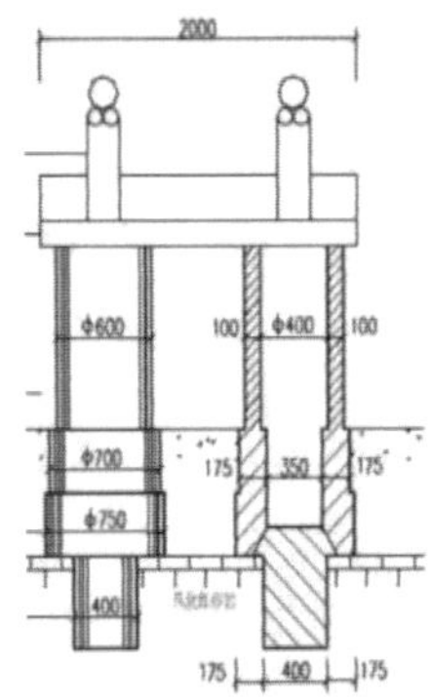

Figure 3. Pipe-column foundation of Taoyuan Yuanshui Bridge[6]

2.3. Brief summary of column foundation engineering application

Super-large diameter column foundation is widely used in bridge engineering in China, and it has good prospects for development. The application of super-large diameter column foundation is summarized as shown in table 1.

Table 1. Column foundation of super-large diameter bridge

Bridge Name	Construction Time	Bridge Type	Maximum Span /m	Bridge Width /m	Type of Foundation	Pile Diameter /cm	Pile Length /m
Jiujiang Bridge	1988	cable-stayed bridge	160	16	bored pile-column foundation without cap	ϕ 300/250/200	70
Xiangjiang Second Bridge	1992	continuous beam	90	20	bored pile-column foundation without cap	ϕ 500/350	18
Yuanshui Bridge	1991	continuous beam	140	16	bored pile-column foundation without cap	ϕ 400/350	12
Yabadu Bridge	1993	continuous beam	30	12	buried pipe-column foundation without cap	ϕ 380/300	35
Nanhuadu Bridge	1994	cable-stayed bridge	50	17	buried pipe-column foundation without cap	ϕ 300/250	35
Shiguishan Bridge	1994	continuous rigid frame	80	12	buried pipe-column foundation without cap	ϕ 500/400	40
Qiushui	1994	double curved arch	54	12	digged pipe-column	ϕ 600/300	12

Bridge		bridge			foundation without cap		
Lugu Bay Bridge	1994	stone slab ribbed arch bridge	48	21	digged pile-column foundation without cap	Φ 510/500	24
Taoyuan Yuanshui Bridge	1995	arch bridge	108	17	digged pile-column foundation without cap	Φ 750/600	20
Tongling Yangtze River Bridge	1995	cable-stayed bridge	432	23	bored pile-column foundation without cap	Φ 460/400/280	108
New Bayi Bridge	1997	cable-stayed bridge	160	26	buried pipe-column foundation	Φ 450/400	17
Guanyin Bridge	1998	stone slab ribbed arch bridge	52	21	digged pipe-column foundation without cap	Φ 500/400	10
Xiangjiang South Bridge	1998	continuous beam	85	24	digged pile-column foundation without cap	Φ 350/280/180	18
Poyang Lake Bridge	1999	cable-stayed bridge	318	27.5	buried pipe-column foundation	Φ 500/400	37
Tianma Bridge	2000	continuous beam	75	26	digged pile-column foundation without cap	Φ 800/600	17
Wuliting Bridge	2003	tied arch bridge	120	30	buried pipe-column foundation without cap	Φ 400/560/350	11
Tianxinzhou Bridge	2009	cable-stayed bridge	504	27	buried pipe-column foundation without cap	Φ 450/400	60

3.Construction technology of super-large diameter column foundation

The construction technology of column foundation of bridge in China has made a series of breakthroughs in construction machinery and methods since 1950s. For the super-large diameter column foundation, some construction technologies have been developed, such as artificial digging, embedding prestressed pipe, sinking and digging pipe, super-large diameter bored pile.

3.1. Technology of artificial digging

The cost of artificial digging method is low, which does not need large-scale machinery and equipment. It not only can be easy to guarantee the construction period and construction quality, but can meet the construction requirements of complex stratum environment[7, 8].

In 1997, the main pier of Xiangjiang South Bridge in Changsha adopted two variable cross-section large diameter column foundations by digging[8, 9]. At present, the largest diameter of artificially excavated column foundation of bridge in China is the super-large diameter pipe-column foundation of Shenzhen Bridge with a diameter of 15 m, as shown in figure 4 [10].

3.2. Technology of embedding prestressed pipe

The super-large diameter drilled-embedded single pipe-column foundation eliminates the large volume of cap concrete which is difficult to construct., which combines the advantages of bored pile and prefabricated pile. It can not only effectively prevent the collapse of holes in the process of bored pile formation, but also avoid the difficulty of driving prefabricated pile or the excessive deviation of piling. Segmental prefabrication of pipe string can realize factory construction.

As the representative of the super-large diameter drilled-embedded single pipe-column foundation, the diameter of the single pipe-column foundation of the navigation hole of Shiguishan Bridge in Changde, completed in 1994, reaches 5m, as shown in figure 5[6].

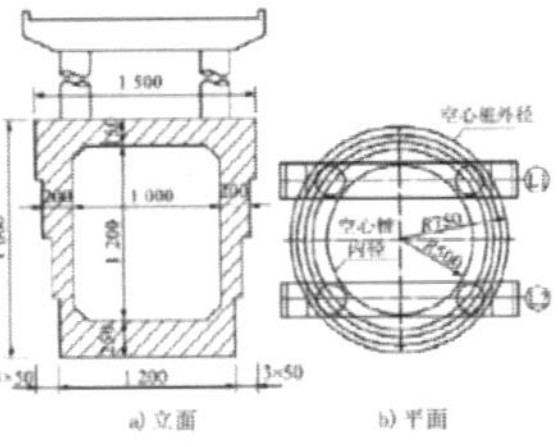

Figure 4. Super-large diameter pipe-column foundation of Shenzhen Bridge[10]

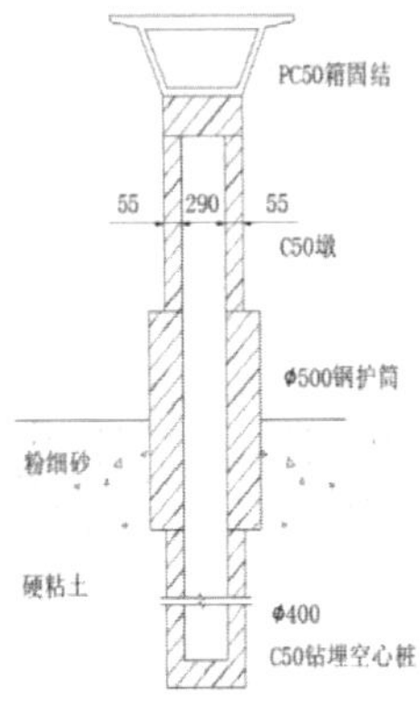

Figure 5. Single pipe-column foundation of Shiguishan Bridge

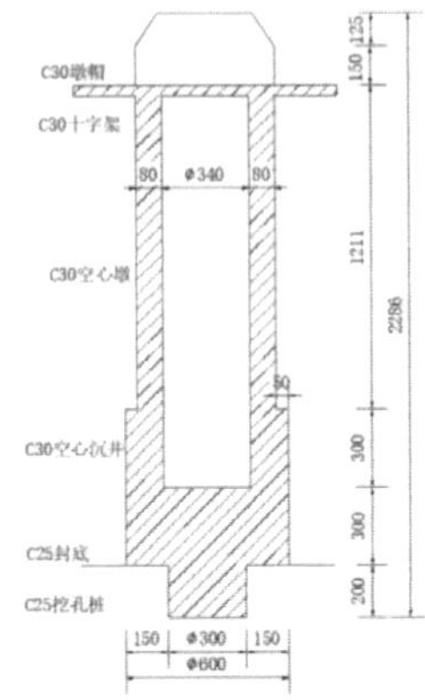

Figure 6. Vertical profile of sinking pipe foundation of Qiushui Bridge

3.3. Technology of sinking and digging pipe

The technology of sinking and digging pipe is a bridge foundation construction technology which combines the traditional construction methods of caisson and hole digging. It has the characteristics of simple technology, short construction period and material saving, which is suitable for all kinds of long-span bridge foundations[11]. This technology was applied in the Guanyin Bridge in Zhangjiajie in 1995 at first, and has been developed on many cross-river bridges. The foundation design of the overbank platform of the Qiushui Bridge completed in 1994 adopts the super-large diameter light circular caisson with a height of 6 m, an outer diameter of 6 m and an inner diameter of 3.4 m. The caisson connects with the super-large diameter pipe-column, which outer diameter is 5 m, forming a variable cross-section large-diameter sinking-pipe-column foundation, as shown in figure 6[12].

3.4. Technology of bored pile

Super-large diameter bored pile-column foundation has the advantages of large bearing capacity of single pile, short construction time, safety and reliability. The main pier of Tongling Yangtze River Bridge in 1995 was designed as a bored pile-column foundation with a diameter of 4.6m/4.0m/2.8m. Wuliting Bridge, which has a span of 120 m and a width of 30 m, adopts the super-large diameter and variable cross-section bored pile-column foundation of 4 m/5.6 m/3.5 m, as shown in figure 7[13].

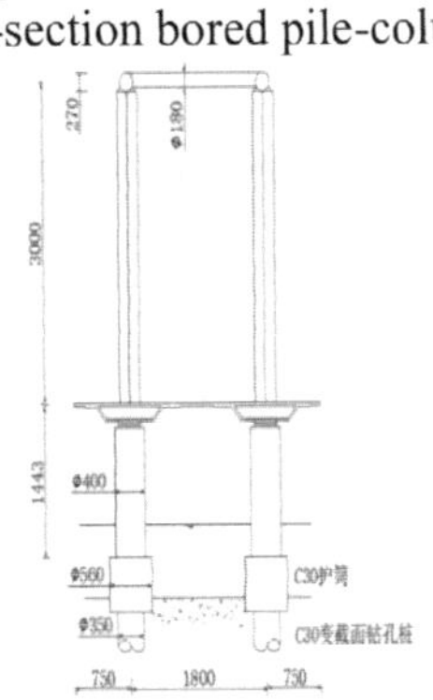

Figure 7. Super-large diameter bored pile foundation of Wuliting Bridge

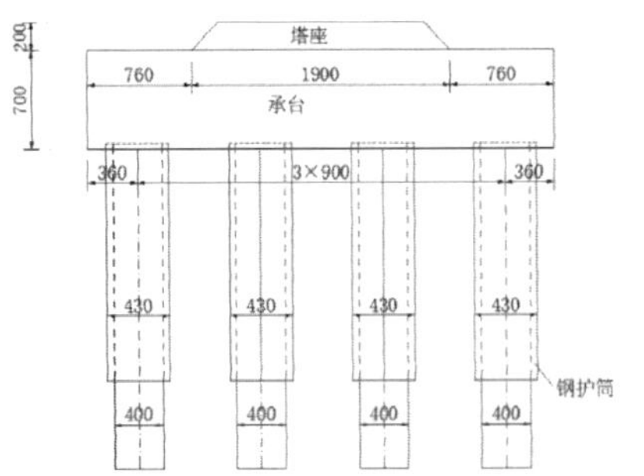

Figure 8 .Main pier foundation of Puqian Bridge[14]

4.Application of super-large diameter column foundation in cross-sea bridges

With the renewal of construction equipment and the improvement of construction methods, the application of super-large diameter column foundation has gradually developed from bridges in land and rivers to deep-water bridges in the ocean, which the diameter and length are also increasing. The

IOP Conf. Series: Materials Science and Engineering **758** (2020) 012021 doi:10.1088/1757-899X/758/1/012021

bored pile-column foundation not only has a good adaptability to marine hydrological and geological environment, but has a wide application range and good seismic effects. The main pier of the Puqian Bridge in Hainan Province in the complex marine environment of strong wind and corrosiveness adopts group-pile foundation, under which 16 variable cross-section super-large diameter bored pile-column foundations of 4.3m are connected, as shown in figure 8[14].

In recent years, it has been widely used in large-scale cross-sea bridges and has broad prospects for development. The approach bridge in water area of Jiashao Bridge, which was completed in 2013, adopts the structure form of super-large diameter bored single-column foundation, as shown in figure 9[15-18]. Its piers and piles are directly connected without caps or tied beams, mainly in order to adapt to the complex hydrological conditions in the bridge site, to reduce the water resistance of piers and the risk of ship collision. It also try not to damage the landscape of tidal in Qiantang River.

The Yushan Bridge, a branch project of the main channel of Zhoushan Port in Ningbo, has 153 single-column piles with super-large diameter and variable cross-section. The diameter of the piles varies from 2.2m to 5m. The substructure of non-navigable holes of Yushan Bridge is shown in figure 10. Pile-column foundation of Yushan Bridge is mostly composite steel pipe, which means when bored cast-in-place pile is constructed, steel retaining barrel is not only a hole-forming protective measure, but also a part of pile foundation to bear load. Composite steel pipe was proposed in the construction of Hong Kong-Zhuhai-Macao Bridge, which has a good flexural capacity and large bearing capacity with simple construction technology. In recent years, it has been widely used in pile foundation engineering of cross-sea bridge and deep-water port[19].

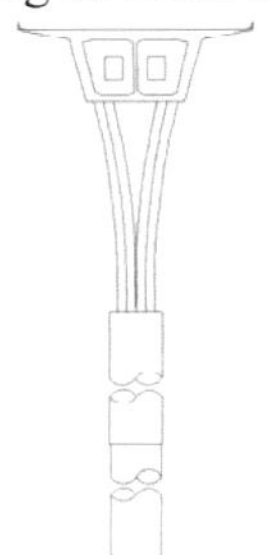

Figure 9. General structural sketch of single-column pier of Jiashao Bridge

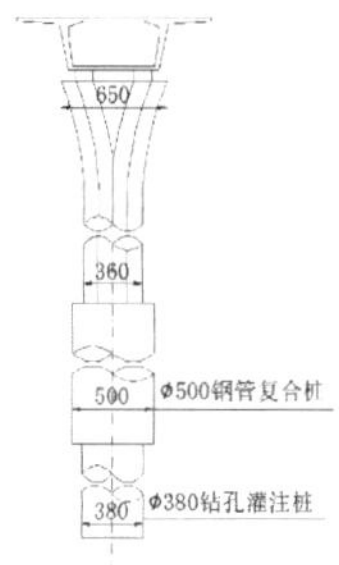

Figure 10. Non-navigable hole lower structure sketch of Yushan Bridge

Super-large diameter column foundation has also been successfully applied in overseas cross-sea bridge projects, among which the Federal Bridge in Canada and the Rion-Antirion Bridge in Greece are the most representative. The Federal Bridge, completed in 1997, is the longest bridge in Canada and the longest bridge in ice water in the world. The bridge has 65 single-tubular foundations with a diameter of 10m, as shown in figure 11. The foundation is assembled by segmental prefabrication, as shown in figure 12[20]. Prefabricated assembling not only avoids site construction in bad environment, but can guarantee construction quality and shorten construction period. Because the bridge located in ice water, it is easy to be affected by drift ice. The column foundation is designed with a conical steel shield below the average sea level of 4m to 2.6m above, in order to reduce the impact of drift ice[21].

Rion-Antirion Bridge in Collins Bay, Greece, opened to traffic in 2004, is located in the earthquake area with magnitude 7. Because the water depth of the bridge site is 65m, and the seabed soil condition is poor, the whole bridge adopts floating concrete structure. The seabed is strengthened by 200 hollow steel pipe networks with the length of 30 m and the diameter of 2 m. The foundation is isolated from the seabed and becomes a sliding foundation, as shown in figure 13[22]. Rion-Antirion Bridge adopts super-large diameter hollow single-tubular column foundation. Four high-strength reinforced concrete columns at the main tower incline toward the center, which are connected at the top of the main tower to provide sufficient stiffness to withstand asymmetric operation load and seismic force. Rion-Antirion super-large diameter hollow column foundation is shown in figure 14[23].

IOP Conf. Series: Materials Science and Engineering **758** (2020) 012021 doi:10.1088/1757-899X/758/1/012021

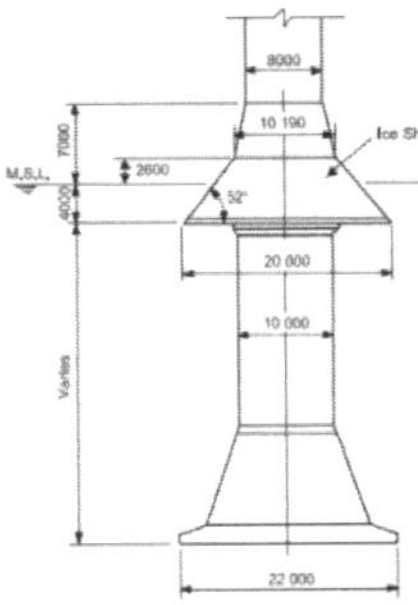

Figure 11. Single-tubular foundation of Federal Bridge

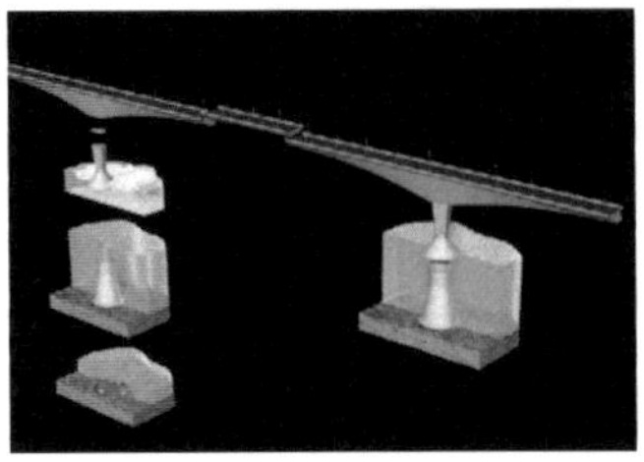

Figure 12. Prefabricated assembly diagram of foundation of Federal Bridge[20]

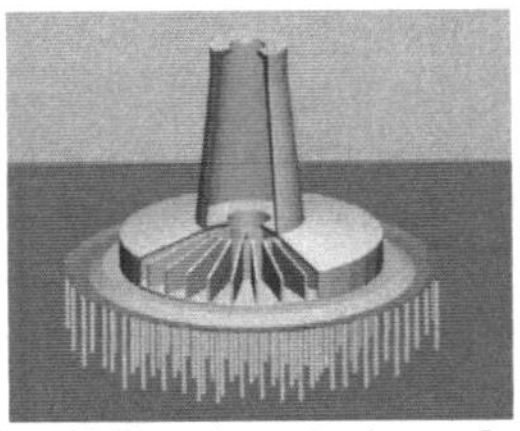

Figure 13. Sliding foundation of Rion-Antirion Bridge

Figure 14. Super-large diameter hollow single-tubular column foundation of Rion-Antirion[22]

5.Development of super-large diameter column foundation

Super-large diameter column foundation is widely used in land and cross-river bridges, which construction technology is relatively mature. For deep-water cross-sea bridges, the natural environment and geological conditions are more complex and harsher. More harmful factors, such as ship collision and earthquake, need to be considered. Super-large diameter single column foundation structure is novel, simple and beautiful. It has strong adaptability to geological conditions, large flexural rigidity and high stability. It cancels complex cap construction and reduces underwater volume, thus reducing the risk of ship collision. At present, the bridge foundation is developing towards a larger, deeper and more economical direction. In recent years, the super-large diameter single column foundation and bored cast-in-place composite steel pipe foundation, which are used in the cross-sea bridge engineering, have good prospects for development.

Root foundation is a new type of variable cross-section foundation. It reserves jacking holes in pile foundation or sidewall of caisson, then pushes prefabricated roots from reserved holes into soil layer, and forms a kind of bionic foundation similar to tree roots after ensuring the close consolidation of root keys and sidewall[24]. Because of the compaction and stress diffusion of the push-in key, the bearing capacity of surrounding soil and the bearing capacity of foundation is greatly improved[25]. After the root foundation was put forward, experiments and applications are carried out in Huai River Bridge, Ma'anshan Yangtze River Bridge, Wangdong Yangtze River Highway Bridge and Chizhou Bridge. The anchorage foundation of Chizhou Yangtze River Bridge is composed of 16 large diameter hollow root-piles and caps. The outer diameter of the hollow root-pile is 5m, the length of the pile is 26m and the wall thickness is 0.9m. The construction schematic of root-pile and root-key is shown in figure 15. Compared with the traditional pile-column foundation, the root foundation improves the bearing capacity of the foundation. However, the pile, the root-key and the soil constitute a more complex interaction system[26], which requires engineers to carry out research in depth.

With the increase of power generation of offshore wind turbines in China, the application of large diameter single-pile foundation in offshore wind turbines is becoming more and more common, which has great reference significance and reference value for the design and construction of pile-column foundation of deep-water cross-sea bridges. The common foundation forms of offshore wind turbines are gravity foundation, cylindrical foundation, single-pile foundation, three-pile foundation, jacket foundation and so on[27]. Among them, the super-large diameter single-pile foundation structure is simple in form, clear in force, and fast in construction, which accounts for more than 65% of wind farms at home and abroad[28]. Single-pile foundation in wind turbines is suitable for many kinds of soil seabed conditions with water depth less than 25m. For example, the surface silt in Bohai Sea and some parts of East China Sea is thinner, where steel pipe-column or bored pile-column foundation can be used. The diameter of a single-pile in China's offshore wind turbines can reach 7m, and the burial depth is generally 20m-40 m[29]. In 2016, the design of Rudong 150 MW offshore wind turbines in Zhongguang Nuclear Power Station is based on single-pile foundation structure. The diameter of steel pipe is between 6 m and 6.7 m, and the length is between 75 m and 94 m, as shown in figure 16. The construction of super-large diameter single-pile foundation of Rudong 150 MW offshore wind turbines in Zhongguang Nuclear Power Station is shown in figure 17[30]. The sliding foundation of Rion-Antirion Bridge is reinforced by hollow steel pipe network, which is similar to the tubular foundation of offshore wind power structure to some extent. Thus, the development of single-pile foundation in design and construction of offshore wind turbines can provide innovative ideas and feasible schemes for super-large diameter column foundation of deep-water bridges.

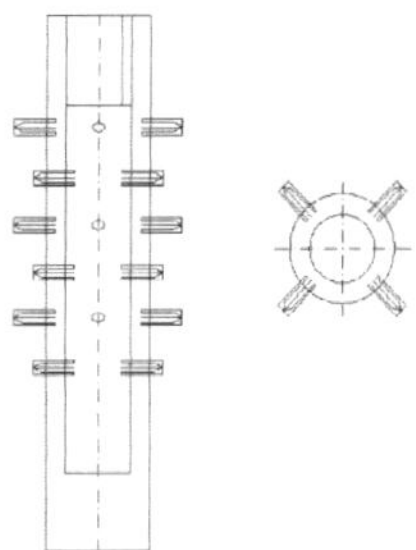

Figure 15. Diagram of root-pile and root-key

Figure 16. Foundation of Rudong 150 MW offshore wind turbine

Figure 17. Construction of single-pile foundation[30]

6.Problems to be solved for super-large diameter column foundation

Although super-large diameter column foundation is more and more widely used in engineering, its theoretical and experimental research is still in the initial stage. Its design and calculation methods and construction methods are not perfect. Especially in the application of super-large diameter column foundation in cross-sea bridges, there are many problems to be solved.

With the increase of diameter and length of pile, construction equipment and technology are facing many difficulties and challenges. For example, in the process of super-large diameter pile foundation, with the increase of diameter, the unstable factors of wall increase. The pore-forming resistance increases in series, which puts forward higher requirements for technology[31]. Complex natural conditions such as tides, strong winds and waves have brought many new problems to construction of bridge foundation. Up to now, there is no complete sets of key technologies for design and construction of steel-pile foundation in marine environment[14].

For the super-large diameter column foundation in cross-sea bridges, considering the long-term action of wind, wave and other loads, the horizontal cyclic loading characteristics are remarkable, which will produce cyclic cumulative deformation and cause adverse effects on the foundation[32]. However, there is no unified recognition of the response characteristics of super-large diameter piles under horizontal cyclic loading, and there is a lack of a widely accepted design calculation method[33].

Super-large diameter single-column pile in cross-sea bridges need to consider the damage caused by strong tide, drifting ice or ships, which requires high strength of pile. In recent years, composite steel pipe is a better foundation form. However, the research at home and abroad does not involve the role of mud, anti-corrosion coatings and shear rings embedded in steel pipes. The existing research results are not fully applicable to the complex pile foundation engineering of cross-sea bridges[34]. In addition, the problem of anti-corrosion and anti-collision of steel casing needs further study by researchers.

When the seabed is cohesionless soil, the soil around the column foundation is prone to scour under the action of wave and current, which greatly weakens the bearing capacity of pile foundation[29].

In recent years, the impact of scouring on the horizontal bearing capacity of pile foundation has gradually attracted the attention of engineering. The specific stress conditions of super-large diameter column foundation need to be further studied by engineers.

7.Conclusion

Bridge engineering needs continuous innovation in order to have better progress and development. Super-large diameter column foundation, especially single column foundation, has its unique advantages for deep-water cross-sea bridges with complex environment. It has been widely concerned by academia and engineering, and there are many successful application cases at home and abroad. The development of construction methods such as technology of sinking pipe and technology of super-large diameter bored pile have further promoted the application of super-large diameter single column foundation. In the future, engineers can also get inspiration from the offshore wind turbine and seek new development ideas. Generally speaking, the super-large diameter column foundation has become one of the best options for deep-water bridges and has broad prospects for development.

Acknowledgments

This work was supported by the National Natural Science Foundation of China: A Study of Performance Design Method of Extra-Large Diameter Piers with Composite Cross Section Against Multi-Hazards Action for Sea Crossing Bridge (No. 51778498).

References

[1] Zhang Jiuchu and Cai Xiangyang 1997 *Construction of boreholes and caps for large diameter piles of the main pier of the New Bayi Bridge in Nanchang [J]* Hunan Communications Science and Technology **2): 48-9**.

[2] Zhang Hong 1998 *Construction of main pier cap of New Bayi Bridge in Nanchang [J]* East China Highway **4): 3-6**.

[3] Wang Yinlong 2011 *Construction supervision of freezing method for East Tower Foundation of Poyang Lake Bridge [J]* Transportation Research **9): 126-30**.

[4] Wu Tongao and Xu Licheng 1996 *Large diameter pile foundation with variable cross-section without cap* Papers of the 12th Annual Conference of the Society of Bridge and Structural Engineering Chinese Society of Civil Engineering (Volume 2) F [C].

[5] Zhao Guoqiang 1994 *Large diameter pile foundation of Xiangjiang Second Bridge in Xiangtan [J]* Hunan Communications Science and Technology **4): 1-5**.

[6] Jiang Wei and Diao Xinhong 2008 *Super Large Diameter Bridge Composite Engineering Pile* National Bridge Academic Conference of Bridge and Structural Engineering Branch of China Highway Society F [C].

[7] Lu Keqin, Zhang Jiantao and Zhang Xuechen 2016 *Construction and mechanical properties of super-large diameter manual excavation pile by multi-guide method [J]* Railway construction technology **6): 107-10**.

[8] Yang Zunjian, Zhao Tongsheng and Liu Decheng 1998 *Manual excavation of large diameter pile foundation [J]* Hunan Communications Science and Technology **1): 50-3**.

[9] Hu Yi and Yan Xueyong 2006 *Construction Technology of Large Diameter Excavated Pile for Xiangjiang South Bridge in Changsha [J]* Western Prospecting Project **18 (7): 63-4**.

[10] Zhu Dongsheng, Yang Yunan and Song Ruibin 2014 *Application of 15m super-large diameter hollow pile in complex karst strata [J]* Waterway Engineering **2): 180-4**.

[11] Chen Bolin and Li Zhigang *Construction of large diameter caisson hollow pile foundation of Yuanshui Bridge in Taoyuan in 1997 [J]* Hunan Communications Science and Technology **1): 36-40**.

[12] Wei Wenyang 1998 *Design and construction of Li Ye Qiushui Bridge [J]* Hunan Transportation Science and Technology **3): 36-9**.

[13] Liu Hui, Xiang Kunshan and Zhao Rudong 2006 *Design and Construction of Wuliting Bridge [J]* Highway and Motor Transportation **3): 149-51**.

[14] Sun Pingkuan 2016 *Design and Construction Technology of Pile-based Steel Barrel for Bridges in Marine Environment [D]* Chang'an University.

[15] He Chenghai, Peng Linlin and He Chenghai 2015 *Key Technologies for Construction of 4.1m Super Large Diameter Steel Sheath of Jiashao Bridge [J]* China Harbour Construction **1): 55-8**.
[16] Song Weiguo, Fu Shousheng and Cheng Delin 2010 *Design of single-pile single-column pier of approach bridge in water area of Jiashao River-Crossing Bridge [J]* Bridge construction **s1): 15-7**.
[17] Wang Rengui, Cao Zongyong and Fu Shousheng 2013 *Study on the structural system of approach bridge in water area of Jiashao Bridge [J]* Highway **5): 226-9**.
[18] Yu Zhengquan and Zhou Aibin 2013 *Key technology of construction of bored pile with 3.8m diameter for Jiashao Bridge [J]* Highway **5): 216-20**.
[19] Fu Shousheng, Fu Yaohua and Wang Rengui 2013 *Study on anti-collision facilities of approach bridge in water area of Jiashao Bridge [J]* Highway **5): 202-4**.
[20] Campbell D. 2013 *Effects of Ice Loads on the Confederation Bridge [J]* Proto-Type,
[21] Lau D T, Brown T and Cheung M S 2004 *Dynamic modeling and behavior of the Confederation Bridge [J]* Canadian Journal of Civil Engineering **31 (2): 379-90**.
[22] Combault J 2011 *The Rion-Antirion bridge - when a dream becomes reality [J]* Frontiers of Architecture & Civil Engineering in China **5 (4): 415-26**.
[23] Combault J *The Rion-Antirion Bridge: Concept, Design and Construction; Structures Congress* F, 2005 [C].
[24] Gong Weiming, Wang Lei and Yin Yonggao 2015 *Application and experimental study of root foundation in thick overburden area [J]* Journal of Civil Engineering **S2: 69-74**.
[25] Yin Yonggao 2007 *Conception of Root Foundation and Root Anchorage Scheme [J]* Highway **2): 46-9.**
[26] Gong Weiming, Hu Feng and Tong Xiaodong 2008 *Experimental study on vertical bearing capacity of root foundation [J]* Journal of Geotechnical Engineering **30 (12): 1789-95**.
[27] Yu Yiming 2012 *Design and Research of Large Diameter Single Pile Foundation for Offshore Wind Power [D]* Tianjin University.
[28] Wang Guojie, Wang Wei and Yang Min 2011 *Design and Analysis of Single Pile Foundation of 3.6MW Marine Fan [J]* Journal of Geotechnical Engineering **33 (s2): 95-100**.
[29] Qi Wengang and Gao Fuping 2016 *Effect of scouring on horizontal bearing capacity of single pile foundation of offshore wind turbine [J]* Chinese Science: Physics, Mechanics and Astronomy **46 (12): 124710**.
[30] Sun Yongxin 2016 *Test and numerical analysis of horizontal bearing characteristics of super large diameter single pile of offshore wind turbines [D]*; Zhejiang University.
[31] Li Mancheng, Yang Bin and Mo Jianjun 1999 *Difficulties and Countermeasures in construction of large diameter piles for bridges [J]* Transportation Science and Engineering **1): 80-3**.
[32] Luo Ruping, Li Weichao and Yang Min 2016 *Accumulative Deformation Behavior of Large Diameter Single Pile at Sea under Horizontal Cyclic Loading [J]* Geotechnical Mechanics **s2: 607-12**.
[33] Dong Aimin 2017 *Study on horizontal bearing capacity of wind power pile foundation [D]*.
[34] Zhang Min 2014 *Study on the working characteristics of steel pipe composite piles under complex loading conditions [D]* Southwest Jiaotong University.

Dynamics of phase transformation of Cu-Cr-Zr-Ag Alloy

Dong Liang[a,b,c], Xujun Mi[b]*, Lijun Peng[b], Haofeng Xie[b], Guojie Huang[b], Zhen Yang[b]

a State Key laboratory of Nonferrous Metals and Processes, GRINM Group Co., Ltd, Beijing 100088, China.
b GRIMAT Engineering Institute Co., Ltd., Beijing 101407, China.
c General Research Institute for Nonferrous Metals, Beijing 100088, China.

*Corresponding author: Tel: +86 010 60662668, address: No. 11 Xingkedong Str., Huairou District, Beijing, 101417, China

Email: sklcopper1967@163.com
Abstract: Dynamic kinetic of phase transformation of Cu-Cr-Zr-Ag alloy has been studied according to the relationship between electrical conductivity and the volume fraction of phases during the aging. Based on above experiments, the Avrami empirical equation and electrical conductivity equation can be respectively identified at different aging treatment. S-curve of the dynamic kinetic during the aging process and C-curve of isothermal transformation for the Cu-Cr-Zr-Ag alloy are obtained.
Keywords: Cu-Cr-Zr-Ag alloy; Electrical conductivity; Microstructure; Phase transformation; Dynamic kinetic

1. Introduction

The Cu-Cr-Zr-Ag alloy has been widely used for many applications such as railway contact wire [1], lead frame and electrodes cap [2-5], because of their good mechanical properties as well as high electrical conductivity [2,6]. The fine Cr-rich and Zr-rich phases are precipitated during aging process attribute to the excellent strength[6,7], whereas the high conductivity is owe to the extremely low concentration of Cr and Zr in Cu matrix at room temperature [8-11].

In Cu-Cr-Zr-Ag quaternary alloy, the process that solute atoms Cr and Zr are precipitated from supersaturated solid solution to form new phases is a complex phase transformation process. Moreover, the volume fraction, size, morphology and distribution of precipitated phases are closely related to the properties of the alloy, and there are many factors affecting the phase transformation process. Therefore, it is almost impossible to systematically analyze the phase transformation process from the perspective of kinetics. However, in the phase transformation process of Cu-Cr-Zr-Ag alloy, the precipitation of the second phase has a great influence on the electrical conductivity [12]. So, by measuring the transformation of electrical conductivity of Cu-Cr-Zr-Ag alloy during aging, this chapter obtains the conversion rate of the second phase during aging, determines the Avrami equation of phase transformation kinetics, and draws the temperature-time-transformation (TTT) curve of Cu-Cr alloy, so as to provide a theoretical basis for experimental research and industrial production of the alloy.

2. Experiment

A cold rolled Φ5mm Cu-0.74Cr-0.12Zr-0.095Ag (wt. %) bar was put under solid solution treatment at 1000℃ for 1h. It was then aged at 400℃, 450℃, 500℃ and 550℃ for 5 min to 24 hours. The electrical conductivity of the samples after different treatments were measured by the TH2512B intelligent DC low resistance tester. The samples used for TEM observation were cut from the bar aged for different time durations and then prepared by the electro-polishing method with the mixture electrolyte of 50ml HNO3 and 150ml CH3OH at -30°C. Most of the thin foils for this study was studied by using acceleration voltages of 200kV in JEM 2100 LaB6 microscope.

3. Experimental results and Discussion

3.1 Calculation of volume fraction of the new phase in the aging process of Cu-Cr-Zr-Ag Alloy

In the aging process of Cu-Cr-Zr-Ag alloy, the supersaturated solid solution is precipitated and decomposed, and a great deal of solute atoms are precipitated from the matrix, forming a new phase. Even if phase equilibrium can be reached inside the alloy, not all solute atoms of Cr, Zr and Ag in the alloy can precipitate out. In this case, the volume conversion f of the new phase β can be expressed as [13]:

$$\mathrm{f} = \mathrm{V}^{\beta} / \mathrm{V}_{\mathrm{p}}^{\beta} \tag{1}$$

Wherein, Vp^{β} is the equilibrium volume of the new phase formed in unit volume, and V_{β} is the volume of the new phase formed in unit volume at a certain time point in the aging process. The moment aging is started, $V^{\beta}=0$, f=0, and the electrical conductivity is marked as initial electrical conductivity σ_0; When the alloy reaches the phase equilibrium state at this aging temperature and the electrical conductivity almost does not change any more, the electrical conductivity can be considered to have reached its maximum value σmax, i.e. $V^{\beta}=Vp^{\beta}$, f=1.0.

Because the concentration of solute atoms in Cu-Cr-Zr-Ag alloy is quite low, the resistivity of the alloy, according to Matthissen-Fuliminge Law [14], can be expressed as:

$$\rho_{\mathrm{S}} = \rho_0 + \mathrm{a}\rho \tag{2}$$

Wherein, ρ_0 is the resistivity of pure copper, a is the atomic fraction of solute, and ρ is the variation of resistivity induced by solute atoms. Formula (2) shows that there is a linear relationship between the solid solution atomic fraction and the resistivity of copper alloys. Therefore, it can be concluded that the volume conversion f of the new phase also have linearly relation with the alloy's electrical conductivity σ, as shown in the following equation:

$$\sigma = \sigma_0 + \mathrm{Af} \tag{3}$$

When phase transformation comes to an end, it can be confirmed that $\sigma=\sigma_{max}$, f=1, and $A=\sigma_{max}-\sigma_0$. Therefore, the volume conversion rate of the new phase can be calculated only by measuring the alloy's electrical conductivity under a certain aging regime.

The influence of aging time at 400~550℃ on the electrical conductivity of Cu-Cr-Zr-Ag alloy shows in Figure.1. Table 1 shows the relationship between the volume fraction of new phase and Cu-Cr-Zr-Ag quaternary alloy's electrical conductivity of solid solution during the aging process at 400~550℃.

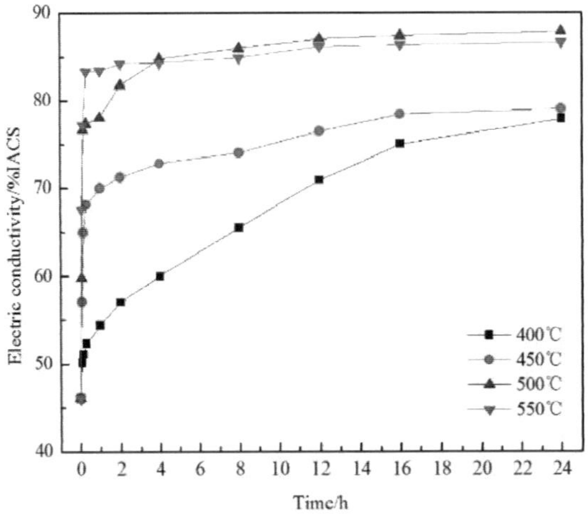

Figure. 1 Electrical conductivity curves of Cu-Cr-Zr-Ag alloy under different aging regimes

Table 1 Volume fraction of new phase and electrical conductivity of Cu-Cr-Zr-Ag alloy under different aging regimes

t/min	400℃		450℃		500℃		550℃	
	σ/IACS%	f/%	σ/IACS%	f/%	σ/IACS%	f/%	σ/IACS%	f/%
0	46.1	0	46.1	0	46.1	0	46.1	0
5	50.1	13.42	57	33.07	59.8	32.83	67.6	53.02
15	53.8	25.84	65	57.35	76.68	73.27	77.23	76.77
30	56.4	34.56	68.16	66.92	77.39	74.99	83.37	91.89
60	58	39.93	70.01	72.56	78.09	76.65	83.44	92.07
120	60.2	47.32	71.24	76.27	81.81	85.57	84.23	94.03
240	62	53.36	72.82	81.07	84.75	92.61	84.31	94.21
480	66.5	68.46	74.04	84.78	85.97	95.55	84.85	95.56
960	71	83.56	76.49	92.21	87	98.02	86.15	98.77
1440	73	90.27	78.45	98.15	87.4	98.97	86.38	99.32

3.2 Precipitation kinetics equation and electrical conductivity equation

The relationship of aging time t with the volume fraction f of Cr phase and Zr-rich phase in the aging process of Cu-Cr-Zr-Ag alloy satisfies the kinetic equation of phase transformation - Avrami empirical equation [14]:

$$f=1-\exp(-bt^{n}) \tag{4}$$

Wherein, b and n are constants. To calculate the two constants, Equation (4) can be converted into:

$$1-f=\exp(-bt^{n}) \tag{5}$$

Taking quadratic logarithm for both sides of Equation (5), we can obtain:

$$\ln(\ln(1/(1-f)))=\ln b+n\ln t \tag{6}$$

According to Equation (6), in a coordinate system where ln(ln(1/(1-f))) is the ordinate and lnt is the abscissa, the relationship between the two should be linear, the value of n is equal to the slope of the straight line, while the intercept is lnb, so constants n and b in Equation (4) can be calculated.

Based on the data provided in Table 1, the curves of relationship between ln(ln(1/(1-f))) and lnt are drawn, as shown in Figure. 2, where the former is the ordinate and the latter is the abscissa. It can be seen from Figure. 2 that after linear regression, the curve becomes a straight line.

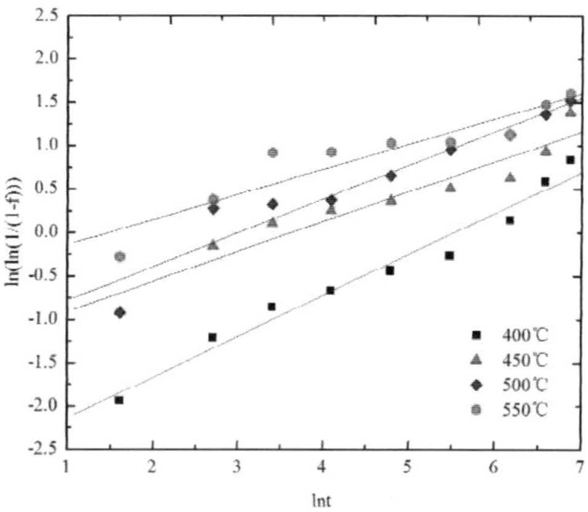

Figure. 2 Relationship between volume fraction and transition time under different aging regimes of Cu-Cr-Zr-Ag alloy

From Equation (6) and Figure.3, we can know the slope and intercept of the straight line, thus obtaining the values of n and b in the equation. These values and the corresponding equations (4) of quaternary alloy at different aging temperatures are shown in Table 2.

Table 2 Equation Coefficients A, n, B and the equations of phase transformation obtained under different aging regimes of Cu-Cr-Zr-Ag alloy

Temperature	Cu-Cr-Zr-Ag			
	A	n	b	Phase transformation equation
400℃	29.8	0.47	0.08	$f=1-\exp(-0.07t^{0.47})$
450℃	32.9	0.34	0.29	$f=1-\exp(-0.29t^{0.34})$
500℃	41.7	0.39	0.31	$f=1-\exp(-0.31t^{0.39})$
550℃	40.6	0.29	0.65	$f=1-\exp(-0.65t^{0.29})$

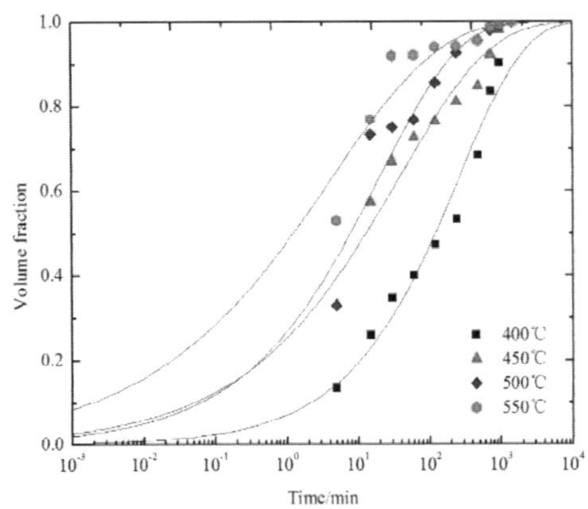

Figure.3 Kinetic curves of phase transformation and transition time of the quaternary alloy under different aging regimes

According to the equations listed in Table 2, we can draw the TTT curves of the precipitate of the alloy aged at 400~550℃. Using Equations (3) and (4) in combination, we can obtain the electrical conductivity equations of Cu-Cr-Zr-Ag alloy isothermally aged at 400~550℃, as shown in Table 3.

Table 3: Electrical conductivity equations of Cu-Cr-Zr-Ag alloy at 400~550℃

Temperature	Cu-Cr-Zr-Ag
400℃	σ=46.1+29.8 (1-exp(-0.07$t^{0.47}$))
450℃	σ=46.1+32.9(1-exp(-0.29$t^{0.34}$))
500℃	σ=46.1+41.7 (1-exp(-0.31$t^{0.39}$))
550℃	σ=46.1+40.6(1-exp(-0.65$t^{0.29}$))

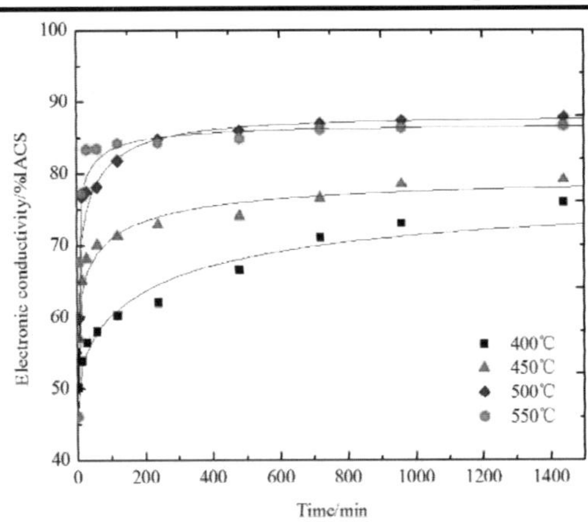

Figure.4 Electrical conductivity curves of Cu-Cr-Zr-Ag alloy under different aging regimes

Based on the equations given in Table 3 and experimental data, the alloy's electrical conductivity curves under different aging regimes are plotted, as shown in Figure. 4. The figure clearly shows the difference between the actual measured value of the alloy's electrical conductivity and the calculated result of the equation under each aging regime. Comparing calculated value (curve, CC) and the measured value (point, TC) of the electrical conductivity of Cu-Cr-Zr-Ag alloy at the aging temperature of 450℃, we see that the difference between the two values is basically controlled within 5%, which shows that the conductivity equation can satisfactorily reflect the change of conductivity.

Table 4 Comparisons between the values calculated from the conductivity equation and the actual measured values of electrical conductivity of quaternary alloy at the aging temperature of 450℃

t/min	Cu-Cr-Zr-Ag	
	TC /IACS%	CC /IACS%
0	46.10	46.10
5	57.00	59.07
15	65.00	63.12
30	68.16	65.91
60	70.01	68.76
120	71.24	71.49
240	72.82	73.93
480	74.04	75.91
720	76.49	76.82
960	78.45	77.35
1440	79.06	77.94

3.3 Temperature-time-transformation (TTT) curve

In this paper, the isothermal aging of Cu-Cr-Zr-Ag inly carried out in the range of 400℃~550℃. According to formula (7), the time that precipitation transformation start and end at each aging temperature can be obtained based on Equation (7). In this paper, the time when 10% of the new phase has appeared is defined as the starting time of precipitation reaction, and the time when 90% of the new phase has been formed is defined as the ending time [15]. According to this definition, the starting time and ending time of the alloy at 400℃~550℃ can be obtained, as shown in Table 6.

$$t = \exp\{\frac{1}{n}\ln[-\frac{\ln(1-f)}{b}]\} \tag{7}$$

Table 5 Calculated precipitation time (min) of Cu-Cr alloy

Alloy	Temperature/℃	400	450	500	550
Cu-Cr-Zr-Ag	t_{start}	2.196	0.053	0.064	0.002
	t_{end}	1544	459	174	78

The TTT curve of Cu-Cr-Zr-Ag alloy plotted based on the data in Table 6 is shown in Figure. 5.

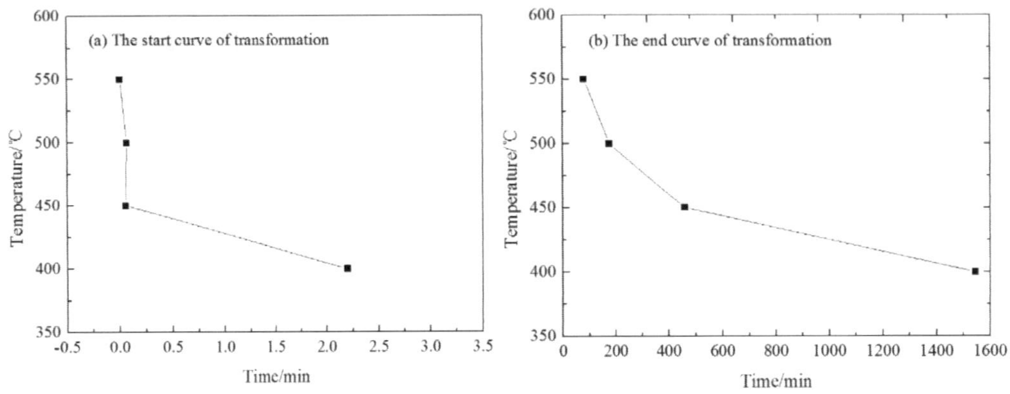

Figure.5 TTT curve of Cu-Cr-Zr-Ag alloy

As shown in Figure. 5, the two TTT curves of the quaternary alloy are both C-shaped, and the starting time of precipitation of the new phase becomes shorter as the aging temperature rises.

3.4 Microstructure observation

Figure.6 shows the Microstructure TEM patterns of Cu-Cr-Zr-Ag alloy aged at 450°C for different aging duration. When the Cu-Cr-Zr-Ag alloy aged at 450°C for 5 min, G.P zones can be found in the matrix (Figure. 6(a)), which leads to the increase of electrical conductivity at the early stage of aging. When the alloy is aged for 4 h, a great deal of Cr phases are found in Cu matrix (Figure. 6(b)).With the extension of aging time, the G.P zones transform to the Cr precipitates gradually and a large number of solute atoms precipitate out from the matrix at the same time, which decreases the contents of the solute atom in the matrix and as a result of a continuous increase in conductivity during the aging treatment. Aged for 24 h, due to the size of Cr precipitates grow larger (Figure. 6(c)) and solute atoms cannot precipitate out any more from Cu matrix, the electrical conductivity becomes stable at the later stage of aging.

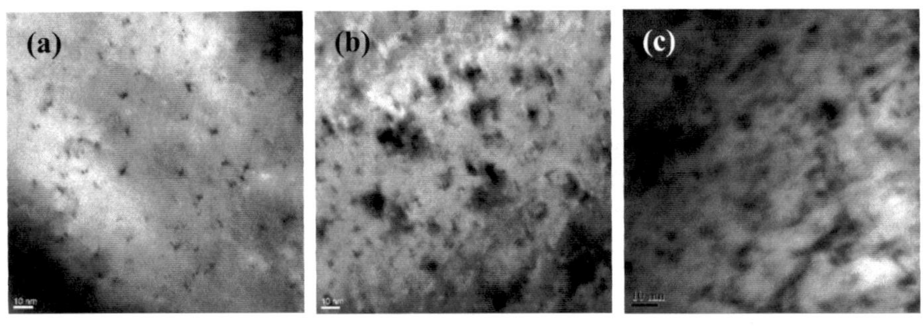

Figure. 6 Microstructure of Cu-Cr-Zr-Ag alloy aged at 450℃ （a）aged for 5 min （b）aged for 4 h （c）aged for 24 h

4. Conclusions

(1) Under different aging regimes, the curve of relationship between the electrical conductivity and aging time of Cu-Cr-Zr-Ag alloy rises rapidly at first, then rises slowly, and finally tends to be flat.

(2) The Avrami kinetic equation of phase transformation and electrical conductivity equation of Cu-Cr-Zr-Ag alloy at different aging temperatures are deduced and calculated.

(3) The starting time and ending time of transformation of the second phase are calculated, the TTT curve of Cu-Cr-Zr-Ag alloy is drawn, and the curve is C-shaped.

Acknowledgments

This work was supported by the National Key R&D Program of China (No. 2016YFB0301400), the technological innovation 2025 & Major special project of NingBo (No. 2018B10030) and the Natural Science Foundation of China (No. 51601017).

5. References

[1] Liu Q, Zhang X, Ge Y, Wang J,. Cui J.H. 2006 *Metall Mater Trans*, 37A, pp. 3233-3238.
[2] Batra I.S, Dey G.K, Kulkarni U.D, Banerjee S. 2001 *J Nucl Mater*, 299, pp. 91-100.
[3] Holzwarth U, Stamn H. 2000 *J Nucl Matter*, 279, pp. 31-45.
[4] Edwards D.J, Singh B.N, Tahtinen S. 2007 *J Nucl Matter*, 367-370, pp. 904-909.
[5] Liu P. Kang B.X. Gao X.G.1999 *Mater Sci Eng ,A* 265, pp. 262-267.
[6] Huang F.X, Ma J.S, Ning H.L, Geng Z.T, Lu C. Guo S.M, Yu X.T, T Wang, H Li, Luo H. F.2003 *Scr Mater*, 48, pp. 97-102.
[7] Chbihi A, Sauvage X, Blavette D. 2012 *Acta Mater*, 60, pp. 4575-4585.
[8] Su J.H, Liu P, Li H.J, Ren F.Z, Dong Q.M. 2007 *Mater lett*, 61, pp. 4963-4966.
[9] Tang N.Y, Taplin D.N.R, Dunlop G.L. 1985 *Mater Sci technol,* 1, pp. 270-275.
[10] Zeng K.J, Hamalainen H, Lilius K. 1995 *Scr Metall Mater*, 32, pp. 2009-2014.
[11]. Batra I.S, Dey G.K, Kulkarni U.D, Banerjee S. 2002 *Mater Sci Eng A,* 356, pp. 32-36.
[12] Su J. H., Liu P., Li H. J. 2007 *Mater Lett,* 61, pp: 4963-4966.
[13] Nagarjuna S., Balasubra K. 1999 *J Mater Sci*, 34, pp: 2929-2942.
[14] Fernee H., Nairn J. 2001 *J Mater Sci*, pp: 362711-2719.
[15] Li Y, Yi D Q. 2010 *Transact Mater Heat Treatment*, 31, pp: 49-52.

Axial compression capacity calculation of slenderness CFST filled with DCLs

Huifeng Zhang and Xue Bai

Department of civil Engineering, Shandong Modern University , Jingshi east Road ,Jinan , P.R. China
Email: zhf.cla.y@163.com

Abstract. The demolished concrete lumps (DCLs) has been demonstrated to replacement partial coarse aggregates when casting concrete. The experimental studied involved 12 Concrete-Filled Steel Tube (CFST) columns. Each columns diameter was 159mm, the length of specimens was 2000, 2200 and 2400 mm. The Diameter-to-thickness (D/t) ratio was 79, 53 and 40. The replacement ratio of FC by DCLs was 0, 20, 40 and 60%. The DCLs had slightly affect on the mechanical performances of slender CFST columns. the code CECS 28:2012 was used to calculate the bearing capacity of slender CFST columns filled with DCLs under axial compression

Keyword: CFST DCLs Axial compression Calculation

1. Introduction

Concrete-Filled Steel Tube (CFST) columns has been widely used in the civil engineering fields[1]. With the acceleration of global urbanization and the rapid development of civil engineering construction, a large number of abandoned concrete had been produced while building, rebuilding and expanding all kinds of buildings on a large scale, which seriously pollutes the environment. In addition, the rapid increase in aggregate demand had resulted in excessive development, which had led to the exhaustion of resources and the great destruction of the ecological environment, so that the contradiction between the sustainable development of concrete and aggregate demand has become increasingly prominent [4-5].

In order to reduce the cost of producing recycled aggregate (RA), The demolished concrete lumps (DCLs) had been demonstrated to replace some coarse aggregates when casting concrete. [5-8]. This method could reduce energy consumption.

However, few studies have focused on the sensitivity of factors affecting mechanical behavior of slender circular steel columns filled with DCLs and FC under axial compression. This research was carried out on the basis of Bo Wu's study. The Orthogonal design method was adopted in this experiment. This study aims to investigate the sensitivity of factors affecting mechanical behavior of slender circular steel columns filled with DCLs and FC under axial compression..

2. Experimental program

2.1. Materials and specimen details

The experimental study involved 12 CFST columns. Each column diameter was 159mm, the length of specimens was 2000, 2200 and 2400 mm. The thickness of the steel tubes was 2, 3 and 4mm, so the

D/t ratio was 79, 53 and 40. The replacement ratio of FC by DCLs was 0, 20, 40 and 60%. 1Aa0, 5Bb0 and 9Cc0 were control specimens that were designed without DCLs in FC. The D/t ratio in these tests met most of the codes' requirements. According the EC4, the slenderness ratio (λ=4L/D) was obtained. The slenderness ratio was 50, 55 and 60. Strain gauge was placed in the mid-height of specimens to measure the strain. The details of the specimens are shown in Table 1.

Table 1. Detail of the test specimens Formatting sections

Specimen	L / mm	D / mm	t / mm	D / t / %	λ	η / %	N / kN
1Aa0	2 000	159	2	79	50	0	890
2Aa1	2 000	159	2	79	50	20	890
3Ac2	2 000	159	4	40	50	40	1 266
4Ab3	2 000	159	3	53	50	60	1 081
5Bb0	2 200	159	3	53	55	0	1 042
6Bb1	2 200	159	3	53	55	20	1 042
7Ba2	2 200	159	2	79	55	40	857
8Bc3	2 200	159	4	40	55	60	1 220
9Cc0	2 400	159	4	40	60	0	1 178
10Cc1	2 400	159	4	40	60	20	1 178
11Cb2	2 400	159	3	53	60	40	1 006
12Ca3	2 400	159	2	79	60	60	827

Note: L, D and t are respectively the length, diameter and thickness of the specimens. λ and η are respectively the slenderness ratio of specimens and the replacement ratio of DCLs.

2.2. Material properties

.

The DCLs were produced from RC beams. Before concrete casting, the DCLs were pre-wetted by pouring water. The DCLs physical properties were list in Table 3.

Table 2. Mechanical properties of steel tubes

Thickness /mm	f_y / MPa	f_u / MPa	E_s /MPa
2	345.6	414.0	2.04×10^5
3	314.2	401.3	2.08×10^5
4	332.7	407.6	2.17×10^5

Table 3. DCLs physical properties

Apparent density kg /m^3	Packing density kg /m^3	Clay percentage /%	Water absorption /%
2957	2130	2.1	7.03

3. Test resultsand discussion

3.1. Calculation

Accompanied by the slight sound of concrete and steel tube separation, the steel tube and concrete separation would produce stress concentration phenomenon. The load ratio of concrete and steel tube varied here. The specimens entered the plastic working stage, the load-bearing capacity of the column decreased rapidly, the deformation increased rapidly, the deformation of the slender column also developed greatly, and there was a little iron pin powder on the surface of the mid-height of the specimens. The maximum lateral deflection was located in the mid-height of the specimens, and the two sides distribute symmetrically along the mid-height of the specimens. The failure mode of the specimens was arched. The failure mode of the specimens belonged to the instability failure.
To calculate the ultimate bearing capacity of the CFST slender columns with DCLs, design codes AIJ, AISC, CE4, CECS 28: 2012, DL/T 5085-1999 were used to predicted the ultimate bearing capacity. According to code AIJ, the predicted ultimate bearing capacity of CFST column can be expressed as follows

$$N_{\mathrm{cr}} = sN_{\mathrm{cr}} + cN_{\mathrm{cr}}$$

Where $sN_{\mathrm{cr}}, cN_{\mathrm{cr}}$ are the bearing capacity of steel tube and concrete.
In code AISC, the ultimate bearing capacity is expressed as follows

$$N_{\mathrm{u}} = F_{\mathrm{cr}} A_{\mathrm{s}}$$

$$\lambda_{\mathrm{c}} = \frac{KL}{r_{\mathrm{m}}\pi}\sqrt{\frac{F_{\mathrm{my}}}{E_{\mathrm{m}}}}$$

Where λ_{c} is the slenderness ratio of the specimens; when $\lambda_{\mathrm{c}} \le 1.5$, $F_{\mathrm{cr}} = (0.658^{\lambda_{\mathrm{c}}^2})F_{\mathrm{my}}$, $\lambda_{\mathrm{c}} \ge 1.5$ $F_{\mathrm{cr}} = \left(\frac{0.877}{\lambda_{\mathrm{c}}^2}\right)F_{\mathrm{my}}$; A_{s} is area cross of steel; L is the length of specimens;

CE4 also provides a mode for calculating the bearing capacity of CFST with DCLs.

$$N_{\mathrm{u}} = k_1 N$$

$$k_1 = \frac{1}{\theta + \sqrt{\theta^2 - \lambda^2}}$$

$$\theta = 0.5\left[1 + 0.21(\lambda - 0.2) + \lambda^2\right]$$

$$\lambda = \sqrt{\frac{N_{\mathrm{k}}}{N_{\mathrm{e}}}}$$

$$N_{\mathrm{k}} = \eta_2 f_{\mathrm{y}} A_{\mathrm{s}} + (1 + \eta_1 \frac{t f_{\mathrm{y}}}{D f_{\mathrm{c}}^{'}})$$

$$N_e = \pi^2 (0.6 E_c I_c + E_{\mathrm{s}} I_{\mathrm{s}}) / L^2$$

Where λ is the slenderness ratio of the specimens; f_{y} is the yield strength of steel; A_{s} is area cross of steel; η_1 and η_2 are coefficient; E_c is elastic modulus of concrete; E_s is elastic modulus of steel; L is the length of specimens; I_c is moment of inertia of concrete; I_s is moment of inertia of steel; $f_{\mathrm{c}}^{'}$ is compressive strength of concrete prism, D is the diameter of steel tube.
According to code CECS 28: 2012, the predicted ultimate bearing capacity of CFST column can be expressed as follows

$$N_{\mathrm{u}} = \varphi f_{\mathrm{sc}} A_{\mathrm{sc}}$$

$$f_{\mathrm{sc}} = (1.212 + \eta_{\mathrm{s}}\xi_0 + \eta_{\mathrm{c}}\xi_0^2) f_{\mathrm{c}}$$

$\eta_s = 0.1759f / 235 + 0.794$

$\eta_c = -0.1038f_c / 20 + 0.0309$

Where f_{sc} is design Value of Axial Compressive Strength of Concrete Filled Steel Tubular Composite; ξ_0 is design value of hoop coefficient; f_c is strength Design Value of Core Concrete in Steel Tube; A_{sc} is section Area of Composite Members; φ is margin of stability.

According to code DL/T 5085-1999, the predicted ultimate bearing capacity of CFST with DCLs column can be expressed as follows

$\xi = A_s f_y / A_c f_c$

$N_c = \varphi N_0$

$N_0 = 0.9 f_c A_c (1 + \theta + \sqrt{\theta})$

Where A_s , A_c is the area of steel and concrete; f_y is the yield strength of steel; f_c is the cylinder strength of the concrete; φ is bearing capacity reduction factor considering slenderness ratio; θ is hoop Coefficient of Concrete Filled Steel Tubular; $[\theta]$ is limit value of hoop coefficient relating to concrete strength grade.

Comparisons of the predicted ultimate bearing capacity of CFST columns with DCLs obtained using five code calculation methods with the measured results are presented in Table 4 and Table 5.

Table 4 Comparison of ultimate bearing capacity

	Measured value / Calculated value				
Specimens	AIJ	AISC	CE4	CECS 28：2012	DL / T5085—1999
1Aa0	0.978	1.210	1.056	0.998	0.942
2Aa1	1.011	1.098	1.342	0.976	1.121
3Ac2	0.942	1.125	1.206	1.102	0.876
4Ab3	0.960	1.064	1.102	1.006	0.956
5Bb0	1.107	1.066	1.064	0.959	0.987
6Bb1	1.021	1.209	1.267	0.978	1.105
7Ba2	1.004	1.145	1.472	0.943	0.996
8Bc3	1.100	1.557	1.068	1.097	1.153
9Cc0	1.045	1.321	1.342	1.064	1.142
10Cc1	0.909	1.427	1.495	0.964	0.972
11Cb2	1.207	1.502	1.341	0.945	0.943
12Ca3	1.064	1.466	1.025	1.062	0.989

Table 5 Statistical characteristics

Statistical characteristics	AIJ	AISC	CE4	CECS 28：2012	DL / T5085—1999

Average value	1.029	1.266	1.231	1.008	1.015
Variance	0.006	0.029	0.026	0.003	0.008
Standard deviation	0.077	0.170	0.161	0.055	0.089

From Table 4 and Table 5, it could be seen that the bearing capacity calculated by American AISC Code and European CE4 Code was less than the ultimate bearing capacity obtained by test. It showed that the calculation capacity of AISC code in the United States and CE4 code in Europe were conservative and had a large safety reserve. From variance and standard deviation, the results of AIJ and DL/T5085-1999 were more discrete than those of CECS 28:2012. Therefore, it was suggested that the code CECS 28:2012 was used to calculate the bearing capacity of slender CFST columns filled with DCLs under axial compression.

4. Conclusion

The mechanical behavior of CFST filled with DCLs under axial compression was presented. The following conclusions are drawn from the test results:

(1) The bearing capacity of 12 specimens was calculated by the domestic and foreign codes. Through mean and variance analysis, it was found that the domestic CECS 28:2012 codes were in good agreement with the test results and could be applied to the calculation of the ultimate bearing capacity of slender CFST column with DCLs.

(2) All the 12 specimens were destroyed by instability failure, and the ultimate failure form was arch. Through comparative analysis of load-strain relationship curve and load-displacement relationship curve, the following basic rules can be obtained: compared with CFST slender columns without DCLs, the rising and falling sections of the curve were steeper and the ultimate bearing capacity was slightly lower for the specimens with DCLs.

Acknowledgments

Authors wishing to acknowledge assistance or encouragement from colleagues, special work by technical staff or financial support from organizations should do so in an unnumbered Acknowledgments section immediately following the last numbered section of the paper.

5. References

[1] B. González-Fonteboa, F. Martínez-Abella, J. Eiras-López, et al., Effect of recycled coarse aggregate on damage of recycled concrete, Mater. Struct (10) (2011) 1759–1771.

[2] B. González-Fonteboa, F. Martínez-Abella, M.F. Herrador, et al., Structural recycled concrete: behaviour under low loading rate, Constr. Build. Mater (1) (2012) 111–116.

[3] N.Y. Ho, P.K. Yang, W.F. Lim, et al., Efficient utilization of recycled concrete aggregate in structural concrete, J. Mater. Civ. Eng. 25 (3) (2013) 318–327

[4] S.I. Ivan, B.M. Snez ˇana, M.M. Zoran, et al., Flexural behavior of reinforced recycled aggregate concrete beams under short-term loading, Mater. Struct (6) (2013) 1045–1059.

[5] J.F. Lamond, Significance of Tests and Properties of Concrete and Concrete-Making Materials, ASTM International, 2006.

Mechanical behaviour of slenderness CFST filled with DCLs under axial compression

Huifeng Zhang and Xue Bai

Department of civil Engineering, Shandong Modern University, Jingshi east Road ,Jinan , P.R. China

Email: zhf.cla.y@163.com

Abstract. The demolished concrete lumps (DCLs) has been demonstrated to replacement partial coarse aggregates when casting concrete. The experimental studied involved 12 CFST columns. Each columns diameter was 159mm, the length of specimens was 2000, 2200 and 2400 mm. The Diameter-to-thickness (D/t) ratio was 79, 53 and 40. The replacement ratio of FC by DCLs was 0, 20, 40 and 60%. Test results indicated that the slenderness ratio was the most sensitive factor on the bearing capacity of slender columns under axial compression, followed by D/t ratio, and the DCLs replacement ratio was smallest factor. The ultimate bearing capacity decreased slightly with increasing DCL replacement ratio.

Keyword: DCLs CFST Axial compression slenderness

1. Introduction

Concrete-Filled Steel Tube (CFST) columns has been widely used in the civil engineering fields[1]. With the acceleration of global urbanization and the rapid development of civil engineering construction, a large number of abandoned concrete had been produced while building, rebuilding and expanding all kinds of buildings on a large scale, which seriously pollutes the environment. In addition, the rapid increase in aggregate demand had resulted in excessive development, which had led to the exhaustion of resources and the great destruction of the ecological environment, so that the contradiction between the sustainable development of concrete and aggregate demand has become increasingly prominent [4-5].

In order to reduce the cost of producing recycled aggregate (RA), The demolished concrete lumps (DCLs) had been demonstrated to replace some coarse aggregates when casting concrete. [5-8]. This method could reduce energy consumption.

However, few studies have focused on the sensitivity of factors affecting mechanical behaviour of slender circular steel columns filled with DCLs and FC under axial compression. This research was carried out on the basis of Bo Wu's study. The Orthogonal design method was adopted in this experiment. This study aims to investigate the sensitivity of factors affecting mechanical behaviour of slender circular steel columns filled with DCLs and FC under axial compression..

2. Experimental program

2.1. Materials and specimen details

The replacement ratio of FC by DCLs was 0, 20, 40 and 60%. 1Aa0, 5Bb0 and 9Cc0 were control specimens that were designed without DCLs in FC. The D/t ratio in these tests. According the EC4, the slenderness ratio (λ=4L/D) was obtained. The slenderness ratio was 50, 55 and 60. The details of the specimens are shown in Table 1.

Table 1. Detail of the test specimens Formatting sections

Specimen	L / mm	D / mm	t / mm	D / t / %	λ	η / %	N / kN
1Aa0	2 000	159	2	79	50	0	890
2Aa1	2 000	159	2	79	50	20	890
3Ac2	2 000	159	4	40	50	40	1 266
4Ab3	2 000	159	3	53	50	60	1 081
5Bb0	2 200	159	3	53	55	0	1 042
6Bb1	2 200	159	3	53	55	20	1 042
7Ba2	2 200	159	2	79	55	40	857
8Bc3	2 200	159	4	40	55	60	1 220
9Cc0	2 400	159	4	40	60	0	1 178
10Cc1	2 400	159	4	40	60	20	1 178
11Cb2	2 400	159	3	53	60	40	1 006
12Ca3	2 400	159	2	79	60	60	827

2.2. Material properties

The DCLs were obtained from one batch beams, The cube (150 × 150 ×150 mm) compressive strength of original beams was 34.6. The approximate size of DCLs were range from 30mm to 80mm. The DCLs physical properties were list in Table 3. The cube (150 × 150 ×150 mm) compressive strength of DCLs was listed as Table 4 and Table 5.

Table 2. Mechanical properties of steel tubes

Thickness /mm	f_y / MPa	f_u / MPa	E_s /MPa
2	345.6	414.0	2.04×10^5
3	314.2	401.3	2.08×10^5
4	332.7	407.6	2.17×10^5

Table 3. DCLs physical properties

Apparent density kg / m^3	Packing density kg / m^3	Clay percentage /%	Water absorption /%
2957	2130	2.1	7.03

Table 4. The mixture proportion

DCLs replacement ratio (%)	Cement (kg.m^{-3})	sand (kg.m^{-3})	Fine aggregate (kg.m^{-3})	DCLs (kg.m^{-3})	Water (kg.m^{-3})
0	380	794	1 096	0	180
20	380	794	219	877	180

40	380	794	438	658	180
60	380	794	658	438	180

Table 5. Concrete compressive strengths

	Compressive strength (MPa)	Standard deviation (MPa)	Coefficient of variation (%)
28-day	35.1	0.8	2.9
Testing day	40.5	0.6	1.8

3. Test resultsand discussion

3.1. General behavior and failure mode

In order to reduce the influence of eccentricity on the specimen, the upper surface of the specimen was polished smoothly before loading. At the beginning of loading, the deformation of the specimen was not obvious, it was in the stage of elastic work, and the increase of deflection was approximately proportional to the load. With the increase of load, the specimens showed slight deformation, but not very obvious. When the specimen was loaded to the ultimate bearing capacity of 60%~80%, the lateral deflection at mid-height of the specimen increased rapidly, and the specimen entered the stage of elastic-plastic work. The steel tube restrained the concrete, the core concrete in the state of three-dimensional compression.

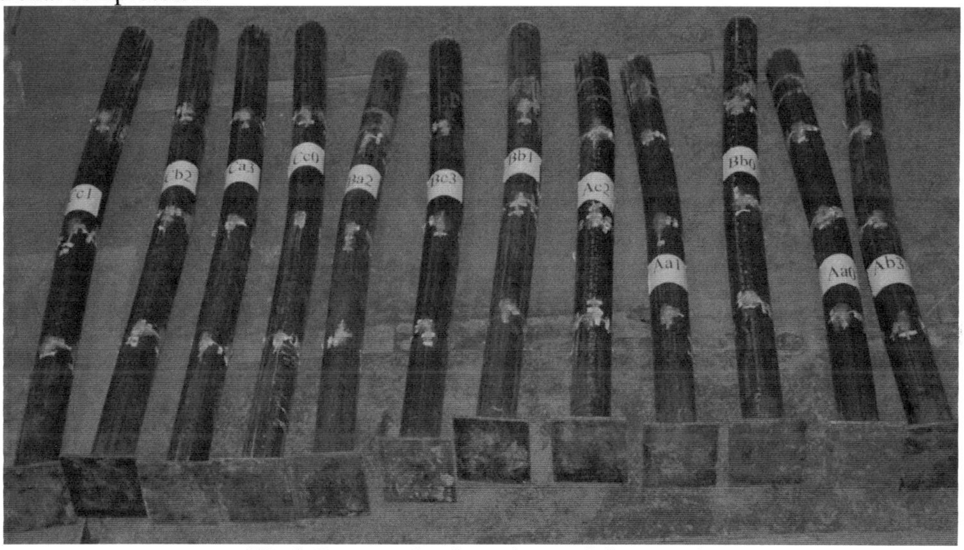

Fig.1. Potograph of specimens failure mode

3.2. Load strain response

With the increased of load, the convex side and concave side strains of the specimens were increased, and the convex side strain was larger than the concave side strain. When the applied load was about 60%~80% of the ultimate bearing capacity, the specimens began to enter the elastic-plastic stage. The increase of convex side and concave side strains were accelerated, and the slope of load-strain curve decreased, and the concave side deformation of the specimen increased. The convex side and concave side strains were large at mid-height of the specimens. After reaching the ultimate bearing capacity, the specimens entered the plastic stage, and the stress fall, the strain increased sharply, and the bearing capacity decreased rapidly.

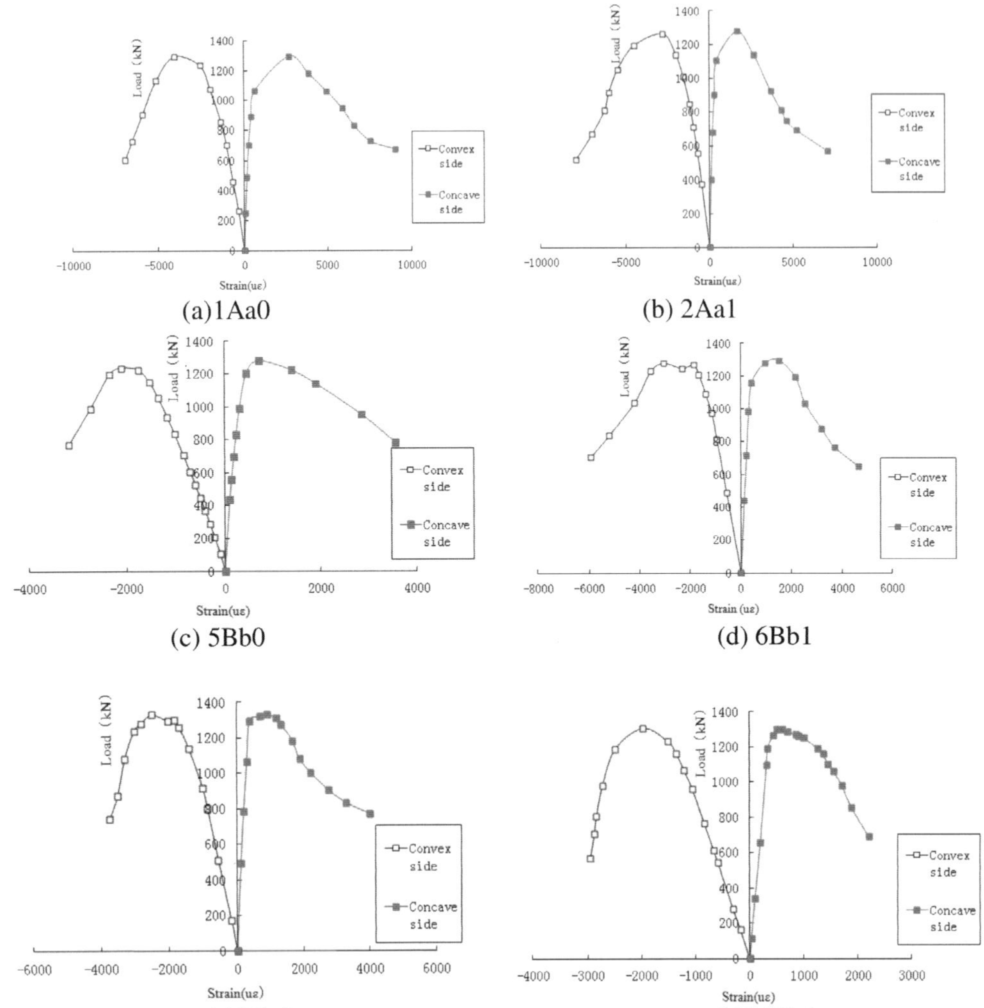

(a)1Aa0 (b) 2Aa1 (c) 5Bb0 (d) 6Bb1 (e)9Cc0 (d)10Cc1

Fig.2 Load strain response of the specimens

Compared with column 1Aa0 and 2Aa1, the slope of load-strain relationship curve of column 1Aa0 ascending section was about 1.3, while column 2Aa1 was about 1.6, thus column 2Aa1 ascending section was steeper. Similarly, the descending section of column 2Aa1 was steeper than that of column 1Aa0. The ultimate bearing capacity of column 2Aa1 was about 98% of column 1Aa0. When the FC was replaced by DCLs, the rising and falling sections of the load-strain relationship curve were steeper, and the ultimate bearing capacity did not change significantly.

4. Conclusion

The mechanical behavior of CFST filled with DCLs under axial compression was presented. The following conclusions are drawn from the test results:

(1) Through comparative analysis, it could be seen that whether there were DCLs had no obvious effect on the mechanism and failure mode of the slender columns, and the failure of the specimens went through three stages: elastic stage, elastic-plastic stage and plastic stage. All the 12 specimens were destroyed by instability failure, and the ultimate failure form was arch. Through comparative analysis of load-strain relationship curve and load-displacement relationship curve, the following basic rules can be obtained: compared with CFST slender columns without DCLs, the rising and falling sections of the curve were steeper and the ultimate bearing capacity was slightly lower for the specimens with DCLs.

(2) The bearing capacity decreased by 7% with DCLs replacement ratio increased from 20% to 40%, , and the effect was not obvious.

5. References

[1] B. González-Fonteboa, F. Martínez-Abella, J. Eiras-López, et al., Effect of recycled coarse aggregate on damage of recycled concrete, Mater. Struct (10) (2011) 1759–1771.
[2] B. González-Fonteboa, F. Martínez-Abella, M.F. Herrador, et al., Structural recycled concrete: behaviour under low loading rate, Constr. Build. Mater (1) (2012) 111–116.
[3] N.Y. Ho, P.K. Yang, W.F. Lim, et al., Efficient utilization of recycled concrete aggregate in structural concrete, J. Mater. Civ. Eng. 25 (3) (2013) 318–327
[4] S.I. Ivan, B.M. Snez ˇana, M.M. Zoran, et al., Flexural behavior of reinforced recycled aggregate concrete beams under short-term loading, Mater. Struct (6) (2013) 1045–1059.
[5] J.F. Lamond, Significance of Tests and Properties of Concrete and Concrete-Making Materials, ASTM International, 2006.
[6] X. Li, Recycling and reuse of waste concrete in China: part I. Mate behavior of recycled aggregate concrete, Resour. Conserv. Recy. 53 (1) (2008) 36–44.
[7] C.H. Liu, J.Y. Fu, Y.L. Pi, et al., Influence of demolished concrete blocks on mechanical properties of recycled blend concrete, Constr. Build. Mater. (2017) 329–347.
[8] S. Lotfi, M. Eggimann, E. Wagner, et al., Performance of recycled aggregate concrete based on a new concrete recycling technology, Constr. Build. Mater 95 (2015) 243–256.
[9] B. Wu, X.Y. Zhao, J.S Zhang, Cyclic behavior of thin-walled square steel tubular columns filled with demolished concrete lumps and fresh concrete. J Constr Steel Res 2012;77(10):69–81.

Sensitive behaviour of slenderness CFST filled with DCLs under axial compression

Huifeng Zhang and Junping Cong

Department of civil Engineering, Shandong Modern University , Jingshi east Road ,Jinan , P.R. China

Email: zhf.cla.y@163.com

Abstract. The demolished concrete lumps (DCLs) has been demonstrated to replacement partial coarse aggregates when casting concrete. The experimental studied involved 12 Concrete-Filled Steel Tube (CFST) columns. Each columns diameter was 159mm, the length of specimens was 2000, 2200 and 2400 mm. The Diameter-to-thickness (D/t) ratio was 79, 53 and 40. The replacement ratio of FC by DCLs was 0, 20, 40 and 60%. Test results indicated that the slenderness ratio was the most sensitive factor on the bearing capacity of slender columns under axial compression, followed by D/t ratio, and the DCLs replacement ratio was smallest factor. The ultimate bearing capacity decreased slightly with increasing DCL replacement ratio..

Keyword: DCLs CFST Axial compression

1. Introduction

Concrete-Filled Steel Tube (CFST) columns has been widely used in the civil engineering fields[1]. With the acceleration of global urbanization and the rapid development of civil engineering construction, a large number of abandoned concrete had been produced while building, rebuilding and expanding all kinds of buildings on a large scale, which seriously pollutes the environment. In addition, the rapid increase in aggregate demand had resulted in excessive development, which had led to the exhaustion of resources and the great destruction of the ecological environment, so that the contradiction between the sustainable development of concrete and aggregate demand has become increasingly prominent [4-5].

In order to reduce the cost of producing recycled aggregate (RA), The demolished concrete lumps (DCLs) had been demonstrated to replace some coarse aggregates when casting concrete. [5-8]. This method could reduce energy consumption.

However, few studies have focused on the sensitivity of factors affecting mechanical behavior of slender circular steel columns filled with DCLs and FC under axial compression. This research was carried out on the basis of Bo Wu's study. The Orthogonal design method was adopted in this experiment. This study aims to investigate the sensitivity of factors affecting mechanical behavior of slender circular steel columns filled with DCLs and FC under axial compression..

2. Experimental program

2.1. Materials and specimen details

The experimental study involved 12 CFST columns. Each column diameter was 159mm, the length of specimens was 2000, 2200 and 2400 mm. The thickness of the steel tubes was 2, 3 and 4mm, so the D/t ratio was 79, 53 and 40. The replacement ratio of FC by DCLs was 0, 20, 40 and 60%. 1Aa0, 5Bb0 and 9Cc0 were control specimens that were designed without DCLs in FC. The D/t ratio in these tests met most of the codes' requirements. According the EC4, the slenderness ratio (λ=4L/D) was obtained. The slenderness ratio was 50, 55 and 60. Strain gauge was placed in the mid-height of specimens to measure the strain. The details of the specimens are shown in Table 1.

Table 1. Detail of the test specimens Formatting sections

Specimen	L / mm	D / mm	t / mm	D / t / %	λ	η / %	N / kN
1Aa0	2 000	159	2	79	50	0	890
2Aa1	2 000	159	2	79	50	20	890
3Ac2	2 000	159	4	40	50	40	1 266
4Ab3	2 000	159	3	53	50	60	1 081
5Bb0	2 200	159	3	53	55	0	1 042
6Bb1	2 200	159	3	53	55	20	1 042
7Ba2	2 200	159	2	79	55	40	857
8Bc3	2 200	159	4	40	55	60	1 220
9Cc0	2 400	159	4	40	60	0	1 178
10Cc1	2 400	159	4	40	60	20	1 178
11Cb2	2 400	159	3	53	60	40	1 006
12Ca3	2 400	159	2	79	60	60	827

Note: L, D and t are respectively the length, diameter and thickness of the specimens. λ and η are respectively the slenderness ratio of specimens and the replacement ratio of DCLs.

2.2. Material properties

For the steel tube, the yield strength, the ultimate strength , and the elastic modulus are shown in Table 2. To measure the displacement of the columns, three LVDTs were placed on the specimens. Through the hydraulic testing machine, the jack act on the top of the column to load step by step. Load grade was directly controlled by hydraulic testing machine. The geometric center of the specimen before loading coincided with the connection of the geometric center of the upper and lower loading plates of the testing machine. Loading can be divided into two stages: preloading and formal loading. In the pre-loading stage, the applied load value was about 1/20 of the ultimate load. The purpose was to test whether the data collected by strain gauges and displacement meters were accurate and effective, and to make the specimen in good contact with the loading device, and then entered the formal loading stage.

In the elastic range, the loading value of each stage was 1/10 of the design ultimate load; when the applied load was about 70% of the design ultimate load, then the loading value of each stage was 1/20 of the design ultimate load, and the loading time of each stage was about 5 minutes. When the applied load was about 95% of the design ultimate load, the lateral deflection at mid-height of the column increased by 1 mm as the control index, and the lateral deflection at mid-height of the column increased gradually until the specimen was destroyed. The data in the test process were collected automatically by the instrument.

2.3. The relationship between slenderness and bearing capacity

Fig.1 shows the influence of slenderness ratio on the bearing capacity of slender columns. it could be concluded that the bearing capacity of slender CFST columns filled with DCLs decreased with the increased of slenderness ratio.

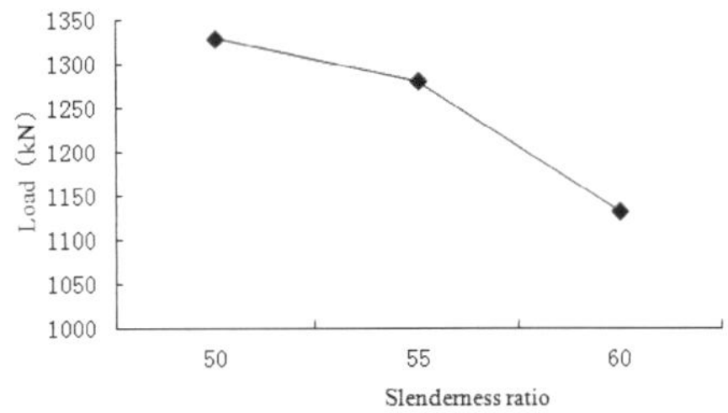

Fig.1. Relationship between slenderness ratio and bearing capacity of specimens

When slenderness ratio increased from 50 to 55, the bearing capacity of the column decreased by about 4%, the slope of the section was about 9.8, and when slenderness ratio increased from 55 to 60, the bearing capacity of the column decreased by about 12%, and the slope of the section was about 29.6. Through comparative analysis, the following basic rules were obtained: the bearing capacity decreased greatly with the increased of slenderness ratio, but the relationship was not linearly. It also illustrated that the elastic-plastic instability failure did not converge with the increase of slenderness ratio.

The bearing capacity decreased greatly with the increased of slenderness ratio. The reason was that the column's deflection did not change significantly at the beginning of loading, and the slender column was in the stage of elastic working. When the load continued to be applied, the lateral deflection began to appear in the mid-height of the slender column. The eccentricity of the specimen occurred, which result in the bending moment and reduced the bearing capacity of the column. When approaching the ultimate load, the column's deflection increased obviously, the eccentricity continued to increased, and the bending moment also increased significantly. At a certain moment, even if the load did not increase, the lateral deflection of the slender column would continue to increase.

2.4. The relationship between D/t ratio and bearing capacity

The relationship between D / t ratio and bearing capacity of specimens is shown in Fig.2.The ultimate bearing capacity increased by about 6%, with the D / t increased from 40% to 53%, the slope of this section was about 22.7%, and the ultimate bearing capacity increased by about 7% with the steel content increased from 53% to 49%, and the slope of this section was about 27.7. Through comparative analysis, the following basic rules were obtained: with the increased of D/t ratio, the increased range of bearing capacity became larger.

The hoop coefficient ($\xi = A_s f_y / A_c f_c$) affected the bearing capacity of the column. The hoop coefficient was introduced to reflect the constraint of steel tube on core concrete. The larger the hoop coefficient was, the stronger the constraint of steel tube on concrete was, the higher the bearing capacity of the column was. D / t ratio indirectly affected the hoop coefficient. For concrete with the same steel and axial compressive strength, the D / t ratio had a linear relationship with the hoop coefficient ($\xi = A_s f_y / A_c f_c$). When the D / t ratio increased, the hoop coefficient increased, the restraint effect of steel tube on core concrete increased, and the bearing capacity of columns increased.

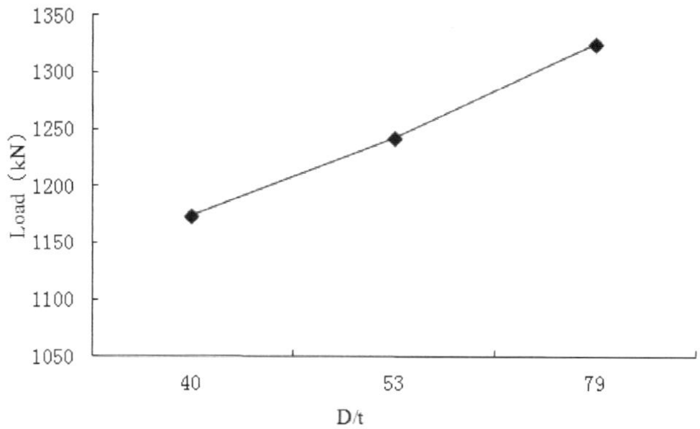

Fig.2. Relationship between D / t ratio and bearing capacity of specimens

2.5. The relationship between DCLs replacement ratio and bearing capacity

The relationship between DCLs replacement ratio and bearing capacity of specimens is shown in Fig.3. The ultimate bearing capacity of long columns decreased by 7% wit hthe replacement ratio of DCLs increased from 20% to 40%, the slope of this section was about 4.6, and the ultimate bearing capacity of slender columns decreased by 3% with the replacement ratio of DCLs increased from 40% to 60%, and the slope of this section was about 1.2. Through comparative analysis, the following basic rules were obtained: the column's bearing capacity slightly decreased with the increased of DCLs replacement rate.

The reason why the DCLs reduced the bearing capacity was that when concrete was poured, because of the small cross-sectional area of the steel tube, there were voids between the DCLs and the steel tube in the process of casting, the concrete was not as dense as expected. In addition, the DCLs were placed unevenly, which reduced the bearing capacity compared with the specimens without DCLs.

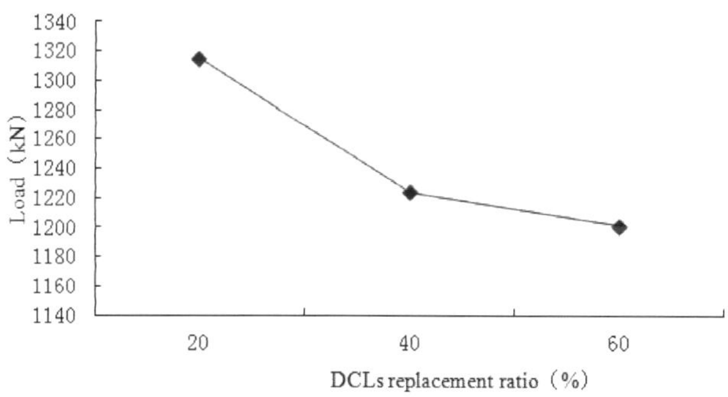

Fig.3. Relationship between DCLs replacement ratio and bearing capacity of specimens

3. Conclusion

The mechanical behavior of CFST filled with DCLs under axial compression was presented. The following conclusions are drawn from the test results:

(1) The influence of slenderness ratio, D / t ratio and DCLs replacement ratio on ultimate bearing capacity and sensitivity of specimens were analyzed by orthogonal test results. The slenderness ratio was the most sensitive factor on the bearing capacity of slender CFST columns, followed by D / t ratio, and the smallest was DCLs replacement ratio.

(2) The bearing capacity decreased by 12% with slenderness ratio increased from 55 to 60,

the bearing capacity increased by 7% with D / t ratio increased from 8% to 11%, , and the effect was not obvious. The bearing capacity decreased by 7% with DCLs replacement ratio increased from 20% to 40%, , and the effect was not obvious. Through the orthogonal test, the influence of various parameters on the bearing capacity was analyzed. The column bearing capacity slightly decreased with the increased of DCLs replacement rate. The column bearing capacity increased with the increased of D / t ratio. The column bearing capacity decreased with the increased of slenderness ratio.

Acknowledgments
Authors wishing to acknowledge assistance or encouragement from colleagues, special work by technical staff or financial support from organizations should do so in an unnumbered Acknowledgments section immediately following the last numbered section of the paper.

4. References

[1] B. González-Fonteboa, F. Martínez-Abella, J. Eiras-López, et al., Effect of recycled coarse aggregate on damage of recycled concrete, Mater. Struct (10) (2011) 1759–1771.
[2] B. González-Fonteboa, F. Martínez-Abella, M.F. Herrador, et al., Structural recycled concrete: behaviour under low loading rate, Constr. Build. Mater (1) (2012) 111–116.
[3] N.Y. Ho, P.K. Yang, W.F. Lim, et al., Efficient utilization of recycled concrete aggregate in structural concrete, J. Mater. Civ. Eng. 25 (3) (2013) 318–327
[4] S.I. Ivan, B.M. Snez ˇana, M.M. Zoran, et al., Flexural behavior of reinforced recycled aggregate concrete beams under short-term loading, Mater. Struct (6) (2013) 1045–1059.
[5] J.F. Lamond, Significance of Tests and Properties of Concrete and Concrete-Making Materials, ASTM International, 2006.
[6] X. Li, Recycling and reuse of waste concrete in China: part I. Mate behavior of recycled aggregate concrete, Resour. Conserv. Recy. 53 (1) (2008) 36–44.

IOP Conf. Series: Materials Science and Engineering **758** (2020) 012026 doi:10.1088/1757-899X/758/1/012026

A Mobile Robot Path Planning Scheme for Dynamic Environments

Kene Li*, Qiaoliang Mo, Zeng Zhang, Bei Liu

School of Electrical and Information Engineering, Guangxi University of Science and Technology, Guangxi, China
E-mails: likene@163.com, 523915089@qq.com

Abstract. This paper proposes a four-direction search method for obstacle avoidance of mobile robots, and the collision energy of obstacles is modeled based on the neural network. For comparison and discussion purposes, the different invariant step lengths are tested in the static environment. To take the advantages of the large and small step lengths effectively, a variable step length method is adopted to improve the performance, such as less iteration number and lower total energy. The variable step length method is also applied to the dynamic environment to explore the real-time performance for path planning. Simulation results demonstrate the effectiveness and practicability of the presented scheme.

1. Introduction

In the field of robotics, one of the hot research issues is how to make a mobile robot move to the goal point as quickly as possible with obstacle-avoidance in an environment [1-10]. Generally speaking, the obstacle-avoidance problem solving of a mobile robot can be transformed into path planning. The set of a series of points or lines interconnecting from the initial point to the goal point is called a path, and the decision to form the path is called path planning. Main problem to be solved in the process of path planning can be described as generating a path from the initial point to the goal point, using algorithms to make a mobile robot avoid obstacles in the path, and making the generated path as smooth as possible [1]. The path planning problem of a mobile robot can be divided into global path planning and local path planning, which depends on the environment of the mobile robot. In global path planning, the environment around the mobile robot is known, and environmental information is static and known. In local path planning, the environment surrounding the mobile robot is unknown or partially known, and then, environmental information can be seem dynamic [2].

In the field of path planning, many scholars have conducted a series of studies. For path planning in a static environment, Priyanka Sudhakara applied an enhanced artificial potential field method to improve the oscillations in navigable trajectories. However, the final generated path was not smooth enough [3]. For the dynamic environment situation, Huckleberry Febbo put forward a method, in which a hard constraint formulation was proposed to solve the moving obstacle avoidance problem of large, high-speed autonomous ground vehicles in an unstructured environment. Nevertheless, after the sensors detected obstacles, it cannot build a suitable mathematical model in the map [4]. Mohamed Elhoseny presented a modified genetic algorithm based on Bezier Curve for path planning in a dynamic environment, in which the local optima may be considered further [5]. In this paper, we propose a mobile robot path planning scheme for a dynamic environment, which can obtain a smooth optimal path and avoid local optima.

2. Problem Formulation

In this section, the algorithm for path planning based on neural network is described for discussion.

2.1. Formulation of Collision Energy Function

Normally, if we have information about the location and shape of the obstacle, we can reformulate the obstacle as a stationary convex polygon or a circle for a known two-dimensional environment [6-8]. For the path planning problem, the collision energy $G(i)(x_i, y_i)$ of every coordinate point is obtained by using neural network [6-8]. In the neural network structure, the input layer is set as the environmental coordinate point. The first hidden layer $\boldsymbol{U}$ is determined by the shape of the environmental obstacles. Based on the neural network, every convex polygon can be represented as a number of closed line segments [6-8], and the line can be described as

$$w_x x + w_y y + \sigma = 0 \tag{1}$$

where w_x, w_y and σ are the line function coefficients. We reformulated the j-th obstacle as a rectangle and we have the information of its four sides as follows:

$$\begin{cases} w_{xj1}x_i + w_{yj1}y_i + \sigma_{uj1} \geq 0 \\ w_{xj2}x_i + w_{yj2}y_i + \sigma_{uj2} \geq 0 \\ w_{xj3}x_i + w_{yj3}y_i + \sigma_{uj3} \geq 0 \\ w_{xj4}x_i + w_{yj4}y_i + \sigma_{uj4} \geq 0 \end{cases}$$

where σ_{ujk} is its threshold coefficient, and w_{xjk}, w_{yjk} are weight values of the input neurons x_i, y_i to the jk-th hidden layer neuron with $k = 1,2,3,4$. For the k-th side of j-th obstacle, the first hidden layer neuron function is set as

$$u_{jk}(x_i, y_i) = \frac{1}{1 + e^{-(w_{xjk}x_i + w_{yjk}y_i + \sigma_{ujk})/T}} \tag{2}$$

where T is the positive design parameter. In addition, If the j-th obstacle is circular, r is its radius, and C_{jx} and C_{jy} are the center coordinates, then we can set its first hidden layer neuron function as

$$u_j(x_i, y_i) = r - (x_i - C_{jx})^2 - (y_i - C_{jy})^2 \tag{3}$$

The second hidden layer neuron function for the j-th obstacle consisted of k sides is taken as

$$o_j(u_{j1}, u_{j2}, .., u_{jk}) = \frac{1}{1 + e^{-(\sum_{i=1}^{k} u_{ji} - \sigma_{oj})/T}} \tag{4}$$

where $\sigma_{oj} = k - 0.5$ (for the circular obstacle, $\sigma_{oj} = 0.5$). For an environment with n obstacles, we can obtain the energy of a point as

$$g_i(o_1, o_2, ..., o_n) = \sum_{j=1}^{n} o_j \tag{5}$$

It is worth mentioned that the distributed shape of collision energy by Eqs. (2) and (4) is influenced by the design parameter *T*. When parameter *T* is larger, the shape is more smooth.

In view of the above-mentioned discussion, we can obtain the energy distribution associated with obstacles in the environment.

2.2. Four-direction search scheme

In this section, we propose a four-direction search method for mobile robot path planning. we connect the start point to the goal point as the initial path, and then divide the initial path into many equally-long line-segments as shown in Figure 1(a). Different from the search algorithm of Refs. [6,7], we present a four-direction search method for mobile robot avoidance obstacle, shown in Figure 1(b).

In Figure 1(b), the search step length is λ, and thus, for the i - th node, we can obtain the four search direction as follows:

$$\begin{cases} x_i^{(k+1)} = x_i^{(k)} + l \cdot \cos(j \cdot \pi / 2) \\ y_i^{(k+1)} = y_i^{(k)} + l \cdot \sin(j \cdot \pi / 2) \end{cases} \tag{6}$$

where the superscript $(k+1)$ denotes the $(k+1)$-th iteration, and $j = 0,1,2,3$. We link the new four nodes with the $(i-1)$-th and $(i+1)$-th nodes, and calculate their obstacle energy. According to the k-th iteration result, we can obtain the path with the minimum energy. Furthermore, in order to reduce the path search computing again, for a path with n nodes, we present the variable step length $\hat{l}$.

$$\hat{l} = \max(G) \cdot l$$

where $G = [g_1, g_2, ..., g_n]$, $\max(\cdot)$ selects the maximum of a vector, and l is an initial invariant step length.

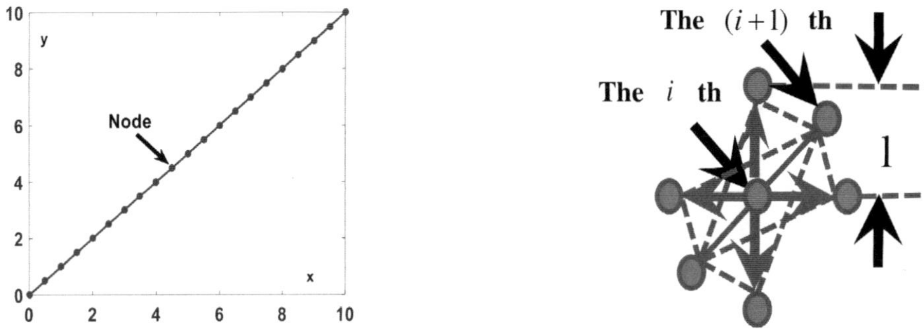

(a) Initial path (b) Four-direction search

Figure 1. The initial path and four direction search

2.3. Formulation of Path Energy Function

Obviously, if we only consider the path energy, the mobile robot will move towards the place where the energy becomes as small as possible, but the final path may be very long. In order to solve this problem, we must consider the obstacle energy as well as the path energy. According to the above-mentioned discussion, we introduce the square of the difference between two adjacent nodes as the path segment energy to solve this problem.

$$L = \sum_{i=1}^{n} [(x_{i+1} - x_i)^2 + (y_{i+1} - y_i)^2] \tag{7}$$

where L denotes the total path energy, and x_i and y_i denote the coordinates of i-th path-node (x_0 and y_0 denote the coordinates of start path-node).

3. Simulation studies

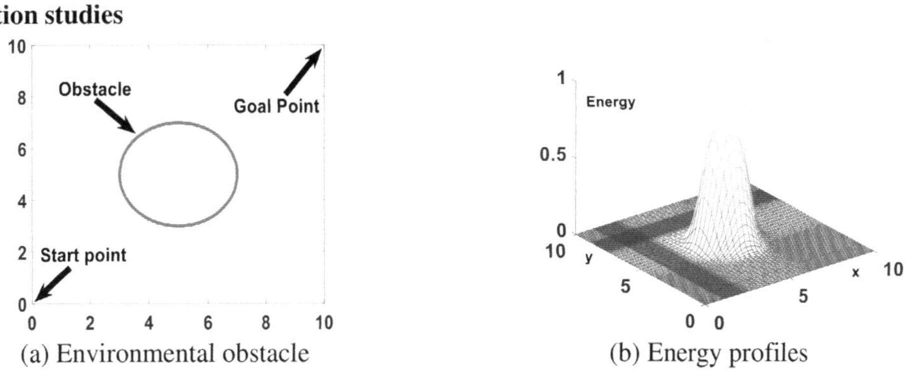

(a) Environmental obstacle (b) Energy profiles

Figure 2. Environmental obstacle and energy profiles

In this section, the simulation is conducted based on the presented four-direction search method with invariable step length in the static environment, and the mutation point problem arising in the simulation is discussed and improved. To enhance the performance of the four-direction search method, a variable step length is applied to reduce the iteration times and make the generation path more smooth. Finally, the four-direction method is tested in the dynamic environment. The simulations verify the effectiveness of the presented four-direction method.

3.1. Invariable search step length situation

In this subsection, obstacle-avoidance is studied and discussed with an invariant search step length for a single circular stationary obstacle.

We set up a simulation environment (i.e., a planar space with $x \in [0,10]$, $y \in [0,10]$), which has a circular obstacle with {(5,5), r=2}, as shown in Figure 2(a). Moreover, the energy distribution of the obstacle is modeled and shown in Figure 2(b). According to the predefined requirements, the mobile robot requires to move from start point (0, 0) to the goal point (10, 10).

In order to make the experiment easy to discuss, we use the invariant search step length $1 = 0.05$ for the four-direction scheme in the first test. The simulation results are shown in Figure 3. We can see that the final path can meet the experimental requirement. However, the final path has mutation points and is not smooth enough. Known from the simulation tests, the appearing mutation points result from the premature convergence caused by the larger search step length. In addition, the total energy (i.e., sum of the path energy and the collision energy) decreases from 15.51 of the initial path to 0.7424 of the final path, and the iteration number is 74. To improve premature convergence, we consider a smaller search step length, e.g., $\lambda = 0.03$. The corresponding simulation results are shown in Figure 4. Comparing with the final energy 0.7424 of Figure 3, the total energy has reduced to 0.7346 in the 97th iteration, and then continue converging to 0.6015, which implies that the reduced search step length can improve the phenomenon of premature convergence. In addition, with reduction of the iteration step length, the generated path is more smooth and the path mutation point is also improved. However, it is worth mentioning that the smaller step length makes the iteration times increase apparently to 118, which may decrease the computation efficacy of this scheme.

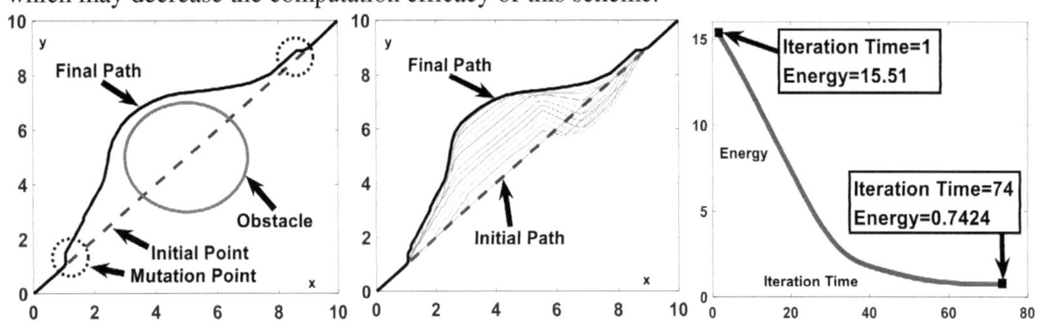

(a) Final path (b) Path profiles (c) Total energy profiles

Figure 3. Path planning and paths synthesized by FDS scheme with $1 = 0.05$

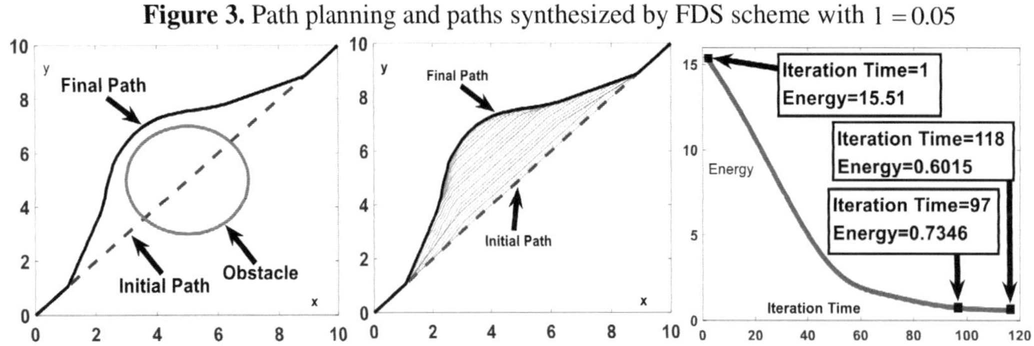

(a) Final path (b) Path profiles (c) Total energy profiles

Figure 4. Path planning and paths synthesized by FDS scheme with $1 = 0.03$

IOP Conf. Series: Materials Science and Engineering **758** (2020) 012026 doi:10.1088/1757-899X/758/1/012026

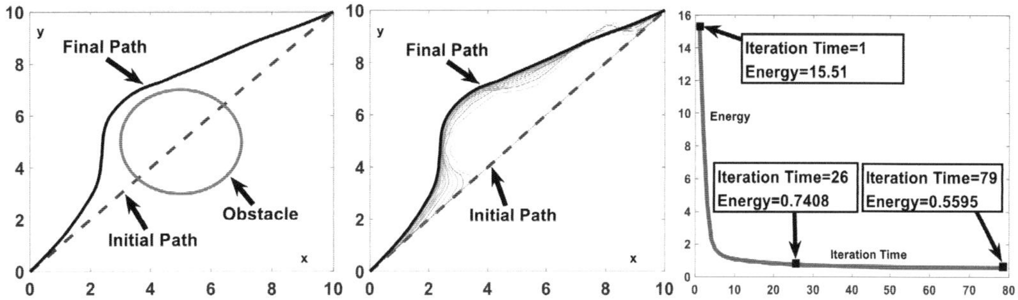

(a) Final path (b) Path profiles (c)Total energy profiles

Figure 5. Path planning and Paths synthesized by FDS scheme with $\hat{\imath} = \mathrm{Max}(G)$

3.2. Variable search step length situation

In order to take the advantages of the large step length (i.e., less iteration number) and small step length (i.e., lower total final energy), a variable search-step-length method is proposed and adopted, inspired by the above analysis and discussion. Therefore, we replace the invariant step-length with variable step-length $\hat{\imath} = \mathrm{Max}(G) \times 1_0$, with 1_0 denoting the initial step length ($\lambda_0 = 1$ for this test). The simulation results are shown in Figure 5. For the comparison purpose, we can see that the total energy of the 26th iteration has been reduced to 0.7408, which is better than 0.7424 of the 74th iteration in Figure 3. Moreover, the final energy converges to 0.5595 after 79 iterations, which shows that the variable step-length method can find the optimal path with lower final energy and lower iteration number. The simulation results demonstrate the variable step-length has better performances than the invariant one, and the scheme can obtain a satisfactory solution.

3.3. Path planning in dynamic environment

Path planning in the dynamic environment is a challenging issue. Due to its real-time requirement, the FDS scheme with the variable step-length can be a feasible alternative. For the test purpose, we set up a map (i.e., a planar space with $x \in [0,2]$, $y \in [0,2]$), and a circular obstacle with {(2, 0), r=0.2} is assumed to move for the start point (2,0) to the goal point (0,2) along the diagonal straight-line path. The variable step length for the dynamic test is set as $\hat{\imath} = \mathrm{Max}(G) \times 0.02$. The simulation results are shown in Figure 6, and corresponding iteration numbers and total energies are shown in Table 1. From Figure 6, we can see that the generated path can avoid the obstacle. For details, Figure 6(a)-(c) show the critical moments, and we can see that the mobile robot moves towards the lower-energy place, but the subsequent planning path will bypass the obstacle automatically. As a comparative result, the final path and the initial path are shown in Figure 6(d). From Table 1, we can see that the FDS scheme works well, and the total energies decrease at the critical moments. This simulation results substantiate that the proposed FDS scheme is effective for obstacle avoidance in a dynamic environment. Note that in this simulation design parameters T is set as 0.28.

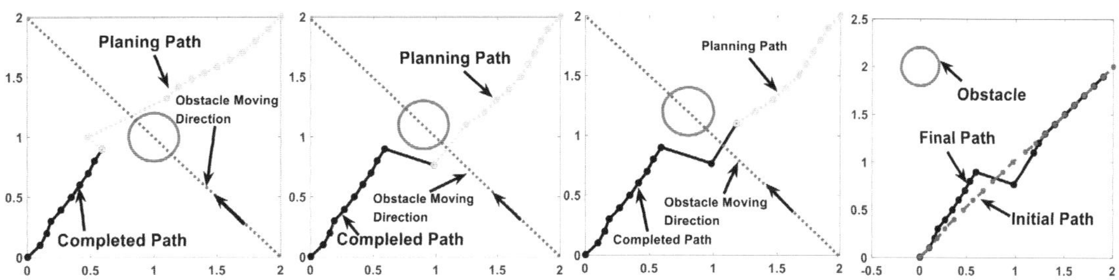

(a) Node 9 (b) Node 10 (c) Node 11 (d)Initial and final paths

Figure 6. Path planning of FDS scheme for dynamic environment

Table 1. Iteration number of FDS scheme for dynamic environment.

#	Iteration number					Initial energy					Final energy				
1-5	1	104	160	173	116	0.0611	0.0991	0.1881	0.3594	0.6287	0.0611	0.0724	0.0813	0.0961	0.1535
6-10	83	33	18	12	6	0.9707	1.3137	1.568	**1.6637**	**1.5746**	0.2723	0.5111	0.8963	**1.2984**	**1.2880**
11-15	5	2	1	1	1	**1.3246**	0.9807	0.6329	0.3541	0.1727	**1.1337**	0.9541	0.6329	0.3541	0.1727
16-19	1	1	1	1		0.0753	0.0309	0.013	0.0058		0.0753	0.0309	0.0130	0.0058	

4. Conclusions

This paper focuses on the path planning method based on the neural network for collision energy building in the map, and a four-direction search method is presented and tested for the problem in the static and dynamic environment. Moreover, the proposed variable step length can improve the search performance apparently. The simulation results verify the effectiveness and practicability of the improved FDS scheme. Future work can consider to improve the generated path in the dynamic environment, and to realize proposed scheme physically.

References

[1] D. Zhu, and M. Yan, Survey on technology of mobile robot path planning, Control and Decision, 25(7), pp.961-967, 2010.
[2] Q. Zhang, B. Chen, X. Liu, and H. Yang, Ant colony optimization with improved potential field heuristic for robot path planning, Transactions of the Chinese Society for Agricultural Machinery, pp. 1-15, 2019.
[3] H. Febbo, J. Liu, P. Jayakumar, J. L. Stein, and T. Ersal, Moving obstacle avoidance for large, high-speed autonomous ground vehicles, American Control Conference, pp.5568-5573, 2017.
[4] P. Sudhakara, V. Ganapathy, B. Priyadharshini, and K. Sundaran, Obstacle avoidance and navigation planning of a wheeled mobile robot using amended artificial potential field method, Procedia Computer Science, 133, pp. 998-1004, 2018.
[5] M. Elhoseny, A. Tharwat, and A. E. Hassanien, Bezier curve based path planning in a dynamic field using modified genetic algorithm, Journal of Computational Science, 25, pp. 339-350, 2017.
[6] J. Yu, V. Kroumov, and H. Narihisa, Path planning algorithm for car-like robot and its application, Chinese Quarterly Journal of Mathematics, 17(3), pp.98-104, 2002.
[7] G. Wei, and M. Fu. An algorithm based on neural network for mobile robot path planning, Computer Simulation, 27(7), pp.112-116, 2010.
[8] K. Li, C. Yuan, J. Wang and X. Dong, Four-Direction Search Scheme of Path Planning for Mobile Agents, Robotica , doi:10.1017/S0263574719000821, pp.1-10, 2019.
[9] B. K Patle, A. Pandey, A. Jadadeesh, and D. R. Parhi, Path planning in uncertain environment by using firefly algorithm, Defence Technology, 14, pp.691-701, 2018.
[10] A. K. Guruji, H. Agarwal, and D. K. Parsediya, Time-efficient A* algorithm for robot path planning, Procedia Technology, 23, pp.144-149, 2016.

Acknowledgments

This work was supported by National Natural Science Foundation of China (NSFC) under Grant No. 61663003.

IOP Conf. Series: Materials Science and Engineering **758** (2020) 012027 doi:10.1088/1757-899X/758/1/012027

Modification of the prediction model of the tensile strength of short fiber reinforced polymer composites

Z H Zhu [1], N Zhang[1], Y Zhang [1], M Y Hao [2] and H W Wu [2]

1 School of advanced manufacturing technology, Guangdong Mechanical & Electrical Polytechnic, Guangzhou, China;
2 The Key Laboratory of Polymer Processing Engineering, Ministry of Education, South China University of Technology, Guangzhou, China
Email: 184823174@qq.com

Abstract: Based on the defect of COX model, the parameters related to the interface properties were introduced, and the fitting results were modified to be closer to the measured values.

1. Introduction

Fiber reinforced composite is a kind of multiphase material and its mechanical properties and failure laws depend on many factors. Researchers have proposed many models to predict the tensile properties of fiber reinforced polymer composites, and the most commonly used model is based on the linear superposition principle.

$$F = F_1(x_1V_1) + F_2(x_2V_2) + F_3(x_3V_3) + \cdots \quad (1)$$

x_1、x_2、x_3 refers to the different components of composite materials. V_1、V_2、V_3 is corresponding to the volume fraction of each component. Based on the above principles, Weibull proposed a formula for calculating the tensile strength of single-fiber reinforced polymer composites [1]

$$\sigma_c = \sigma_f V_f + \sigma_m(1 - V_f) \quad (2)$$

σ_c、σ_f and σ_m are the tensile strength of the composite fiber and polymer matrix respectively. V_f is the volume fraction of fiber.

Cox introduced fiber length factor and proposed Cox model for prediction of tensile strength of short fiber reinforced composites. [2]

$$\sigma_c = \eta_{LE}\sigma_f V_f + \sigma_m(1 - V_f) \quad (3)$$

η_{LE} is the effective factor of fiber length and its calculation formula is given as follows.

$$\eta_{LE} = 1 - \frac{tanh\cdot(\beta L/2)}{\beta L/2} \quad (4)$$

$$\beta = \sqrt{\frac{2G_m}{E_f r_f^2 ln(R/r_f)}} \quad (5)$$

L is the fiber length, r_f is the fiber radius, G_m is the shear modulus of matrix, R is half the distance of the adjacent fibers which can be obtained by equation 6:

$$R = \frac{r_f}{2} \cdot \sqrt{\frac{\pi}{V_f}} \tag{6}$$

If polymer matrix is isotropic material, the shear modulus G_m of which can be calculated as follows.

$$G_m = \frac{E_m}{2(1+v_m)} \tag{7}$$

E_m, v_m are the young's modulus and Poisson's ratio of the polymer matrix. On the basis of Cox model, Krenchel considered the effect of fiber orientation, introduced fiber orientation distribution factor C_0 , and modified the prediction model of Equation 3-3 to Equation 8. [3]

$$\sigma_c = C_0 \eta_{LE} \sigma_f V_f + \sigma_m (1 - V_f) \tag{8}$$

Different composites molding process can get different value C_0. In general, the more oriented the fiber, the greater the value C_0. In this work, Sisal fibers were randomly distributed in PLA matrix and the value C_0 is 0.125 which was calculated by Fukuda when the orientation is randomly distributed in the three-dimensional direction. [4]

Through the single fiber pull-out experiment in our previous work, it was known that the interface strength of PLA/USFs and PLA/ASFs are different, and thus the interface strength of composites will certainly change due to the different proportion of USFs and ASFs.[1] Therefore, Untreated sisal fibers (USFs) and Alkali-treated sisal fibers (ASFs) are mixed into Hybrid SFs (HSFs) with different proportions, and then HSFs are mixed with PLA to prepare different PLA composites in this work. Our purpose is to consider introducing the interface strength into equation 3-8 to obtain a new modified model.

2. Experiment

2.1. Materials

PLA 3051D, Nature Works ™, was purchased by Shenzhen Bright China Industrial Company (China). SF bundles (GB/T 15031-94) were provided by Guangxi Sisal Company of Guangxi Province, China. Other chemical reagents were commercial products.

2.2. Samples Preparation

SFs were submerged into 10% NaOH solution to prepare ASFs for 3 h. Then, ASFs were washed with distilled water and dried in the open air for 48 h. USFs were mixed with ASFs to fabricate HSFs in a mass ratio of 1:9、2:8、3:7、4:6、5:5、6:4、7:3、8:2、9:1. After that, PLA, USFs, ASFs and HSFs were dried at 100 degrees for 4 h. Then, PLA was compounded with USFs, ASFs and HSFs with a mass ratio 4:1 respectively using a two-roll plastic mill at 180 degrees. Eleven kinds of samples were made, as shown in Table 1. Finally, they were compressed into sheets by a hot press. The compression molding temperature and pressure were 190 degrees and 10 MPa, respectively.

Table 1 Mass ratios of USFs and ASFs

Composites	C1	C2	C3	C4	C5	C6	C7	C8	C9	C10	C11
USFs:ASF	0:10	1:9	2:8	3:7	4:6	5:5	6:4	7:3	8:2	9:1	10:0

2.3 Mechanical test

An Instron Universal Testing Machine (model 5566, USA) was used to perform the tensile test, the crosshead speed of which was 5 mm min^{-1}.

3. Results and discussion

3.1. Tensile property of composites with different SFs

The tensile properties of pure PLA and elven PLA composites were shown in Figure 1, which could be observed that the tensile properties of PLA/USFs composite and PLA/ASFs composite were worse than those of pure PLA. It could also be seen that the addition of HSFs could improve the tensile properties of PLA effectively.Tensile strength and tensile modulus of PLA/HSFs are higher than those of PLA/USFs and PLA/ASFs except to C9 and C10 composites. C5 has the best tensile properties, whose tensile strength and tensile modulus were 53.9 and 2526.7 MPa, respectively, which improved by 41.5% and 48.1% compared with PLA/USFs, and 10.0% and 27.1% compared with PLA/ASFs. Those indicated that HSFs as reinforcement could improve the tensile properties of PLA composites more significantly, which mainly because that HSFs can give full play to the advantages of USFs in tensile properties and the advantages of ASFs in interfacial compatibility with PLA matrix. There is a synergistic effect between USFs and ASFs when they are mixed in polymer composites.

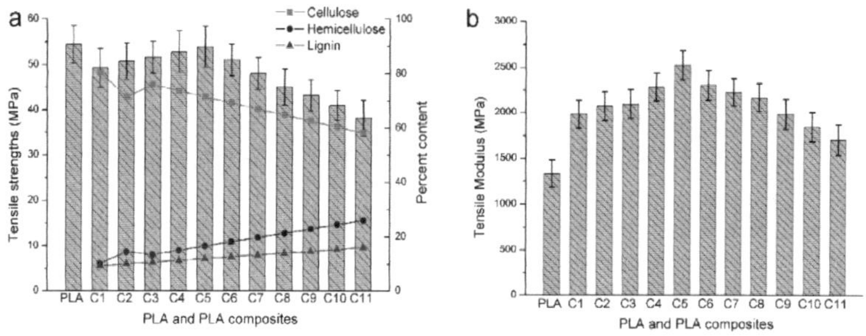

Fig. 1 Tensile properties of PLA and PLA composites

3.2 Discussion of the prediction model of the tensile strength of short fiber reinforced polymer composites

By substituting the data of USFs and ASFs into equation (8), it is surely found that the theoretical value is not equal to the measured data, and there is a considerable difference. Tensile failure of short fiber reinforced polymer composites involves fiber, matrix and interface which is a random process of cumulative microscopic damage, such as fiber fracture, matrix failure, interface debonding, etc. Therefore, a defect of Cox model lies in that it does not pay attention to the impact of interface properties on the failure process and strength, thus affecting the accurate judgment of tensile strength of composite.

The strength of the short fiber reinforced composite is mainly affected by three aspects, which are the fiber strength, the matrix strength and the interface strength. Based on the structure of composite, we can take a single fiber and its surrounding matrix as a unit, as shown in Figure 2.

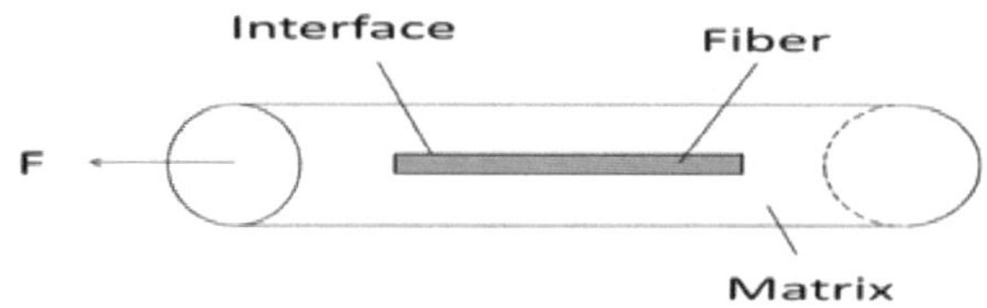

Fig. 2 Schematic diagram of pulled-out single fiber from PLA matrix

The following stress analysis is performed for this model. When the composite is subjected to external forces, the matrix is the first to bear the force, and the force is transferred to the fiber as the reinforcement through the interface. When the adhesion between the fiber and the matrix is strong enough, external forces can be effectively transferred to the fiber through the interface, and the fiber thus plays a strengthening role. The mechanical analysis can be expressed as follows.

$$F = F_m + F_f + F_i \tag{9}$$

However, the interface between the matrix and the fiber was too weak due to too much USF after the mixing ratio was greater than 7:3, and the external force on the matrix could not be effectively transmitted through the interface. The fiber can not give full play to its strengthening function, but also causes the stress concentration in the composite due to the fact that the fiber itself is easy to aggregate. At this time, the prediction of the strength of composite is more complex, and there is no great significance and need to discuss because the fiber has no real strengthening effect.

The destruction of the composite is a random process, and there are various failure modes and thus F_i and σ_i exists nonlinear correlation. For short fiber reinforced polymer composites, Equeation 3-8 can be corrected as Equation 10.

$$\sigma_c = C_0\eta_{LE}\sigma_f v_f + \sigma_m(1 - v_f) + f(\sigma_i) \tag{10}$$

For hybrid fiber reinforced polymer composites, Equeation 3-10 can be corrected as Equation 11.

$$\sigma_c = C_0\eta_{LE}(\sigma_{asf}v_{asf} + \sigma_{usf}v_{usf}) + \sigma_m(1 - v_f) + f(\sigma_i) \tag{11}$$

Then

$$f(\sigma_i) = \sigma_c - C_0\eta_{LE}(\sigma_{asf}v_{asf} + \sigma_{usf}v_{usf}) - \sigma_m(1 - v_f) \tag{12}$$

Then the tensile strength of hybrid fiber reinforced polymer composites can be calculated by Equation 13.

$$\sigma_i = \lambda\sigma_{iasf} + (1 - \lambda)\sigma_{iusf} \tag{13}$$

λ is the mass ratio of the ASF to the hybrid fiber.

The relationship beteween $f(\sigma_i)$ and σ_i is shown on Figure 3 from which it could be observed that $f(\sigma_i)$ and σ_i have nonlinear negative correlation.

$$f(\sigma_i) = \varepsilon - \xi\sigma_i^t \tag{14}$$

ε, ξ and t are parameters related to interface properties of composites.

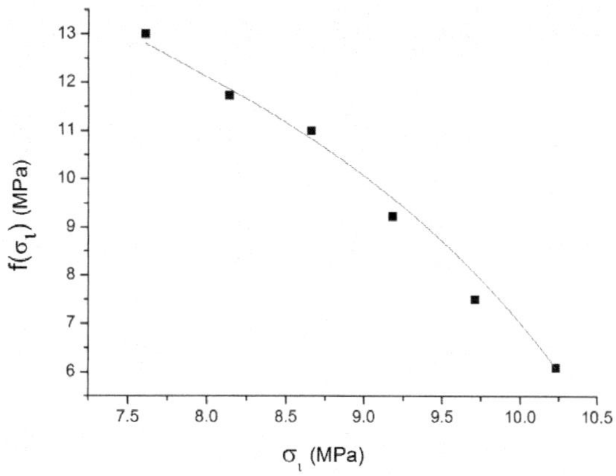

Fig. 3 Nonlinear curve fitting between $f(\sigma_i)$ and σ_i

By substituting equation (14) into equation (11), the model of short-fiber reinforced PLA composites can be obtained as follows.

$$\sigma_c = C_0\eta_{LE}\sigma_f v_f + \sigma_m(1 - v_f) + (\varepsilon - \xi\sigma_i^t) \quad (15)$$

For hybrid fiber reinforced polymer composites

$$\sigma_c = C_0\eta_{LE}(\sigma_{asf}v_{asf} + \sigma_{usf}v_{usf}) + \sigma_m(1 - v_f) + (\varepsilon - \xi\sigma_i^t) \quad (16)$$

Through the fitting function of Origin software, values of ε, ξ and t can be obtained.

$$\varepsilon = 21.7469$$

$$\xi = 0.342$$

$$t = 1.5763$$

4. Conclusion

Tensile strength of short fiber reinforced polymer composite is mainly affected by three aspects which are fiber strength, the matrix strength and the interface strength. Based on the defect of COX model, the parameters related to the interface properties were introduced in this work. It can be seen that the fitting results were modified to be closer to the measured values.

Reference

[1] Zhu Z, Hao M, Zhang N, et al. Journal of Natural Fibers, 2018: 1-12.
[2] Waloddi W. Journal of Applied Mechanics-Transactions of the Asme, 1951, 18 (2): 293-297
[3] Cox H L. British Journal of Applied Physics, 1952, 3(3):72
[4] Krenchel H. Alademisk forlag, 1964:110-115
[5] Fukuda H, Chou T W. Journal of Materials Science, 1982, 17 (4): 1003-1011

Acknowledge

This research was supported by Key Research Platforms and Projects of Universities of Guangdong Province (grant no. 2018GKTSCX111) and Funding Project for Scientific Research of High-level Talents of Guangdong Mechanical & Electrical Polytechnic (grant no. Gccrcxm-201903).

IOP Conf. Series: Materials Science and Engineering **758** (2020) 012028 doi:10.1088/1757-899X/758/1/012028

Improvement in microstructure and properties of MgO-C refractory by addition of Fe

Wanting Liang, Shijie Deng, Zhihong Zhou, Xiaoyong Li and Lei Xu*

School of Metallurgy, Northeastern University, 3-11 Wen-Hua Road, Shenyang 110819, China
*E-mail: xul@smm.neu.edu.cn

Abstract. In order to achieve self-repairing of MgO-C refractory in oxidizing atmosphere at high temperatures, Fe powder, serving as both antioxidant and direct-bond former, was introduced into this refractory. The effect of Fe addition on the microstructure, properties and oxidation resistance was investigated. The coexistence of $(Mg,Fe)O_{ss}$ and $Mg(Fe_{1.2}Al_{0.8})O_4$ was identified for the Fe-containing MgO–C refractory after graphite was oxidized, by which a network was formed in the matrix. As a consequence, both densification and direct-bond in the oxidized layer were greatly enhanced compared with the conventional MgO-C refractory, solving the issue of strength degradation during high temperature oxidation. An excellent oxidation resistance was obtained owing to the above improvements.

1. Introduction

It is well known that MgO–C refractory played important role in the steelmaking industry by using as the linings of converters, electric arc furnaces and steel ladles. The excellent thermal shock and slag corrosion resistance owing to the presence of graphite is the key reason for the success of this refractory [1]. However, a major disadvantage regarding the incorporation of graphite is its high oxidation potential at high temperatures, accompanied by the formation of a porous structure and decrease of mechanical strength. To solve this issue, many technologies, such as introduction of antioxidants [2-4], graphite surface modification [5] and attachment of anti-oxidation coating [6], were developed. Besides, the structure stability of the decarburized layer is also of utmost importance to the oxidation resistance, which has been confirmed by the successful application of slag-splashing technology in converters. On the other hand, the presence of a decarburized layer with considerable thickness can also decrease the contact area between graphite and liquid, slowing down or even stopping the carbon pick-up in molten steel. However, very few studies have focused on this point.

Fe_xO is an effective sintering aid to enhance the densification and mechanical properties of MgO based ceramics and/or refractories [7]. Similarly, improvement in structure and properties of the porous layer can be achieved if Fe^{2+}/Fe^{3+} were present. Moreover, other benefits such as anti-oxidation and toughening effects could be attained if metal Fe is introduced into MgO–C refractory. To date, the role of Fe in the properties of MgO–C refractory is still unknown. In this work, therefore, Fe-containing MgO–C refractories were prepared, and their microstructure, oxidation resistance and mechanical properties were studied by comparing with the conventional one.

2. Experimental procedures

2.1. Preparation

High purity fused magnesia, flake graphite and aluminum powder were used as the raw materials in this work, and the liquid phenol resin was used as binder. An initial mixture was prepared based on the composition of the refractories used as the linings of steelmaking converters, as shown in Table 1. Then 4, 8 and 12 wt% high purity Fe powders were added into this mixture, respectively. The final products were prepared by the following procedure: mixing (30 min) – aging (25 °C for 12 h) – pressing (250 MPa to cylindrical samples) – curing (200 °C for 24 h) – firing (1500 °C for 3 h in air atmosphere).

Table 1. Chemical composition of the designed MgO–C refractories.

Samples	MT14-0	MT14-4	MT14-8	MT14-12
Fused magnesia	86	86	86	86
Graphite	14	14	14	14
Phenolic resin[a]	4	4	4	4
Aluminum[a]	2	2	2	2
Fe powder[a]	0	4	8	12

[a] with external addition.

2.2. Characterization and analysis

The bulk density and apparent porosity were measured by Archimedes principle according to GB/T 2997-2015 standard. The cold crushing strength was evaluated by hydraulic universal test machine (Model YEW-2000). 6 valid samples were used for each test. The phase assemblage in different layers was identified by X-ray diffraction (XRD; X'pert PRO, PANalytical, Netherlands). Microstructure observation as well as chemical analysis in different zones was performed by field-emission scanning electron microscopy (FE-SEM; Model Ultra Plus, Carl Zeiss, Germany) attached with energy dispersive X-ray analyzer (EDX).

3. Results and Discussion

Figure 1 shows the typical cross-section photographs of the four refractories after firing at 1500 °C in air for 3 h. Two distinct regions, i.e. unoxidized layer and oxidized layer (or decarburized layer), with clear reaction interface were observed for all the samples. For the MT14-0 sample, the thickness of the decarburized layer was around 4 mm after oxidizing. This value decreased apparently with increase of Fe content, which was only 2 mm after 12 wt% Fe was introduced into this refractory. Smooth cut surfaces also were observed for the Fe-containing refractory, indicating that the bonding strength of the decarburized layer was greatly improved by Fe addition.

Figure 2 shows the XRD patterns of the oxidized layers for different samples. The major diffraction peaks were periclase (MgO) and spinel ($MgAl_2O_4$) for the MT14-0 sample after the graphite was oxidized. Besides these phases, another spinel of $Mg(Fe_{1.2}Al_{0.8})O_4$ was identified when 4 wt% Fe powder was added into the refractory. The peak intensity of this new phase increased obviously with increasing Fe content, while an opposite trend was observed for that of $MgAl_2O_4$. These changes were caused by the following reactions: Fe powder began to oxidize after the oxidation of carbon, after which the produced FeO and Fe_2O_3 both can react with periclase based on the MgO-Fe_xO system, forming $(Mg,Fe)O_{ss}$ and $MgFe_2O_4$, respectively [8]. Both $MgFe_2O_4$ and $MgAl_2O_4$ belong to the minerals of spinel [9], and thus isostructural $Mg(Fe_{1.2}Al_{0.8})O_4$ can be obtained by substituting Al^{3+} for Fe^{3+}. Moreover, the ionic radius of Fe^{3+} (0.55 Å) is similar to Al^{3+} (0.53 Å), which favours the formation of continuous solid solution. As a consequence, only MgO and $Mg(Fe_{1.2}Al_{0.8})O_4$ were detected for the Fe-containing refractories with higher Fe content.

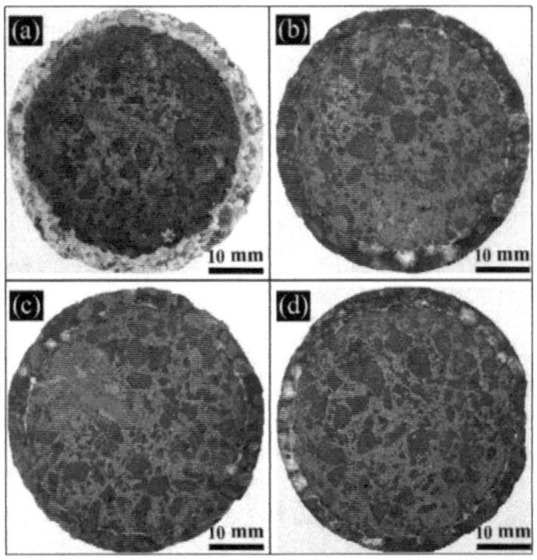

Figure 1. Photographs of the refractories after firing at 1500 °C for 3 h: (a) MT14-0, (b) MT14-4, (c) MT14-8, (d) MT14-12.

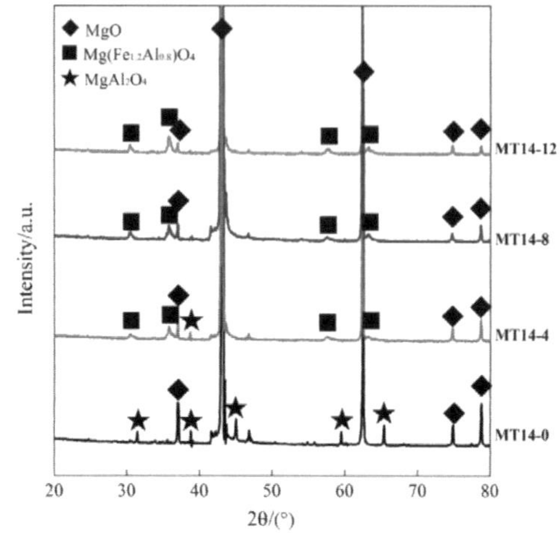

Figure 2. XRD patterns of the oxidized layers for different samples fired in air at 1500 °C for 3 h

Figure 3 displayed the representative microphotographs of the two samples after firing at 1500 °C for 3 h. High porosity and big pores (50-200 μm) were clearly observed in the matrix structure of the sample MT14-0, and interconnected pores were left after graphite was oxidized. What is worse, the direct-bond between neighbouring grains was rather poor. However, the pore sizes were obviously decreased and the direct-bond was greatly improved after Fe powder was introduced, as shown in Figure 3(b). On the one hand, the large formation of magnesio-wustite $(Mg,Fe)O_{ss}$ by replacing Mg^{2+} with Fe^{2+} enhanced the sintering of the magnesia matrix [9], as shown on the right of Figure 3(b). On the other hand, owing to the formation of $Mg(Fe_{1.2}Al_{0.8})O_4$ network by the interdiffusion of Fe^{3+} and Al^{3+}, as shown in Figure 4, the isolated periclase grains can be effectively rebonded. Undoubtedly, both bonding strength and oxidation resistance can be greatly enhanced by the formation of the complicated network and the tortuous pores.

Figure 3. Microstructure comparison of the two samples after oxidation: (a) MT14-0, (b) MT14-8. +2: $(Mg,Fe)O_{ss}$.

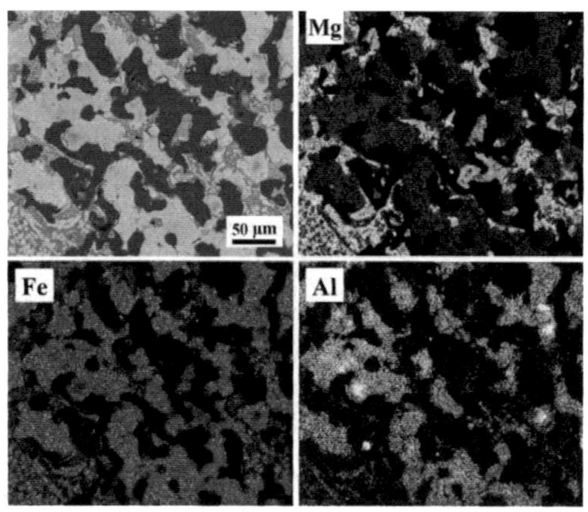

Figure 4. Map scanning results in the matirx of the sample MT14-8 after firing at 1500 °C for 3 h.

Figures 5 and 6 show the effects of Fe addition on the physical and mechanical properties, respectively. For the apparent porosity, it was about 5% after curing for all the refractories, which is accordant with the theoretical values [10]. The porosity then rose to ~10% after coking at 1500 °C due to the pyrolysis of the phenol resin. After firing in oxidizing atmosphere, this value further increased to around 16%

for the sample MT14-0, which is consistent with the microstructure observation, as shown in Figure 3(a). Nevertheless, the increase of porosity after oxidation was much less pronounced for the Fe-containing samples. The volume expansion owing to Fe oxidation and $Mg(Fe_{1.2}Al_{0.8})O_4$ formation and effective sintering owing to solid solution formation were the key reasons for this change. Similar change rule was observed for the cold crushing strength. It can be seen that about 45% strength was lost after firing at 1500 °C for the conventional MgO–C refractory. However, the strength was kept in a constant value for the samples MT14-8 and MT14-12, suggesting that the self-repairing was effectively achieved for the porous and weak decarburized layer by addition of Fe powder. The main reasons for this improvement can be described as follows: (i) the increase of densification (ii) the promotion in sintering by accelerating the ions migration; (iii) the enhancement of direct-bond owing to the formation of $Mg(Fe_{1.2}Al_{0.8})O_4$ network.

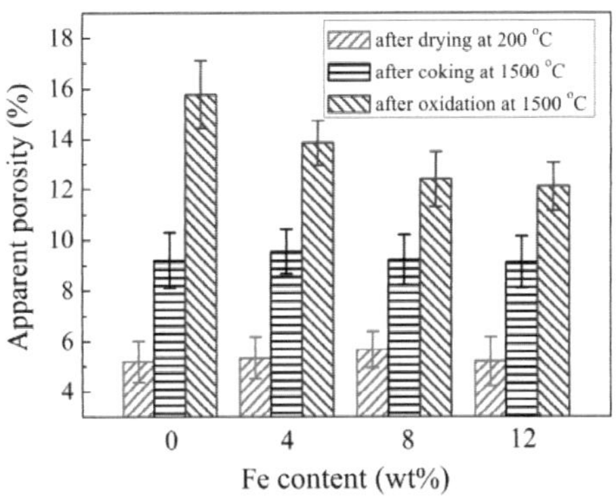

Figure 5. Effect of Fe addition on the apparent porosity.

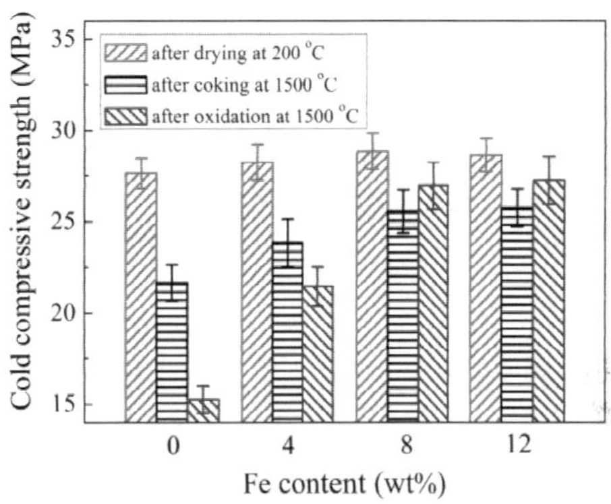

Figure 6. Effect of Fe addition on the cold crushing strength.

4. Conclusions

The microstructure, physical and mechanical properties, and oxidation resistance of Fe-containing MgO–C refractory were studied by comparing the conventional one. Based on the above results and discussion, the following conclusions can be obtained:

(1) $(Mg,Fe)O_{ss}$ and $Mg(Fe_{1.2}Al_{0.8})O_4$ were formed for the Fe-containing refractory by the reaction between the produced FeO and Fe_2O_3, periclase and $MgAl_2O_4$ in the oxidized area, by which the neighbouring matrix particles were firmly connected.

(2) Porous structure with poor direct-bond was observed for the conventional MgO–C refractory. However, the densification and bonding degree were greatly improved after Fe powder was introduced by the formation of network structure in the matrix.

(3) The oxidation resistance of MgO–C refractory was greatly improved by addition of Fe. The key factors were revealed as follows: (1) the improvement in densification; (2) the increase in the diffusion resistance of gases by the $Mg(Fe_{1.2}Al_{0.8})O_4$ network.

(4) The issue of strength degradation during high temperature oxidation for the MgO–C refractory was solved by addition of Fe powder. This is of great importance to the stability of the decarburized layer, thus further inhibiting the oxidation of graphite.

5. References

[1] Ewais E M M 2004 Carbon based refractories *J. Ceram. Soc. Jpn.* **112** 517-32

[2] Zhang S, Marriott N J and Lee W E 2001 Thermochemistry and microstructures of MgO-C refractories containing various antioxidants *J. Eur. Ceram. Soc.* **21** 1037-47

[3] Gokce A S, Gurcan C, Ozgen S and Aydin S 2008 The effect of antioxidants on the oxidation behaviour of magnesia–carbon refractory bricks *Ceram. Int.* **34** 323-30

[4] Ghasemi-Kahrizsangi S, Dehsheikh H G and Boroujerdnia M 2017 Effect of micro and nano-Al_2O_3 addition on the microstructure and properties of MgO-C refractory ceramic composite *Mater. Chem. Phys.* **189** 230-36
[5] Cho G H, Kim E H, Li J, Lee J H, Jung Y G, Byeun Y K and Jo C Y 2014 Improvement of oxidation resistance in graphite for MgO−C refractory through surface modification *Trans.* Nonferrous *Met. Soc.* China **24** 119-24
[6] Aslanoğlu Z 2011 Performance of a ceramic frit anti-oxidation coating on a MgO–C refractory brick *Ceram. Int.* **37** 3419-23
[7] Gómez Rodríguez C, Das Roy T K, Shaji S, Castillo Rodríguez G A, García Quiñonez L, Rodríguez E, González J O and Aguilar-Martínez J A 2015 Effect of addition of Al_2O_3 and Fe_2O_3 nanoparticles on the microstructural and physico-chemical evolution of dense magnesia composite *Ceram. Int.* **41** 7751-58
[8] Honda T and Kaneko S 2003 Phase relations in the MgO-rich region of the system MgO-Fe_2O_3 *J. Ceram. Soc. Jpn.* **111** 841-47
[9] Kingery W D 1976 *Introduction to Ceramics* 2nd ed (New York: John Wiley & Sons)
[10] Hashemi B, Nemati Z A and Faghihi-Sani M A 2006 Effects of resin and graphite content on density and oxidation behavior of MgO-C refractory bricks *Ceram. Int.* **32** 313-19

6. Acknowledgments

This work was supported by National Training Program of Innovation and Entrepreneurship for Undergraduates (Grant no. 201910145040) and the Fundamental Research Funds for the Central Universities (Grant no. N182410001).

IOP Conf. Series: Materials Science and Engineering **758** (2020) 012029 doi:10.1088/1757-899X/758/1/012029

Surface Roughness Prediction in Grinding Process of the SKD11 Steel by using Response Surface Method

N Hong Son[1], D Duc Trung[2,*], Nhu-Tung Nguyen[2]

1 Mechanical Engineering Center, Hanoi University of Industry, No. 298, Cau Dien Street, Bac Tu Liem District, Hanoi, Vietnam
2 Faculty of Mechanical Engineering, Hanoi University of Industry, No. 298, Cau Dien Street, Bac Tu Liem District, Hanoi, Vietnam
E-mail: doductrung@haui.edu.vn

Abstract. This study was performed to to develop a surface roughness model in the grinded process by using the response surface method (RSM). The experimental research was performed in the grinding the SKD11 steel by using CBN grinding wheel (symbol of HY-180x13x31.75-100#). The experimental matrix was developed basing on the Box - Behnken plan with 15 experiments. Experimental results were analyzed by Minitab 16 statistical software to develop a surface roughness model. The proposed model is a quadratic function of the grinding parameters such as workpiece velocity, feed rate, and depth of cut. The model of surface roughness was verified by experimental with very consistent results. The proposed model can be used to calculate the surface roughness values of the machined part in machining process the SKD11 steel by using CBN grinding wheel.

1. Introductions

The grinding surface roughness of the parts is an important parameter that greatly affects on the workability of the parts. This parameter is often chosen as an indicator for the evaluation of the grinding process. Most of the studies were performed to reduce the machine adjustment and the test machining time as well as to ensure that the machining surface roughness is a small value or an appropriate value. According on this principle, many researches developed the models to predict the machining surface roughness in the grinding processes.

Lal and Shaw [1] conducted a model to predict surface roughness when grinding basing on the analyzing the model of the cut thickness; In the studies of K. Nakayama et al [2], K. Sato [3], and C. Yang et al [4], the authors built a model to predict the surface roughness when grinding with assuming the grinding grains were evenly distributed on the surface of the grinding wheel; A model was proposed to predict the surface roughness when grinding by determining the average cut thickness into the machining surface of the grinding grains [5]; Depending on the probability analysis of the cutting process of the grinding grains into the workpiece surface, a model was proposed to predict the surface roughness in grinding processes [6-11].

There are many the advantages of the prediction of surface roughness when grinding based on the theoretical background analysis of the cutting process, Besides, this approach can be applied to many different cases. However, in almost those studies, the impact of many factors on the surface roughness have been not considered during grinding processing. Therefore, the accuracy of the predicted results was limited. In order to ensure quite high accuracy in predicting, one of the modelling methods to predict the output parameters of the machining processes is the response surface method. This method has been applied by many authors to build a model for predicting the surface roughness of a part in

IOP Conf. Series: Materials Science and Engineering **758** (2020) 012029 doi:10.1088/1757-899X/758/1/012029

many different machining processes. Some studies that have applied this method including: P. Krajnik et al [12] developed a prediction model of surface roughness when centerless grinding the 9SMn28 steel; A. Arunpremnath et al [13] developed a model for predicting the surface roughness when milling the hybrid aluminum composites; A model was developed for predicting the surface roughness when cylindrical grinding the AISI 4140 steel [14]; A model was developed to predict the surface roughness when turning aluminum 6063 T6 [15]; Radhakrishnan B et al developed a model to predict the surface roughness when milling the aluminum 6063 [16]; N. Ganesh et al proposed a model to predict the surface roughness when turning the EN 8 steel by using Cemented Carbide cutter [17].
In this study, the response surface method was applied to develop a surface roughness model for the grinding process when machining the SKD11 steel by using a CBN grinding wheel.

2. Experimental research

2.1. Material, Tool, and Equipment
Experiments of grinding process was applied to machine the SKD11 steel which is the steel grade representing the high alloy steel group. This steel is widely used to manufacture the mechanical parts that require high precision and high surface gloss by using grinding technology. The equivalent symbols of these steel grades for some countries are presented in Table 1. The dimensions of the workpiece are 50 x 40 x 10 (mm). The compositions of the main elements of SKD11 steel is listed in Table 2.

Table 1. Symbols equivalent to SKD11 steel grade of some countries [18]

Japan	Russia	America
SKD11	X12M	D2

Table 2. Composition of the main elements of SKD11 steel

Element	C	Mn	Si	Cr	V	Mo	Ni
%	1.5	0.3	0.25	11.5	0.25	0.3	0.35

The experiments were performed in the Toyoda - Taiwan surface grinding machine (Fig. 1).
A CBN grinding wheel (Korea), HY-180x13x31.75-100# was using in this study, the outer diameter size x thickness x the inner diameter of the stone are 180mm x 13mm x 31.75mm, respectively.

Figure 1. Experimental machine

The surface roughness values were measured by a surface roughness tester (TESA RUGOSURF 10 Roughness Gauge tester). For each experiment, the surface roughness was measured at least 3 times, surface roughness value at each experimental point was the average value of successive measurements.

2. 2. Response surface method

RSM is a combination of statistical theory and mathematical model, which is very useful in the modelling and analysing the technical problems. According to B. Radha Krishnan et al. [14], Raymond H. Myers et al. [19], the main objective of RSM is to determine the optimum value of the target surface affected by many different initial parameters. Furthermore, RSM also allows control of input parameters to ensure the surface reaches a certain value. In RSM, the relationship between desired response and the input parameters is expressed in the following form.

$$Y = F(v, f, t) \quad (1)$$

For the specific case of this study, Y is the surface roughness value of the part; F is the response function; v, f, t are the workpiece velocity, the feed rate, and the depth of cut, respectively. Depending on the studied B. Radha Krishnan et al. [14], Raymond H. Myers et al [19], in engineering, most of the relationship between the target surface roughness and the input parameters can be expressed and represented by a second order model. This model works quite well across the entire range of input variables. Consequently, the expression (1) is written in the following form.

$$Y = \beta_0 + \sum_{i=1}^{k} \beta_i x_i + \sum_{i=1}^{k} \beta_i x_i^2 + \sum_i \sum_j \beta_{ij} x_i x_j + \varepsilon \quad (2)$$

In which: $'Y'$ is corresponding response; x_i is (i^{th}) value of the input parameters; the quantities β are regression coefficients; ε is residual measure.

2.3. Experimental design

Experimental plans including the number of experiments and the sequence of experiments are formulated in the form of Box-Behnken plans. According to Raymond H. Myers et al. [18], this is a type of experimental plan commonly is used in the optimization of machining processes. Box-Behnken testing plan includes 3 input parameters that were the workpiece velocity, depth of cut, and feed rate. Throughout the experiment, each parameter received 3 levels of coding values, the actual value of each parameter at the coding levels as shown in Table 3. Experimental matrix of 15 experiments is shown in Table 4.

Table 3. Value of input parameters at the coding levels

Parameter	Value at the coding level		
	-1	0	1
$v(m/\min)$	5	10	15
$f(m/stroke)$	3	4	5
$t(mm)$	0.01	0.015	0.02

Table 4. Experimental matrix

Run	Cutting mode						$R_a(\mu m)$
	Real value			Coding value			
	$v(m/min)$	$f(mm/stroke)$	$t(mm)$	v	f	t	
1	5	3	0.015	-1	-1	0	0.46
2	15	3	0.015	1	-1	0	0.75
3	5	5	0.015	-1	1	0	0.82
4	15	5	0.015	1	1	0	0.68
5	5	4	0.010	-1	0	-1	0.59
6	15	4	0.010	1	0	-1	0.66
7	5	4	0.020	-1	0	1	0.82
8	15	4	0.020	1	0	1	0.80
9	10	3	0.010	0	-1	-1	0.55
10	10	5	0.010	0	1	-1	0.65

IOP Conf. Series: Materials Science and Engineering **758** (2020) 012029 doi:10.1088/1757-899X/758/1/012029

11	10	3	0.020	0	-1	1	0.62
12	10	5	0.020	0	1	1	0.66
13	10	4	0.015	0	0	0	0.54
14	10	4	0.015	0	0	0	0.52
15	10	4	0.015	0	0	0	0.55

Besides, the workpiece velocity, the feed rate, and the depth of cut that were adjusted for each experiment orderly as shown in Table 4, the testing process was carried out with the value of other parameters as follows:

- Grinding wheel velocity: 26 m/s.
- Before each experiment, the grinding wheel was dressed with a depth of 0.01 mm and the feed rate when dressing of 150 mm/min.
- Cool irrigation technology: the experimental was conducted by the irrigation method, the flow rate of the coolant irrigation is 4.6 litters/minute. The coolant irrigation that was used is the Emulsion 10% oil.

3. Results and Discussions

The experiments were carried out orderly as shown in the Table 4. The surface roughness values of each experiment are also stored in the Table 4. The analysis of variance of the experimental results are listed in Table 5.

Table 5. Variance analysis of the experimental results

Source	Sum of Squares	DF	Mean Square	F value	P-value Prob>F	
Model	0.1655	9	0.0184	6.11	0.030 < 0.05	significant
Linear	0.0534	3	0.0178	5.92	0.042 < 0.05	
v	0.0050	1	0.0050	1.66	0.254	
f	0.0231	1	0.0231	7.68	0.039 < 0.05	
t	0.0253	1	0.0253	8.41	0.034 < 0.05	
Square	0.0629	3	0.0209	6.98	0.031 < 0.05	
v^2	0.047	1	0.0524	17.43	0.009 < 0.05	
f^2	0.0011	1	0.0017	0.58	0.482	
t^2	0.0140	1	0.0140	4.67	0.083	
Interaction	0.0491	3	0.0163	5.45	0.049 < 0.05	
$v * f$	0.0460	1	0.0462	15.37	0.011 < 0.05	
$v * t$	0.0020	1	0.0020	0.67	0.449	
$f * t$	0.0009	1	0.0009	0.30	0.608	
Residual	0.0150	5	0.0030			
Lack of Fit	0.0145	3	0.0048	20.82		
Pure Error	0.0005	2	0.0002			
Total	0.1806	14				
R^2						0.9167

The results in the Table 5 showed that the depth of cut has the greatest influence on the surface roughness, the second and third factors which influence on the surface roughness were the feed rate and the workpiece velocity. In the interaction effect between the experimental factors on the surface roughness, the interaction factor between the workpiece velocity and the feed rate has the largest degree of the influence on the surface roughness, the second interaction factor that influenced on the surface roughness was the interaction between the cutting speed and the depth of cut. The interaction between the feed rate and the depth of cut has the third interaction factor impacting on surface roughness.

- In order to increase the accuracy of predicted model, the surface roughness model was modelled by a function of all factors such as workpiece velocity, feed rate, depth of cut and the interaction factors of these three factors. The surface roughness model was regressed by a quadratic function with the defined coefficient, $R^2 = 0.9167$, that is very close to 1. This proved that the model has a great degree of compatibility to the experimental data.

$$R_a = 0.53667 + 0.02500 * v + 0.05374 * f + 0.05625 * t + 0.11917 * v^2 + 0.02167 * f^2 + 0.06167 * t^2 - 0.10750 * v * f - 0.02250 * v * t - 0.01500 * f * t \quad (3)$$

The compared results of the predicted and experimental values of surface roughness was described in Fig.2. From the results in Fig.2, the surface roughness value that were predicted by Eq. (3) is very close to the measured surface roughness values. So, the Eq. (3) can be used to calculate the surface roughness when grinding the SKD11 steel by using CBN grinding wheel.

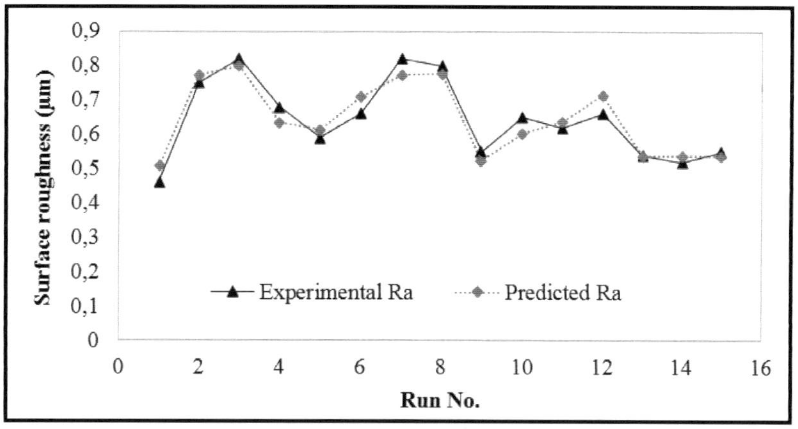

Figure 2. Comparison of the predicted and experimental values of surface roughness

4. Conclusions

Depending on the analysed results, the conclusions of this study can be drawn as follows.

- In this study, the RSM method was be used to model and predict the surface roughness in grinding process. The predicted results were very close to the experimental data. Using this proposed model, the time and cost can be reduced during grinding processes.
- The cutting depth had the greatest impact on the machining surface roughness, the feed rate was the second factor that influenced on the surface roughness, and the third factor that influenced on the surface roughness was the workpiece velocity.
- The interaction factor had the different effect on the machining surface roughness, the interaction factor between the workpiece velocity and feed rate had the largest degree effect on the surface roughness, the interaction factor between the cutting speed and the cutting depth was the second interaction factor that influenced on the surface roughness. The interaction between the feed rate and the depth of cut has least impact on the surface roughness.

5. References

[1] G K Lal and M C Shaw 1975 *The role of grain tip radius in fine* (Journal of Engineering for Industry August) p 111

[2] K Nakayama and M C Shaw 1968 *Study of finish produced in surface grinding* (part 2, Proceeding of the Institution of Mechanical Engineers, No182)

[3] K Sato 1955 *On the surface roughness in grinding*, (Technology Reports, Tohoku University, No20) p 59

[4] C Yang and M C Shaw 1955 *The grinding of titanium alloys* (Transactions of ASME, No77) p 645

[5] X. Zhou and F. Xi, 2002, *Modelling and predicting surface roughness of the grinding process* (International Journal of Machine Tools and Manufacture, No42) p 969

[6] Rogelio L Hecker and Steven Y Liang 2003 *Predictive modeling of surface roughness in grinding*, (International Journal of Machine Tools & Manufacture 43) p 755

IOP Conf. Series: Materials Science and Engineering **758** (2020) 012029 doi:10.1088/1757-899X/758/1/012029

[7] Sanjay Agarwal and P Venkateswara Rao 2005 *A probabilistic approach to predict surface roughness in ceramic grinding* (International Journal of Machine Tools & Manufacture 45) p 609

[8] A. Sanjay and P. V. Rao *2005 Surfacce roughness prediction model for ceramic grinding, (*ASME International Mechanical Engineering Congress and Exposition, November 5-11, Orlando, Florida USA) p 1

[9] Sanjay Agarwal and P Venkateswara Rao 2005 *A new surface roughness prediction model for ceramic grinding* (Proceedings of the Institution of Mechanical Engineers, Part B: Journal of Engineering Manufacture 219) p 811

[10] K.K. Sanchit and A. Sanjay 2015 *Predictive modelling of surface roughness in grinding* (Procedia CIRP 31) p 375

[11] Krishna Kumar Saxena, Sanjay Agarwal and Raj Das, 2016, *Surface Roughness Prediction in Grinding: a Probabilistic Approach*, (MATEC Web of Conferences 82, 01019) p 1

[12] P. Krajnik, J. Kopac, and A. Sluga, 2005, *Design of grinding factors based on response surface methodology* (Journal of Materials Processing Technology 162–163) p 629

[13] A Arunpremnath, T Alwarsamy, T Abhinav and C Adithya Krishnakant 2012 *Surface Roughness Prediction by Response Surface Methodology in Milling of Hybrid Aluminium Composites* (Procedia Engineering, Vol 38) p 745

[14] B Radha Krishnan, R Aravindh, M Barathkumar, K Gowtham and R Hariharan 2018 *Prediction of Surface Roughness (AISI 4140 Steel) in Cylindrical Grinding Operation by RSM* (International journal for research & development in technology, Vol 9, Issue-3) p 702

[15] B. Radhakrishman, P. Ramakrishnan, S. Sarankumar, S. Tharun Kumar and P. Sankarlal 2017 *Optimization of CNC machining parameters for surface roughness in turning of aluminium 6063 T6 with response surface methodology* (SSSG international journal of mechanical engineering – (ICCRESt 17), Specia issue) p 23.

[16] B. Radhakrishnan, T. Sathish, T.B. Siva Subramanian, N. Tamizharasan and E. VarunKarthik, 2017, *Optimisation of Surface Roughness in CNC Milling Process Using RSM (*SSRG International Journal of Mechanical Engineering-Special issue) p5

[17] N. Ganesh, M. U. Kumar, C. V. Kumar, B. S. Kumar, 2014, *Optimization of cutting parameters in turning of EN 8 steel using response surface method and genetic algorithm* (International Journal of Mechanical Engineering and Robotics Research, Vol 3, No 2)

[18] Tran Van Dich, Ngo Tri Phuc 2006 *World steel handbook* (Science and Technics Publishing House) Ha Noi.

[19] Raymond H Myers, Douglas C Montgomery, and Christine M Anderson-Cook 2009 *Response Surface Methodology* (Process and Product Optimization Using Designed Experiments, John Wiley & Sons, Inc, 3rd Edition).

IOP Conf. Series: Materials Science and Engineering **758** (2020) 012030 doi:10.1088/1757-899X/758/1/012030

Study on Temperature Related Factors of Asphalt Pavement Based on APRIORI

W Y Lin[1], Z H Zhang[2], X Qiu[3], W Y Xu[3] and H J Hong[3]

1 Jinhua Traffic Engineering Supervision Station, Jinhua, 321013, China
2 Zhejiang Zhengda Testing Technology Co., Ltd., Jinhua, 321000, China
3 Road and Traffic Engineering Research Center, Zhejiang Normal University, Jinhua, 321004, China
E-mail: zjnuhhj@outlook.com

Abstract. Asphalt is a typical temperature-sensitive material. In order to explore the correlation between surface temperature of asphalt pavements and meteorological factors, the association rules mining algorithm (Apriori) was applied to identify the key factors affecting the surface temperature of asphalt pavement. Firstly, errors and missing data in the meteorological dataset were cleaned. Then, Apriori was applied to identify the key factors affecting the asphalt pavement temperature. The results indicate that Apriori would perform an excellent ability to analyze the correlation rules between meteorological factors with the minimum confidence of 0.8 and the key meteorological factors which affect the temperature change of asphalt pavements are excavated including air temperature, air pressure, dew point temperature and relative humidity. The research would serve as a technical support for the machine learning algorithms applied in the field of the association rule analysis.

1. Introduction

Asphalt is a common material which was widely used in pavement due to its properties, but asphalt is a typical temperature-sensitive material, and the change of road surface temperature and state will directly affect the road traffic safety level and traffic efficiency, so meteorological conditions are important factors affecting road surface temperature change. Therefore, how to accurately identify the key meteorological factors affecting the pavement temperature has important theoretical and practical significance for road traffic safety.

Now, researchers have conducted extensive research on the variation law of road surface temperature. In 1975, Barder [1] first used the heat conduction equation to establish a model for the highest temperature prediction of the road surface and performed statistical analysis. With the gradual improvement of the prediction model of asphalt pavement temperature field, researchers in various countries began to pay attention to the selection of input parameters of the model. Hermansson [2] and Grimbacher [3] selected the appropriate meteorological factors to establish the model to predict the temperature field of the road surface. As the amount of data increases, the process of model establishment becomes complicated, also the precision of the prediction model is hard to be guaranteed. In addition, the input parameters of the model set by artificial can reduce the prediction accuracy. Therefore, how to accurately identify the relevant factors is the research emphasis.

The emergence of machine learning method has broken the limitations of traditional methods. Also, its excellent performance has been shown in association rule mining and model ensemble prediction. Association rules mining is an unsupervised learning algorithm for machine learning, and Apriori is its representative algorithms which have been proposed by Agrawal [4] in 1994 for the first time. Apriori is a frequent itemsets mining algorithm for extracting association rules using frequent itemsets obtained by loop iteration. Apriori algorithm was originally used for market basket analysis. Then numerous scholars such as Borgelt [5] began to apply it to the analysis of association rules of data. The result demonstrates that Apriori has a greater advantage than Eclat in mining the closed itemsets association rules. Guo [6] also found that association rules drew by Apriori have powerful ability in handling the forecasted wind speed values correction. However, with the exponential increase in data volume, low efficiency of Apriori operations becomes prominent, thus the improved form of Apriori algorithm has developed rapidly. For example, in order to improve the efficiency of association rule mining, Park [7] reduced the size of candidate itemsets by using an appropriate hash function. Similarly, by improving the operation process of Apriori algorithm, Han [8] proposed a method called FP-growth (frequent pattern growth) that succeeded in eliminating candidate generation. As can be seen from the foregoing, Apriori algorithm is widely used in association rule mining and has a good application effect in the field of association analysis. Therefore, in this paper, hash-based Apriori algorithm is used to mine the correlation between meteorological factors and asphalt pavement temperature.

This paper, in the light of related research, selects the winter meteorological observation data from December 2015 to January 2018 in Jinhua area of Zhejiang Province of China. The research goal is to identify key meteorological factors affecting the road temperature based on the association rules mined by Apriori algorithm. The study would serve as a technical support for the accurate prediction of winter asphalt pavement temperature.

2. Data Preprocessing

2.1. Data Cleaning

Data cleaning is intended to handle anomalous data in data sets, standardize dataset forms, and provide numerically accurate, formal-standard datasets for data mining. In this study, python platform was used to organize the data, and 3261 sets of valid data are generated.

2.2. Data Discretization

The meteorological data used for association analysis mainly include: air temperature (T), dew point temperature (D), air pressure (P), relative humidity (H), wind speed (W), precipitation (R), visibility (V) and asphalt pavement temperature (APT). Before the Apriori algorithm analyzes the correlation of meteorological data, it is necessary to discretize each data [6]. And the discretization of meteorological data is usually based on national grading standards of meteorological data, and the degree of grading of each factor is reduced by enlarging the grading interval to improve the operational efficiency of the Apriori algorithm. In this paper, the Short-term Weather Forecast (GB/T 21984-2008) [9] is used as the basis for the classification of meteorological factors, and the classification of meteorological factors is given in Table 1.

Table 1. Classification of Meteorological Factors.

Factors	Level 1	Level 2	Level 3	Level 4
T (℃)	T≤0	0<T≤13.9	13.9<T≤21.9	21.9<T
D (℃)	D≤-9.9	-9.9<D≤0	0<D≤9.9	9.9<D
P (Hpa)	P≤1010	1010<P≤1020	1020<P≤1030	1030<P
H (%)	H≤25	25<H≤50	50<H≤75	75<H
W (m/s)	W≤3.4	3.4<W≤10.7	10.7<W≤20.7	20.7<W
R (mm/h)	R≤2.6	2.6<R≤8.1	8.1<R≤16	16<R
V (m)	V≤50	50<V≤500	500<V≤2000	2000<V
APT (℃)	APT≤0	0<APT≤10	10<APT≤20	20<APT

Data in the meteorological database are subjected to value conversion according to the scale quantization standard and get the set of transaction dataset for association rule mining. The level of 8 meteorological factors within an hour is taken as an itemset I, which can be expressed as I = {Ti, Di,

Pi, Hi, Wi, Ri, Vi, APTi}, where i is the level of each meteorological factor. For example, the data item at 8:00 on December 5, 2015 is {6.4, 1.9, 1020.9, 73, 2, 0.2, 94, 8.1}, and each data in item corresponds to {temperature, dew point temperature, air pressure, relative humidity, wind speed, Precipitation, visibility, asphalt pavement temperature}, Referring to the meteorological factor classification of Table 1, the data items can be converted into itemset as I = {T2, D3, P3, H3, W1, R1, V2, APT2}. According to the conversion rule, all the data items in the database can be converted into an itemset, and then the transaction set for association rule mining can be obtained.

3. Association Rules Mining

3.1. Association Rule Mining Process

Apriori is a two-stage recursive algorithm. The first step is to find frequent itemsets from the transaction dataset that satisfy the minimum support, and the second step is to derive association rules from frequent items that satisfy the minimum confidence [5]. Therefore, it is necessary to preset minimum support and minimum confidence before the association rules are mined. The equation for support and confidence is as follows:

$$\begin{aligned} support(A \rightarrow B) &= P(A \cup B) \\ confident(A \rightarrow B) &= P(A \cup B)\ /\ P(A) \end{aligned} \tag{1}$$

Among them, support indicates the probability that A and B appear simultaneously, and confidence indicates the probability that *A* and *B* appear simultaneously on the premise of *A* has occurred.

Apriori algorithm has two steps in the process of finding frequent itemsets: Join step and Prune step. In the join step, Apriori scans the database and searches for *K*+1 itemset depend on *K* itemset, finally extract all the frequent itemsets that satisfy the minimum support. The prune step is interleaved with the join step, and the purpose is to discard the itemset that does not satisfy the minimum support and retain the frequent itemset.

After all frequent itemsets are found by Apriori algorithm, derived association rules from the frequent itemset, discard all association rules that do not meet the minimum confidence, and the retained association rule is the mining result of the Apriori algorithm.

3.2. Association Rule Mining Results

The weather with a road surface temperature below zero is rarely in Jinhua area of Zhejiang Province. For the sake of the comprehensiveness and reliability of the association rule mining results, it is necessary to pay attention to the association rules of asphalt pavement temperature below zero. Therefore, the Apriori algorithm is used to perform a large number of trials on the transaction set to determine the minimum support and minimum confidence. In association rule mining analysis, in order to make the extracted association rules have higher accuracy, the minimum confidence value is generally greater than 0.75. So, the minimum confidence is set to 0.8 in this study. On this foundation, the trial results are presented in Figure 1. It shows that the two-dimensional association rule with zero asphalt pavement temperature is stable to three when the support degree is no more than 0.003. Therefore, the minimum support is set to 0.003.

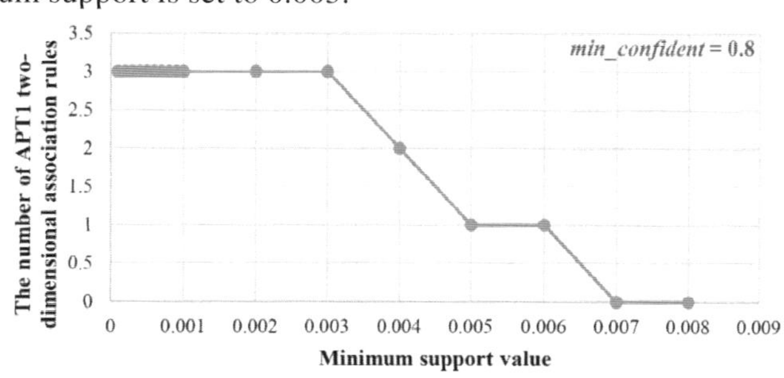

Figure 1. Apriori algorithm trial results.

Apriori algorithm mines 7 association rules for APT1; 28 association rules for APT2; 22 association rules for APT3; and due to the fact that the asphalt pavement temperature in the APT4 grade is rarely in winter, so only 1 two-dimensional association rule is obtained for the mining result. By analyzing the relationship between the association rules of each dimension, it is found that the Three-dimensional association rules can cover all meteorological factors, and a large number of invalid association rules appear from the four-dimensional association rules. Therefore, the analysis focuses on the analysis of one, two and three-dimensional association rules. It is worth mentioning that the existence of invalid association rules will directly affect the accuracy of the analysis results and should be eliminated in the analysis process. For example, in the analysis of the three-dimensional association rule, if the confidence of the two-dimensional association rule {H3, P4}→APT1 is 0.8333, and also the confidence of the three-dimensional association rule {H3, W1, P4}→APT1 is 0.8333, then the three-dimensional association rule is determined to be an invalid association rule, and needs to be discarded.

3.3. Association Rule Analysis

Figure 2 to Figure 4 show the statistical analysis results of the two-dimensional association rules. The analysis results show that in the two-dimensional association rules of different asphalt pavement temperature grades, two meteorological factors, dew point temperature (D) and air pressure (P), appeared in antecedent, and both grades change with the change of asphalt pavement temperature grade. It can be known that the dew point temperature and air pressure are 2 key factors affecting the temperature change of asphalt pavement.

Figure 5 to Figure 7 show the statistical analysis results of the three-dimensional association rules. The analysis results are demonstrated that the two factors of relative humidity (H) and air temperature (T) appear in different asphalt pavement temperature grade correlation rules, and all change with the pavement temperature grade. The result is indicating that relative humidity and temperature are also two key factors affecting asphalt pavement temperature change. In addition, there is no similar pattern of wind speed (W), visibility (V) and precipitation (R).

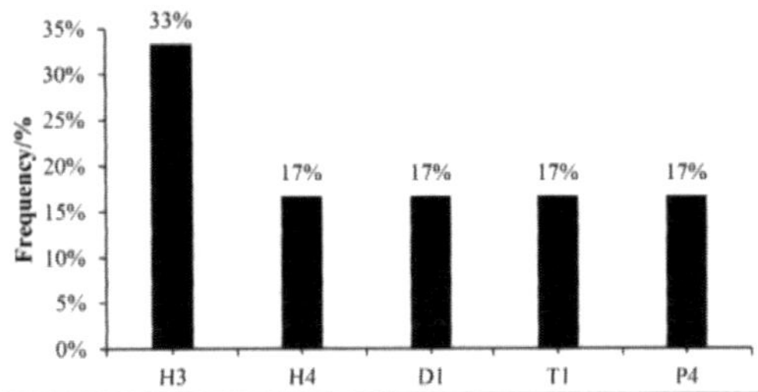

Figure 2. Statistics frequency for APT1 two-dimensional association rules..

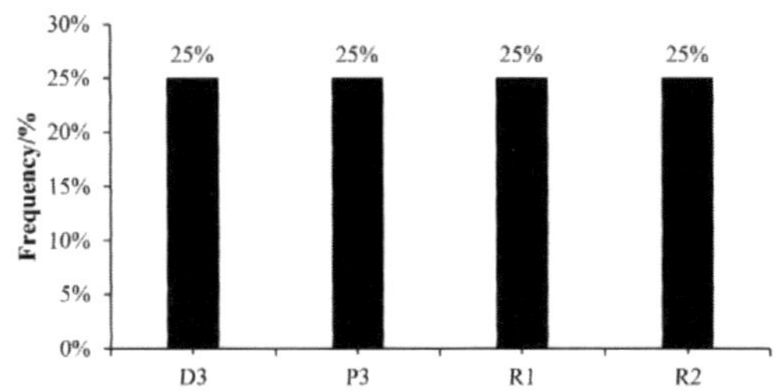

Figure 3. Statistics frequency for APT2 two-dimensional association rules.

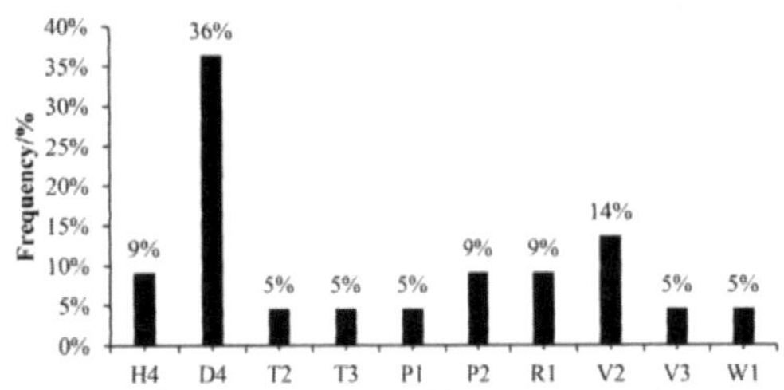

Figure 4. Statistics frequency for APT3 two-dimensional association rules.

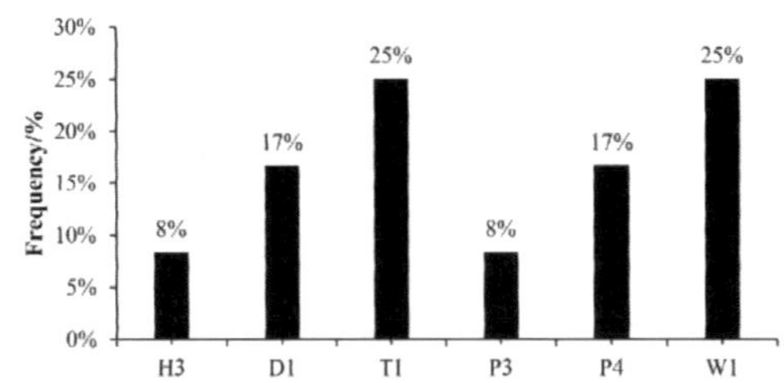

Figure 5. Statistics frequency for APT1 three-dimensional association rules.

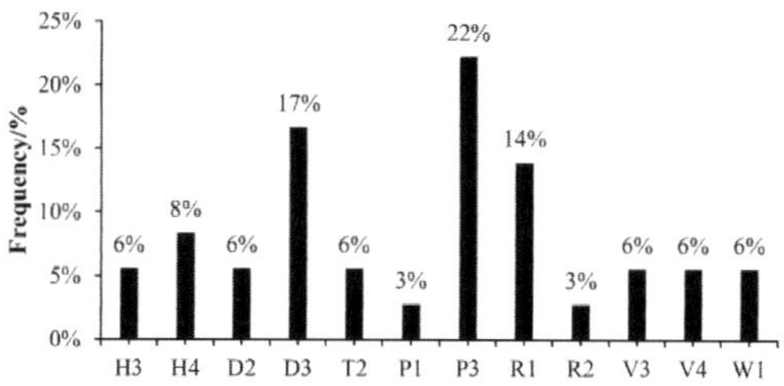

Figure 6. Statistics frequency for APT2 three-dimensional association rules.

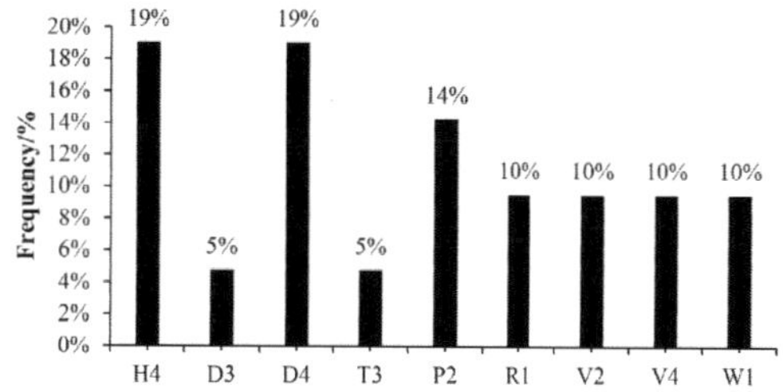

Figure 7. Statistics frequency for APT3 three-dimensional association rules.

According to the analysis results of the above association rules, there are four fundamental meteorological factors affecting the temperature change of asphalt pavement: dew point temperature, air pressure, relative humidity and air temperature.

4. Conclusion

The following major findings were achieved from this study:

(1) Apriori algorithm preforms well in terms of mining association rules, and compared with the linear fitting methods, it has the advantage of obtaining the correlation between different factors without drawing figures or calculating the correlation coefficient, which greatly improves the efficiency of the correlation analysis, especially dealing with big data.

(2) According to the mining results of Apriori algorithm, there are four key meteorological factors affecting the temperature change of asphalt pavement in Jinhua area of central Zhejiang Province: dew point temperature, air pressure, relative humidity and temperature. Also, the meteorological factors collected in this paper are relatively small, so in the further study the number of meteorological factors can be further increased.

Acknowledgements

The authors are grateful to the financial supports from the Natural Science Foundation of Zhejiang Province (LY18E080020).

References

[1] Barber E S 1957 Calculation of Maximum Pavement Temperatures from Weather Reports Highway Research Board Bulletin

[2] Hermansson 2001 Mathematical Model for Calculation of Pavement Temperatures: Comparison of Calculated and Measured Temperatures *Transportation Research Record Journal of the Transportation Research Board*

[3] Grimbacher T and Schmid W 2005 Nowcasting Precipitation, Clouds, and Surface State in Winter. *Atmospheric Research,*. **77** pp 378-387

[4] Agarwal R and Srikant 1994 R Fast Algorithms for Mining Association Rules. P*roc. of the 20th VLDB Conference* p 487-499

[5] Borgelt C 2003 Efficient Implementations of Apriori and Eclat *Proc.ieee Icdm Workshop on Frequent Item Set Mining Implementations. ceur* Workshop Proceedings **90**

[6] Guo Z, Chi D and Wu J 2014 A New Wind Speed Forecasting Strategy Based on the Chaotic Time Series Modelling Technique and the Apriori Algorithm *Energy Conversion & Management* **84** pp 140-151

[7] Park J S, Chen M S and Yu P S 1995 An Effective Hash-based Algorithm for Mining Association Rules *ACM* p 175-186

[8] Han J, Pei J and Yin Y 2000 Mining Frequent Patterns Without Candidate Generation *ACM SIGMOD International Conference on Management of Data* p 1-12

[9] GB/T 21984-2008. Short-term Weather Prediction.

IOP Conf. Series: Materials Science and Engineering **758** (2020) 012031 doi:10.1088/1757-899X/758/1/012031

Surface Temperature Prediction of Asphalt Pavement Based on GBDT

X Qiu[1], W Y Xu[1], Z H Zhang[2], N N Li[1] and H J Hong[1]

1 Road and Traffic Engineering Research Center, Zhejiang Normal University, Jinhua, 321004, China
2 Zhejiang Zhengda Testing Technology Co., Ltd., Jinhua, 321000, China
E-mail: zjnuhhj@outlook.com

Abstract. Asphalt is a temperature sensitive material, distribution characteristics and vary rules of asphalt pavement temperature have an important impact on the bearing capacity and performance of pavement, which is a concern of domestic and foreign researchers. The objective of this study was to explore the correlation between pavement temperature of asphalt pavements and meteorological factors and implement an accurate trend prediction of the asphalt pavement temperature. First, errors and missing data in the meteorological dataset were cleaned. Then, the three kinds of temperature prediction models of asphalt pavements in winter were established by Gradient Boosting Decision Tree (GBDT), Random Forest (RF) and Linear Regression (LR). The results indicate that GBDT would perform an excellent ability on prediction. The mean-square-error of the GBDT predicting results has a lower value of 1.5 when compared with the Random Forest and Linear Regression owing to the high robustness and the good generalization ability, which reflects the GBDT model has a good applicability in the field of prediction. The research would serve as a technical support for the machine learning algorithms applied in the field of the application of prediction problems.

1. Introduction

In terms of the properties of asphalt, it's a typical temperature-sensitive material, and asphalt pavement temperature is one of the important factors affecting road traffic safety. Changes in road surface temperature and state will have a direct impact on the safety and efficiency of road traffic, while meteorological conditions are the most precise causes affecting the change of pavement temperature. Therefore, how to effectively predict the change law of pavement temperature has important theoretical and practical significance for road traffic safety.

Now, domestic and foreign researchers have conducted extensive research on the temperature change of asphalt pavement, while their research direction is primarily focussed on the prediction model of pavement temperature field. In 1975, Barder [1] first used the heat conduction equation to establish a model for the highest temperature prediction of the road surface and performed statistical analysis. Subsequently, Huber [2] conducted a regression analysis of local temperature and road surface temperature. Then a simple regression prediction model is established for predicting the maximum temperature of the road surface. Then, Krsmanc [3] proposed a purely statistical approach for forecasting road surface temperature is based on stepwise linear regression analysis with appropriate selection of the input parameters. In summary, the current research method of asphalt pavement temperature still stays in statistical analysis and linear regression analysis. These methods are relatively simple with a high accuracy in most studies. However, as the amount of data increases, the process of model establishment becomes complicated, also the precision of the prediction model is

IOP Conf. Series: Materials Science and Engineering **758** (2020) 012031 doi:10.1088/1757-899X/758/1/012031

hard to be guaranteed. Therefore, how to establish a high accuracy asphalt pavement temperature prediction model is the research emphasis.

Gradient Boosting Decision Tree (GBDT) belongs to decision tree-based models in machine learning which has been introduced by Professor Firedman [4] for the first time. The algorithm trains each of the decision trees in a sequential manner, and each iteration fits a decision on the residuals left by the previous one and then the prediction is accomplished by combining the trees [5-6]. In recent years, more and more scholars have begun to choose to improve the accuracy of predictive models in an iterative way, making GBDT become the new direction of forecasting [7]. For example, GBDT is applied to short-term subway ridership prediction in Ding's study, and the findings suggest that the GBDT model has considerable advantages in improving the short-term subway ridership forecast [8]. Similarly, Yang also proposed a short-term traffic prediction method based on gradient boosting machine, concluded that the prediction errors of GBDT are smaller than SVM and BPNN [9]. It's prediction effect is excellent. Then, Gradient boosting decision trees, an ensemble learning method, is proposed to make asphalt pavement temperature prediction based on the meteorological data in this study. And compared with the temperature prediction model built by random forest (RF) and linear regression (LR) to verify the applicability and performance of GBDT in the field of asphalt pavement temperature prediction.

This paper, in the light of related research, selects the winter meteorological observation data from December 2015 to January 2018 in Jinhua area of Zhejiang Province of China. The study analyzes the relationship between meteorological factors and road surface temperature and provides technical support for the accurate prediction of winter road surface temperature.

2. Data Preprocessing

2.1. Data Cleaning

The amount of meteorological data is huge, and it is inevitable that data is missing and wrong during the process of collection and storage. Data cleaning is intended to handle anomalous data in data sets for data mining. The study reads meteorological data through the Pandas function in the Python platform and uses the dropna function to delete the data set where the missing data are located. Finally, 3261 sets of valid data are generated.

2.2. Key Factor Determination

In order to determine the key factors affecting the pavement temperature, the relationship between meteorological factors and asphalt pavement temperature was linearly fitted to calculate the Pearson coefficient, and the calculation formula is as shown in the Equation (1). The meteorological data used for association analysis mainly include: air temperature, dew point temperature, air pressure, relative humidity, wind speed, precipitation, visibility and asphalt pavement temperature.

$$Pearson(x,y)=\frac{\sum(x-\overline{x})(y-\overline{y})}{\sqrt{\sum_{i=1}^{n}(x_i-\overline{x})^2\cdot\sum_{i=1}^{n}(y_i-\overline{y})^2}} \tag{1}$$

Where, x and y are two independent variables; $\overline{x}$ and $\overline{y}$ are the mean of two independent variables; n is the number of variables; i is the variable number.

Table 1 shows the results of Pearson coefficient, it can be known that the dew point temperature, air pressure, relative humidity and air temperature have a great influence on the temperature change of the asphalt pavement.

Table 1. Pearson coefficient of meteorological factors

Meteorological factors	Air temperature	Dew point temperature	Air pressure	Relative humidity	Wind speed	Precipitation	Visibility
Pearson coefficient	0.79	0.37	-0.38	-0.34	0.07	-0.11	0.06

3. Asphalt Pavement Temperature Prediction

In this part, key factors obtained by linear fitting are used as input variables and asphalt pavement temperature is used as the output variable to establish the prediction model of asphalt pavement temperature based on gradient boosting decision tree (GBDT). Two algorithms, random forest and linear regression, were introduced for comparative analysis.

3.1. Predictive Model Building Process

Taking the winter meteorological observation data in Jinhua area of central Zhejiang Province as the data sample of the model, using the train_test_split function in the Python platform, 3246 sets of data are randomly selected from 3261 sets of meteorological data to set up the asphalt pavement temperature prediction model, and the remaining 15 sets of data are used as test data to verify the prediction effect of the model.

Gradient Boosting Decision Tree is an integrated machine learning algorithm, which consists of Classification and Regression Tree (CART), Gradient Boosting and Learning Rate. During the establishment of the model, there are 5 main parameters that need to be tuned, such as learning rate, loss function, the number of trees, max-depth and max-features. The default parameter of the model learning rate is 0.1, in order to slow down the learning rate of the model and improve the accuracy of the model, the learning rate in this paper is set to 0.01; The loss function is used to calculate the residual of the prediction result. As shown in Equation (2), here the mean square loss function is used in study; Since there are only 4 features in the training sample, and all of them are related parameters, so the max-features is set to None which indicating that all features will be used for model building. On this basis, the max-depth of GBDT is first set as the default value, and the value of the number of trees can be set by multiple trial calculations. As is shown in Figure 1, when the number of trees reaches 1000, the mean absolute error of the GBDT prediction results tends to be stable, so the number of trees of the model is set to 1000. Similarly, the max-depth of GBDT is calculated. It can be observed in from Figure 2 that when the depth of the tree is 4, the average error of prediction result is the smallest, so the max-depth is set to 4.

$$L(y, f(x)) = \sum_{i=1}^{n} (y_i - f(x_i))^2 \tag{2}$$

Where, x are the input variables, y are the corresponding labels, $f(x)$ are the prediction results, and i is the iteration number of the prediction model.

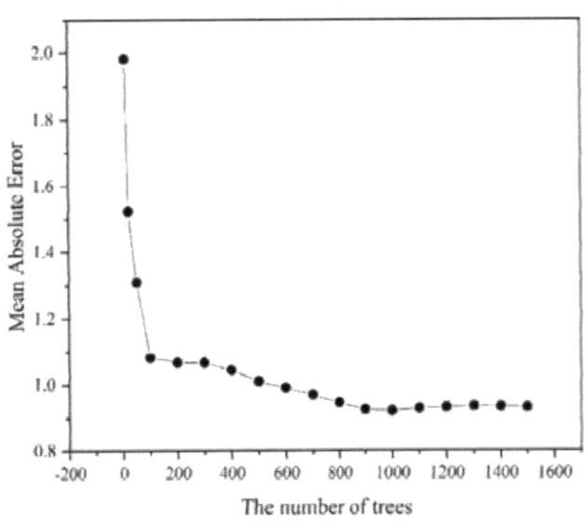

Figure 1. Prediction errors varying with the number of trees.

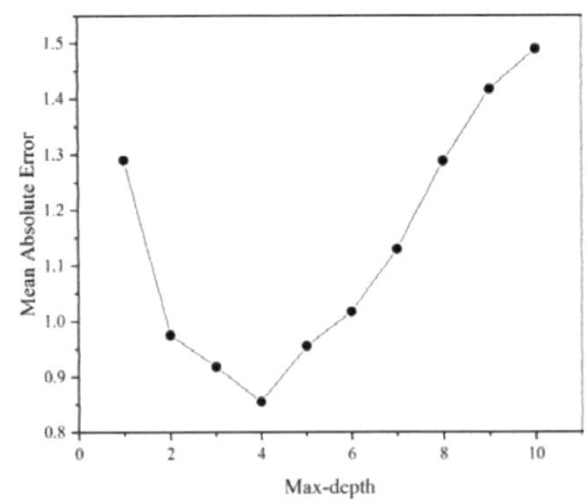

Figure 2. Prediction errors varying with max-depth.

Random Forest is an integrated machine learning algorithm, which consists of decision trees and bagging. During the establishment of the model, there are 3 main parameters that need to be tuned, such as max-depth, max-features and the number of trees. Since each tree in the random forest is independent of each other and the prediction results do not affect each other, so max-depth can be set to None which indicating that the depth of CART is not limited. The default setting parameter for max-features is auto, same as GBDT. Here the max-features is set to None. On this basis, the value of the number of trees can be set by multiple trial calculations. As is shown in Figure 3, when the number of trees reaches 12, the mean absolute error of the random forest prediction results tends to be stable, so the number of trees of the model is set to 12.

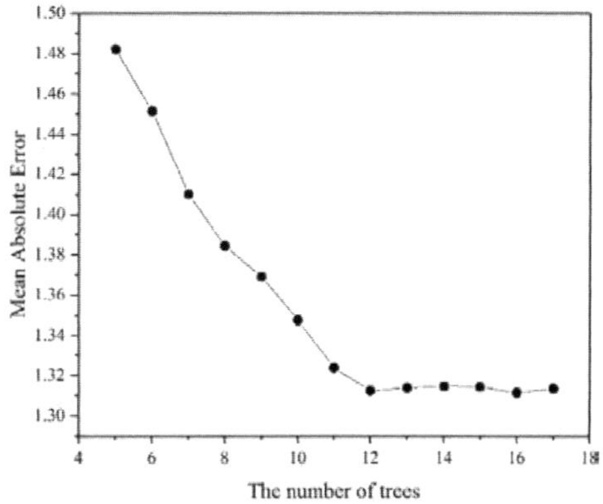

Figure 3. Random Forest predictive model building process.

Linear regression is a method to establish the relationship between dependent variable and multiple independent variables by using a linear model. The linear regression model finds the corresponding regression coefficient according to the input parameters and multiplies the regression coefficient by the corresponding independent variable to accumulate the dependent variable.

3.2. Model Prediction Result Analysis

The prediction results of each model are shown in Figure 4. In the figure, Test represents the test data, GBDT represents the gradient boosting decision tree prediction result, RF represents the random forest prediction result, and LR represents the linear regression prediction result.

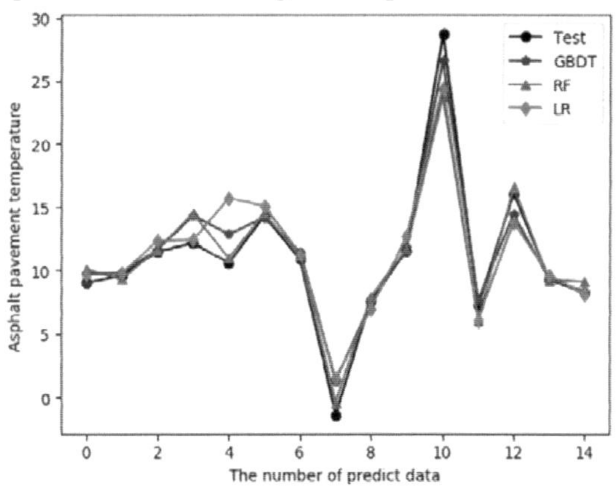

Figure 4. Asphalt pavement temperature prediction model test results.

In order to verify the accuracy of each asphalt pavement temperature prediction model, the Mean Square Error (MSE) of the prediction results of GBDT, random forest and linear regression models was calculated. The calculation formula is as shown in Equation (3), and the MSE variation trend of each temperature prediction model is shown in Figure 5.

$$MSE = \frac{1}{N}\sum_{i=1}^{n}(f(x_i) - y_i)^2 \quad (3)$$

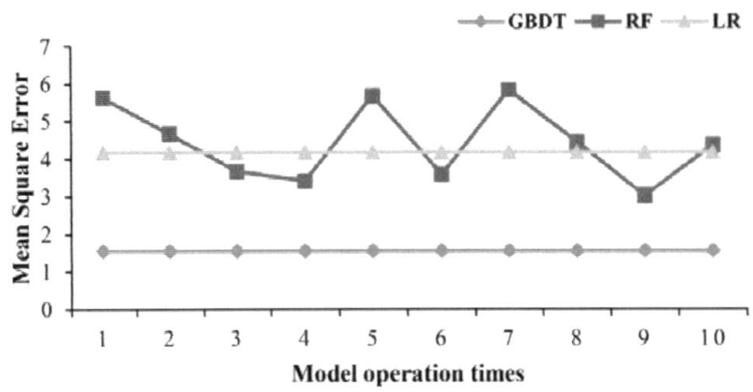

Figure 5. MSE variation trend for each temperature prediction model.

The analysis results show that the mean square error of GBDT prediction results is the smallest and the robustness is high, which has great advantages in asphalt pavement temperature prediction; the mean square error of random forest prediction result is always in the state of fluctuation; the prediction model built by linear regression is as stable as the GBDT model, but has little advantage in prediction accuracy. In conclusion, the model built by GBDT has good prediction effect, and has good generalization ability and robustness. It has good applicability in the prediction of asphalt pavement temperature.

4. Conclusion

The following major findings were achieved from this study:

(1) Both GBDT and Random Forest belong to the decision tree integration algorithm. In the case that the training samples have constraints, the GBDT has higher prediction accuracy than the random forest, so GBDT is more suitable for establishing the prediction model.

(2) The asphalt pavement temperature prediction model established by GBDT has good generalization ability and robustness, can predict the temperature change of asphalt pavement more accurately, and has certain feasibility for the prediction of pavement temperature. The study would serve as a technical support for the application of machine learning algorithms in prediction.

Acknowledgements

The authors are grateful to the financial supports from the Natural Science Foundation of Zhejiang Province (LY18E080020).

References

[1] Barber E S 1957 Calculation of Maximum Pavement Temperatures from Weather Reports Highway Research Board Bulletin

[2] Huber G A 1993 Weather Database for the Superpave™ Mix Design System Binders

[3] Krsmanc R, Slak A S and Demsar J 2013 Statistical Approach for Forecasting Road Surface Temperature Meteorological Applications **20** pp 439-446

[4] Friedman J H 2001 Greedy Function Approximation: A Gradient Boosting Machine Annals of Statistics **29** pp 1189-1232

[5] Friedman J H 2002 Stochastic Gradient Boosting Elsevier Science Publishers B. V. pp 367-378

[6] Liu L, Ji M and Buchroithner M 2017 Combining Partial Least Squares and the Gradient-Boosting Method for Soil Property Retrieval Using Visible Near-Infrared Shortwave Infrared Spectra. Remote Sensing **9** pp 1299-1317

[7] Hong T, Pinson P and Fan S 2016 Probabilistic Energy Forecasting: Global Energy Forecasting Competition 2014 and Beyond International Journal of Forecasting **32** pp 896-913

[8] Ding C, Wang D and Ma X 2016 Predicting Short-Term Subway Ridership and Prioritizing its Influential Factors Using Gradient Boosting Decision Trees Sustainability **8** pp 1100-1116

[9] Yang S, Wu J and Du Y 2017 Ensemble Learning for Short-term Traffic Prediction Based on Gradient Boosting Decision Trees Journal of Sensor **2017** pp 1-15

IOP Conf. Series: Materials Science and Engineering **758** (2020) 012032 doi:10.1088/1757-899X/758/1/012032

Kinematic synthesis of a reconfigurable robot manipulator with lattice morphing mechanisms based on enhanced QPSO

Bo Yin , Zhenyu Zhao, Haibing Xiao, Mingjun Liu and Wen Hu

School of intelligent manufacturing and equipment, Shenzhen Institute of Information Technology, 2188 Longxiang Road, Shenzhen 518172, China

Corresponding author's e-mail:zhaozy@sziit.edu.cn

Abstract. A kinematic synthesis approach of a modular reconfigurable robot manipulator with lattice morphing mechanisms is proposed to find the optimal robot configuration with the minimum weight. First, the structure features and types of lattice mechanisms are introduced. Then, a synthesis approach using an enhanced QPSO (Quantum-behaved Particle Swarm Optimization) to get the optimal robot configuration with the minimum lattice is presented. Finally, a computation example is given to show the optimization process. The obtained results demonstrate the feasibility of this approach.

1. Introduction

Reconfigurable robot manipulators have the potential to exceed specialized robots in multi-functionality, flexibility, and robustness. However, reconfigurable robots have too much redundant mass to perform more deftly than specialized robots. Hence, finding an optimal robot configuration with minimum weight by kinematic synthesis becomes essential for reconfigurable robots. Finistauri provided guidance on the topological reconfiguration of a fully reconfigurable parallel robot to achieve task-based reconfiguration[1]. Valsamos introduced an approach for the determination of the best configuration of a reconfigurable manipulator based on the location of the task[2]. Zhao proposed a universal approach for configuration synthesis of reconfigurable robots based on fault tolerant indices[3]. Baca proposed a heterogeneous reconfigurable robot design approach based on three types of modules for fast response to a diversity of tasks[4]. Tarkian presented a multidisciplinary design optimization framework for automated design of a modular industrial robot[5]. Yin proposed a configuration optimization approach including a kinematic description of joints for a lattice distortable reconfigurable robot and an optimization method based on Genetic Algorithm[6]. In this paper, inspired by the above approaches, the kinematic synthesis model is built for the reconfigurable robot manipulator with lattice morphing mechanisms[7]. Different from the above methods, an enhanced QPSO is selected to solve the optimization problem because of its fast convergence.

2. Types of lattices

2.1. Features of the module in lattices

As shown in Figure.1, each module has six connectors and can connect other modules on its six faces.

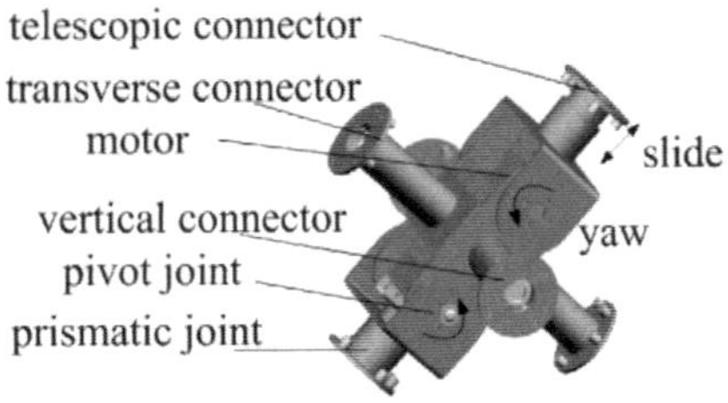

Figure 1. The structure of a single module. A single module has two actuators driven by two pivot joints that their rotation axes are parallel each other. Two motors placed in the module body provide the driving torque to yaw. In the end of each pivot joint, there is a prismatic joint.

2.2 .The morphing rule of the lattice morphing mechanism

Four modules can configure a planar lattice morphing mechanism. When the joint of each module of a lattice rotates a same angle, the telescopic connector on one side of the lattice is extruded with a corresponding length, and then the shape of the lattice is morphed to an isosceles trapezoid from a square, shown in Figure 2.

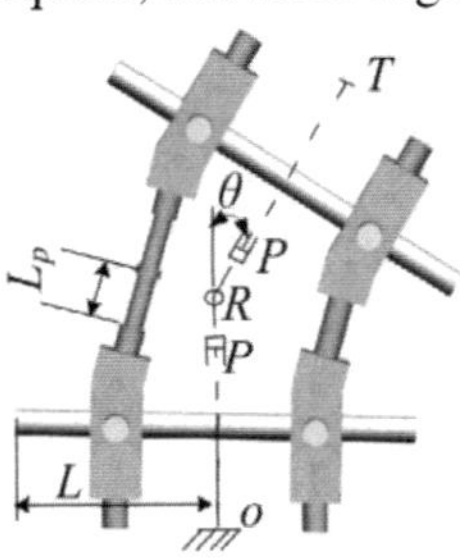

Figure 2.The illustration of the lattice morphing rule. A morphing lattice can be simplified to a "Prismatic–Rotation -Prismatic (PRP) " mechanism, which has 3 degrees of freedom including one rotation and two translations. L=0.156m is the length of a module; θ is the rotation angle of the lattice morphing mechanism. L_p is the sliding length of each prism and its maximum sliding length is 0.039m.

2.3. Types and kinematic models of lattices

Based on their role in the configuration of a robot manipulator, lattices can be distinguished into five types as shown in Table 1. In this paper, the D-H formulation is used to describe the kinematic model of lattice mechanisms.

Table 1. Types of lattices

type	Rotation axis or Connection direction	Transformation matrix	model
P-R-P joint	Y-axis	$T^{jy}=\begin{bmatrix}\cos\theta & 0 & \sin\theta & (L+L_p)\sin\theta\\ 0 & 1 & 0 & 0\\ -\sin\theta & 0 & \cos\theta & (L+L_p)(1+\cos\theta)\\ 0 & 0 & 0 & 1\end{bmatrix}$	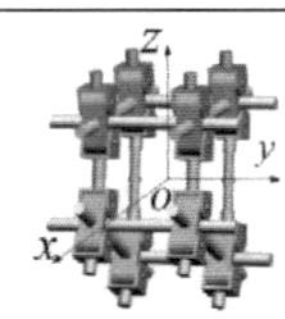
P-R-P joint	X-axis	$T^{jx}=\begin{bmatrix}1 & 0 & 0 & 0\\ 0 & \cos\theta & \sin\theta & (L+L_p)\sin\theta\\ 0 & -\sin\theta & \cos\theta & (L+L_p)(1+\cos\theta)\\ 0 & 0 & 0 & 1\end{bmatrix}$	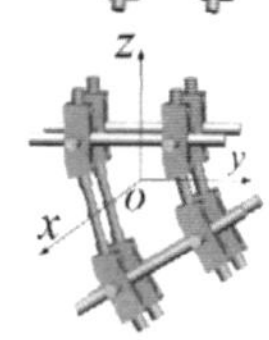
Straig ht link	Z-direction	$T^{z}=\begin{bmatrix}1 & 0 & 0 & 0\\ 0 & 1 & 0 & 0\\ 0 & 0 & 1 & L\\ 0 & 0 & 0 & 1\end{bmatrix}$	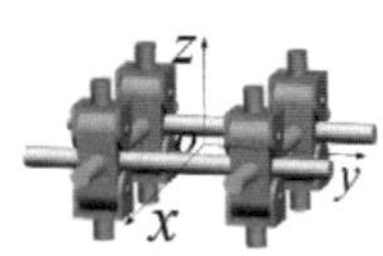

Elbow link	ZY-direction	$T^{y}=\begin{bmatrix}1&0&0&0\\0&0&1&L/2\\0&1&0&L/2\\0&0&0&1\end{bmatrix}$	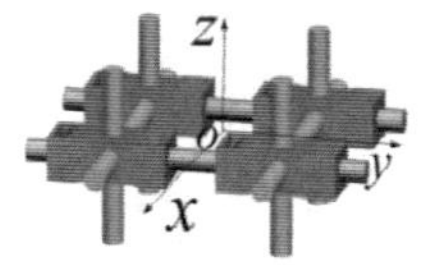
Elbow link	ZX-direction	$T^{x}=\begin{bmatrix}1&0&0&L/2\\0&0&1&0\\0&1&0&L/2\\0&0&0&1\end{bmatrix}$	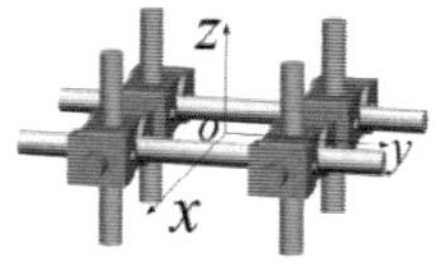

3. Kinematic synthesis for a manipulator

3.1. Optimization model

The goal of the synthesis problem is to determine the types of joints and the dimensions of links in a manipulator that can position a tool held by its end effector in a given set of task positions. The synthesis equation can be written in the form:

$$E_i=\left\|\mathbf{P}_i(\boldsymbol{\theta})\text{-}\mathbf{P}_{Ti}\right\| \quad i=1,2,\cdots n \tag{1}$$

Where $\boldsymbol{\theta}$ is the vector of rotation angle of PRP joints; $\mathbf{P}_i(\boldsymbol{\theta})=[x_i,y_i,z_i]^{\mathrm{T}}$ is the ith work position point of the end actuator; $\mathbf{P}_{Ti}$ is the identified ith task position point; E_i is the acceptable bias between the work position of the actuator and the identified task position.

In order to count easily, a lattice with "PRP joint + straight link + elbow link + straight link" can be regarded as a lattice group as shown in table 2. $x_1\in\{0,1\}$ represents the state of PRP joint, 0- no PRP joint, 1- a PRP joint; $x_2\in\{0,1,2,3\}$ and $x_4\in\{0,1,2,3\}$ represent the length of straight link, 0- no straight link; 1- a straight link; 2- two straight links; 3- three straight links; $x_3\in\{0,1,2\}$ represents the state of elbow links, 0- elbow link; 1- a Z-Y elbow link, 2- a Z-X elbow link.

Table 2. A lattice group

State of PRP chain	Length of straight link	State of Elbow link	Length of straight link
x_1	x_2	x_3	x_4

Then, the kinematic model T_{LG} of a lattice group can be represented as:

$$T_{LG}=\begin{bmatrix}1-x_1+x_1\cos\theta & -x_1x_3\sin\theta & 0 & x_1L_p\sin\theta+x_1x_2L\sin\theta+x_1x_3L\sin\theta/2\\ 0 & 1-x_3 & 1 & x_3L/2+x_4L\\ -x_1\sin\theta & x_1x_3-x_3-x_1x_3\cos\theta & 0 & x_1(L+L_p+L_p\cos\theta)+x_2L(1-x_1+x_1\cos\theta)+x_3L(1-x_1+x_1\cos\theta)/2\\ 0&0&0&1\end{bmatrix} \tag{2}$$

To minimize the weight of the manipulator, the number of PRP joints and links must be the least.

$$\min\ weight\rightarrow\min\ number(x1,x2,x3,x4)=\sum_{j=1}^{m}(sign(x1_j)+x2_j+x3_j+x4_j) \tag{3}$$

Where m is the minimum number of lattice groups; $x1_j$ is the state of the PRP joint in the jth lattice group; $x2_j$ and $x4_j$ are the states of straight links in the jth lattice group; $x3_j$ is the state of elbow links in the jth lattice group.

Here, the kinematic synthesis problem is converted to an optimization problem. States of PRP joints, straight links and elbow links are regarded as design variables. The best configuration with the least lattice that can perform the given task accurately is the aim. The optimization model is:

$$\begin{aligned} & find\ x1, x2, x3, x4 \\ & \min f(x1,x2,x3,x4) = \sum_{i=1}^{n} \left\| \mathbf{P}_i(\boldsymbol{\theta}) - \mathbf{P}_{Ti} \right\| + \sum_{j=1}^{m} (sign(x1_j) + x2_j + x3_j + x4_j) \\ & s.t.\ \ x1 = x2 = 0,1,2 \\ & \quad\ \ x2 = x4 = 0,1,2,3 \\ & \quad\ -\pi/12 < \theta < \pi/12 \end{aligned} \tag{4}$$

3.2. Optimization algorithm

Quantum-behaved Particle Swarm Optimization (QPSO) can search the solution quickly and efficiently [8]. Therefore, QPSO is adopted to solve the optimization problem.

3.2.1. Quantum-behaved Particle swarm optimization. The particle of QPSO algorithm moves according to the following equation:

$$mbest = \frac{1}{M}\sum_{i=1}^{M} P_{i,best} \tag{5}$$

$$P_i = \phi \cdot P_{i,best} + (1-\phi) \cdot G_{best} \tag{6}$$

$$X_i(t) = P_i \pm \alpha \cdot \left| mbest - X_i(t-1) \right| \cdot \ln(1/u) \tag{7}$$

Where *mbest* is the mean best position among the particles; M is the population size; P_i is the local attractor of the ith particle and is a stochastic point between $P_{i,best}$ and G_{best}; ϕ is a random umber distributed uniformly on [0,1]; u is another uniformly-distributed random number on [0,1]; α is a parameter of QPSO that is called Contraction-Expansion Coefficient.

$$\alpha = (\alpha_{\max} - \alpha_{\min}) \cdot (Miteration - t) / Miteration + \alpha_{\min} \tag{8}$$

Miteration is the max iteration number; in general, α_{max}=1, α_{min}=0.5.

3.2.2. Enhanced QPSO for kinematic synthesis of manipulators. To avoid falling into a local optimal solution, we set up a judgment rule in the optimization process: if the function values are same in three continuous iterations, then the algorithm reset the initial random particles and reiterate from the beginning. The process include two stages: one stage is to calculate the difference between the reachable workspace of every population and the task points; the other stage is to search the best configuration with the least lattices.

4. Computation example

The task trajectory is a spatial V-type polyline which starts from point A, crosses through point B, and ends at point C. The coordinates of these points are (0.376, 0.482, 0.313),(0.293, 0.478, 0.467),(0.211, 0.471, 0.572), unit: m. The base of the manipulator is located at the origin of the task coordinate system. According to the coordinates of the task points, the max distance in x direction is 0.165m, and then at least three joint lattices and one elbow link are needed in the configuration, because the max translation distance of one joint lattice is 0.08m. The max distance in y direction is 0.011m, and then there exists one joint lattice in the configuration. The max distance in z direction is 0.259m, and then there are two joint lattices in the configuration. Considering the above cases, the required configuration must have at least three groups and at least one elbow link. Then, the initial number of lattice groups N_{LG}=3.The other parameters of the enhanced QPSO are listed in Table 3.

Table 3. Parameters of QPSO for kinematic synthesis

Parameter	Value
maximum contraction-expansion coefficient α_{max}	1
minimum contraction-expansion coefficient α_{min}	0.5
Allowable position tolerance $E1$	0.1×10^{-6}m
population size M	60
maximum generation $Miteration$	20

The computing result is obtained as shown in Figure 3. The output is $x1$=(1 1 1), $x2$=(0 0 0), $x3$=(0 1 2), $x4$=(0 0 1). The configuration of the manipulator is shown in Figure 4.

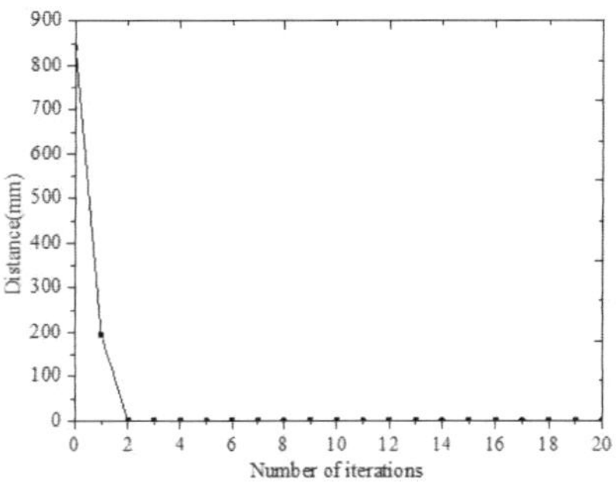

Figure 3. Optimal result of the enhanced QPSO.

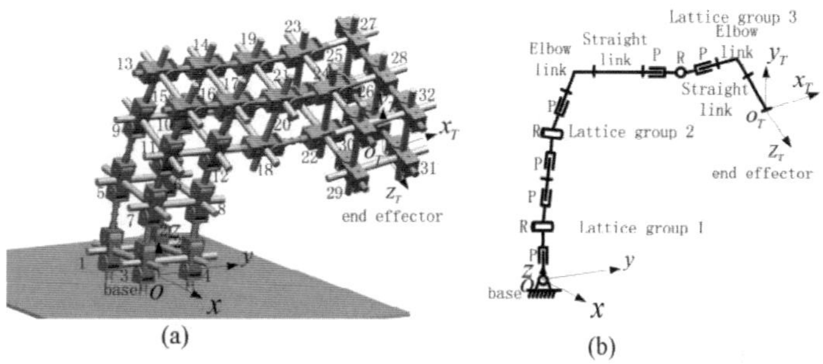

Figure 4. The best configuration. (a) The manipulator and the number of modules; (b) The equivalent kinematic chain.

5. Conclusion

The enhanced QPSO developed in this paper is valid to solve the kinematic synthesis of manipulators. The computing time of the algorithm can be reduced by analyzing the characteristic and the size of the given task and setting suited parameters.

Acknowledgments

Project supported by Science and Technology Program of Shenzhen (JCYJ20170817112440533, JCYJ20170817114441260)，Guangdong Province Higher Vocational Colleges & Schools Pearl River Scholar Funded Scheme(2018), Guangdong University Innovation Team Project(2018GKCXTD001), School-enterprise cooperation project (SZIIT2019KJ015).

References

[1] Finistauri AD and Xi FJ 2013 *J Mech Robot.* **5** 041002
[2] Valsamosa C, Moulianitisb V and Aspragathosa N 2012 *Robot Cim-Int Manuf.* **28** 517
[3] Li Q and Zhao J 2012 *Industrial Robot: An International Journal.* **39** 69
[4] Baca J, Ferre M and Aracil R 2012 *Robot Auton Syst.* **60** 522
[5] Tarkian M, Persson J and Feng X 2012 *J Mech Design.* **134** 124502
[6] Yin B, Liang ZN, Dai X, Mo JQ and Wang SG 2016 *P I Mech Eng C-J Mec.* **230** 1532
[7] Yin B, Liang ZN, Hu W, Xu W and Wang SG 2014 *Adv Robotics.* **28** 895
[8] Xu ZY, Du L, Wang HP, Deng ZC 2019 *Appl Math Mech-Engl.* **40** 111

Study on dynamic behaviors of a breakwater triggered by earthquakes

Jie Lai[1], Yun Liu[1,2], Wei Wang[1], Chen-qiang Gao[1], Zi-rong Niu[1], and Hai-bo Zhu[1]

1 Xi'an Research Institute of High-Tech, Xi'an 710025, China
2 Faculty of Architechtural and Enviromental Engineering, Chongqing industry polytechnic college, Chongqing 401120, China
E-mail: 513516059@qq.com

Abstract. To ensure the safety requirement of an artificial island in the construction period, the modified Hardin-Drnevich dynamic constitutive model and strength reduction method was used to analyze the dynamic response of a breakwater under the action of earthquakes. Research indicates the following: (1) the modified Hardin-Drnevich can effectively simulate the soil stress-strain relationship under dynamic action, and obtain strain hysteresis, and get permanent deformation of material; (2) the deeper the buried depth of breakwater, the greater soil pressure; (3) dynamic pressure response is more obvious on both sides of the breakwater at peak time in earthquakes.

1. Introduction

Artificial islands are located at sea, and as such represent complex environments, subject to the potentially adverse effects such as waves, ocean currents, wind, ice, and earthquakes. Breakwaters are important enclosure structures designed to protect artificial islands, and an instability of a breakwater will have catastrophic consequences for the artificial island it is built upon.

China's rapid economic development has brought strain upon land transportation networks. This, when coupled with its long coastline and numerous islands, means the development of coastal traffic has important political and economic significance. Coastal breakwaters in China are near the Circum–Pacific belt; hence, earthquakes happen frequently. These conditions can create numerous problems in engineering, so research in this area is of great significance for the construction of stable, effective breakwaters. As large-scale breakwater projects are in the initial development stages, the related research into breakwaters mainly focuses on construction technologies such as sandy soil liquefaction and static stability analysis [1-5]. WANG [6] applied the nonlinear constitutive model of a multiple shear mechanism type and calculated the effect of rotation of principal stress axis direction.

Matsuda, et al. [7] proposed the damage mechanism of breakwaters during earthquake–tsunami events, in which a simulated breakwater was destabilized by not only wave pressure, but also long-acting tsunami seepage flow and overflow.

However, the dynamic stability of breakwaters has not been fully investigated, which may mean the stability requirements are not being met in engineering practices. Based on this concern, a breakwater project in the East China Sea is analyzed in this paper by a strength reduction approach, taking into account the effects of seepage, earthquakes, pore water pressure, and other factors. These research results would provide an important reference for engineering construction.

Further, the dynamic characteristics and stability of this breakwater are studied by considering the coupled actions of seepage, earthquakes, and waves. This research provides an important reference for engineering construction.

2. Soil-fluid interaction

Based on Biot's consolidation theory [8], we can consider the soil-fluid interaction using Darcy's law. The response equation for the pore fluid can be formulated as

2.1. Seepage equation [9]

$$-\frac{\partial q_i}{\partial x_i}+q_v=\frac{1}{M}\frac{\partial P}{\partial t}+a\frac{\partial \varepsilon}{\partial t}-\beta\frac{\partial T}{\partial t} \tag{1}$$

Here, q_i is the seepage vector; q_v is the volumetric fluid source intensity; M is the Biot modulus; p is the pore water pressure; a is the Biot ratio; ε is the volumetric strain; T is the temperature; t is time and β is the coefficient of thermal expansion for the solid-fluid interaction. As the artificial island is at sea, its temperature change is relatively small, and has little effect on the stability, deformation, and structural forces. This article considers only the coupled actions of seepage, waves, and earthquakes, without considering the influence of the temperature; hence, $\partial T / \partial t = 0$

2.2. Fluid-solid coupling equation [9]

$$\delta\sigma_{ij}+a\frac{\partial P}{\partial t}\delta_{ij}=H(\sigma_{ij},\varepsilon_{ij}-\varepsilon_{ij}^{T},\kappa) \tag{2}$$

Here, δ_{ij} is the Kronecker delta factor; σ_{ij} is the stress; $\delta\sigma_{ij}$ is the stress increment; ε_{ij} is the total strain; H_{ij} is from the given constitutive model; κ is a history parameter and ε_{ij}^{T} is the strain rate determined by temperature variation, for this scenario $\varepsilon_{ij}^{T}=0$.

3. Strength Reduction Method

Zheng [10, 11] used a new method, referred to as the strength reduction method, to calculate the slope's dynamic stability and consider the slope's dynamic characteristics under earthquake conditions. This method can fully describe soil tensile-shear failure during earthquakes. Traditional static slope stability analysis is based on the limit equilibrium method, and the safety coefficient w relating the anti-sliding force s and sliding force t is expressed as follows:

$$w = \frac{s}{t} = \frac{\int_0^l (c + \tan\varphi)dl}{\int_0^l \tau dl} \tag{3}$$

Here, c is the cohesion, φ is the angle of internal friction and τ is the dynamic shear stress. In the strength reduction method, the slope is made to approach the limit equilibrium state by reducing the shear strength of the soil as

$$c' = c / \omega, \quad \varphi' = \arctan(\tan\varphi / \omega) \tag{4}$$

where ω is the reduction parameter, which is the safety factor. c' and φ' are the values of c and φ after reduction, respectively.

However, slope failures under earthquake excitations are the result of both tensile and shear damage; hence, the tensile strength of the rock and soil mass should be considered. When using the strength reduction method, the tensile strength should also be reduced:

$$\sigma^{t'} = \sigma^{t} / w \tag{5}$$

Here, σ^t is the tensile strength and $\sigma^{t'}$ is the numerical value of σ^t after reduction.

4. **Test case**

4.1. Basic breakwater set-up

The environment of this breakwater features smooth ocean floors and a simple underwater landscape, which is composed of clay soil, backfill sand, and compaction sand, proportionately. The breakwater structure is composed of a plug-in steel cylinder structure. The steel cylinder is 40 m long, its thickness is 16 mm and its diameter is 22 m. The elevation of the cylinder head is 3.5 m, and the elevation of the cylinder base is -36.5 m. The top level of the soil layer is -8.5 m, and the water level is 2.57 m, as shown in Figure 1.

The numerical model is 190 m long, with a height of 80.5 m. The water level for the breakwater is 2.57 m, and the artificial pumping water level is -12.5 m on the inland side. To improve the calculation efficiency and reduce the grid size, a typical cross section was selected in calculation. The cross-sectional element was chosen to simulate the interactions between the cylinder and the surrounding soils. The backfill sand as well as compaction sand are considered as liquefiable soils, and the parameters are shown in Table 1.

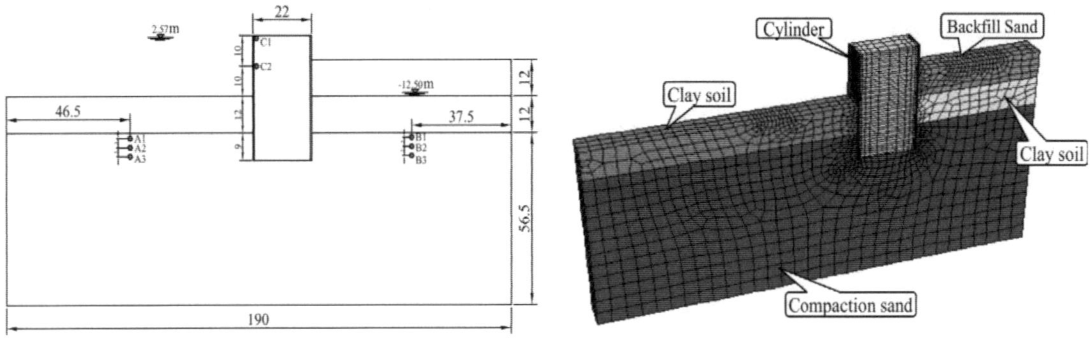

(a)Model dimensions (b) Calculation model

Figure 1. calculation model for the breakwater

Table 1. Geotechnical material parameters

Materials	Dry density kg/m^3	Elastic modulus /Mpa	Poisson's ratio	Angle of internal friction	Cohesivity /kPa
Backfill Sand	1380	40	0.35	30	4
Clay soil	1350	30	0.38	13	22
Compaction sand	1380	45	0.35	32	2
Cylinder	3500	2000	0.17	--	--

4.2. Input earthquake wave

The typical earthquake intensity is category VII in the project area, but considering the importance of its strategic location and economic status, category VIII is decided upon for the anti-seismic design. The peak amplitudes of the Kobe earthquake (Japan, 1995) were between 0.2 g and 0.3 g in numerical calculations. The input acceleration amplitude of 0.2 g was used to calculate the stress, acceleration response and stability of the breakwater. The input amplitudes from 0.2 g to 0.3 g were applied to analyze the liquefaction of the artificial island. The input acceleration waveform is shown in Figure 2.

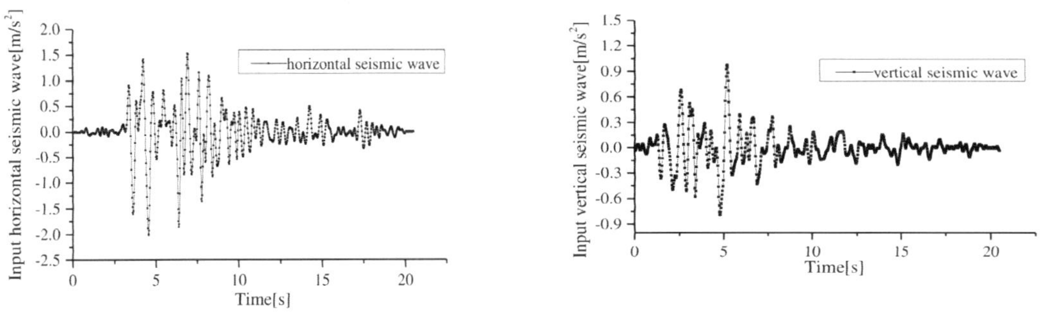

(a) Horizontal direction (b) Vertical direction

Figure 2. Put earthquake waves (Kobe wave)

4.3. Dynamic earth pressure

Monitoring points were set up at 3 m intervals on both sides of the steel cylinder to record the time-history curve of the dynamic earth pressure. The distribution of the dynamic earth pressure is shown in Figure 3.

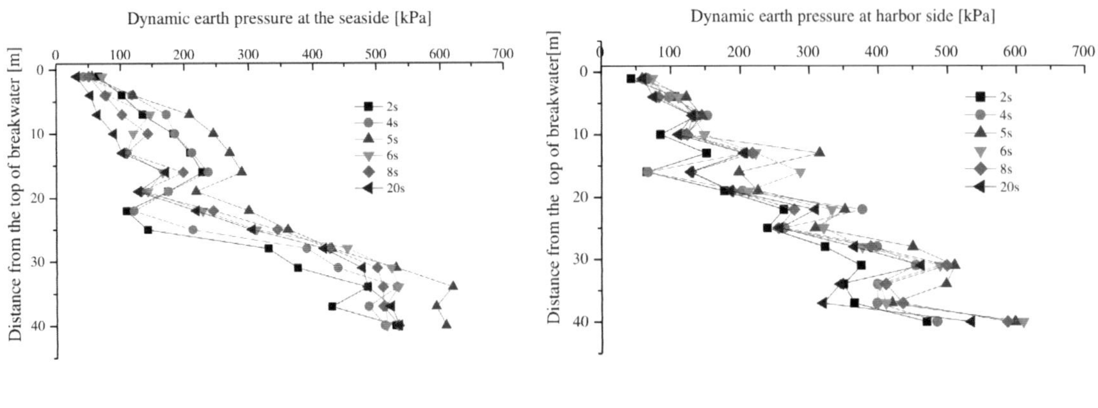

(a)Left-hand side of the breakwater (b) Right-hand side of the breakwater

Figure 3. Dynamic earth pressure distribution on both sides of the breakwater when triggered by an earthquake

Based on the dynamic earth pressure distribution on both sides of the breakwater under earthquake excitations (see Figure 3), it can be concluded that the deeper the buried depth of the breakwater, the greater the dynamic earth pressure. The pressure response on both sides of the breakwater is more notable between 4s and 6s under earthquake conditions.

4.4. Stability of the breakwater

The strength reduction method was used to find the dynamic stability of the artificial island. In this method, the artificial island was made to approach the limit equilibrium state by reducing the shear strength of the soil. When the reduction factor $\omega = 1.39$, the displacements of monitoring points C1 and C2 converged after the earthquake, so the breakwater was stable. When the reduction factor $\omega = 1.40$, the displacements of monitoring points C1 and C2 did not converge and would continue to increase after the earthquake, so the breakwater was not stable and entered the limit equilibrium state (as shown in Figure 4).

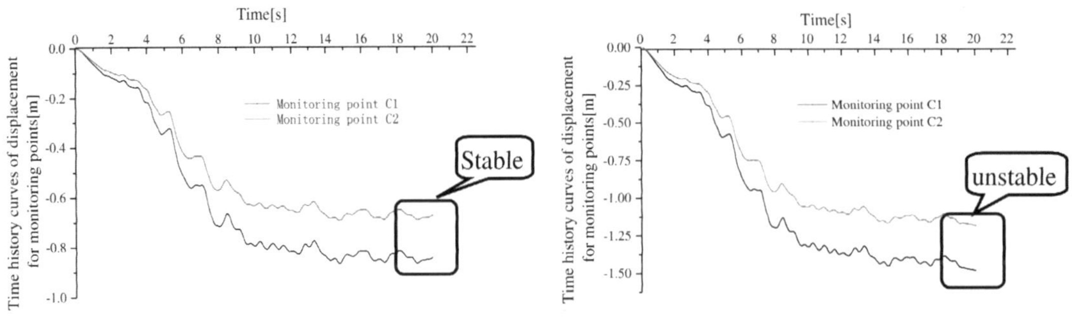

(a) Reduction factor of 1.39 (b) Reduction factor of 1.40

Figure 4. Horizontal displacement time-history curves of monitoring points under different reduction factors

From Figure 5, it can be seen that a displacement mutation occurs when the reduction factor $\omega = 1.40$, which shows the difference in stability occurring with a small change in reduction factor.

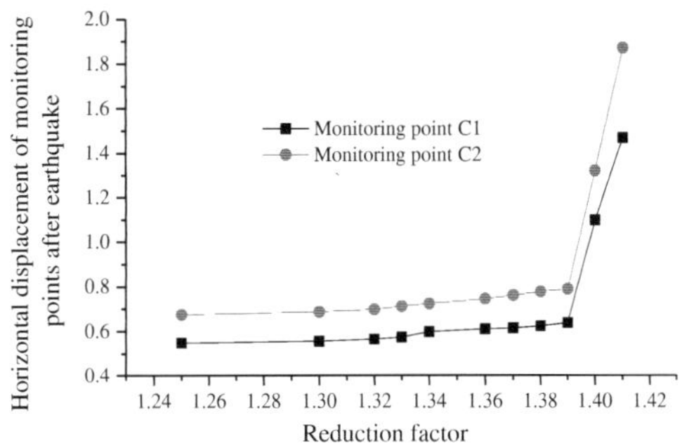

Figure 5. Reduction factor - Horizontal displacement curve

When the reduction factor $\omega = 1.40$, the plastic shear zones of the artificial island become connected, and the final failure surface of the artificial island can be seen as the red line shown in Figure 6. Based on the above analysis, the dynamic safety factor of the breakwater is found to 1.39.

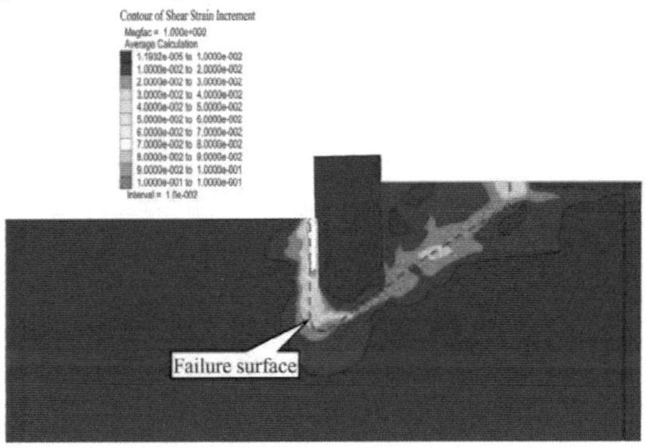

Figure 6. Plastic shear zone

5. Conclusions

(1) When the water level was increased to the design height, the dynamic safety factor of the breakwater calculated with the strength reduction method was 1.39, which meets the safety requirements. The final failure surface appeared at the breakwater-soil interface.

(2) The deeper the buried depth of the breakwater, the greater the dynamic earth pressure will be. The dynamic earth pressure response is more notable on both sides of the breakwater from 4 to 6 s under earthquake conditions.

(3) The dynamic safety factor of the breakwater, calculated using the strength reduction method, concerns the water levels. When the water level is pumped to the design height, it can meet the safety requirements. The ultimate failure surface appeared at the breakwater-soil interface.

Acknowledgements

This research was supported by the National Natural Science Foundation of China (51178457), National Natural Science Foundation of China (51378496), Natural Science Foundation of Chongqing City of China(CSTC2013jcyjys0002), Graduate Student Scientific Research Project of Chongqing City of China(CYB15108), and the Youth Foundation of Force Engineering University (2018QNJJ003).

References

[1] Wang Y L and Wu Z S 2013 *Risk Assessment for the Construction Drawing Design of the East and West Artificial Islands of Hong Kong-Zhuhai-Macao Bridge* CONSTRUCTION TECHNOLOGY 42 pp 64-68.

[2] Mnasri C, Hafsia Z, Omri M and Maalel K 2010 *A moving grid model for simulation of free surface behavior induced by horizontal cylinders exit and entry* Engineering Applications of Computational Fluid Mechanics 4 pp 260–275.

[3] Williams A N and Li W 2000 *Water wave interaction with an array of bottom-mounted surface-piercing porous cylinder* Ocean Eng. 27 pp 841-866.

[4] Bricker J D 2013 *Turbulence Model Effects on VOF Analysis of Breakwater Overtopping during the 2011 Great East Japan Tsunami* In Proceedings of the 35th IAHR World Congress, Chengdu, China.

[5] Mnasri C, Hafsia Z, Omri M and Maalel K 2010 *A moving grid model for simulation of free surface behavior induced by horizontal cylinders exit and entry* Engineering Applications of Computational Fluid Mechanics 4 pp 260–275.

[6] WANG L Y, JIANG P M and LIU H L 2011 *Mechanism analysis of residual liquefied deformation of breakwater during earthquake*, Rock and Soil Mechanics 31 pp 3556-2562.

[7] Matsuda T, Maeda K, Miyake M, Miyamoto J, Sumida H & Tsurugasaki K 2016 *Instability of a Caisson-Type Breakwater Induced by an Earthquake–Tsunami Event* International Journal of Geomechanics 16 pp C4016003.

[8] BIOT M A 1956 *General Solution of the Equation of Elasticity and Consolidation for a Porous Material* Journal of Applied Mechanics 23 pp 91-96.

[9] FLAC3D Fast Lagrangian Analysis in Three Dimensions User's Guide. Itasca Consulting Group, Minneapolis 2013.

[10] Lai J, Zheng Y R, Liu Y, Li X D and Liu H W 2016 *Seismic design of multiple anti-slide piles by strength reduction dynamics analysis method* Electronic Journal of Geotechnical Engineering 181 pp 1-10.

[11] ZHENG Y R, ZHAO S Y and ZHANG L Y 2002 *Slope stability analysis by strength reduction FEM* Engineering Science 4 pp 57-61.

IOP Conf. Series: Materials Science and Engineering **758** (2020) 012034 doi:10.1088/1757-899X/758/1/012034

Properties of AlN-MWCNTs/epoxy Nanocomposites

FangXie*, Guiqiang Diao, Na Qiang and Dingshu Xiao

School of Chemistry and Materials Engineering, Huizhou University, Huizhou, Guangdong, 516007, China

*Corresponding author's e-mail:xiefang4498@126.com

Abstract. Synergistic effect of AlN in combination with MWCNTs in AlN-MWCNTs/epoxy nanocomposites was found. Thermal conductivity of AlN-MWCNTs/epoxy nanocomposites increased obviously with the increase of AlN content. The added AlN can be filled in the networks of MWCNTs to form a more dense three-dimensional thermal conductivity network and provide more thermal conductivity channels. In addition to that, mechanical and flame-retardant properties of the nanocomposites were improved also. AlN-MWCNTs/epoxy nanocomposites presented optimum mechanical properties with the addition of 5~7.5%wt AlN powder. With the addition of AlN powder in AlN-MWCNTs/epoxy nanocomposites, its LOI value is increased from 30.8% to 43.4% and UL 94 test achieves V-0 ranking.

1. Introduction

In electronic devices, thermally conductive polymer composites have been widely applied [1]. To improve the thermal conductivity of polymers, an effective way is to incorporate the thermally conductive fillers [2,3]. Carbon nanotubes (CNTs) possess excellent properties in thermal conductivity and have been used as the fillers to fabricate polymeric composites materials. [4]. The conductivity of a single multiwalled carbon nanotubes (MWCNTs) is about 3000W/mK, for a single-walled carbon nanotubes (SWCNTs) is about 2000W/mK [5,6].

In order to improve the thermal conductivity of CNTs-nanocomposites, various filler combinations have been investigated, such as boron nitride/CNTs, hydrotalcite clay/ CNTs and graphite nanoplatelets/CNTs, etc [7~9]. In this work, aluminium nitride (AlN) and MWCNTs were incorporated into the epoxy resin to prepare AlN-MWCNTs/epoxy nanocomposites. Thermal conductivity, mechanical and flame-retardant properties of the above nanocomposites were investigated. The synergetic effects of AlN and MWCNTs were also discussed.

2. Experimental

2.1. Materials

A EP828 epoxy resin (Diglycidyl ether of bisphenol-A, Epoxy equivalent weight (EEW) of 185~195 g/mol) was obtained from Momentive Specialty Chemicals Inc., Columbus, USA. 4,4'-diaminodipheny ethane was purchased from Aladdin Reagents (Shanghai) Co., Ltd. The carbon nanotubes used were multi-walled carbon nanotubes (MWNTs, Shenzhen Nanotech Port Co., Ltd.) and their properties are listed in Table I. The particle size of AlN powder (Shanghai Chaowei Nano Technology Co. Ltd.) is about 50 nm with purity over 99.9%.

IOP Conf. Series: Materials Science and Engineering **758** (2020) 012034 doi:10.1088/1757-899X/758/1/012034

Table 1. Properties of MWNTs

Diameter	Length	Purity	Ash	Amorphous carbon	Special surface area
10~30nm	5~15μm	≥95%	≤0.2%	< 3%	40~300 m^2/g

2.2. Preparation of AlN-MWCNTs/Epoxy nanocomposites

In order to minimize the effects of moisture, MWNTs and AlN powder were dried at 120°C in vacuum for 24 hours at first. With the aid of solvent of acetone, the epoxy resin, MWCNTs and AlN powder were added in a vessel and put in ultrasonic bath sonicator for 2 hours at 60°C to achieve good distribution of MCNTS and AlN powder in epoxy resin. Then, the acetone in the mixture was removed under vacuum condition. After that, DDM was added to the mixture by stoichiometry and it was stirred further for 5 minutes at 90°C. The above mixture was degassed in a vacuum oven. To prepare AlN-MWCNTs/epoxy nanocomposites samples, the mixture was cast into the a mould and cured for 3 hours at 95°C, for 2 hours at 115°C, and for 2 hour at 145°C, then heated for 2 hours at 175°C. After that, to avoid stress cracking, the samples were cooled to the ambient temperature. The ratios of AlN:MCNTs:epoxy resin for AlN-MWCNTs/epoxy nanocomposites are 2.5:3:94.5, 5:3:92, 7.5:3:89.5, 10:3:87 and 12.5:3:84.5, respectively.

2.3. Characterization

Thermal conductivity of AlN-MWCNTs/epoxy nanocomposites was measured by a thermal constants analyzer (Hot Disk TPS2200, Sweden). According to the ASTM D2863-1997 standard, limiting oxygen index (LOI) measurement was conducted on an oxygen index instrument manufactured by Fire Testing Technology Co.Ltd., UK. The dimension of the sheets is 150×6.5×3.2 mm^3. According to ANSL UL 94-1985, a UL 94 flammability meter (Fire Testing Technology Co.Ltd., UK) has been used to carry out vertical burning tests, the dimension of the sheets is 130×13×3.2 mm^3. According to ASTM standards D 790-07e and D 638-08, a universal testing machine (Z010, Zwick/Roell Company, Germany) has been used to measure flexural and tensile properties. Six specimens for each set were tested with a 5 mm/min crosshead speed and the average value was calculated.

3. Results and discussion

3.1. Thermal conductivity of AlN-MWCNTs/epoxy nanocomposites

Fig.1 shows the thermal conductivity of AlN-MWCNTs/epoxy nanocomposites.

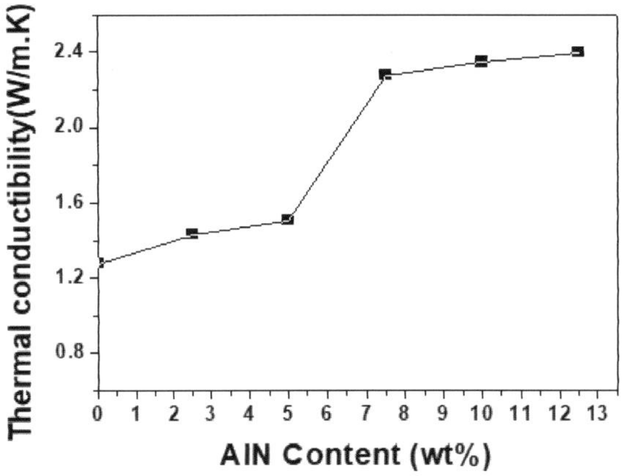

Figure 1. Thermal conductivity of AlN-MWCNTs/epoxy nanocomposites

According to Fig.1, it can be found that the thermal conductivity of AlN-MWCNTs/epoxy nanocomposites increased obviously with the increase of AlN content. When AlN content exceeds 7.5wt%, thermal conductivity of the nanocomposites is over 1.75 w/m.k.
Fig.2 shows the structural models of the nanocomposites.

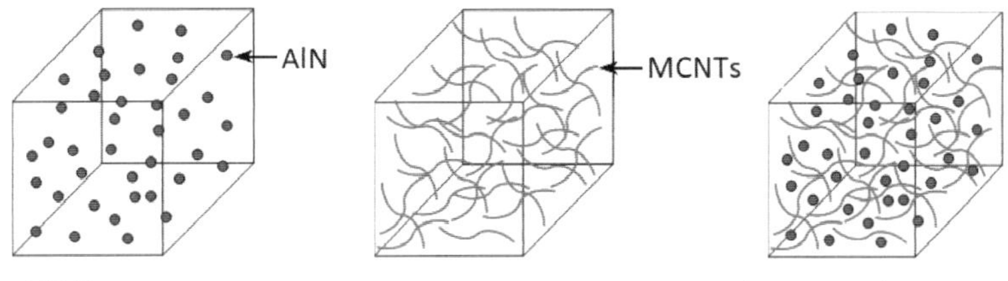

Figure 2. Structural models of AlN/epoxy, MWCNTs/epoxy and AlN-MWCNTs/epoxy nanocomposites

It is well known that the thermal conductivity of polymer composites with conductive fillers has been dominated by thermal conductivity channels and networks formed with fillers in the matrices [10]. As showed in Fig.2, the contact and overlap of dispersed MWCNTs can form thermal conductivity channels and networks. The AlN added can be filled in the networks of MWCNTs, thus to form a more dense three-dimensional thermal conductivity network and provide more thermal conductivity channels, which improved the thermal conductivity of AlN-MWCNTs/epoxy nanocomposites significantly.

3.2. Mechanical properties of AlN-MWCNTs/epoxy nanocomposites

Carbon nanotubes present the excellent mechanical properties, for example, high tensile strength and elastic modulus etc., which make them the most ideal and promising reinforce material in enhancing the mechanical properties of carbon nanotubes/polymer composites [11]. Table 2 presents the measurements of tensile strength and flexural strength of AlN-MWCNTs/epoxy nanocomposites.

Table 2. Mechanical properties of AlN-MWCNTs/epoxy nanocomposites

AlN:MCNTs:epoxy(wt)	Tensile strength(Mpa)	Flexural strength(Mpa)
0:0:100	58.9±2.1	122.3±1.5
0:3:97	76.7±1.9	171.3±3.0
2.5:3:94.5	78.2±1.7	176.5±2.7
5:3:92	81.9±2.3	182.1±2.9
7.5:3:89.5	80.3±2.7	185.2±1.8
10:3:87	73.6±2.6	176.7±2.6
12.5:3:84.5	67.3±3.1	142.8±2.9

According to Table 2, the addition of 3%wt MWCNTs results in a significant improvement of epoxy resin. The measurements in Table 1 also revealed that a synergetic effect with the incorporation of AlN in MWCNTs/epoxy resin. AlN-MWCNTs/epoxy nanocomposites presented optimum mechanical properties with the addition of 5~7.5%wt AlN powder. At the same time, the content of MWCNTs in nanocomposites is reduced. It can be inferred that the addition of a certain amount of AlN can reduce the agglomeration and winding of MWCNTs in AlN-MWCNTs/epoxy nanocomposites and improve the dispersion of MWCNTs, thus enabling the AlN-MWCNTs/epoxy nanocomposites to exhibit better tensile strength and flexural strength.

3.3. Flame-retardant properties of AlN-MWCNTs/epoxy nanocomposites

To characterize the flame-retardant properties of AlN-MWCNTs/epoxy nanocomposites, UL-94 and LOI tests were performed. The results are listed in Table 3.

Table 3. Flame-retardant properties of AlN-MWCNTs/epoxy nanocomposites

AlN:MCNTs:epoxy(wt)	LOI(%)	Total burning time (s)	Drop	Classification
0:0:100	22.4	67	No	UL94 V-2
0:3:97	30.8	50	No	UL94 V-1
2.5:3:94.5	32.1	46	No	UL94 V-0
5:3:92	35.9	39	No	UL94 V-0
7.5:3:89.5	38.3	31	No	UL94 V-0
10:3:87	40.6	23	No	UL94 V-0
12.5:3:84.5	43.4	22	No	UL94 V-0

It can be found that the LOI value of epoxy composites increased from 22.4% to 30.8% with the addition of 3wt% MWCNTs. Three-dimensional network structure in epoxy matrix can be formed by adding a certain amount of MWCNTs. During the combustion of MWCNTs/epoxy nanocomposites, a protective char network layer covering the sample surface is formed that acts as insulative barrier and resulting in a significant increase in the flame retardant properties.

In addition to that, combination of AlN and MWCNTs in epoxy nanocomposites shows a positive effect on both LOI and UL94 test: LOI value is increased from 30.8% to 43.4% and UL 94 test achieves V-0 ranking with the further addition of AlN powder. Compared with individual MWNTs component filled sample, the coexistence of MWNTs and AlN can reduce the flammability of the nanocomposites more efficiently, which indicated that the coexistence of AlN and MWNTs in the composites can form a more effective network structure, which is responsible for the improved flame retardancy for AlN-MWCNTs/epoxy nanocomposites. Similar synergetic effect has also been found in ABS/clay/MWCNTs nanocomposite [12].

4. Conclusion

The introduction of AlN and MWCNTs to the epoxy resin matrix is an effective methods for improving thermal conductivity, mechanical and flame-retardant properties of AlN-MWCNTs/epoxy nanocomposites. Synergistic effect on improving the related properties was found between AlN and MWNTs. The thermal conductivity of AlN-MWCNTs/epoxy nanocomposites increased obviously with the increase of AlN content. The AlN added can be filled in the networks of MWCNTs, thus to form a more dense three-dimensional thermal conductivity network and provide more thermal conductivity channels, which improved the thermal conductivity of AlN-MWCNTs/epoxy nanocomposites significantly. At the same time, mechanical and flame-retardant properties of the nanocomposites were improved also. AlN-MWCNTs/epoxy nanocomposites presented optimum mechanical properties with the addition of 5~7.5%wt AlN powder. With the addition of AlN powder in AlN-MWCNTs/epoxy nanocomposites, its LOI value is increased from 30.8% to 43.4% and UL 94 test achieves V-0 ranking.

Acknowledgments

This research was financially supported by the Natural Science Foundation of Guangdong Province, China (2014A030313642).

References

[1] Pernot G, Stoffel M, Savic I, Pezzoli F, Chen P, Savelli G 2010 *Nat.Mater.* 9 491-495
[2] Chen H, Ginzburg V, Yang J, Yang Y, Liu W, Huang Y, Du L, Chen B 2016 *Prog.Polym.Sci.* 59 41-85
[3] Burger N, Laachachi A, Ferriol M, Lutz M, Toniazzo V, Ruch D 2016 *Prog.Polym.Sci.* 61 1-28
[4] Yang X, Liang C, Ma T. Guo Y, Kong J, Gu J, Chen M, Zhu J 2018 *Adv.Compos.Hybrid.Mater.* 1 207-230
[5] Kim P, Shi L, Majumdar A, McEuen PL 2001 *Phys.Rev.Lett.* 87 215502/1-4

[6] Yu C, Shi L, Yao Z, Li D, Majumdar A 2005 *Nano.Lett.* 5 1842-6
[7] Xu M, Lei Y, Ren D, Chen S, Chen L, Liu X 2018 *Nanomaterials.* 8 997-1010
[8] Vittorio R, Carlo N, Liberata G, Luigi V 2015 *Polym.Compos.* 36 1118-1123
[9] Yu A, Ramesh P, Sun X, Bekyarova E, Itkis M.E, Haddon R.C 2008 *Adv.Mater.* 20 4740-4744
[10] Ngo I.L， Prabhakar V, Byon C 2016 *Int.J.Heat.Mass.Transf.* 102 713-722
[11] Arash B, Wang Q, Varadan V.K 2014 *Sci.Rep.* 4 64791-8
[12] Li H, Liu H, Li J 2016 *Sci.Eng.Compos.Mater.* 23 29-35

IOP Conf. Series: Materials Science and Engineering **758** (2020) 012035 doi:10.1088/1757-899X/758/1/012035

An improved pretreatment method for the fabrication of Cu-Ag core-shell nanoparticles

Q Y Zou[1], G N Yang[1], P Y Wang[1], Y Zhang[1], C Q Cui[1], K Zhang[1*]

1 State Key Laboratory of Precision Electronic Manufacturing Technology and Equipment, School of Electromechanical Engineering, Guangdong University of Technology, Guangzhou 510006, China
E-mail: zhangkai@gdut.edu.cn

Abstract. A pretreatment method of nitric acid pickling is applied in the fabrication of Cu-Ag core-shell nanoparticles. By tuning the pH value and the concentration of Cu ions after acid pickling, the cleaning and separation process can be eliminated. This simplified pretreatment process can prevent the oxidation and particle loss. The as-prepared Cu@Ag nanoparticles show promising coating quality, oxidation resistance and quasi-spherical core-shell microstructure. Such a pretreatment method provided by this study could be promising in further rapid fabrication of Cu-Ag core-shell nanoparticles.

1. Introduction

With the development of the third generation of semiconductor power device, the copper nanoparticles have attracted wide attention due to their excellent electronic/thermal conductivity, mechanical properties, low sintering temperature, high temperature stability and low cost[1]. However, the poor oxidation resistance of copper nanoparticles is still an main restriction for their further application[2]. Fabrication of copper nanoparticles with Ag shell becomes a feasible solution to overcome such a problem[3].

At present, a main fabrication method for Cu-Ag core-shell nanoparticles is wet chemical reduction, in which the original Cu particles is mixed in Ag solution [4,5]. The Ag is coated onto the Cu particles through reduction reaction. In this method, the size of the Cu@Ag nanoparticles can be tuned by the size of the original Cu nanoparticles[6]. Consider that the original Cu nanoparticles have been partly oxidized, a pretreatment of acid pickling is required to remove the oxide layer on the Cu nanoparticles[5]. Dilute sulfuric acid is conventionally used in different studies. However, some drawbacks should also be considered for its application in the fabrication of Cu nanoparticle. The sulfate and Ag ions can form the less soluble Ag_2SO_4, which will influence the coating quality. After the acid pickling, the nanoparticles should be separated from the solution and cleaned by deionized water, which will also introduce oxidation and particle loss.

In this study, consider the good compatibility between the nitrate and silver ion, we applied another nitric acid to the pretreatment process in the fabrication of Cu-Ag core-shell nanoparticles. By tuning the pH value and the concentration of Cu ions in the solution after acid pickling, the particle cleaning and separation process can be eliminated. And the chemical coating process can proceed directly. The results show good coating quality of the as-prepared Cu-Ag core-shell nanoparticles. The study provides a promising pretreatment method for the rapid fabrication of Cu-Ag core-shell nanoparticles.

IOP Conf. Series: Materials Science and Engineering **758** (2020) 012035 doi:10.1088/1757-899X/758/1/012035

2. METHODS

Cu nanoparticles (the perpetration method is shown in Ref.[7]) with an average diameter of 55 nm are dispersed in deionized water, and then mixed with dilute nitric acid for acid pickling. KOH is dripped into the solution to tune the pH to the range of 8~11. Then the solution is mixed with ammonia. Another solution of silver nitrate and ammonia is prepared and mixed with the original solution for chemical coating. Ultrasonic agitation is applied during the whole process. After the chemical coating, nanoparticles are separated from the solution by centrifugation and cleaned by deionized water for 3 times and alcohol for 3 times. The morphology, composition and element distribution of the as-prepared nanoparticles are inspected through scanning electron microscope (SEM), X-Ray Diffraction (XRD) and energy dispersive spectrometer (EDS), respectively.

3. RESULTS AND DISCUSSIONS

Figure 1 and Figure 2 show the morphologies of the original Cu nanoparticles and the as-prepared Cu@Ag nanoparticles. Both the two kinds of nanoparticles are in globular shape. The morphologies of the nanoparticles before and after the chemical coating are very similar. The Cu@Ag nanoparticles are slightly smaller than the original Cu nanoparticles, which probably results from the surface dissolution of Cu nanoparticles during the acid pickling process. Size distribution of the nanoparticles are analyzed by ImageJ software. It shows that the size of the Cu nanoparticles are in the range of 26~102 nm, with an average diameter of 55nm. The size of the Cu@Ag nanoparticles are in the range of 21~89nm, with an average diameter of 50nm.

Figure 1. SEM image of the original Cu nanoparticles.

Figure 2. SEM image of the as-prepared Cu@Ag nanoparticles.

IOP Conf. Series: Materials Science and Engineering **758** (2020) 012035 doi:10.1088/1757-899X/758/1/012035

Figure 3 shows the XRD spectra of the as-prepared Cu@Ag nanoparticles and the nanoparticles after exposure in air for 24 days. Sharp peaks of face-centered-cubic (FCC) Cu crystals can be observed at the positions of 2θ=43.3°, 50.4°, and 74.1°, which relate to the crystal planes of {111}, {200}, and {220}, respectively. Sharp peaks of FCC Ag crystals can be observed at the positions of 2θ=38.1°, 44.3°, 64.4°, and 77.4°, which relate to the crystal planes of {111}, {200}, {220}, and {311}, respectively. No peak of oxide can be detected on the spectra. The spectrum of the nanoparticles exposed in air for 24 days barely shows any difference compared to that of the as-prepared nanoparticles, demonstrating a promising oxidation resistance of the Cu@Ag nanoparticles fabricated by this method.

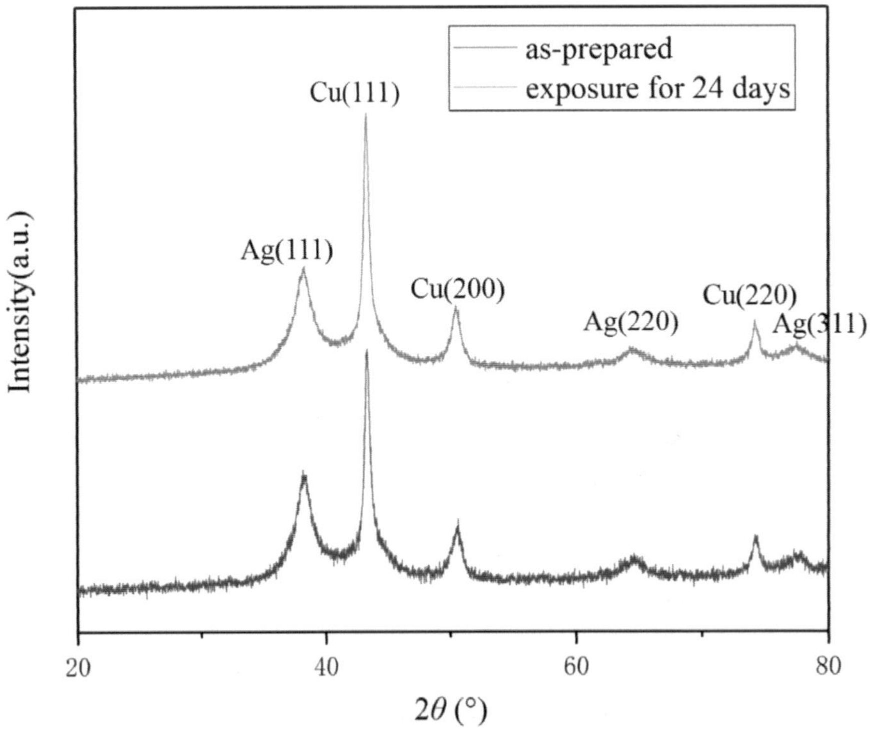

Figure 3. XRD spectra of the as-prepared Cu@Ag nanoparticles and the nanoparticles after exposure in air for 24 days.

To confirm the core-shell microstructure of the as-prepared Cu@Ag nanoparticles, the element distributions of the nanoparticle are inspected by EDS. Figure 4 shows the element distribution of an as-prepared Cu@Ag nanoparticle along its diameter direction, obtain by line sweeping of EDS. The signal intensity of Cu reaches the maximum at the center of the particle, and decreases dramatically near the edge of the particle. The signal intensity of Ag is high near the edge of the particle and is relatively stable in the positions inside the particle. The results certificate that the Cu@Ag nanoparticles are in core-shell microstructure.

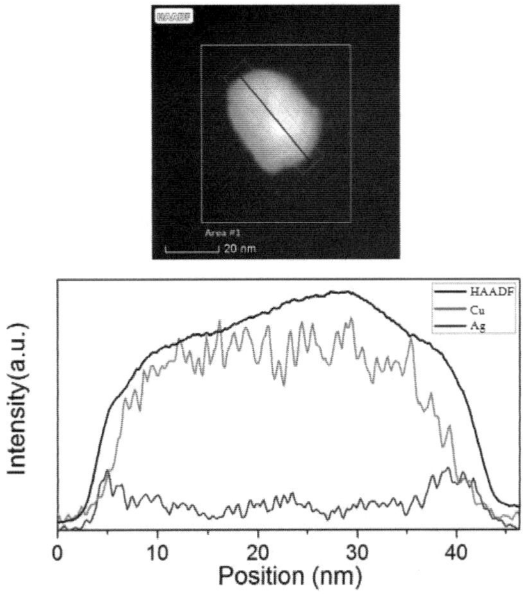

Figure 4. The element distribution of an as-prepared Cu@Ag nanoparticle along its diameter direction, obtain by line sweeping of EDS.

4. CONCLUSIONS

This study applied nitric acid to the acid pickling pretreatment process in the fabrication of Cu@Ag nanoparticles. Compared to the conventional pretreatment using sulfuric acid, the substitution of nitric acid simplifies the pretreatment process by eliminating the cleaning and separation process after acid pickling, and therefore can prevent oxidation and particle loss. The morphologies of the nanoparticles before and after the chemical coating are well consistent. The as-prepared Cu@Ag nanoparticles show promising coating quality, oxidation resistance and quasi-spherical core-shell microstructure. The pretreatment method provided by this study could be applied in the rapid fabrication of Cu@Ag core-shell nanoparticles in the future.

5. References

[1] Yang W, Wang C, and Arrighi V 2019 *J. Mater. Sci. - Mater. Electron* **30** 11607-11618
[2] Wang JQ and Shin SH 2017 *Rsc Advances* **7** 21607-21617
[3] Milon B, Mousumi P, Debanjan D, Riya M and Tarakdas B 2018 *Nanotechnology* **29** 325102
[4] Jin X, Mao AQ, Ding ML, Ding PP, Zhang TC, Gu XL, Xiao W and Yuan J 2016 *Appl. Spectrosc* **70** 1692-1699
[5] Yanhong T, Zhi J, Chenxi W, Su D, Jiayue W, Zhiquan L and Chunqing W 2016 *Rsc Advances* **6** 91783-91790
[6] Xing Y, Junjie L, Tielin S, Chaoliang C, Guanglan L, Jinhu F, Tianxiang L and Zirong T 2017 *J. Alloys Compd.* **724** 65-372
[7] Lai T, Zhang Y, Cui CQ, Zhang K, Chen T, Chen X, Chen X, Gao J , He YB, Tang H and Chen Y 2018 *presented at the 19th International Conference on Electronic Packaging Technology* 1702-0705

Acknowledgments

This work is supported by the National Natural Science Foundation of China (61874155, 61704033, U1601202 and 61874035) and Science and Technology Planning Project of Guangdong Province, China (2017A050501053 and 2017A010106005).

Research Status of Ultrasonic Vibration Assisted Plastic Forming Process

Zhanbin Chen[1,a],Lianfa Yang[2,b*],Kunyan Zhang[3,c],Jingyu Jiang[4,d],Pengju Zong[5,e]

[1,2,3,4,5]Faculty of Mechanical and Electrical Engineering, Guilin University of Electronic Technology, Guilin,Guangxi, 541004, China
E-mail:[a]C_zhanbin@163.com,[b]y-lianfa@163.com,[c]445969897@qq.com
[D]779227042@qq.com,[e]452667657@qq.com

Abstract: Ultrasonic vibration assisted plastic forming is a process method for assisting the forming of metal materials by means of ultrasonic vibration devices. It is proved that the ultrasonic vibration could reduce the forming force, improve the friction environment, and enhance the formability of the metal material. However, there are some differences in forming characteristics for different materials under different vibrational plastic forming processes.In this paper, the forming principles, the vibration application modes and the forming characteristics of the ultrasonic vibration assisted the wire drawing, equal channel angular pressing (ECAP), upsetting, drawing of the sheets and hydroforming of the tube are reviewed. It emphasizes on the various processes' improvement of the forming efficiency and the quality of the formed parts under the aid of ultrasonic vibration. Finally, the future development trend of ultrasonic vibration assisted plastic forming was prospected.

1. Introduction

On the basis of the conventional plastic forming, adopting the ultrasonic vibration device to the blank or the tool with an adjustable ultrasonic vibration can obtain a better forming effect. There are two main effects on the mechanism of ultrasonic vibration [1]: one is the volume effect that acts on the inside of the material and affects the flow stress of the material, which mainly explains by the two theories of thermal softening effect and mechanical superposition principle related to crystal dislocation[2]. The other is the surface effect acting on the outside of the material which affects the contact condition between the material and the die, which mainly explains by the reduction of friction and the reversal of the friction vector [3].

The ultrasonic vibration apparatus is mainly composed of an ultrasonic generator, a transducer, a horn and a blank or a die [4, 5], as shown in Fig1. The function of the ultrasonic generator is to convert the alternating current into an electrical oscillating signal. The output power is usually 1~2 kW, and the vibration frequency is generally 20 kHz. It mainly provides the energy for generating reciprocating vibration. The function of the transducer is to convert the electrical oscillation signal into mechanical vibration, and the amplitude is generally less than 10 μm. The function of the horn is to amplify the amplitude generated by the transducer to suit different working conditions, and the amplitude after amplification is up to 100~150 μm. The blank or die is usually connected to the horn by screws, and the whole apparatus is connected to the device through the flange on the horn to provide high frequency ultrasonic vibration[6].

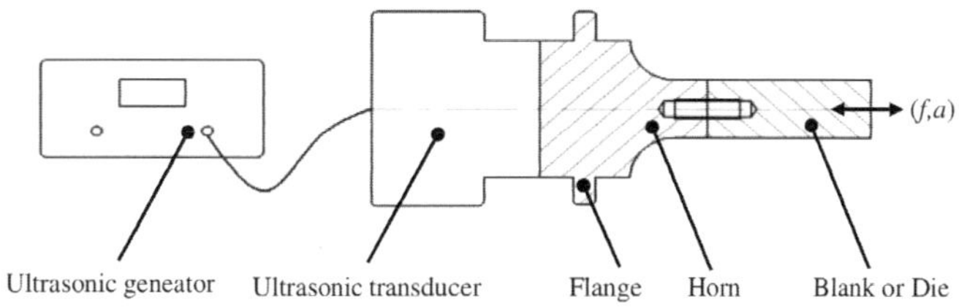

Fig 1. Ultrasonic vibration apparatus

Based on the brief introduction of the ultrasonic vibration apparatus and the mechanism of action above, the following research situation is reviewed from the of ultrasonic vibration's forming principles, the vibration application method and the forming characteristic, such as the assisted drawing, the equal channel angular pressing (ECAP), the upseting, the drawing of the sheet and the hydroforming of the tube and so on, which provides a new reference method and ideas for the application and promotion of ultrasonic vibration in the field of plastic forming.

2. Ultrasonic vibration assisted wire drawing

Ultrasonic vibration wire drawing is based on the traditional drawing, applying vibration to the wire drawing die for vibration drawing, as show in Fig 2. For easy-to-machine materials such as pure aluminum and stainless steel, radial vibration is superior to axial vibration in improving tensile speed and improving surface forming quality [7], but radial vibration is not as good as axial vibration in reducing drawing force [8]. And the critical speed at break is 14.9% higher than that of conventional wire drawing [9]. For the difficult-to-machine material of titanium, Yang et al.[10] found in the drawing test in 2016 that the longitudinal-torsional composite vibration drawing force was reduced by 15.6% compared with the longitudinal vibration, but the friction was slightly increased and the surface quality was poor. In 2017 and 2018, Liu et al.[11,12] found that compared with conventional drawing, large amplitude longitudinal vibration and longitudinal-torsion composite vibration could improve the surface quality, and large amplitude longitudinal vibration was more effective in reducing the drawing force. In addition, in the wire drawing test of the longitudinal vibration of two passes, the drawing force was reduced by 50%, and the effect of increasing the wire speed and improving the surface quality was better than increasing the amplitude.

It can be seen that ultrasonic vibration exhibits good characteristics for easy-to-machine materials and difficult-to-machine materials in wire drawing. Moreover, in the vibration state, the drawing pass of the blank can be reduced, and multiple passes can be performed, which significantly improves the processing efficiency and forming performance of the material.

3. Ultrasonic vibration assisted ECAP

ECAP is an important technical for achieving ultra-fine grain structure, and the vibration application way is shown in Fig 3. In 2014, Ahmadi et al. [13] constructed a simulation model of rigid and plastic pressing punch, and found that the simulation error of the plastic pressing punch was only 0.6% while the simulation error of the rigid body pressing punch was 7.3%. The plastic model was closer to the actual machining environment. In 2015, Ahmadi et al. [14] found that the grain refining efficiency increased by 25.8% after the vibration pressing of pure aluminum, and microstructure were more uniform by increasing the amplitude. In 2016, Bagherzadeh et al. [15] applied longitudinal or transverse ultrasonic vibration directly to the deformation zone in the vibrational equal channel angular pressing of pure aluminum. The pressing force was reduced by about 30%, the hardness was increased by 50% to 52.9%. Moreover, the effect in refining the grains, reducing the pressing force, strengthening the hardness, and the longitudinal vibration is better.

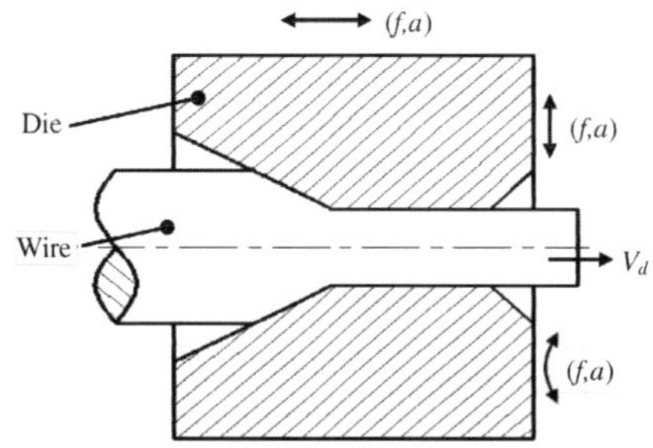

Fig 2. Ultrasonic vibration wire drawing process

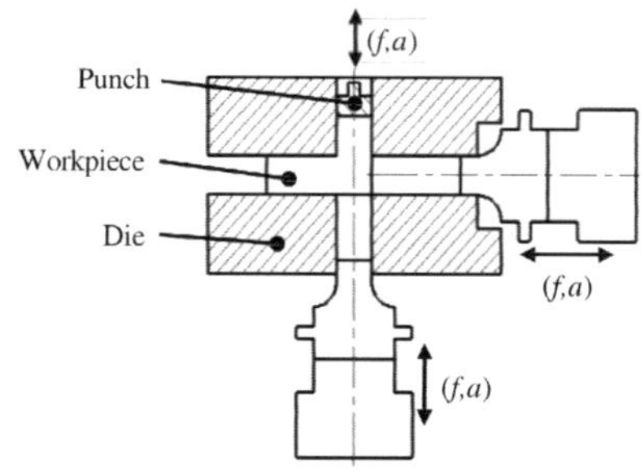

Fig 3. Ultrasonic vibration ECPA

It can be seen that in the extrusion forming, the ultrasonic vibration can reduce the extrusion force, and have higher grain refining efficiency and better hardness strengthening effect. At the same time it make the microstructure of the forming component more uniform. Moreover, in the simulation study, it is very important to construct a reasonable simulation model to improve the accuracy of the simulation study.

4. Ultrasonic vibration assisted deep drawing

Nowadays, ultrasonic vibration deep drawing is mainly used for drawing the thin sheet barrel, and the vibration is applied as shown in Fig 4, and the thick-direction vibration of the blank holder or the die was more obvious for the increase of the limiting drawing ratio[16]. In 2014, Huang et al. [17] found that LDR increased by 8.5%~8.9% and the punching force decreased by 24%~38% in the vibration micro-draw test of stainless steel sheet, effectively overcoming the problem of work hardening. During 2016-2018, CAO et al. [18-20] performed a soft punching of the AZ31B magnesium alloy sheet at a high temperature by longitudinal vibration of the die, and found that the vibration reduced the optimum forming temperature by 10.3%~17.2%. Moreover, the punching force was lowered by about 55.2% at 260℃, the LDR was increased by 22.1%. In addition, in the experimental study on the friction characteristics of the sheet under ultrasonic vibration, it was found that the friction coefficient was reduced by about 40%.

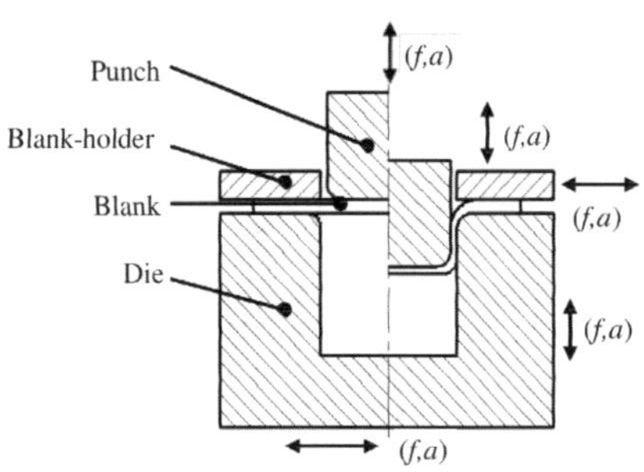

Fig 4. Ultrasonic vibration deep drawing process

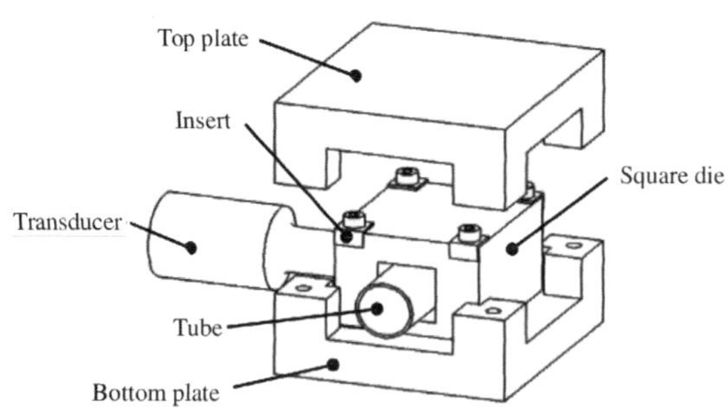

Fig 5. The apparatus of ultrasonic vibration hydroforming of the tube

It can be seen that in the deep drawing, the ultrasonic vibration reduces the punching force, improves the forming limit of the sheet, significantly reduces the friction coefficient during the processing, and overcome the problem of the work hardening of the smaller parts. In addition, for the magnesium alloy sheet which is not easily formed at normal temperature, the optimum forming temperature is effectively lowered, and the processing process is improved.

5. Ultrasonic vibration assisted hydroforming of the tube

At present, the application of ultrasonic vibration in the hydroforming process of tube has gradually gained attention and achieved some results. In 2008, Bauget et al. [21] developed a set of ultrasonic vibration hydroforming apparatus, as shown in Figure 5. In the experimental study of the annealed copper alloy tube, ultrasonic vibration was found to reduce the angular radius of the formed section by 7.7%.In 2015, Shahri [22] changed the position of the gasket on the die based on Bauget, and constructed a two-dimensional rigid square die model. The results showed that the thickness distribution of the formed section was more uniform after applying ultrasonic vibration. The angular filling rate was increased from 55.8% to 92.93%. In 2017, Zarei et al. [23] also used the Bauget's method to fix the die directly without using a gasket. The two-dimensional deformed square die model was used for simulation analysis. It was found that there was a high frequency contact-separation state between the inner surface of the die and the tube. This state improved the friction between the tube and the inner surface of the die, so that the angular radius of the formed component was reduced, the thickness distribution was uniform, and no extreme thinning occured at the straight side and the rounded corner.In the same year, the team [24] applied ultrasonic vibration along the axial direction in the simulation study of tube hydro forming with axial feeding, and found that increasing the vibration frequency or vibration amplitude can reduce the axial compression load and increase the bulging height.

It can be seen that the ultrasonic vibration in the hydroforming of the tube improves the filling property of the pipe, the moldability and the uniformity of the wall thickness, improves the forming limit of the pipe.But at present, it is basically in the stage of simulation research. Although some scholars have carried out experimental research on the forming properties, when using the forming device developed by the researcher, the volumetric effect of the ultrasonic vibration on the tube was negligible, and it is more suitable for the related research of the surface effect.

6. Summary and forecast

In summary, in the plastic forming process, Ultrasonic vibration plays a very good role in the formation of easy-to-machine materials such as copper, aluminum, stainless steel, and difficult-to-machine materials such as titanium and titanium alloys. It also effectively overcomes the work hardening problem caused by the small size of the small piece, Nevertheless, due to the further research and verification of the mechanism of ultrasonic vibration plastic forming, there is still much work to be done in the future. For example, The different characteristics of friction in different processes and different temperatures, and the depth mechanism need further study. In the simulation study of ultrasonic vibration plastic forming, the establishment of more accurate theoretical models, material models and friction models can significantly improve the accuracy of research. Moreover, the research and development of high-power ultrasonic vibration devices and forming equipment is crucial for the application and promotion of ultrasonic vibration in the field of plastic forming.

Acknowledgments

This work was financially supported by the National Natural Science Foundation of China

(51564007), Natural Science Foundation of Guangxi Province (2017GXNSFAA198133), and Guangxi Key Laboratory of Manufacturing System & Advanced Manufacturing Technology (14-045-15-005Z).

References

[1] Dnwson G R, WinspeT C E, Sansome D H 1996 *J.Metal Forming*,**231**,234-238.

[2] Wen T, Li W,Xia C and Chunlei P 2011 *J.International Journal of Minerals, Metallurgy and Materials*,**18**,70-76.

[3] Storck H, Littmann W, Wallaschek J and Mracek M 2002 *J.Ultrasonics*, **40**,379-383.

[4] Junwen Z,Shusen W,Youwu M and An P 2008 *J. Materials Review*,**S1**,189-193.

[5] Yanxiong L and Lin H 2015 *J.Journal of Plasticity Engineering*,**22**,8-14.

[6] Shendun Z,Yonfeng L and Shuqin F 2013 J. China Mechanical Emgineering,2013,24,835-840.

[7] Murakawa M and Jin M 2001 J.Journal of Materials and Processing Tcchnology,113,81-86.

[8] Hayashi M, Jin M, Thipprakmas S, Murakawa M, Hung J C,Tsai Y C, Hung C H 2003 J.Journal of Materials Processing Technology,140,30-35.

[9] Haiqun Q, Xiaobiao S and Tao X 2009 J. Chinese Journal of Mechanical Engineering, 22,580-586.

[10] Yang C Q,Shen X B and Xie T 2016 *J.The International Journal of Advanced Manufacturing Technology*,2016(1/4):645-655.

[11] Liu S,Shen X B,Guo K, Yang Y Cand Xie T 2017 *J. Ultrasonics*,**83**,60-67.

[12] Liu S,Yang Y C,Xie T and Shen X B 2018 Experimental Study on Fine Titanium Wire Drawing with Two Ultrasonically Oscillating Dies *J. IEEEAccess.*

[13] Ahmadi F,Farzin M 2014 *J.Journal of Mechanical Engineering Science*, **228**,1859-1868.

[14] Ahmadi F,Farzin M,Meratian M,Loeian S M and Forouzan M R 2015 *J.The International Journal of Advanced Manufacturing Technology*, **79**,503-512.

[15] Bagherzadeh S,Abrinia K and Han Q Y 2016 *J.MaterialsLetters*,**169**,90-94.

[16] Jimma T,Kasuga Y,lwaki N, Miyazawa O, Mori E,Ito K and Hatano H 1998 *J.Journal of Materials ProcessingTechnology*,406-412.

[17] Huang Y M,Wu Y S and Huang J Y 2014 *J.The International Journal of Advanced Manufacturing Technology*, **71**,1455-1461.

[18] Cao Miayan,Li Jianchao,Yang Zhuoyun,Zhao Changcai and BI Jiang 2016 *J.Transactions of Nonferrous Metals Society of China*, **26**,2118-2127.(In Chinese)

[19] Cao M Y,Li J C,Yuan Y N and Zhao C C 2017 *J.Trasaction of Nonferrous Metals Society of china*, **27**,163-171.

[20] Miaoyan C,Jianchao L,Yanyang L,Yaning Y, Changcai Z and Guo-jiang D *J.Journal of Central South University*, **25**,1879-1887.

[21] Bunget C 2008 Mechanics of Ultrasonic Tube Hydroforming *D.North Carolina State University.*

[22] Shahria S E E, Boroughanib S Y A, Khalilib K and Kang B S 2015 *J. Procedia Technology*,90-97.

[23] Zarei M, Faghani G R,Farzin M and Mashayekhi M 2017 *J.Journal of Applied and Computational Mechanics*, **3**,178-184.

[24] Zarei M,Faghani G R, Farzin M and Mashayekhi M 2017 *J.Journal of Applied and Computational Mechanics*,**3**,251-257.

IOP Conf. Series: Materials Science and Engineering **758** (2020) 012037 doi:10.1088/1757-899X/758/1/012037

Experimental Study on Continuous Laser Polishing Die Steel S136D

Zhen Yu Zhao[1,*], Zhen Jiu Zhang[1], Bo Yin[1], Xiao Mei Xu[1]

1School of Intelligent Manufacturing and Equipment, Shenzhen Institute of Information Technology, Shenzhen, 518172, China
Corresponding author's e-mail: yxpzzy01@163.com

Abstract. With the traditional polishing method the mold material will cause problems such as difficulty in processing the micro-area surface and damage to the surface of the workpiece. Laser polishing technology is an advanced and new surface processing technology that has emerged in recent years. Because laser polishing has the characteristics of good polishing quality and low thermal effect, it is very suitable for high-precision polishing of mold steel, titanium alloy and other materials. By properly controlling the process parameters such as laser energy density, scanning speed, scanning pitch and defocusing amount during laser polishing, the workpiece can be polished with high precision and the surface roughness can be quickly reduced. The processing examples in the paper further illustrate the superiority of laser polishing.

1. Introduction

In industrial production, mold automatic polishing technology has a variety of more mature technology, basically meet the needs of industrial manufacturing, such as mechanical polishing, electrochemical polishing, thermal chemical polishing, fluid polishing and ultrasonic polishing, etc., and widely used in the field of mechanical equipment, electronic instrumentation, medical machinery equipment, precision optical components. However, there are some shortcomings in these polishing methods [1], and such polishing techniques are faced with the problem of post-processing [2]. For example, electropolishing, electrochemical polishing, grinding, etc., can result in rounded corners or some areas cannot be processed. For chemical polishing, electrolytic polishing, and ultrasonic polishing, the surface needs to be cleaned after processing, and the electrolyte and the grinding suspension need to be frequently replaced and processed. For manual polishing, etc., the abrasive is consumed, and the ablated particles need to be disposed of in time during processing. Therefore, it is not feasible for an irregular mold surface. If you use precision CNC machining, you will face programming difficulties. When polishing with conventional mechanical equipment, different areas of the same plane cannot be polished differently. When polishing an anisotropic workpiece, it is necessary to select a polishing path depending on the crystal orientation. These polishing methods are easy to impact the workpiece and may affect the mechanical properties and material properties of the workpiece [3]. Mold surface finishing is a difficult problem in mold processing that has become a bottleneck in mold processing.

Laser polishing is an emerging manufacturing process which improves the roughness of components through controlled melting of a microscopic surface-layer in industrial production in Figure 1. The laser polishing technique is a new surface processing technology. Therefore, with the development of precision machinery industry, laser polishing is attracting more and more attention and is widely used in industrial production, and produces products of high quality produced, reduces product costs and has multi functions.

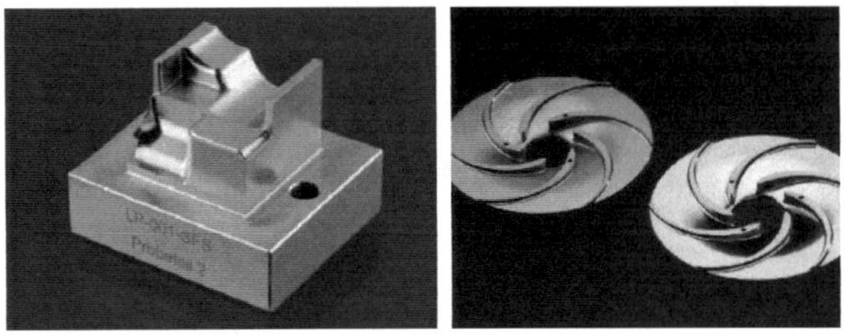

Figure.1 Laser polishing applications

In 2002, Ramos J.A et al. [4] used Nd:YAG laser and CO_2 laser to polish the selected parts of the laser sintered part, reducing the surface roughness Ra from 7.0-9.0um to 1.4-2.1um. In 2006, Lamikiz et al. [5] also polished selected laser sintered parts using CO_2 lasers. It is noteworthy that Lamikiz and other methods have not only achieved good results in surface polishing, but also achieved good results with three-dimensional polishing, and the roughness Ra was reduced from 7.5um to 1.2-1.3 um around. In addition, Lamikiz et al [6] performed a metallographic analysis on the polished parts, and found that the heat-affected zone has no defects, and is more uniform than the original sintered molded part, and the hardness of the metal surface is also improved. Ukar et al [6] found that under the optimal parameters, both lasers achieved remarkable results, the surface roughness was reduced by 80%, and Ra was less than 0.5um; but the semiconductor laser was compared with CO_2. Compared with the laser, the polishing efficiency is higher. In 2012, Temmler.A et al. [7] used a five-axis linkage manipulator combined with a three-dimensional scanning system and a fiber laser to study the laser polishing of the tool steel 1.2343 metal free-form surface. It is believed that the polishing process is mainly the remelting of the thin metal surface. After the laser irradiates the workpiece, the molten metal at the peak flows to the trough under the action of the surface tension, and there is no material removal during the polishing process, but the melt is redistributed. In 2012, Hafiza A. M. K. [8] and others studied two consecutive laser beam laser polishing AISIH13 tool steels to study the effects of surface quality. Under the conditions of argon protection, laser polishing irradiation of the same energy density was applied at the same time, and the 95% overlap ratio gave the best result, and the surface roughness was reduced from 1.35 μm to 0.18 μm. Pfefferkorn FE [14]studied the improvement of metal surface finish by thermal capillary flow during pulsed laser polishing. In the hot capillary state, the longer melt duration was used during pulsed laser polishing, resulting in a surface tension gradient and causing the Marangoni flow. The flow moved outward from the centre and caused a rise at the edge of the weld pool. The results showed that the combination of capillary and hot capillary was more effective than the single use, and polishing on Ti-6Al-4V samples, achieving a surface roughness reduction of up to 72%. Hafiz, Abdullah M. Khalid [15] studied the applicability of picosecond lasers in laser micro-polishing of Ni-based superalloy Inconel 718 (IN718). The research contents include: determining the melting state, establishing the concept of polish ability of the initial surface topography space content, and experimenting with the potential micro-polishing process capability of the ultrashort picoseconds (ps) laser. During 1D laser micro-polishing, the ps pulsed laser polishing reduced the Ra value of the material Inconel 718 from 0.50 μm to 0.24 μm, which was reduced by about 52%. It significantly improved the surface quality of the workpiece.

Laser polishing is a new material surface treatment technology that has emerged with the development of laser technology. It uses a laser beam of a certain energy density and wavelength to radiate on the workpiece to melt or evaporate a thin layer of material on the surface to obtain a smooth surface. Laser polishing solves the problems of conventional polishing methods. For surfaces with very complex topography that are difficult or impossible to polish, laser polishing enables rapid polishing of these materials. Through the numerical control system, the laser polishing liquid realizes the automatic processing of the workpiece. Therefore, it is a very promising new material processing technology [8].

2. Laser polishing mechanism

When the laser acts on the surface of the workpiece, due to the different wavelengths and absorption modes of the laser, or the chemical bond vibration of the surface material of the workpiece is intensified, the laser energy is absorbed, converted into heat, the surface melts or evaporates, or the chemical bond of the surface material lattice is broken. The surface material is removed [9]. Laser polishing is performed by the above process, through reasonable selection of processing parameters to obtain a smoother surface. Although the principle of laser polishing is not conclusive, most researchers believe that laser polishing of metallic materials is based on laser irradiation, which causes a thin layer of the surface of the workpiece to melt. Due to the influence of surface tension, the convex portion will be flattened. After the molten surface solidifies, the overall roughness of the surface decreases, and the molten pool continues to move forward as the beam moves, and the above process is repeated, as shown in Figure 2.

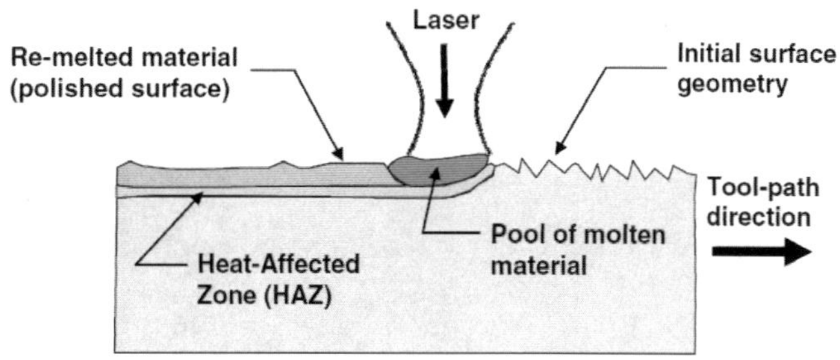

Figure.2 Laser polishing principle

The laser polishing process is essentially the interaction between the laser and the material surface, which follows the general rule of laser and material action. The interaction between laser energy and the area being processed can be divided into two main types of processes: thermal effect and photochemical decomposition. Depending on the laser and material interaction process, laser polishing can be easily divided into two types: one for thermal polishing and the other for cold polishing.

Thermal polishing generally uses a continuous long wavelength laser as a laser source. For example: a YAG laser with a wavelength of 1.06 um and a CO_2 laser with a wavelength of 10.6 um. When a continuous long-wavelength laser beam is irradiated onto the surface of the material, the surface is heated by absorbing laser photon energy. When the temperature of the active region reaches the melting point of the material to be polished, the surface active region of the material begins to melt and evaporate. Since the surface melting portion has different curvature radii, the material in the molten state flows toward a place where the curvature is small, and finally the curvature tends to be uniform everywhere. At the same time, the solid-liquid interface of the surface material solidifies at a speed of several meters per second, and the final metal material obtains a smooth and smooth polished surface [10-12].

The cold action of the laser and the material is photochemical decomposition, also known as photochemical action. It refers to the action of the surface of the material after laser illumination, because the surface of the "single photon absorption" or "multiphoton absorption" laser photon energy, the chemical bond of the surface layer of the material is destroyed, or its lattice structure is damaged. After photochemical decomposition, the resulting product expands at an active rate, eventually exploding itself and detaching from the area, while taking away excess energy. The process of action is also known as photolysis stripping [13]. The polishing process is performed by removing the material of the surface layer of the workpiece by the above-mentioned photo-disbonding action, which is called laser cold-polishing.

3. Experimental study

This experiment was carried out on a self-developed five-axis laser polishing machine. As shown in Figure 3, the selected material for polishing was S136D mold steel, as shown in Figure 4. In this experiment, a continuous laser is used to process the surface of the mold steel, and the continuous laser is polished according to a given numerical control program. Under the control of the CNC motion control system, the continuous laser can achieve high-precision polishing of the surface of the die steel S136D. Continuous laser polishing of the surface of the mold steel, the processing parameters set: 1000w continuous laser, power: 210W, polishing speed is V = 160mm / s, scanning spacing: 0.15mm. The initial surface roughness of the mold steel before polishing was 7.245 um, as shown in Figure 5. After continuous laser polishing, the surface roughness Ra of the material was 0.782 um, and the surface roughness of the workpiece was reduced by 89.2%, as shown in Figure 6. There is a lot of undulation on the surface of the original mold steel S136D. The principle of continuous laser polishing is that the laser beam irradiates the surface of the workpiece, so that the thin surface of the metal surface is remelted and then solidified rapidly. During this process, the surface of the remelted surface is convex, and the surface tension is Redistribution under the influence, so that the surface of the mold steel is flat, which further demonstrates that continuous laser polishing has the effect of reducing the surface roughness of the mold.

Figure.3 Laser polishing machine

Figure.4 Die steel before and after laser polishing

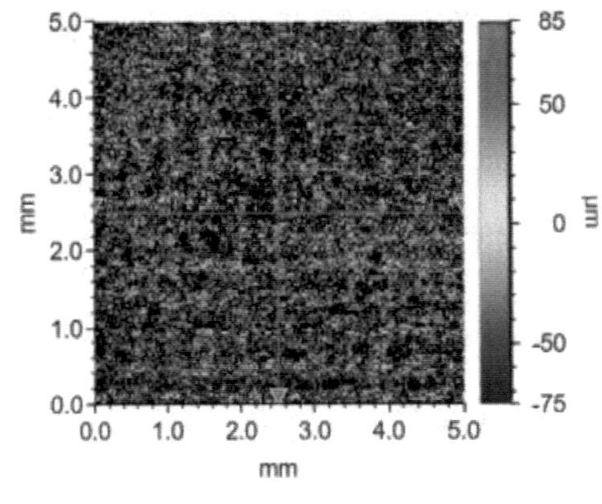

Analytical Results

Labe	Value	Units
Sa	7.245	μm
Sku	3.319	
Sp	76.813	μm
Sq	9.085	μm
Ssk	-0.071	
Sv	-68.645	μm
Sz	145.458	μm

Figure.5 Surface roughness before laser polishing

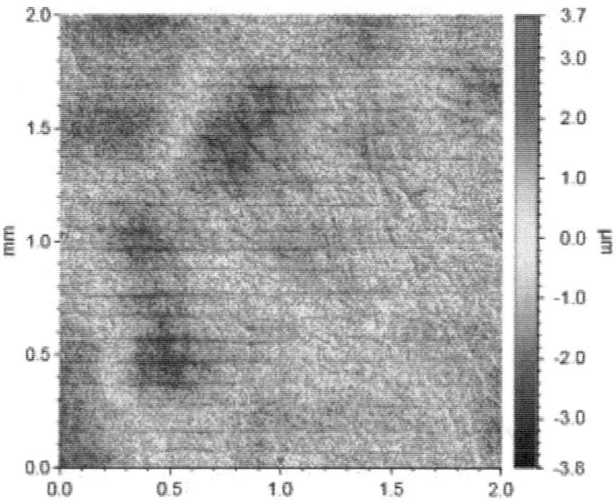

Analytical Results

Label	Value	Units
Average	0	nm
Data Points	4187291.904	
Percent Data Points	99.94	%
Ra	0.782	µm
Rp	3.714	µm
Rq	0.958	µm
Rt	7.549	µm
Rv	-3.835	µm

Fig.6 Surface roughness after laser polishing

4. Conclusion

Traditional polishing methods have certain limitations, and laser polishing technology is a new type of surface processing technology. Because laser polishing has great advantages, it is widely used in high-precision polishing of materials such as die steel and titanium alloy. In the paper, the continuous laser is used to process the die steel S136D, and the polishing effect is obtained. The surface roughness is reduced by 89.2%. The example further demonstrates that laser polishing can significantly reduce the surface roughness of the material.

Acknowledgement

Funding: The research was funded by Shenzhen Science and Technology Plan (No. JCYJ20170817112440533, No. GGFW2017041209483817 and No. GJHZ20180416164715805), Guangdong University Innovation Team (No. 2018GKCXTD001), Guangdong Pearl River Scholars Funded Scheme 2018, Research Funds of Shenzhen Institute of Information Technology (No. ZY201708), and School-enterprise cooperation project (SZIIT2019KJ015).

References

[1] Ji RL2010 *Mould Industry*. **36** 71
[2] Jia NZ 2011 *Mold manufacturing*. **5** 85
[3] Temmler A, Willenborg E and Wissenbach K 2012 *LAMOM*. **7** 1
[4] Ramos J A, Bourell D L 2002 *Rapid Prototyping of Materials*. **36** 191
[5] Lamikiz A, Sanchez J A and Lopez L N 2006 *Material Science Forum*. **526** 217
[6] Ukar E, Lamikiz A and De Lacalle 2010 *International Journal of Machine Tools and Manufacture*. **50** 115
[7] Hafiz A M K, Bordatchev E V 2012 *Journal of Manufacturing Processes*. **14** 425
[8] Jiang C, Wang Y Q and Hu S L 2002 *Laser Technology*. **26** 421
[9] Laguarta and Ferran L 1997 *Optical Soc of America*. **56** 352
[10] Yukon H, Cheng H L and Xiao H 2009 *Microsystems Technol*. **15** 1045
[11] Chang C S, Chung C K and Lin J F 2016 *Journal of Materials Processing Technology*. **234** 177
[12] Zhou Y Q, Zhao Z Y and Zhang W 2019 *Coatings*. **9** 324
[13] Zhao FH, Wang QY and Ren FH 1994 *Journal of Natural Science of Jilin University*. **2** 79
[14] Pfefferkorn F E, Duffie N A, Morrow J D 2014 *CIRP Annals-Manufacturing Technology*. **63** 237
[15] Hafiz A M K, Bordatchev E V 2014 *The International Journal of Advanced Manufacturing Technology*. **70** 1963

IOP Conf. Series: Materials Science and Engineering **758** (2020) 012038 doi:10.1088/1757-899X/758/1/012038

Improvement of physical, mechanical and strength behavior of cohesive soils with natural pozzolana and brick dust

E Chang[1], J Villalta[1], C Fernandez[2], G Duran[2]

1 Bachelor, Civil Engineering Program, Universidad Peruana de Ciencias Aplicadas, Lima, Peru
2 Full Professor, Civil Engineering Program, Universidad Peruana de Ciencias Aplicadas, Lima, Peru. Email: gary.duran@upc.edu.pe

Abstract: This research project seeks to improve soil properties through experimentation with geotechnical purposes. For this, will be used natural volcanic pozzolana in 5%, 10%, 15% and brick dust in 10% giving it a second reuse. The soil improvement will be analyzed with the proposed additions and its influence on the results. It is concluded that the addition improves the behavior of the soil by decreasing its plasticity index, increases the compaction index and improves the geotechnical parameters.

1. Introduction

Nowadays there are many methods for the improvement and stabilization of soils, whose objectives are aimed at improving the performance and behavior of the properties. These methods are related to the improvement of physical and chemical mechanisms, which are widely related to the various fields of civil engineering. According to R. J. Gallardo [1], they investigated about the behavior of high plasticity soils with the addition of residue material in brick making. The addition of 10% brick significantly decreased the plasticity index and the optimum humidity was reduced by 14.29%. The CBR value only increased by 3%. On the other hand, S. Aref al and H. Ibrahim [2] carried out their research in order to know the effect of the addition of natural pozzolana on clayey soils stabilized with lime. They used a clay of high plasticity (CH), consisting of kaolinite, montmorillonite and ilite with percentages of 23%, 7% and 14% respectively, with 95% of its passant particles of the mesh No. 200, plasticity index was 28.8% and CBR of 2.89%. They obtained a decrease in the plasticity index by adding the natural pozzolana, moving from a clayey soil CH to one of type ML. The CBR increased by 90% compared to the soil with lime without pozzolana.

This article presents the experimental evaluation of the effect on the physical and mechanical properties of a clayey soil improved with natural pozzolana and brick dust.

2. Soil

The geology that characterizes the samples extraction zone is constituted by red silty clays interspersed with white sandstones and gray limestones. According to the SUCS, it is a low plasticity clay (CL) that has 70% of passant material of the mesh No. 200.

3. Natural pozzolana

Natural pozzolans come from volcanic eruptive rocks such as rhyolites, traquites, andesites and basalts. Natural pozzolans mainly consist of a vitreous mass that cements fragments of pumice, slags, small crystals of augite, mica, pyroxenes, etc. [3].

The extracted natural pozzolan has the chemical composition shown in Table 1. Due to the high contents of silica (SiO_2) and alumina (Al_2O_3), it is sought that they react resulting in the formation of crystals that when occupying the voids of the soil increase the shear strength of itself [4]. It is worth mentioning that natural pozzolana has lime which enhances the chemical reaction.

Table 1. Chemical composition of natural pozzolana (Ingemet 2016)

Chemical element	Al_2O_3	CaO	Fe_2O_3	K_2O	MgO	MnO	NaO	P_2O_5	SiO_2	TiO_2	LOI*
Percentage (%)	14.26	1.29	1.34	4.8	0.77	0.08	2.47	0.05	60.03	0.18	6.14

4. Brick dust

The brick dust is classified as an artificial pozzolana, for this investigation the material was collected from the residue of the distributor trucks of craft bricks.

5. Experimental methodology

For the analysis of physical and mechanical properties, granulometric distribution tests were performed. Limits of Atterberg, Modified Proctor and shear strength. Likewise, mixtures with percentages of 5%, 10%, 15% natural pozzolana (PN) and 10% brick dust addition (PL) were prepared. Table 2 presents the symbology of the mixtures and the clayey soil.

Table 2. Symbology of the mixtures

MATERIAL	ABBREVIATION
Soil	S100%
Pozzolana and brick dust	S+PN5%+PL10%
	S+PN10%+PL10%
	S+PN15%+PL10%

6. Analysis and results

6.1. Atterberg limits

Table 3 presents the results obtained. The plasticity index is reduced to a greater amount of natural pozzolana.

Table 3. Liquid limit, plastic limit and plasticity index

TESTS	LL	LP	IP
S100%	36.3	18.9	17.4
S+PN5%+PL10%	34.7	22.18	12.52
S+PN10%+PL10%	27.7	21.35	6.35
S+PN15%+PL10%	29.22	22.885	6.335

6.2. Modified Proctor

The maximum dry density of the soil and the optimum moisture content were determined. According to Figure 1, it is observed from the compaction curves that when the percentage of natural pozzolana increases the dry density decreases and the optimum humidity content increases.

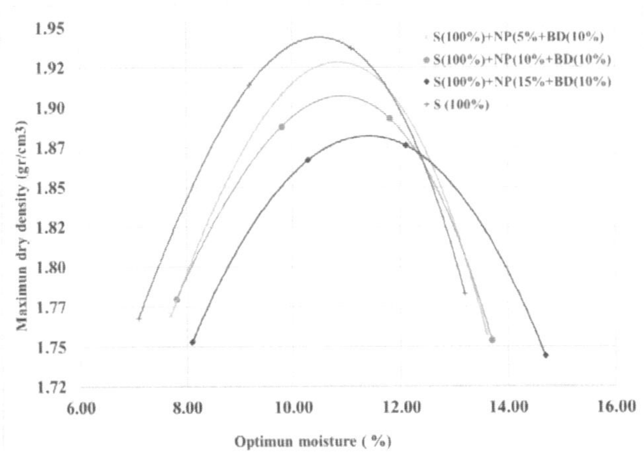

Figure 1: Compaction curves.

6.3. Shear Strength

The samples tested were prepared with the optimum humidity of the Modified Proctor, were cured for seven days in an airtight container, Table 4 presents the strength parameters obtained for the clayey soil and mixtures. The mixture S + PN5% + PL10% has the highest cohesion value (1.15 kg/cm2) with respect to the clayey soil and the other mixtures. Although the friction angle of S + PN5% + PL10% decreases with respect to the clayey soil, the shear strength is higher for the confinement levels analyzed. The higher the level of confinement, the shear strength of the clayey soil and the mixture with greater cohesion are closer, establishing a clear influence of the confinement efforts on the shear strength of the clayey soil improved with the natural pozzolana and the brick dust.

Table 4. Friction angle and cohesion

Description	Internal friction angle (Φ)	Cohesion (c´) kg/cm^2
S100%	31.5	0.51
S+PN5%+PL10%	22.2	1.15
S+PN10%+PL10%	26.7	0.92
S+PN15%+PL10%	31	0.69

6.3.1. Dilatant behavior

In the graphs of Figures 2, 3, 4 and 5 show the behavior of the shear effort and of the dilatation with respect to the horizontal displacement during the direct cutting test performed on the clayey soil and the mixtures S + PN5% + PL10%, S + PN10% + PL10% and S + PN15% + PL10%, respectively. In all specimens, a contractive behavior can be seen in the initial part of the test, together with an increase in the shear effort necessary to achieve horizontal displacement at constant speed. The contractive behavior changes towards a dilating behavior before reaching 1 mm of horizontal displacement, this for confinement efforts 0.5 and 1.0 kg / cm2. On the other hand, for the 2.0 kg / cm2 confinement effort, the dilating behavior occurs after 1 mm of horizontal displacement. The growth ratio, of the shear strength, decreases when the 2 mm horizontal displacement is reached coinciding with a purely dilating behavior. All the mixtures have a greater shear strength than the clayey soil, with the difference that the level of

IOP Conf. Series: Materials Science and Engineering **758** (2020) 012038 doi:10.1088/1757-899X/758/1/012038

dilatation of the sample S + PN5% + PL10% is lower than that of the other mixtures, may be the explanation that it reaches greater shear strength.

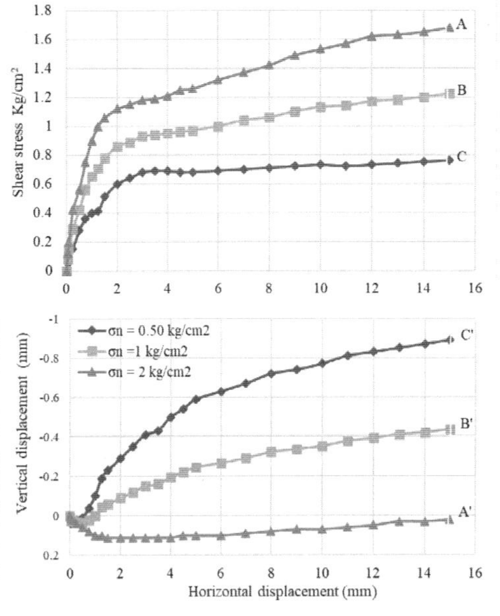

Figure 2. Results of shear stress for clayey soil (S100%)

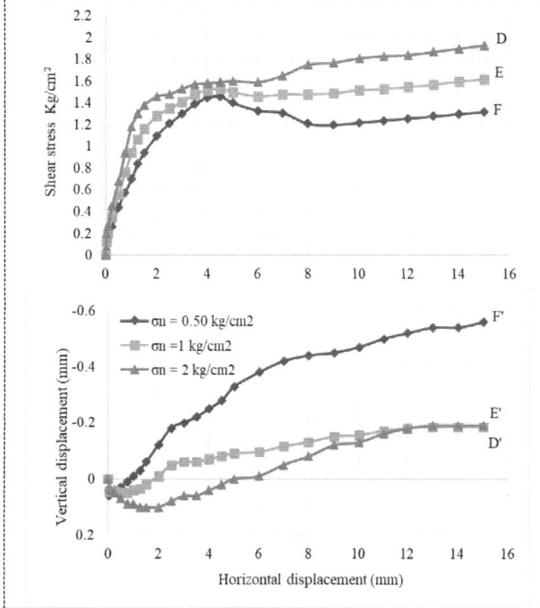

Figure 3. Results of shear stress for (S+NP5%+BD10%)

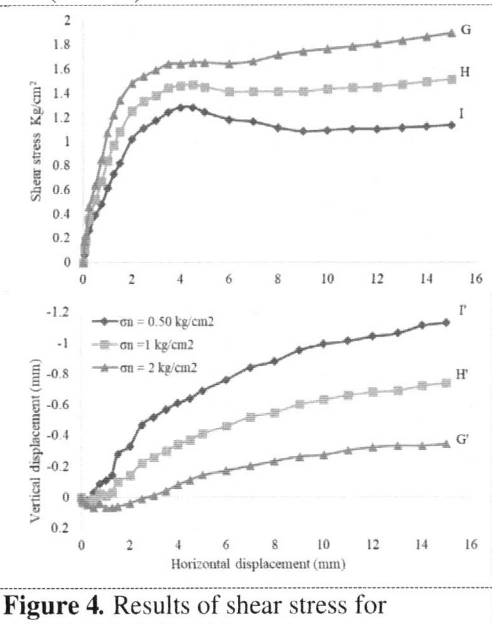

Figure 4. Results of shear stress for (S+NP10%+BD10%)

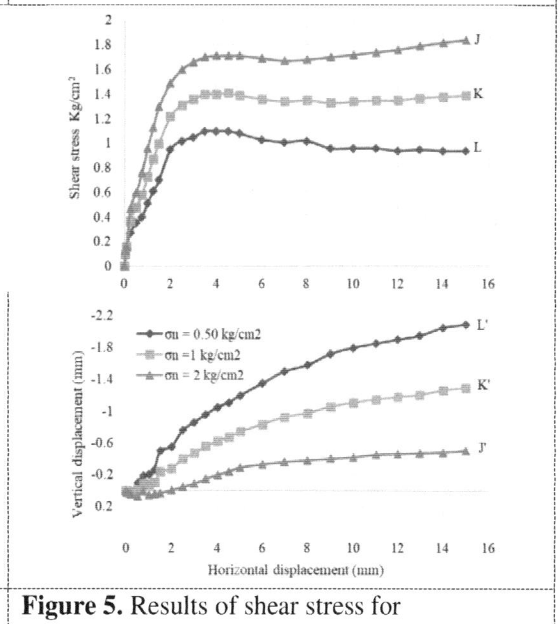

Figure 5. Results of shear stress for (S+NP15%+BD10%)

6.3.2. Cohesion and friction angle

Figures 6 and 7 confirm the variation of resistance parameters as a function of horizontal displacement level. Both the cohesion and the friction angle have a constant value trend after 8 mm of displacement, coinciding with a constant dilatation ratio.

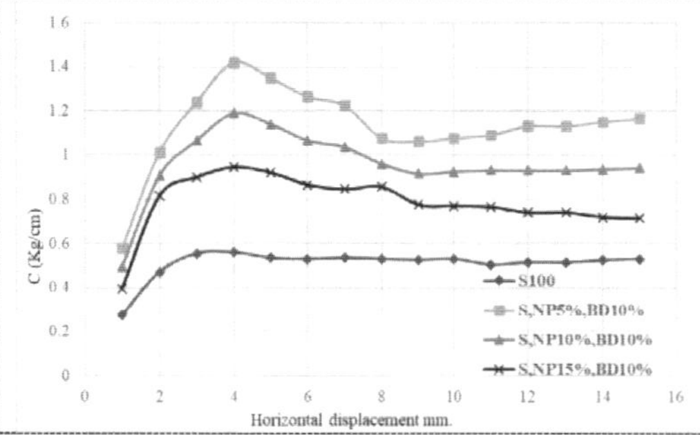

Figure 6. Cohesion variation at every 1 mm of horizontal displacement.

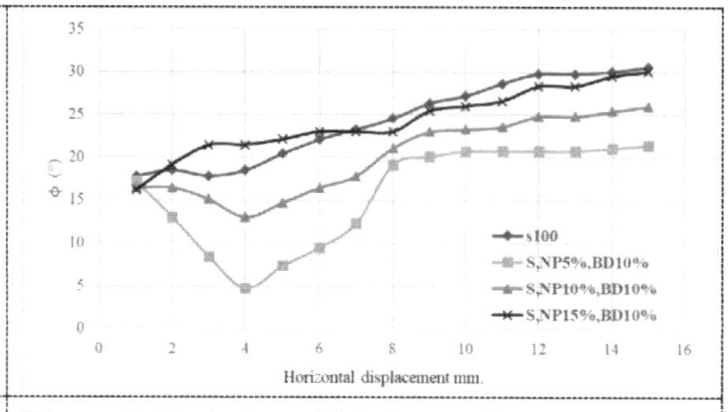

Figure 7. Variation of friction angle at every 1mm of horizontal displacement.

7. Conclusions

- The maximum dry density of the soil decreases with the increase of the natural pozzolana content. The lowest dry density was 1.88g / cm3 with the addition S + PN15% + PL10%.
- All mixtures have greater shear strength than clayey soil, being that for confinement levels greater than 2.0 kg / cm2, the difference decreases.
- The sample called S + PN5% + PL10% shows the best performance compared to clayey soil and the other mixtures. This sample also dilates less than the samples S + PN10% + PL10% and S + PN15% + PL10%, this may be the reason for reaching higher values of shear strength.
- The cohesion and friction angle vary depending on the level of horizontal displacement, being that for 4 mm these resistance parameters reach maximum and minimum values, respectively.
- Cohesion and friction angle have a tendency not to vary from 8 mm of horizontal displacement, coinciding with a constant dilatation ratio.

8. References

[1] R. J. Gallardo Amaya, O. A. Cuanalo Campos, L. J. Quintero Lemus y A. A. Muñoz, «*Análisis del comportamiento de suelos de alta plasticidad con la adición del material de residuo en la fabricación de ladrillo cerámico*,» Revista Colombiana de Tecnologías de avanzada, vol. 1, nº 29, pp. 105-110, 2017.

[2] S. Aref al y H. Ibrahim, «Effect of adding natural pozzolana on geotechnical properties of,» Journal of Rock Mechanics and, vol. 8, pp. 714-725, 2016.

[3] INGEMMET, «*Estudio de Recursos de Rocas y Minerales Industriales para la inclusion Económica Social y Desarrollo en la Región Huancavelica.*,» INGEMMET, Boletín Serie D: Estudios Regionales, nº 31, pp. 29-31, 2016.

[4]Ghrici M., Kenai S. y Said Mansour M., «Mechanical properties and durability of mortar and concrete containing natural pozzolana and limestone blended cements,» Cement and Concrete Composites , vol. 29, pp. 542-549.

IOP Conf. Series: Materials Science and Engineering **758** (2020) 012039 doi:10.1088/1757-899X/758/1/012039

Use of recycled broken bricks as *Partial Replacement Coarse Aggregate* for the Manufacturing of Sustainable Concrete

S Pinchi[1], J Ramírez[1], J Rodríguez[2] and C Eyzaguirre[2]

1 Bachelor, Civil Engineering Program, Universidad Peruana de Ciencias Aplicadas, Lima, Perú
2 Full Professor, Civil Engineering Program, Universidad Peruana de Ciencias Aplicadas, Lima, Perú

Abstract. The bricks are one of the primary materials required for construction of homes that no used completely when executes all the walls due, the excess purchase, the cutting to be settle, the breaking for their transfer and its fixed dimensions; this situation requires monitoring on work site the order, cleanliness and accidents. A common practice is these bricks and/or waste are included in the clearing construction before being deposited or eliminated in dumps or sanitary landfills, with their early clogging and shortening them to ther design lifespan. An important alternative to reduce this waste, is to recycle them and reuse them as a concrete component material, due to their high absorption percentage that allows them to keep the water inside of them and then use it in the cement hydration process as internal curing of the concrete. In the present investigation, the effect of crushed clay brick as a replacement for coarse aggregate in concrete processing is studied. The results indicate that with 21 % replacement brick, the plastic contraction decreases, and the compressive strength and flexural strength increase.

1. Introduction

Nowadays one of the main concerns around the world is environmental pollution, where the construction industry is the one that produces a high impact on the atmosphere and natural resources, due to the high consumption of raw materials, being one of the main waste sources of building materials [1]. This waste, in the case of brick in the construction of multifamily buildings, is estimated in Britain in 380000 tons and in the United States in 120000 tons [2]. Studies by [3] in Hong Kong shows that the waste brick unused in the walls and masonry is between 4-8%. As a mitigation measure, it is proposed to associate the use of brick waste in the elaboration of a lightweight aggregate to later form part of the mixture, and thereby promote the sustainability of concrete. Lightweight aggregates are generally porous and can absorb a considerable amount of water when immersed in it [4]. The clay brick has a high water absorption capacity due to its capillarity, which is directly related to its internal pores that make it variable depending on its quantity, size and shape; being this, an important physical property of the brick to be taken into account in the internal curing of the concrete. [5] performs tests with crushed brick as coarse aggregate and finds that this material has an absorption percentage of 9.90%. In this sence, here, [6] proposes the use of crushed brick for its effectiveness as a means of internal curing, based on their experiments suggest using 20% of crushed brick as replacement thick aggregate to achieve a resistance to the compression greater. Some researchers have experimented with lightweight aggregates and [5] evaluates different concrete mixtures with 10%, 20, 30% replacement of broken brick and obtains a decrease from 2% - 6% of the resistance the compression and concluded that their results are similar to control concrete and that it is feasible to use the broken brick. Regarding the flexural strength, it shows a decrease in resistance between 0% - 2%, which represents a behaviour similar to the control concrete. Similarly, Rajamanickam [7] studies light expanded clay (LECA) as a replacement for natural aggregate, finding that the compressive strength of concrete with LECA is less than 7 days than control concrete. In the case of 28 days, with a 15% replacement of LECA, it is obtained that the compressive strength is 1.84% more than the control concrete due to the presence of an adequate moisture content in the aggregate; for a percentage greater than 15% the resistance decreases due to the weaker nature or the very pronounced water absorption characteristics of the aggregate. In this paper we study the influence of crushed brick clay as partial replacement of coarse aggregate in making concrete by plastic shrinkage, the compressive strength and flexural strength.

2. Materials y methods

2.1. Materials
The crushed brick used has a Ø 3/4", the cement is the Portland Type I brand Sol, the aggregates used are coarse sand and natural stone of Ø 3/4", the water was drinkable.

2.2. Method
The mix design was made for a particular f'c= 280 Kg/cm2 according to the Fineness Modulus of the Combination of Aggregates [8], the brick ground is saturated for 24 hours. Table 1 shows the designs of the mixtures carried out and Table 2 shows the tests performed and the number of samples used.

Table 1. Dosage for concrete designs in a cubic meter.

Components	Control	Design 1 brick 15%	Design 2 brick 21%	Design 3 brick 27%
a/c ratio	0.50	0.50	0.50	0.50
Water	178.30	193.10	193.10	193.10
Cement	390.00	390.00	390.00	390.00
Coarse aggregate	918.10	780.40	725.30	670.21
Fine aggregate	750.60	750.60	750.60	750.60
Crushed brick	0.00	113.40	192.80	247.88

Table 2. Tests performed

Essays	Applicable standard	N°. tests	Cure age (days)
Plastic shrinkage	ASTM C 1579 [9]	3	1
Compressive strength	ASTM C39 [10]	9	7, 21, 28
Flexural strength	ASTMC 78 [11]	3	28

3. Results and Analysis

3.1. Plastic shrinkage
In the Table 3 the influence of crack initiation time of crack width for design 2 and control mixture shown, and the reduction ratio of fissure (CRR). In it you can see that as the drying time increases the crack width and CRR grow, reaching for Design 2 the maximum value of 0.508 mm, which represents 38% less with respect to the control mixture for a time of 1440 min; achieving for this maximum value a CRR ratio of 37.50%. [12], [13] relate this behaviour to the degree of saturation of the aggregate,

IOP Conf. Series: Materials Science and Engineering **758** (2020) 012039 doi:10.1088/1757-899X/758/1/012039

which in the case of recycled aggregates for the condition of superficially dry saturates, the crack start time and the crack width increase.

Table 3. Crack width, drying time and CRR factor (Cracking reduction ratio)

Fissure initiation time (min)	Crack Width (mm)		
	Control	Design 2	CRR (%)
0	0.00	0.00	0.00
30	0.00	0.00	0.00
60	0.058	0.058	0.00
90	0.2932	0.1524	25.00
120	0.3048	0.2540	16.67
150	0.4572	0.3556	22.22
180	0.6604	0.4064	38.46
1440	0.8128	0.508	37.50

3.2. Compressive strength

The influence of the test age on the compressive strength of concrete for 0%, 15%, 21% and 27% of crushed brick is shown in Figure 1. In it can see that as the test age varies, the compressive strength increases, this being maximum for 21% of crushed brick at the age of 28 days, reaching the value of 355.55 kg/cm2, which represents a 4.07 % more with respect to the control sample. [6] study the clay splinter with replacements of 10%, 20% and 25%, obtaining that the specimens cured 7 days in water and 21 days in the laboratory without plastic coverage for a 20% partial replacement reach a resistance of 25% more than the control mixture, then decreases for a value of 30% partial replacement. [2] they study the clay splinter with replacements of 10%, 20% and 25%, obtaining that the specimens cured for 28 days in the laboratory without plastic coverage and with a 20% partial replacement reach a resistance greater than 4.6% with respect to the control mixture, from there the resistance drops for a 30% partial replacement. These growth trends are like those obtained in the study [14] indicates that this behaviour is due to the continuous hydration of the mixture at older ages, promoted by the extra water stored in the light aggregate.

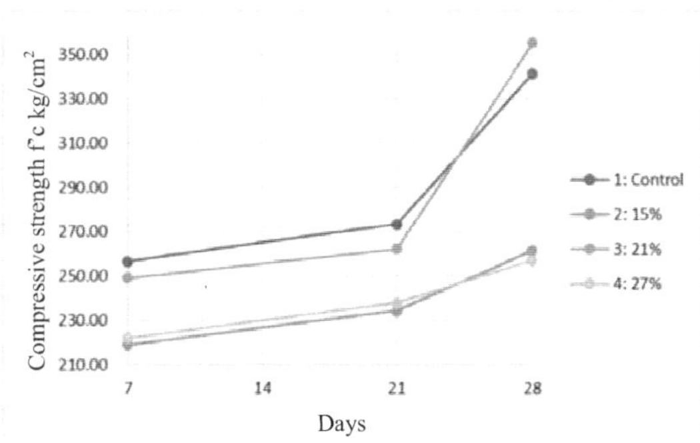

Figure 1. Influence of the test age on crushed brick compressive strength

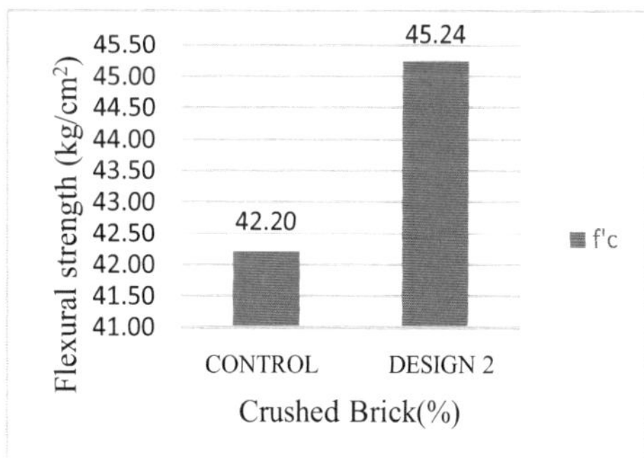

Figure 2. Influence of the percentage of on the resistance to bending

Table 4. Unit price analysis of control and design mix 2 per m^3

Materials	Unit	Control design			Design 2		
		Metrado	U.P.	Partial	Metrado	P.U.	Partial
Water	m^3	0.20	2.83	0.57	0.20	2.83	0.57
Cement	bls	9.18	22.20	203.80	9.18	22.20	203.80
Coarse aggregate	kg	884.10	0.14	123.77	725.30	0.14	101.54
Crushed brick	kg	0.00	0.00	0.00	158.80	0.23	36.52
Fine aggregate	kg	750.60	0.15	112.59	750.60	0.15	112.59
Mix workforce	hh	2.00	15.00	30.00	3.00	15.00	45.00
Trompo (70 lt)	hm	1.00	4.02	4.02	1.00	4.02	4.02
Workforce (cured)	hh	0.50	15.00	7.50	0.00	0.00	0.00
Water (cured)	m^2	1.00	2.83	2.83	0.00	0.00	0.00
Maintenance crack	m^3	1.00	76.00	76.00	0.00	0.00	0.00
TOTAL				561.08			504.04

3.3. Flexural strength
Figure 2 shows the influence of the percentage of crushed brick on the flexural strength at 28 days. In it you can see that with design 2 a value of 45.24 Kg/cm2 is reached, this being 7% higher than the control mixture. [15] studies the brick with recycled fine aggregate and finds that with 55% recycled fine aggregate 17.9% more is obtained than the control brick.

3.4. Cost evaluation
Table 4 shows the unit price analysis of the control mix and design 2 for 1 m^3 of concrete. As it can be seen, the variation in cost for design 2 is S/. 57.04 ($ 17.04 US dollars) of lower cost, which represents 10.17%, so design 2 is a cheaper alternative as a replacement for coarse aggregate with crushed brick. [2] they concluded that using partial replacement of coarse aggregate by saturated clay chips, is achieved decreases go using water and coarse aggregate, causing the unit price of concrete is reduced and is a good cost - effective alternative.

4. Conclusions
- The mixture with 21% crushed clay brick as a replacement for coarse aggregate has an average resistance of 5% and a lower cost that makes it possible to prepare a sustainable concrete.
- With the processed concrete, a reduction of cracks of 47% of the control sample is achieved because the good absorption capacity of the clay brick facilitates self-curing.
- The reuse of clay bricks from homes, and / or their waste, in the preparation of concrete contribute to the improvement of cement hydration.

5. References

[1] Khairulzan, Y., & Halim, B. (2010). Quantifying Environmental Impacts and Eco-costs from Brick Waste. Architectural Engineering and Design Management, 189-206.
[2] Iffat, S., Manzur, T., & Noor, M. A. (2016). Durability Performance of Internally Cured Concrete Using Locally Available Low Cost LWA. KSCE Journal of civil Engineering , 1256-1263.
[3] Tam, V., Tam, C., & Ho, S. (2000). MATERIAL WASTAGE IN CONSTRUCTION ACTIVITIES – A HONG KONG SURVEY. Department of Buildin g & Real Estate, The Hong Kong Polytechnic University, 125-131.
[4] Zhang, J., Hou, D., & Sun, W. (2010). Experimental study on the relationship between shrinkage and interior humidity of concrete at early age. Mag. Concrete Res., 191-199.

[5] Perez Rojas, A. V. (2012). Uso de triturado de ladrillo reciclado como agregado grueso en la elaboracion de concreto. Revista Ingenium, 116-125
[6] Iffat, S., Tanvir, M., Sayedur, R., Ahmed, N. M., & Nur, Y. (2017). Optimum Proportion of Masonry Chip Aggregate for Internally. International Journal of Concrete Structures and Materials, 513-524.
[7] Rajamanickam, G., Vaiyapuri, R. (2016). Sel compacting self curing concrete with lightweight aggregates. Gradevinar 68, 279-285.
[8] Torre, Ana (2004). *Basic course of technology of concrete.* Lima, Perú: Universidad Nacional de Ingenieria.
[9] ASTM Designation: C 1579 "Standard Test Method for Evaluating Plastic Shrinkage Cracking of Restrained Fiber Reinforced Concrete". 2013
[10] ASTM Designation: C 39 "Standardized Test Method for Compressive Strength of Concrete Cylindrical Specimens".
[11] ASTM Designation: C 78 "Standard Test Method for Flexural Strength of Concrete (Using Simple Beam with Third-Point Loading)".
[12] Salgue, M., Souche, J.-C., Devillers, P., & Garcia-Diaz, E. (2016). Influence of initial saturation degree of recycled aggregates on fresh cement paste characteristics: consequences on recycled concrete properties. European Journal of Environmental and Civil Engineering.
[13] Souchea, J.-C., Bendimeradb, A. Z., Roziereb, E., Salguesa, M., Devillersc, P., Garcia-Diaza, E., & Loukilib, A. (2017). Early age behaviour of recycled concrete aggregates under normal and severe drying conditions. Journal of Building Engineering, 244-253.
[14] Lura, P. (2003). Autogenous Deformation and Internal Curing of Concrete (Doctoral thesis). Delft University of Technology
[15] Pinchi S., Ramirez J. (2019) Proposal for the application of the self-curing method by adding crushed brick to the coarse aggregate to reduce surface cracks and increase the compressive strength of concrete in warm areas (North Lima).(Undergraduate thesis) Universidad Peruana de Ciencia Aplicadas
[16] Khalid, F. S., Herman, H. S., Azmi, N. B., & Juki, M. I. (2017). Sand Cement Brick Containing Recycled Concrete Aggregate as Fine-Aggregate Replacement. MATEC Web of Conferences.

IOP Conf. Series: Materials Science and Engineering **758** (2020) 012040 doi:10.1088/1757-899X/758/1/012040

Analysis of high plasticity clayey soil improvement at subgrade level through Portland cement added to decrease volumetric change

M Castro[1], J Navarro[1], G Aybar[2] and G Duran[2]

1 Bachelor, Civil Engineering Program, Universidad Peruana de Ciencias Aplicadas, Lima, Peru
2 Full Professor, Civil Engineering Program, Universidad Peruana de Ciencias Aplicadas, Lima, Peru. Email: gary.duran@upc.edu.pe

Abstract. This research includes the potential for resistance and the expansion that the soil presents, this evaluation was carried out through CBR tests. The soil cement technique was used to improve the physical and mechanical characteristics; this process consists in mixing the material with Portland cement type I. That combination forms soil cement 10%, 15% y 20%, which present an increase of the CBR (max: 138.7% and min: 91.9%) achieving a type of extraordinary subgrade to resist the structure of the pavement and a reduction of 7.18% in the expansion of the samples.

1. Introduction

In Peru, as in much of Latin America, there is a large deficit in road infrastructure, to a greater extent the absence of paved routes for regional connections is highlighted, reaching a total of 141,603.00 km of unpaved road network [1]. One of the most frequent problems, especially in road construction, is the presence of highly cohesive soils, which represents a challenge for civil and geotechnical engineers around the world. By improving the mechanical characteristics of these soils it is possible to enable an unpaved road.
Various investigations conducted in cohesive soils, such as high plasticity clays [3] [5] and soft clays [4] [6] [7] were studied by adding various additives, slag-based superfine cement, cement mixed with polypropylene fibers, lime and Portland cement. The improvements were seen with the increase in unconfined compression [3] [4] [5], the reduction in settlements [6] and the increase in CBR [7].

The previously mentioned studies provide an overview of the soil-cement technique to achieve increased shear strength and its importance for the improvement of subgrade or support soil. Under this background, the aforementioned technique is adopted to stabilize the clayey soil of high plasticity in a regional route of the Peruvian jungle, in order to resist the vehicular load and reduce the volumetric changes.

2. Materials and methods

Clayey soil of high plasticity (CH) extracted in unaltered samples from a regional road in the province of Oxapampa, Pasco, Peru was evaluated. Portland cement Type I was used.

IOP Conf. Series: Materials Science and Engineering **758** (2020) 012040 doi:10.1088/1757-899X/758/1/012040

Figure 1. Unaltered soil sample.

Figure 2. Sample cement soil 15%.

Figure 3. Sample cement soil 20%.

The tests carried out were: the limits of Atterberg, Compaction and CBR. The mixtures evaluated were those containing 10%, 15%, and 20% cement in relation to the dry weight of the soil. Figures 1, 2 and 3 present the natural soil in an unaltered state and the mixtures of 15 and 20%, respectively.

3. Results

The results of the programmed tests are shown in Tables 1, 2, 3, 4, 5 and 6, together with Figures 4, 5, 6, 7, 8 and 9.

3.1. Natural soil

Table 1. Characterization tests of natural soil.

	Atterberg limits				Classification		Modified Proctor	
	Liquid limit	Plastic limit	Plasticity index	Humidity content	SUCS	AASHTO	OCH	MDS (gr/cm3)
Natural soil	66.10%	26.90%	39.20%	29.37%	CH	A-7-6 (45)	14.30%	1.827

Table 2. Maximum dry density (MDS)

Penetration	2.54 mm (0.1")	5.08 mm (0.2")
100% MDS	3.30%	3.00%
95% MDS	2.60%	2.30%

Table 3. Natural soil expansion

N° Hits	Expansion	Average expansion
57	7.21%	
25	7.00%	8.12%
12	10.14%	

3.2. Cement soil

Table 4. Modified proctor test on cement soil.

	Modified Proctor	
Cement_soil	OCH	MDS (gr/cm3)
10%	14.60%	1.831
15%	15.00%	1.832
20%	15.20%	1.833

Table 5. CBR test on cement soil.

	Cement soil 10%		Cement soil 15%		Cement soil 20%	
Penetration	2.54 mm (0.1")	5.08 mm (0.2")	2.54 mm (0.1")	5.08 mm (0.2")	2.54 mm (0.1")	5.08 mm (0.2")
100% MDS	130.80%	105.80%	116.70%	111.80%	153.60%	146.30%
95% MDS	105.40%	91.90%	108.00%	104.70%	138.70%	129.30%

3.3. Compaction curves

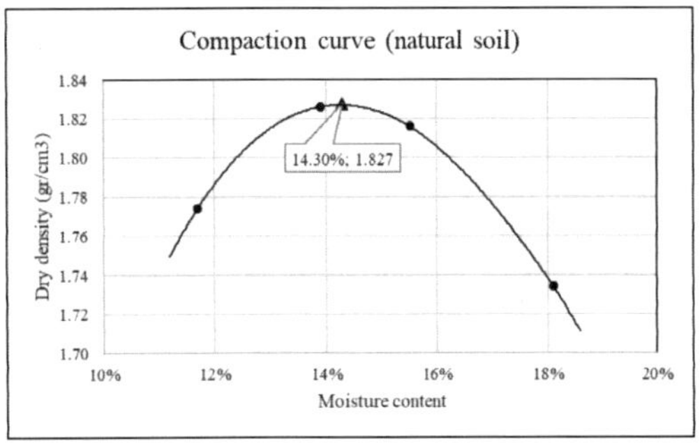

Figure 4. Natural soil compaction curve.

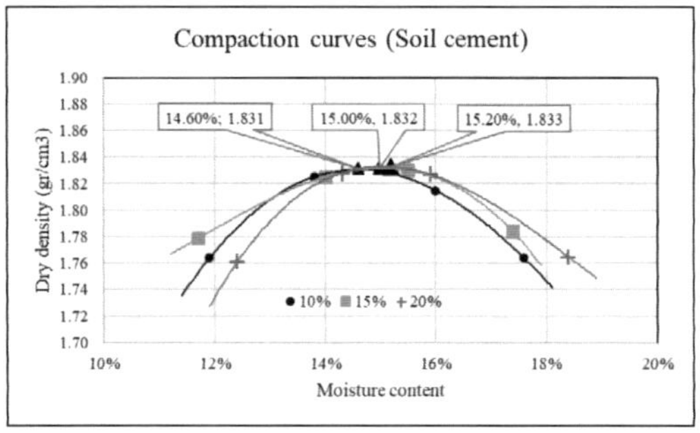

Figure 5. Soil cement compaction curves.

3.4. Comparative

3.4.1. Modified Proctor

The curve in Fig. 7 shows the behavior of the optimum humidity content (O.C.H.) according to the percentage of Portland cement in the cement soil mixture.

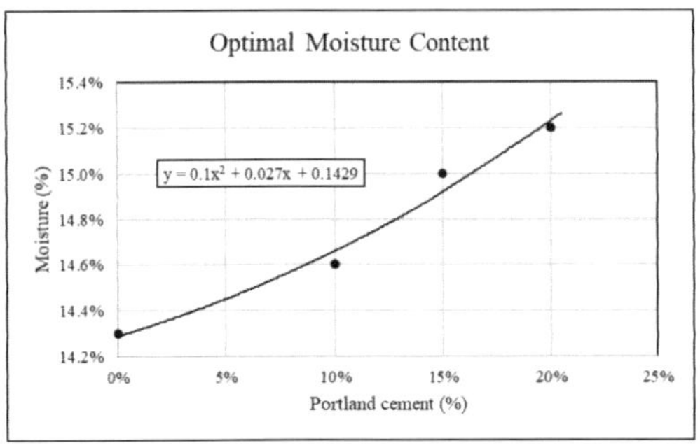

Figure 6. Portland cement v. Humidity.

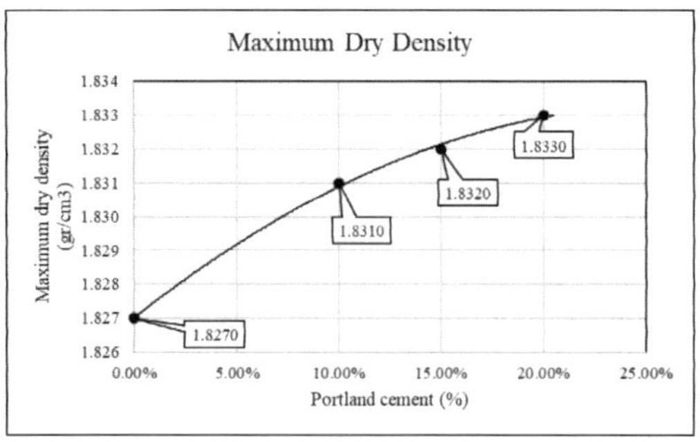

Figure 7. Portland cement v. MDS.

3.4.2. CBR

For the curve of Fig. 8, the CBR values corresponding to 95% of the MDS were related to the percentage of Portland cement added.

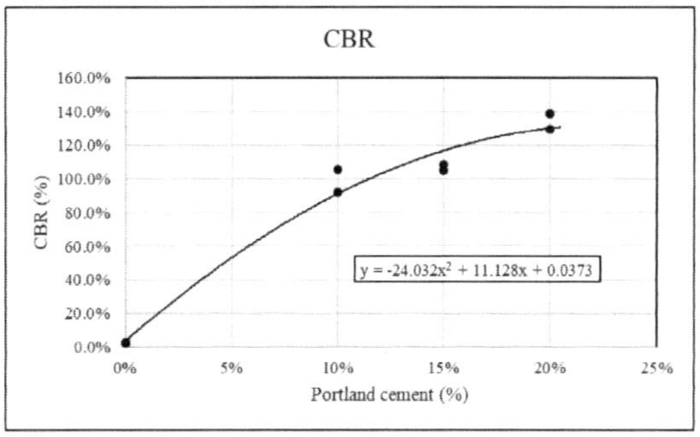

Figure 8. Portland cement v. CBR.

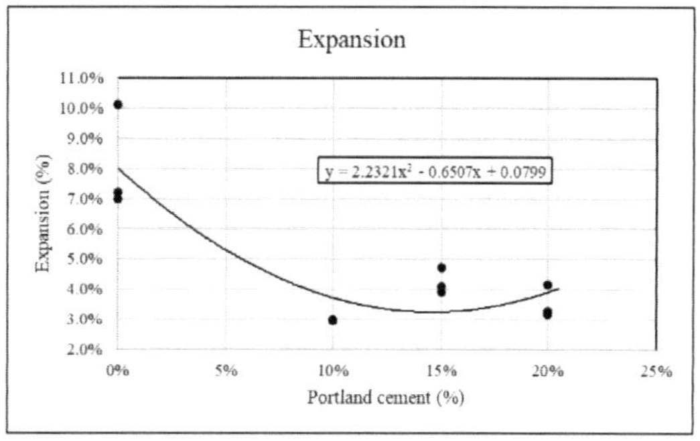

Figure 9. Portland cement v. Expansion.

4. Discussion

It is observed in Fig. 7 that the optimum humidity content varies almost linearly, presents a little magnitude change in the humidity value, the lowest value being 14.3% when the soil is in natural conditions, and reaches to 15.2% maximum when 20% cement is added to the soil. Therefore, a variation of 0.9% humidity between the samples is obtained and indicates that the greater the amount of cement, the greater the water requirement will be to achieve the maximum density of the sample. The additional humidity necessary to develop maximum consistency is due to the hydration of the cement, action necessary for the components react and forge occurs. The more cement is present in the cement soil, the greater the amount of water. The amount of water in the mixture is important to achieve greater resistance in the improved soil. Likewise, a comparison of the MDS in relation to the percentage of Portland cement added is shown in Fig. 8. As a result, it is observed that the value of the MDS remains almost unchanged as Portland cement increases to the mixture. On the other hand, Fig. 9 shows the variation of the CBR due to the percentage of cement added. A quadratic trend line with direct proportionality between the variables is presented. The more cement is added to the soil, the higher the CBR of the cement soil.

The natural soil being a clay of high plasticity has a low CBR value (2.3%) which, according to the above, does not meet the resistance necessary to support the pavement structure or the vehicle load. Portland cement together with the clayey soil achieves very high CBR values, from 91.9% with 10% cement to 129.3% with 20% cement, becoming characterized as extraordinary subgrade. However, there is a growth that is decelerating as the percentage of cement added to the clayey soil increases, that is, a limit of resistance contribution by the cement is reached. Finally, the variation of the expansion with respect to the percentage of cement added is shown in Fig. 10. This curve shows an inversely proportional relationship, where it can be seen that, the greater the amount of cement in the mixture, the lower the expansion of the material during the four days under saturated conditions. It is possible to reduce the expansion of the material up to 7.18% by adding Portland cement. However, it should be noted that the expansion of cement soil samples remain within the same range that is not related to the amount of cement, this range ranges from 2.96% to 4.72%, that is, the volumetric change that the clay presents decreases by the presence of Portland cement and not by the amount of it. In the same way that was exposed about the CBR limit that is reached by the cement soil, it is emphasized that the expansion reaches a limit in which the volumetric change remains relatively unchanged.

5. Conclusion

According to the results obtained from the compaction tests and CBR, the high plasticity clayey soil used when combined with Portland cement is improved.

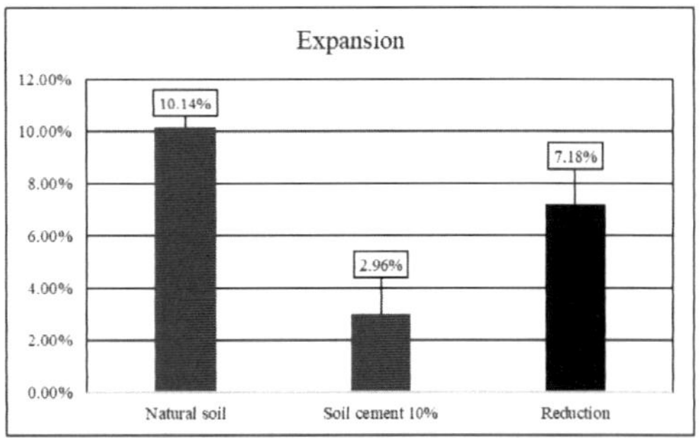

Figure 10. Samples expansion in CBR test.

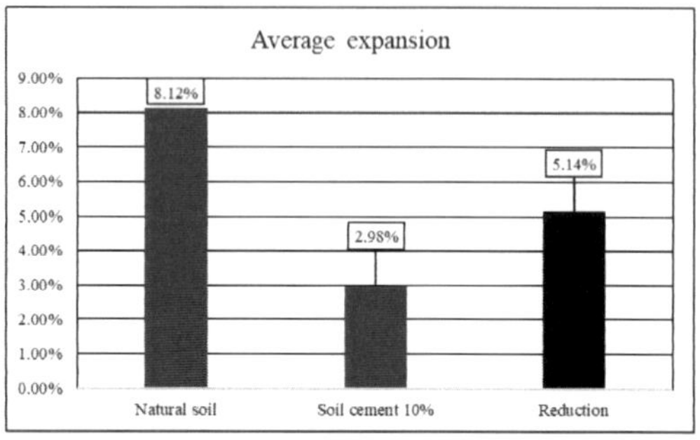

Figure 11. Average Samples expansion in CBR test.

The largest reduction in expansion is 7.18% and occurs when the mixture contains 10%. Likewise, the highest average of volumetric change reduction also occurs in cement soil 10%, reducing from a value of 8.12% expansion in natural soil to 2.98% in cement soil, so the variation is equivalent to 5.14%. The reduction of the volumetric change could be due to the Portland cement particles reacting with the water molecules and as a consequence the contact of said molecules with the clayey soil particles of high plasticity is avoided, causing the clayey part of the mixture to be less reactive The decrease of the expansion in clayey soils of high plasticity is of great importance to reduce settlement and swelling in the subgrade. By using the cement soil technique, the expansion is reduced, so that settlements and swellings in the road are avoided and as a consequence the condition of the pavement and the serviceability of it are preserved. The cement soil technique greatly increases the value of the CBR, achieving an extraordinary subgrade from the 10% cement soil. Finally, according to the tests carried out, it is observed that the behavior of the cement soil adopts a tendency to increase the value of the CBR by containing more cement. This parameter maintains a relationship with the volumetric change which indicates that, while the value of the CBR is greater, the expansion of the sample will be reduced and both the water requirement, expressed in the optimum humidity content, as the MDS will go increase.

6. References

[1] Ministry of Transportation and Communications - General Office of Planning and Budget 2015.

[2] Chindris L, et al. 2017 "Expansive Soil Stabilization - General Considerations" *Proceedings of the International Multidisciplinary Scientific GeoConference SGEM* 17 (3–2) pp 247–55.

[3] Mollamahmutoglu M and Avcı E 2018 "Engineering Properties of Slag-Based Superfine Cement-Stabilized Clayey Soil" *ACI Materials J.* **115** issue 4 pp 541–48.

[4] Shen Z, Cao Y and Fang L 2017 "Experimental Investigation of Rapid Stabilization of Soft Clay Soils Using Chemical Admixtures" *Soil Mechanics and Foundation Engineering* **54** issue 3 pp 202–10.

[5] Chen M, et al 2015 "Laboratory evaluation on the effectiveness of polypropylene fibers on the strength of fiber-reinforced and cement-stabilized Shanghai soft clay", *Geotextiles and Geomembranes* **43** issue 6 pp 515–23.

[6] Pancar E and Akpinar M 2016 "Comparison of Effects of Using Geosynthetics and Lime Stabilization to Increase Bearing Capacity of Unpaved Road Subgrade" *Advances in Materials Science & Engineering* pp 1–8.

[7] Olivia M et al. 2019 "Jambi and Palembang clay soil stabilization for pavement matrix in road construction by using Portland cement type I" *MATEC Web of Conferences* 276 p N.PAG.

IOP Conf. Series: Materials Science and Engineering **758** (2020) 012041 doi:10.1088/1757-899X/758/1/012041

Stabilization of clayey soil for subgrade using rice husk ash (RHA) and sugarcane bagasse ash (SCBA)

F. Hidalgo[1], J Saavedra[1], C Fernandez[2] y G Duran[2]

1 Bachelor, Civil Engineering Program, Universidad Peruana de Ciencias Aplicadas, Lima, Peru
2 Full Professor, Civil Engineering Program, Universidad Peruana de Ciencias Aplicadas, Lima, Peru. Email: gary.duran@upc.edu.pe

Abstract. This document studies the stabilization of the soil used as a subgrade, by adding locally available materials such as rice husk ash (RHA) and sugarcane bagasse ash (SCBA). These aggregates were added to the soil in substitution by weight between 5%, 7.5% and 10%. By adding these, the expansiveness is reduced while the maximum dry density increases, in addition the tendency of CBR is increasing and then tends to decrease proportionally to the addition of the aforementioned aggregates. This indicates a peak in CBR and expandability. The best result obtained from CBR was 33.75% with the 5% replacement mixtures.

1. Introduction

One of Peru's first economic activities is agriculture, about 3 million tons of rice cultivation is produced each year, which provides about 900 thousand tons of rice husk. For every 1000 kg of ground rice crops, about 300 kg (30%) of rice husk is produced, and when it is burned, about 45 kg (15%) of rice husk ash (RHA) is generated. On the other hand, sugar production is estimated at 9.4 million tons per year, La Libertad-Peru, is the department with the highest production around 4.5 million tons, representing almost 50% of the country's annual production.

Kumar Yadab et al. [1] conducted various studies by adding RHA to evaluate the physical and mechanical properties of clayey soils. This author proposed replacement percentages that vary from 2.5% to 12.5%. Through this replacement it was observed that the optimum humidity content and CBR increase proportionally to the addition of ash, while the maximum dry density decreases.

Delgado López et al. [2] evaluated the effect of adding alkaline activated sugarcane bagasse ash (SCBA) to a clayey soil. Performed tests with additions that varied from 10 to 20% of addition with respect to the weight of the soil, resulting in an optimal mixture of 15% where the soil improved by 20% in compression with respect to the unaltered sample.

Castro (2017) [3] conducted trials by mixing RHA, lime and clayey soil. The RHA additions proposed by the author varied between 10 and 40% and 3% lime. The mixture with 20% RHA showed a CBR improvement of 5% to 38.5%. Finally, Castro (2017) concludes that this mixture can be used as a subgrade of pavements.

2. Materials and methods

Mixtures with different percentages of each ash and natural soil (S100 + A0 + B0) were prepared, the addition was 5% (S90 + A5 + B5), 7.5% (S85 + A7.5 + B7.5) and 10 % (S80 + A10 + B10). At the time of immersion for 96 hours of the mixtures the reaction of the silica with the calcium of the soil is activated resulting in the formation of calcium silicates, subsequently the crystallization of these components is generated, giving greater resistance to the soil in its quality as subgrade.

IOP Conf. Series: Materials Science and Engineering **758** (2020) 012041 doi:10.1088/1757-899X/758/1/012041

2.1. Physical and chemical properties of stabilizing agents

Rice husk ash was obtained at the "El Rey" mill, located in the department of San Martin-Peru. The rice husk was burned at temperatures between 600-700 ° C in open fire. In this part of the process silica is amorphous. The sugarcane bagasse burned in a temperature range of 800-1000 ° C.

2.1.1. Stabilization through the use of RHA

Rice husk ash contains silica, an element that with the usual characteristics in clayey soils such as calcium generates calcium silicates that harden the mixture and allows the increase of the soil density and its resistance characteristics.

2.1.2. Stabilization through the use of SCBA

Additionally, the SCBA contains a high percentage of silica that borders 57% of its entire composition, this contributes to its ability to be a pozzolanic material. Similarly, other compounds such as calcium oxide and ferric oxide that together with silica comprise more than 70% of SCBA are large pozzolana enhancers.

2.2. Laboratory Investigations

The soil was classified as clay of low plasticity CL. The natural soil particle distribution test was carried out by granulometric analysis as shown in Figure 1.

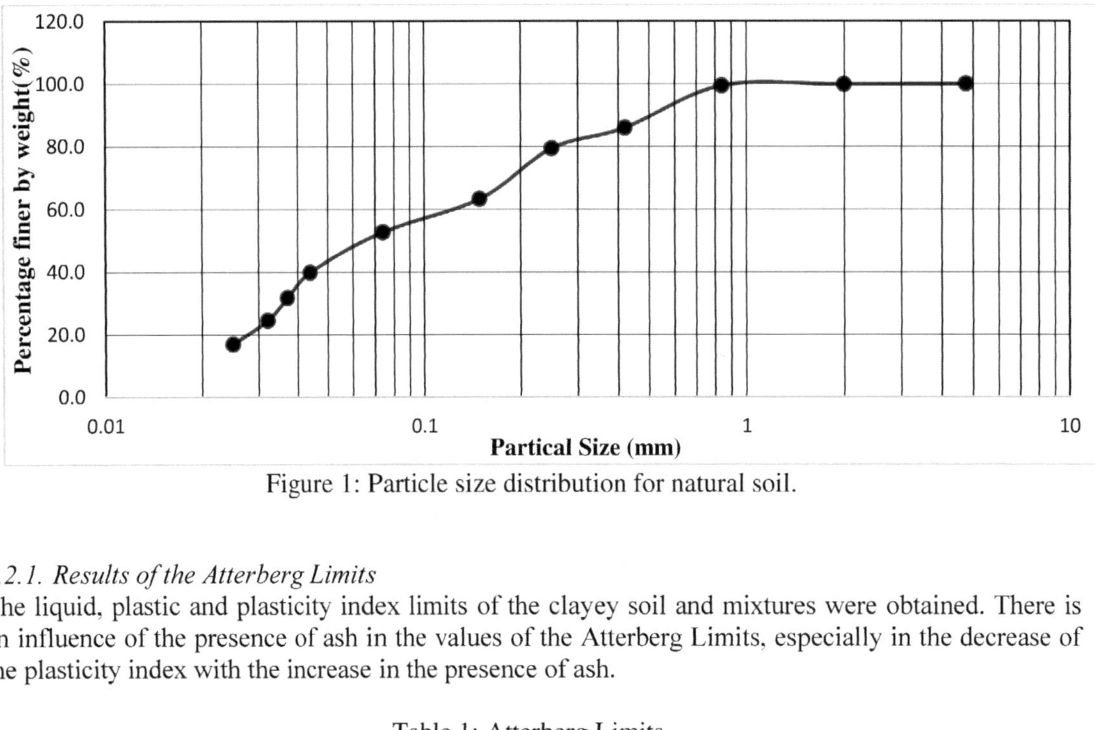

Figure 1: Particle size distribution for natural soil.

2.2.1. Results of the Atterberg Limits

The liquid, plastic and plasticity index limits of the clayey soil and mixtures were obtained. There is an influence of the presence of ash in the values of the Atterberg Limits, especially in the decrease of the plasticity index with the increase in the presence of ash.

Table 1: Atterberg Limits.

Name	Mixture Soil: RHA: SCBA	LL	LP	IP
S100+A0+B0	100: 0:0	30.69	18.75	11.94
S90+A5+B5	90: 5:5	35.10	24.48	10.62
S85+A7.5+B7.5	85: 7.5:7.5	34.59	27.76	6.83
S80+A10+B10	80: 10:10	34.10	27.87	6.23

IOP Conf. Series: Materials Science and Engineering **758** (2020) 012041 doi:10.1088/1757-899X/758/1/012041

2.2.2. Compaction Results

The Proctor Standard test was performed according to ASTM-D-1557 to determine the optimum humidity content and maximum dry density of clayey soil and mixtures.

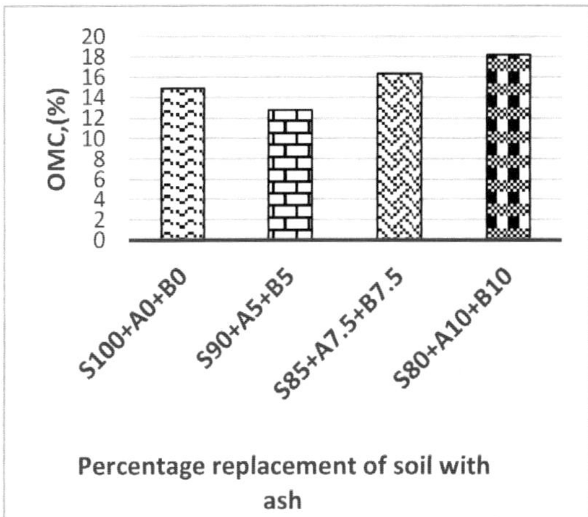

Figure 2: Effect of ash on OMC.

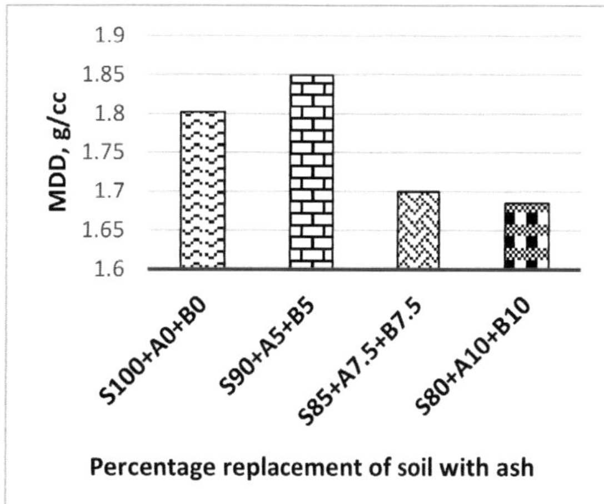

Figure 3: Effect of ash on MDD.

Figure 2 shows that increasing the presence of ash in the mixture increases the optimum humidity content, except for the S90 + A5 + B5 mixture, this is consistent with the increase of the liquid limit and plastic limit. In Figure 3 it can be seen that the maximum dry density of the S90 + A5 + B5 mixture is greater than that of the clayey soil, then for the mixtures with greater ash presence the maximum dry density decreases.

2.2.3 CBR Results

The CBR test for natural soil and stabilized soil was performed according to ASTM-D-1883 in immersed conditions to assess its carrying capacity. The results are shown in Figure 4.

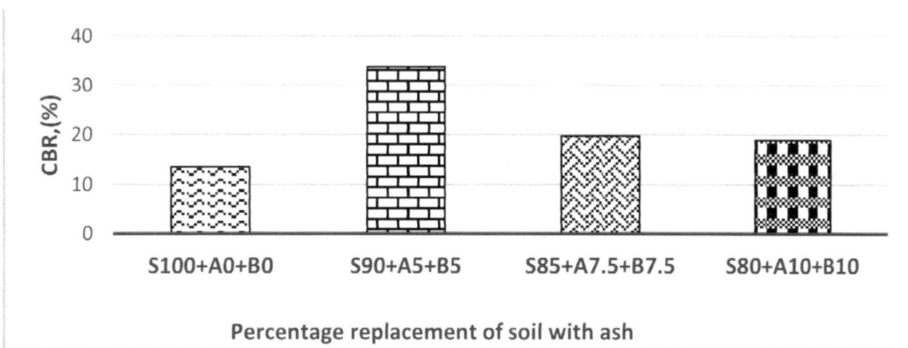

Figure 4: Effect of ash on CBR.

All mixtures achieved CBR values greater than that of clayey soil, being the mixture S90 + A5 + B5 the highest CBR value, 34.75%. The higher the presence of ash in the mixture, the CBR value decreases. The increase in CBR occurs due to the formation of calcium silicates after the reaction of silica from the ash and calcium of the soil that reacted during the 96-hour immersion. For the mixture S85 + A7.5 + B7.5 and S80 + A10 + B10 the CBR values were 19.73% and 18.90%, respectively.

Finally, with the S90 + A5 + B5 mixture, it manages to exceed the minimum value of CBR, 30%, a requirement for a subgrade to be considered as "excellent", according to the materials manual of the Ministry of Transportation and Communications of Peru.

3. Discussion of the results:

- The S90 + A5 + B5 mixture manages to achieve an optimum humidity content lower than that of the clayey soil and a maximum dry density greater than the other mixtures. The greater presence of ash in the mixture, before the pozzolanic reaction, helps to reduce the density due to the lower specific weight of the ash and dramatically modifies the granulometric distribution, directly affecting the compaction results. On the other hand, the lower amount of ash in the mixture, before the pozzolanic reaction, could contribute to positively modify the granulometric distribution, so that during compaction, the fine grains of the ash settle between the grains of the natural soil, thus achieving a higher dry density in relation to mixtures with a higher percentage of ash.

- The CBR value stands at 33.75 for the 5% mixture and this produces the best result. In the same way, during the compaction test, the highest value (1,849 g / cm3) and the lowest optimum humidity content (12.78%) are obtained. For larger ash additions, the CBR decreases, this could be due to the amount of ash that fails to react and produce crystals. So instead of having soil grains, you have unreacted ash grains that do not contribute to soil resistance.

- The results obtained during this investigation showed that the soil stabilized with rice husk ash and sugarcane bagasse brings favorable changes that would make it possible to use it as a subgrade. Due to the high percentage of silica present in these ashes, it converts them into potential primary or secondary stabilizing agents.

- The proposal generates a possible solution to an environmental problem of the disposal of these wastes, mitigating the effects of pollution such as the generation of respiratory diseases, among others.

4. Conclusions

- The addition of rice husk ash (RHA) and sugarcane bagasse (SCBA) increases the value of CBR in a clay of low plasticity. The mixture with 5% RHA and 5% SCBA reaches a CBR value of 33.75%, being that of the clayey soil is about 12%.

- The mixtures compared to the natural soil sample show significant increases in the plastic limit value, this could be explained by the greater absorption of water from the ashes in relation to the natural soil.

- In comparison with the other mixtures, the mixture S90 + A5 + B5 obtained the highest value in CBR and MDD 33.75% and 12.78% respectively; and, the lowest value of OCH (12.78%).

- Finding new uses for RHA and SCBA, entails adding value to these wastes and reducing the negative impact on the environment and the generation of respiratory or other diseases.

5. REFERENCES

[1] Anjani Kumar Yadav, Kumar Gaurav, Roop Kishor, S.K. Suman, (2017): “Stabilization of alluvial soil for subgrade using rice husk ash, sugarcane bagasse ash and cow dung ash for rural roads” International Journal of Pavement Research and Technology Vol 10, 27/02/2017, p254-261. 8p

[2] Delgado López, Carol Ashlly; Mendoza Meléndez, Ivette Anamile, (2018): "Influence of the percentage of alkaline-activated sugarcane bagasse ash on the effective tension in soils susceptible to liquefaction." (Undergraduate thesis). National University of Trujillo, Trujillo, Peru.

[3] Castro, Axel, (2017): “Stabilization of clay soils with rice husk ash for subgrade improvement”. (Undergraduate thesis). National University of Engineering, Lima, Peru.

IOP Conf. Series: Materials Science and Engineering **758** (2020) 012042 doi:10.1088/1757-899X/758/1/012042

Productivity improvement of tower crane in tall buildings

A Manrique[1] J Saman[1] S Rodriguez[2] and K Melendez[2]

[1] Bachelor School of Civil Engineering, Universidad de Peruana de Ciencias Aplicadas, Lima Peru.
Email: u201410477@upc.edu.pe, u201312421@upc.edu.pe
[2] Professor of School of Civil Engineering, Universidad de Peruana de Ciencias Aplicadas, Lima Peru.
Email: sandra.rodriguez@upc.edu.pe, pcipkmel@upc.edu.pe

Abstract. The tower crane is an electromechanical equipment that is used for the vertical transport of materials in a construction project and together with the two riggers form the work team to carry out this task. One of the main problems in the construction of multifamily buildings corresponds to the use of the tower crane because vertical transport causes non-contributory times, which is, dead times and waits above expectations. This research analyzes the current vertical transport process and proposes its optimization through some management tools with the aim of improving the productivity of the use of the tower crane by reducing non-contributory times. To this end, the productivity of the work team is recorded in several projects with similar characteristics, then the main problems are selected to analyze them and finally the process is optimized. The results determined that non-contributory times can be reduced by 10% if there is an orderly and continuous process.

1. Introduction

The tower crane, being a heavy equipment in construction, must justify its cost by optimizing its use; Therefore, it should be used optimally in construction works in order to make construction processes fast and safe, to obtain an improvement in productivity.

The literature on the productivity and operation of the tower crane is extensive. Ratios are used to measure the productivity of the tower crane, such as: the necessary space of the tower crane in action, the emptying time using the tower crane and the vertical transport time of construction materials. [1] Construction information modeling (BIM) is used to position the tower crane at construction sites with repetitive operations. BIM modeling optimizes the process through the user, which selects an appropriate optimization approach and formulates the positioning problem to minimize the total cycle duration. [2] Advance planning is used to plan daily operations and to provide a reliable daily schedule of the tower crane. The results show that overlapping activities and coupling advance planning balances the demand for work and reduces the duration of the tower crane process. [3] An agent-based simulation model is proposed to quantitatively assess the effect of the conflict over time and the overall costs of tower crane operations. The proposed model compares several tower crane design combinations to achieve an optimal solution. [4] Tower cranes operate in environments that are sensitive to unavoidable external disturbances, such as having buildings in much of the perimeter. An adaptive monitoring controller is proposed to deal with the uncertainties of external disturbances, which is implemented through experimental tests. [5]

This research focuses on analyzing the productivity of the tower crane in the vertical transport of materials during the construction of multi-family buildings, and then redesign the process through management tools.

2. Method

To initiate the investigation it is necessary to define the characteristics of the projects to be evaluated, so that buildings with residential use between 15 and 20 floors are included, building height between 30 and 60 m, land area between 800 and 1200 m2, evaluation at the stage of structures and use of a tower crane. Three projects with the characteristics defined for this research are evaluated:

Table 1. Characteristics of the project evaluated

	Project 1	Project 2	Project 3
Number of floors	15	20	20
Project height (m)	50	58	58
Land area (m2)	840	920	1200
Project stage	Structures	Structures	Structures
Number of tower cranes by project	1	1	1

For the analysis of the vertical transport process it is necessary to use some management tools, such as: balance sheet that is used for the measurement of each activity in a period of time whose objective is to analyze the efficiency of the construction method used, rather than the staff efficiency; Pareto diagram used to detect failures caused during the process whose objective is to graphically select the process problems; Ishikawa diagram that is used to analyze the causes of the main problems previously identified; and the flow chart that is used to present the new process, including those responsible at each control point.

In order to meet the objective of this study and present the information and results in an orderly manner, the following research process is developed:

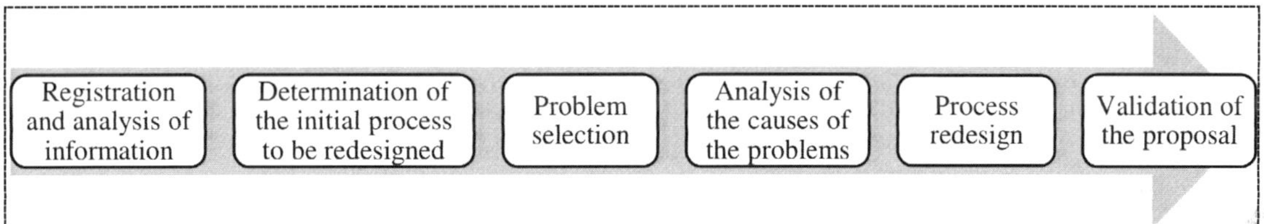

Figure 1. Research methodology

As a first step, the information of the measurements made in the selected projects must be recorded and analyzed. It is important to take into account that the work team for the general process of vertical transport of materials with tower crane is composed of rigger 1 in the free areas of the project, rigger 2 in the production area of the work and the tower crane.

Subsequently, the initial process must be determined, which presents the similar activities carried out within the data collection of each project. The general process presents four activities: movement, mooring, transport and unmooring. Movement and transportation are activities executed by the tower crane; while mooring and unmooring are activities executed by rigger 1 and rigger 2, respectively. The movement is defined as the maneuver of the tower crane without material; while transport is defined as the maneuver of the tower crane with material.

Once the activities are defined, each one must be classified by type of work PW - Productive work, CW - Contributory work and NCW - Non-contributory work.

Table 2. Activities for type of work

	Movement	Tie up	Transport	Untie
Resource	tower crane	rigger 1	tower crane	rigger 2
Type of work	CW	PW	PW	PW

Then the main problems must be selected through Pareto statistics, with the aim of subsequently analyzing the causes of these problems until the root cause is identified. Finally, taking into account the previous information, the redesign of the process is proposed and applied in a study case to validate the research proposal.

3. Results

3.1. Registration and analysis of information

The recording of the information of the work team measurements is made for a period of five days for each project. The measurements are made at the structures stage, specifically at construction levels that exceed the zero level. Once the information is registered, the quantity of the main materials transported in the three projects must be classified to compare them.

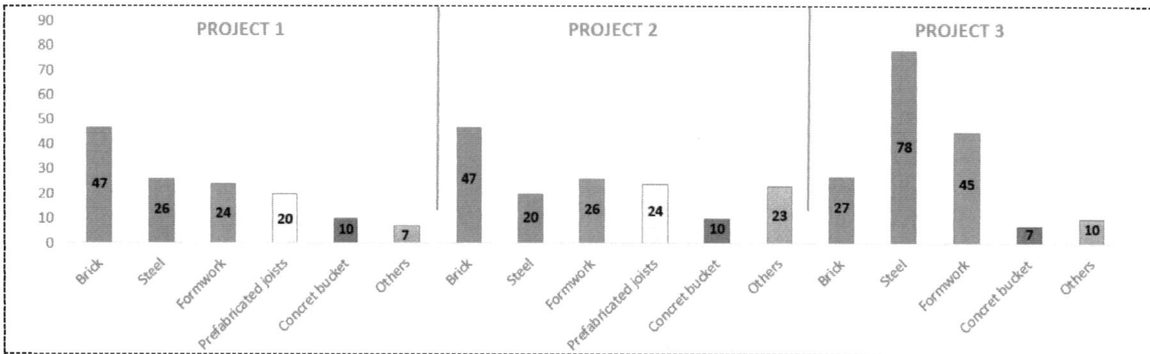

Figure 2. Quantity of transported materials by project

In project 3, prefabricated joists were not used because the construction system was made of solid slabs, so that steel is the most transported material. In the rest of the projects lightened slabs with prefabricated joists were used to optimize the construction processes.

All recorded information is analyzed and the overall results of the productivity of the three projects are classified according to the type of work.

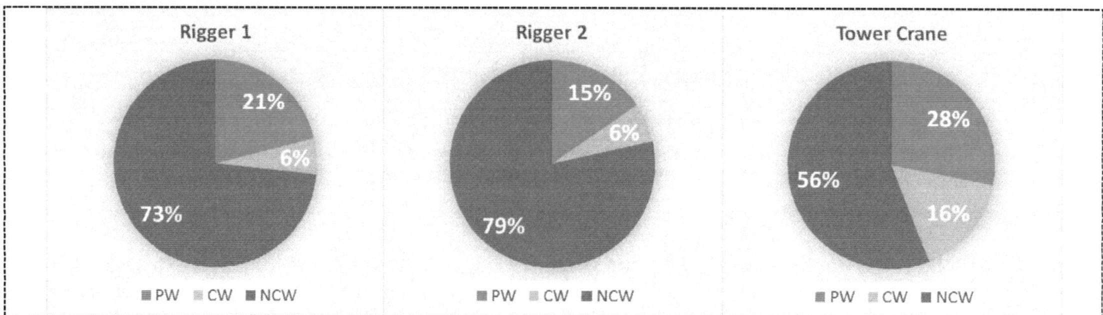

Figure 3. Overall productivity results

For the riggers, 21% and 15% of TP is the mooring and unmooring of the materials. 6% of CT are the use of the radio to guide the tower crane in the transport and unloading of the material and the use of the beep. 73% and 79% of TNC are the waiting times caused by the work done by the other rigger and the crane.

For the tower crane, 28% of TP is the transport of all types of materials to any point of the work. 16% of TC is the movement that the tower crane executes without material, it includes the movement of the empty brick cell. The 56% of TNC are the waiting times caused by the coordination of material with the foremen and by the work of mooring and unmooring of each rigger.

Within the TNC of each team member, the dead times (19%) are empty in which none of the members performs any activity; Research is not focused on these times.

3.2. Determination of the initial process to be redesigned

The common activities that together represent the initial process of vertical transport of materials with tower crane, observed in the measurements of the three projects, are identified.

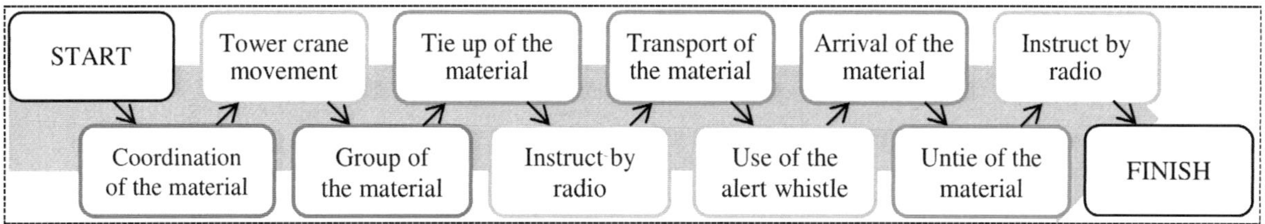

Figure 4. Initial vertical transport process

3.3. Problem Selection

Pareto statistics are applied to the activities corresponding to the TC and TNC, with the objective of identifying the activities that are within 80%. The following nomenclature is presented:

- Waiting for the rigger 1 : E1 - Waiting for the crane : EG - Movement : M
- Waiting for the rigger 2 : E2 - Dead times : U - Radio use : R

	Quantities (min)	%	Quantities accum (min)	% accum
W2	5,375	32.0%	5,375	32.0%
W1	4,666	27.7%	10,041	59.7%
WT	3,439	20.5%	13,480	80.1%
D	2,024	12.1%	15,504	92.2%
M	793	4.7%	16,297	96.9%
R	523	3.1%	16,820	100.0%
TOTAL	16,820	100.0%		

Table 3. Calculations and percentages of the activities

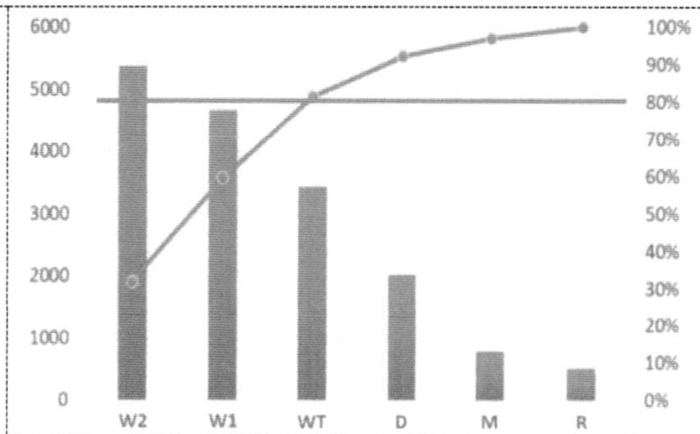

Figure 5. Pareto's diagram

3.4. Analysis of the causes of the problems

The main problems identified are analyzed. An analysis is carried out with the six categories to identify the causes, the sub causes and the root cause of the problem.

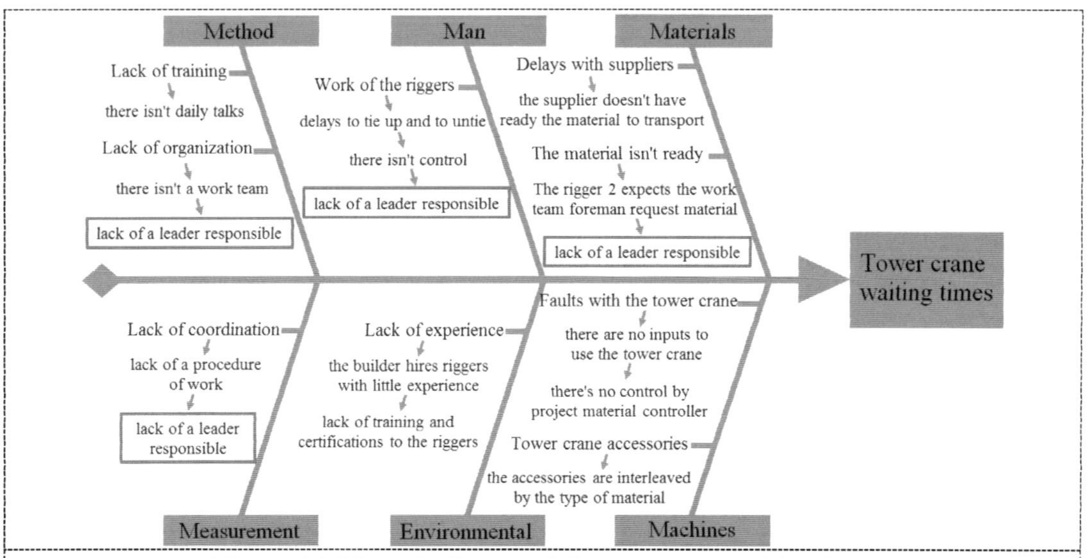

Figure 6. Ishikawa's diagram of the tower crane waiting times

IOP Conf. Series: Materials Science and Engineering **758** (2020) 012042 doi:10.1088/1757-899X/758/1/012042

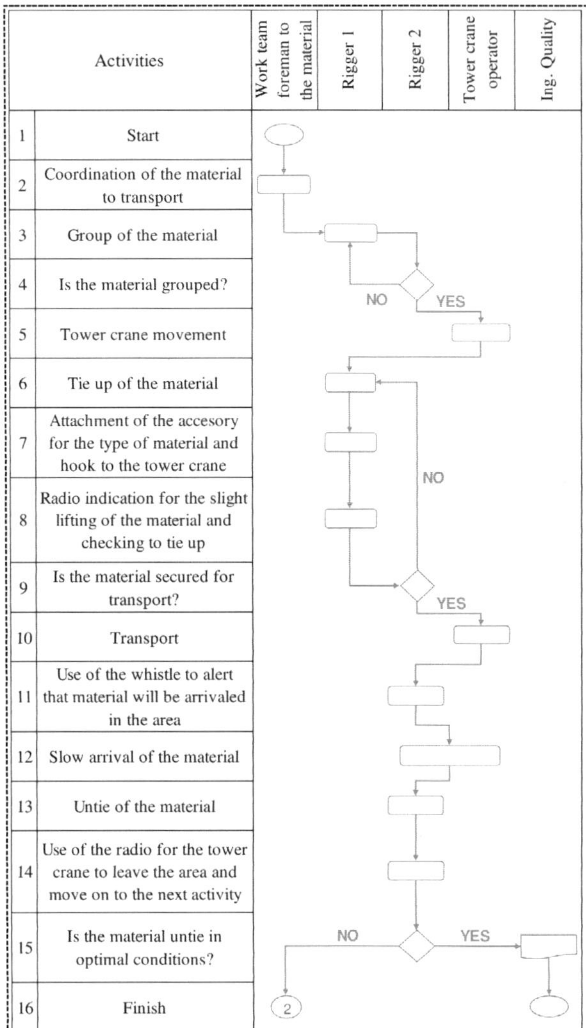

Figure 7. Redesigned vertical transport process

3.5. *Process redesign*

The objective of the redesign is to ensure that the general process is carried out consistently with an established sequence, complying with safety standards.

With respect to the initial process, control points have been added before the operator performs the respective maneuver and at the end of the process, in which the Quality Engineer is added for documentation through an inspection protocol.

3.6. *Validation of the proposal*

The proposal, applied to a project with similar characteristics, obtains a 10% productivity improvement in the TP of the tower crane product of an orderly process in which the rigger 1 has the grouped materials and makes the respective moorings in the times in which the rest of the work team performs its functions; also generating an 11% increase in CT.

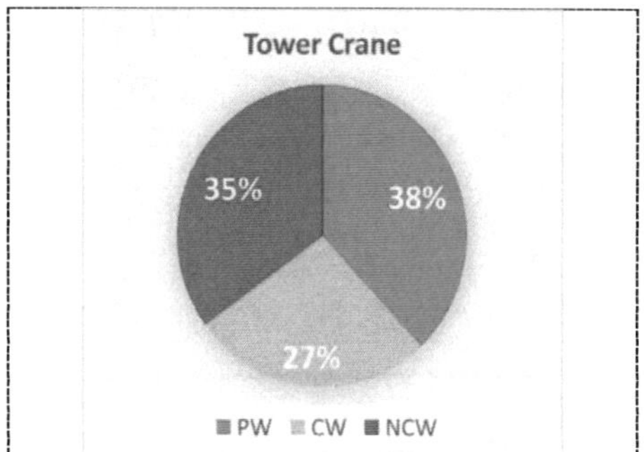

Figure 8. Productivity results

4. Conclusions

The proposal focuses on the team members working in a more integrated and intelligent way. For this, if you have an organization and programming you get to increase the productivity of the overall process. The organization is relevant because each member knows their roles and responsibilities within the process. The programming helps to have well defined schedules on the materials to be transported during the day. On the other hand, the reduction of the tower crane's TNC from 56% to 35% means that the tower crane is in operation for 1.2 more hours per day, taking into account that 8 hours per day are worked. Considering that the monthly rent of a tower crane is \$ 10,000 on average, producing the 1.2 hours is equivalent to a saving of \$ 60 per day, which is equivalent to \$ 1,500 per month, which for a one-year project of the tower crane's operation equals \$ 18,000 savings.

5. References

[1] Sudie S 2017 *Periodica Polytechnica Architecture* **48** 87
[2] Dasović B, Galić M and Klanšek U 2019 *Buildings* **9** 21
[3] Hattab M, Zankoul E and Hamzeh F 2017 *Journal of Computing in Civil Engineering* **31**
[4] Younes A and Marzouk M 2018 *Automation in Construction* **93** 348
[5] Chen H, Fang Y and Sun N 2019 *Mechanical Systems and Signal Processing* **123** 426
[6] Ngabesong R and Yilmaz M 2019 *M. Int. J. Dynam. Control* **7** 430
[7] Chen H, Fang Y and Sun N 2017 *Chinese Automation Congress* 3609-14

IOP Conf. Series: Materials Science and Engineering **758** (2020) 012043 doi:10.1088/1757-899X/758/1/012043

Numerical simulation of the plunge stage of friction surfacing of AA5083 aluminum alloy

Hongjun Li, Zhuang Xu, Zhewei Zhou,

Zhejiang Sci-Tech University, Hangzhou, 310018, China
E-mail: lihongjun@zstu.edu.cn

Abstract. Plunge stage is crucial to the success of whole friction surfacing process. To investigate the effect of parameters, the CEL modelling method in the finite element software model was used to simulate the plunge stage of friction surfacing. The results showed that the higher rotation speed would result a smaller axial force but higher the center point temperature and frictional dissipation; Besides, a higher feed rate, who can cause higher peek value of axial force and the center point temperature rises faster, is inversely proportional to the frictional dissipation, and the final axial force and the center point temperature tended to be in the same range.

1. Introduction

Friction surfacing is a solid phase joining technique that can be used to produce metal coatings. Schematic diagram of friction surfacing is shown in Figure 1. Under the axial force, the high-speed rotating consumable rod was rubbed against the substrate and the end of the consumable rod was heated. With the relative movement of the consumable rod and the substrate, the consumable rod was continuously transferred to the substrate to form a continuous deposited layer [1]. The welding process parameters, including the feed rate, the rotating speed and the welding speed would affect the thermal coupling between the surfacing layers, and consequently the welding quality of the surfacing layer [2].

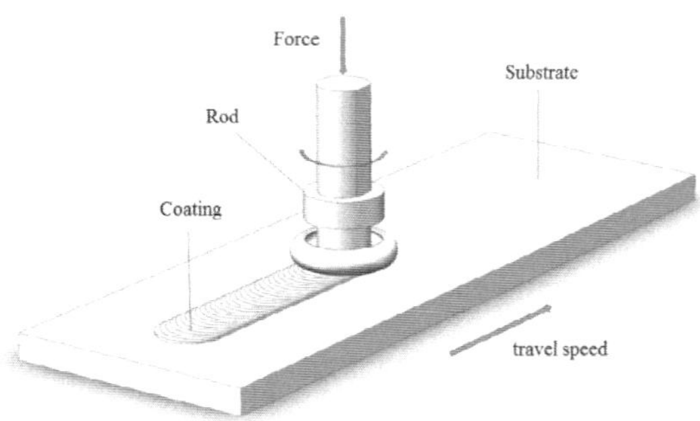

Figure 1. Schematic diagram of friction surfacing

At present, the research on the mechanism of friction surfacing is mainly based on deposition process of steel over steel [3-4], aluminum alloy over aluminum alloy [5-6], but there are few studies on aluminum/steel surfacing in the literature [7-11]. By means of numerical analysis, the material plastic deformation, temperature change of the consumable rod, formation mechanism of the deposited layer can be obtained. Liu et al. [12] applied the finite element method to the coupling calculation of the friction welding process. But due to the limitations of the Lagrangian method in dealing with large

IOP Conf. Series: Materials Science and Engineering **758** (2020) 012043 doi:10.1088/1757-899X/758/1/012043

nonlinear deformation, it was impossible to simulate the stable surfacing stage. Other numerical simulation studies [13-15] only analyzed the temperature field using a simplified heat source model. So the simulation study of the surfacing process is still very scarce.

Deposition process of aluminum over steel has a wide range of requirements in shipbuilding, automotive and aerospace [16]. In this paper finite element software was used to establish a thermal mechanical coupled friction surfacing model and to simulate the compression stage and then to obtain the deformation of consumable rod in the compression stage.

2. Modelling

2.1 Model geometry

The consumable rod undergo large plastic deformation during the surfacing process, was defined as Eulerian body, and the substrate was defined as rigid body. The consumable rod was cylindrical, with diameters and heights of 20 mm and 100 mm, respectively. The Eulerian body included two regions, 'full' and 'void'. The outer region 'void' contained no initial AA5083 material properties. The substrate diameter and height are 15 mm and 3 mm. As shown in Figure 2 and 3:

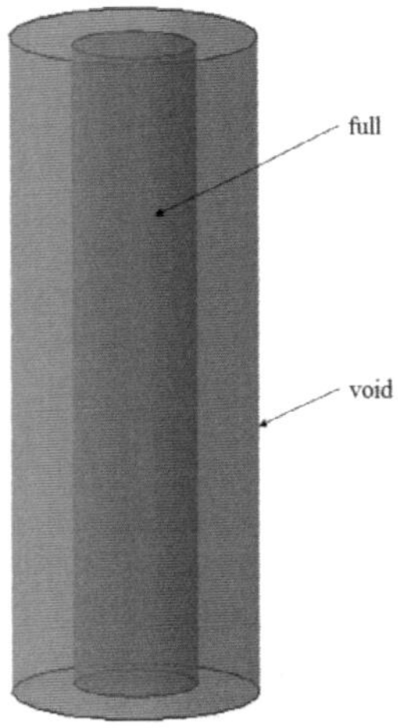

Figure 2. Rod geometrical model

Figure 3. Substrate geometrical model

2.2 Material properties

The consumable rod are AA5083 aluminum alloy, that belongs to Al-Mg alloy, which has high strength, corrosion resistance and excellent welding performance [17]. The physical properties of AA5083 are shown in Tables 1:

Table 1. Temperature dependent properties of AA5083 steel [18]

Temperature (℃)	Thermal conductivity ($W·m^{-1}·K^{-1}$)	Specific heat capacity ($J·Kg^{-1}·℃^{-1}$)	Density ($Kg·m^{-3}$)
-20	112.5	924.1	2673.9
80	122.7	984.2	2642.7
180	131.6	1039.6	2629.4
280	142.3	1081.2	2611.5
380	152.5	1136.6	2589.3
480	159.5	1178.2	2567.0
580	177.2	1261.4	2549.2

The Johnson-Cook plasticity model was used in the finite element simulation of friction surfacing with high strain rate deformation [19]. The parameters of the AA5083 aluminum alloy Johnson-Cook are shown in Table 2:

Table 2. Johnson-Cook Plasticity Coefficients for AA5083 [20]

A(GPa)	B(GPa)	n	C	m	T_m(℃)	T_r(℃)
0.13789	0.21673	0.4845	0.002	1.225	660	20

2.3 Boundary conditions

The simulation mainly studied the influence of process parameters on the friction surfacing process. Since only the compression stage of the friction surfacing was simulated, the rotating speed and the feed rate were merely considered in the experiment. Five sets of simulated parameters were tested, and the feed was 7 mm. The specific process parameters were shown in Table 3:

Table 3. Process parameters of simulation

Test	Rotating speed（$r\cdot min^{-1}$）	Feed rate（$mm\cdot min^{-1}$）
1	700	20
2	900	20
3	1100	20
4	900	15
5	900	25

3. Results and discussion

3.1 Axial force

When the feed rate was 20 mm/min, the effects of the rotational speeds of 700, 900 and 1100 r/min on the axial force *P* are shown in Figure 4. Due to the instability of plastic deformation during frictional surfacing. The axial force was in a fluctuating state. The higher the rotational speed, the more heat was generated by the consumable rod, and the material at the bottom of the consumable rod was more softer, so the lower the axial force; when the rotation speed was 900r/min, the influence of the feed rate of 15, 20 and 25 mm/min on the axial force is shown in Figure 5.The peak value of the axial force P_{max} (25 mm/min) > P_{max} (20 mm/min) > P_{max} (15 mm/min), but when the surfacing is carried out, the downforce tended to be in the same range.

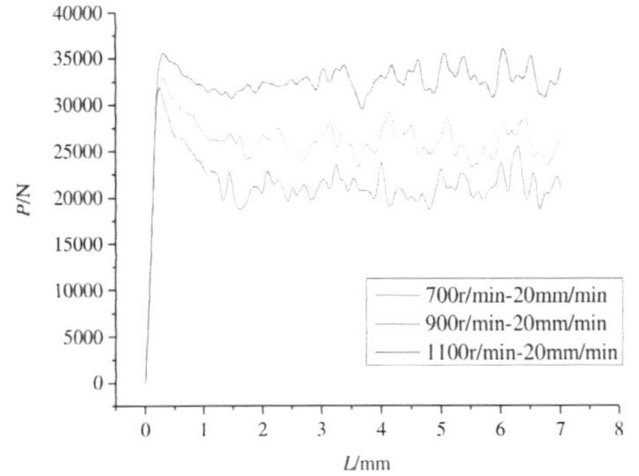

Figure 4. Effect of rotational speed on axial force

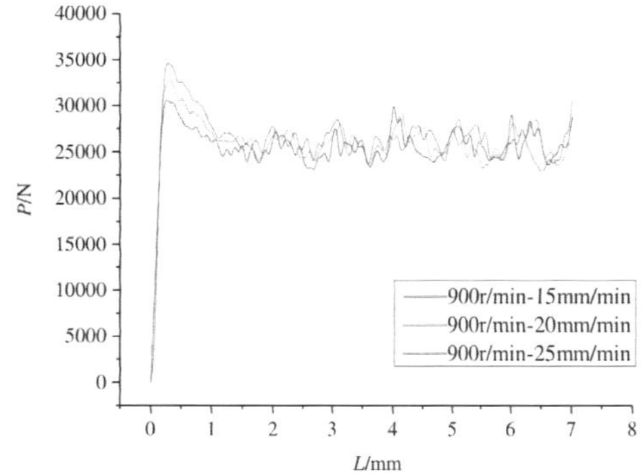

Figure 5. Effect of feed rate on axial force

3.2 Temperature of the center point of the contact surface

When the feed rate was 20 mm/min, the effects of the rotational speeds of 700, 900 and 1100 r/min on the contact point center temperature are shown in Figure 6. The higher the speed, the higher the center point temperature rises fast, and the center point temperature T (1100 r/min) > T (900 r/ min) > T (700 r/min), when the rotation speed was 900 r/min, the feed rate was 15, 20 and 25 mm/min on the center point temperature is shown in Figure 7. The faster the feed rate, the more friction between the consumable rod and the substrate, the faster the temperature rises at the center point, and due to the same displacement, the center point temperature tended to be in the same range as the friction surfacing continued.

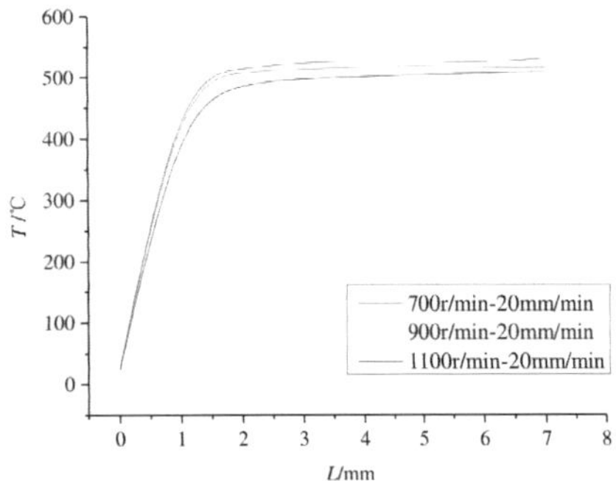

Figure 6. Effect of rotational speed on temperature

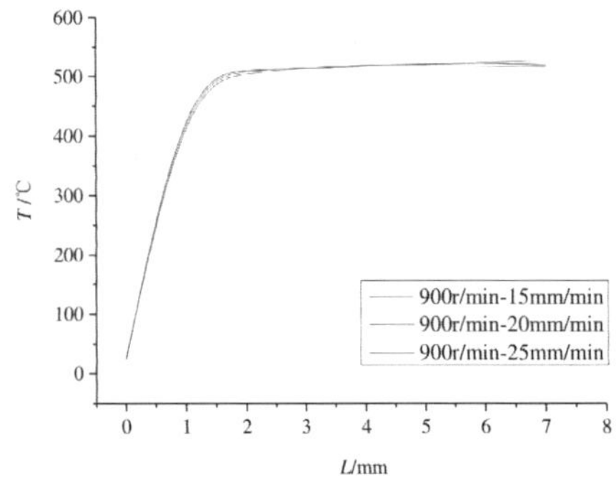

Figure 7. Effect of feed rate on temperature

3.3 Frictional dissipation

When the feed rate was 20mm/min, the effects of the rotational speeds of 700, 900 and 1100 r/min on the material frictional dissipation Q_f are shown in Figure 8. The higher the rotation speed, the greater the frictional dissipation. When the rotation speed was 900 r/min, the effect of the feed rate of 15, 20 and 25 mm/min on Q_f is shown in Figure 9. The higher the feed rate, the shorter friction surfacing time and the smaller the frictional dissipation.

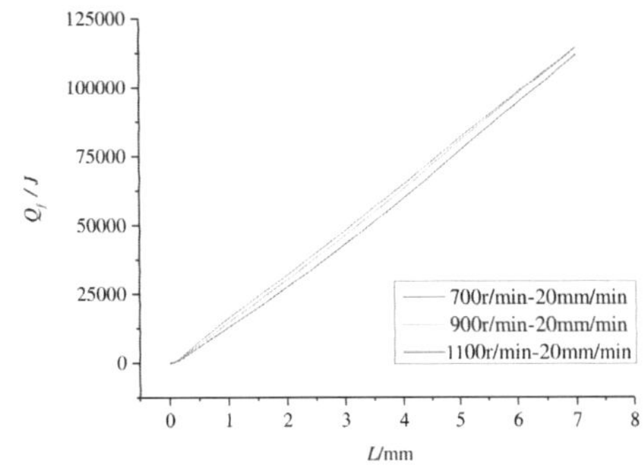

Figure 8. Effect of rotational speed on frictional dissipation

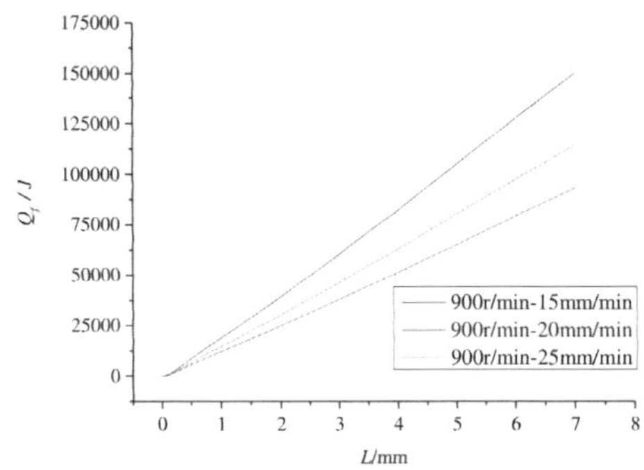

Figure 9. Effect of feed rate on frictional dissipation

4. Conclusions

1) The higher the rotation speed, the smaller the axial force; the higher the feed rate, the greater the peak value reached, but eventually axial force tended to be in the same range.
2) The higher the rotation speed, the higher the center point temperature; the higher the feed rate, the faster the center point temperature rises, but finally center point temperature tended to be in the same range.

3) The higher the rotation speed, the greater the frictional dissipation; the higher the feed rate, the smaller the frictional dissipation.

References

[1] Badheka K and Badheka V 2017 Friction surfacing of aluminium on steel: An experimental approach *Materials Today: Proceedings* **4**(9) pp 9937-9941
[2] Padhy G K, Wu C S and Gao S 2018 Friction stir based welding and processing technologies-processes, parameters, microstructures and applications: A review *Journal of Materials Science & Technology* **34**(1) pp 1-38
[3] Rafi H K , Ram G D J, Phanikumar G, Rao and K P 2011 Microstructural evolution during friction surfacing of tool steel H13 *Materials & Design* **32**(1) pp 82-87
[4] George S N R , Mohanty B S and Sathish R 2018 Friction surfacing of AISI 316 over mild steel: A characteriation study. *Defence Technology* **14**(4) pp 306-312
[5] Galvis J C, Oliveira P H F, Hupalo M F, Martins J P and Carvalho A L M 2017 Influence of friction surfacing process parameters to deposit AA6351-T6 over AA5052-H32 using conventional milling machine *Journal of Materials Processing Technology* **245** pp 91-105
[6] Krohn H, Hanke S, Beyer M and Dos Santos J F 2015 Influence of external cooling configuration on friction surfacing of AA6082 T6 over AA2024 T351 *Manufacturing Letters* **5** pp 17-20
[7] Chandrasekaran M , Batchelor A W and Jana S 1997 Friction surfacing of metal coatings on steel and aluminum substrate *Journal of Materials Processing Technology* **72**(3) pp 446-452
[8] Li, H.; Qin, W.; Galloway, A.; Toumpis, A. 2019 Friction Surfacing of Aluminium Alloy 5083 on DH36 Steel Plate. *Metals*, **9**, 479
[9] Sugandhi V and Ravishankar V 2012 Optimization of friction surfacing process parameters for aa1100 aluminum alloy coating with mild steel substrate using response surface methodology (RSM) technique *Modern Applied Science* **6** pp 69–80
[10] Badheka K and Badheka V 2017 Friction Surfacing of Aluminium on Steel: An Experimental Approach *Materials Today: Proceedings* **4**(9) pp 9937-9941
[11] Kumar B V, Reddy G M and Mohandas T 2015 Influence of process parameters on physical dimensions of AA6063 aluminium alloy coating on mild steel in friction surfacing *Defence Technology* **11**(3) pp 275-281
[12] Liu X M, Zou Z D, Zhang Y H, Qu S Y and Wang X H 2008 Transferring mechanism of the coating rod in friction surfacing *Surface & Coatings Technology* **202**(9) pp 1889-1894
[13] Liu X, Yao J, Wang X, Zou Z and Qu S 2009 Finite difference modeling on the temperature field of consumable-rod in friction surfacing *Journal of Materials Processing Technology* **209**(3) pp 1392-1399
[14] Vitanov V I and Javaid N 2010 Investigation of the thermal field in micro friction surfacing *Surface & Coatings Technology* **204**(16) pp 2624-2631
[15] Jaworski B, Voutchkov I I, Vitanov V I and Hughes V 2000 Modelling of the friction surfacing process for turbine blade reclamation *Proceedings of the 33rd International MATADOR Conference* Springer London pp 307-312
[16] Ravisekhar S, Das V C and Govardhan D 2017 Friction surfaced deposits for industrial applications *Materials Today: Proceedings* **4**(2) pp 3796-3801
[17] Ahmad B, Galloway A and Toumpis A 2018 Advanced numerical modelling of friction stir welded low alloy steel *Journal of Manufacturing Processes* **34** pp 625-636
[18] Gupta M S N, Balunaik B and Murti K G K 2012 Finite element modeling and thermo-mechanical analysis of friction stir welded Al/Cu bimetallic lap joints *International Journal of Mechanical Engineering and Robotics Research* **2**(17) pp 165-173
[19] Johnson GR and Cook WH 1983 A constitutive model and data for metals subject to large strains, strain rates and high temperature. In: proceedings of 7th Int. Sym. Ballistics,Editors. Seventh International Symposium on Ballistics Hague-Netherlands pp 541-547
[20] Polyzois I and Polyzois I 2010 Finite element modeling of the behavior of armor materials under high strain rates and large strains

IOP Conf. Series: Materials Science and Engineering **758** (2020) 012044 doi:10.1088/1757-899X/758/1/012044

Eco-Concrete for *Hydraulic Structures* with Addition of Colloidal Nano-Silica

C Salguero[1], L Castañeda [1], J Rodríguez[2] E Carrera[2]

1 Bachelor, Civil Engineering Program, Universidad Peruana de Ciencias Aplicadas, Lima, Perú
2 Full Professor, Civil Engineering Program, Universidad Peruana de Ciencias Aplicadas, Lima, Perú
Email: U201613500@upc.edu.pe

Abstract. In the construction of buildings and infrastructures, high resistance materials are used due to current design requirements, concrete being one of the main materials used in the execution of these projects whose cement content is limited to obtaining an economic concrete and of minimum retraction. This limitation requires the use of new additions such as Nano Silica (NS), which due to its nanometric structure is used as a partial replacement for cement, producing an increase in strength in concrete. The present investigation studies the partial replacement of the NS in the cement to determine its behavior in compressive strength, diametric compressive strength, water permeability coefficient. The results indicate that with an addition of 0.225% of NS the compressive strength and splitting tensile strength are increased and the water permeability coefficient decreases, all of them compared to a conventional concrete.

1. Introduction

The conventional concrete is a material that has been developed approximately 150 years ago to imitate natural stone while providing less labor-intensive methods of shaping the material (casting rather than hewing and carving) [1], its components are Portland cement, sand, stone and water [2]. Cement is the main element that contributes to its durability and strength; it must be designed with an economical amount because it is more expensive than aggregates [3]. Important studies have been carried out in recent years on the use of mineral additions in concrete, ranging from natural pozzolans, through fly ash and blast furnace slags to micro silica; the latter being used, in one of its forms, as a partial replacement for Portland cement [4]. Pozzolans have had extensive uses as a component of Portland cement concrete, particularly for marine and hydraulic structures [5]. The hydraulic structures have different pathologies caused by the presence of external agents, with sulphates causing the greatest damage when they come into contact with the concrete, because they react with the cement components, expanding the hardened mixture and causing its progressive detachment. The canals are responsible for transporting large amounts of water, which causes damage by abrasion, erosion and cavitation on its surface; because the liquid they carry carries sediments and particularly sulphates, which wear away the concrete [6]. On the other hand, the wetting of the soils by irrigation or rain seepage can activate the soluble salts that would produce a chemical attack on the concrete of the foundations or of the structure that is in direct contact with the ground, such as the canals [7].

Recent advances in the field of Nanotechnology and cement indicate that the NS as a partial replacement of cement produces an increase in the density of its microstructure in the concrete mix [8]. This NS in low percentages increases the speed of the hydration process from its chemical reaction and fills the empty spaces of the mixture by its size [9]. Today, the NS is an attractive cement replacement solution because it reduces the greenhouse effect and CO2 emission, and because its high specific surface area / volume ratio provides efficient chemical reactivity. [10] studying the compressive strength of concrete for different ages from 3 to 180 days, obtaining that with 10% replacement of NS the compressive strength is greater than the CS, indicating that the NS acts as a filler which increases the density of the microstructure and as an activator in the hydration reaction. On the other hand, [11] studying mortar mixtures at the age of 28 days for a 3% and 6% replacement of NS, finding that the compressive strength increases 19% and 44% with respect to the conventional mortar. Similarly, [12] evaluates different properties of the mortar, finding that with 1% replacement of NS the compressive strength and diametric compressive strength at 28 days, increase by 12% and 25% compared to the mortar conventional, and the water absorption coefficient evaluated 38.5% compared to the standard mortar. In the present work, study the effect of the addition of 0.225% of nanoparticles as a replacement for cement in compressive strength, splitting tensile strength and water permeability coefficient in order to elaborate an eco-concrete to be used in the construction of hydraulic channels.

2. Materials and Methods

2.1. Materials

The cements used are Portland Type I and Portland Type HS; the aggregates were from the quarry of Cieneguilla, the sand particles of 0.15-9.50 mm and the stone particles of 2.36-12.50 mm; the water was drinkable; and the additives used were Gaia Nanosilica and the Sika CEM superplasticizer.

2.2. Method

The present work has been developed in the Laboratory of Concrete Technology of the UPC-Monterrico. The mix design was made for a concrete f'c = 280 Kg / cm^2 and according to [13]. Table 1 shows the mixing designs made for Conventional Concrete (CS) and Concrete with Nano Silica (CNS), with Superplasticizer (SP) and Nano Silica (NS) and in Table 2 the tests and numbers of specimens used in the experimental development of the present investigation. The specimens for the tests of compressive strength, splitting tensile strength were cylindrical 6"x12" and for the test of water permeability also cylindrical 4 "x 8", both specimens shown I Figure 1 y Figure 2; which were cured in the laboratory until 28 days according to [14], then they were demoulded and tested. As a previous part of the investigation, a CS with slump of 3” to 4” was designed and since the slump values for the CNS1 and CNS3 mixtures were outside the allowed range, the mixtures were discarded and the CNS2 mixture was considered with a percentage of addition of NS of 0.225% in relation to the weight of the cement, which is called CNS.

Table 1. Mix design for concrete f'_c= 280 Kg/cm^2 units Kg/m^3

Code	Cement	Water	Aggregate	Sand	SP	NS	Slump (in.)
CS	480[1]	216	954.38	615.44	5.65	0	3.20[1]
CNS 1	480[1]	216	954.38	613.12	5.65	0.96	2.70[1]
CNS 2	480[1]	216	954.38	612.86	5.65	1.08	3.30[1]
CNS 3	480[1]	216	954.38	612.54	5.65	1.20	5.50[1]
CS-HS	480[2]	216	954.38	615.44	5.65	0	3.40[2]
CNS-HS	480[2]	216	954.38	612.86	5.65	1.08	3.10[2]

[1] Portland Cement Type I
[2] Portland Cement Type HS

Table 2. Tests performed

Concrete Tests	Methods	N° Specimens		Age (days)
		CS[a]	CNS[a]	
Compression strength	ASTM C39[15]	3	3	7, 28
Splitting tensile	ASTM			

Figure 1. Specimens concrete 6"x12"

Figure 2. Specimen in water permeability test

3. Analysis and Results

3.1. Compression strength

Figure 3 shows the effect of curing time on the compressive strength of concrete with addition of 0.225% nanosilica. It shows that the compressive strength increases with the cure time, obtaining the values of 424.14 and 487.41 kg / cm^2 at the ages of 7 and 28 days for the CNS that represent an increase of 5.79% and 6.28% with respect to the CS. [18] studies the compressive strength at 28 days with an addition of 0.25% nanosilica, and obtains a 20-25% increase over conventional cement paste. Similarly, [19] for an addition of 2% of NS finds that the compressive strength increases at 28 days by 7.30% with respect to the CS. These growth trends are similar to those obtained in the present study. [20] indicates that this behavior is due to the good dispersion of the silica nanoparticles in the concrete, which generates a denser microstructure and produces a decrease in voids and weak areas, which can be cracking points.

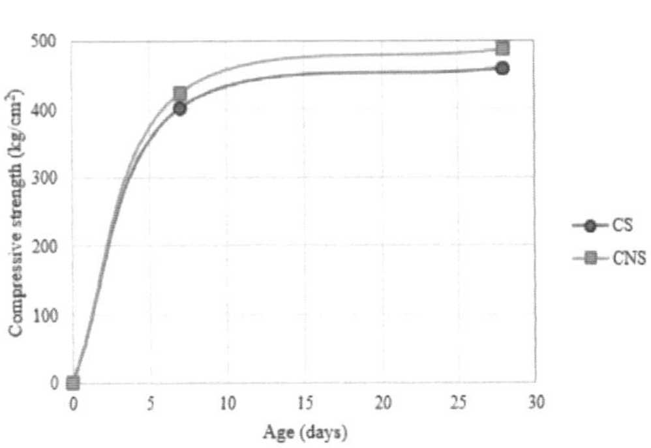

Figure 3. Effect of the test age on the compressive strength

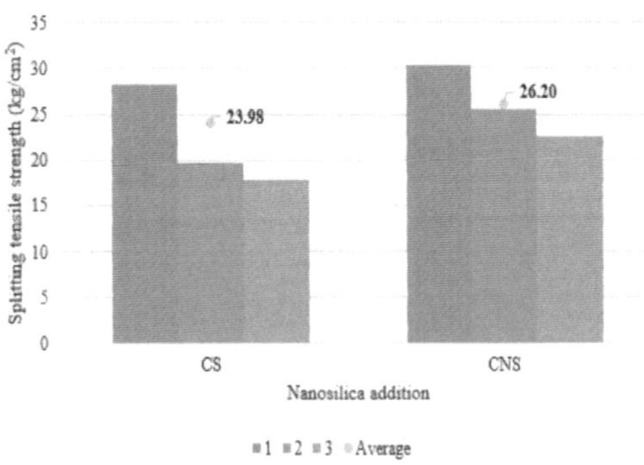

Figure 4. Effect of the addition of 0.225% on the splitting tensile strength

3.2 Splitting tensile strength

Figure 4 shows the effect of the addition of 0.225% nanosilica on splitting tensile strength at the age of 28 days. It shows that the individual values and the average obtained for the CNS are greater than the values reached by the CS, the average value being 26.20 Kg / cm2 19.47% more than the 21.93 Kg / cm2 of the CS. [18] studies the increase in splitting tensile strength for an addition of 1% and 2% NS with cure at 28 days; from this investigation it obtains a 17% and 24% increase, respectively, compared to the CS. [21] in his research, he made concrete mixtures with additions of NS of 1.5% and 3% in concrete mixtures with different water-cement ratios (0.412, 0.435 and 0.44 for 1.5% of NS; and 0.374, 0.42 and 0.434 for 3% of NS), obtaining that with 1.5% of NS and water-cement ratio of 0.44 the average resistance by splitting tensile strength reached at 28 days is 3.83 MPa, this being lower than the CS in 8.8%, which can due to poor particle dispersion. [20] indicates that this behavior is due to the existence of a greater specific surface area in the concrete, which causes a dispersion of nanoparticles in the matrix that produces a denser and more homogeneous CSH gel, promoting a better zone of interfacial transition between aggregates. and cement paste.

3.3 Water Permeability Coefficient

Table 3 shows the coefficient of permeability (K) at 28 days for an addition of 0.225% nanosilica. It shows that the water permeability coefficient for CNS reaches a value of 8.47x10-12 which gives it a low permeability level [17] and represents 50.12% more than the CS that has a medium permeability level [17]. [18] in their study, considering a 0.30% percentage of NS addition, the CNS water permeability coefficient increases 56% more than the CS. Likewise, as noted [22] in his investigation, he obtained a water permeability coefficient of 2.65x10-13 for an addition of 9% nanosilica, which compared to CS decreased by 69.60%. These K growth trends are similar to those obtained in the present study. [23] indicates that this behavior is due to the fact that the pozzolanic reaction between silica (of NS) with calcium hydroxide in solution forms a super saturated CSH gel, producing a reduction in the pore structure of the CSH microstructure that decreases the fibril and the porous structure of CSH to a denser gel. It should be noted, as mentioned [22] in his research, the lower the water permeability coefficient of the concrete, the greater the compressive strength, this relationship indicates that the addition of nanosilica improves the microstructure of the concrete and, together with it, the mechanical properties.

Table 3. Water permeability coefficient at 28 days with 0.225% nanosilica

Type	CS	CNS
K	$1.69E^{-11}$	$8.47E^{-12}$
Level	Medium	Low

4. Conclusions

- The partial replacement of cement with the addition of 0.225% of NS allows the development of an eco-concrete that contributes to the reduction of greenhouse gases.
- The compressive and splitting tensile astrength improved with the addition of NS because its nanoparticles fill the voids of the cement microstructure producing a more resistant concrete
- The water permeability coefficient decreases because the pores in the cement matrix are reduced, contributing to the development of a less porous and durable concrete.

Acknowledgements

The authors acknowledge the company ULMEN S.A. for donating HP-300 product for carrying out tests in this investigation.

IOP Conf. Series: Materials Science and Engineering **758** (2020) 012044 doi:10.1088/1757-899X/758/1/012044

5. References

[1] Hansson C 1997 *Non conventional concrete technologies* (USA: National Academy Press) pp 1-125.

[2] Mehta P, Monteiro P 2014 *Concrete: Microstructure, Properties and Materials* (USA: McGraw Hall Education) 4th ed. pp 1-239.

[3] Neville A M 2011 *Concrete Technology* (England: Pearson Education) 5th ed. pp. 1-442.

[4] ACI Committee 234R-06 2000 *Guide for the use of silica fume in concrete* pp 1-51.

[5] Shetty M. 2000 *Concrete Technology Theory and Practice* (India: S. Chand & Company Ltd. s.a) pp 1-647.

[6] Gómez A, Schnabel S y Lavado F 2011 *Processes, factors and consequences of erosion by gullies; works developed in the Iberian Peninsula* (Spain: Boletín de la Asociación de Geógrafos Españoles) n 55, pp 59 - 80.

[7] Heinzen R y Arulanandan K 1977 *Dispersive Clays, Related Piping, and Erosion in Geotechnical Projects* ed J L Sherard and R S Decker (USA: ASTM) pp 202-217.

[8] Baena O 2007 Portland cement blended with nanoparticles (Colombia: Elsevier) Nro 152, pp 277-291.

[9] Hendriks C F y Pietersen H S 2000 Sustainable raw materials: construction and demolition waste (RILEM Publications SARL) pp 216

[10] Saloma, Nasution A, Imran I y Abdullah M 2015 *Improvement of concrete durability by nanomaterials* (Indonesia: Elsevier) vol 125, pp 608-612.

[11] Joshaghani A y Amin M 2018 Evaluating the effects of sugar cane bagasse ash (SCBA) and nanosilica on the mechanical and durability properties of mortar (USA: Elsevier) vol 152, pp 818– 831.

[12] Lee H, Balasubramanian B, Gopalakrishna G, Kwon S J, Karthick S y Saraswathy V 2018 Durability performance of CNT and nanosilica admixed cement mortar (India: Elsevier) vol 159, pp 463-472.

[13] ACI Committee 211.1 2009 Standard practice to select the proportion of concrete of normal, heavy and massive weight.

[14] ASTM C192/C192M-18 2018 Standard practice for preparation and curing of concrete specimens for laboratory testing.

[15] ASTM C39/C39M-18 2018 *Standardized test method for compressive strength of cylindrical concrete specimens.*

[16] ASTM C496/C496M-17 2017 *Standard test method for determination for splitting tensile strength.*

[17] NTC 4483 1998 Test method to determine the permeability of water concrete.

[18] Balapour M, Joshaghani A, y Althoey F 2018 *Nano-sio2 contribution to mechanical, durability, fresh and microstructural characteristics of concrete: A review* (USA: Elsevier) vol 181, pp 27-41

[19] Singh L, Karade S, Battacharyya S, Yousuf M y Ahalawat S 2013 *Beneficial role of nanosilica in cement-based materials – A review* (India: Elsevier) vol 47, pp 1069-1077.

[20] Serag M, Yasien A, Muhammad S y Elkady H 2017 *Effect of nanosilica on concrete bond strength modes of failure* (International Journal of GEOMATE) vol 12, pp 73- 80.

[21] Gopinath B, Mouli P, Murthy A, Iyer N y Maheswaran S 2012 *Effect of nano silica on mechanical properties and durability of normal strength concrete* (India: Archives of Civil Engineering) vol 58, pp 433- 444.

[22] Haruehansapong S, Pulngern T y Chucheepsakul S 2017 *Effect of nanosilica particle size on the water permeability, abrasion resistance, drying shrinkage and repair work properties of cement mortar containing nano-sio*$_2$ ed Berthod P (Thailand: Hindawi) vol 2017, pp 1- 11.

[23] Belkowitz J, Belkowitz W, Nawrocki K y Fisher F 2015 *Impact of Nanosilica size and surface area on concrete properties* (USA: ACI Materials Journal) n 112, pp 419- 428.

IOP Conf. Series: Materials Science and Engineering **758** (2020) 012045 doi:10.1088/1757-899X/758/1/012045

Microstructure and mechanical characterization of high strength low alloy steel welded joint by hybrid laser arc welding

Liang Liang Bao, Yong Wang and Tao Han

1 School of Materials Science and Engineering, China University of Petroleum, Tsing Dao, China
E-mail: baoliangfighting@163.com

Abstract. Hybrid laser arc welding was developed to joint EQ70 steels. Welds with no visible defects were obtained. Investigations on microstructure and mechanical properties showed that: the welded metal (WM) consisted of acicular ferrite (AF), granular bainite (GB) and M-A constituents. The arc zone and laser zone possessed semblable HAZ microstructures. Coarse grained heat affected zone (CGHAZ) and fine grained HAZ (FGHAZ) were both composed of lath martensite (LM), GB and polygonal ferrite (PF). The intercritically HAZ (ICHAZ) consisted of martensite and M-A constituents. The microstructure of the subcritically HAZ (SCHAZ) was tempered sorbite. Partly cracked blocky M-A constituents were found at the grain boundary of intercritically reheated CGHAZ (ICCGHAZ). Some M-A constituents was slender and distributed in the grain of ICCGHAZ. The arc zone and laser zone had similar hardness and micro-shear strength distributions. The highest hardness and micro-shear strength of the backing and double-pass welded joints appeared near CGHAZs, reheated WM and ICCGHAZ. Slightly soften showed up at the SCHAZs. SCHAZ, CGHAZ and ICCGHAZ showed a reduction of micro-shear toughness. -40℃ sub-size Charpy impact testing showed that FL specimens possessed the lowest impact values. The coarse CGHAZs and cracked blocky M-A constituents of ICCGHAZ may be the reason caused the toughness deterioration.

1. Introduction

Thick section high strength low alloy (HSLA) steels are widely used in ocean engineering due to good combination of strength and toughness [1]. Arc welding processes have a broad use in shipbuilding industry because of the high-quality stable process and cheap equipment. However the efficiency of arc welding processes are extremely low for the welding of thick steels owing to the low penetration depth. Multi-pass welding procedure is usually applied which inevitably increases residual stress and distortion. High energy beam welding processes such as laser welding, electron beam welding have great welding depth and efficiency. But the costs of high power special apparatuses are especially large. There is an urgent need to develop an efficient welding process suitable for the welding of thick section steels. Hybrid laser arc welding (HLAW) is a promising process which can compensate the disadvantages of laser welding and arc welding by utilizing both features [2]. With the synergistic interaction of laser and arc, HLAW allows increased gap tolerances compared with laser welding, deepen penetration and faster welding speed than arc welding [3]. HLAW is receiving extensive and continuous attention as an efficient welding technique suitable for thick section steels.

X. Cao et al. [4] developed a single-pass HLAW process to weld 9.3 mm thick HSLA-65 steel. Compared with the MIG welds, the HLAW process resulted in a less distortion welds. Ivan Bunaziv et al. [5] realized a stable double-side butt weld process of 45 mm thick high strength steel HLAW. M.

Wahha et al. [6] reported a one-pass weld of 25 mm thick plates in I-butt configuration using HLAW. The HLAW was successfully applied to double-side welding of 50 mm thick joints. No detailed information was presented about the mechanical properties of HLAW joints. Most researchers' attentions were focused on how to realize the HLAW process on thick section steels [7], the studies on microstructure and mechanical properties of the HLAW welded joints were relatively less. Yanbin Chen et al. [8] studied the microstructure and properties of a four-pass double-side HLAW welded joint and found that welded joint possessed higher strength and lower toughness than the base metal. S. Gook et al. [9] investigated the possibilities of HLAW on longitudinal welds of high strength X80 pipe steels. 23.4 mm thick section X80 steel was jointed by HLAW. The hardness and impact toughness of HLAW WM met the requirements. The microstructure and properties of HAZ were not considered in these studies. Most of the researchers are primarily concentrated on the WM of HLAW joints [10, 11], studies on HAZs were relatively few. A. E. Nolting et al. [12] joined 9.1 mm thick HSLA-65 steel plates by HLAW. The sub-size Charpy impact test found that specimens notched in the HAZ were less tough than the base metal and the HAZ impact values fluctuated greatly. In this investigation, HSLA EQ70 steel plates were welded by HLAW. The microstructure and mechanical property of the WM and HAZ were evaluated in detail.

2. Experimental procedure

2.1. Hybrid laser arc welding experiments

EQ70 steel plates and ER100S-G filler wire were used. The chemical compositions are listed in Table 1. Shielding gas of 80% Ar, 18% CO_2 and 2% O_2 with a flow rate of 20 L/min was used. Butt joint of 45° V-groove configuration with a 6 mm root face was prepared. An IPG photonics corporation (YLS-10000) fiber laser system and a Fronius (TPS-4000) welding source were used. Laser leading arc mode was selected because incomplete fusion often occurs under arc leading laser mode. The distance between laser beam and filler wire was 10 mm. The laser defocus distance was +6 mm. Laser power is the most important parameter [3]. The various laser powers in the trials were 4.5 kW, 5.5 kW and 6.5 kW. The steels were welded under different welding speed of 0.6 m/min, 0.8 m/min and 1.0 m/min.

Table 1 Chemical compositions of EQ70 and ER100S-G (wt-%).

Elements	C	Si	Mn	S	P	Cu
EQ70	0.12	0.25	1.09	0	0.004	0.28
ER100S-G	0.08	0.47	1.62	0.004	0.015	0.11
Elements	Cr	Mo	V	N	B	Al
EQ70	0.58	0.53	0.04	0.0039	0.0009	0.074
ER100S-G	0.22	0.25	/	/	/	/

2.2. Metallographic and mechanical characterization

Metallurgical specimens were etched with 4% nital solution. Lepera etchant was used to reveal M-A constituents. The microstructures were characterized by optical microscope (OM), scanning electron microscope (SEM), electron backscattered diffraction (EBSD) and transmission electron microscope (TEM). Sub-size Charpy V-notch impact specimens (55 mm×10 mm×5 mm) were tested at -40 ℃. The Vickers hardness of the arc zone (Line 1 in figure 2), laser zone (Line 2 in figure 2) and overlapping zone (Line 3 in figure 3) were measured with a force of 10 kgf (HV10) and 10 s dwell time. Micro-shear test was conducted to illustrate the variation tendency of mechanical properties. Specimens with dimensions of 2 mm×2 mm×12 mm were cut and grinded into 1.5 mm×1.5 mm×12 mm from the welded joints as shown in figure 2 and figure 3. The test was conducted at room temperature with a loading speed of 0.5 mm/min. The curves of shear load P versus the displacement of shear cutter S were recorded. The micro-shear strength τ_b and micro-shear toughness W_s are estimated referring to X. P. Zhang [13].

3. Results and Discussions

3.1. Weld formation

Figure 1 shows the weld formations under different laser power and welding speed. Lack of penetration was observed under 4.5 kW laser power. Severe collapse of molten pool occurred, when laser power reached 6.5 kW. Complete fusion was achieved under 5.5 kW laser power. 1.5 kW laser power was used in the subsequent welding passes, because high laser power may cause the turbulence of the molten pool. Joints with sufficient penetration and no visible defects were achieved under 0.6 and 0.8 m/min welding speed. Lack of penetration appeared when the welding speed was 1.0 m/min. Relative lower 0.6 m/min welding speed was selected for the backing welding pass to maintain sufficient penetration. The welding speed of subsequent welding passes was 0.8 m/min to reduce the heat input and improve welding efficiency. To match the laser powers and the welding speeds, the arc currents varied from 260 A to 280 A. The optimized welding parameters were determined in Table 2.

The backing welded joint was typical “wine-cup” shape with upper arc zone and lower laser zone as shown in figure 2. The arc zone had a wider WM than the laser zone, however the arc zone HAZ was narrower than laser zone HAZ. A high laser power of 5.5 kW and a slow welding speed of 0.6 m/min result in a higher heat input and a wider laser zone HAZ. To study the influence of subsequent bead on the backing bead, a double-pass HLAW welded joint with 15 mm penetration depth was achieved, as shown in figure 3.

Table 2 Optimized welding parameters.

Welding Pass	Laser power [kW]	Arc current [A]	Arc voltage [V]	Welding velocity [m/min]
Backing	5.5	260	26.0~26.4	0.6
Subsequent	1.5	280	25.0~26.2	0.8

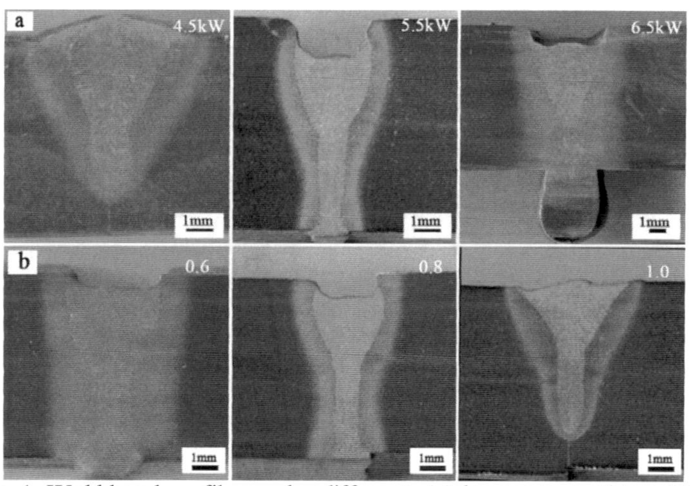

Figure 1. Weld bead profiles under different: (a) laser power, (b) welding speed.

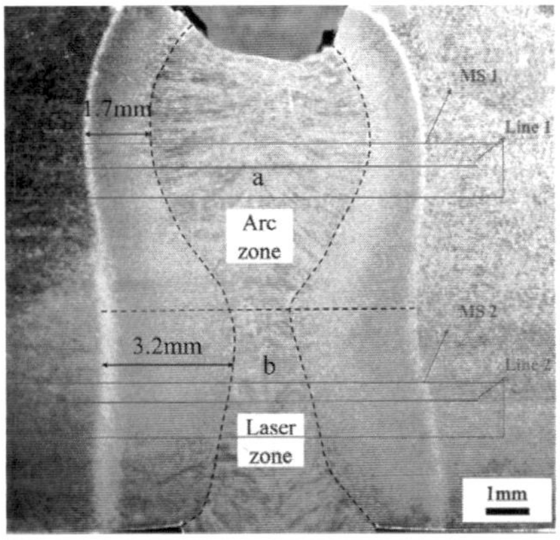

Figure 2. Weld bead profile of the backing welded joint.

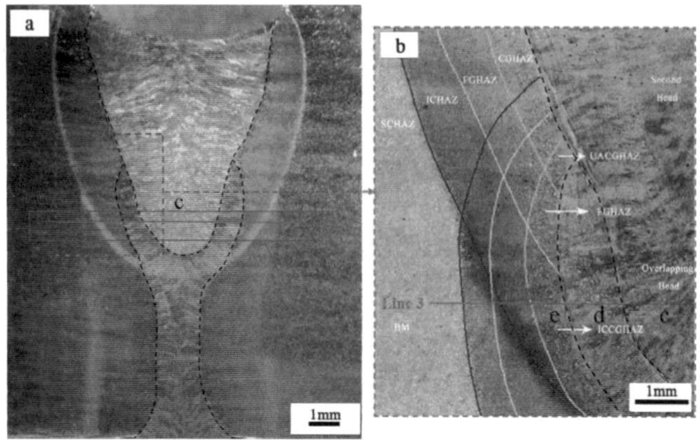

Figure 3. Weld bead profile of the double-pass welded joint.

3.2. Microstructures

Figure 4 shows the microstructures of the WM. The WM of the arc zone and laser zone were mainly composed of AF. Some GB and M-A constituents were also found in the WM. M-A constituents (reveals white and bright) were mainly scattered within the grain. The grain size of the laser zone WM was much smaller than that of the arc zone WM, relative fast cooling rate led to the small grains. The double-pass welded joint became further complicated due to the thermal cycles overlapping effect. Part of the arc zone WM went through a remelting process (point c in figure 3a). The remelt region was composed of AF, GB and a bit of LM as shown in figure 4e. Part of the arc zone WM experienced a reheating process (point d in figure 3b). Compared with the remelting WM, more GB and LM were formed in the reheat weld as shown in figure 4f.

The arc zone and laser zone possessed semblable HAZ microstructures. Figure 5 shows the arc zone HAZ microstructures. The CGHAZ and FGHAZ were both mainly composed of LM, some GB and polygonal ferrite (PF) also can be found. The ICHAZ consisted of martensite and M-A constituents, but the lath structure of martensite disappeared. The microstructure of the SCHAZ was tempered sorbite and the lath structure was visible. Reheated by the second thermal cycle, the CGHAZ was divided into unaltered CGHAZ (UACGHAZ), FGHAZ, ICCGHAZ and subcritically reheated CGHAZ (SCCGHAZ) as shown in figure 3b. The toughness deterioration always occur at the ICCGHAZ [13]. Detailed microstructure characterization of ICCGHAZ were conducted by OM, SEM as shown in figure 6. Most of the M-A constituents was blocky and distributed on the grain boundary, some M-A constituents was slender and distributed in the grain. The blocky M-A constituents was 1~2 μm. The ICCGHAZ inherited the coarse grain size of the CGHAZ. The blocky M-A constituents continuously connected with each other at the boundaries. And some of the blocky M-A constituents cracked already as shown in figure 7c and 7d.

3.3. Hardness and micro-shear properties

Figure 7 shows the Vickers hardness and micro-shear testing results. The variation tendency of the micro-shear strength was well matched with the hardness. The arc zone and laser zone of the backing welded joint had similar hardness and micro-shear strength distributions. The highest hardness and micro-shear strength of the arc zone and laser zone both appeared near the CGHAZs. Slightly soften showed up at the SCHAZs with relative lower hardness and micro-shear strength compared with base metal, because tempered sorbite was formed in SCHAZs. The hardness and micro-shear strength of the laser zone were higher than that of the arc zone, especially for the WM. The grains in the laser zone was much smaller than that of the arc zone. The small grain mains more grain boundaries per unit volume. It is generally recognized that grain boundaries can inhibit the dislocation movement and increase the hardness and strength. As for the double-pass welded joint, the reheated WM (Part 2 in figure 3) and ICCGHAZ owned the highest hardness and micro-shear strength with values of 432 HV10 and 1100 MPa which were basically equal to that of the arc zone CGHAZ. The LM and M-A constituents in the ICCGHAZ and reheated WM led to the hardness increase. The remelt WM (Part 1 in figure 3) of the double-pass joint has a hardness of 302 HV10 and micro-shear strength of 700 MPa which were much lower than that of the backing welded joint WM. The reduction of the harden LM caused the decrease. The lowest micro-shear toughness of the arc zone and laser zone both presented at SCHAZs with values of 0.70 J/mm^2 and 0.82 J/mm^2, respectively. The low micro-shear strength of SCHAZ accounted for the low toughness. Reduction of micro-shear toughness also appeared at the arc zone and laser zone CGHAZs with average values of 0.94 J/mm^2 and 0.98 J/mm^2 despite of high micro-shear strength. The lowest micro-shear toughness of the overlapping zone presented at SCHAZ (0.74 J/mm^2) and ICCGHAZ (0.72 J/mm^2). Obviously, the subsequent thermal cycle deteriorated the CGHAZ micro-shear toughness. The SCHAZ, CGHAZ and ICCGHAZ were the local brittle zones.

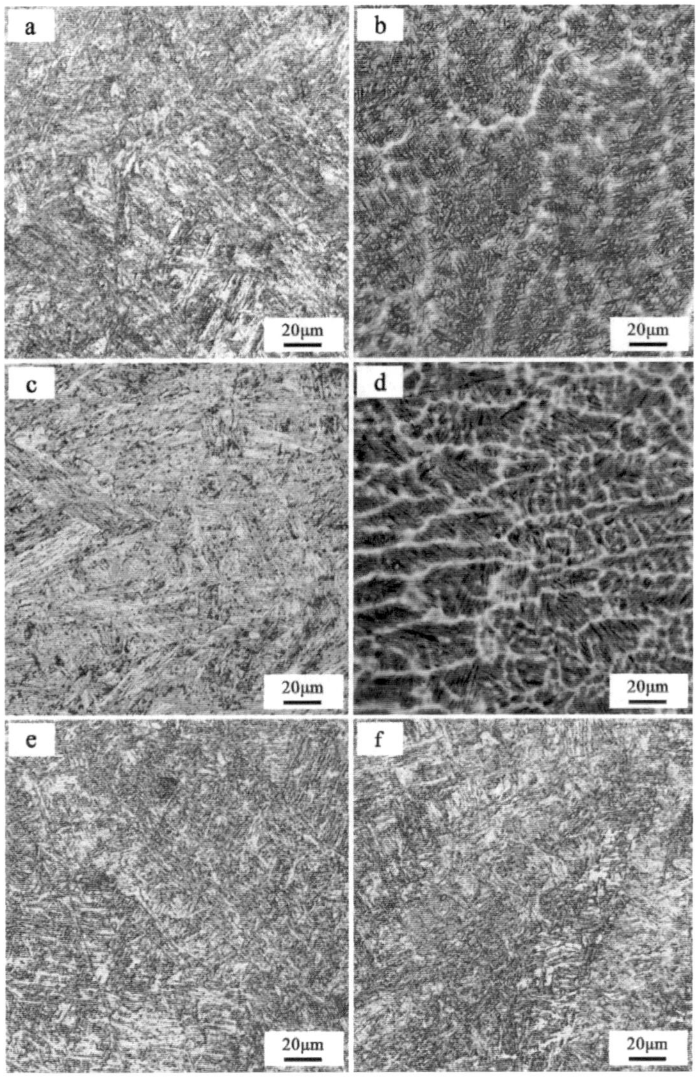

Figure 4. Optical microstructures of WM: (a), (b) are arc zone at point a in figure 2, (c), (d) laser zone at point b in figure 2, (e) overlapping zone at point c in Figure 3b, (f) overlapping zone at point d in figure 3b (a, c, e, f were etched with 4% nital solution and b, d were etched with Lepera etchant).

IOP Conf. Series: Materials Science and Engineering **758** (2020) 012045 doi:10.1088/1757-899X/758/1/012045

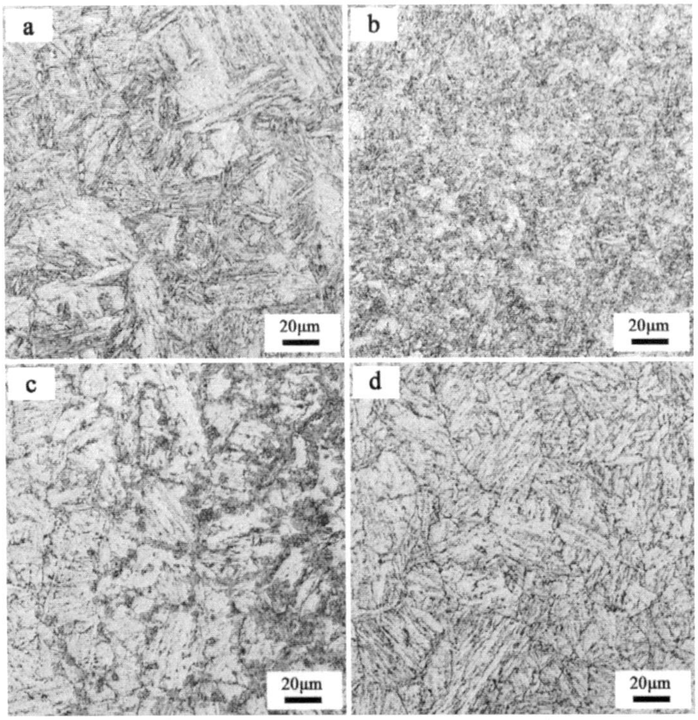

Figure 5. Microstructures of the arc zone: (a) CGHAZ, (b) FGHAZ, (c) ICHAZ, (d) SCHAZ.

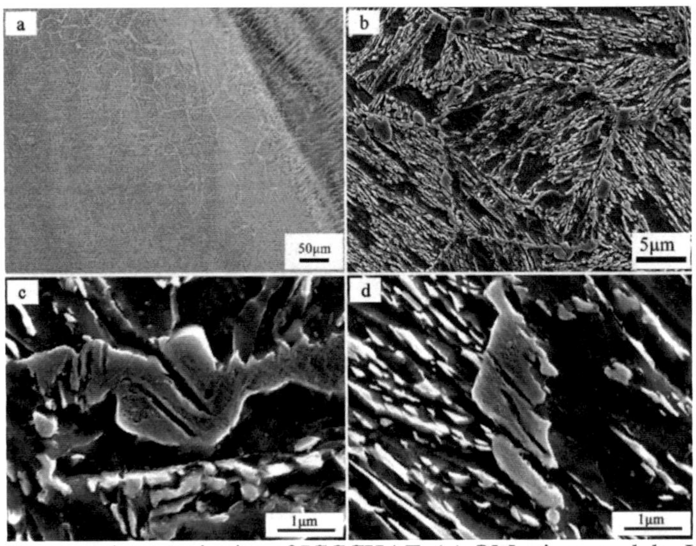

Figure 6. Microstructure characterization of ICCGHAZ: (a) OM micrograph by Lepera etchant (b) SEM micrograph, (c) enlarged SEM micrograph of region 1, (d) enlarged SEM micrograph of region 2.

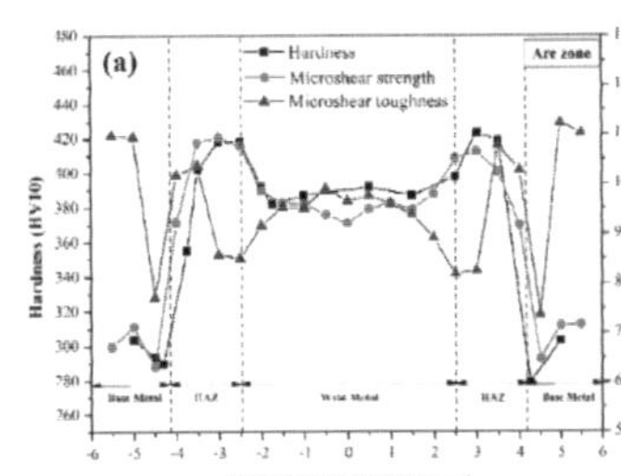

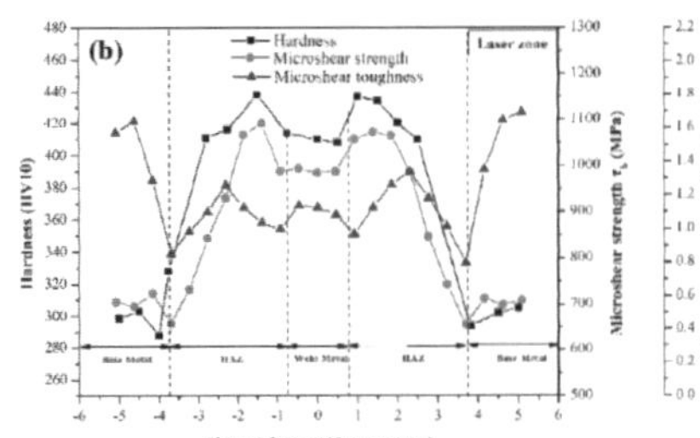

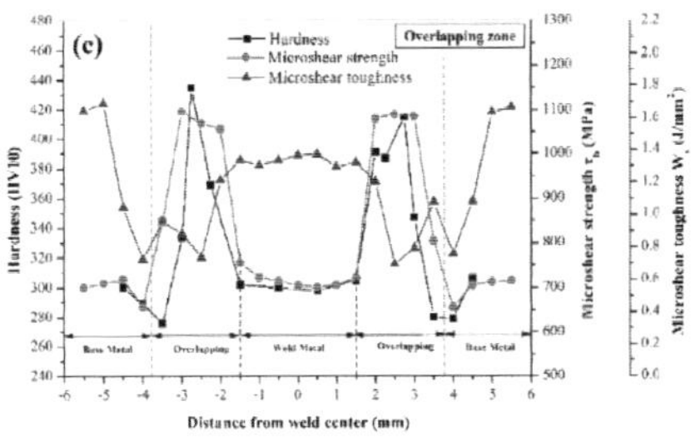

Figure 7. Vickers hardness and micro-shear properties of: (a) arc zone (MS 1), (b) laser zone (MS 2), (c) overlapping zone (MS 3).

3.4. -40 ℃ Charpy impact toughness

The sampling location are shown in figure 8. The V-notch was placed at the welded metal center (WM), fusion line (FL) and fusion line +2 mm (FL+2). The results were shown in figure 9. The average impact values of base metal was 61.3 J. The FL specimens possessed of the lowest impact values. The average impact values of the arc zone and laser zone FL specimens were 35.7 J and 32 J, respectively. The overlapping zone FL specimen owned an average impact value of 27.3 J. To identify the microstructures that caused toughness deterioration, the macro-fracture and micro-fracture morphologies of the base metal and FL specimens were characterized by stereomicroscope and SEM.

As shown in figure 10, the macro-fracture surface of the base metal was dark and uneven, the shear lip nearly accounted for half of the fracture surface. The micro-fracture surface was a typical dimple ductile fracture with large dimples linked by small dimples as shown in figure 11a. As for the arc zone and laser zone FL specimens, the notches crossed the WM and CGHAZs. Despite of the large percentage of shear lip similar to the base metal specimen, the fibrous zone was very small and most of the macro-fracture surface was radiative zone characterized by bright and even surface. Figure 12b and figure 12d are the micro-fracture morphologies of point b (in Figure 10b, arc zone CGHAZ) and point d (in figure 10.c, laser zone CGHAZ). The surface was quasi-cleavage fracture featured with fracture facet and tear ridges. The fracture facets varied from several to dozens of microns, some dimples were distributed around the tear ridges. Figure 11c and figure 11e are the micro-fracture morphologies of point c (in figure 10b, arc zone WM) and point e (in figure 10c, laser zone WM). The morphologies were both typical dimple fracture surface. The dimples at point c were uniform sized but smaller than that of the point e. With overall consideration of macro-fracture and micro-fracture morphologies, it can be inferred that the CGHAZs caused the toughness deterioration of the arc zone FL and laser zone FL specimens. The coarse CGHAZs may be the reason caused the toughness deterioration. As for the overlapping FL specimen, the shear lip was the smallest only 20% of the fracture surface and its macro-fracture surface was bright white and flat. There was basically no fibrous zone. Figure 11f and figure 11g are the micro-fracture morphologies of point f (in figure. 10d, ICCGHAZ) and point g (in figure. 10d, reheated WM). The micro-fracture morphology at reheated WM was typical dimple. The micro-fracture morphology at ICCGHAZ was quasi-cleavage fracture and the fracture facets were larger than that of the CGHAZs (figure 11d and figure 11d). The ICCGHAZ led to the toughness reduction of the overlapping FL specimen.

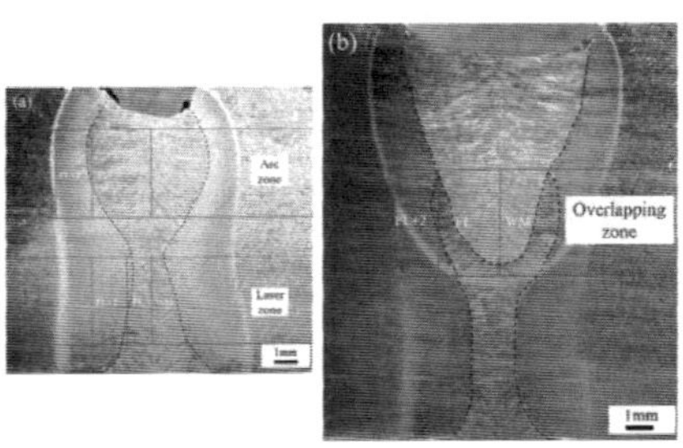

Figure 8. Impact sampling location for: (a) single-pass welded joint, (b) double-pass welded joint.

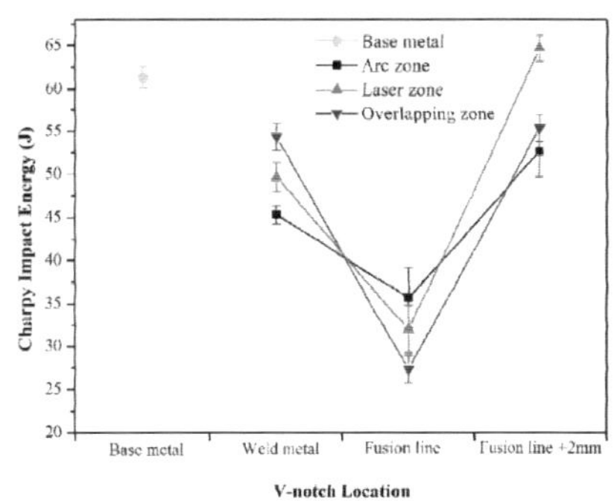

Figure 9. Charpy impact testing results.

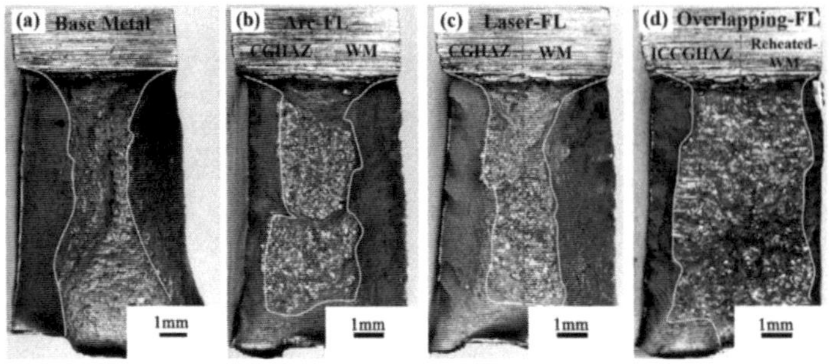

Figure 10. Macro-fracture morphologies of: (a) base metal, (b) arc zone FL, (c) laser zone FL and (d) overlapping zone FL.

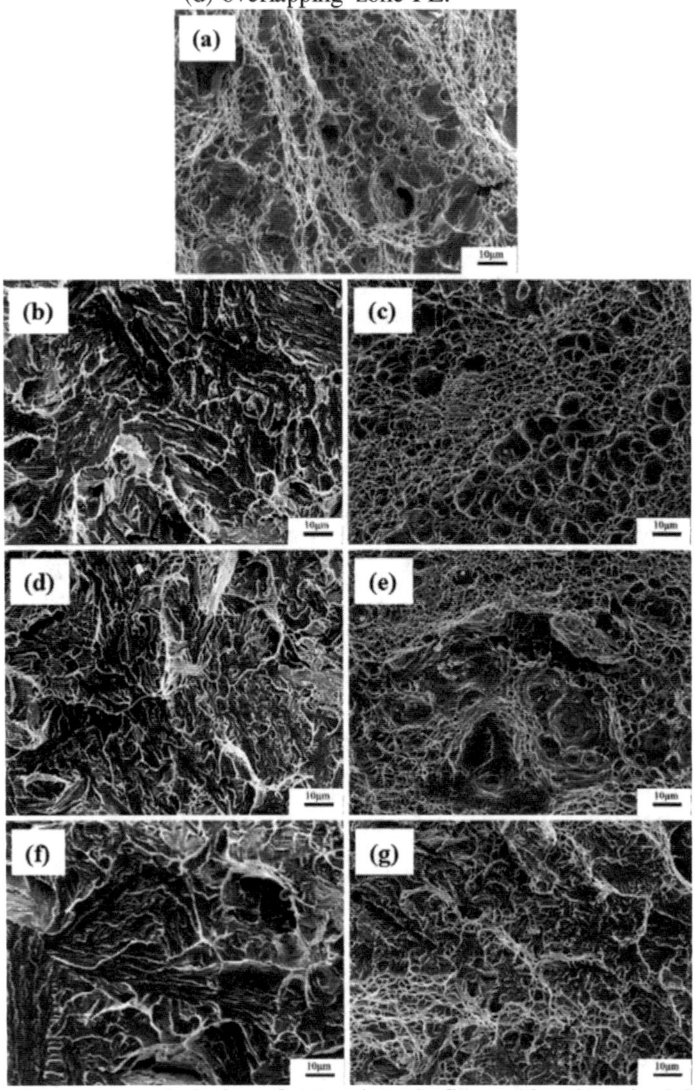

Figure 11. Micro-fracture morphologies of: (a) point a in figure 10a, (b) point b in figure 10b, (c) point c in figure 10b (d) point d in figure. 10c, (e) point e in figure 10c, (f) point f in figure 10d, (g) point g in figure 10d.

4. Conclusion

The backing and double-pass HLAW welded joints with no visible defects were obtained under optimised parameters. Arc zone WM, laser WM and overlapping zone WM both consisted of AF, GB and M-A constituents. Arc zone and laser zone HAZs had similar microstructures. CGHAZs were composed of LM, GB and PF. ICCGHAZ inherited the coarse grain size of CGHAZ and cracked blocky M-A constituents were found at grain boundary. The variation tendency of the micro-shear strength was well matched with the hardness. The highest hardness and micro-shear strength appeared near CGHAZs, reheated WM and ICCGHAZ. Slightly soften showed up at SCHAZs. SCHAZ, CGHAZ and ICCGHAZ showed a reduction of micro-shear toughness. -40℃ sub-size Charpy impact testing showed that FL specimens possessed the lowest impact values. The coarse CGHAZs and cracked blocky M-A constituents of ICCGHAZ may be the reason caused the toughness deterioration.

5. References

[1] Blekkenhorst F, Ferrari GM, Wekken CJVD, et al. 1986 *Br Corros J* **21(3)**: 163-176
[2] Flemming Ove Olsen 2009 *Hybrid Laser-Arc Welding* (UK: Woodhead Publishing Limited)
[3] Acherjee B. 2018 Opt Laser Technol **99** 60-71
[4] Cao X, Wanjara P, Huang J, et al. 2011 *Mater Des* **32(6)** 3399-3413
[5] Bunaziv I, Frostevarg J, Akselsen O M, et al. 2018 *J Mater Process Technol* **102** 34-44
[6] Wahba M, Mizutani M, Katayama S 2016 *Mater Des* **97** 1-6
[7] Gu X, Li H, Jiang X, et al. 2017 *Int J Adv Manuf Technol* **89(9-12)** 1-11
[8] Chen Y, Feng J, Li L, et al. 2013 *Mater Sci Eng A* **582** 284-293
[9] Gook S, Gumenyuk A, Rethmeier M. 2014 *Sci Technol Weld Joi* **19(1)** 15-24
[10] Chen Y, Yang C, Chen H, et al. 2015 *Int J Adv Manuf Techno* **78(1-4)** 457-464
[11] Zhang X, Mi G, Li S, et al 2017 *Int J Adv Manuf Techno* **3** 1-12
[12] Nolting AE, Munro C, Cao XJ, et al.2012 *Can Metall Quart* **51(3)** 336-345.
[13] P. Zhang X, Dorn L 1998 *Int J Pres Ves Pip* **75** 37-42

Acknowledgments

This work was supported by the National Natural Science Foundation of China (No.51874345), Natural Science Foundation of Shandong Province (No.ZR2018MEE012) and Key Research & Development project of Shandong Province (No.2017CXGC0812).

Short Review of Polymer Composites for 3D Printing

Z H Zhu [1], N Zhang [1], T Wang [1] and M Y Hao [2]

1 School of advanced manufacturing technology, Guangdong Mechanical & Electrical Polytechnic, Guangzhou, China;

2 The Key Laboratory of Polymer Processing Engineering, Ministry of Education, South China University of Technology, Guangzhou, China

Email: 184823174@qq.com

Abstract

3D printing technology is a rapid prototyping technology, which can use different materials to prepare real objects through layer by layer printing based on three-dimensional digital model. At present, the material with the largest amount of application and the most extensive molding method is polymer material。The research of ABS, PLA, PCL, TPU, photosensitive, resin polymer hydrogel and their composites are reviewed in this paper.

1. Introduction

3D printing technology, also known as additive manufacturing technology, is a new manufacturing technology made by computer-aided design (CAD), which is similar to ordinary printers and directly prints products. The forming process is adopting 3d drawing tools such as UG Pro/E Solid Works to conduct 3d modeling of the products to be printed and export them into STL format firstly. Then, 3d slicing software was used to conduct language segmentation and slicing processing on the established model and generate gcode file which is transferred to the 3D printer. Finally, the 3D printer is used to deposit the material layer upon layer on the working platform and finally print the whole product.

A 3D printing system can realize the printing of different materials. For example, metallic,stone polymer materials can be applied to 3D printing among which the polymer materials have the largest amount of application and the most extensive molding method. Several typical polymer materials for 3D printing including ABS, PLA, PCL and TPU photosensitive resin polymer hydrogel and their composite materials are introduced in this paper firstly. Then the application of materials for 3D printing in biomedical electrical aerospace and other fields was discussed.

2. Typical materials for 3D printing

2.1 ABS resin

ABS resin is a thermoplastic resin composed of three components. ABS material is suitable for 3D printing technology with the following characteristics. The first is the moderate detail performance. The second is the high strength. The third is suitable for product functional verification. However, the unsaturation of butadiene makes ABS have poor heat resistance and ozone resistance and warping deformation and dimensional shrinkage will occur under the action of thermal stress, which limit the development of pure ABS in the field of 3D printing technology.

Quill et al. [1] obtained BN (boron nitride) /ABS composites with thermal conductivity and electrical insulation by FDM and injection molding respectively. The study showed that the thermal conductivity increased significantly and the impact strength of bending strength decreased with the increase of BN content.Tekinalp et al. [2] applied staple fiber of 0.2 ~ 0.4 mm to reinforce ABS as the raw material for 3D printing. It was found that the tensile strength and modulus of 3D printed samples were improved by 115% and 700% respectively, and the fiber orientation was up to 91.5 than traditional technologies. However, samples prepared by FDM have high porosity. Therefore, improving the adhesion and compatibility of two phases through surface treatment of fibers or using appropriate additives is the main direction of future research.

Nikzad M et al. [3-4] applied iron powder and copper powder particles to modify ABS which was used them in FDM 3D printing. The results show that the thermodynamic and mechanical properties of ABS substrate have been significantly improved. When the iron content of ABS/Fe composite is high, the compression strength and energy storage modulus of FDM products are greatly improved and the rigidity of the composite is significantly enhanced at the same time. In addition, some scholars [5] blended glass fiber and carbon nanofiber with ABS for FDM 3D printing. It was found that the mechanical and machining properties of the printed products have been improved obviously. However, the molten ABS smells bad during the printing process and thus people don't like to use ABS as printing material.

2. 2 PLA

PLA is mainly derived from plant starch which has good biocompatibility and biodegradability. PLA can be used for 3D printing, the main advantages of which are low melting point, low viscosity and excellent mechanical properties, but high brittleness, transparent materials and high gloss.

Qin [6] improved the crystallization rate and toughness of PLA by blending polycarbonate with PLA, and also retained the degradability of PLA. Song [7] prepared WF/PLA wood-plastic composite with eucalyptus powder (WF), which was a degradable environment-friendly plastic. When the content of wood powder was 50% (mass fraction), the tensile strength of composite increased by 10 MPa compared with that of pure PLA. Tekinalp [61] prepared staple fiber reinforced ABS composite through fused deposition molding and die casting molding respectively to study the mechanical properties and structural characteristics of the composites. The results showed that the tensile strength and elastic modulus of FDM products have been improved, and the FDM fibers have a higher orientation in the composites. Although the porosity of FDM printed composites is higher, its tensile strength and tensile modulus are higher than those of traditional die-casting composites, which are determined by the orientation and dispersion of fibers in ABS.

Y Tao [66] prepared 3D printing wire of wood powder reinforced PLA composite and successfully applied it in FDM molding. The tensile properties, thermal properties, crystallinity and surface morphology of composites were tested and characterized by scanning electron microscope (SEM) thermogravimetric analysis (TGA), differential scanning calorimetry (DSC) and X-ray diffraction (XRD). The results showed that the initial bending resistance of the composite was enhanced, the initial thermal degradation temperature of the composite decreased slightly, and the melting temperature was hardly affected. Chizari [10] prepared PLA nanocomposite with 40% (mass fraction) carbon nanotubes (CNT). It was found that the conductivity of the composite was up to 3 800 S /m. Thstudy explored the application of 3D printed conductive composites in smart sensors in textile or electronic fields.

In the field of tissue engineering, Lactic acid produced by degradation of PLA scaffold can cause inflammation and allergic phenomenon. Blending PLA with CS (chitosan) can improve the toughness of PLA and alleviate the acid produced by the degradation of PLA. The PLA/CS scaffold has been successfully applied in the field of neural tissue engineering as a three-dimensional system to increase the differentiation of neurotic cells. [11]

2. 3 Polycaprolactone (PCL)

PCL is widely used as medical biodegradable materials and drug control release system , with good biodegradability, biocompatibility and non-toxicity. As PCL has the ability to store and restore deformation, it can adapt to the rapid development of 4D printing. At present, researchers mainly improve the mechanical strength of PCL through blending and copolymerization modification.

Visser et al. [12] developed a 3D printing technology, called melt electrospinning writing, which could print a filament with a diameter of 5 microns, which can be blended with methylacrylamide to prepare high-strength composite scaffold. Lee et al. [13] mixed chitosan (CS) obtained by electrospinning with PCL nanofibers to prepare CS /PCL chains by 3D printing and produced artificial blood vessels successfully. Qu, etc. [14] prepared PCL/nHA nanocomposite scaffolds based on electrohydrodynamic printing. The average diameter of the scaffold is (8. 85±12)μm, which is close to the size of living cells. In addition, scaffolds have good biocompatibility which can realize the multi-scale regulating cell microenvironment and promote tissue regeneration.

2. 4 Thermoplastic polyurethane (TPU)

TPU has excellent wear resistance, corrosion resistance, radiation resistance, oil resistance, oxygen resistance and ozone resistance which is widely used in the field of 3D printing. Akbarian et al. [15] prepared short aramide fiber reinforced polyurethane composite and found that the short fiber had serious axial splitting and fracture fibrification during its blending with the polyurethane.

In recent years, biodegradable natural fibers with low cost have gradually become a research hotspot. El-shekeil et al. [15-16] strengthened thermoplastic polyurethane with Kenaf fiber and Cocoa Pod Husk Fibers. It was found that the tensile and flexural properties of the composites were improved with the load increasing. Polyurethane is widely used in conductor field. Kim et al. [17] invented a 3D multi-axial force sensor manufactured by FDM, whose structure part USES thermoplastic polyurethane (TPU) as filament, and the sensing parts use carbon nanotube (CNT) /TPU nanocomposite filament to realize the customization, rapid manufacturing and economic feasibility of multi-axial sensor. The nano porous PMMA/TPU composite prepared by Wang et al. [18] has higher mechanical properties than ordinary honeycomb materials with a thermal conductivity of 24.8mk/mK, which has potential applications in the field of 3D printing. Chen et al. [19] prepared biocompatible TPU/PLA/GO nanocomposite elastomer, which is a kind of good tissue engineering scaffold material. Its unique shape memory function, antibacterial effect and excellent electro-thermal performance can be applied in advanced scientific and technological fields.

2.5 Light cured solid material

Research on 3D printing of photocurable materials at home and abroad focuses on two aspects. The first is that different prepolymer photoinitiators were used to optimize the parameters to obtain the resin meeting the performance requirements. The second is to optimize the 3D printing equipment. Wang et al. [20] prepared epoxy resin matrix with continuous through-hole by 3D printing technology, and then prepared CNTs/epoxy resin composites with excellent electrical properties and electrical shielding properties by dipping process, which provided necessary reference for the preparation of composites with electromagnetic shielding properties.

Tumbleston[21] adopts the continuous liquid interface production (CLIP) technology to enable the parts to be formed within 10 minutes, and improves the accuracy and speed of SLA technology, which is 25 ~ 100 times faster than traditional printing technology.

Comptom et al. [22] used epoxy resin and other raw materials to prepare a lightweight honeycomb composite with excellent mechanical properties, which is a 3D printing honeycomb composite.

3. Conclusion

3D printing is a creative technology which are not widely used in industry. There are still some limitations to overcome in the research of polymer materials of 3D printing. Firstly, the orientation and content of fibers are important factors affecting the properties of materials. At present, the nozzle will be blocked obviously when the fiber content is greater than 40% (mass fraction). Therefore, the fiber content should be further improved and the materials should be processable. Secondly, the nanoparticles should be treated with chemical surface treatment or modified by adding compatibilists in order to avoid agglomeration of nanoparticles and ensure a good interface between nanoparticles and polymers. Thirdly, how to eliminate the gap in printing process and ensure the good interface between matrix and reinforcement is the main content of future research.

Reference

[1]. Quillt J, Smith M K, Zhou T, et al. Thermal and mechanical properties of 3D pirnted boron nitride-ABS composites. Applied Composite Materials, 2018, 25: 1205-1217.

[2]. Tekinalp H L, Kunc V, Velez Garcia G M, et al. Highly oriented carbon fiber-polymer composites via additive manufacturing. Composites Science and Technology, 2014, 105: 144-150.

[3]. Nikzad M, Masood S H, Sbarski I. Thermo-mechanical properties of a highly filled polymeric composites for Fused Deposition Modeling. Materials & Design, 2011, 32(6):3448-3456.

[4]. Nikzad M, Masood S H, Sbarski I, et al. ThermoMechanical Properties of a Metal-filled Polymer Composite for Fused Deposition Modelling Applications. Journal of Periodontology, 2007, 77(12):2084-9.

[5]. Zhong W, Li F, Zhang Z, et al. Short fiber reinforced composites for fused deposition modeling. Materials Science & Engineering A, 2001, 301(2):125-130.

[6]. Qin S X, Zhao L F, Liu J. A PLA composite suitable for 3D printing with high strength and high toughness and its preparation method. China patents, CN104629280A.

[7]. Song L X, Yao N N, Song Y Z, et al. Study on the properties of wood powder/poly (lactic acid) degradable composites. Functional Materials, 2014, 45(5) : 5037-5040.

[8]. Tekinalp H L, Kunc V, Velez-Garcia G M, et al. Highly oriented carbon fiber–polymer composites via additive manufacturing. Composites Science & Technology, 2014, 105:144-150.

[9]. Tao Y, Wang H, Li Z, et al. Development and Application of Wood Flour-Filled Polylactic Acid Composite Filament for 3D Printing. Materials, 2017, 10(4): 339.

[10]. Chizari K, Daoud M A, Ravindran A R, et al. 3D printing of highly conductive nanocomposites for the functional optimization of liquid sensors. Small, 2016, 12(44) : 6076-6082.

[11]. Hoveize E, Tavakol S, Ebrahimi-Barough S. Neuroprotective effect of transplanted neural precursors embedded on PLA/CS scaffold in an animal model of multiple sclerosis. Molecular neurobiology, 2015, 51(3) : 1334-1342.

[12]. Visser J, Melchels F P W, Jeon J E, et al. Reinforcement of hydrogels using three-dimensionally printed microfibers. Nature Communications, 2015, 6: 6933.
[13]. Lee S J, Heo D N, Park J S, et al. Characterization and preparation of bio-tubular scaffolds for fabricating artificial vascular grafts by combining electrospinning and a 3D printing system. Physical Chemistry Chemical Physics, 2015, 17(5) : 2996-2999.
[14]. Qu X L, Xia P, He J K, et al. Microscale electrohydrodynamic printing of biomimetic PCL /nHA composite scaffolds for bone tissue engineering. Materials Letters, 2016, 185: 554-557.
[15]. Akbarian M, Hassanzadeh S, Moghri M. Short twaron aramid fiber reinforced thermoplastic polyurethane. Polymers for Advanced Technologies, 2008, 19(12) : 1894-1900.
[16]. El-shekeil Y A, Sapuan S M, Abdan K, et al. Influence of fiber content on the mechanical and thermal properties of Kenaf fiber reinforced thermoplastic polyurethane composites. Materials & Design, 2012, 40: 299-303.
[17]. El-shekeil Y A, Sapuan S M, Algrafi M W. Effect of fiber loading on tensile properties of cocoa pod husk fibers reinforced thermoplastic polyurethane composites. Appied Mechanical Materials, 2014, 606: 25-28.
[18]. Kim K, Park J, Suh J, et al. 3D printing of multiaxial force sensors using carbon nanotube (CNT) /thermoplastic polyurethane (TPU) filaments. Sensors and Actuators A: Physical, 2017, 263: 493-500.
[19]. Wang G, Zhao J, Mark L H, et al. Ultra-tough and super thermal-insulation nanocellular PMMA/TPU. Chemical Engineering Journal, 2017, 325: 632-646.
[20]. Chen Q, Mangadlao J D, Wallat J, et al. 3D printing biocompatible polyurethane /poly (lactic acid) /graphene oxide nanocomposites: anisotropic properties. ACS applied materials &interfaces, 2017, 9(4) : 4015-4023.
[21]. Wang Y, Fan Z W, Zhao J, et al. Carbon nanotubes/epoxy resin electromagnetic shielding composites prepared by 3D printing. Journal of composite materials, 2019, 1: 1-6.
[22]. Tumbleston J, Shirvanyants D, Ermoshkin N, et al. Additive manufacturing. Continuous liquid interface production of 3D objects. Science, 2015, 347(6228) : 1349-1352.
[23]. Comptom B G, Lewis J A. 3D-printing of lightweight cellular composites. Advanced Materials, 2014, 26: 5930-5935.

Acknowledge

This research was supported by Key Research Platforms and Projects of Universities of Guangdong Province (grant no. 2018GKTSCX111) and Funding Project for Scientific Research of High-level Talents of Guangdong Mechanical & Electrical Polytechnic (grant no. Gccrcxm-201903).

IOP Conf. Series: Materials Science and Engineering **758** (2020) 012047 doi:10.1088/1757-899X/758/1/012047

Methodology for determining optimized traffic light cycles based on simulation

G Rivera[1], L Velasquez[1], A Bravo[2] and P Miñano[2]

1 Research Student. Faculty of Civil Engineering, UPC, Lima, Peru
2 Full Professor. Faculty of Civil Engineering, UPC, Lima, Peru
E-mail: aldo.bravo@upc.pe

Abstract. In large urbanized cities, a major problem that affects the economy and health of all citizens is vehicular congestion. This is because the traffic light cycles are not adequate. In the present study, we seek to optimize traffic light cycles based on simulation, in order to improve vehicle flow. For this, the PTV Vissim 9.0 software was used as a simulator and the Synchro 10.0 software to determine the initial optimal traffic light cycle. Through several runs and having as variables the length of queues, delay times and the average speed, the optimal traffic light cycle could be found for the study area. The results obtained reflect a 14% reduction in delay times and 10% in queue lengths. On the other hand, the average vehicle speed increased by 10.56%. All this represents an improvement in the service level of the study intersections.

1. Introduction

The growth of the motor park in the major cities of the world brings with it the increase in vehicular congestion, which has an impact on the economy and health of all citizens. This is due to a delay in the management of urban circulation systems that causes various problems that affect traffic [1]. Currently finding a solution to these problems requires that the investment of significant resources is aimed at improving existing infrastructure.

With the rise of the motor park in modern cities, the problem remains unresolved and even worse is compounded every year. There are several solution proposals to reduce this problem. The first is to modify the existing road infrastructure (rails, bridges, roundabouts or underground passages). This solution can be implemented in non-metropolitan areas, but it is very expensive. Another concerns the restriction of mobilization of private vehicles in high-density flow times. The most promising approach is the smart traffic light system that can adapt to real-time traffic conditions. This proposal is the most cost-effective because it only requires the integration and updating of existing systems, without imposing restrictions on the movement of the vehicle [2].
Programs used for the investigation of traffic control systems are divided into two types. The first concerns Adaptive Traffic Control Systems (ATCS), which consist of changing the duration of the traffic light cycle based on sensors located on the tracks which detect queued vehicles in real time. However, managing this system requires a high operating cost and qualified personnel. This real-time traffic information allows drivers to plot an optimal route to reach their destination [3]. The second approach is to use highly modern simulators [4], to simulate real-world scenarios cost-effectively and test them by iteration parameters (queue lengths or speeds) microscopic or macroscopic traffic views [5].

Within this context, the work done consists specifically of the topic of optimizing simulation-based traffic light cycles.

2. Case study

For this study, two intersections located in the city of Lima will be evaluated. According to the Dutch GPS company "TomTom" [6], it is ranked third in the world ranking of congested cities. Javier Prado Avenue is one of the main roads used so it is necessary to have an optimal vehicular flow. That is why the assessment of the intersections of the Av. Javier Prado – Ca. Las Flores y la Av. Javier Prado – Ca. Las Palmeras.

3. Methodology

The methodology used in this study is to generate an initial optimal cycle with Synchro 10.0 software. To do this, information (vehicle count, traffic light cycles, geometry and signaling) will need to be collected from the evaluated intersection. From this, through the simulation with the PTV Vissim 9.0 software will modify the traffic light cycles, taking into account variables such as tail lengths and average speeds (vehicles). This process is iterative so you can determine that the traffic light cycle is optimized when the tail length, decrease; and average speed; increase; in the best possible expression.

3.1 Data collection and analysis

To optimize traffic light cycles based on simulation, information was required from high vehicle density intersections in which the following were performed:

i. A vehicular count of the intersections was carried out for three days.
ii. Information was acquired on the geometry and signage of the study area.
iii. The "peak hour" (period where there is higher vehicular density) at intersections was determined.

3.2 Modeling and calibration

A. PTV Vissim 9.0

The Vissim 9.0 program is a microscopic traffic simulation model: each vehicle is explicitly modeled, has its own route and moves individually, through a route including vehicle tracking logic and lane change [7].

B. Synchro 10.0

Synchro 10.0 is a program for modeling, optimizing, managing and simulating traffic systems. The generation of optimal time plans consists of optimizing the cycle duration, partial times, compensation and phase sequence to minimize driver stops and vehicle delay [8].

For modeling and calibration, the sketchup and Vissim 9.0 programs were used, which was performed as follows:

i. The Sketchup program included geometry, signaling, and all users involved in intersections.
ii. This file was exported to the Vissim 9.0 software in (skp) format.
iii. The vehicle composition (cars, buses, trucks and motorbikes) and pedestrian composition were defined. In addition, the assignment of vehicle and pedestrian routes with priority rules, and control signals. The current traffic light cycles of the intersections studied can be seen in Table 1.
iv. The parameters of the model worked in the Vissim 9.0 program (speeds, times and queue lengths) were compared with the results obtained in the field. To do this, calibration was performed with the travel times of vehicles and pedestrians collected in the field.

Table 1: Current traffic light cycles present at intersections

	Av. Javier Prado & Ca. Las Flores			**Av. Javier Prado & Ca. Las Palmeras**		
Time (s)	**Green**	**Amber**	**Red**	**Green**	**Amber**	**Red**
W-E/E-W	169	3	150	248	3	177
Exclusive Turn	15	3	135	20	3	157
S-N	75	3	244	85	3	340
N-S	63	3	254	75	3	350

3.3 Validation and Simulation

Another dataset (aphores collected during a different day than the evaluated one) was used for model validation. On the other hand, the simulation was performed in time intervals of 15 min.

i. The validation process consisted of analyzing the model with another dataset (vehicular and pedestrian capacity). The analysis was carried out on vehicles and pedestrians in the stretch with the highest demand.
ii. 15 runs were performed in each 15-minute time interval with the current traffic light cycle present at both intersections to assess traffic indicators (speeds, delay times and queue lengths); it was observed that the speed of vehicles at each access varies depending on the time set in each traffic light cycle, so at each interval, the time it takes to cross vehicles was measured by a distance of 235 m and then calculated the average speed per access. The above mentioned, can be seen in Table 2.
iii. The proposal was to simulate the model with the optimized cycles obtained from the Synchro 10.0 program and as the parameters of tail lengths and vehicle speeds vary at each access, more optimized traffic light times were obtained, which are in Table 3.
iv. In Table 4, it is noted that through changes in traffic light cycles higher average speeds were obtained, for each time interval, which will optimize the vehicular flow.

Table 2: Simulation of current parameters

	Av. Javier Prado & Ca. Las Flores			Av. Javier Prado & Ca. Las Palmeras		
Access	**Time (s)**	**Distance (m)**	**Speed (km/h)**	**Time (s)**	**Distance (m)**	**Speed (km/h)**
W-E/E-W	58	235	14.58	70	235	12.09
S-N	40	235	21.13	46	235	18.39
N-S	45	235	18.79	50	235	16.92

Table 3: Optimized traffic light cycles every 15 minutes

	7:15 - 7:30 a.m.			7:30 - 7:45 a.m.			7:45 – 8:00 a.m.			8:00 - 8:15 a.m.		
Time (s)	**Green**	**Amber**	**Red**	**Green**	**Amber**	**Red**	**Green**	**Amber**	**Red**	**Green**	**Amber**	**Red**
W-E/E-W	155	3	122	158	3	137	160	3	128	155	3	130
Exclusive Turn	25	3	97	30	3	107	25	3	103	25	3	105
S-N	45	3	232	50	3	245	48	3	240	50	3	235
N-S	55	3	222	60	3	235	58	3	230	58	3	227

	8:45 - 9:00 a.m.			9:00 - 9:15 a.m.			9:15 – 9:30 a.m.			9:30 - 9:45 a.m.		
Time (s)	**Green**	**Ámber**	**Red**	**Green**	**Ámber**	**Red**	**Green**	**Ámber**	**Red**	**Green**	**Ámber**	**Red**
W-E/E-W	235	3	142	240	3	142	238	3	147	240	3	147
Exclusive Turn	20	3	122	15	3	127	20	3	127	20	3	127
S-N	70	3	307	70	3	312	75	3	310	70	3	317
N-S	55	3	322	60	3	322	55	3	330	60	3	327

Table 4: Simulation of optimized parameters

	Av. Javier Prado & Ca. Las Flores			Av. Javier Prado & Ca. Las Palmeras		
Access	**Time (s)**	**Distance (m)**	**Speed (km/h)**	**Time (s)**	**Distance (m)**	**Speed (km/h)**
W-E/E-W	50	235	16.92	57	235	14.84
S-N	36	235	23.50	42	235	20.14
N-S	40	235	21.15	46	235	18.39

4. Discussion of the results

4.1 Tail lengths

Figure 1 shows the results of the simulation in the PTV Vissim 9.0 program by 15-minute intervals. The results indicate that the tail lengths decrease by 10%. This reduction is related to the optimization of traffic light cycles.

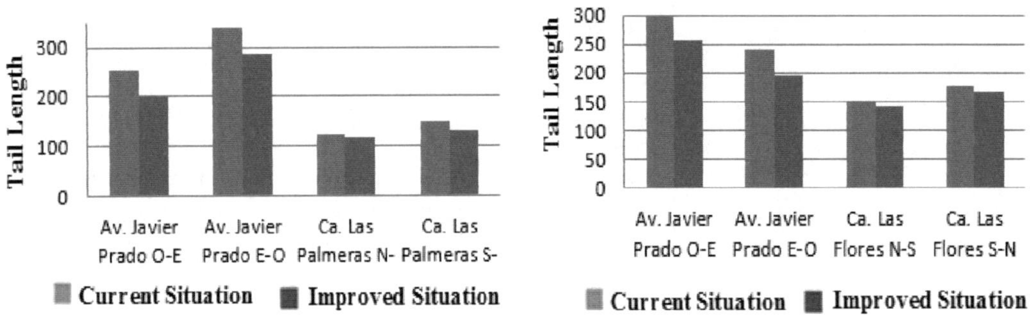

Figure 1: Comparison of queue lengths at both intersections, with current and optimized traffic light cycles.

4.2 Delay times and average speed

Table 5 shows travel delay times per vehicle. This parameter is based on the quality and operation of the intersections under study. The results obtained indicate that the delay times decrease by 14% and with respect to the average speed of vehicles per access increases by 10.56%.

Table 5: Parameter comparison in both scenarios

	Current traffic lights			Optimized traffic lights		
Access	**Delay time (s)**	**Service level**	**Current average speed (km/h)**	**Delay time (s)**	**Service level**	**Current average speed (km/h)**
O-E/E-O	89.7	F	14.58	77.1	E	16.92
S-N	98.7	F	21.13	78.6	E	23.50
N-S	95.1	F	18.79	76.6	E	21.15

Current traffic lights	Optimized traffic lights

Access	Delay time (s)	Service level	Current average speed (km/h)	Delay time (s)	Service level	Current average speed (km/h)
O-E/E-O	91.9	F	12.09	78.8	E	14.84
S-N	100.3	F	18.39	79.7	E	20.14
N-S	98.5	F	16.92	79.1	E	18.39

5. Conclusions

- Average vehicle speed increased by 10.56%, which represents an increase in the volume of traffic that flows through the intersection for each traffic light cycle.
- Optimized traffic light cycles every 15 minutes improve the average speed of vehicles during the first 235 meters of their journey.
- Tail length decreases by 10%, while delay times by 14%, which represents an improvement in the service level of both intersections.

6. References

[1] Quintero J.& Prieto L (2015) “Sistemas inteligentes de transporte y nuevas tecnologías en el control y adminsitración del transporte” *Revista digital de la Facultad de Ingeniería, Universidad Pedagógica y Tecnológica de Colombia sede Tunja*, pp. 53-54.

[2] D.Panovski and T.Zaharia, "Simulation-based vehicular traffic lights optimization" *Consumer Electronics (ICCE) 2019 IEEE International Conference on*, pp.258, 2019.

[3] Molina Navarro. A., Zamora Castro. S., Remess Pérez. M. & Lagunes Lagunes E. (2016). Los semáforos inteligentes en la logística urbana sustentable. *Revista Aplicaciones de la Ingeniería*, 28-30

[4] D. Zhao, S. members, Y. Dai, and Z. Zhang, "Computational Intelligence in Urban Traffic Signal Control: A Survey," *IEEE Trans. Syst. Man, Cybern.*, vol. 42, no. 4, pp. 485-494, 2012.

[5] C. Tolba, D. Lefebvre, P. Thomas, and A. The Moudni, "Continuous and Dated Petri Networks for Macroscopic and Microscopic Traffic Flow Modeling," *Simul. Model. Practy. Theory*, vol. 13, no. 5, pp. 407-436, 2005.

[6] TomTom (2019). World ranking of congested cities. Ámsterdam, Países Bajos

[7] PTV VISSIM (2016). Getting Started VISSIM 9. Karlsruhe.

[8] TRAFFICWARE, Engineered by N Naztec. Synchro 8 Studio Analysis, optimization, and simulation applications. Fecha de consulta: 10 de Abril del 2015. Disponible en: http://www.trafficware.com/products.

IOP Conf. Series: Materials Science and Engineering **758** (2020) 012048 doi:10.1088/1757-899X/758/1/012048

A 3-D Analysis on the Loading Behaviours of an Under-reamed Anchor in Cohesion-less Soils

Shih-Tsung Hsu[1], Xiang-Yang Chen[2]and Yi-Fan Wang[3]

1Professor and Chairman, Department of Construction Engineering, Chaoyang University of Technology, Taichung, TW
2 Lecturer, School of Civil Engineering and Architecture, Nantong Vocational University, Nantong, CN
3 Master student, Department of Construction Engineering, Chaoyang University of Technology, Taichung, TW
E-mail:sthsu@cyut.edu.tw

Abstract. This study adopts a constitutive model for sand and a 3-D numerical software, named "FLAC^{3D}" to examine the anchorage behaviours of an under-reamed anchor in sandy soil. A series of pull-out test on model anchors is performed to confirm the results obtained numerical analyses. Test results reveal that the load-movement curves computed numerically are get close to those estimated experimentally. For under-reamed anchors with dimensions of D=0.075m, L/D=4.5, and installed in sandy soil of D_r=70% (D_r: relative density, D: diameter of anchor, L/D: fixed end/diameter ratio), the shape of load-movement relationship on a deep anchor differs from that of a shallow anchor, especially the shape of end resistance-movement curve. However, the friction force, end resistance and total load do not attain peak values at same anchor movement. The end resistance of a shallow anchor displays obviously peak value. However, the peak value the end resistant cannot treasured from a deep anchor. The friction load presents peak and residual values. The mainly pull-out load of a shallow anchor is donated by the friction load, meanwhile the end bearing governs the pull-out load of a deep anchor; overburden depth of 12D is the shallow/deep anchor demarcation. According to the definition of end bearing coefficient, it is categorised to a deep anchor as the overburden surpasses 12D, it is a shallow anchor if the overburden is smaller than 7.5D, and as the overburden equal to 7.5D~12D, the anchor can be classified as a transit anchor.

1. Introduction

Foundations of diverse structures, such as submerged pipelines and transmission towers, are frequently applied huge lift forces. The application of ground anchors as part of a foundation system looks to be a feasible choice for preserving the stability and increasing a vertical uplift capacity of the structure. Additionally, vertical anchors are commonly applied to offset eccentric forces that load on foundation mats, to fix basements against uplift pressure resulting from groundwater, or to resist seismic force.

The top stratum of the Taipei Basin comprises the Sung-Shan formation, depending on the local geological characteristics, the thickness of which ranges from 30 to 60 m. Nearly all of the fixed anchor lengths of the anchors installed in the Taipei Basin are in the Sung-Shan formation, which contents of alternating layers of clay and sand. The fixed lengths are naturally located in sandy layers. In some places, however, the thickness of the sand that provides shear strength is restricted. As a result, an anchor with both of high anchorage load and short fixed length is necessary in this scenario.

To achieve both of high anchorage load and short fixed length in an anchor, Hsu and Liao [1] recommended the under-reamed anchor, which has an equal diameter in its fixed length. As displayed in Figure 1(a) is the under-reamed anchor, the fixed anchor is formed by the enlarged blade system. Initially, the device with four blades remains closed. When the drilling rod drills to the pre-set depth, all four blades are opened and turn around concurrently to expand the fixed length. The diameter can be determined by the length and rotating rate of the blades. After enlargement, the drilling rod is substituted by the strand system coupled to the swivel. A grout with the water/cement ratio of W/C=0.45~0.55 is then pressurized at 1 MPa to produce the under-reamed length. In accordance with the mechanism of load transfer within the fixed length, the anchor is categorized as a kind of compression anchor.

Studies on anchorage behavior of anchors are various, and can be categorized into several fields, such as field tests [2], model tests [3], limit analyses [4] and numerical analyses [5]. For field tests, most tested anchors are not stressed until failure; thus, their ultimate loads cannot be estimated accurately. The limit analysis cannot realistically examine the pre- and post-peaks of soil behavior. Regarding numerical analyses, these seldom use a fair stress-strain relationship for soil that includes strain softening to research the anchorage behavior, a complete load-movement curve of an anchor cannot be calculated precisely.

Therefore, a precisely numerical model and an analyzed method are demeaned to investigate the behavior of ground anchor, and the loading behaviors of under-reamed anchors are necessary to be researched further.

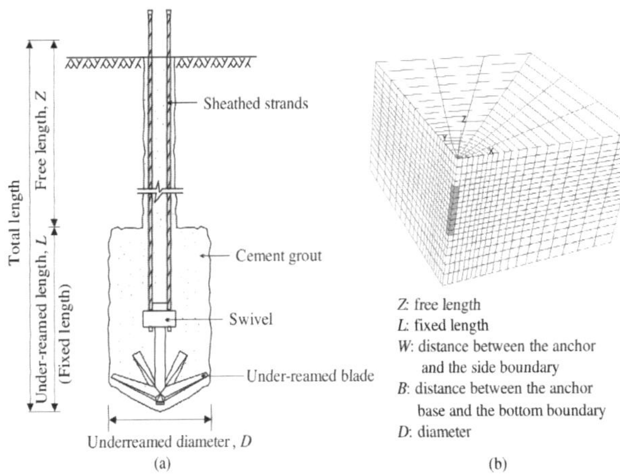

Figure 1. Illustration of an under-reamed anchor and represnetative meshes for numerical analysis (Z/D=2.5 、L/D=4.5).

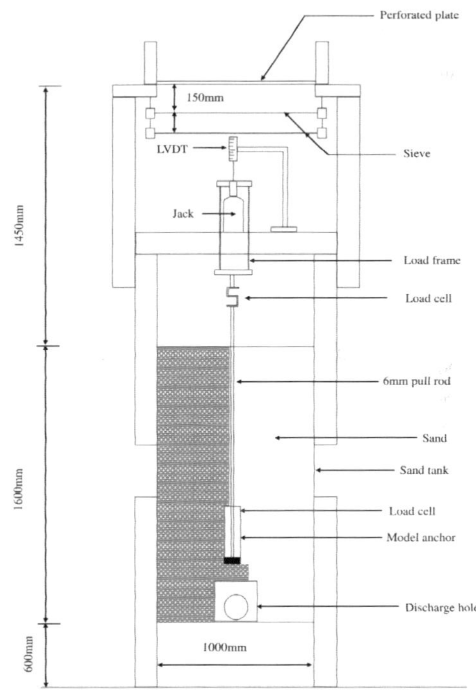

Figure 2. The experimental model anchor test's setups.

2. Numerical Model of Sand and Analytical Method

Hsu and Liao[1] recommanded a numerical model that consists both the strain hardening/softening and volumetric dilatancy, called *SHASOVOD* model. The model had been adopted with the software $FLAC^{2D}$ to study the anchorage behaviours of several types of anchors in sandy soil. The results revealed that the numerical results were consistent with those results obtained in-situ. Therefore, this work applies above model and the commerical software FLAC^{3D} to research the anchorage behavior of an under-reamed anchor in sandy soil.

The sandy specimen is taken from Yi-lan, it is classified as the *SP*-type soil, its specific gravity G_s of 2.69, which with a minimum dry unit weight of 13.70 kN/m^3, and a maximum dry unit weight of 17.34 kN/m^3.

The numerical model is related to the non-associated flow rule of plasticity theory. The yield function *f* can be expressed as

$$f = \sigma_1 - \sigma_3 \frac{1+\sin\phi^*}{1-\sin\phi^*} \tag{1}$$

where ϕ^* is the mobilized friction angle, which is related to accumulative plastic strain ε^p; and σ_3 denotes the minor principal stress, σ_1 represents the major principal stress. The plastic potential function g is

$$g = \sigma_1 - \sigma_3 \frac{1+\sin\psi^*}{1-\sin\psi^*} \tag{2}$$

where ψ^* resprents the mobilized dilatancy angle, and related to accumulative plastic strain ε^p. To illustrate the reliability of the numerical model, the stress-strain relations of sand with a relative density of 70%, calculated using the proposed model, are used to compare with those evaluated from triaxial test. As depicted in Figure 3, the triaxial test results get perfectly close to the results from numerically calculated results. Hsu and Liao [6] elucidated the relevant details of the model.

The assumptions for numerical analyses are listed as followings

- The fixed anchor (under-reamed length) is elastic and homogenous.
- The load is uniformly applied to the bottom of the anchor, because the under-reamed anchor is a compression anchor; and
- To reduce calculated consumption, one-quarter symmetric meshes are used for the under-reamed anchor.

Figure 1(b) illustrates the representative numerical meshes and details adopted to depict an anchor.

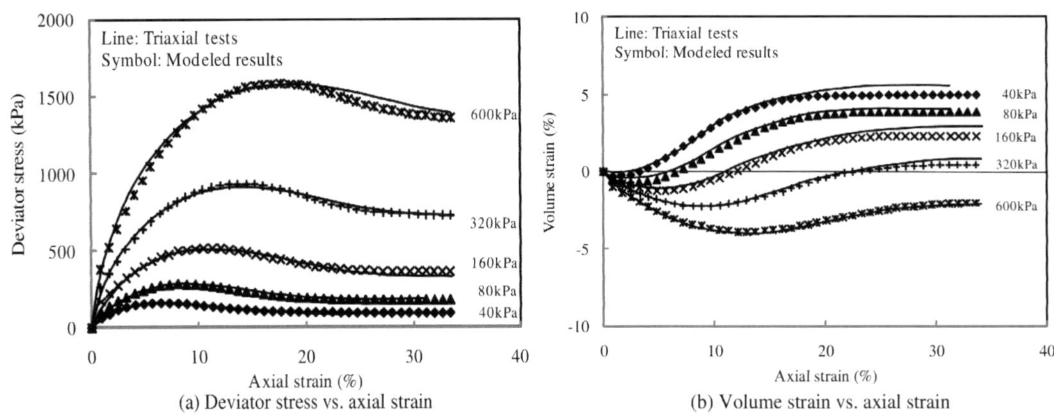

Figure 3. Tested and calculated deviator stress-axial strain-volume strain curves (D_r=70%).

3. Verifications of Numerical results

3.1. Experimental Model Tests

In engineering practice, the length/diameter ratios, L/D for under-reamed anchors can be ranged from 2.5~5.0. However, an under-reamed anchor with an L/D ratio of 4.5 for generating both shaft friction and end bearing is adopted for the model tests and numerical analyses in this paper.

The model anchor employed for the experimental test is made of aluminum tube, the diameter is 75 mm, length is 338 mm, and L/D is 4.5. The surface of the tube is corrugated in a pitch of 1.6 mm.

As shown in Figure 2, the anchor is installed in the center of a container. Sand particles fall freely into the container from a punctured plate with holes' diameter of 8 mm and throughout two sieves of opening dimension of 3 mm, the processes can produce a relative density of around 70%. When the sand surface reached the top of the container, the sand raining is stopped. Finally, the model anchor is pulled vertically during the whole pull-out processes at rate of 1mm/min.

3.2. Verifications

Hsu et al. [6] suggested that an under-reamed anchor placed in sandy soil of D_r=70% is classified as a shallow anchor, when overburden depth Z smaller than 8D. Else, the anchor is categorized as a deep anchor.

Figure 4 presents the experimental and numerical load-movement relationships of under-reamed anchors with overburden depth ratios of Z/D=2.5 and 10. Figure 4 reveals that the numerically calculated load-displacement curves are consistent with those measured by laboratory tests. Therefore, this study uses the numerical method to investigate the behaviors under-reamed anchor with various overburden depths in sand.

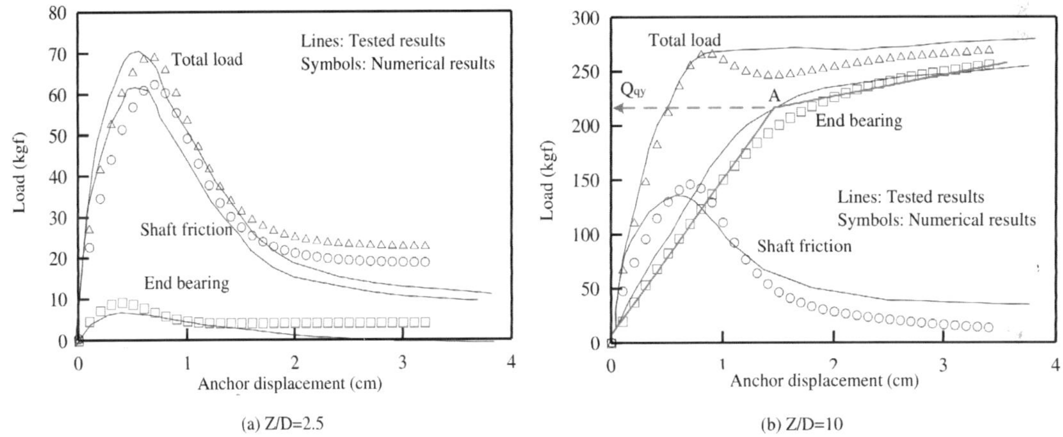

Figure 4. Numerical and experimental load-displacement (anchor movement) behaviours of two under-reamed anchors with Z/D=2.5 and Z/D=10.

4. Discussion on the experimental and numerical results

4.1. Load-displacement (anchor movement) relationships of the anchor

Figure 4 depicts the load-movement of anchor with Z/D=2.5 and 10. Figure 4(a) demostrates the total load and friction force raise with the anchor movement until the peaks take place; after that, decline obviously from peak to residual values. The end resistant exhibits a pre-peak and post-peak as well, indicating that the anchor is a typical shallow anchor of Z/D=2.5.

The friction force similarly behaves the phenomenon of pre-peak and post-peak for a deep anchor of Z/D=10 (Figure 4(b)). The end resistance of a deep anchor increases with anchor movement constantly without a peak. As a result, the total load originally declines and then rises when the anchor movement increases in the post-peak stage. For simplicity of assessment with the ultimate load of the anchors, this research defines the first peak load as the anchorage capacity for a deep anchor.

In brief, the ultimate anchor load is not the summation of peak friction load and peak end resistance (yield end bearing), because the peak total load, peak shaft friction and peak (yield) end bearing occurs in different anchor movements. Additionally, a greater overburden depth ratio Z/D indicates a greater corresponding anchor movement of the ultimate end bearing; and when the anchors are installed in the relative densities of 70%, with overburdens that exceeds 7.5D, the end bearings increase without peaks. Because the anchor's overburden exceeds the critical overburden depth H, no peak can be found on end bearing [6]. To decide the amount of the end resistance depicted formerly, an artificial recommendation by Trautmann and Kulhawy[7] is adopted. Figure 4(b) is the numerical end resistance-movement relations, Point A is obtained from the intersection of the initial tangent line and the latter half section of an end resistance-movement curve, and the ordinate that corresponded to Point A is the yielding end resistance Q_{qy} of an under-reamed anchor that had an end bearing without a peak.

4.2. Differentiating a shallow anchor from a deep anchor by coefficient of end bearing

This study applied the coefficient of end bearing N_q to classify an anchor as a shallow anchor or deep ancho. For anchors installed in cohesion-less soil, the coefficient of end bearing N_q can be expressed as

$$N_q = \frac{Q_q}{\gamma\, Z\, A_q} \tag{3}$$

where Q_q denotes the end bearing of an anchor, γ is the unit weight of the silty sand, Z represents the overburden (free length) of an anchor, and $A_q$is the cross-area of the fixed end.

Based on Equation (3), the relationship between the coefficient of an end bearing and the overburden/diameter ratio is plotted and is shown in Figure 5. The curve depicted in Figure 5 appears to increase and is initially upwardly concave. Consequently, the curve increases and becomes downwardly concave. The point of inflection in respect to the overburden is labelled as the critical overburden depth of shallow anchor H, as suggested by Vermeer and Sutjiadi [8]. The $H$is 7.5D. When the overburden depth is shallower than the critical overburden depth H, an obvious peak is found on end bearing-displacement curves, and the yielding soil around the under-reamed length develops to the ground surface, the anchor is a shallow anchor.

When the overburden depth exceeds 12D for anchors in sand, the coefficient N_q decreases with Z/D, the maximum in the curve corresponding to the overburden is the critical depth of transited anchor H_l. When the overburden of an anchor exceeded H but was smaller than H_l, the anchor belonged to a transited anchor. The anchor's end bearing-displacement does not have a peak, but the yielding soil around the anchor developed to the ground surface. However, when the overburden depth of an anchor exceeds H_l, the anchor is categorized as a deep anchor, and the end bearing-displacement curve does not behave a peak, the yielding soil around the under-reamed length does not develop to the ground surface.

Three types of anchors are found based on the relationship between the coefficient of end bearing N_q and overburden Z: shallow, transited, and deep anchors. These have behaviors that are similar to those of general shear failure, local shear failure, and punching shear failure in shallow foundations.

4.3. Dividing an end-bearing anchor from a frictional anchor

Anchorage capacity comprises the end bearing and friction load. Therefore, it is worthwhile to differentiate whether an anchor is an end-bearing or friction type. Research results indicated that the proportions of end bearing and friction load changed when the overburden and fixed length/diameter ratio were changed. When the fixed length remains unchanged, the proportion of friction decreases with the increase in the overburden/diameter ratio Z/D, the proportions of the end bearing and friction remains approximately constant until the Z/D exceeds 16. Figure 6 shows the percentages of the end bearing and friction load that dominates the anchorage capacity versus the Z/D for an anchor with a L/D of 4.5. Figure 6 demonstrates that when the Z/D is less than 10, the anchor is categorized as an end-bearing type; otherwise, it is a friction anchor.

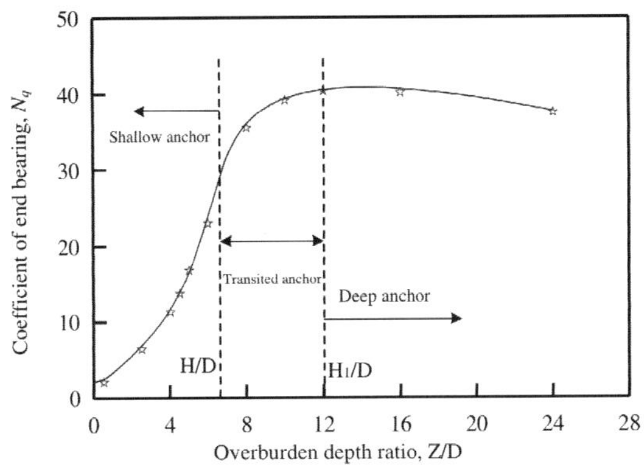

Figure 5. Relationship between coefficient ofend bearing and overburden depth/diameter ratio.

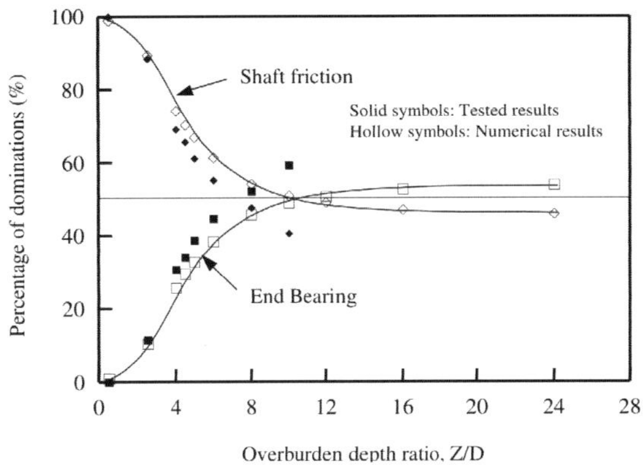

Figure 6. Percentages of end bearing and friction that dominate the loading capacity when peak load is applied.

5. Conclusion

This study adopted both the laboratory model test and numerical investigation to examine behaviours of under-reamed anchors in dense sand with a verity of overburden/diameter ratio Z/D. The results pilot the conclusion as following

- Research results demonstrated that the load-movement relationships evaluated numerically are in an agreement with those obtained experimentally. Hence, the numerical model together with FLAC^{3D} software could evaluated the behaviour of under-reamed anchor perfectly.
- The shapes of load-movement relationships on deep anchors were different from those of shallow anchors, especially the shapes of end resistance-movement curves.
- The frictional force dominated the ultimate load of a shallow anchor; meanwhile the end resistance is the main part of total load for a deep anchor. The demarcation overburden depth was 12D.
- According to the categorization of end bearing coefficient, if an overburden less than 7.5D, it was a shallow anchor; when an overburden exceeded 12D, it was classified to a deep anchor; and as an overburden depth was between 7.5D~12D, the anchor belonged to a transit anchor.

6. Reference

[1] Hsu S T and Liao H J 1998 Uplift behavior of cylindrical anchors in sand *Canadian Geotechnical Journal* vol 35 no1 p 70-80
[2] Liao H J Ou C D and Hsu S T 1996 Anchorage behavior of shaft anchors in alluvial soil *Journal of Geotechnical Engineering Division* (ASCE) vol 122 no 77 p 526-533.
[3] Hanna T H Sparks R Yilmaz M 1972 Anchor behavior in sand *Journal of the Soil Mechanics and Foundation Division* (ASCE) vol 98 no 11 p 1187-1208
[4] Ganesh R Khuntia S and Sahoo J P 2018 Seismic uplift capacity of shallow strip anchors: A new pseudo-dynamic upper bound limit analysis *Soil Dynamics and Earthquake Engineering* vol 109 p 69~75
[5] Wang D Hu Y Randolph M F 2010 Three-dimensional large deformation finite-element analysis of plate anchors in uniform clay *Journal of Geotechnical and Geoenvironmental Engineering*, (ASCE) vol 136 no 2 p 355–365
[6] Hsu S T Wang C C Wu S Tung H C 2016 Computer simulation on the uplift behaviors of arrayed underreamed anchor groups in sand *Systems and Informatics* (IEEE) p 534-539
[7] Trautmann C H and Kulhawy F H 1988 Uplift Load-Displacement Behavior of Spread Foundations *Journal of Geotechnical Engineering* vol 114 no 2 p 168-184

IOP Conf. Series: Materials Science and Engineering **758** (2020) 012049 doi:10.1088/1757-899X/758/1/012049

Control of fissures generated by the retraction in rigid pavements, applying synthetic fibers of recycled polypropylene.

V Torres1, K Chirinos2 and C Cuervo3

1 Civil Engineering, Peruvian University of Applied Sciences, Lima, Peru
2 Civil Engineering, Peruvian University of Applied Sciences, Lima, Peru
3 Civil Engineering, Peruvian University of Applied Sciences, Lima, Peru
victor.torres@upc.pe

Abstract. The retraction affects the setting process and the useful life of the concrete with the appearance of fissures; in last year's studies and methods have been generated to mitigate and control it with the use of different products and applications. The development of road infrastructure with the use of concrete as a rolling folder, requires methods to guarantee the durability and reduce the effects of the efforts incorporated by the use, climatic conditions, support base and restrictions of movement of the structure. To evaluate the effects of recycled synthetic polypropylene fibers in plastic retraction tests (ASTM C 1579), 3 mix designs were prepared with different ratios 58 gr., 116 gr., and 176 gr. of recycled and virgin synthetic fibers; the most significant and positive result to reduce fissures without affecting the resistance of concrete by bending and compression, was 0.50 mm without addition fibers, 0.10 mm and 0.15 mm with 176 gr. of virgin and recycled synthetic fibers. Finally, it can be concluded that adding a ratio of 4 kg per m3 allows good workability, in addition, the costs of the fibers are not representative compared to the high costs for future repairs.

1. Introduction.

One of the main problems of the world population in constant growth, is that every year more waste is generated due to the increase in consumption rates (Hoornweg et al., 2013). These rates directly affect the depletion of raw materials and environmental pollution due to the lack of correct recirculation flows (Koppelaar and Wiekard, 2013). [1] (Gilardino, Rojas, Mattos, Larrea and Vásquez, 2017).

To improve the durability of airport pavement concrete, four types of reinforced synthetic fibers in concrete were investigated in this document. Modified polyester synthetic fiber concrete, Polypropylene monofilament reinforced with synthetic concrete fiber, reticular synthetic polypropylene concrete reinforced with fiber and concrete reinforced with synthetic polyacrylonitrile fiber, the resistance, impermeability and resistance to the frost of these were investigated. The results show that synthetic fiber improves the flexural strength of concrete, but has little effect on compressive strength. [2] (Chen, Cen and Cui, 2018).

Macro recycled plastic fiber offers significant environmental benefits over virgin plastic fiber and steel reinforcement. However, as there is limited research on the performance of recycled plastic fiber in concrete, it has not yet been widely adopted by the construction industries. In this investigation, *post-cracking* studied the performance of different types of recycled polypropylene fibers from industrial wastes. Compared with virgin polypropylene fiber in concrete. The recycled fibers during the performance of the test demonstrated a good balance between tensile strength, elastic modulus and concrete bonding, thus producing brilliant post cracking performance. This research demonstrated the feasibility of using recycled fibers as reinforcement in concrete pavement [3] (Yin, et al., 2015).

The basic guidelines for the design of a concrete reinforced with polypropylene-based fiber on road pavement, as it is applied in a real test section that rests inside a tunnel of the Marche-Umbria Quadrilatero, road empowerment project, Italy. The results of a six-month follow-up with the presence of real traffic loads, such as feedback to the design stage. The supervision covers the direct measurement of the level of tension inside the mold, as well as the acoustic measurement, it was a design solution for roads, especially within tunnels [4] (Nobili, Lanzoni and Tarantino, 2013).

2. Materials and Design.

The materials used in the design of the concrete and preparation of the mixture are sand, crushed stone, cement, recycled synthetic fibers of polypropylene "figure 1" and virgin synthetic fibers of polypropylene "figure 2", plasticizing and accelerating additives, necessary to prepare the slab in accordance with the mold proposed in the Norma ASTM C 1579, "figure 3", cylinders and rectangular beams for compression and flexural tests.

Figure 1. Recycled polypropylene fibers

Figure 2. Virgin polypropylene fibers

Figure 3. Norma ASTM C 1579 Metal Mold

The mixture design was prepared by the method of combining aggregates for an amount of 0.044 m3, enough amount to prepare the mold and the test tubes for resistance tests, also proportions of 58 gr, 116 gr. and 176 gr. were added of recycled and virgin fibers respectively, plasticizer additive for ease of handling and accelerating additive and achieve settings and contraction of concrete in less time, allowing visualize the fissures generated.

IOP Conf. Series: Materials Science and Engineering **758** (2020) 012049 doi:10.1088/1757-899X/758/1/012049

In the process of setting and contraction of concrete, fissures of 0.50mm, "figure 4", were evident in the standard mold without fibers; Then an amount of 58 gr. was added to the same proportion of concrete of recycled fiber registering a fissure of 0.40 mm, "figure 5".

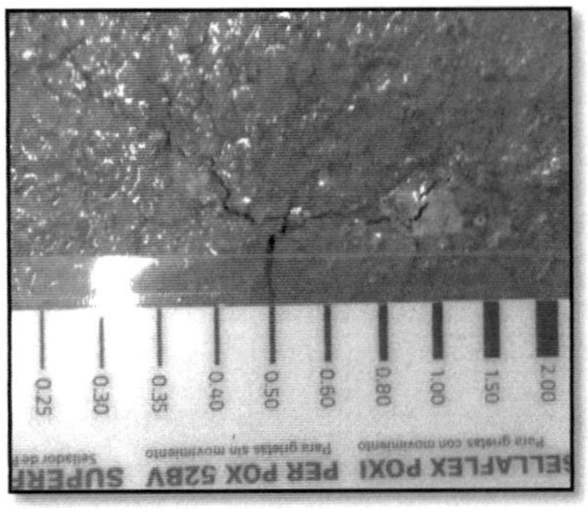

Figure 4. 0.50mm fissure.

Figure 5. 0.40 mm fissure.

Then using the same proportion of concrete 116 gr. was added of recycled fibers, and 0.35 mm. fissures were observed, "figure 6"; The procedure is repeated and changed for 116 gr. of virgin fibers, obtaining fissures of 0.25 mm; "Figure 7".

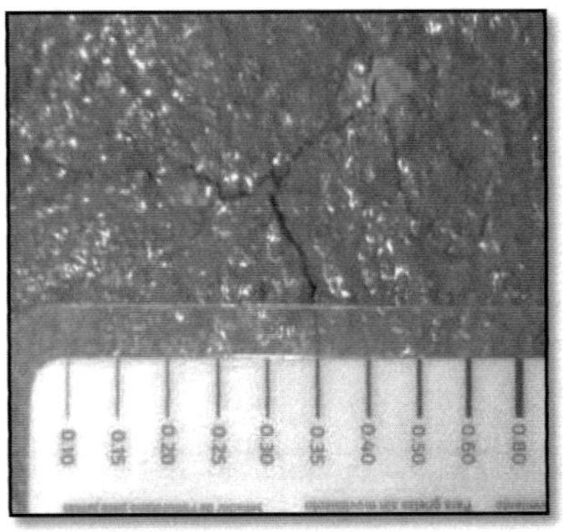

Figure 6. 0.35 mm fissure.

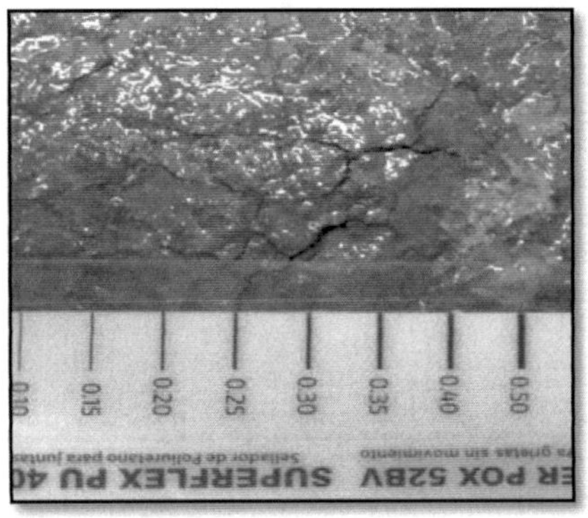

Figure 7. 0.25 mm fissure.

Finally, the amount of fibers is modified by 176 gr. from recycled fibers, 0.15 mm fissures were obtained, "figure 8"; and finally, with 176 gr. of virgin fibers, fissures of 0.10 mm were obtained, "figure 9".

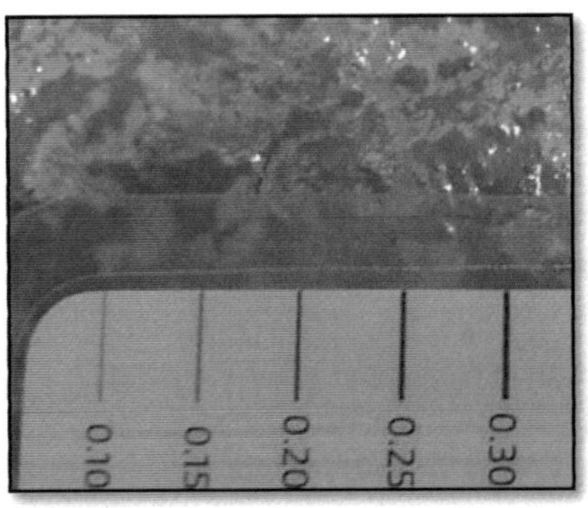

Figure 8. 0.15 mm fissure.

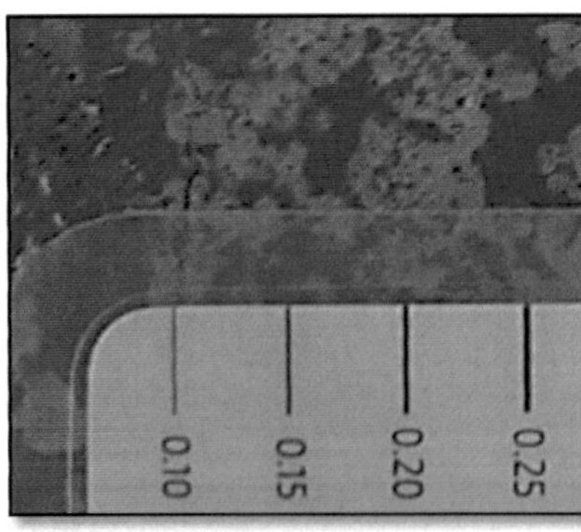

Figure 9. 0.10 mm fissure.

3. Analysis of results

After compiling the data of the different tests carried out, the results obtained from the different dosages were compared with and without the addition of recycled and virgin synthetic fibers; in addition, the samples were subjected to tensile tests, bending and fissures monitoring to verify the impact on resistance, "figure 10" and "figure 11"; The results were positive as there were no effects on the physical characteristics of the concrete. This allows us to affirm that the use of recycled synthetic polypropylene fibers is a reliable alternative for the control of fissures and their use can occur without any restrictions.

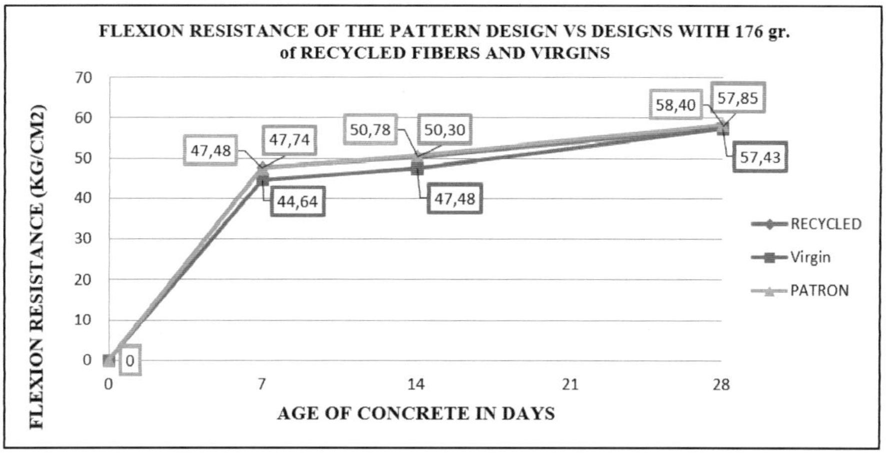

Figure 10. Flexural strength pattern design, with 176 gr. of recycled and virgin fibers

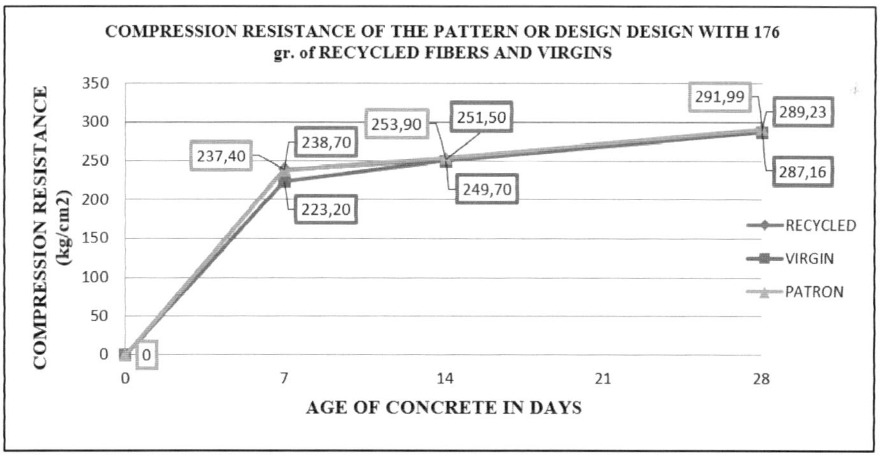

Figure 11. Compressive strength pattern design, with 176 gr. of recycled and virgin fibers

4. Conclusions

1. It was observed in the experimental analysis of concrete with the addition of recycled polypropylene fibers, which, by increasing the amount of fibers in the concrete, decreases the size of the fissures due to the effect of retraction; in the standard mold, fissures of 0.50 mm thick were found, with the recycled fibers fissures of 0.35 and 0.15 mm were found with

respect to 0.25 and 0.10 mm of virgin fibers in amounts of 116 and 176 gr respectively, observing a better control of fissures.

2. The variation of 210 Kg / cm2 concrete mechanical properties (compression and flexion at 28 days) was verified with the addition of recycled and virgin polypropylene synthetic fibers for a volume of (V = 0.044m3). With the addition of 116 and 176 grams of fibers, it was evident:

- For a proportion of 116 gr, of recycled fibers there is a higher compressive strength of 285.32 kg / cm2 compared to a concrete with virgin fiber of 279.45 kg / cm2; on the other hand, the flexural strength of 57.06 kg / cm2 was verified in comparison to a concrete with virgin fiber of 55.89 kg / cm2.
- For a proportion of 176 gr, the recycled fibers obtain a greater compressive strength of 289.23 kg / cm2 compared to a concrete with virgin fiber of 287.16 kg / cm2; on the other hand, the flexural strength is 57.85 kg / cm2 compared to a concrete with virgin fiber of 57.43 kg / cm2.

3. The economic analysis to use synthetic fibers of recycled polypropylene ($ 88.90 m3), demonstrates that they are a technically and economically viable alternative with respect to virgin fibers ($ 91.70 m3), because it allows a decrease in costs; In addition, it helps in reducing environmental pollution caused by polypropylene.
4. Finally, with the results and analysis obtained it was determined that the proportion of 176 gr. of recycled polypropylene synthetic fibers in the volume of concrete tested, represents an addition of 4 kg / m3; concrete maintains its characteristics and provides advantages in reducing and controlling the size of fissures in concrete pavements. The quantities of materials for 1 m3 are; 414.15 kg of cement, 193 liters of water, 821.51 kg coarse aggregate and 869.97 kg fine aggregate.

5. REFERENCES

[1] Gilardino A, Rojas J; Mattos H; Larrea G; y Vázquez I (2017) Combining operational research and Life Cycle Assessment to optimize municipal solid waste collection in a district in Lima (Perú) *Journal Cleaner of Production,* doi: 10.1016/j.jclepro.2017.04.005

[2] Yin, S., Tuladhar, R., Sheehan, M; Combe, M., y Collister T. (2016). A life cycle assessment of recycled polypropylene fibre in concrete footpaths. *Journal Cleaner of Production*, 112, 2231-2242.

[3] Chen, Y., Cen, G., y Cui, Y. (2018). Comparative study on the effect of synthetic fiber on the preparation and durability of airport pavement concrete. *Construction and Building Materials*, 184, 34-44.

[4] Yin, S., Tuladhar, R., Collister, T., Combe, M, Sivakugan, N, y Deng, Z. (2015). Post-cracking performance of recycled polypropylene fibre in concrete. *Construction and Building Materials*, 101, 1069-1077.

[5] Nobili, A., Lanzoni, L., y Tarantino, A. (2013). Experimental investigation and monitoring of a polypropylene-based fiber reinforced concrete road pavement. *Construction and Building Materials*, 47, 888-895.

Re-liquefaction of Sand in Shaking Table

Dikshit Babu Nepal[1], Jianliang Deng[1, 2], Jinjian Chen[1], Tshegofatso Maakoe[1]

1 State Key Laboratory of Ocean Engineering, Department of Civil Engineering, Shanghai Jiao Tong University, Shanghai 200240, China
2 Corresponding author: dengjianliang@sjtu.edu.cn; Tel: +8613916341178

Abstract. Subsequent aftershocks can induce reliquefaction of sand. In the shaking table experiment, the settlement, the excess pore water pressure and the acceleration response of the sand in the process of reliquefaction were measured. It was found that: (1) the liquefaction resistance of sand was the lowest in the second liquefaction, not in the first liquefaction; (2) Liquefaction resistance is depth-dependent, and the soil layer near the surface of soil is most likely to be liquefied; (3) In the model test, the attenuation of seismic shear wave acceleration is not only related to distance but also to the vibration frequency and to the liquefied layer; (4) The amplitude of the acceleration response increases with the number of the shaking. This result indicates that: (1) the phenomenon of re-liquefaction of sand induced by earthquake is not same as that of the first liquefaction. Therefore, the conventional strategy for liquefaction resistance needs to be improved before it can be implemented for analysis of reliquefaction. (2) The attenuation of seismic shear waves in liquefied soil layer has some properties different from the normal soil layer; thus the traditional attenuation calculation method needs improvement in the evaluation in the analysis of site liquefaction.
Keywords: Re-liquefaction, Shake table, Excess pore water pressure (EPWP), Sand resistance

1. Introduction

Reliquefaction of sand has been reported extensively in the recent earthquake research and reconnaissance [4, 7] while prehistoric liquefaction [10,14,17] illustrated that the sand deposits once liquefied might still get liquefied again in the future by a subsequent earthquake similar or even smaller than the previous one which is very alarming issue. Finn et al. [2] conducted the triaxial shear test and observed that the reliquefaction resistance of sand decreases during the subsequent earthquake. Following this pioneering work, Ishihara et al. [5] and Suzuki et al. [11] also investigated the reliquefaction resistance of the sand using the triaxial test and found that the sand reliquefies more easily despite the increase in density in the second earthquake. Additionally, Oda et al. [9] explained that the destruction of aged soil structures by large shear straining makes the liquefied sand to act like a fresh deposit following a post-liquefaction consolidation which may reduce the post shaking liquefaction resistance [12]. Olson et al. [8] further reinforced that the re-liquefaction resistance of the sand may not increase despite the increase in density due to post-liquefaction consolidation. Similarly, by conducting triaxial test, Wang et al. [15] revealed that the re-liquefaction resistance of silt also decreases.

On the other hand, some studies show that the reliquefaction resistance of sand can increase in the subsequent earthquake. Based on the laboratory test of dynamic shear modulus after primary consolidation, and field measurement of penetration resistance after ground densification, Mesri et al. [6] explained the resistance of the sand in re-liquefaction may increases only after the significant increase in relative density of the sand. In contrary, Ye et al. [19] argued and stated the reliquefaction resistance of the sand remains constant, but the dissipation time of excess pore water pressure (EPWP) decreases with increasing density. Yamada et al. [16] concluded that the reliquefaction resistance of the sand may increase or decrease depending upon the level of anisotropy developed before the shaking. After performing shaking table test in sand, Ha et al. [3] and Ye et al. [18] concluded that the reliquefaction resistance of sand decreases during the second shaking and starts increasing after the second shaking. Despite of intensive research, the results observed in the sand reliquefaction test is inconsistent. In this study, a series of shaking table tests were conducted successively in a sandy model to investigate the liquefaction resistance of sand by measuring the settlement in the first and the subsequent shakings. Furthermore, the response of EPWP and acceleration in multiple layers were also studied.

2. Materials and Methods

Four pore water transducers and four accelerometers were fixed in the model box designed and fabricated for the shake table test, as shown in **Figure 1**. Three accelerometers (A_2, A_3, and A_4) were buried in the sand, and one accelerometer (A_1) was set on the shaking table near the base of the model to measure the input acceleration. The linear variable differential transformer (LVDT) of 50mm capacity was used to measure the ground subsidence, which was extended with a plate (3cm×3cm×0.2cm thick) to prevent the tip from plunging in the sand. In this experiment, yellow sand within the liquefiable range as reported by Tsuchida [13] was used. The grain size distribution and the index properties of the sand used for the experiment is as shown in **Figure 2** and **Table 1** respectively. The model box was filled with water, and oven-dried sand was allowed to fall from a narrow tip funnel so that the sand lies above the prior sand particles by gravity. The excess water at the surface was removed by siphoning. Finally, out of many experimental trials, analysis was performed for the sample with the least relative density (22.9%). The water content, void ratio, and degree of saturation of the thus prepared sample were obtained to be 28.4%, 0.754, and 98.1% respectively.

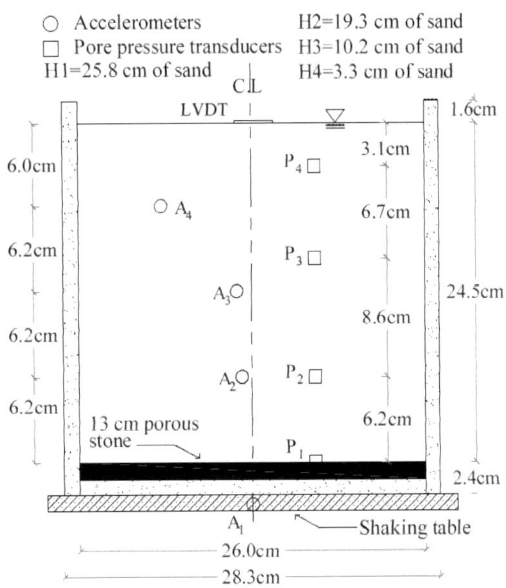

Figure 1. Schematic model and layout.

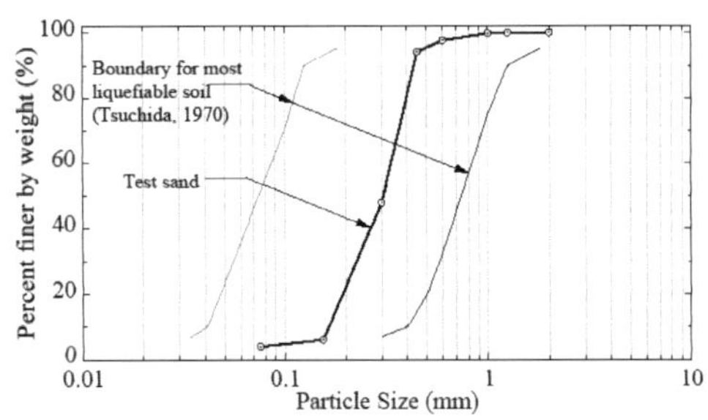

Figure 2. Grain size distribution of sand

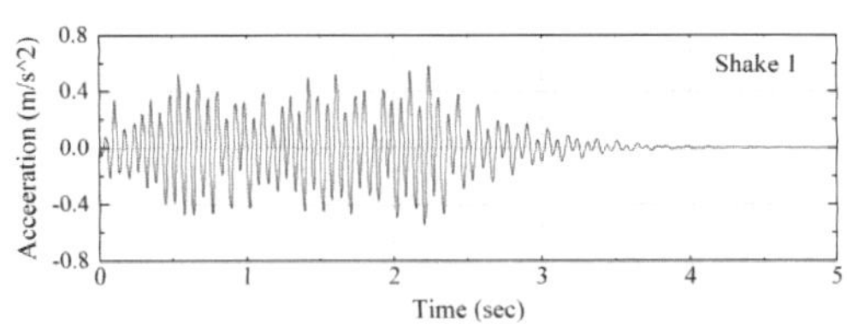

Figure 3. Acceleration time history during for first shaking event.

Table 1. Index and gradation property of experimental sand

Grain size (mm)				Coefficient of uniformity (C_u)	Specific gravity (G_s)	Dry unit weight (kN/m^3)		Void ratio	
D_{10}	D_{30}	D_{50}	D_{60}			Min.	Max.	Min.	Max.
0.17	0.24	0.31	0.34	2.10	2.60	14.30	17.00	0.53	0.82

The sand model was subjected to input acceleration of similar PGA till the 6th shaking as shown in **Figure 3**. The subsequent shakings were applied only after the dynamic pore water pressure were completely dissipated. The ground settlement was measured before and after each excitation. Based on the settlement data, the relative density (D_r) and void ratio (e) after each test were evaluated.

3. Result and Discussion

3.1. PWP response and reliquefaction mechanism

Liquefaction occurs when EPWP (Δu_x) reaches the initial vertical effective stress (σ'_{vo}), i.e., excess pore water pressure ratio ($r_u = \Delta u_x/\sigma'_{vo}$) is unity. However, the directly measured EPWP has significant fluctuation due to the irregular input acceleration. Therefore, all the Δu_x histories have been smoothened by using 1-D median filtering and smooth function in MATLAB to show the trend of Δu_x and were compared to $r_u = 1$ condition for respective sand layer. Here, the sand layer H4 has been omitted due to its inconsiderable height which imparts insignificant response. In some cases (Figure 4), r_u corresponding to the liquefied height is obtained slightly higher than 1. Ecemis et al. [1] explained this phenomenon to be either a result of the settlement of the piezometers or the rise in the water table elevation.

The time required to reach the condition $r_u = 1$ were significantly varying for the different shaking events. For liquefied sand layers ($r_u = 1$), N_L and N_R are event-dependent parameters whose value depends upon the product of dominant frequency and time; where, N_L and N_R are the numbers of cycles required to trigger liquefaction during the *1st* and following shakings, respectively. However, for some layers where sand couldn't reliquefy ($r_u < 1$), the corresponding maximum r_u for non-liquefied layers are listed in Table 2, and **Figure 5** graphically represents its data.

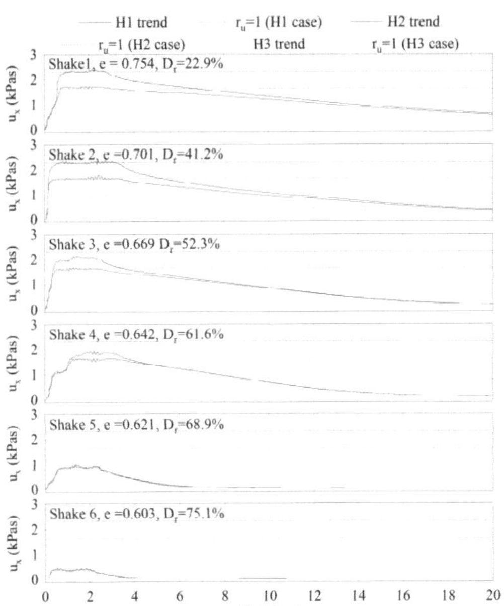

Figure 4. Excess PWP time history in each of the shaking events.

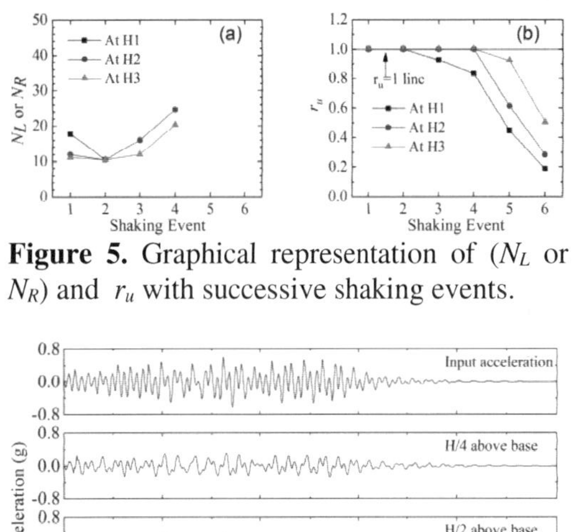

Figure 5. Graphical representation of (N_L or N_R) and r_u with successive shaking events.

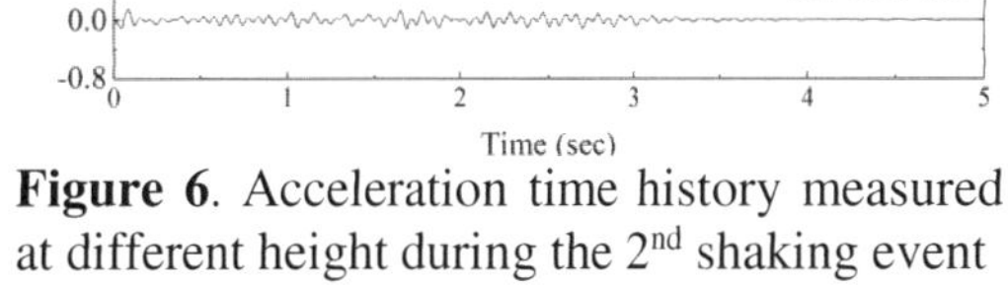

Figure 6. Acceleration time history measured at different height during the 2nd shaking event

Table 2. Number of cycles required to trigger liquefaction/reliquefaction during each of the shaking events.

Height	Event 1	Event 2	Event 3	Event 4	Event 5	Event 6
H1	N_L=18	N_R=10	r_u= 0.93	r_u= 0.83	r_u= 0.41	r_u= 0.19
H2	N_L=12	N_R=10	N_R=16	N_R=25	r_u= 0.61	r_u= 0.28
H3	N_L=11	N_R=10	N_R=12	N_R=20	r_u= 0.92	r_u= 0.50

Figure 5 shows that the values of N_L in H1 ~ H3 for the first shaking are significantly larger than the corresponding values of N_R for the second shaking event. Therefore, the resistance to reliquefaction in the second shaking event decreases relative to the first shaking event despite of the increase in the density. The destruction of initial anisotropic sand fabrics during the first shaking could possibly be the reason for the rate of increase in pore water pressure during the second shaking. However, N_R exhibits the upward trend during the subsequent shaking events, meanwhile, reliquefaction resistance also starts to increase with density which makes the second event most vulnerable to reliquefaction phenomenon. The value of N_R in the second shaking event has been reported by Ha et al. [3] and Ye et al. [18] to be minimum which validates the results observed in this study. As indicated in Figure 4, all the sand layers in H1, H2, and H3 heights were liquefied in both the *1st* and the *2nd* shaking events, but reliquefaction was only observed for H2 and H3 heights in *3rd* and *4th* shaking events indicating that H2 and H3 are more likely to re-liquefy than H1. For *5th* shaking event, none of the sand height re-liquefied, however, the value of r_u for H3 height was 0.92 (implies closely re-liquefied) which means that the shallow layer (e.g. H3 layer) is most susceptible to re-liquefaction. Therefore, reliquefaction behaviors are depth-dependent. Similarly, none of the sand layers were re-liquefied even in the *6th* shaking event ($r_u \leq 0.5$). Finally, due to the significant reduction in the value of r_u, the need for further investigation was irrelevant. According to Ha et al. [3], the small increases in effective confining stress and relative density with depth contributes to this depth-dependent reliquefaction behavior. Since the effective confining stress and the relative density could not be measured precisely, more studies are needed to further investigate the underlying mechanism of depth-depending reliquefaction.

3.2. Acceleration response and mechanism

The amplitude of acceleration response attenuates with the propagation distance from the point of excitation, while the rate of attenuation is related to the frequency. The seismic waves were still under effect even after the corresponding input excitations were completely ceased. Thus, the input acceleration time history for 3 seconds for all the shakings were subjected to fast Fourier transform (FFT) to obtain the dominant frequency. A typical acceleration time history and its FFT analysis, are shown in **Figure 6** and **Figure 8**, respectively. That FFT normalization was based on mean square amplitude with a sampling interval of 0.001 seconds. Thus obtained first dominant frequency was nearly 16 Hz for all the input shakings, whereas, the second dominating frequency was 7 Hz. Besides, other sub and super harmonics were also observed during those periods.

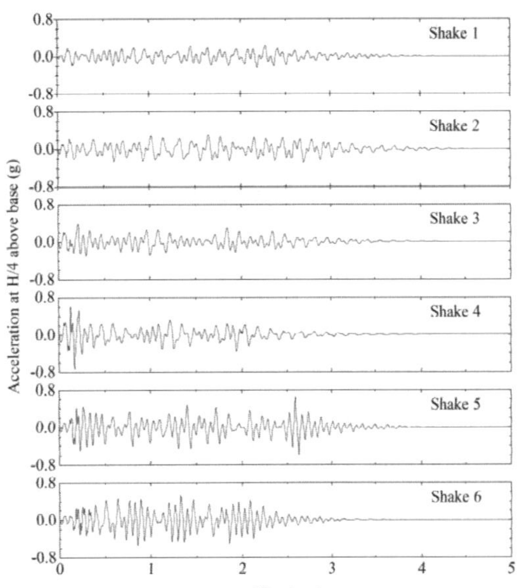

Figure 7. Acceleration response measured by accelerometer A_2 placed H/4 above the base.

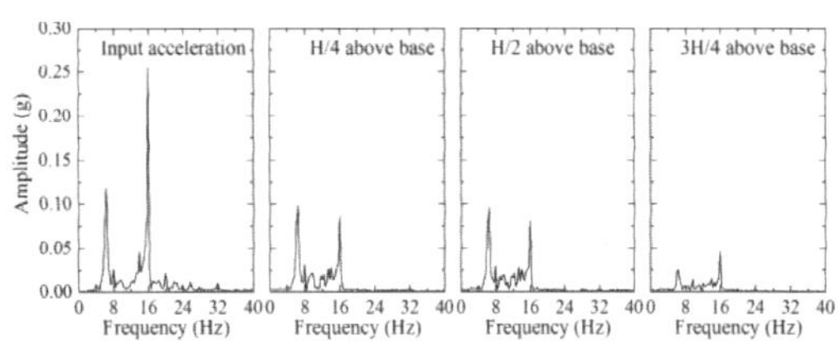

Figure 8. FFT for acceleration measured at different height during the 2nd shaking event.

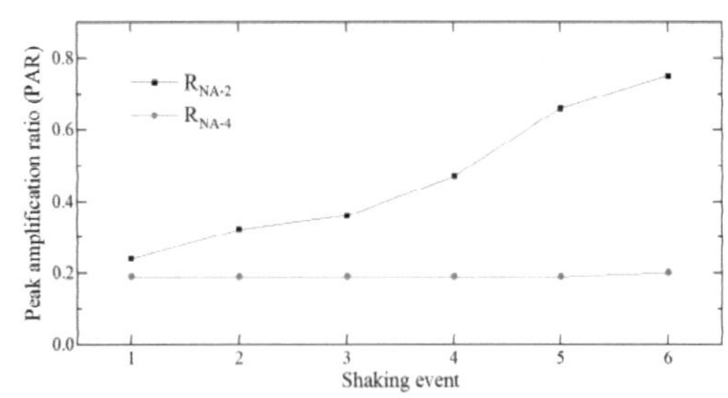

Figure 9. R_{NA} for the sand layer at H/4 and 3H/4 above the base.

It should be noted that the attenuation rate with lower frequency can be much smaller than that of higher frequency: For the acceleration of 16 Hz, it attenuates from 0.25g to 0.07g when it propagates from the table (point of excitation) to the height H/4 (from the base); however, for the acceleration of 7Hz, it marginally attenuates from 0.12g to 0.1g (Figure 9). The mechanism of this different attenuation rate may be dependent on the natural frequency of the sand layer, hence, further studies are necessary for its deeper understanding. The amplitude of acceleration response might increase with the number of shakings during which the densification of the sand layer increases correspondingly. **Figure 7** shows the accelerations measured by accelerometer A_2 at the height H/4 from the base. The peak values of acceleration with frequency 16 Hz (FFT spectrum based on measured accelerations) during each shaking event was obtained.

The normalized acceleration response R_{NA-2} (or R_{NA-4}) is the ratio of the peak acceleration measured by A_2 (or A_4) to that measured by A_1. R_{NA-2} and R_{NA-4} thus calculated in each shaking event is presented in **Figure 9**, where R_{NA-2} increases significantly with the number of shaking events. This shows that the attenuation of acceleration decreases with the number of shaking events. The reason for the decrease in attenuation is possibly related to the increase in relative density of the sand layer. However, R_{NA-4} increases marginally with the number of shaking events. This small increase in R_{NA-4} is possibly due to the strong mitigation effect of re-liquefied layer below A_4 layer (during the shear wave propagation). This study found that the attenuation of seismic shear wave acceleration is related to the liquefied layer. The reasons for the small R_{NA-4} (about 0.2) is not due to the small distance (20 cm from the shaking table to the position A_4), but probably due to the liquefied layer below A_4 where the shear module might be very small. Similarly, the value of R_{NA-2} is 0.2 ~ 0.3 at the first two shaking events which could be due to the liquefied lowest layer. To date, the effect of number of shaking and the position of the liquefied layer on attenuation of seismic shear wave has been overlooked. As the attenuation of seismic shear wave is very crucial factor for the evaluation of site liquefaction, the traditional attenuation calculation method needs its improvement in the future practice.

4. Conclusion

A series of shaking table tests were conducted to investigate the multiple liquefaction phenomenon in layered sand via its acceleration response and the excess pore water pressure, and following main conclusions have been derived:

(1) Besides the relative density of sand layer, other factors such as the structure or the fabrics of grains determine the re-liquefaction behaviour of sand because the lowest liquefaction resistance was observed in the second shaking event (not in the first shaking event where the relative density was the largest one among the all events). This is possibly due to the destruction of initial anisotropic sand fabrics during the first shaking leading to the increase in rise of PWP during the second shaking. Therefore, the relative density of sand layer does not uniquely determine the reliquefaction behaviour.

(2) Shallow sand layers are highly susceptible to multiple liquefaction in comparison to the deeper sand layers. Therefore, reliquefaction phenomenon is influenced by the depth of the targeted layer. However, further intensive research needs to be conducted to understand the depth-dependent behaviour considering the density and confining stress of that particular layer.

(3) Distance is not only the influential parameter for the attenuation of seismic shear wave acceleration, but is also affected by the number of shaking events, the vibration frequency and the position of liquefied layer. To be specific, the attenuation rate with lower frequency can be much smaller than that of higher frequency. Additionally, normalized acceleration response is influenced by the position of liquefied layer rather than the distance from the point of excitation. Hence, for evaluating the site liquefaction, the conventional method of calculation needs to be improvised.

5. References

[1] Ecemis N, Demirci H E and Karaman M 2014 Influence of consolidation properties on the cyclic re-liquefaction potential of sands *Bulletin of Earthquake Engineering* **13** 1655–73
[2] Finn L, Bransby P L and Pickering DJ 1970 Effect of strain history on liquefaction of sand *Journal of the Soil Mechanics and Foundations Division* **96** 1917-1934
[3] Ha I-S, Olson S M, Seo M-W and Kim M-M 2011 Evaluation of reliquefaction resistance using shaking table tests *Soil Dynamics and Earthquake Engineering* **31** 682–91
[4] Hashash Y M A *et al.* 2015 *Geotechnical Field Reconnaissance: Gorkha (Nepal) Earthquake of April 25, 2015 and Related Shaking Sequence* (Berkeley Electronic Press)
[5] Ishihara K and Okada S 1978 Effects of stress history on cyclic behavior of sand. *Soils And Foundations* **18** 31–45
[6] Mesri G, Feng T W and Benak J M 1990 Postdensification Penetration Resistance of Clean Sands *Journal of Geotechnical Engineering* **116** 1095–115
[7] Moss R E S *et al.* 2015 Geotechnical Effects of the 2015 Magnitude 7.8 Gorkha, Nepal, Earthquake and Aftershocks *Seismological Research Letters* **86** 1514–23
[8] Olson S M, Green R A and Obermeier S F 2005 Geotechnical analysis of paleoseismic shaking using liquefaction features: a major updating *Engineering Geology* **76** 235–61
[9] Oda M, Kawamoto K, Suzuki K, Fujimori H and Sato M 2001 Microstructural Interpretation on Reliquefaction of Saturated Granular Soils under Cyclic Loading *Journal of Geotechnical and Geoenvironmental Engineering* **127** 416–23
[10] Sims J D and Garvin C D 1995 Recurrent liquefaction induced by the 1989 Loma Prieta earthquake and 1990 and 1991 aftershocks: Implications for paleoseismicity studies *Bulletin of the Seismological Society of America*
[11] Suzuki T and Toki S 1984 Effects of preshearing on liquefaction characteristics of saturated sand subjected to cyclic loading. *Soils And Foundations* **24** 16–28
[12] Terzaghi K and Peck R B 1967 *Soil mechanics in engineering practice* (New York, NY: John Wiley & Sons)
[13] Tsuchida H 1970 *Prediction and counter means against the liquefaction in sand deposits* (Abstract of the seminar in the Port and Harbor Research Institute, Yokohama, Japan)
[14] Wakamatsu K 2012 Recurrent Liquefaction Induced by the 2011 Great East Japan Earthquake *Journal of Japan Association for Earthquake Engineering* **12**
[15] Wang S, Yang J and Onyejekwe S 2013 Effect of Previous Cyclic Shearing on Liquefaction Resistance of Mississippi River Valley Silt *Journal of Materials in Civil Engineering* **25** 1415–23
[16] Yamada S, Takamori T and Sato K 2010 Effects Of Reliquefaction Resistance Produced By Changes In Anisotropy During Liquefaction *Soils And Foundations* **50** 9–25
[17] Yasuda S and Tohno I 1988 Sites of reliquefaction caused by the 1983 Nihonkai-Chubu Earthquake. *Soils And Foundations* **28** 61–72
[18] Ye B and Hu H 2018 Investigation on the Reliquefaction Behaviors of Sand Using Shaking Table Tests *Proceedings of GeoShanghai 2018 International Conference: Advances in Soil Dynamics and Foundation Engineering* 300–7
[19] Ye B, Ye G, Zhang F and Yashima A 2007 Experiment And Numerical Simulation Of Repeated Liquefaction-Consolidation Of Sand *Soils And Foundations* **47** 547–58

IOP Conf. Series: Materials Science and Engineering **758** (2020) 012051 doi:10.1088/1757-899X/758/1/012051

Effect of End-to-End Connectors in Elastic GFRP Gridshell

Kedar Baral[2], Bin Cheng[1,2*], Sheng Xiang[2] and Suraksha Sharma[2]

[1]State Key Laboratory of Ocean Engineering, Shanghai Jiao Tong University, Shanghai 200240, China.
[2]Department of Civil Engineering, Shanghai Jiao Tong University, Shanghai 200240, China.
[*]Corresponding author's e-mail: cheng_bin@sjtu.edu.cn

Abstract. The presence of end-to-end connectors for lengthening the glass fiber reinforced polymer (GFRP) pipes in elastic gridshell causes imperfections in the structure. Such geometrical imperfections alter the structural properties of members and affect the overall performance of the structure. Hence, for a comprehensive analysis of such structures, the aforementioned imperfections need to be acknowledged. In this study, an equivalent rigid part (ERP) for the connector was proposed for taking into account the effect of connectors in the structural analysis. The applicability of ERP for predicting the deformation of structural members was first verified by bending beam models. A typical GFRP elastic gridshell was further utilized for investigating the effects of connectors (with two different arrangement) in the shape and stress of the structure. Results show that proposed ERP is practicable for simplifying the end-to-end connectors in the structural analysis model. Significant differences in results were found between the analysis with and without the consideration of end-to-end connectors, and therefore the effect of connectors should be considered during structural analysis of elastic gridshell.

1. Introduction

Elastic gridshells are long-span shell-like[1] structures which derives its stiffness from its double curvature[2]. They have simple construction techniques and employ the principle of active bending[3]. To achieve the target shape, regular flat quadrangular grids which are arranged in superimposed layers[4] are elastically deformed.

Elastic gridshell came into existence when Frei Otto, Edmund Happold and W. I. Liddell achieved famous Mannheim Multihalle in 1975[5]. This structure is an inspiration in both form and aesthetic to modern gridshell like Weald and Downland gridshell. Traditionally wooden laths were used in elastic gridshells because of their high elastic limit strain[6]. However, in recent decades, the attention of the scientific community has shifted towards the use of new innovative materials that possess better mechanical properties than wood, have simpler connection systems, and are economical. GFRP, as suggested by Ashby, has higher elastic limit strain and Young's modulus compared to wood[6-8]. Moreover, industrial production of GFRP ensures more reliability on mechanical properties[9].

Realizing the benefits of GFRP, Navier laboratory constructed a GFRP elastic gridshell in 2011 for Solidays[7] and Ephemeral Cathedral of Créteil in 2013[10]. In both gridshells, nodes are formed with the help of commercially available swivel couplers applied at the member intersections[7, 10]. On the other hand, gridshell members are lengthened by using proper end-to-end connection system. The connection system must transfer normal stress effectively and also have comparable bending stiffness as that of the connecting members to confirm the continuity of the global shape[11]. Thus, an specific end-to-end connector should be employed in the gridshells for member lengthening[12]. In the Solidays gridshell, two swivel couplers are used as end-to-end connectors to connect members and observations showed that such treatment does not provide sufficient stiffness[7]. In Ephemeral Cathedral gridshell, a specially designed end-to-end connection system consisting of two steel tube and a linked threaded rod is adopted[13].

The aforementioned connectors provide geometrical imperfection in structural members. They might affect the global shape and have negative impact on the structure's performance. Therefore, such imperfections need to be integrated for predicting the response of the gridshell. Even though the advancement in finite element analysis and numerical modelling have aided comprehensive structural analysis of structures, integrating such geometric imperfections are still very complicated. In addition, limited research has been conducted focusing on the effect of connectors in gridshell. Therefore, in this study, efforts have been made to understand the impact of the presence of end-to-end connectors in gridshell by evaluating the equivalent rigid part (ERP) for the connector. The ERP is then included as a connector in the finite element model of the gridshell for analysis. Moreover, connectors are placed in two different ways in the gridshell to investigate the effect of the arrangement of connectors in the structure.

In this paper, finite element method is described in Section 2 where ERP is worked out and validated. In addition, a typical gridshell is briefly described and analyzed in this section. Similarly, in Section 3, results and discussions are mentioned, followed by Section 4 where few conclusions are drawn.

2. Finite Element Model

The finite element analysis was performed in two stages. In the first stage, ERP was worked out and validated as mentioned in Section 2.1.2. Moreover, the flexural behavior of the GFRP pipe was analyzed for three cases: with connector, without connector and ERP. To represent connector, respective sectional bending stiffness values were assigned throughout the length of the connector. Similarly, in the second stage, the effect of presence of the connectors in the gridshell was investigated by comparative analysis. The analysis was performed by comparing the gridshell, with and without connectors.

To model the GFRP pipes in ABAQUS, deformable wire feature and thin-walled pipe section were chosen. Moreover, to simulate the swivel connector at grid nodes, a hinge connector in an interaction module was used. Geometric non-linear solver of ABAQUS was employed so that large deformation behaviors during the erection can be accurately analyzed. The element type chosen for the analysis of the structure was B32, which is a Timoshenko beam element that considers all the axial forces, shear forces and bending moments. To incorporate end-to-end connectors in the gridshell, the ERP was used. The ERP derived in Section 2.1.1 was placed in all the locations of the end-to-end connectors in the gridshell for considering the effects of such connectors.

2.1 Equivalent rigid part

2.1.1 Formula derivation. This section aims to calculate the ERP for the connector. The ERP provides a simple technique to integrate the connectors in the gridshell. For simplification, effect of bolts and tube tolerance were not considered.

Figure 1 depicts the symmetric half portion of the member with connector and ERP. Figure 1(a) shows the connection system similar to Ephemeral Cathedral gridshell[11]. The connection system consists of steel sleeve of diameter (d_4) 20 mm and a steel tube having external diameter (d_3) 48 mm with a wall thickness of 3 mm. Similarly, the ERP along with the member is showcased in Figure 1(b). The GFRP pipe has an external diameter (d_2) of 42 mm, and a wall thickness of 3.5 mm.

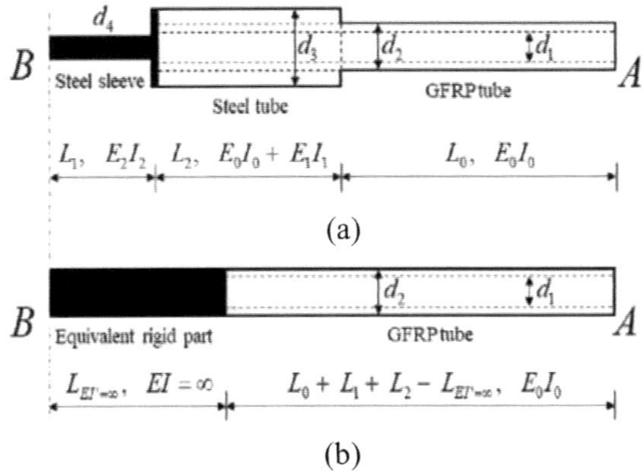

Figure 1. Connector assembly and the corresponding ERP.

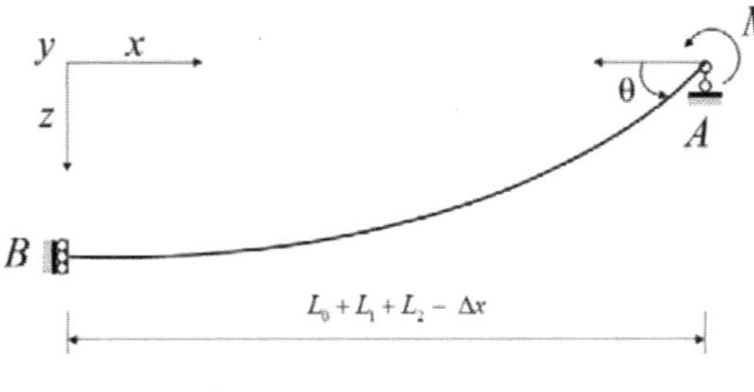

Figure 2. Deflection of member BA under external loading.

Here, EI denotes bending stiffness and L_0, L_1, L_2, and $L_{EI'=\infty}$ are the length of GFRP pipe, steel sleeve, steel tube, and ERP, respectively. Similarly, E_0 and E_1=E_2 denotes Young's modulus of GFRP and steel, respectively and ERP has infinite bending stiffness. The second moment of inertia (I_z) of a pipe or tube is given by $\frac{\pi(d_e^4 - d_i^4)}{64}$ while that of rod is given by $\frac{\pi d^4}{64}$ where, d_e is the external diameter, d_i is internal diameter of the tube, and d is diameter of the rod.

Similarly, Figure 2 depicts the bending of a member BA under pure bending, M. End A is roller supported while another end B is collar supported. In order to calculate $L_{EI'=\infty}$, it is assumed that the slope at end A due to pure bending, M, must be equal in both cases. To calculate the slope at the end A, widely used moment area theorem is used. In the figure, Δx is the horizontal translation of the end A.

Equating slope at end A for both cases, Equation (1) was derived, which provides the length of the ERP of the connector.

$$L_{EI'=\infty} = (1-\mu)L_1 + \left(\frac{1}{1+\beta}\right)L_2 \tag{1}$$

where $\beta = \frac{E_0 I_0}{E_1 I_1} = \frac{E_0(d_2^4 - d_1^4)}{E_1(d_3^4 - d_2^4)}$, and $\mu = \frac{E_0}{E_1}\left(\frac{d_2^4 - d_1^4}{d_4^4}\right)$.

2.1.2 Validation of ERP. Figure 3 depicts the bending of a 9 m long GFRP pipe under different circumstances. Figure 3(a) (or Case A) represents a pipe deformed under the influence of gravity load alone, while Figure 3(b) (or Case B) depicts buckling of a GFRP pipe by displacing support A by 4 m. Case B is the typical case of large curvature bending.

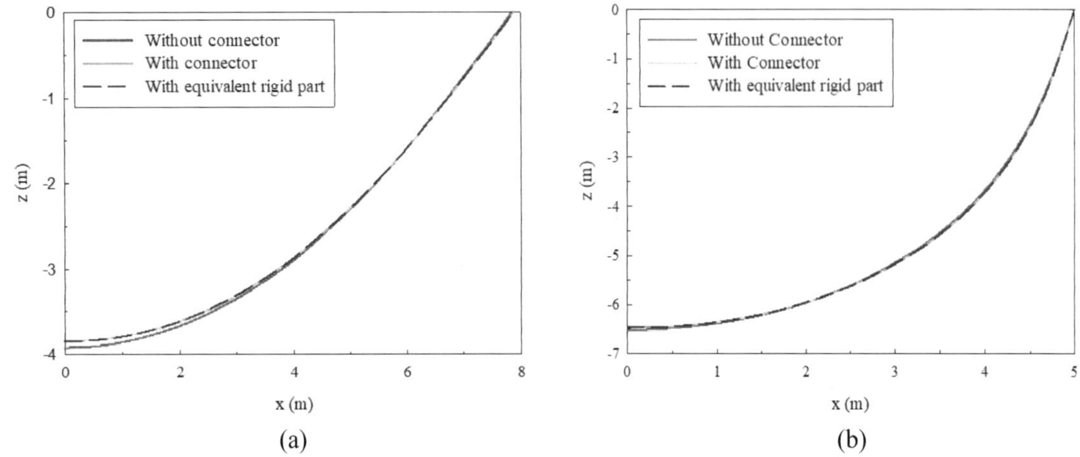

Figure 3. Deflection of gridshell member due to (a) gravity load (b) translation of support A ($\Delta x = 4$ m).

The coinciding graphs (with connector and the ERP) from Figure 3(a) and Figure 3(b) justifies the convenient use of ERP instead of using a connector for finite element analysis. The deflections and stresses developed in pipe sections are recorded in the Table 1. The difference of the deflection at end B (crown) is found to be minute. Also in both load cases, the stresses developed in the gridshell member, with connector and ERP, are almost equivalent. This suggests that the ERP can predict the influence of the connector in the structure with great precision. Furthermore, Figure 3(b) ensures that even at larger curvature, ERP has the capability to measure the influence of connector in a member. Moreover, the effect of connector in the member was also analyzed. From comparative study of Table 1, it can be observed that the presence of connector has minor but noticeable effect on flexural behavior of the member having smaller curvature. Meanwhile, for larger curvature member, the effect is comparatively larger.

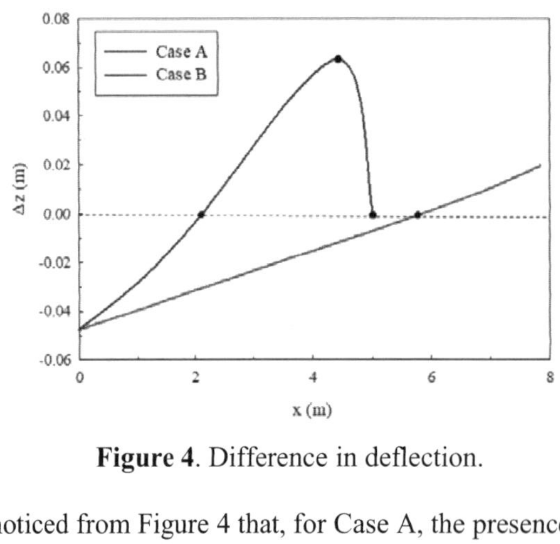

Figure 4. Difference in deflection.

Additionally, it can be noticed from Figure 4 that, for Case A, the presence of the connector has caused a small but noticeable decrease in vertical deflection at crown and a linear decrement away from the crown but it again starts increasing from inflection point highlighted by black dot. While in Case B, the connector has caused a significant flattening of the curve. In this case, two inflection points, one near to the middle and other near to the end, have appeared in the member. Overall, the result suggests that the adoption of

IOP Conf. Series: Materials Science and Engineering **758** (2020) 012051 doi:10.1088/1757-899X/758/1/012051

ERP can accurately predict the effects of the end-to-end connectors in a lengthened member with large deformation.

Table 1. Comparisons of deflection and stress for two load cases.

	Deflection at B, m			Max. von Mises Stress, MPa		
	Without connector	With connector	With ERP	Without connector	With connector	With ERP
Case A	-3.920	-3.844	-3.843	63.000	63.350	63.360
Case B	-6.516	-6.470	-6.467	131.110	133.830	133.810

2.2Elastic gridshell

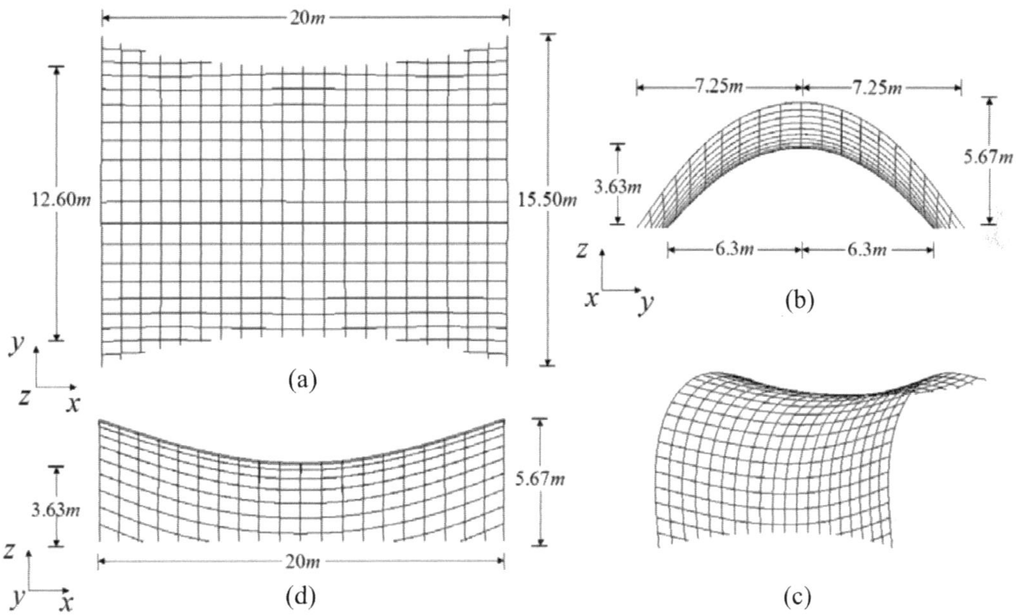

Figure 5. Gridshell (a) Plan view (b) Side view (c) Elevation view (d) Perspective view.

Figure 5 illustrates a simple elastic gridshell of grid size of 1m with all the corresponding views and dimensions considered in the study. Similarly, Figure 6 demonstrates two schemes of the connector arrangements taken into account. The connectors were applied considering the length of the member to be limited to 9 m. For Arrangement I as shown in Figure 6(a), the connectors were symmetrically placed in the plane grid. For y-directional members, the connectors are arranged at the waist of the structure, i.e., nearly at a distance of L/4 from the member ends. Connectors in the crown regions, i.e., L/2 of the members are avoided intentionally. On the other hand, for Arrangement II as shown in Figure 6(b), connectors were placed at all the possible locations of the members. The arrangement of the connectors is centrosymmetric so that the placement of the connectors have no or minor effect on the prescribed shape of the structure. It should be noted that, for the y-directional members of Arrangement II, some connectors are placed at the crown region of the structure where the stress is high during deformation of the grids.

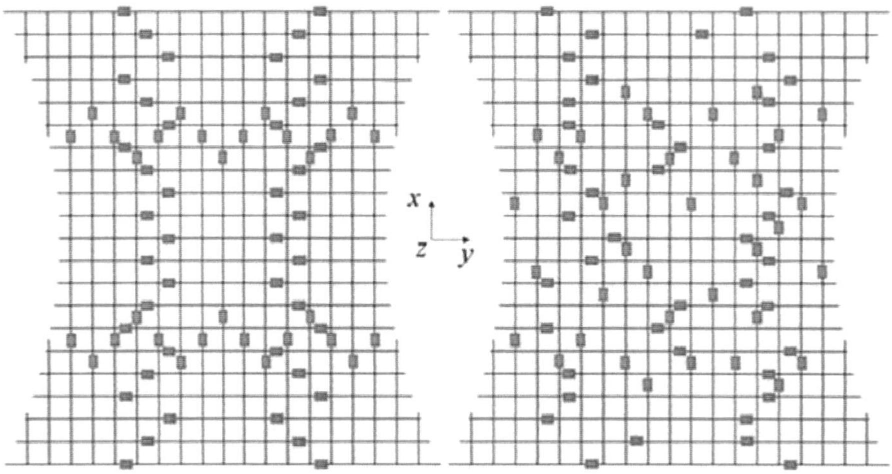

(a) (b)

Figure 6. Flat grids with connectors (a) Arrangement I (b) Arrangement II

3. Results and Discussions

Figure 7 and Figure 8 represent the stress distribution in gridshell members and deformation of the gridshell after erection. Figure 7(a) and Figure 8(a) depict the stress in pipe sections and deformation for the gridshell without connectors. Similarly, Figure 7(b) and Figure 8(b) are for gridshell with connectors employing Arrangement I while Figure 7(c) and Figure 8(c) considers proposed Arrangement II.

Figure 7 indicates a significant rise in the stress level due to the presence of connectors in member sections. The maximum stress developed in the member sections without connectors was found to be 160.89 MPa. Whereas, analysis of sections with connectors in gridshell showed an evident increase in stress level in few portions, specifically in the crown region as highlighted in Figure 7(b) and (c). The stress in the highlighted regions of the gridshell with connectors exceeded the estimated maximal stress (160.89 MPa) from gridshell without connectors. These regions are susceptible to failure and member breakage may occur due to underestimation of the sectional stress. Figure 7(b) shows that 11 members were over-stressed for Arrangement I while Figure 7(c) shows that 12 members were over-stressed for Arrangement II. Moreover, Figure 7(b) shows that the over-stressed regions are localized in the central regions of the structure. It can also be observed from Figure 7(b) and (c), more beam sections in the structure with connectors employing Arrangement I exceeded the stress (160.89 MPa) compared to the structure with connectors using Arrangement II. But the intensity of stress was higher in the latter case. It can also be noticed that the arrangement of the connectors caused significant rise in the stress, which cannot be neglected in the design. For Arrangement I, the stress rose by 6.51 MPa, whereas for Arrangement II, the stress rose by 20.55 MPa. A surplus stress of 14.04 MPa was caused due to the poor arrangement of the connectors (i.e., the Arrangement II) in the structure. In the gridshell considered in this research, y-directional members had large curvature compared to the x-directional members. Therefore, presence of connectors in middle part of the y-directional members in Arrangement II caused significant rise in maximal stress in member sections.

Similarly, it is apparent from Figure 8 that presence of connectors does not cause any significant change in the global shape whatsoever. The distribution of deformation is almost similar in all the cases and same in case of both type of connector arrangements. An increase in maximum deformation of 0.01 m was caused by the presence of connectors which is negligible compared to the global dimension of the structure.

IOP Conf. Series: Materials Science and Engineering **758** (2020) 012051 doi:10.1088/1757-899X/758/1/012051

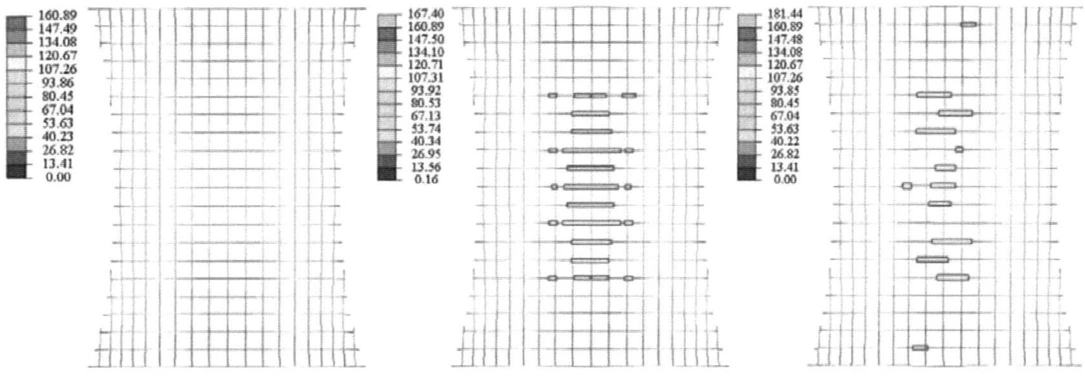

(a) (b) (c)

Figure 7. Stress after erection (a) Without connector (b) Arrangement I (c) Arrangement II (MPa).

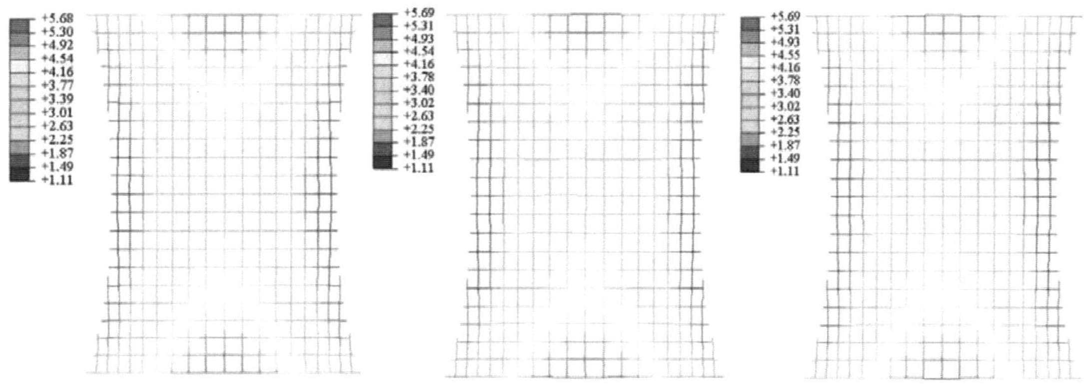

(a) (b) (c)

Figure 8. Deformation after erection (a) Without connector (b) Arrangement I (c) Arrangement II (m).

4. Conclusion

A concept of equivalent rigid part (ERP) for simplifying the modeling of end-to-end connector in the structural analysis of GFRP elastic gridshell is proposed in this paper. The ERP is further adopted for evaluating the effects of the connectors in the GFRP elastic gridshell by utilizing a typical structure, where the deformations and stresses are considered. The major conclusions drawn from the study are as follows:

- It is evident from the study that the ERP can predict the effect of connector to a greater precision.
- The effect of connector in flexure of a gridshell member is found to be minimal. Moreover, no significant effect on global shape was observed. Therefore, the flexibility of the member shall be restored as long as possible to attain the prescribed form.
- The presence of connector increased the member stress significantly. Hence, including the connectors while designing gridshell structures, aids to predict the true member stress and prevents failure of structural members before attaining the form.
- It is observed that the connector arrangement has significant effect on the member stress. Therefore, extra attention must be given to the proper arrangement of the connectors in the gridshell while designing and analyzing the structure.

IOP Conf. Series: Materials Science and Engineering **758** (2020) 012051 doi:10.1088/1757-899X/758/1/012051

Therefore, for designing GFRP elastic gridshell, where the end-to-end connectors are involved, it is suggested to incorporate the connectors in the analysis for better prediction of structural behaviors. For this purpose, the presented concept of ERP could be an option for simplification instead of connectors in the whole model for analysis.

References:

[1] Paoli C 2007 *Past and Future of Grid Shell Structures* [Masters](Massachusetts Institute of Technology)

[2] Peloux LD, Baverel O, Caron J-F and Tayeb F 2013 From shape to shell a design tool to materialize freeform shapes using gridshell structures *Des. Model. Symp.* (*Berlin*)

[3] Lienhard J, Alpermann H, Gengnagel C and Knippers J 2013 Active Bending, a Review on Structures where Bending is Used as a Self-Formation Process *Int. J. Sp. Struct.* **28** 187-96

[4] Hernández EL, Baverel O and Gengnagel C 2013 On the Design and Construction of Elastic Gridshells with Irregular Meshes *Int. J. Sp. Struct.* **28** 161-74

[5] Happold E and Liddell WI 1975 Timber lattice roof for the Mannheim Bundesgartenschau *Struct. Eng.* **53**

[6] Tayeb F, Lefevre B, Baverel O, Caron J-F and Peloux Ld 2015 Design and realisation of composite gridshell structures *J. Int. Assoc. Shell Spat. Struct.* **56** 49-59

[7] Baverel O, Caron J-F, Tayeb F and Peloux LD 2012 Gridshells in composite materials: Construction of a 300 m2 forum for the solidays' festival in Paris *Struct. Eng. Int.* **22** 408-14

[8] Hernández EL, Gengnagel C, Sechelmann S and Rörig T 2011 On the materiality and structural behaviour of highly-elastic gridshell structures *Computational Design Modelling* ed Gengnagel C, Kilian A, *et al.* (Berlin: Springer) p 123-35

[9] Douthe C, Baverel O and Caron J-F 2006 Form-finding of a grid shell in composite materials *J. Int. Assoc. Shell Spat. Struct.* **47** 53-62

[10] Du Peloux L, Tayeb F, Caron J-F and Baverel O 2015 The Ephemeral Cathedral of Créteil: a 350m2 lightweight gridshell structure made of 2 kilometers of GFRP tubes *CIGOS 2015: "Innovations in Construction", Proc. of CIGOS 2015* (*Chanchan, France*)

[11] Du Peloux L, Tayeb F, Baverel O and Caron J-F 2016 Construction of a large composite gridshell structure: a lightweight structure made with pultruded glass fibre reinforced polymer tubes *Struct. Eng. Int.* **26** 160-7

[12] Quinn G and Gengnagel C 2014 A review of elastic grid shells, their erection methods and the potential use of pneumatic formwork *Mobile and Rapidly Assembled Struct. IV* **136** 129

[13] THINkSHELL 2014 *Temporary Cathedral in Paris 2013* [20-Aug-2019] Available from: http://thinkshell.fr/gridshell-cathedral-2013/

Photocatalytic Activity of Cu-based Nanoparticles Synthesized by Glycine-Nitrate Combustion Technique

Hsuhui Cheng[1]*, Liangwei Jang[2], Shiaoshing Chen[2]* and Huamin Chen[1]

1School of Environmental and Chemical Engineering, Zhaoqing University, Zhaoqing, 526061, China
2 Institute of Environment Engineering and Management, National Taipei University of Technology, 1, Sec.3 Chung-Hsiao E. Rd., Taipei 10643, Taiwan, China

E-mail: hhcheng1126@gmail.com; f10919@ntut.edu.tw

Abstract. The copper based nanoparticles was synthesized by glycine-nitrate process (GNP), using copper nitrate trihydrate [$Cu(NO_3)_2 \cdot 3H_2O$] as main starting materials and glycine [$C_2H_5NO_2$] as complexant and incendiary agent. The as-prepared powders were characterized by X-ray diffraction (XRD), and scanning electron microscopy (SEM) techniques. Results of the photocatalytic degradation of Nonylphenol polyethoxylates (NP9EO) in a custom-made photoreactor indicated that the maximum degradation (more than 94% and 70% TOD removal) of NP9EO occurred with $CuO+Cu_2O$ composite catalyst (dosage of 0.3 g/L) when a combination of ultraviolet (UV) irradiation for 600 min, and a heterogeneous system was used.

1. Introduction

Nano-sized of transition metal oxides (TMO) has attracted great attention due to their unique applications in various fields, such as chemical, catalysts [1], photo-catalysts [2], solar cells [1], and gas sensors [3] etc. Preparation of high quality nanostructures of defined, controllable size and morphology are the main determinant of the catalytic performance [4]. Among them, transition metal oxides (TMOs)-based catalysts offers an active sites with low activation potential for hydrogen evolution systems [3], especially noble metals (such as gold, silver, and platinum) are widely employed optimize the catalytic efficiency of these. On the other hand, high cost of noble metals are still major obstacles in developing stable highly efficient catalysts. Consequently, reducing the amount of noble metals without compromising their catalytic performance is an urgent task.

Compared with noble metal catalysts, being cheap, quite abundant and safe, copper has recently drawn the attention as co-catalyst for photoreforming [5]. The Cu oxides (Cu_xO_y, where x:y = 1:1 or 1:2) is as TMO catalysts, is a p-type semiconductor having monoclinic structure with a bandgap of 1.7–2.17 eV [6], and high theoretical photocurrent density of −14.7 mA/cm^2 [6-8], which makes it a promising material in application of catalysis. In this paper, we present a glycine nitrate process (GNP) methods with varying G/N ratio conditions, it is possible to obtain Cu_xO_y and metallic copper catalysts. Meanwhile, this study aimed to investigate the heterogeneous copper oxides (including metallic copper) catalyst degradation of nonylphenol polyethoxylates (NP9EO) under ultraviolet (UV) irradiation in a custom-made photoreactor.

IOP Conf. Series: Materials Science and Engineering **758** (2020) 012052 doi:10.1088/1757-899X/758/1/012052

2. Experimental

2.1. Catalyst Preparation

To synthesize Cu based materials by using the GNP route, copper nitrate trihydrate [$Cu(NO_3)_2 \cdot 3H_2O$] (Merck, USA, 99.5% purity) and glycine [$C_2H_5NO_2$] (Merck, USA, 99.7% purity) were dissolved in distilled water. The glycine (fuel) was added into the metal nitrate (oxidizer) solution at the various glycine-nitrate (G/N) ratio (G/N=0.3, 0.5, 0.7). The glycine-nitrate solution was then mixed and heated overnight on a hot plate stirrer (Corning, USA) at 105°C to form clear, homogeneous and viscous gel-like solutions. The gel was then placed in a ceramic bowl, and place it in the oven, further heated to the temperatures of 200°C until the gel was self-ignited, producing a catalyst ash.

2.2. Characterization

The crystal structure of the samples was determined by powder X-ray diffraction (XRD) using Cu Kα radiation (λ = 1.5418 Å) on a Rigaku DMAX 2200VK/PC diffractometer (Tokyo, Japan). All peaks measured using XRD were assigned by comparison to those of the Joint Committee on Powder Diffraction Standards (JCPDS) data. The morphology of the synthesized samples were analyzed by scanning electron microscope (SEM, Hitachi S-4800) coupled with EDS (energy dispersive X-ray spectroscopy) measurements.

2.3. Photocatalytic experiments

NPEOs (reagent grade) with an average of 9 EOs (Sigma-Aldrich, USA, purity ≥ 99.7%) was chosen as model compound. The photocatalytic reaction was carried out in a custom-made photoreactor (Panchum Scientific Co., PR-2000) as illustrated in Fig. 1. The photocatalytic activities of Cu oxides samples under UV irradiation were evaluated based on the degradation rate of nonylphenol polyethoxylates (NPEOs) in a quartz reactor (120/115 mm OD/ID, Height: 240 mm) containing 0.3 g of Cu oxides (G/N=0.3) and 1 L of a 100 mg/L NPEOs aqueous solution. The system was open to the atmosphere, and the cylindrical quartz reactor was surrounded by a total of 14 light tubes. The UV light tubes were germicidal lamps with a wavelength of 254 nm (Sankyo Denki Co., Ltd.). The photo-degradation rates of NPEOs solutions was determined by high performance liquid chromatography (HPLC) (Agilent 1100 series; Agilent Technologies, USA) with Wondasil-C18 (4.6 × 150 mm, 5 μ m), and a 2487-UV detector at a wavelength of 277nm.

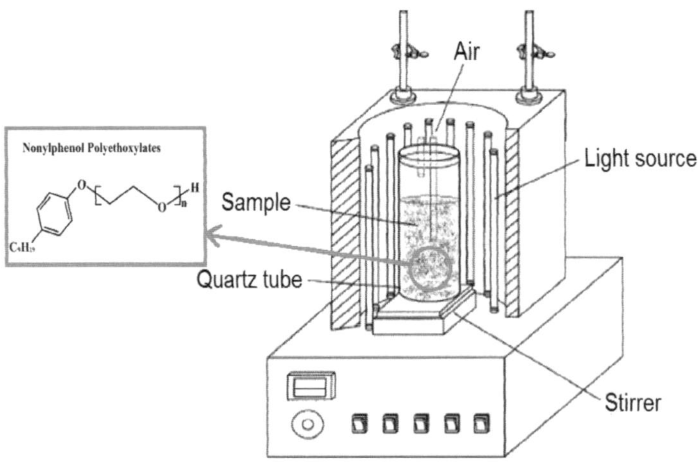

Figure 1. Schematic drawing of the custom-made photoreactor

3. Results and discussion

3.1 Characterization of Cu-Based nanoparticles catalyst

Fig. 2 exhibits the XRD pattern of the as-synthesized copper based nanoparticles by GNP method with different G/N ratio conditions. At the stoichiometric G/N ratio of 0.5 (see Sample (b) in Fig. 2), intense reflections are observed with 2θ values of 29.5°, 36.8°, 38.7°, 42.3°, 43.4°, 50.5°, 61.4°, and 74.2° . The peaks at 38.7° correspond to the (111) plane of divalent copper (II) oxide (CuO) whereas the reflections at 29.5°, 36.8°, 42.3° and 61.4°corresponding to(110), (111), (200), and (220) planes of cuprite, respectively (JCPDS File No. 01-078-0428), indicate the formation of cubic copper (I) oxide (Cu_2O) nanocrystals (JCPDS File No. 01-077-0199). In addition, the zero-valent copper (Cu^0) peaks at 50.5°, and 74.2° correspond to the (200), and (220) peaks were observed, respectively. This obviously illustrates that the Cu^0 nanoparticles formed in the chemical reduction stage go through decomposition due to limited stability of Cu [9,11], and Cu_2O might be formed by oxidation [10-11]. When the G/N ratio was increased to 0.7, broad reflections due to a highly dispersed Cu^0 phase was shown in sample (c) of Fig. 2 The XRD pattern (111), (200), and (220) of Cu^0planes at 43.4°, 50.5°, and 74.2°, respectively, which are in good agreement with the bulk copper crystallographic data (JCPDS File No. 01-085-1326). The results clearly show that the formation of Cu^0 as a favoured product with increased glycine (fuel) content in the combustion synthesis [12]. When the G/N ratio was decreased to 0.3, the sample (a) contains a large number of primary CuO phases and a few number of Cu_2O phases (about 9.5%) were observed. This indicates that the nanocrystalline CuO reduces to a stable Cu_2O phase rather than forming metallic copper directly [13].

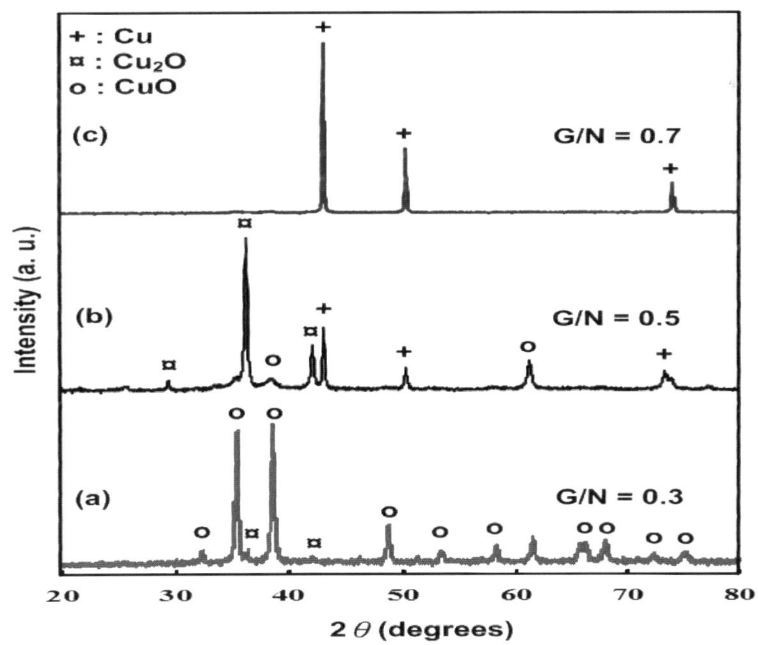

Figure 2. The XRD pattern of combustion synthesized products of copper-based materials with various glycine-nitrate ratio (G/N) = 0.3(a), 0.5(b), and 0.7(c)

Fig. 3 shows the scanning electron microscope (SEM) with EDS images of prepared Cu based nanoparticles at various G/N ratios. The typical EDS spectrum from the particles in Figure 5 indicates that the particles were composed of copper (Cu), and oxygen (O). From SEM photographs, the catalysts formed a sponges-like aggregates containing some small nanoparticles with the broad-sized distribution (~5–100 nm), and some irregular granules with a high degree of shape anisotropy. It could be noticed that prepared Cu based nanoparticles were not single-domain particles. Obviously, the particles shape and size in homogeneity was typical for combustion syntheses as well as the presence of sponges or flake-like structures. The results indicate that the high amount of gas evolution during the combustion process affects the porosity of solid products [14]. Moreover, high combustion temperature promotes nanoparticles agglomeration and crystal grains growth [15].

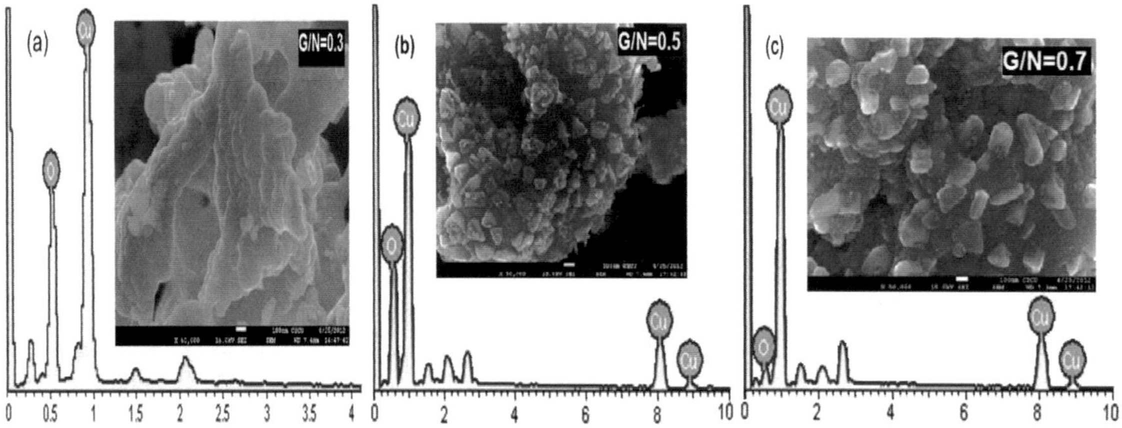

Figure 3. The SEM with EDS images of prepared Cu based nanoparticles at glycine-nitrate ratio (G/N) = 0.3(a), 0.5(b), and 0.7(c).

3.2 Degradation of NP9EO in batch photoreactor

Fig. 4 shows the photocatalytic NP9EO degradation curves for copper/oxygen catalysts synthesized at different G/N ratios. The photocatalytic activity evolves as follows: G/N=0.5 > G/N=0.7> G/N=0.3. The G/N=0.5 catalysts shows the highest photocatalytic activity and leads to 94.6% NP9EO degradation in 600 min. This clearly suggests that G/N=0.5 condations, catalysts have a higher photocatalysts for the photodegradation of NP9EO. The Cu_2O (bandgap of around 2.2 eV) has superior electric conductivity to CuO (bandgap of around 1.6 eV) [6], and thus a better removal effect in a photocatalytic system. Besides, Cu_2O and CuO are more stable at a higher potential for photoelectrochemical (PEC)/hydrogen evolution reaction (HER) because a high potential restrains the photoreduction of Cu_2O (or CuO) to metal copper [16]. Consequently, for the copper-based catalysts for heterogeneous system, the best photocatalytic performance with a glycine/nitrate ratio of 0.5. Furthermore, in our results, TOC is consistently lower than the removal efficiency of NP9EO, indicating other than mineralization to CO_2, some of the intermediates were also generated [17]. The main reason for NP9EO cannot, therefore, be effectively mineralized.

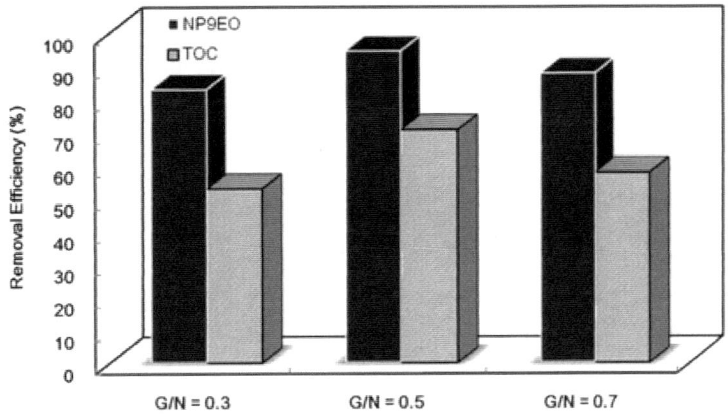

Figure 4. NP9EO degradation curves for copper/oxygen catalysts synthesized at different G/N ratios

4. Conclusions

Nanocrystalline Cu-based heterogeneous catalyst was successfully synthesized using combustion synthesis with glycine as the fuel and nitrate as the oxidizer. For the Cu-based catalysts for heterogeneous system, the best photocatalytic performance with a glycine/nitrate (G/N) ratio of 0.5, which leads to 94.6% NP9EO degradation and 70.5% TOC reduction were obtained after 600 min. Furthermore, in our results, TOC is consistently lower than the removal efficiency of NP9EO, indicating other than mineralization to CO_2, some of the intermediates were also generated.

References

[1] Lim, H.H., Chua, P.N., Mun, H.P., Horri, B.A. *Int. J. Adv. Sci .Eng. Inf. Technol.* **2018**, *6*, 2321-8991
[2] Jiao, F., Frei, H. *Energy Environ. Sci.* **2010**, *3*, 1018-1027
[3] Li, W. Y., Xu, L. N., Chen, J. *Adv. Funct. Mater.* **2005**, *15*, 851-857
[4] Tran, T. H., Nguyen, V. T. *Int. Sch. Res. Notices.* **2014**, *2014*, 1-14
[5] Clarizia, L., Spasiano, D., Di Somma, I., Marotta, R., Andreozzi, R., Dionysiou, D.D. *Int. J. Hydrogen Energy* **2014**, *39*, 16812-16831
[6] Bandara, J., Udawatta, C. P. K., Rajapakse, C. S. K. *Photochem. Photobiol. Sci.* **2005**, *4*, 857-861
[7] Janczarek, M., Kowalska, E. *Catal.* **2017**, *7*, 317-342
[8] Abd-Elkader, O.H., Deraz, N. *Int. J. Adv. Sci .Eng. Int. J. Electrochem. Sci.* **2013**, *8*, 8614-8622
[9] Aslam, M., Gopakumar, G., Shoba, T.L., Mulla, I.S., Vijayamohanan, K., Kulkarni, S.K., Urban, J., Vogel, W. *J. Coll. Interf. Sci.* **2002**, *255*, 79-90
[10] Feng, L., Zhang, C., Gao, G., Cui, D. *Nanoscale Res. Lett.* **2012**, *7*, 276-286
[11] Khan, A., Rashid, A., Younas, R., Chong, R. *Int. Nano. Lett.* **2016**, *6*, 21-26
[12] Rao, G.R., Mishra, B.G., Sahu, H.R. *Mater. Lett.* **2004**, *58*, 3523-3527
[13] Pike, J., Chan, S. W., Zhang, F., Wang, X. *Appl. Catal. A-Gen.* **2006**, *303*, 273-277
[14] Patil, K.C., Hegde, M.S., Rattan, T., Aruna, S.T. *World Scientific Publishing Co. Pte. Ltd.* 2008, Singapore.
[15] Podbolotova, K. B., Khorta, A. A., Tarasovb, A. B., Trusovc, d, G. V., Roslyakovd, d, S. I., Mukasyan, A. S. *Combust. Sci. Technol.* **2017**, 189, 1878-1890
[16] Yang, Y., Xu, Di., Wu, Q., Diao, P. *Sci. Rep.* **2016**, *6*, 35158-35161
[17] Chen, S. S.; Hsu, H. T.; Tsui, H. J.; Chang, Y. M. *Desalination Water Treat.* **2013**, *51*, 1678–1684

IOP Conf. Series: Materials Science and Engineering **758** (2020) 012053 doi:10.1088/1757-899X/758/1/012053

Mathematical modeling and analysis of energy conversion efficiency of particle jet accelerating nozzle

Tiancheng Fang[1] , Fushen Ren[2*], Xiaoze Cheng[3], Qing Hu[2] and Tiexin Tong[1]

1 School of Civil Engineering and Architecture, Northeast Petroleum University, Daqing, Heilongjiang, China
2 School of Mechanical Science and Engineering, Northeast Petroleum University, Daqing, Heilongjiang, China
3 The China National Petroleum Corporation, Beijing, China
E-mail: renfushen@126.com

Abstract. The impact kinetic energy of particle jet determines rock-breaking effect and velocity, and as acceleration device of particle jet, the energy conversion efficiency of nozzle determines rock-breaking efficiency. According to three different types of nozzles of the straight-taper nozzle, the pressure drop model and energy conversion efficiency of particle jet was established, and the energy conversion efficiency under different type nozzles was studied. And then, the acceleration ability of different nozzles for was analyzed with simulation. The results of model calculation and simulation show that constant-acceleration nozzle has the best acceleration ability and energy conversion efficiency. The rock-breaking experiment was carried out with the adopted nozzle, and comparison analysis between experimental results and numerical calculation prove that the applicability of mathematical model.

1. Introduction

Particle jet impact rock breaking technology is a rather efficient non-contact rock-breaking method which can solve the problem of difficult drilling in deep-well and hard formation. Its working principle is that drilling fluid and steel particles are mixed at certain ratio, and then the acceleration action with nozzle would form high-velocity particle jet to destroy and impact rock [1]. In particle jet impact rock-breaking system, the nozzle plays a role of energy conversion [2], which can convert pressure of the system into kinetic energy of particle water jet, thus particle jet would have high velocity and can greatly improve drilling speed of deep well and hard formation. Therefore, as the key component of the particle jet impact rock breaking device, the acceleration capability and energy conversion efficiency of the nozzle determine the drilling speed and rock breaking effect.

At present, the majority of researchers changed the nozzle cone angle and length to diameter ratio to improve the acceleration capability of nozzle [3]. Shen [4], Huang [5] et al. analyzed the relationship between cone angle and nozzle diameter on jet velocity. Liu et al. researched the relationship between energy loss and flow coefficient and verified experimentally [6]. Ren et al. studied the acceleration effects of four different nozzle structures [7].

The studies on energy conversion efficiency in acceleration process of nozzle are less at present, so it is necessary to select the appropriate nozzle type, so as to improve energy conversion efficiency and rock-breaking efficiency. In this paper, the effect factors of energy conversion efficiency can be analyzed with the establishment of mathematical model of energy conversion efficiency and CFD simulation, so as to optimize the nozzle type of flow channel. Based on adopted nozzle type, the comparative analysis of experiment and numerical calculation could be used to verify the correctness of mathematical model.

2. Modeling and analysis

2.1. The accelerating mechanism of the nozzle

Particle jet impact rock-breaking technology is inspired by projectile impact rock-breaking technology, and it is proposed to add a certain proportion of steel particles into high-speed mud, and through the acceleration of drill pipe and nozzle to form high-speed particle jet. The high-speed particle jet impacts on rock surface to form hole or grove, which could release the compaction effect at the bottom of deep-well and reduce rock strength, and then quickly break rock under the grinding effect of drilling bit. Debris and steel particles formed by rock-breaking are carried to surface separation equipment by drilling fluid for particle recovery and separation. The process of rock-breaking under particle jet impact is shown in figure 1.

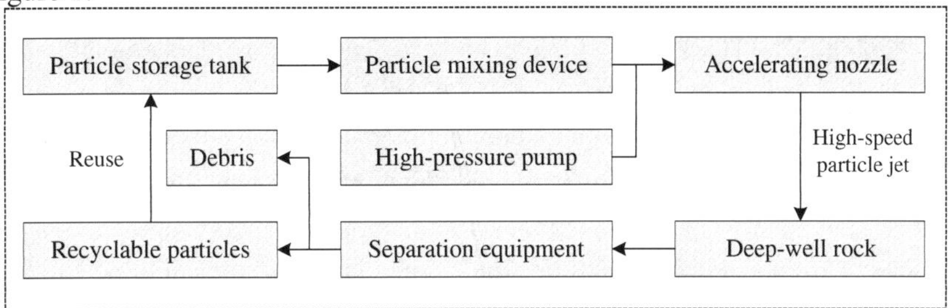

Figure 1. The process of rock-breaking under particle jet impact.

According to the rock-breaking mechanism of particle jet, the rock-breaking effect depends on impact kinetic energy of particle jet, and the acceleration of particle jet mainly depends on the nozzle. Therefore, it is of great significance to study energy conversion efficiency of particle jet in different type of nozzles.

2.2. Energy conversion efficiency model

The structural model of nozzle refers to the shape and geometry of the flow channel. At present, the widely used high-efficiency nozzles are composite nozzles, which are composed of contraction section and cylinder section. In this paper, the flow channel shape is mainly studied of straight-taper nozzle, constant-acceleration nozzle and streamline nozzle.

When drilling fluid flows through nozzle, firstly, it will be accelerated by contraction section, and then flows cylinder section. Fluid mechanics is applied to analyze the particle jet accelerating process [8]. The resistance of drilling fluid flowing through nozzle mainly consists of contraction resistance and on-way resistance, namely:

$$h_w = h_j + h_f \tag{1}$$

Where, h_w is resistance of drilling fluid; h_j is contraction resistance; h_f is on-way resistance.

The contraction resistance refers to the resistance when fluid flows through contraction section. The calculation formula is as follows:

$$h_j = \zeta_j \frac{u_f^2}{2g} \tag{2}$$

Where, ζ_j is local resistance coefficient; g is gravity acceleration.

Because local resistance coefficient varies with the shape of contraction section, and its value is determined by experimental measurement and empirical formula [9].

The flow state of particle water-jet is turbulent, and the energy loss is mainly due to overcoming the on-way resistance. The on-way resistance mainly includes the on-way resistance of contraction section and cylinder section, and the expression is as follows:

$$h_f = \int \lambda \frac{L_1}{2y} dx \frac{u_f^2}{2g} + \lambda \frac{L_2}{d_2} \frac{u_{f1}^2}{2g} \tag{3}$$

Where, λ is on-way resistance factor, which is related to nozzle roughness and fluid pattern[8]; L_1 is length of contraction section; L_1 is length of cylinder section; d_2 is outlet diameter of nozzle.

The expression of total resistance coefficient when particle jet flows through bit nozzle could be calculated:

$$\zeta = \zeta_j + \int \lambda \frac{L_1}{2y} dx + \lambda \frac{L_2}{d_2} \tag{4}$$

Where, ζ is total resistance coefficient.

According to energy equation of fluid flow, the relationship between flow coefficient of nozzle and total resistance coefficient C can be deduced:

$$C = \sqrt{1/(1+\zeta)} \tag{5}$$

The water power at the nozzle inlet and outlet could be obtained:

$$N_i = Q\nabla P \tag{6a}$$

$$N_o = \frac{1}{2}\rho Q u_{f1}^2 \tag{6b}$$

Where, N_i is water power at the nozzle inlet; N_o is water power at the nozzle outlet; ∇P is pressure drop of nozzle.

When drilling fluid flows through nozzle, the pressure difference between inlet and outlet of nozzle will generate, which called pressure drop. It is an important parameter to determine kinetic energy of particle water-jet, and the expression is:

$$\nabla P = \frac{\rho_f Q^2}{2nC^2 A_1^2} \tag{7}$$

Where, ρ_f is fluid density; ρ_f is flow rate; n is the number of nozzle; A_1 is sectional area of inlet.

As energy conversion element in acceleration of particle jet, the performance of nozzle can be expressed by energy conversion efficiency:

$$\eta = \frac{N_o}{N_i} = C^2 = 1/(1+\zeta) \tag{8}$$

Where, η is energy conversion efficiency.

According to established mathematical model, the energy conversion efficiency of nozzle is directly proportional to the square of flow coefficient.

3. Simulation analysis of acceleration effect

In order to simulate the acceleration and pressure change of particle jet, the computational fluid dynamics (CFD) method is used to simulate the internal flow field of different types of nozzles, as shown in figure 2. In simulation process, the boundary conditions are velocity inlet and outflow [10]. The velocity of particles and drilling fluid at the inlet is 14 m/s, and the volume fraction of particles is 0.2%. The standard *k-ε* equation and Eulerian model are used in simulation.

According to simulation results, the acceleration of particles is obviously different. The acceleration process runs through contraction section and cylinder section, and the acceleration effect of constant-acceleration nozzle is the best.

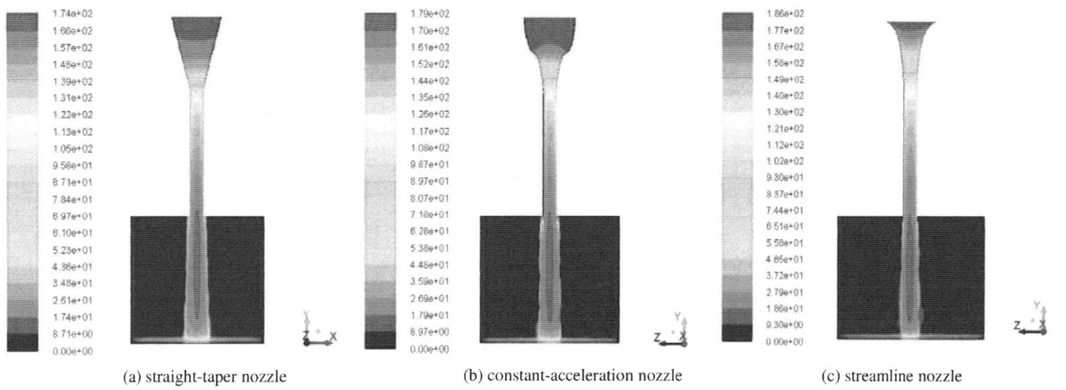

Figure 2. The velocity nephogram of particle with different nozzles.

4. Results and analysis

According to the established mathematical model, the relationship between energy conversion efficiency of different nozzles and contraction section is shown in figure 3. With increase of contraction section length, the energy conversion efficiency would gradually decrease. When the length of contraction section is fixed, energy conversion efficiency of constant-acceleration nozzle is the best. Because pressure drop is inversely proportional to energy coefficient, the nozzle with the highest energy conversion efficiency should be selected at the same time to meet the rock breaking requirements, that is, in particle jet impact rock-breaking process, it is adopted with constant-acceleration type.

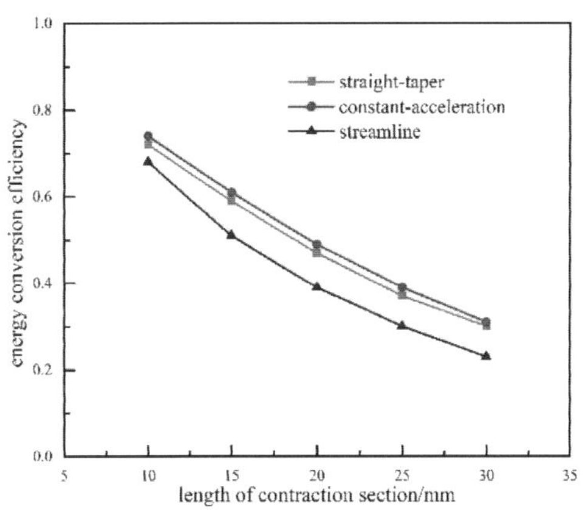

Figure 3. Energy conversion efficiency and the length of contraction section.

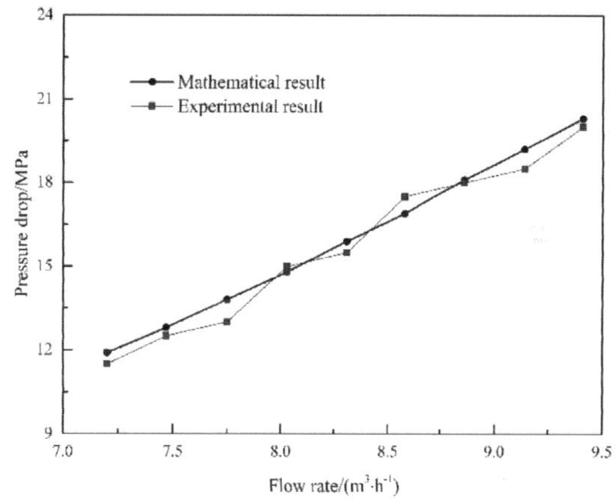

Figure 4. The comparison of pressure drop between experimental data and mathematical model.

Particle jet impact rock-breaking tests were carried out with constant-acceleration nozzle. The comparison between experimental data and mathematical model is shown in figure 4. According to comparison results, experimental value is basically consistent with mathematical calculation, which proves the applicability of mathematical model. The results are of great significance to study of energy conversion of nozzles and rock breaking efficiency.

5. Conclusion

(1) The mathematical model of energy conversion efficiency of particle and water jet in nozzle is established, and the influence of different types of nozzle on energy conversion efficiency is analyzed
(2) The type of nozzle is determined by model calculation and simulation, and the applicability of mathematical model is verified by comparison of experiment and numerical simulation. The results are of great significance to study of energy conversion of nozzles and rock breaking efficiency.

6. References

[1] Thomas Hardisty 2007 *American Oil & Gas Reporter* **50** 86
[2] Zhou W, Huang J, Li L 2018 *IOP Conference Series Earth and Environmental Science* **108** 1-6
[3] Huang Z, Xie Y 2011 *Mechanical Design* **28** 62-5
[4] Shen C M, Lin B Q, Meng F W 2011 *Advanced Materials Research* **228** 1001-6
[5] Huang Z, Xie Y 2011 *Applied Mechanics and Material* **84** 368-72
[6] Liu C, Li Z 2000 *Drilling & Production Technology* **23** 1-3
[7] Ren F, Ma R, Cheng X 2014 *Material Mechanical and Manufacturing Engineering* **II** 475-8
[8] Li S. 1980 Engineering fluid mechanics *China Machine Press* 79-156
[9] Nie B, Hui W, Li L 2011 *Procedia Engineerin* **26** 48-55
[10] Ren F, Fang T, Cheng X 2019 *Journal of Petroleum Science and Engineering* **183** 106419

Acknowledgments
The study was funded by National Natural Science Foundation of China (11972113), the Post-doctoral Research Start-up Fund Project of Heilongjiang Province (LBH-Q15018), and the Guiding Innovation Fund Project of Northeast Petroleum University (2019YDL-07).

IOP Conf. Series: Materials Science and Engineering **758** (2020) 012054 doi:10.1088/1757-899X/758/1/012054

Investigation of process for producing carbide-containing welding fused fluxes using plasma granulation

A O Artemov[1], S V Naumov[2], M N Ignatov[3], D S Belinin[4]
1,2,3,4 Perm National Research Polytechnic University, 29 Komsomolsky prospekt, Perm, 614990, Russia
E-mail: artemoff87@mail.ru

This article proposes a technology for producing fused welding fluxes using plasma granulation. The developed technology enables producing welding flux granules with carbide-containing additives. This is possible due to high dynamic and temperature effects of plasma arc on mineral raw materials that stops refractory additives from melting and thus remaining in welding materials composition. Using petrographic analysis, it was found that silicon carbide particles remain unchanged in granules of fused welding fluxes composition. Plasma granulation technology makes it possible to obtain both finished products such as welding fluxes and a charge ready for manufacturing cored wire and coated electrodes.

1. Introduction

Expanding the use of mineral raw materials, especially gabbroid and basaltoid groups of the Ural region becomes an urgent task, since there are sufficiently large reserves of both natural and technogenic raw materials in the Ural region. One of the directions for obtaining new types of products from mineral raw materials is producing welding materials (fused flux, cored wire, coated electrodes) using plasma granulation technology that helps to reduce energy and material costs and increase welding properties in general [1-3].

Welding materials are an integral part of modern welding production. Therefore, using high-quality welding materials with high welding-technological properties is required during welding and repairing (surfacing) critical and building structures of general purpose. To this end, the chemical composition of welding materials is improved for arc burning stability and the favorable course of physical and chemical reactions during welding (surfacing). Mineral-based welding materials with various additives (e.g. carbides) are in particular demand.

In this regard, the aim of this scientific work is to obtain welding materials on oxide-silicate basis with the addition of silicon carbide. The main task of the study on the identification of silicon carbide in welding fused flux granules composition is to carry out a petrographic analysis of welding flux granules obtained by plasma granulation.

2. Experimental research

The experiments on plasma granulation of welding flux from mineral raw materials with carbide-containing additives were carried out on modern HyperTherm Powermax 45 air-plasma cutting equipment. This equipment allows to programmatically control heat input into the material by changing the arc parameters, as well as changing the speed and trajectory of the plasmatron.

Rocks of oxide-silicate base of the gabbroid group of the Ural region were used to study plasma granulation process. Aforesaid mineral raw materials have all the necessary components for a slag basis of fused flux and a low content of harmful impurities. Such rocks of the gabbroid group as hornblendite possess good fusibility, required melting point, and homogeneity in chemical and phase composition. Silicon carbide (SiC with a grain size of F60, 0.2-0.3 mm) was used as an additive that increased arc burning stability and ensured high deposition coefficient and fluidity of the bath [4].

Sections from synthesized flux granules obtained under action of high dynamic and temperature effects of the plasma arc underwent petrographic analysis using a Nikon Eclipse E 600 POL polarizing microscope.

Petrographic analysis showed that composition of the granules is similar in structure to hornblende. Hornblende is characterized by absolute cleavage in two directions, where the cleavage angle is 124 ° and 56 °, but there is no such cleavage observed in this section, since cracks are technogenic in nature [5].

In this section a hornblende has brown and green color with a clear pleochroism (pleochroic from pale green to green or pale brown to dark brown) without an analyzer. With the analyzer, the grains are isotropic, although the hornblende crystallizes in monoclinic system, and it must be anisotropic.

In this section, there are inclusions of silicon carbide, where the grain size is 0.08-0.12 mm. These clusters are located unevenly on the periphery (Figure 1, a) or inside (Figure 2, a) of single hornblende grains. Carbides in the section are colorless (without analyzer Figure 1, 2 b) and have birefringence (with analyzer-photo Figure 1, 2, a), that is characterized by bright pearly interference coloring.

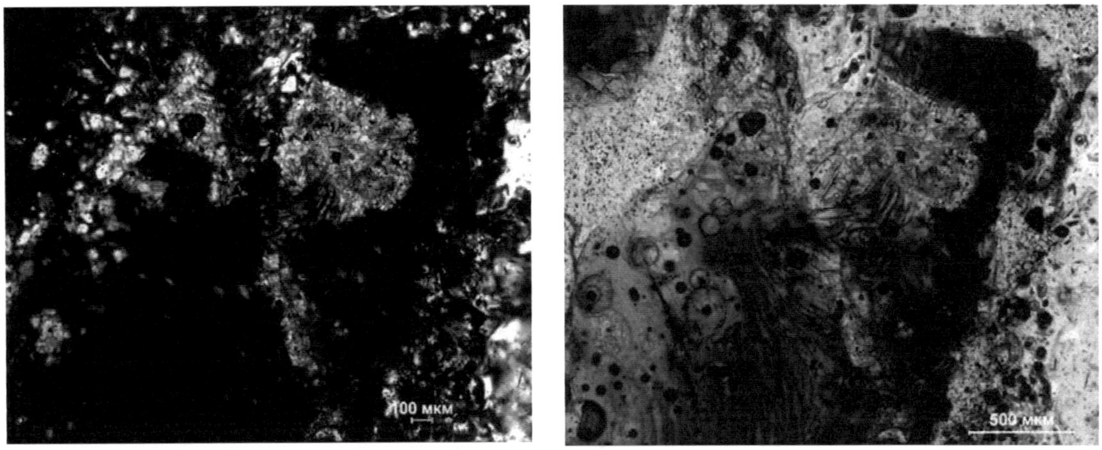

a b

Figure 1. Petrographic analysis of welding fused flux granules with SiC particles on the periphery: a - with analyzer-photo, 50x, b –without analyzer, x50.

It is possible that reduction in particle sizes of silicon carbide is associated with thermal decomposition. The oxide-silicate base affects the particles of silicon carbide, since an acidic medium causes thermal decomposition and a decrease in concentration and size of particles in welding flux granules. To prevent a decrease in concentration of carbides in the oxide-silicate base, it is possible to use a calcium fluoride base [6].

Using a calcium fluoride base, silicon carbide will be less susceptible to thermal decomposition. Fluorite can be such a calcium fluoride base. Fluorite-based granules obtained by plasma granulation are shown in Figure 3.

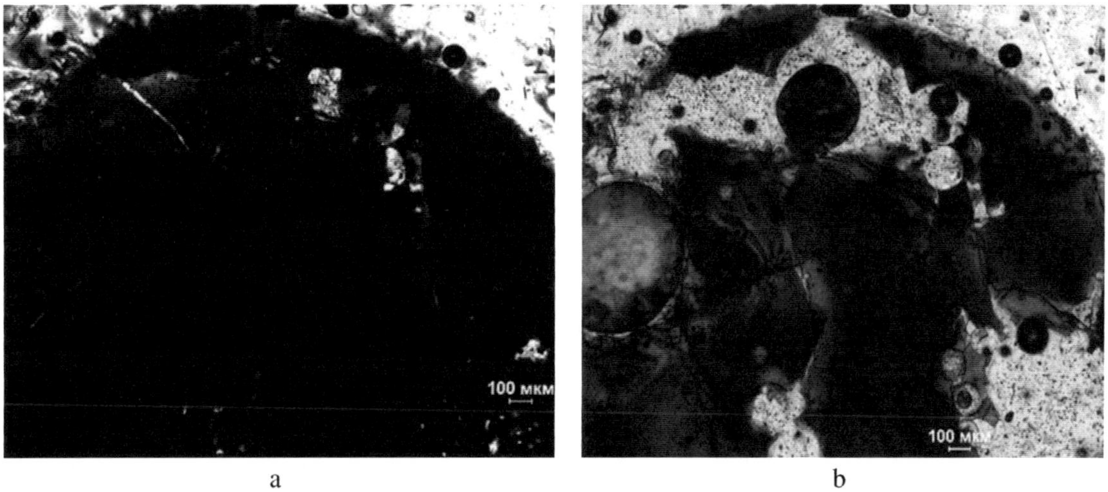

Figure 2. Petrographic analysis of welding fused flux granules with SiC particles inside the section: a - with analyzer-photo, 50x, b –without analyzer, x50.

Fluorite-based granules have absolute cleavage in four directions; therefore, the crystal can split along the octahedron. The figure shows two out of four directions, the angle between the cleavage is ~ 90 ° (Figure 3, a). Partial cleavage is also observed in the granule part due to its technogenic origin (Figure 3, b). Since cubic system is characteristic of fluorite, it lacks one of the main optical properties - birefringence. Under a microscope in a section, fluorite without an analyzer is most often colorless, with an analyzer it is isotropic.

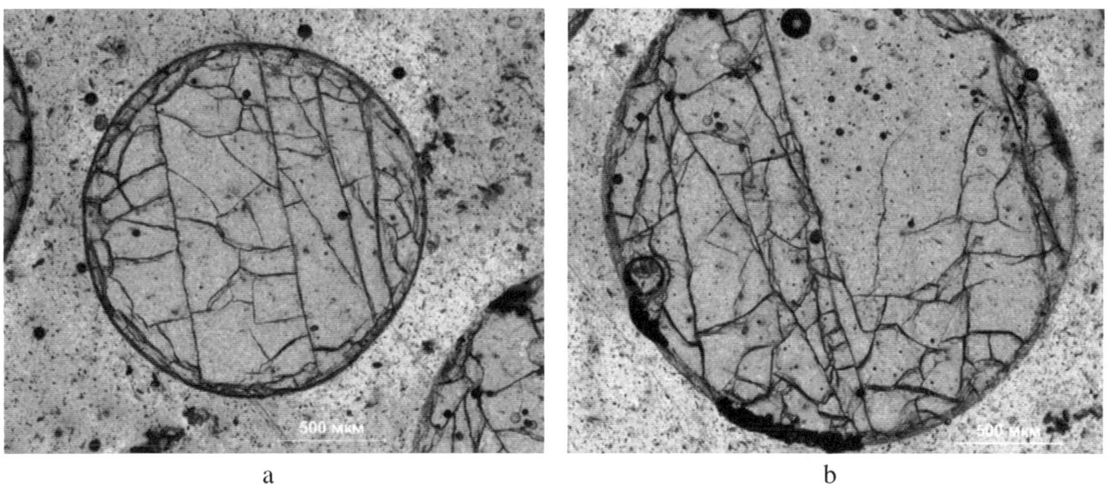

Figure 3. Petrographic analysis of welding fused flux granules on fluorite base: a – granule with absolute cleavage, x50, b –granule with anisotropic cleavage, x50.

It is assumed that during formation of such granules after plasma granulation carbides will be located along the cleavage boundaries of fluorite that indicates uniform distribution of such particles.

3. Conclusion

Plasma granulation of fused welding flux technology is proposed that allows to obtain granules from mineral raw materials of the Ural region with the addition of silicon carbide, that later in the welding process participate in metallurgical processes that favorably affect modification of the weld. It has been established that silicon carbide particles up to 0.12 mm in size are located both inside and on the periphery of the granules. The synthesis of such composite granules will expand the range of welding materials.

Acknowledgements
The reported study was funded by RFBR according to the research project № 18-33-00599.

References

[1] S.V. Naumov, A.O. Artemov, K.I. Belousov. Research on Plasma Electric Arc Granulation Processes of the Ural Region Mineral Raw Materials to Welding Materials for Submerged Arc Welding. *Materials Science Forum.* 946 (2019) 945-949
[2] S.V. Naumov, M.N. Ignatov, M.A. Sheksheev. Technology of mineral raw materials granulation by electric arc for manufacturing of welding fused flux. *Solid State Phenomena.* 265 (2017) 290-295
[3] Takaya Kawabe, Takeshi Hayashi. Plasma physics and plasma technology. *Physica Scripta.* 75 (1998) 72-74
[4] S.V. Naumov, M.N. Ignatov, A.M. Ignatova, A.O. Artemov. Development of slag base for welding fluxes from man-made mineral formations of ural mining and smelting companies. *Key Engineering Materials.* 743 (2017) 406-410
[5] E.A. Gorbatova, E.G. Ozhogina, A.N. Lebedev, E.A. Emelyanenko, A.S. Kharchenko, V.N. Selivanov. Justifying the integration of mineralogical analysis methods when studying metallurgical slags. *Vestnik of Nosov Magnitogorsk State Technical University.* 15 (2017) 31-39
[6] S.G. Parshin, RU Patent 2544317 (2015)

IOP Conf. Series: Materials Science and Engineering **758** (2020) 012055 doi:10.1088/1757-899X/758/1/012055

Obtaining welding materials of various structures, fractional and chemical composition using plasma granulation

S V Naumov[1], A O Artemov[2], M F Kartashev[3] and M A Sheksheev[4]

1,2,3 Perm National Research Polytechnic University, 29 Komsomolsky prospekt, 614990, Perm, Russia
4 Nosov Magnitogorsk State Technical University, 38, Lenin Avenue, 455000, Magnitogorsk, Russian Federation
E-mail: NaumovStanislav@yandex.ru

Abstract. The article proposes the technology for producing fused welding fluxes grains of various structures, fractional and chemical composition using a highly concentrated heat source. Interaction processes between the highly concentrated heat sources and mineral raw materials are presented in the article. The necessary components for obtaining welding flux with various granule structures are determined. The proposed technology allows to obtain welding flux granules with a range from 0.1 to 3.0 mm that meets regulatory documentation requirements. It was found that during the process of plasma granulation there is a decrease in the concentration of alkali metals to the required content for welding fluxes. At the same time, there is an increase in the concentration of basic elements such as silicon, calcium, aluminum and other metals.

1. Introduction

Currently there is a number of studies to obtain welding flux granules with a spherical shape and a vitreous grain structure using a plasma arc. Expanding the possibility of obtaining granules of different morphology and grain structure using highly concentrated heat sources is an urgent task [1].
In this regard, the aim of this work is to develop technological foundations for obtaining fused welding fluxes with different grain structures, morphology and chemical composition produced by plasma-arc method.
These aims are achieved by choosing the right source of mineral raw materials, manufacturing samples for the granulation process and the subsequent production of fused welding flux using plasma granulation.

2. Plasma granulation technology

Plasma granulation process was carried out on modern welding equipment for plasma cutting. Due to the fact that compressed air was used as a plasma-forming gas in the direct-action plasmatron, a copper electrode was used for a stable granulation and arc burning process. The scheme and process of granulation of fused welding flux by plasma-arc method is presented in Fig 1.
Rocks of the gabbroid group of the Ural region were selected as mineral raw materials for producing welding materials. Hornblendite contains all the necessary elements for a slag base and chalk makes it possible to increase slag basicity to the desired value, silicon carbide increases stability of arc burning and provides high surfacing coefficient and fluidity of the bath [2-4].
For granulation process samples were made in the form of plates (Fig. 2, a), using a technology similar to manufacturing of coated electrodes and agglomerated or ceramic fluxes, where hornblendite and

chalk charge was mixed with sodium liquid glass to produce plates [5]. The surface of the plate is porous. This is possible due to the chemical reaction of the components with liquid glass (Fig. 2, b).

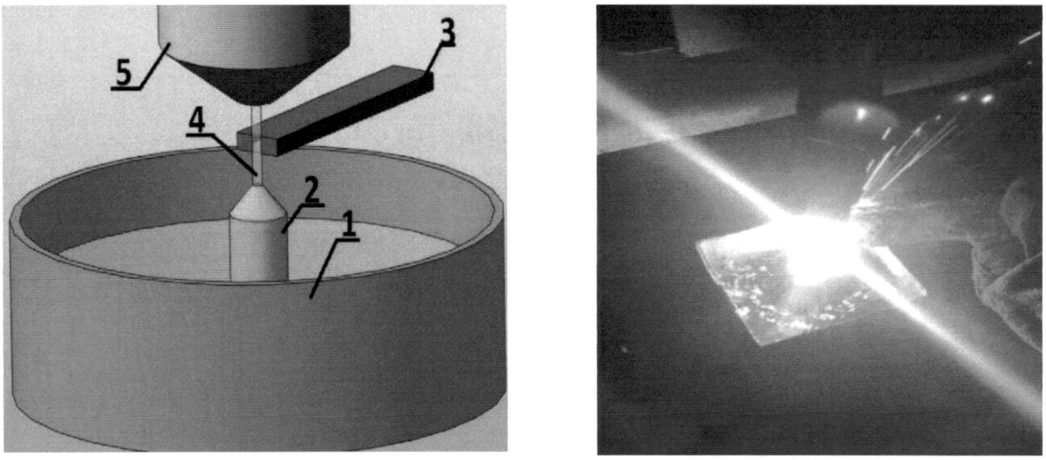

a b

Figure 1. The scheme (a) and granulation process (b) of fused welding flux by plasma-arc method: 1 – granulation pool; 2 – copper electrode; 3 – granulated plate; 4 – plasma arc; 5 – plasmatron.

a b

Figure 2. Samples for plasma granulation: a - general view of samples for plasma granulation; b - surface of samples, x150

Plasma arc 4 was ignited between the plasmatron 5 and a copper electrode 2 immersed into the granulation pool 1. Then, plates 3 were fed into the arc burning zone. Under dynamic and temperature action of the compressed arc the edges of the plates melted and fell into the granulation pool in the form of granules.

3. Experimental research

As a result, welding flux granules were obtained (Fig. 3), the morphology and size of these granules were examined using scanning electron microscopy with a ZEISS microscope. It was found that the granules have a particle size of 0.1 to 3 mm that corresponds to fractional composition for welding fluxes.

Figure 3 «a» shows general view of fused welding flux granules that have both spherical and fragmental form. The surface of the granules is porous because of the fact that chalk and partially silicon carbide form CaO, SiO_2, CO_2 under the conditions of an oxidizing medium and therefore

IOP Conf. Series: Materials Science and Engineering **758** (2020) 012055 doi:10.1088/1757-899X/758/1/012055

bubbles and pores on the surface of the granules are formed [6]. The formation of a pumiceous grain structure occurs more intensively when granules fall into heated water.

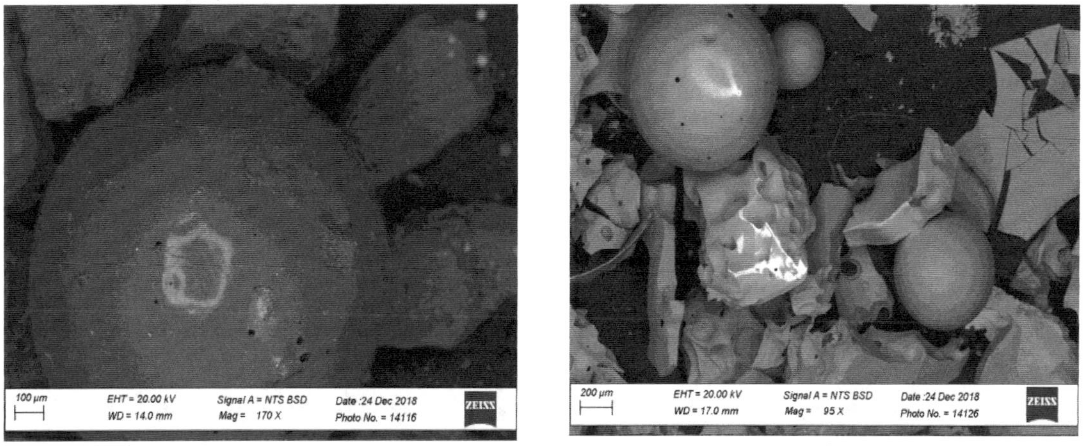

a b

Figure 3. Fused welding flux granulated by plasma arc, a – general view of the granules, b – general view of glass-pumiceous grain structure flux granules.

Plates produced from the charge with a greater amount of added chalk led to formation of a fused welding flux with a glass-pumiceous grain structure during the granulation process (Fig. 3 b). Formation of such structures is explained by the fact that vitreous fluxes mainly consist of Ca^2+, Mg^2+, Fe^2+, Na+, K+ cations and complex silicon-oxygen or aluminum-oxygen anions.

Some examples of how your references should be listed are given at the end of this template in the 'References' section which will allow you to assemble your reference list according to the correct format.

Chemical composition of mineral raw materials before and after plasma arc granulation is shown in Table 1.

Table 1. Chemical composition of mineral raw materials before and after plasma granulation.

Material	Elemental chemical composition, mass. %									
	C	O	Na	Si	Ca	Fe	Al	Mg	K	Ti
Raw material	12.80	52.52	27.59	1.57	4.13	1.18	0.21	-	-	-
Formed granules	0.54	60.01	3.40	15.22	9.00	2.61	3.67	4.86	0.20	0.49

It was found that during plasma processing alkali metals are partially burned to the required concentration for welding fluxes, also there is an increase in the concentration of basic elements such as silicon, calcium, aluminum and other metals. In addition, proposed chemical composition and method of granulation of fused welding flux makes it possible to obtain granules of the required chemical and fractional compositions, with a degree of basicity of 0.7, as well as the required grain structure that corresponds to regulatory documentation [6].

4. Conclusion

In conclusion, it can be noted that grain structure, fractional and chemical composition during granulation of mineral raw materials of the Ural region using plasma-arc method is affected by changing chemical composition of the charge and methods of granules formation. The processes leading to a structure change, fractional and chemical composition are well studied and can be predicted in the development of welding materials for various purposes.

IOP Conf. Series: Materials Science and Engineering **758** (2020) 012055 doi:10.1088/1757-899X/758/1/012055

Acknowledgements
The reported study was funded by RFBR according to the research project № 18-33-00599.

References

[1] A.O. Artemov, S.V. Naumov, M.N. Ignatov. Technology and Equipment for Plasma Electric Arc Granulation of Fused Welding Flux. *Materials Science Forum.* 946 (2019) 389-394
[2] S.V. Naumov, M.N. Ignatov, A.M. Ignatova, A.O. Artemov, Development of slag base for welding fluxes from man-made mineral formations of Ural mining and smelting companies. *Key Engineering Materials.* 743 (2017) 406-410
[3] A.O. Artemov, M.N. Ignatov, A.M. Ignatova, S.V. Naumov, Composition development and production technology of stone casting silicate materials and items. *Key Engineering Materials.* 743 (2017) 401-405
[4] N.K. Skripnikova, V.V. Shekhovtsov, O.G. Volokitin, R.E. Gafarov. Low-temperature plasma in the manufacturing process of glassceramic materials. *Bulletin of Science and Practice.* 4. №2 (2018) 204-208
[5] S.A. Shamin, RU Patent 2257988 (2003)
[6] I.P. Ershov, E.A. Sergeeva, L.A. Zenitova, I.Sh. Abdullin. The effect of plasma treatment on the surface properties of fiberglass. *Bulletin of the Technological University.* 4 (2013) 97-99

Combined opto-mechanical measurements with the transparent indenter's tip

I Maslenikov[1] and A Useinov[2]

1 Research Fellow, Technological Institute for Superhard and Novel Carbon Materials, 7a Centralnaya str., Troitsk, Moscow, 142190, Russia
2 First Deputy Director, Technological Institute for Superhard and Novel Carbon Materials, 7a Centralnaya str., Troitsk, Moscow, 142190, Russia

E-mail: useinov@mail.ru

Abstract. Transparent indenter's tip proposed elsewhere allows to observe sample surface before, after and during an indentation measurements. Such an ability allows not only to select the indentation with the real-time optic image and observe residual imprints, but also conduct spectroscopic measurements. Current work shows an examples of the tip application, in particular the possibility to observe a cracks during a scratching.

1. Introduction
Instrumented indentation technique allows to measure surface local values of hardness and local elastic modulus from the load-depth data without the need to use an optical microscope. The method has been used for several decades. Modern equipment allows to combine this method with high-temperature measurements [1], conduct an experiments within electron microscopes [2], detect acoustic emission [3], use an indenters along with another analytical equipment, in particular combine raman spectroscopy with an indentation. However, in most of the work the collection of raman spectra were performed only during the indentation of a transparent spectra in the location close to the sample edge, which allowed to perform indentation and spectroscopy measurements in the perpendicular directions. The transparent indenter – objective suggested elsewhere [6, 7] allows to overcome the limitation.

2. Indenter – objective scheme
Detailed description of the construction was given in the work [6]. Indenter-objective is made out of the diamond cylinder that has both ends faceted in the shape of the Berkovich pyramid in the way that each face on one end has a parallel face on the opposite end. For the Berkovich-shape pyramid this parallelity constrain means that the pyramids on the ends are rotated 60 degrees versus each other (Fig.1). This way faceted indenter has the following property. Parallel light beam incident along the indenters axis undergoes double refraction and maintains its parallelity after passing through the described tip. However, propagated beam is split into three sectors which are lineary shifted vs their original line of propagation. The visible through the indenter picture also split into three parts. An example of the testing structure TGZ-2 surface reconstructed image is shown on fig.2.

Analogous measurements of the sample surface can be conducted not only during the indenter's positioning but as well as during an indentation and after it. That gives a possibility to measure indentation contact area, which in turn allows to determine hardness directly according to the residual imprint, without need to perform additional movement beneath an optical microscope.

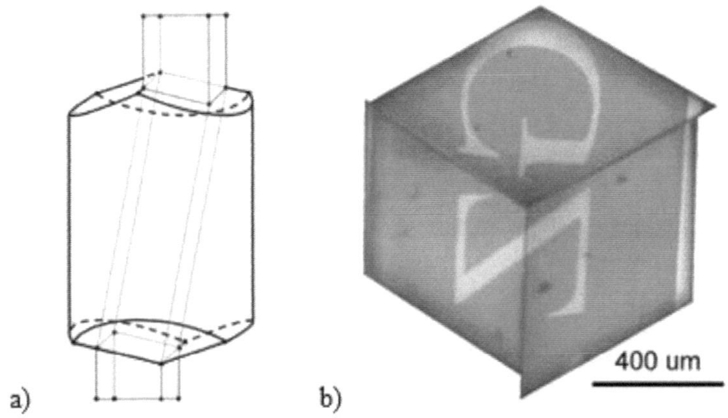

Figure 1. a) Transparent indenter-objective and beam optical path b) reconstructed image of the TGZ testing grating surface

Another example of the transparent indenter-objective application is an imaging of the surface during the scratch testing. Figure 2 shows an example of picture taken during the scratching made over the explosion-weld metals border. This case corresponds to the plastic surface deformation and scratch width can be used to calculate sample's hardness.

Figure 2. Image of explosion welded metals boarder taken during the scratching with a 300 mN load.

Scratching on the fused silica can show a different situation. At small load scratch appears to be mostly plastic, but after the load reaches a critical value cracks reach the surface so that the surface "explodes". Figure 3 shows corresponding set of pictures that were taken during the scratching of fused silica surface with an increasing load from 1mN up to 1N. As it can be seen from the figure 2 a) when load is relatively low on can observe plastic deformation on the sample's surface, as the load increases crack reaches the surface scratch explodes and further deformation, which is shown on the figure 2c, occurs in a brittle manner.

IOP Conf. Series: Materials Science and Engineering **758** (2020) 012056 doi:10.1088/1757-899X/758/1/012056

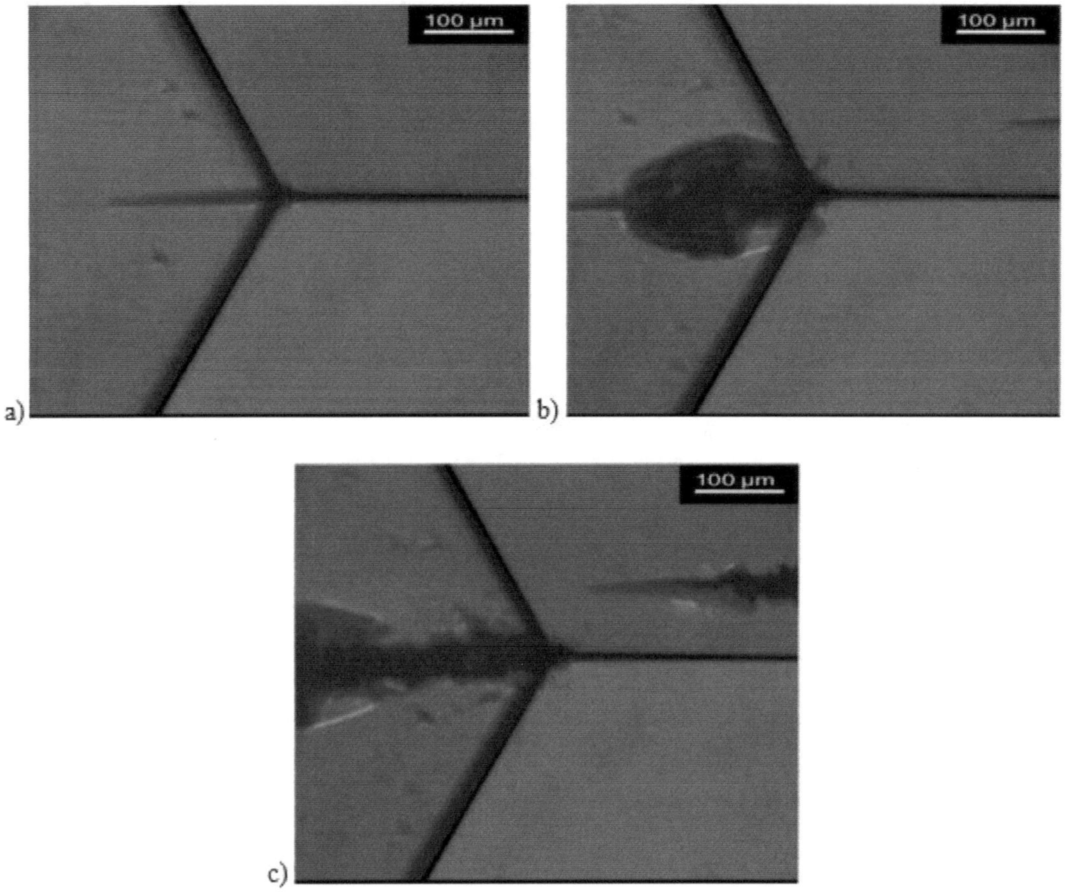

Figure 3. (a) plastic deformation of fused silica surface (b) fused silica surface at the moment of cracking

Such kind of in-situ surface optical measurements can be particularly useful for the adhesion measurements. In these measurements the determination of coating delamination moment plays an important role as it defines the critical load, which is used for further adhesion energy evaluation [8].

3. Raman spectroscopy during indentation measurements

An ability to obtain optical images during indentation directly through the tip leads to possibility to combine mechanical testing with optical spectroscopy measurements, in particular obtain Raman spectra simultaneously during an indentation [7]. Corresponding measurements were performed with the help of the special module designed to fit in the Raman spectrometer. The module allows applying and measure the load up to the force of 50 N. Measurements were performed on silicon sample's (100) surface. Two spectra collected on free and 50N loaded surfaces are shown on the figure 4.

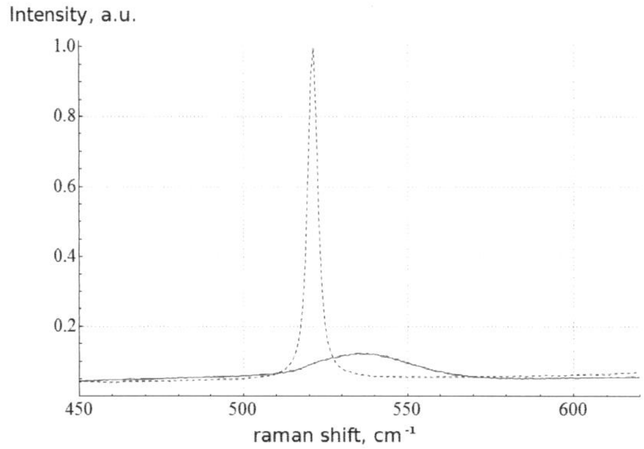

Figure 4. Shift of Si-I line under the load of 50N.

As it can be seen from the figure 3, when the load is applied Si-I line is shifted right to 20 cm-1. 50N spectra also have a peak at 376 cm-1, which are transform to series of peaks corresponding to the Si-III phase.

4. Conclusions

Transparent indenter-objective allows to obtain optical image of the surface, define deformed area surface during an indentation and afterwards. The indenter allows to conduct simultaneous indentation measurements and collect Raman spectra of the indented surface during its loading and unloading cycle.

Acknowledgments

This study was supported by the Ministry of Science and Higher Education of the Russian Federation within the framework of the Agreement no. 14.577.21.0274 (the unique identifier of the project is RFMEFI57717X0274).

References

[1] Gibson J S K L, Roberts S G and Armstrong D E J 2015 High temperature indentation of helium-implanted tungsten *Mater. Sci. Eng. A* **625** 380

[2] Rzepiejewska-Malyska K A, Mook W M, Parlinksa-Wojtan M, Hejduk J and Michler J 2009 In situ scanning electron microscopy indentation studies on multilayer nitride films: Methodology and deformation mechanisms *J. Mater. Res.* **24** 1208

[3] Dyjak P and Singh R P 2006 Acoustic emission analysis of nanoindentation-induced fracture events *Exp. Mech.* **46** 333

[4] Gerbig Y B, Michaels C A, Forster A M, Hettenhouser J W, Byrd W E, Morris D J and Cook R F 2012 Indentation device for in situ Raman spectroscopic and optical studies *Rev. Sci. Instrum.* **83** 125106

[5] Manimunda P, Hintsala E, Asif S and Mishra M K 2017 Mechanical Anisotropy and Pressure Induced Structural Changes in Piroxicam Crystals Probed by In Situ Indentation and Raman Spectroscopy *JOM* **69** 57

[6] Maslenikov I I, Reshetov V N, Useinov A S and Doronin M A 2018 In Situ Surface Imaging Through a Transparent Diamond Tip *Instruments Exp. Tech.* **61** 719

[7] Maslenikov I I, Reshetov V N and Useinov A S 2019 Raman Spectroscopy through the Indenter Working as an Optical Objective *Mater. Trans.* **60** 1433

[8] Malzbender J, Den Toonder J M J, Balkenende A R and De With G 2002 Measuring mechanical properties of coatings: A methodology applied to nano-particle-filled sol-gel coatings on glass *Mater. Sci. Eng. R Reports* **36** 47

IOP Conf. Series: Materials Science and Engineering **758** (2020) 012057 doi:10.1088/1757-899X/758/1/012057

Experimental study of the mechanical effect of a clayey soil by adding rubber powder for geotechnical applications

N Alvarez[1], J Gutierrez[2], G Duran[3] and L Pacheco[4]

[1,2] Bachelor, Peruvian University of Applied Sciences, Lima, Peru.
[3] Full time profesor, Peruvian University of Applied Sciences, Lima, Peru.
[4] Part time profesor, Peruvian University of Applied Sciences, Lima, Peru.
E-mail:[1]u201513703@upc.edu.pe, [2]u201610652@upc.du.pe, [3]gary.duran@upc.edu.pe, [4]pccilpac@upc.edu.pe

Abstract. At present, worrying quantities of tires are discarded due to the growth in demand for vehicles in the world, which has a direct impact on the deterioration of the environment since they normally go to landfills. Based on the background found, the use of this material for geotechnical applications can help reduce the pollution they generate and improve the physical and mechanical properties of soils. Therefore, this research seeks to evaluate a greater shear strength and capacity of support to the penetration of the clayey soil by means of the addition of 1.5%, 2.5% and 3.5% of rubber powder recycled. For this, the Atterberg limits analysis, the modified proctor compaction test, shear box and CBR were performed. For the shear box test, the results reflect that the cohesion of the mixture increased and the angle of internal friction decreased with respect to the natural soil, resulting in the sum in an increase of shear strength. On the other hand, the percentage of CBR increased, this means that the rubber helped the soil to be more rigid and have a greater resistance to penetration. These mixtures could be used in different projects within geotechnical engineering, as it presents an improvement in shear strength and an acceptable support index value (CBR).

1. Introduction

During the last decades Thailand is the leading producer of rubber worldwide, this material serves as a raw material for the manufacture of tires through an industrial process, which increases the emission of greenhouse gases that damage the environment [1]. In addition, the final disposal of this material often ends up being landfills and then burned. Experimental studies show that this process generates CO and SO_2 in high quantities, which exceeds the allowed limits [2]. It is also known that clayey soils have a deficit in their physical and mechanical properties. Therefore, there are several investigations on the improvement of these parameters making use of recycled rubber particles from disused tires, with the aim of having a soil suitable for building embankments, roads, etc. In Turkey, the influence of the addition of recycled tire rubber fibers and silica fume to clayey soils was investigated, with the aim of improving shear strength. The results showed that cohesion and friction angle increased 1.1 and 1.7 times respectively more than the clayey soil sample [3]. In the same way, the modification of clayey soils was investigated by adding tires and synthetic fibers with the aim of finding reinforced samples subjected to a shear strength. The results showed that the parameters of the shear strength increased 1.2 times to the samples of the non-reinforced soil [4]. The improvement of the California Bearing Ratio (CBR) of a clayey soil with the addition of polished rubber and lime was then investigated in order to stabilize it and thus be used as a material for road construction. In the results it was observed that the addition of this mixture increased the percentage of CBR and managed to decrease the thickness of the

IOP Conf. Series: Materials Science and Engineering **758** (2020) 012057 doi:10.1088/1757-899X/758/1/012057

pavement [5]. In addition, researchs on this last parameter has already been taking place in India, where it was also concluded that recycled rubber does help improve the CBR of a clayey soil, up to 5% addition of rubber improves soil support properties, helping to reduce the compression rate and swelling pressure [6]. This research aims to show the feasibility of using recycled rubber powder to improve the mechanical and physical properties of a clayey soil, and then be applied to the subgrade of a road. Finally, the use of this recycled material will help reduce the environmental pollution caused by this waste in large quantities and highly polluting if they do not have an appropriate end use.

2. Experimental Program

2.1. Materials

2.1.1. Clayey soil

The clayey soil was extracted at a depth of 1.5m from the Callampampa, Cajamarca, Peru (Figure. 1) with a plasticity index of 13%, liquid limit of 34%, plastic limit of 21% and the specific gravity is 2.13. In addition, the granulometric curve shows that this soil has 28.8% thick fraction, 17.9% silt, 24.1% clays and 29.2% colloids (Figure. 3). This clayey soil is classified as CL according to SUCS and A-6(8) according to AASHTO.

2.1.2. Rubber powder

Powdered rubber (Figure. 2) used to improve the physical and mechanical properties of clayey soil, was obtained from out-of-use tires. In the granulometric distribution (Figure. 3) it was obtained that the size of its particles varies between 2mm to 0.075mm and its average size is 1,452mm. This material has a specific gravity of 0.90.

Figure 1. Clayey Soil

Figure 2. Rubber Powder

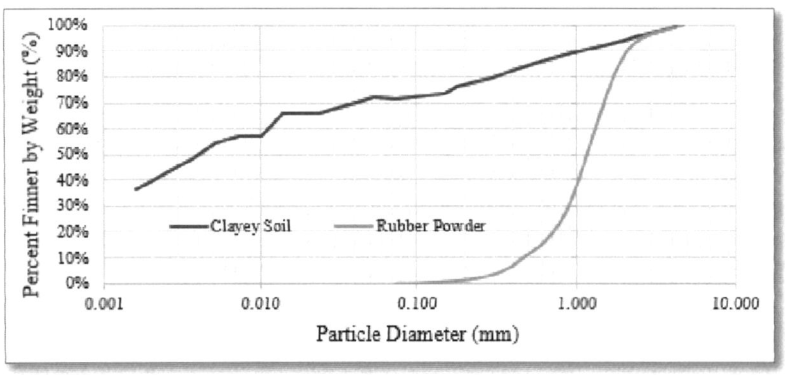

Figure 3. Particle size distribution curve of powder rubber and clayey soil

2.2. Preparation of the mixtures

The clayey soil that was used in this study was dried in an oven at 105 °C, then carefully crumbled in order not to alter the interaction between its particles. With this process a more homogeneous soil was obtained to add rubber powder in 1.5%, 2.5% and 3.5% of the dry weight of the clayey soil, the symbology that was used to denote these mixtures can be seen in Table 1.

Table 1. Symbologies to denote mixtures of clayey soil with rubber powder.

Material	Soil (%)	Rubber Powder (%)	Symbology
Clayey Soil	100	0	S100
Mixture 1	100	1.5	S100/C1.5
Mixture 2	100	2.5	S100/C2.5
Mixture 3	100	3.5	S100/C3.5

2.3. Laboratory tests

2.3.1. Physical characterization
To know the physical properties of clayey soil and mixtures, the following tests were performed:
- ASTM D422 - Standard Test for Particle-Size Analysis of Soils
- ASTM D4318 - Standard Test for Liquid Limit, Plastic Limit, and Plasticity Index of Soils
- ASTM D854 - Standard Test for Specific Gravity of Soil Solids by Water Pycnometer

2.3.2. Mechanical characterization
To know the mechanical properties of clayey soil and mixtures, the following tests were performed:
- ASTM D1557-02 - Modified Proctor Test.
- ASTM D3080 – Standard Test for Direct Shear Test.
- ASTM D1883 - Standard Test for California Bearing Ratio Test (CBR).

The modified Proctor, method A, was carried out with the objective of determining the optimum humidity content and the maximum dry density of the natural clayey soil and of the soil samples improved with rubber powder. These parameters will be used to prepare the test specimens of the Shear Box and CBR tests.
The shear box test was carried, in order to determine the parameters of shear strength of clayey soil and mixtures. For this, we worked with test specimens of 6.323cm in diameter and 2.02cm in height. A tangential travel speed of 0.25 mm/min and a normal load of 49kPa, 98.1kPa and 196.1kPa were applied until the sample fails by shear. In addition, parameters such as cohesion and friction angle were known from the fault envelope.
For the CBR the samples were compacted, then saturated for 4 days and then the penetration test was performed.

3. Analysis and Results
The granulometric curves (Figure 4) represent the distribution of the particles of the thick portion, greater than the mesh No. 200, of the mixtures and the clayey soil. On the other hand, the results of the plasticity index (IP), liquid limit (LL), plastic limit (LP) and the classification by the SUCS and AASHTO method are summarized in Table 2.

Table 2. Physical characterization of clayey soil and mixtures.

Mixtures	Gs	LL (%)	LP (%)	IP (%)	SUCS	AASHTO
S100/C1.5	2.134	33	22	11	CL	A-6 (6)
S100/C2.5	2.142	31	22	9	CL	A-4 (5)
S100/C3.5	2.135	29	21	8	CL	A-4(4)

The maximum dry density of the mixtures will increase with respect to the S100 (Table 3), this means that there will be a lower pore index between the particles as the rubber powder content increases, achieving a more densified soil. The optimum humidity content decreases as the addition of rubber powder increases, this is because the rubber does not absorb water and the mixtures have less clayey soil content.

Table 3. Optimum moisture and maximum dry unit weight of clayey soil and mixtures.

	S100	S100/C1.5	S100/C2.5	S100/C3.5
MDS(g/cm^3)	1.535	1.543	1.56	1.562
OCH (%)	17.4	17.0	16.7	16.4

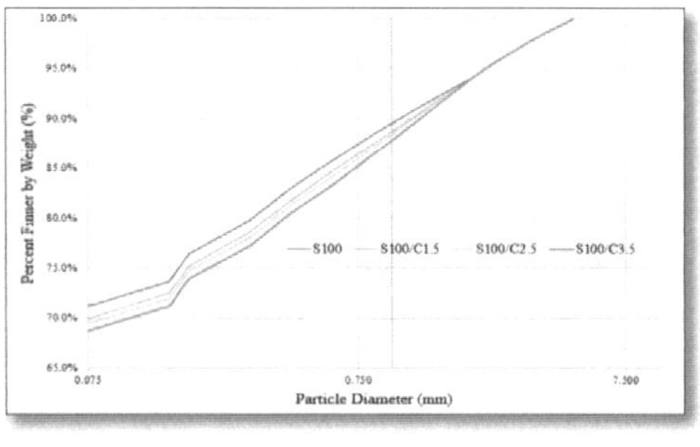

Figure 4 Particle size distribution curves of the thick portion of the clayey soil and mixtures.

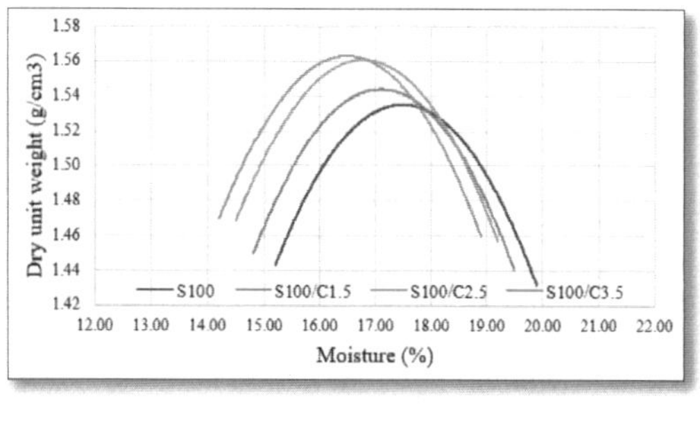

Figure 5. Modified proctor test curves for the clayey soil and mixtures.

Shear box tests showed that the addition of rubber powder improved the resistance to the shear strength of the clayey soil and that the samples tested will be more densified as the confinement effort increases. With the data obtained, the fault envelope was made (Figure. 6) where the cohesion parameters and the friction angle for each sample will be obtained. The cohesion of the low plasticity clayey soil is 0.25kg/cm^2 and the friction angle is 20.9 °, the first parameter increases as the addition of rubber powder increases and the second parameter has a behavior opposite to the first (Figure. 7). On the other hand, the ratio between the average size of the rubber powder particles and clayey soil is 1/300, this helps to improve the interaction between them. This means that a particle of rubber powder will have a considerable amount of clayey soil particles and thus achieve an improvement in the area of influence, which helps to increase cohesion and thus a greater resistance to the shear strength.

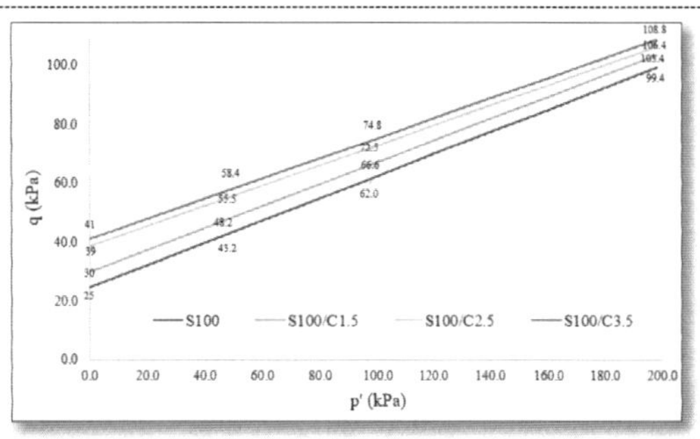

Figure 6. Strength envelopes of clayey soil and mixtures.

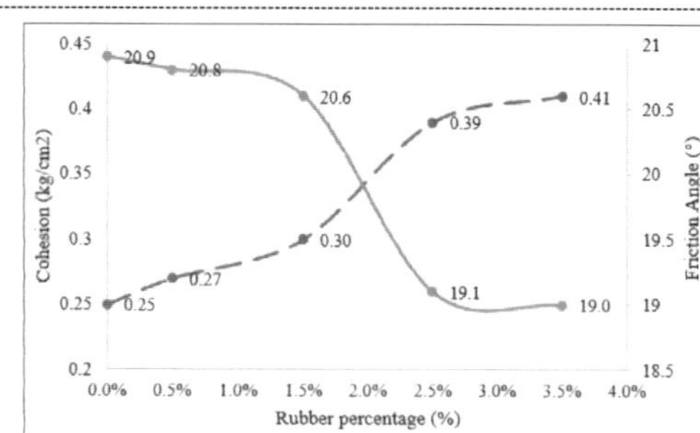

Figure 7. Variation of Cohesion and Friction Angle with respect to rubber percentages.

The CBR made to the natural clayey soil and to the samples of the soil improved with rubber powder was worked at 95% of the maximum dry density and at 2.5mm penetration. Clayey soil (S100) has a CBR of 3.2%, however, this parameter increased as the addition of rubber powder to the clayey soil increased. This is because the amount of voids in the mixtures interact with the rubber powder, which prevents water from entering, creating a larger contact surface. Table 4 shows that mixtures S100 / C2.5 and S100 / C3.5 whose CBR value is 8.7% and 9.4%, meet the conditions for an acceptable subgrade, according to AASTHO 93.

Table 4. CBR at 95% M.D.S and at 5mm penetration of clayey soil and mixtures.

	S100	S100/C1.5	S100/C2.5	S100/C3.5
CBR (%)	3.2	5.0	8.7	9.4

4. Conclusions

- The maximum resistance to the shear strength was presented when the horizontal displacement is 2.5mm. When the normal applied effort is 49kPa, the shear strength of the S100 / C3.5 mixture compared to natural soil (S100) increased from 43.2kPa to 58.4kPa, in the same way when the normal applied effort is 98.1kPa it increased from 62kPa to 74.8kPa and when the normal effort is 196.1kPa it increased from 99.4kPa to 108.8kPa.
- Cohesion increases as the addition of rubber increases, for the soil mixture plus 3.5% rubber (S100 / C3.5) the cohesion value is 0.41 kg/cm^2, which compared to the cohesion of S100 It increased by 64%. On the other hand, the friction angle has an inverse behavior of cohesion, that is, it decreases as the rubber addition increases. The S100 / C3.5 mixture has a value equal to 19 °, which compared to the friction angle of the natural soil decreased by 9.1%.
- For the CBR, it was observed that the mixtures S100 / C1.5, S100 / C2.5, S100 / C3.5 improve by 56%, 172% and 194% respectively compared to S100, with this it was possible to go from one subgrade listed as inappropriate to a regular. This allows the use of rubber powder to improve the subgrade of a road.
- The use of recycled rubber contributes to the conservation of the environment, as it has a strong impact on society, because the vast majority of disused tires are not properly recycled.

References

[1] Petsri S, Chidthaisong A, Pumijumnong N, Wachrinrat C 2013. Greenhouse gas emissions and carbon stock changes in rubber tree plantations in Thailand from 1990 to 2004. *J Clean Pro* **52**, pp 61-70.

[2] Jimoda LA, Sulaymon ID, Alade AO, Adebayo GA 2018. Assessment of environmental impact of open burning of scrap tyres on ambient air quality. *Int J Environ Sci Technol.* **15**(6), pp1323-30.

[3] Kalkan E 2013. Preparation of scrap tire rubber fiber-silica fume mixtures for modification of clayey soils. *Appl Clay Sci.* **80-81,** pp 117-25.

[4] Akbulut S, Arasan S, Kalkan E 2007. Modification of clayey soils using scrap tire rubber and synthetic fibers. *Appl Clay Sci.* **38**(1-2), pp 23-32.

[5] Cabalar AF, Karabash Z, Mustafa WS 2014. Stabilising a clay using tyre buffings and lime. *Road Mater Pavement Des.* **15**(4), pp 872-91.

[6] Yadav JS, Tiwari SK 2018. Influence of crumb rubber on the geotechnical properties of clayey soil. *Environ Dev Sustainability.* **20**(6), pp 2565-86.

IOP Conf. Series: Materials Science and Engineering **758** (2020) 012058 doi:10.1088/1757-899X/758/1/012058

Stabilization of a Subgrade Composed by Low Plasticity Clay with Rice Husk Ash

E Ormeño[1], N Rivas[1], G Duran[2] y M Soto[3]

1 Bachelor, School of Civil Engineering – Peruvian University of Applied Sciences, Lima, Peru. E-mail: u201312433@upc.edu.pe, u201421496@upc.edu.pe
2 Full time professor, School of Civil Engineering – Peruvian University of Applied Sciences, Lima, Peru. E-mail: gary.duran@upc.edu.pe
3 Full time professor, School of Civil Engineering – Peruvian University of Applied Sciences, Lima, Peru. E-mail: pccimsot@upc.edu.pe

Abstract. The construction of road works in the world has always been a challenge for engineering, especially in areas where the conditions and types of soil are not adequate for the execution of this type of projects. The present investigation has as main objective to determine the influence that has the rice husk ash (RHA) to stabilize the subgrade layer of a pavement, composed of a low resistance clayey soil. RHA is a waste and pollutant material for the environment; therefore that its use can be considered as an economic and ecological alternative. Thus, several tests were carried out where it proved the value of CBR increased from 4.30% to 20.70%, by adding a 20% RHA dosage, achieving its optimum value to be considered a very good subgrade. In this way, it is possible to affirm that the addition of RHA improves the geotechnical properties of the soil.

1. Introduction

The paving of a road allows easy access, communication and transport within a country. Depending on their structural composition, pavements can be classified as rigid or flexible [1]. For the effectiveness of a pavement, it must have the ideal dimensions to support the applied loads, provide a suitable surface for comfortable driving, minimal deformation, long service life, among other conditions [2]. The durability and quality of a pavement depend on the strength and mechanical properties of its subgrade, because this layer of the soil fulfills the role of foundation. Thus, in the case of clay subgrade, which does not have adequate resistance, these must be stabilized or improved by means of methods that are economical and that favor environmental care. The most common techniques are based on the addition of cementitious materials to the soil in order to improve their resistance.

In 1992, in Malaysia, Haji et al. [3] studied the geotechnical properties of a clayey soil stabilized with rice husk ash (RHA). The authors classified the soil as a sandy clay (SC), A-7-5 according to the AASHTO standard, with a liquid limit of 73%, a plastic limit of 36% and a CBR value of 2% for saturated samples; also, the rice husk was obtained as a by-product of the rice milling and an ash was obtained with an amount of silica of 90.73%. After the respective laboratory tests, it was determined that at a higher amount of RHA the maximum dry density decreases and the optimum humidity content increases. On the other hand, in 2016, Rukenya et. Al [4] performs CBR tests on an expansive clay in Nairobi, where it is sought to solve the problems of soil resistance. Through the use of RHA with 78.9% silica, dosages of RHA and lime are performed that result in an 800% improvement in the CBR value and a 70% decrease in soil swelling, both in comparison to the soil sample natural. Likewise, rice production in Peru has increased by 19.2% in November 2018 compared to the same month in 2017, with total production of 171,207 tons [5]. Rice husk ash represents approximately 25% of the husk when burned and the latter, in turn, is equivalent to 22% of the total grain [6].

In summary, the research seeks to quantitatively determine the improvement of the geotechnical properties of a soil with poor support capacity against the addition of rice husk ash as a stabilizing agent. This soil is used as a subgrade of an unpaved low traffic road, where the optimum CBR value is 20% according to the manual of the Ministry of Transportation and Communications of Peru. Thus, due to the poor development and investment in innovative methods, in this country, there is not enough information from similar studies to generate standard civilizations. Therefore, laboratory tests are carried out to determine the physical, chemical and mechanical properties of natural and dosed soil at different proportions of rice husk ash to determine an optimal mixture that meets the required CBR value.

2. Experimental program

2.1 Materials

The materials that have been used in the investigation are the clay extracted from the Callampampa farmhouse in the province of Chota, Cajamarca and the rice husk ash, obtained in a mill in the city of Guadalupe, Lambayeque. Both components were taken to various preliminary tests to know their physical, chemical and mechanical properties.

In the first place, a granulometric analysis was carried out on the clay, in its natural state, where 75.06% of material was reached through the # 200 mesh without the presence of gravels. Also, the main physical properties were determined, which are shown in Table 1.

Table 1. Physical properties of the soil.

Soil properies	Tests results
Humidity content (%)	19.00
Liquid limit (%)	26.00
Plastic limit (%)	19.00
Plasticity index (%)	7.00
SUCS Classification	Low Plasticity Clay (CL)
Maximum dry density (kN/m3)	16.61
Optimun humidity content (%)	18.60
Soaked CBR (%)	4.50

In second place, through X-ray fluorescence test was obtained the chemical composition of low plasticity clay and rice husk ash. Through the results obtained in Table 2, we can highlight that RHA is an excellent pozzolan having 90.81% silica (SiO2) in its chemical composition. This characteristic is very important for the stabilization of the clay since, the RHA, when combined in various dosages with the CL, will form a cementitious compound of greater stability.

Table 2. Chemical properties of CL and RHA

Item	Chemical Component	CL	RHA
1	Silicon dioxide (SiO_2)	69.00	90.81
2	Aluminum oxide (Al_2O_3)	14.94	0.21
3	Iron (III) oxide (Fe_2O_3)	3.99	0.61
4	Potassium oxide (K_2O)	1.96	4.62
5	Magnessium oxide (MgO)	0.90	0.59
6	Calcium oxide (CaO)	0.61	0.67
7	Others	9.60	2.49

2.2 Dosage

Through the tests of physical and chemical properties carried out on the samples of natural soil and ash, the dosage of this project was determined. In recent years, many researchers have stabilized clayey soils, of similar characteristics, with an average of 15% RHA as the optimal dosage. In this way, the research uses the dosages shown in Table 3.

Table 3. Dosage and denomination of samples.

Sample	Sample proportion (%)	
	CL	RHA
100CL0RHA	100	0
90CL10RHA	90	10
85CL15RHA	85	15
80CL20RHA	80	20
75CL25RHA	75	25

Figure 1. (a)100CL0RHA (b)90CL10RHA (c)85CL15RHA (d)80CL20RHA (e)75CL25RHA

3. Results and Analysis

3.1 Compaction

For this process, the modified proctor test was carried out in order of determining the optimum humidity content (OMC) and the maximum dry density (MDD). Figure 2 shows the variation that exists in (OMC) according to the dosage of RHA added. By increasing the percentage of ash and, therefore, by decreasing the proportion of soil, the OMC is increasing. This result means that more water will be needed to compact the CL-RHA mixture.

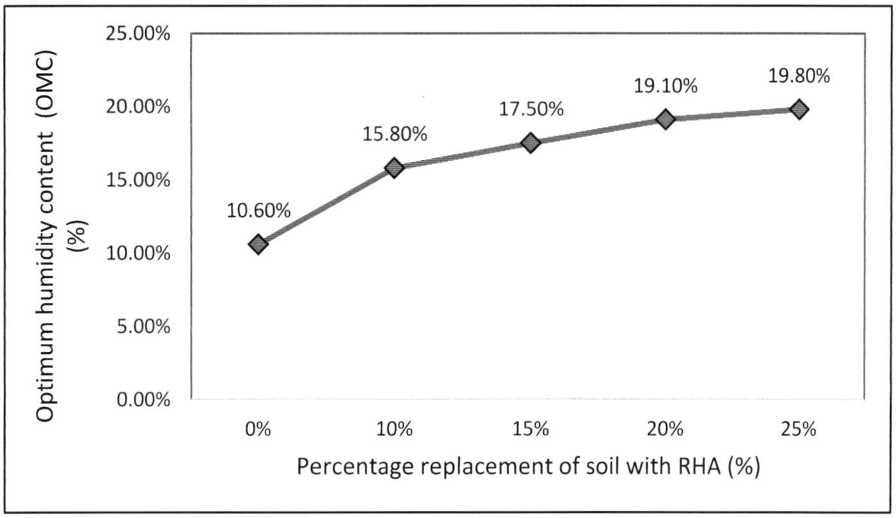

Figure 2. Effect of RHA on OMC.

On the other hand, Figure 3 shows the variation presented by the maximum dry density with respect to the added RHA dosage. Second the graph shows that at a higher percentage of RHA, the MDD decreases. This is because the specific gravity of the ashes is less than that of the clay.

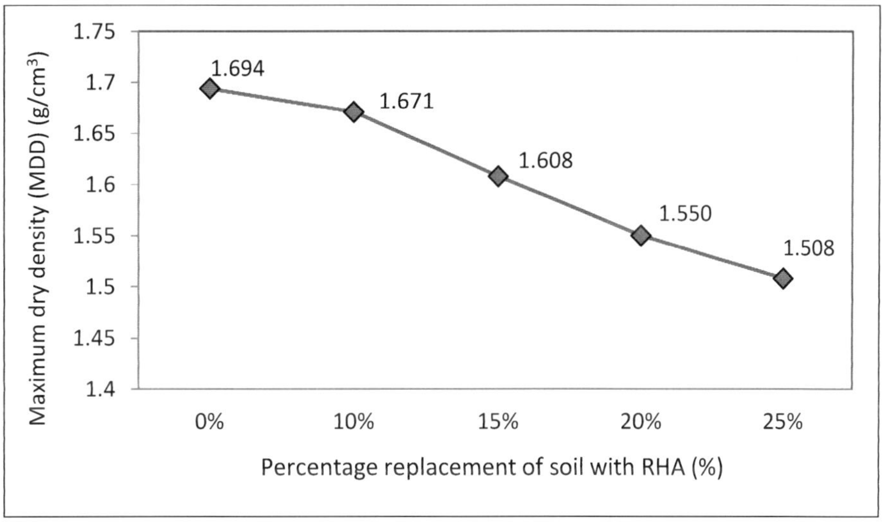

Figure 3. Effect of RHA on OMC.

3.2 CBR test

The CBR was the most important test that was carried out in the investigation since it allowed to know the resistance, both of the clay under study, and of the stabilized mixtures with various proportions of RHA. These analyzes were performed with their established humidity and density parameters, according to preliminary analyzes; as in its most critical condition, soaked.

Figure 4 indicates the effect that RHA has on the parameter of a saturated CBR. It can be determined that, while the RHA is added in greater quantity, the percentage of resistance increases. This is due to the formation of calcium silicates and calcium aluminates, which are very important parameters for the conformation of cementitious elements.

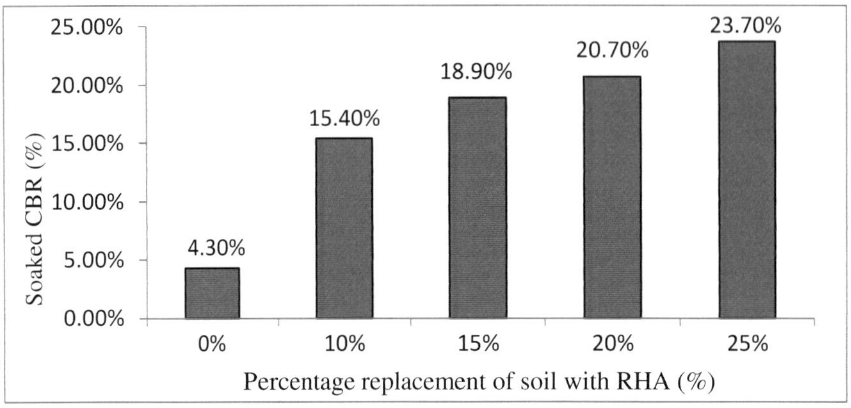

Figure 4. Effect of RHA on soaked CBR.

4. Conclusions

- When performing the X-ray Fluorescence test, the chemical components of rice husk ash and clay are determined. It is observed that the values of silica (SiO2), aluminum oxide (Al2O3) and iron oxide (Fe2O3) exceed 70% of the total components. Therefore, it is concluded that they are materials with a high pozzolanic behavior that can be used as a cement in soil stabilization.
- As the proportion of rice husk ash increases, the optimum humidity content (OMC) increases and the maximum dry density (MDD) decreases. In this way, we deduce that the addition of ash improves the soil and less energy is needed to perform its compaction.
- CBR values increase when more RHA is added. For this reason, to obtain a very good subgrade, a CBR value of 20% must be reached, this value being exceeded with the addition of 20% RHA, which gives a result of 20.70% CBR. Likewise, the different dosages provide various CBR values which could be used, referentially, for the conformation of retaining walls or as stabilizers for embankments.

Acknowledgments
The authors would like to thank the Faculty of Civil Engineering of the "Peruvian University of Applied Sciences" (UPC) and the LabGeo CRVV S.A.C. laboratory for the facilities for conducting soil tests.

References

[1] Tom V. and Krishna K. 2007 Introduction to pavement design *National Programme on Technology Enhanced Learning* p 19.1-19.7.
[2] Muntohar A. 2003 Engineering behaviour rice husk ash blended soil and it's potential as road base construction *Conference of the Australian Road Research Board* **21** 1295
[3] Haji F., Adnan A. and Chew K. 1992 Geotechnical properties of a chemically stabilized soil from Malaysia with rise husk ash an additive *Geotechnical and Geological* Engineering **10** 117-134
[4] Rukenya T., Wanbua J., Kabubo Ch. and Thiong'o 2016 Soil stabilization using rice husk ash and natural lime as an alternative to cutting and filling in road construction *Journal of Construction Engineering and Management* **143** 5
[5] National Institute of Statistics and Informatics (INEI) 2019 *Rice husk production in November 2018* (Lima: https://www.inei.gob.pe/media/MenuRecursivo/noticias/nota-de-prensa-no-009-2019-inei.pdf)
[6] Tutur N. and Noor R. 2016 The potential of rice husk ash (Rha) and coconut fiber (cf) as partial replacement of cement *AIP Conference Proceedings* **2020** 020061-1 – 020061-6

Theoretical design proposal for simulated hot asphalt mixture at a temperature below zero degrees Celsius

H Chávez[1], A Pezo[2], G Llerena[3] y V Torres[4]

1 Civil Engineering, Peruvian University of Applied Sciences, Lima, Peru
2 Civil Engineering, Peruvian University of Applied Sciences, Lima, Peru
3 Civil Engineering, Peruvian University of Applied Sciences, Lima, Peru
4 Civil Engineering, Peruvian University of Applied Sciences, Lima, Peru
Email: victor.torres@upc.pe

Abstract. In the world there are adverse climates, climates that hinder the good construction and paving of roads, generating insecurity among the locals and visitors. This over time affects the economy of a country, as a road boosts tourism, transport and commerce. Therefore, a mixture was designed to mitigate a problem in the placement of hot asphalt mixture at temperatures below zero degrees Celsius. That is, a conventional mix design was proposed, but with different types of filler (lime, Portland cement type I and silica) tested with the Marshall and Lottman method which are governed according to the EG-2013 standards [1] and parameters established in the Asphalt Institute [2]. To find the optimum, it was tested with 5.0%, 5.5% 6.0% and 6.5% asphalt cement. Then with the results obtained a comparative analysis was performed. Finally, specimens without any additives were made, the specimens once prepared at 140°C were subjected to freezing, resulting in the three types of filler, that the hot asphalt mixture with incorporation of Portland cement type I to a 5, 90% of asphalt cement is the optimum since, subject to extreme temperatures below 0ºC they comply with the parameters required in the standards.

Keywords. Marshall methodology, quality tests (Lottman), filler percentage, combination of aggregates, optimum asphalt content (C.A) and EG-2013 (Manual of General Technical Specifications for Construction in Peru).

1. Introduction

A big problem in places where there are climates of temperature below zero, is the decrease of the productivity of the projects during the process of placing hot asphalt mixture, where the temperature can vary in a matter of minutes and remain below 0ºC during all the day, which represents a restriction for the placement of hot asphalt mixture (MAC) in cold climates that is stipulated in the normal EG-2013 [1],

According to the researcher Javier Reyes, who analyzed the variations in the temperature of asphalt mixing observed that the placement of a mixture at temperatures below zero results in segregation, micro fissures, undulating surfaces, tears and especially changes in mechanical and dynamic properties of the mixture [3], thus decreasing the life time of the asphalt folder.

All the development of the research was in the collection of information and laboratory tests that demonstrated the design of optimal hot asphalt mixture that meets the established parameters.

2. Materials and Design

2.1. Materials

- **Selection of coarse and fine aggregates**

Figures 1 and 2 show the aggregates that have been used for the preparation of the asphalt mixture, these have also been washed and dried to perform the tests and are free of impurities to show their optimal behavior.

Figure 1. Crushed sand without humidity.

Figure 2. Thick Aggregate ¾" and Medium Aggregate 3/8" dried

- **Filler selection.**

The three types of filler (Portland cement type I, lime and silica) have been selected for research due to their physical and chemical characteristics in the hot asphalt mixture design. See figure 3.

Figure 3. The filler has been selected according to its composition, lime and portland cement type I for being oxidants and silicas for being mineral with properties similar to lime [1].

- **Asphalt Cement Selection (C.A)**

For the selection of asphalt cement it was verified that they can have an optimum performance and abrasion with the aggregates, therefore, in accordance with the parameters of the standards for the design, asphalt cement 85/100 was used. See figure 4.

Figure 4. Asphalt cement 85/100 was selected according to reference of Standard EG-2013 [1] and for the accessibility of getting on the market.

- **Marshall Methodology. -** It delimited the optimum content, stability, percentage of voids and resistance to compression of asphalt cement for a specific proportion of aggregates.

IOP Conf. Series: Materials Science and Engineering **758** (2020) 012059 doi:10.1088/1757-899X/758/1/012059

- **Lottman test. -** It determines the tensile strength of the specimens, measuring the loss of cohesion of a compacted mixture as a result of the effects of accelerated saturation in water.

2.2. Design

- The procedure consists in selecting the gravel aggregates at 48% (3/4" and 3/8"), sands (Crushed and Fine) at 50% and filler at 2%. This is placed in the oven at a temperature of 140°C.
- In figure 5 it can be seen that the outlet temperature of the asphalt mixture was taken from the oven. Also while the combined aggregates remain in the oven they are beaten evenly at a temperature of 140 ° C.
- Figure 6 shows the procedure of placing MAC (Hot Asphalt Mixture) in specimens.
- The specimens, as shown in Figure 7, have been placed freezing, this allows simulating the climates under 0 ° C to which they can be subjected during the placement of the asphalt folder.

Figure 5. Temperature verification.

Figure 6. MAC is placed left fresh from the oven in the Marshall test specimens, which are struck an average of 75 times around the mold.

Figure 7. The specimens are frozen for 24 hours. Simulating extreme climates where they can be placed [4]

For the Marshall and Lottman tests, the MAC is frozen to simulate cold climates (below 0°C), this abrupt change in temperature so variant (from hot to cold) shows that this type of MAC can be applied since the tests have been frozen for more than 24 hours.

The freezing time of the MAC has been considered for the reason that it is desired to see the physical and chemical behavior of the design at temperatures below 0°C, this generates a simulation in the MAC placement in the project [5].

IOP Conf. Series: Materials Science and Engineering **758** (2020) 012059 doi:10.1088/1757-899X/758/1/012059

3. Results and discussions

3.1. Results

In Table 01, Marshall's tests and the results obtained with the three types of filler are analyzed, only designs made with silica and cement comply with what is established according to the EG-2013 standard [6] and the asphalt institute.

Table 01. Analysis of MAC-2 by Marshall tests

Filler	Silica	Cement	Lime
%AC. Optimal (± 0.20)	5,8	5,9	6,0
% Vacuum with C.A.	78,2	81,6	87,6
Rigidity Index	3792	3840	3949
% of V.M.A	16,4	17,2	15,8
Stability	1371	1125	1385
Unit Weight kg / m3	2,383	2,39	2,419
% Air Vacuum	4,0	3,8	2,2
Flow (in.)	14,0	11,5	14,0

In Table 02 according to Lottman's tests and the results obtained with the three types of filler, only the designs made with cement and lime comply with what is established according to EG-2013 [1] and AASHTO T 283 standard [6].

Table 02. MAC-2 analysis by Lottman tests

Filler	Silica	Cement	Lime
Optimum% of C.A.	5,8	5,9	6,0
% Dry Vacuum	6,89	7,2	6,96
% Saturated Vacuum	6,73	7,3	7,2
Humidity Resistance (kg / cm2)	4,35	3,617	3,43
Dry Resistance (kg / cm2)	7,49	4,37	4,19
Retained Resistance (Adhesion)	58,08	82,75	81,72

3.2. Discussions

- In accordance with the parameters established by the asphalt institute and the EG-2013 standard [1] it is verified that of the materials used as filler in the tests, both silica and lime, do not meet certain restrictions such as lime in the percentage of vacuum with C.A. it has to be 87.6% and the limit according to the EG-2013 standard is 85%, so the percentage of air vacuum is 2.2% and the standard allows us at least 3.0%.

- Silica has had many variations, compared with other results obtained with cement and lime, silica in the different tests (5.0%, 5.5%, 6.0% and 6.5%) the optimum that is has C.A. it is 5.8%, however, in some tests they do not meet the parameters required by the standard, so the optimal percentage of asphalt cement does not only meet the retained resistance (adhesion) which results in 58.08 and the minimum according to AASHTO T 283 [6] is 80.00.

IOP Conf. Series: Materials Science and Engineering **758** (2020) 012059 doi:10.1088/1757-899X/758/1/012059

4. Conclusions

- According to the results obtained and the comparative analysis performed with the different types of fillers, it is obtained that the hot asphalt mixture subjected to temperatures below 0ºC complies with all the parameters established by the EG-2013 standards [1], AASHTO T 283 [6] and the asphalt institute so that a hot asphalt mixture can be placed under these conditions [2].
- The results of the other types of filler (lime and silica), which do not meet certain conditions can be studied for other types of research since, with the use of some additive, they may meet the parameters required for the placement of hot asphalt mixture subjected to temperatures below 0 ° C.
- The results obtained from tables 1 and 2 show that it is possible to place MAC at temperatures below 0ºC, complying with the standards of conventional designs, so that higher production would be obtained, although in EG-2013 standard [1] there are restrictions where it says us that the minimum temperature of placement is at 6ºC in ascent.
- From the results in table 01, it was obtained that the hot asphalt mixture design with lime incorporation does not comply with the parameters mentioned in the EG-2013 standard in the properties of percentage of air voids. Therefore, the result obtained of 2.20% of air voids will generate a greater range of exudation and in turn this may disrupt the structural response to ice and thaw due to the increase in its density.
- In Table 02, it has been verified that the Lottman tests carried out on the asphalt mixture with silica incorporation do not comply with the restriction parameters of ASHTO 283 [6], so that this mixture will cause problems of adhesion and premature wear.
- The asphalt mix, with the incorporation of Portland cement type I as a filler, meets all the parameters established in the EG-2013 standard [1], with the best asphalt cement content of 5.90% the best mechanical behavior of the mixture was achieved, in addition that the briquettes have suffered a freezing of -20ºC, the results were obtained: Stability (1125 kg), Flow (11.5 - 0.01 "), Density (2,338 gr / cm^3), Retained Resistance (82.7%), and Resistance to Compression (3.0 Mpa). (3.0 Mpa).

5. References

[1] Ministry of Transportation and Communications; 2013; Road Manual "general technical specifications for construction"; (EG-2013); Vol. 1; Peru; Pag-1282.

[2] Asphalt Institute (1982); Manual of Principles of Construction of Hot Asphalt Mix Pavements MS-22. Lexington; U.S.

[3] Oscar Javier Reyes Ortiza, Javier Fernando Camacho Tauta, Fredy Reyes Lizcanob; March 5, 2006; Influence of the temperature and energy level of compaction on the dynamic properties of an asphalt mixture; Faculty of Engineering Magazine No. 36. pp. 121-130; Vol. 36; Magazine Faculty of Engineering Universidad Antioquia; Colombia; Page-9.

[4] M.I. Yelitza Ayala del Toro, Dr. Horacio Delgado Alamilla, Dr. Paul Garnica Anguas; April 7, 2016; Effect of Temperature on the Evaluation of Fatigue in Binders Vol. 18; Asphalt; Road Infrastructure Magazine / LanammeUCR / ISSN: 2215-3705; Mexico; Page-8.

[5] M. Sc Ing Lucia del Pilar Saez ALvan, 2015, PT-21 Pavements. Selection of main articles - 4.a; Mechanical Behavior of Asphalt Mixtures at Low Temperatures in the Altitude Regions above 3500 meters; Vol. 04; institute of construction and management; Peru- Lima (ICG); Page 421.

[6] American Association of State Highway and Transportation Officials (AASHTO); 2014; AASHTO T 283.

IOP Conf. Series: Materials Science and Engineering **758** (2020) 012060 doi:10.1088/1757-899X/758/1/012060

Mechanical properties and self-cleaning mortar capacity C/A 1: 5 of Portland cement modified with titanium dioxide (TiO2)

H Flores[1], G Bernuy[1], C Huerta[2]

1 Bachelor, Civil Engineering Program, Peruvian University of Applied Sciences, Lima, Peru
2 Professor, Civil Engineering Program, Peruvian University of Applied Sciences, Lima, Peru
E-mail: carlos.huerta@upc.edu.pe

Abstract. The deterioration of the surfaces of the constructions made with mortar C/A 1: 5 of Portland cement, are produced by being exposed to the emission of toxic gases emanating from the growing automobile fleet, this problem causes alternatives to be sought in order to counteract its effect on buildings and the environment. A new method to deal this problem is the incorporation of the titanium dioxide photocatalyst (TiO2) into the Portland cement mortar, which can develop self-cleaning and air purification properties to be in contact with sunlight. This work seeks to introduce this organic component to the Portland cement mortar, used for the facade charging and structural elements, for this purpose, different percentages (5%, 7.5% and 10%) of titanium dioxide (TiO2) are added and the properties of the modified mortars making use of [1] compression tests, [2] fluidity tests, [3] absorption tests and [4] photocatalytic activity tests with which the self-cleaning capacity was verified. This study concludes that the best percentage of titanium dioxide addition is 5%, with which the Portland cement mortar is granted self-cleaning property without substantially damaging its mechanical properties.

1. Introduction

In South American countries, air pollution is largely due to the disproportionate increase in its car fleet. The gases emitted by public and private transport vehicles (some with more than 15 years old) contain polluting particles, according to the investigations carried out there is a good mortality rate, due to cardiovascular, respiratory and lung cancer problems. Likewise, they produce aesthetic and physical damages to the materials of the buildings, monuments and works of art, which are mostly constructed of concrete, which are located around places with moderate vehicular influx, and which are damaged by the dry or wet deposition of particulate material [5]. That is why the investigations of Titanium Dioxide (TiO2) in the world have concentrated on finding techniques to solve pollution problems, this through the process called photocatalysis that by making sunlight contact with a catalyst such as dioxide Titanium (TiO2) generates self-cleaning properties that not only help keep the material clean, but also give it the property to purify the air. [6] This research, after reviewing an extensive bibliography [7] [8] [9], hopes to introduce a new material that helps not only improve the conservation of cement-based materials but also the quality of life of people. For this purpose, tests are carried out to determine if after adding titanium dioxide to the mortar mixture with Portland cement, the mechanical properties are not altered and if the self-cleaning property is added.

IOP Conf. Series: Materials Science and Engineering **758** (2020) 012060 doi:10.1088/1757-899X/758/1/012060

2. Materials and methods

2.1. Materials

Figures 1, 2 and 3 show the components used to make the portland cement mortar mix, Figure 4 shows the organic component (rhodamine) that will be used to verify if is granted or not the mortar self-cleaning capacity of this investigation.

Figure 1. *Portland cement type I.*

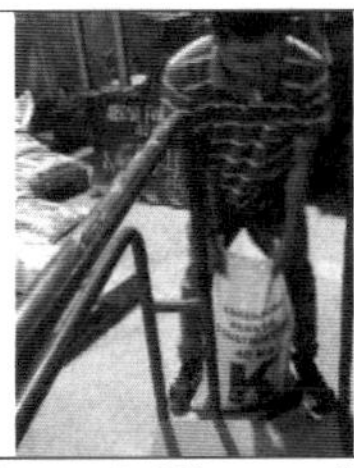

Figure 2. *Fine sand*

Figure 3. *Titanium Dioxide (TiO2)*

Figure 4. *Rhodamine*

2.2. Method

The mortar was mixed according to [10] table 1, which shows the dosage used and the percentage of titanium dioxide addition (TiO2), table 2 summarizes the number of specimens, tests and standards used. All samples had a curing time of 28 days except those used for the compression test which were tested at 3, 14 and 28 days respectively.

Table 1. Mortar dosage C:A 1:5 for a volume of 0.000375m3

Specimen	TiO2		Cement	Fine Aggregate	Water
	%	kg	Kg	Kg	lts
Mortar C/A 1:5	0	0.000	0.177	0.892	0.128
Mortar C/A 1:5 + 5% TiO2	5	0.009	0.164	0.892	0.128
Mortar C/A 1:5 + 7.5% TiO2	7.5	0.013	0.160	0.892	0.128
Mortar C/A 1:5 + 10% TiO2	10	0.017	0.155	0.892	0.128

Table 2. Test chart and specifications

Status	Type of test	Number of specimens	Curing time (days)	Methods used
Plastic	Fluency	12		[10], [2]
Tough	Compression resistance	54	3,14,28	[10], [1]
Tough	Absorption	12	28	[10], [3]
Tough	Self-cleaning capacity	16		[10], [4]

3. Results and analysis

3.1. Mortar fluidity

Table 3 and Figure 5 show the percentages of fluidity that are achieved after incorporating the different replacements (in percentage) of titanium dioxide in the mortar mixture, the results are compared with the corresponding tables to observe their consistency [11]. Thus, the percentages of decrease of this property of the mortar were 8.26%, 14.51%, 21.65% according to the addition of titanium dioxide in percentages of 5%, 7.5% and 10% respectively.

Table 3 Average fluency

Modified mortar	Average fluency (%)	Consistency
Mortar C/A 1:5	109.87	Medium
Mortar C/A 1:5 + 5% TiO2	100.80	Medium
Mortar C/A 1:5 + 7.5% TiO2	93.61	Tough
Mortar C/A 1:5 + 10% TiO2	84.87	Tough

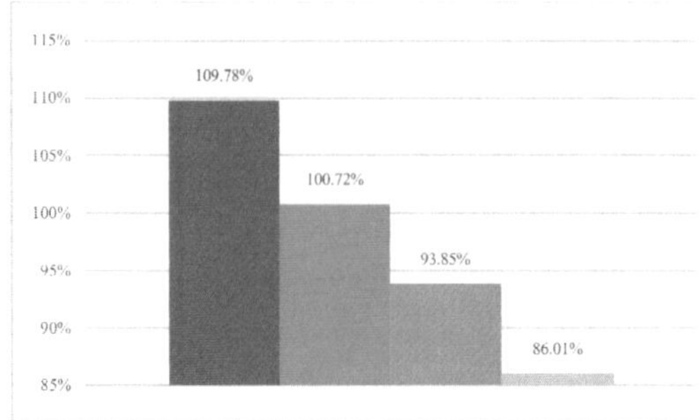

Figure 5. Average fluency

3.2. Compression resistance

Table 4 and Figure 6 show the compression resistance according to the curing days requested by [1] which shows that the higher the percentage of replacement of titanium dioxide, the lower the compression resistance, however the only one set of samples that at 28 days do not exceed the minimum resistance limit for a mortar of C: A 1: 5 as it refers [12] are those that have a 10% replacement. This decrease in resistance is due to the replacement of the weight percentage of the cement with titanium dioxide. It can be noted that the percentages of decrease in mortar compression were 4.76%, 15.24%, 24.76% according to the addition of titanium dioxide in percentages of 5%, 7.5% and 10% respectively.

Table 4 Compression resistance

Modified mortar	Curing days		
	3*	**14***	**28***
Mortar C/A 1:5	47.83	78.63	124.30
Mortar C/A 1:5 + 5% TiO2	44.05	77.72	118.38
Mortar C/A 1:5 + 7.5% TiO2	40.07	70.02	105.35
Mortar C/A 1:5 + 10% TiO2	36.80	62.07	93.52

*The records of this table in kg/cm2

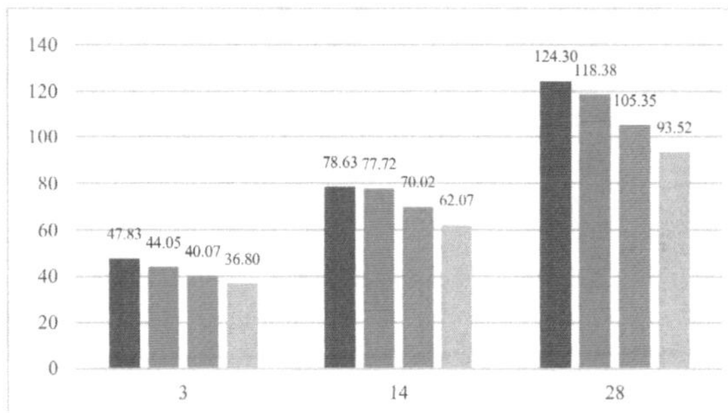

Figure 6. Compression resistance

3.3. Absorption percentage

Table 5 and Figure 7 show the results of the absorption test according to [3] a decrease is observed in comparison with the standard mortar in the range of 13.49%, 16.46% and 17.09% according to the additions of 5%, 7.5% and 10% of titanium dioxide, here the mortar added with 5% presents a slight advantage against the rest of percentages it can be said that the modified mortar has the ability to retain a certain amount of water that can help in the plastering of the facades of ceramic bricks.

Table 5 Absorption percentage

Modified mortar	Test Hours		
	0.25	**1.4**	**24**
Mortar C/A 1:5	15.95	29.79	131.35
Mortar C/A 1:5 + 5% TiO2	12.74	25.69	113.63
Mortar C/A 1:5 + 7.5% TiO2	9.11	23.19	109.72
Mortar C/A 1:5 + 10% TiO2	6.50	20.88	108.91

*The records in this table are in percentage (%)

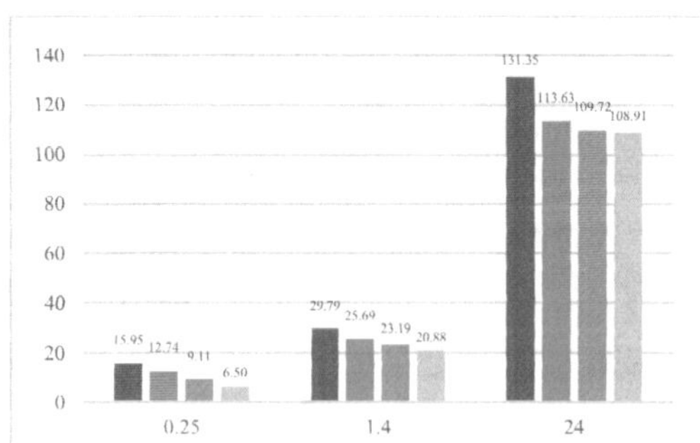

Figure 7. Absorption percentage

3.4. Self-cleaning capacity

Figures 8 and 9 show the specimens prepared following [4] as well as the exposure to UV rays contained in the standard, figure 10 shows the specimens exposed to the sun. The results presented in table 6 and figure 11 show the final condition of the tests, where it is verified that the mortars added with titanium dioxide acquire the photocatalytic capacity since the indexes obtained after the exposure comply with the following condition:

$$R_4 > 20\%\ y\ R_{24} > 50\%$$

Where:
R4 index obtained at 4 hours of exposure to UV rays and material evaluation using a digital colorimeter.
R24 index obtained at 24 hours of exposure to UV rays and material evaluation using a digital colorimeter.

Figure 8. Specimens with rhodamine

Figure 9. Artificial UV exposure

Figure 10. Sun exposure

Table 6 Photocatalytic capacity indexes

Modified mortar	Indexes R4	Indexes R24	Condition
Mortar C/A 1:5	6.94	19.60	It is not photocatalytic
Mortar C/A 1:5 + 5% TiO2	43.95	67.50	It is photocatalytic
Mortar C/A 1:5 + 7.5% TiO2	36.48	82.21	It is photocatalytic
Mortar C/A 1:5 + 10% TiO2	47.04	84.19	It is photocatalytic

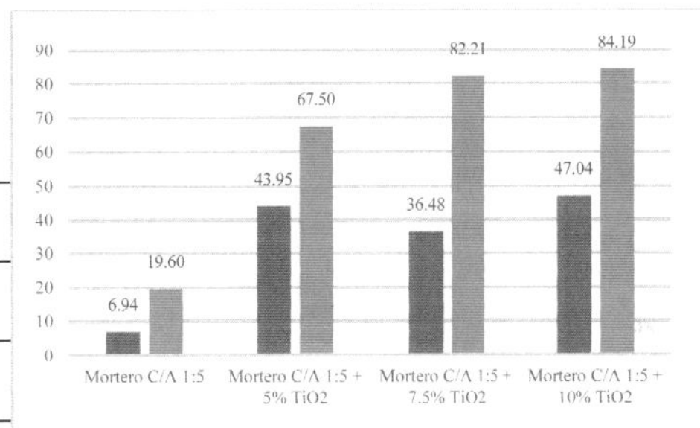

Figure 11. Photocatalytic capacity indexes

4. Conclusions

- The addition of titanium dioxide (TiO2) to the mortar mixture used for charging (C:A 1:5) alters its mechanical properties as fluidity, compression resistance, absorption percentage are reduced according have varied the percentages of addition of the organic component (5%, 7.5% and 10%) in the mixture. Likewise, the modified mortar acquires the photocatalytic capacity since the results of the tests obtained after exposing the samples to UV rays are in the range established by [4].
- By granting photocatalytic capacity to the portland cement mortar, the number of times a building is give maintenance when is exposed to sunlight is reduced.
- This new mortar could and should be applied in the short term on facades of buildings, murals, parking lots, central berms and others.
- Also, this research suggests that improving the appearance of construction works creates a more friendly environment for ordinary citizens.

5. References

[1] N. T. P. 334.051:2013, CEMENTS Test method to determine the compression resistance of Portland cement mortars using 50 mm side specimens., Lima: INACAL, 2018.

[2] N. T. P. 334.057:2011, CEMENTS Test method to determine the fluidity of Portland cement mortars., Lima: INACAL, 2016.

[3] N. T. P. 3. 2010, MASONRY UNITS. Standardized test method for water absorption rate of masonry mortars., Lima: INACAL, 2015.

[4] U. 11259:2016, Photocatalysis - Determination of the photocatalytic process of hydraulic binders - Rodamina Method., Milán: Ente Nazionale Italiano di Unificazione, 2016.

[5] MINAM, « National Air Quality Report 2013-2014, ¨Ministry of Environment¨, Lima, 2016.

[6] TioCem, «Innovative building materials - reduction of pollution with TioCem,» November 2010. [En línea]. Available: http://www.lehighpermanente.com/tiocem-new.

[7] D. Jimenez Cortez y J. J. Moreno Burbano, Effect of replacement of portland cement with titanium dioxide on the mechanical properties of the mortar, Cali: Pontificia universidad javeriana, 2016.

[8] L. E. Lisbona García, Materiales fotocatalíticos y sus aplicaciones en construcción, Cataluña: Universidad Politécnica de Cataluña, 2016.

[9] E. Boonen y A. Beeldens, «Recent Photocatalytic Applications for Air Purification in Belgium,» *Coatings,* vol. 4, pp. 553-573, 2014.

[10] N. T. P. 334.003:2017, CEMENTOS: Procedimiento para la obtención de pastas y morteros de consistencia plástica por mezcla mecánica., Lima: INACAL, 2017.

[11] L. G. d. Lopez, El concreto y otros materiales para la construcción, Cali: Universidad Nacional de Colombia Sede Manizales, 2003.

[12] U. C.-L. Mancha, «Morteros,» de *Cátedra de Ingeniería Rural*, Castilla, 2013.

[13] N. T. P. 399.607:2018, UNIDADES DE ALBAÑILERÍA. Agregados para mortero de albañilería, Lima: INACAL, 2018.

[14] J. Ohama y D. Van Gemmert, Application of Titanium Dioxide Photocatalysis to Construction Materials, EEUU: Springer Netherlands, 2011.

IOP Conf. Series: Materials Science and Engineering **758** (2020) 012061 doi:10.1088/1757-899X/758/1/012061

Finite element analysis and experimental study on laser welding temperature field of dissimilar metal materials

Haibing Xiao[1,a*],SongLing Zhang[2,b],Yongquan Zhou[1,c],XiaomeiXu[1,d] and Bowei Luo[1,e]

[1]Mechanical & Electrical School, Shenzhen Institute of Information Technology, Shenzhen, China

[2]Hymson Laser technology Co.Ltd, Shenzhen, China

*Corresponding author: xiaohb@sziit.edu.cn

Abstract. Aiming at saving defects and cracks of dissimilar metal materials laser welding, the finite element analysis of temperature field and new welding process of stainless steel/Al, stainless steel/Cuwere developed in this work.Taking stainless steel/Al, stainless steel/Cu dissimilar materials laser welding for example, the physical model was established, and the finite element simulation of three-dimensional temperature field of laser welding was carried out based on Abaqus CAE. The results show that the simulation of temperature field provides reference for experimental process parameters, fiber laser welding is applicable to stainless steel/Cu, stainless steel/Al with lower laser welding defect.The process parameters were optimizedto achieve better welding quality and solve welding defects.The accurate prediction of temperature field distribution in dissimilar materials laser welding is great significance for practical process guidance.

Keywords: Dissimilar materials; Laser welding; Temperature field; Finite element analysis; Welding process.

1. Introduction

Dissimilarmaterialsarewidely used inpower batteries, new energy vehicles and aerospace manufacturing. A single metal can no longer meet the requirements of the use of metal engineering materials, Dissimilar materials welding is an effective way to solve the component simultaneously and meet the requirements of various performance, the combination of dissimilarmaterials advantages. Dissimilar materials welding such as stainless steel/Alhave made great progress, but the laser welding process is very complex.

At present, many research institutions such as TWI, Aachen university in Germany, Braunschweig university, 0saka university, EWI in the United States and Technical university in Denmark have developed metal material welding technology[1-3].In 2009, Anawa et al.[4]demonstrated that CO_2 laser can be used for stainless steel AISI316 and stainless steel AISI1008 welding, the welding of ferrite and austenitic steel could be successfully welded by laser, which obtained a small residual stress and heat affected area, and designed experiments to optimize welding parameters. In 2016, Jesper Sundqvist et al.[5, 6]form South Korean investigated welding models of dissimilar materials through finite element analysis. In 2017, Naoyuki Matsumoto et al. [7] used high-power optical fiber laser to conduct welding research on low carbon steel and 6056-T4 and 6016-T6 Al alloy, and established the influence rule of laser power, defocus, welding speed, pulse frequency and shielding gas on steel/Al welding defects. In 2016, Sun daqian et al.[8] systematically studied the characteristics of temperature field, nucleation formation process and microstructure of steel/Al dissimilar metal spot welding, interfacial atomic inter diffusion behavior of steel/Al, interfacial reaction mechanism, intermetallic compound layer growth mechanism, joint crack characteristics and formation mechanism.

In 2018, He xiuli et al.[9] reported the mutual mechanism of welding between stainless steel and nickel heterogeneous metal materials, established the dynamic characteristics of welding pools for these two metals, and established the laser heat transfer and weld pool morphology.

In this study, the finite element simulation of three-dimensional temperature field of dissimilar materials laser welding was carried out to predict the distribution of temperature field based on Abaqus CAE. The physical model was established, and new welding process of stainless steel/Al, stainless steel/Cu were developed in this work.

2. The mechanism of dissimilar materials laser welding

2.1. The mechanism of dissimilar materials laser welding

Mechanism of dissimilar materials laser welding is shown in Figure 1.Dissimilar metal materials laser welding is a complicated process, which contains laser melting and solidification. The theoretical model of heat flow for dissimilar metal materials laser welding is established as follows: the small element is established, and the heat on the element reaches an equilibrium.

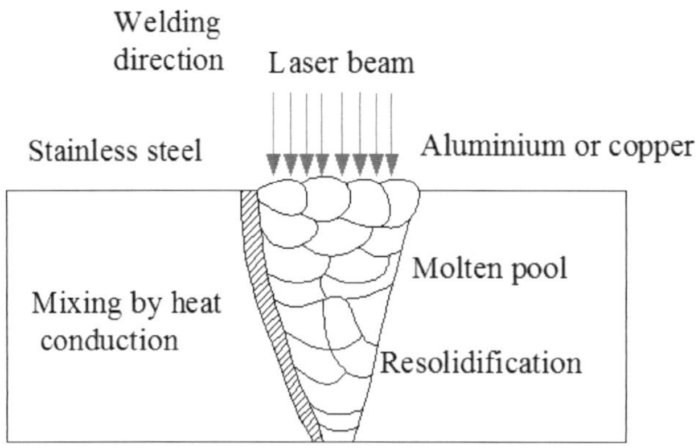

FIG.1.Mechanism of dissimilar materials laser welding

2.2. Lattice heat transfer model of ultrafast laser

According to the size of dissimilar materials laser welding beam size, material size and thermal characteristics, action time, the temperature field can be processed into a three-dimensional model. The temperature field can be divided into stable heat conduction and non-stable heat conduction according to its changing law along with time. The temperature field is described as follows[10]:

$$-\lambda(u\frac{\partial T}{\partial x}+v\frac{\partial T}{\partial y}+w\frac{\partial T}{\partial z})=-\lambda\nabla T=q\ (1)$$

where, q is heat flow vector, ∇T is welding temperature field gradient, λ is coefficient of thermal conductivity, T is temperature of the molten pool. The welding heat conduction equation is described as follows.

$$\frac{\partial}{\partial x}(\lambda\frac{\partial T}{\partial x})+\frac{\partial}{\partial y}(\lambda\frac{\partial T}{\partial y})+\frac{\partial}{\partial z}+(\lambda\frac{\partial T}{\partial z})=\rho C\frac{\partial T}{\partial t}+Q(x,y,z,t)\ (2)$$

where, $\alpha=\lambda/\rho C$ is thermal diffusivity, ρ is density, C is material heat capacity, Q is heat source. Stable heat conduction is simplified to poisson equation.

$$\nabla^2 T+Q/\lambda=0(3)$$

Different laser processes can be produced according to the action spectrum of laser power time. There are three main factors affecting the temperature field of laser heat treatment: power density, action time, pulse width or beam interface/welding speed, etc.

$$S = \frac{0.16}{\rho HL}(H^2 t_v - H^2 t_m) \quad (4)$$

where, S is the fusion depth corresponding to the heat flux, L is Latent heat of fusion, H is material constant.

2.3. Simulation calculation

The melting point of dissimilar metal materials varies greatly. Moreover, stainless steel/Cu,stainless steel/Al have the characteristics of high reflectivity and high heat conduction coefficient, so it is difficult to form keyhole in the welding process, and high energy density is needed in laser welding. The model is composed of three plates of dissimilar metal materials, as shown in Figure 2. The mesh is divided after the model is built, as shown in Figure 3.The mesh is refined near the welding site and roughened away from the welding seam during the division to reflect the temperature gradient at the welding site accurately.

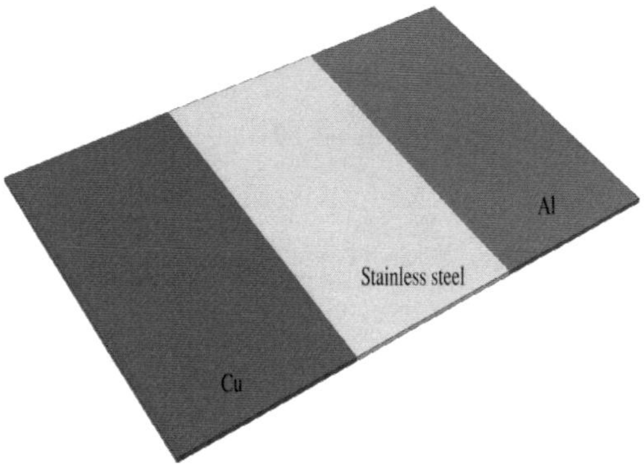

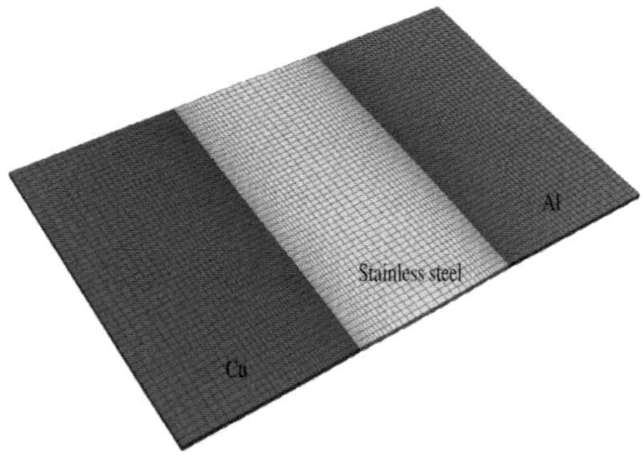

FIG.2.Geometric model of dissimilar materials FIG.3.Three-dimensional mesh model

The temperature field parameters of the flat plate are divided into three types, which are assigned to three plates respectively. Physical and chemical properties of materials, mainly including density, the thermal conductivity and specific heat capacity, are shown in Table 1.

Table 1.Physical and chemical properties of materials

	Density (g/cm^3)	Thermal conductivity (w/m·°C)	Specific heat capacity(J/kg·°C)	Melting point (℃)
Al	2.7	204	0.88	580~740
Stainless steel	7.85	17	0.46	1 399~1 455
Cu	8.96	383	0.39	1084

Figure 4 and Figure 5are the simulated temperature field of stainless steel/Cu, stainless steel/Al, laser welding respectively. Figure 6and Figure 7are the simulated temperature change curve of stainless steel/Cu, stainless steel/Al laser welding respectively .From Figure 4and Figure 6, it can be seen that the highest temperature of the molten pool reached 1375.3°C.From Figure 5 and Figure 7, it can be seen that the highest temperature of the molten pool reached 1489.6°C.

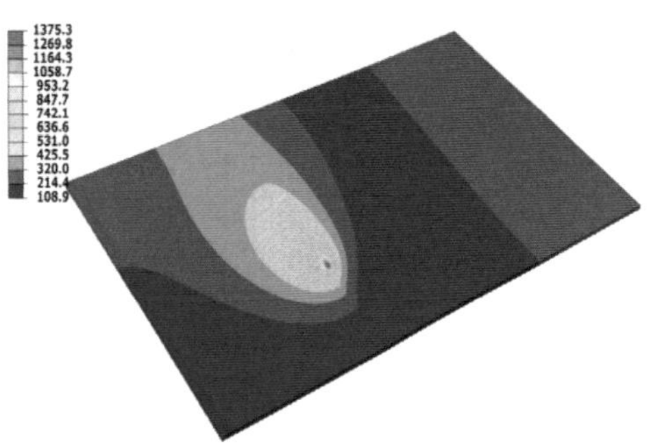

FIG.4.Simulated temperature field of Stainless steel/Cu FIG.5.Simulated temperature field of stainless steel/Al

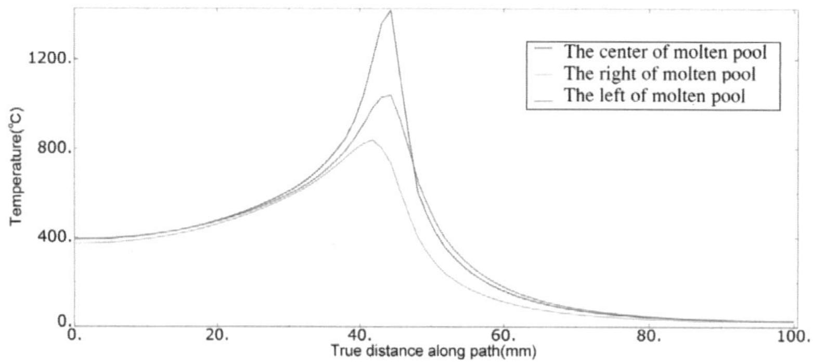

FIG. 6. Welding pool temperature curve of stainless steel/Cu

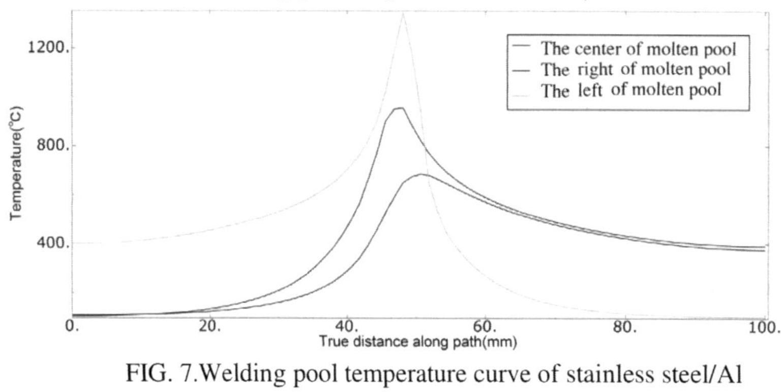

FIG. 7.Welding pool temperature curve of stainless steel/Al

The temperature field of the dissimilar metal materials laser welding pool, that is, the temperature distribution of the pool directly affects its convection, heat and mass transfer, and further affects its solidification process and composition uniformity. Therefore, it is very important for the optimization process parameters.

3. Experimental Analysis

3.1. Experimental Equipment

Fiber laser welding equipment is shown in Figure 8,which mainly consists of fiber laser, control system, vision system, working table, fixtureand cooling system. 1000W IPG fiber laser is the core of the whole system. The laser wavelength of the fiber laser welding system is 1064nm, the average power is 1000W, repetition frequency is 10MHz, the spot diameter is 50μm after focusing through the 3d scanning galvanometer, and the maximum scanning speed of the galvanometer is 20mm/s.

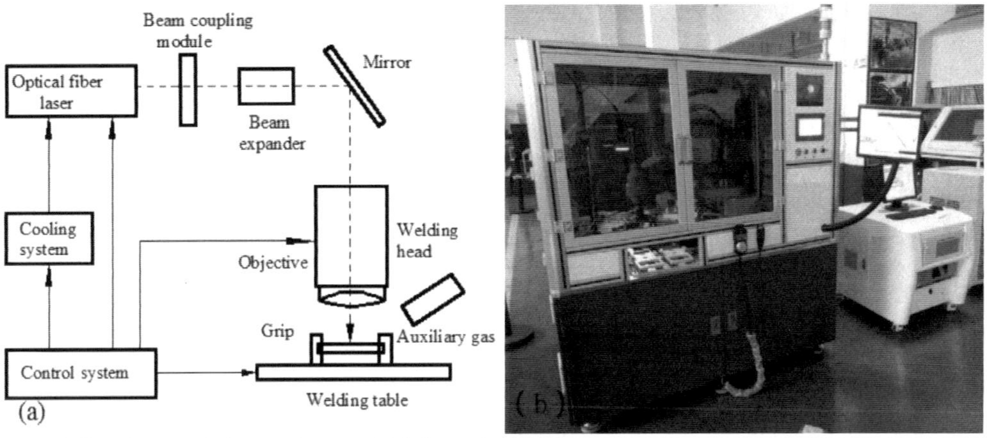

FIG.8.Fiber laser welding equipment (a)Schematic diagram of welding experiment (b)Photograph offiber laser welding equipment

The parameters of the laser welding system are shown in Table 2.

Table 2.The parameters of laser welding system

Parameters	Value	Parameters	Value
Fiber laser power	1000W	Scanning speed	5-20mm/s
Wavelength	1064nm	Focus diameter	50μm
Repetition frequency	10MHz	Positioning accuracy	0.02mm

3.2. Experimental scheme

Test program includes test equipment, test materials, are shown in Figure 9. Through fiber laser welding experiments, the quality of welding is analyzed by tests.

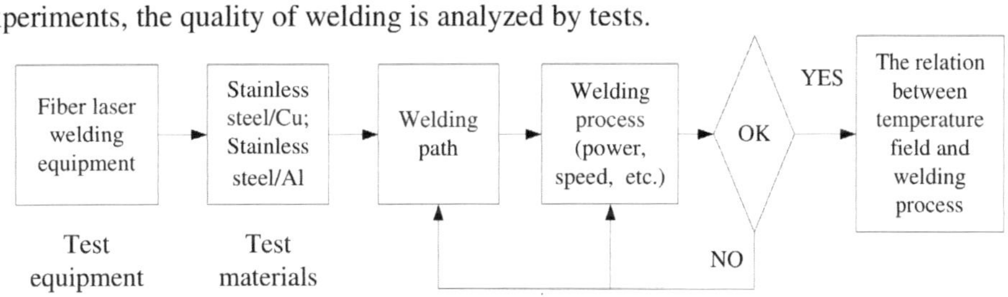

FIG.9.The welding experiment flow diagram

3.3. Experimental results and discussion

To study the influence of process parameters on welding defects, the effects of average power, scanning speed, repetition frequency on the thermal affected area and scanning path are optimized to reduce welding defects. Dissimilar materials welding samples are shown in Figure 10.Figure 10 (a) is stainless steel/Cu, while Figure10(b) is stainless steel/Al under optimum technology parameter.

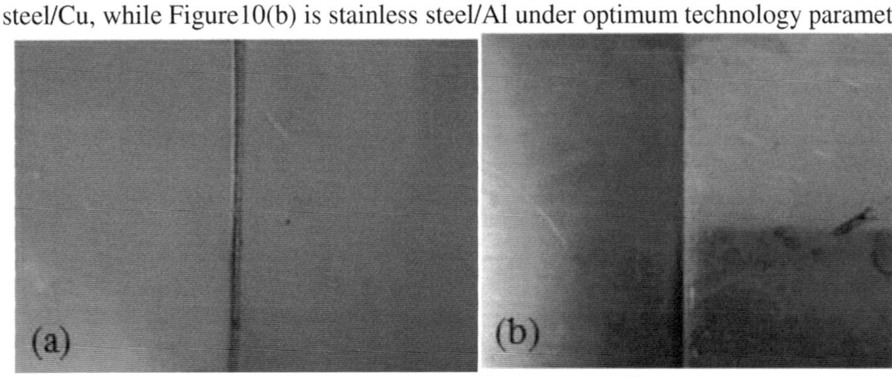

FIG.10.Dissimilar materials welding sample(a) stainless steel/Cu (b) stainless steel/Al

4. Conclusions

(1)The simulated temperature field of laser welding of dissimilar metal materials and its distribution curve along the welding temperature are consistent with the theory. The finite element simulation is used to predict the temperature field to explore the influence rule of welding temperature.

(2)According to the dissimilar metal materials of stainless steel/Cu and stainless steel/Al, the high-power laser welding test is adopted to optimize the laser welding process parameters and establish a new welding process for dissimilar materials. The results show that the optical fiber laser welding is suitable for stainless steel/Cu, stainless steel/Al dissimilar metal materials laser welding defects to achieve a better welding quality by optimizing the process parameters.

Acknowledgments

This work was financially supported by Shenzhen Science and Technology Plan (Grant Nos. JCYJ20170817114441260，GGFW2017041209483817), 2019 Guangdong Special fund for Scientific and Technological Innovation Training for College Student(Grant No.pdjh2019b0864), 2018 Scientific Research Project of General Universities in Guangdong Province(Grant No.2018GKTSCX013),Shenzhen Institute of Information Technology Project (Grant No. SZIIT2019KJ023).

Reference

[1]Yutaka S. Sato, Seung Hwan C. Park, Masato Michiuchi, et al. Constitutional liquation during dissimilar fraction stir welding of Al and Mg alloys, *Scripta Characterization*,2004,52: 49-64.

[2]AlHazaa A, Khan T I, Haq I. Transient liquid phase boding of AL7075 to Ti-6AL-4V ally.*Materials Characterization*,2010,61(3):312-317.

[3]Ghoo B Y, KeumY T, KimY S. Evaluation of themechanical properties of welded metal in tailored steel sheet welded by CO_2laser.*Journal of Materials Processing Technology*, 2015, 113 (1-3) :692-698.

[4]Anawa E M, Olabi A G, Elshukri F A. Modeling and optimization of tensile shear strength of titanium/aiuminum dissimilar welded component.*Journal of Physics: Conference Series*,2009, 181:012033.

[5]Libor Mrna, Martin Sarbort, Simon Rerucha, and PetrJedlicka. Autocorrelation analysis of plasma plume light emissions in deep penetration laser welding of steel.*Journal of Laser Applications*, 2017, (2):012009-1_012009-10.

[6]Jesper Sundqvist, Kyoung-HakKim, Hee-SeonBang, Han-Sur Bang&Alexander F.H.KAPLAN. Numerical simulation of laser preheating of friction stir welding of dissimilar metals.*Science and Technology of Welding and Joining*,2017(10): 1-6.

[7]Naoyuki Matsumoto, YousukeKawahito,KojiNishimoyto, et al. Effects of laser focusing properties on weldability in high-power fiber laser welding of thick high-strength steel plate.*Journal of Laser Application* ,2017,29 (1):012003.

[8]Q Chang,D Sun, X Gu, H Li.Microstructures and mechanical properties of MIG welded joints of aluminum alloy and ultra-high strength steel using Al-Mg and Al-Cu fillers.*Journal of Materials Research*,2017,32(3):666-676.

[9]C.Y. Cui, X.D Li, C. fang, et al. Effects of marangoni convection on the embedding dynamic behavior of SiCnano-particles into the Al molten pool during laser micro-melting.*Materials and Design*,2018(143):256-267.

[10]Zhendong Guan. Laser processing.*Chian metrology press*,2007.

IOP Conf. Series: Materials Science and Engineering **758** (2020) 012062 doi:10.1088/1757-899X/758/1/012062

Cooling performance improvement of circulating internal cooling turning tool by built-in additional spray cooling nozzle

Shengrong Shu[1], Yu Zhang[1], Caiwei Xiao[2], Xiaoning Qi[2], and Leping Liu[1]

1 School of Mechatronics & Vehicle Engineering, East China Jiaotong University, Nanchang 330013, P.R. China
2 Institute of Chemical Materials, China Academy of Engineering Physics, Mianyang 621900, P.R. China
E-mail: shushengrong201@163.com, 976524010@qq.com, xiaocaiwei@caep.cn, 549348812@qq.com, jxllp@163.com

Abstract. Aiming at the demand of high efficiency green cooling cutting of difficult-to-machine materials, a novel composite cooling turning tool design concept based on circulating internal cooling combined with spray cooling is proposed in this paper. Thermal-fluid-solid coupling simulation models were developed based on FLUENT software, which are applied to investigate the cooling performance of two type of composite cooling tool which are single-nozzle type and double-nozzle type, and the circulating internal cooling tool. Simulations results shown that the tool-chip interface temperatures of the three tools are from low to high under the same conditions, followed by the composite cooling turning tool with double-nozzle, the composite cooling tool with single-nozzle and the circulating internal cooling tool. The double-nozzle type of the composite cooling tool exhibits a better cooling and lubrication performance.

1. Introduction

In recent years, in order to meet the requirements for high hardness, high toughness and high wear resistance of the parts in various industries, a number of difficult-to-machine materials such as titanium alloys and nickel-based alloys have been widely used. Heat is easily accumulated in the cutting zone when processing these materials, which results in rapidly tool wear and defects such as scales and burns on the surface of the workpiece. The traditional cooling method by pouring a large amount of cutting fluid in cutting zone not only pollutes the environment, but also causes harm to the human body. It does not conform to the current concept of sustainable green development[1-2]. To reduce or even eliminate the negative-side effects of using the cutting fluid, scholars have conducted in-depth research on cooling methods for green cutting.

Circulating internal cooling allows coolant to flow below the cutting zone through the internal channels of the tool, thereby decreasing cutting temperature and slowing the wear rate of the cutting edge and improving the quality of the workpiece. Vicentin et al.[3] designed an internal cooling turning tool, which is a hollow toolholder with cooling groove and flow channel, and coolant can circulate in the tool to take away heat in cutting process. Another internally cooled tool proposed by Carlo et al.[4] which mainly consists of a hollow cutting insert, a support seat with cooling channels and a toolholder. However, the circulating internal cooling method cannot cool the chips and assist the chip removal because the cooling liquid cannot wash the chips. Spray cooling dissipates heat by forced convection, droplets and thin fluid film evaporation and even nuclear boiling, so it has a large heat dissipation capability and can promote chip evacuation[5]. Therefore, integrating the spray cooling technology into the circulating internal cooling tool would further increase the cooling effect and promote chip removal.

To improve the integration, the flow channels and nozzle for spray cooling can be built in an internal cooling tool to form a composite cooling tool, which not only enables the cooling liquid to fully enter the workpiece-tool-chip interfaces, but also assists chip evacuation.

The CFD method based on Fluent software has been successfully applied to the research of circulating internal cooling and spray cooling in cutting process[6-8], but the research about the influence of the number of spray nozzle on the cooling performance of the composite cooling tool is rarely. In this paper, the composite cooling turning tool was designed and its three-dimensional CFD simulation model was developed. The thermal-fluid-solid coupling simulations were carried out by using Fluent software to investigate the cooling performance of the single-nozzle type and double-nozzle type composite cooling turning tools, and compared with the circulating internal cooling turning tool. This work provides the basis for the application of the composite cooling turning tool in reality.

2. Construction of the tool cooling structure

As shown in Fig.1(a), the circulating internal cooling tool proposed in this paper is a transformation design based on a standard turning tool which composed of the toolholder, insert, support seat and clamp. In order to bring the coolant closer to the cutting zone, the insert thickness is reduced to 1.8mm without seriously affecting the strength of the turning tool. There is an annular groove on top of the support seat and some micro holes are arranged in the toolholder and support seat, and cooling liquid channels can be formed after assembling. The inlet and outlet of the coolant channels are placed on the bottom surface of the toolholder to facilitate the installation of external cooling system.

The design scheme of the composite cooling turning tool is set an additional spray cooling inlet on the bottom surface of the toolholder on the basis of the circulating internal cooling tool. As shown in Fig.1(b) and Fig.1(c) are two types composite cooling turning tool respectively. Due to the limitation of the size of the toolholder head, the single-nozzle located on the top surface of the toolholder head which mainly used for cooling and lubricating the tool-chip interface. For the double-nozzles type composite tool which one nozzle is located on the top surface of the toolholder and another nozzle is located on the lower front of the toolholder, thereby both tool-chip and tool-workpiece interfaces can be cooled and lubricated simultaneously. Because spray cooling needs to realize the function of assist chip removal, the outlet of the spray nozzle on the top of the toolholder should be close to the rake face and perpendicular to the main cutting edge. It is worth noting that the spray cooling channels should not interfere with the circulating internal channels due to the difference between the internal cooling and the spray cooling medium.

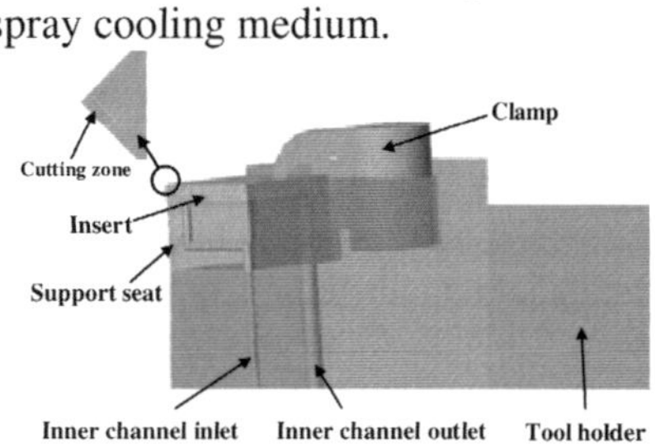

(a) Circulating internal cooling turning tool

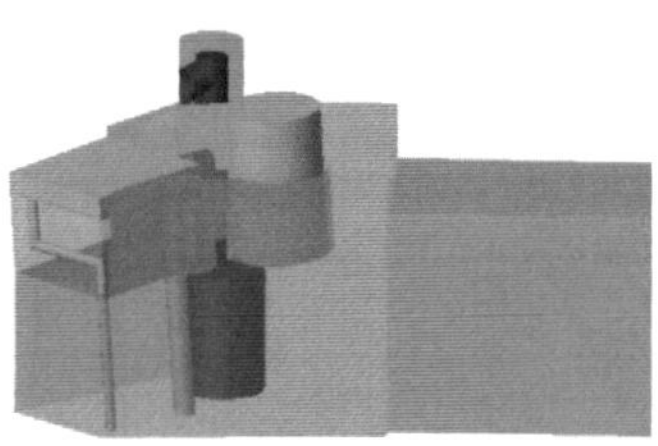

(b) Single-nozzle composite cooling turning tool

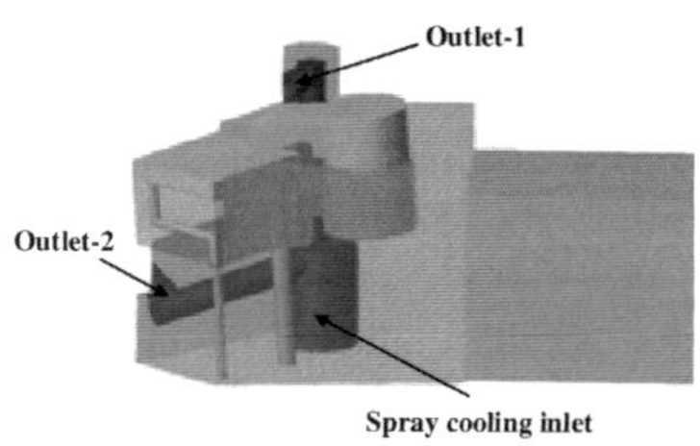

(c) Double-nozzle composite cooling turning tool

Fig.1 The 3-D model of tools

3. CFD modeling of cutting temperature

3.1. Meshing

The tools geometry models were imported into ANSYS Workbench for meshing. The tetrahedral meshing have been used for all the turning tools model. Fig.2(a) shows an example of the mesh model of simulation. In order to make the simulation results closer to the actual situation, a fluid domain with a volume of 95 mm × 80 mm × 70 mm was established for all the analysis models to simulate the flow of air and droplets near the cutting zone and the heat transfer process. The workpiece is added to all the analytical models to simulate the physical process of turning.

As shown in Fig.2(b), the mesh size around the cutting zone was refined, which can better capture changes in the temperature gradient of the tool-chip interface and improve the accuracy of the simulation results. The mesh was also inflated for the interface between solid and fluid, which can be seen as refinement of the mesh elements perpendicular to the interface. Fig.2(c) shows the mesh of the internal cooling channel. Both refinement and inflation can improve the calculation accuracy of the heat transfer between solid domain and fluid domain. Mesh independence test has been performed, and the number of mesh element is about 7.45×106 cells for all simulation models.

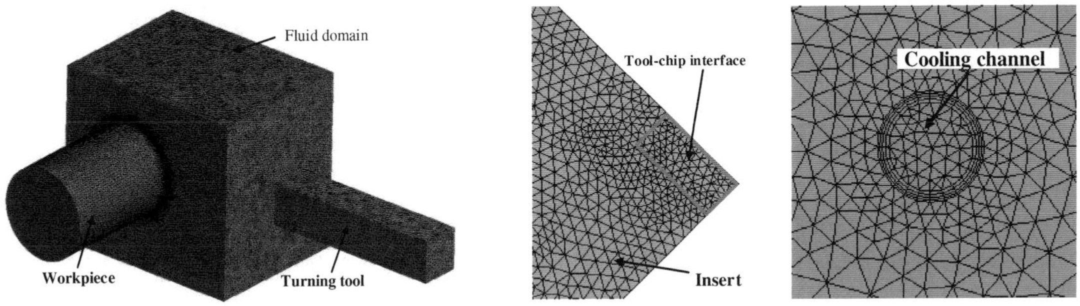

(a) Mesh of the simulation model (b) Mesh of the insert (c) Mesh of the cooling channel

Fig.2 The mesh model of simulation

3.2. Boundary conditions

Boundary conditions applied to the simulation model should be reasonably defined according to the actual cutting process so as to simulate the actual turning process more realistically. The purified water was chosen as the cooling fluid for circulating internal cooling and spray cooling in this study. This is because it has a good heat transfer performance, and it is readily available and harmless to environment and human. The water temperature is set as 20°C, which is equal to the ambient temperature in the cutting process. The inlet of circulating internal cooling is set as velocity-inlet, the inlet velocity is 1 m/s, and the outlet of internal cooling is set as pressure-outlet.

For the spray cooling of the composite cooling tool, the inlet of spray cooling is defined as pressure-inlet, the inlet pressure is 0.3 MPa, the coolant mass flow rate is 1.38639×10^{-5} kg/s and the corresponding flow rate is 50 mL/h, and the fluid domain wall is set as pressure-outlet. The DPM boundary conditions of the inlet and outlet of spray cooling are set to escape. The realizable k- ε turbulence model is used to model the disturbance caused by the fluid impingement on the wall and the rotation of the workpiece. The species transport model is activated in order to model the vaporization and boiling of the coolant droplets and fluid film. The Wall-film model is applied to the interfaces between solid and fluid domain, which can simulate the collision and heat transfer between wall and droplets.

For all turning tools used for simulation, the tool-chip contact area is 0.5 mm×1.5 mm, the heat flow rate conducted into the cutting insert is 20 W/mm^2, and the cutting speed of the turning tool is set to 100 m/s. In this simulation, the workpiece material is steel which adopt from Fluent, and the materials properties of the tool and other parameters used in the simulation can be referred to [6].

4. Simulation results and discussions

In the cutting process, the cutting temperature depends not only on the amount of heat transferred into the tool but also on the cooling method. Different cooling methods have different ways to dissipate heat, and their cooling performance will be different. This paper mainly examines the influence of different cooling methods on the tool-chip interface temperature. During the simulation, the temperature variation of the tool-chip interface was recorded until it was stable, then terminated the calculation and the results were output.

Fig.3 shows the temperature contour of each tool. It can be seen from Fig. 3 that under the same conditions, the maximum temperature of the circulating internal cooling turning tool is 591.4K, and that of the composite cooling turning tool is 568.5K. It is clearly shows that the cooling efficiency can be improved by adding spray cooling to the circulating internal cooling tool. This is mainly because spray cooling increases heat dissipation through convection heat transfer of high-pressure air, evaporation and boiling of liquid droplets, and convection heat transfer, evaporation and boiling of liquid film formed after liquid droplets collide with the wall surface. It is worth noting that the tool-chip interface temperature of the single-nozzle type and double-nozzle type composite cooling tool is almost the same, thus the cooling efficiency of the two cooling methods is almost the same. From the cooling performance point of view, the choice of the two methods is indistinguishable.

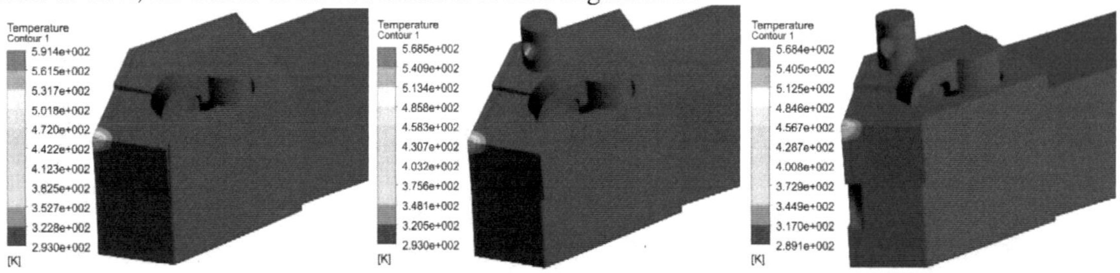

(a) Temperature contour of the circulating internal cooling tool (b) Temperature contour of the single-nozzle composite cooling tool (c) Temperature contour of the double-nozzle composite cooling tool

Fig.3 The temperature contour of each tool

When the liquid droplets flow in the curved channel, some of them will collide with the wall of the channel and adhere to the wall to form liquid film. The accumulated liquid droplets or film on the wall would be re-atomized into bigger particles by high speed air flow and escape from the nozzle outlet to cool the cutting zone. As shown in Fig. 4, it illustrates the velocity and the spatial dispersion contours of the cooling droplets of the two type composite cooling turning tool. It can be seen that the maximum droplet velocity of the double-nozzle type composite cooling tool is slightly larger than that of the single-nozzle type composite cooling tool under the same inlet pressure. It can also be seen that some of the liquid droplets rebound or splash after colliding with the walls due to the high velocity of the liquid droplets after exiting the nozzles. The droplets only flow to the tool-chip interface after they flow out from the single-nozzle composite cooling turning tool, while the droplets flow to both the tool-chip and tool-workpiece interfaces after they flow out from the double-nozzle composite cooling turning tool. Therefore, it can be inferred that the lubrication performance of the double-nozzle composite cooling turning tool is better than that of the single-nozzle composite cooling turning tool. From the point of view of cooling and lubrication, the double-nozzle composite cooling tool has more research value than the single-nozzle composite cooling tool.

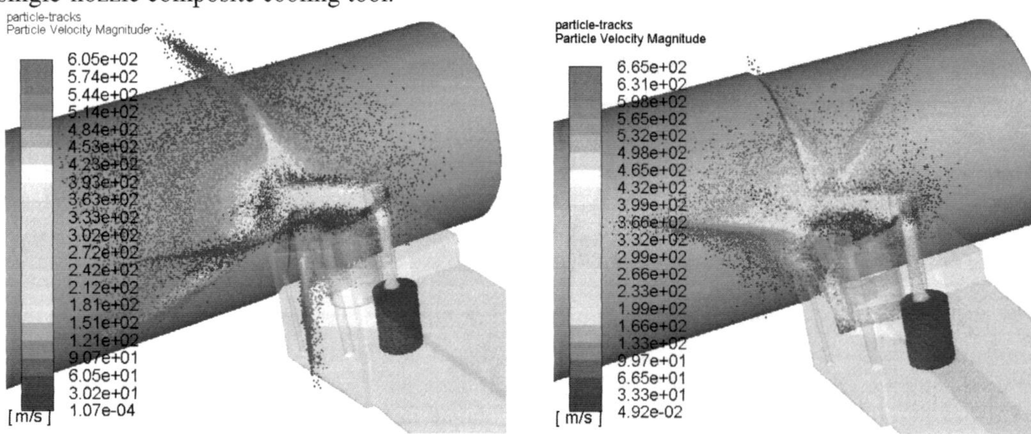

(a) Droplet velocity contour of single-nozzle composite cooling tool (b) Droplet velocity contour of double-nozzle composite cooling tool

Fig.4 The droplets velocity contour of two type composite cooling tool

5. Conclusions

IOP Conf. Series: Materials Science and Engineering **758** (2020) 012062 doi:10.1088/1757-899X/758/1/012062

In this paper, the thermal-fluid-solid coupling simulations were carried out by FLUENT software to investigate the temperature in cutting process between the composite cooling turning tool and the circulating internal cooling turning tool. The composite cooling method based on circulating internal cooling and spray cooling can greatly reduce the temperature in cutting zone, and can lubricate the tool-chip interface thus extend the tool life. The influence of the number of spray nozzle on the cooling capacity of composite cooling tool was also studied. The simulation results show that the cooling performance of single-nozzle type and double-nozzle type composite cooling tool is almost the same, while the lubrication performance of the double-nozzle composite cooling turning tool is better. The cooling structure and nozzle of spray cooling can be further optimized to improve the cooling performance.
From above results, it is found that the composite cooling cutting tool has great potential for machining difficult-to-machine materials such as Titanium, Inconel and chemical explosive material. Furthermore, it is worthwhile to investigate the influence rule of cooling parameters such as inlet pressure, the temperature and flow rate of the cooling liquid on the cooling performance of the composite cooling tool, and to apply it for adaptive machining where abnormal temperature rising of the cutting tool and thus tool wear can be avoided.

6. Acknowledgments
The authors would like to acknowledge the funding support for this research by the National Natural Science Foundation of China (Grant No. 51705153) and the Jiangxi Provincial Natural Science Foundation of China (Grant No. 20192BAB206029).

7. References
[1] BYRNE G, DORNFELD D, DENKENA B. Advancing cutting technology[J]. CIRP Annals-Manufacturing Technology, 2003, 52(2): 483-507.
[2] ZHOU JI. Intelligent Manufacturing - Main Direction of "Made in China 2025"[J]. China Mechanical Engineering, 2015, 26(17): 2273-2284.
[3] VICENTIN G C., SANCHEZ L E A., SCALON V L., et al. A sustainable alternative for cooling the machining processes using a refrigerant fluid in recirculation inside the toolholder[J]. Clean Technologies and Environmental Policy, 2011, 13(6): 831-840.
[4] Carlo F., Timothy M., Saiful B C G., Cheng K. Efficiency in contamination-free machining using microfluidic structures[J]. CIRP Journal of Manufacturing Science and Technology, 2014, 7(2): 97-105.
[5] Hoyne A C., Nath C., Kapoor S G. Characterization of fluid film produced by an atomization–based cutting fluid (ACF) spray system during machining[J]. Journal of Manufacturing Science and Engineering-Transactions of the ASME, 2013, 135(5): 051006.
[6] SHU SR, CHENG K, DING H, et al. An innovative method to measure the cutting temperature in process by using an internally cooled smart cutting tool[J]. Journal of Manufacturing Science and Engineering, 2013, 135(6).
[7] WU T, LI TJ, DING XH, et al. Design of a modular green closed internal cooling turning tool for applications[J]. International Journal of Precision Engineering and Manufacturing-Green Technology, 2018, 5(2): 211-217.
[8] EL-BOURI. W, DEIAB. I, KHANAFER. K, WAHBA. E. Numerical and experimental analysis of turbulent flow and heat transfer of minimum quantity lubrication in a turning process using discrete phase model[J]. International Communications in Heat and Mass Transfer, 2019, 104:23-32.

IOP Conf. Series: Materials Science and Engineering **758** (2020) 012063 doi:10.1088/1757-899X/758/1/012063

Effect of Quartz Sand on Compressive Strength of High Performance Basalt Fiber Concrete

Yao Dongdong[1], Ji Hongru[2], Lu Zhe[2], Li Xinjun[2], Li Jie[1] and Feng Zhengang[2]

[1] Jilin Provincial Transport Scientific Research Institute, Changchun, 130012
[2] School of Highway, Chang'an University, Xi'an 710064

Corresponding author: Feng Zhengang

E-mail: zgfeng@chd.edu.cn

Abstract. The QS with different sizes was used to prepare the HPBFC according to the closing packing theory. The universal mechanical tester was applied to test the compressive strength of HPBFC specimens after different curing periods. Effects of the QS with different sizes and mixing ratios on compressive strength of the HPBFC were compared and the development of compressive strength of the HPBFC during different curing time was analysed. Results show that the QS with small particle size can significantly increase the compressive strength of HPBFC. The HPBFC with small size of QS not only possesses the highest compressive strength at different curing stages, but also has the fastest increasing rate during the standard curing conditions.

1. Introduction

High performance concrete (HPC) has been developed for more than 20 years, which is primarily used in many large structures, especially high-rise buildings and long-span bridges for its characters of high strength, high performance and multi-function [1]. In order to meet the engineering requirements, fiber is often used to improve the crack resistance and toughness of HPC. Basalt fiber, as a silicate fiber, is naturally compatible with cementitious composites and often is used to prepare high performance basalt fiber concrete (HPBFC) which possesses higher tensile and compressive strength than ordinary concrete [2].

Sand, as the largest part of HPBFC, can obviously influence the mechanical property, alkali aggregate reaction, volume stability and durability of HPBFC due to its different particle size, shape and elastic modulus [3].Therefore, the properties of sand cannot be ignored in HPBFC. It is known that the commonly used sand in HPBFC is river sand (RS), dune sand, crushed sand or quartz sand (QS), etc. Compared with other types of sand, QS is harder to be weathered and has better chemical stability under alkaline conditions, which can improve the strength and durability of concrete.

Many experiments have been carried out to study the influence of QS on performance of HPC. Ma et al. found that QS could improve the distribution and compressive strength of concrete and decrease the density of cementitious composites linearly with the rising content of QS [4]. Wang et al. used QS as concrete reinforcement agent and found that QS could improve the impermeability and salt-frost resistance of self-repairing concrete [5]. The results of He et al. showed that quartz powder increased compressive strength of reactive powder concrete [6]. Most of the existing studies investigated the QS with a certain size on properties of HPC. However, few of them studied the QS with different sizes and mixing ratios on the performance of HPC, especially on that of HPBFC.

IOP Conf. Series: Materials Science and Engineering **758** (2020) 012063 doi:10.1088/1757-899X/758/1/012063

In this paper, the QS with different sizes was used to prepare the HPBFC according to the closing packing theory. The universal mechanical tester was applied to test the compressive strength of HPBFC specimens after different curing periods. Effects of the QS with different sizes and mixing ratios on compressive strength of the HPBFC were compared and the development of compressive strength of the HPBFC during different curing time was analysed.

2. Experiments

2.1. Materials

The P.O 52.5 Portland cement was produced in Tangshan Jidong Cement Co., Ltd. The technical index of the cement is shown in Table 1.

Table 1. Technical index of P.O 52.5 Portland cement

		National standard	Measured value
Loss of ignition/%		≤5.0	1.01
Specific surface area/(m^2/kg)		≥300	416
Condensation time/min	Initial setting	≥45	105
	Final setting	≤600	350
3day flexural strength/MPa		≥4.0	6.0
3day compressive strength/MPa		≥23.0	31.1
28day flexural strength/MPa		≥7.0	9.0
28day compressive strength/MPa		≥52.5	59.8

The fly ash was produced in Lingshou Danxu Mineral Products Processing Plant. The technical index of the fly ash is shown in Table 2.

Table 2. Composition of fly ash

Composition	SiO_2	Al_2O_3	Fe_2O_3	CaO	MgO	Na_2O
Measured value /%	58	30	4.3	1.5	2.8	3.2

The silica fume was produced in Lingshou Erping Mineral Products Processing Plant. The technical index of silica fume is shown in Table 3.

Table 3. Technical index of silica fume

Test items	Test result
Average particle size/μm	0.1～0.3
Specific surface area/(m^2/g)	20～28
Refractoriness/℃	≥1600

The short-cut basalt fiber was produced in Wuhan Hengchida New Material Co., Ltd. The technical index of the short-cut basalt fiber is shown in Table 4.

Table 4. Short-cut basalt fiber (for cement concrete)

Test items	Standard requirements	Measured average value
Diameter/μm	9～25	14.09
Fracture Strength/MPa	≥1200	1988
Elongation at break/%	≤3.1	2.4
Density/(g/cm^3)	2.60～2.80	2.64

The polycarboxylate super-plasticiser was produced in Changchun Dehai Building Materials Co., Ltd. The technical index of polycarboxylate super-plasticiser is shown in Table 5.

Table 5. Technical index of polycarboxylate superplasticiser

Product model	Water reduction rate/%	Bleeding rate /%	Air content /%	7day strength ratio/%	28day strength ratio/%	Shrinkage /%	Cement paste fluidity /mm
HK-1	≥25	≤60	≤6.0	≥150	≥140	≤110	≥240

The QS with size less than 1mm and 1-2mm were used in test, which was produced in Lingshou Zhenling Mineral Products Processing Plant. The technical index of QS is shown in Table 6.

Table 6. Technical index of quartz sand

Test items	Standard requirements	Measured value
Silica content/%	>96	99.5
Ferric Trioxide Content/%	<0.06	0.012
Mud content/%	<0.2	0.01
Loss of ignition/%	<0.4	0.1
Density/(g/cm³)	⩾2.55	2.64

2.2. Mix proportion
The QS with particle sizes of less than 1 mm and 1-2 mm were selected for this paper. The total amount of QS was the same as RS(1220 kg/m³), and the ratio of QS with particle size of less than 1 mm and 1-2 mm was 10:0, 8:2, 7:3, 6:4 and 5:5, which was numbered from 1 to 5, respectively. The detailed design mix proportion of HPBFC is summarized in Table 7.

Table 7. Experiment mix proportion of the HPBFC /(kg/m³)

Cement	Fly ash	Silica fume	Basalt fiber	Water	Super plasticiser	Quartz Sand		Number
						<1mm	1-2mm	
760	110	130	2	190	30	1220	-	1
						976	244	2
						854	366	3
						732	488	4
						610	610	5

2.3. Experiment procedures
Firstly, the powder material and RS or QS were dried stirred by planetary mixer at a speed of 140 rpm for 1 minute. Secondly, put basalt fibers in the blender and then the materials were stirred for 1 minute. Finally, after adding water and super-plasticiser in the blender, the materials were wet stirred at a speed of 285 rpm for 2 minutes. The dimension of compressive strength test specimens was 150mm× 150 mm × 150 mm cube. The specimens were cured in the standard curing room for 3 days, 7 days and 28 days respectively. The compressive strength test was carried out according to the cubic compressive strength test method (ASTM C39 / C39M-18).

3. Results and Discussion

3.1. Effect of quartz sand with different particle sizes and mixing ratios on compressive strength of HPBFC
The compressive strength of different HPBFC cured for 3days is shown in figure 1. The specimen of no.1 has the highest compressive strength, which can reach 54MPa, while the compressive strength of no. 2 is the lowest at 47MPa. The overall trend of compressive strength is on the rise as the increasing

ratio of QS with larger size, for example, the specimens of no. 3 and no. 4 are 51MPa and 52.3MPa, respectively.

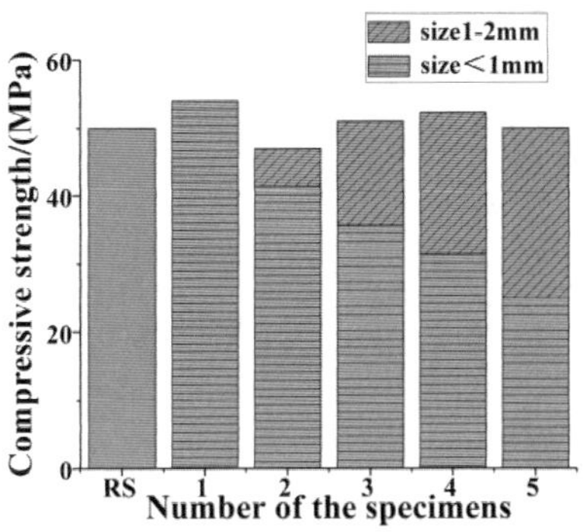

Figure 1. 3 days compressive strength of the HPBFC.

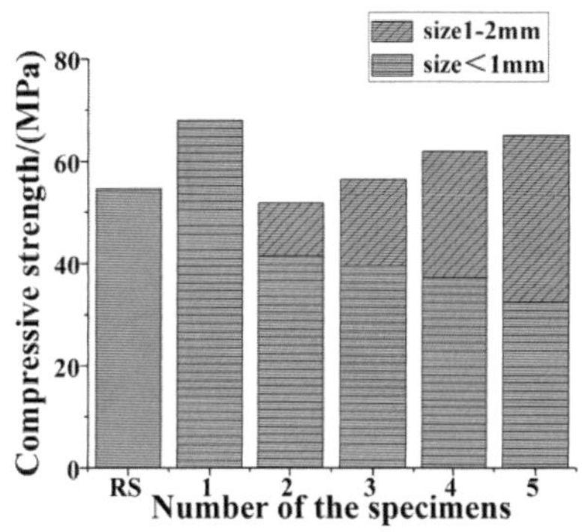

Figure 2. 7 days compressive strength of the HPBFC.

Figure 2 shows the measured compressive strength of HPBFC after cured for 7 days. The compressive strength of HPBFC with RS is 54.6MPa. The specimen of no. 1 has the compressive strength at 68MPa, which is much higher than that of HPBFC with RS. However, after mixed with larger size of QS, the compressive strength of the HPBFC shows a decreasing trend compared with the HPBFC with only small size of QS. The more the larger size of QS is contained, the higher the compressive strength of HPBFC is. The specimen of no. 1 has the highest compressive strength at 54MPa and the specimen of no. 2 owns the lowest compressive strength at 47MPa. Cohesion between of cement and aggregate increase rapidly in the early stage by increasing the specific areas of aggregate, so the QS with particle size less than 1 mm is advantageous to the early strength of concrete.

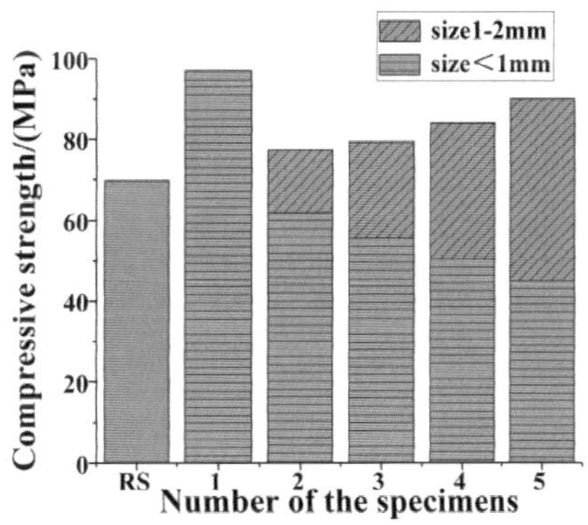

Figure 3. 28 days compressive strength of the HPBFC.

As shown in figure 3, the compressive strength of specimen no. 1 is 97MPa, which is higher than that of the RS control group of 69.7MPa. As the increase of the QS with larger size, the compressive strength of the specimens increases gradually. However, the compressive strength of specimens with larger size of QS are all smaller than that with only small size of QS (less than 1mm), which is in agreement with the results in figures 1 and 2. This indicated that the QS with small particle size can significantly increase the compressive strength of HPBFC.

3.2. Effect of curing time on compressive strength of HPBFC

The effect of curing time on compressive strength of HPBFC by using different proportion of QS is shown in figure 4. As shown in figure 4, the compressive strength of all samples increase with the rise of curing time. However, the increasing rate of compressive strength for different samples is diverse to some extent. Compared with 3 days compressive strength, the 28 days compressive strength of different specimens (RS, no. 1, no. 2, no. 3, no. 4 and no. 5) increase by 39.4%, 79.6%, 64.4%, 55.4%, 60.6%, 73.1%, respectively. This indicates that the HPBFC with small size of QS not only possesses the highest compressive strength at different curing stages, but also has the fastest increasing rate during the standard curing conditions.

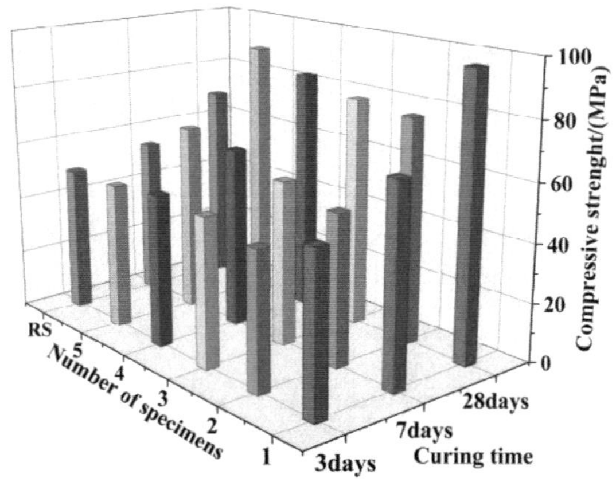

Figure 4. Compressive strength of HPBFC at different curing time.

4. Conclusions

(1) The QS with small particle size can significantly increase the compressive strength of HPBFC, which can be ascribed that the cohesion between of cement and aggregate increase rapidly by increasing the specific areas of aggregate.

(2) The HPBFC with small size of QS not only possesses the highest compressive strength at different curing stages, but also has the fastest increasing rate during the standard curing conditions.

Acknowledgements

This work is supported by the Transportation Science & Technology Project of Jilin Province (2015-1-6, 2017-1-7), the Fundamental Research Funds for the Central Universities (300102219216). The authors gratefully acknowledge their financial support.

References

[1] Xia X D, Ding D J 1994 High performance concrete *Industrial Construction* No.**11** PP 21-6+40

[2] Ayub T, Shafiq N and Nuruddin M. F 2014 Effect of Chopped Basalt Fibers on the Mechanical Properties and Microstructure of High Performance Fiber Reinforced Concrete *Advances in Materials Science and Engineering* Volume: 2014 Article No. **587686** Published: 2014

[3] Huang Z Y, Liu Q 2009 Research of Ultra-High Strength and High Wear-Ear-Resistant Concrete Material *Natural Science Journal of Xiangtan Univrsity* Vol.**31**, No.1 PP 92-7

[4] Ma B G, et al 2013 Effect of mixed quart z sand on the particle size dist ribution of the cement and performance of the mortar *Concrete* No. **6** PP 93-6

[5] Wang L, Bai Y 2017 Study on the effect of high performance repair agent on the permeability and salt resistance of concrete *Structural Engineers* Vol.**33**, No.1 PP 144-8

[6] He F, Huang Z Y 2006 Compressive strength contribution analyses of silica fume and crush quartz in RPC *Concrete* No. **1** PP 39-42

Evaporation processes of alloying components duringwire-arcdeposition of aluminum alloy 5056

E.S. Salomatova[1], M.F. Kartashev[1], D.N. Trushnikov[1], G.L. Permykov[1], T.V. Olshanskaya[1], I.R. Abashev[1], E.M. Fedoseeva and E.G. Koleva[2]
1 Perm National Research Polytechnic University, Russia
2 Institute of Electronics, Bulgarian Academy of Sciences, Sofia, Bulgaria
E-mail: weld-katy@mail.ru

Abstract. Additive manufacturing technologies are developing fast world wide. Never the less, in the machine-building industry, manufacturing of especially large products is required, and common processes, for example, selective laser melting, are not able to satisfy this requirement. Multilayer wire-arc deposition, allows to make high-quality large-scale products. In addition, the productivity of wire-arc deposition is many times higher than the productivity of powder additive manufacturing. Never the less, the metal obtained by this method is likely to lose easily evaporated alloying elements, which is mainly due to excessive heat and high deposition rates. This process leads to reduced mechanical properties of the deposited metal. The paper presents the results of chemical analysis of the sample obtained by CMT deposition of aluminum alloy 5056.It was revealed that during wire-arc deposition there is evaporation of some alloy components, for example magnesium. A nonlinear theoretical model of nonequilibrium processes in the liquid phase of the deposited metal and the processes of evaporation of chemical easily evaporated elements in the zone of influence of the heating source of wire-arc deposition is presented. Verification of the model was carried out by studying the chemical composition of the samples. During X-ray fluorescence analysis, reduced magnesium content of the first deposited layer was revealed. Increase in the magnesium content was in the upper deposited layers.

1. Introduction

The use of additive technologies helps to reduce the preparatory and technological cycle of creating a product in production, which leads to a reduction in the cost of finished products. Application of wire materials of various compositions expands the range of products obtained in the additive production, in comparison with expensive powder materials. When using wire materials and various surfacing technologies (laser, plasma, CMT surfacing) [1] with technological methods (pneumatic hardening of layers [1-2], ultrasonic action in the process of surfacing), it is possible to obtain products with unique properties, which is an undeniable advantage of additive manufacturing with use of wire materials. This article discusses the samples obtained during CMT (Cold Metal Transfer) surfacing in a pulsed mode. The process of CMT surfacing proceeds as follows: when the deposited wire comes into contact with the surface of the substrate, a short circuit occurs, when the first short circuit is detected, the current decreases to the minimum acceptable value, at the same time, the droplet breaks off due to the reverse movement of the deposited wire. Metal transfer occurs at a current value of almost zero and therefore the contribution of heat is very small [3].

In studies [4] and [5], during investigation of the characteristics of controlling the droplet transfer process with CMT technology it was found that CMT requires less current for the same amount of deposited material as compared to conventional GMAW (Gas Metal Arc Welding).In study [6], the possibility of using the CMT process for welding of aluminum alloy was investigated. Pires et al. [7] compared the GMAW, the pulsed GMAW and the CMT process and found that the CMT process reduces the emission of harmful fumes due to low heat generation. The above studies provide a preliminary understanding of the CMT process, but the details of the physical processes are still unclear. A complete understanding of the CMT process cannot be obtained by experimental data alone. Thanks to advances in numerical methods and high power computers, the numerical method provides a convenient way to better understand the mechanisms. In recent years, a number of articles have been published on modeling transport phenomena in the process of arc welding [8-11].However, not enough literature is devoted to the CMT-pulsed welding process - a combination of CMT and pulsed arc, especially for welding aluminum alloys and galvanized steel. This technology has been known and used not so long ago, since 2004, so its comprehensive study, including process modeling, is an important and urgent task.

2. Mathematical model

In a first approximation, the mathematical model is based on the solution of the heat problem in a three-dimensional formulation in the first approximation. Due to symmetry, it is sufficient to include only half of the entire object in the computational domain for the description. The geometry of the computational domain is shown in Fig. 1.

The following simplifications were made when modeling the surfacing process. Firstly, the entire volume of the metal under study is in a solid state, secondly, the heating source is written in the form of a Gaussian distribution, thirdly, the geometric parameters of the bead are associated with the parameters of the CMT process.

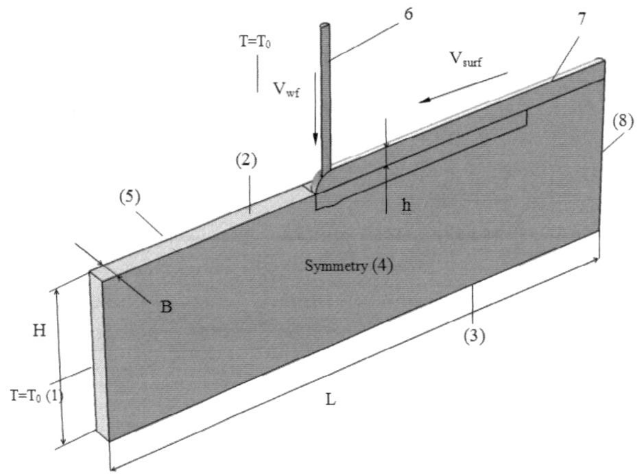

Fig. 1. The design model: 1 – entrance surface; 2 – side surface;3 – bottom surface; 4 – symmetry surface; 5 – top surface; 6 – fused wire; 7 – weld bead; 8 –exit surface

The mathematical model is based on the solution of the differential heat equation (energy transfer) in a three-dimensional formulation

$$- \quad \left(- \quad - \quad - \right) \quad - \quad —— \qquad (1)$$

where T is the temperature, $a = \lambda / C_{eff} \cdot \rho$ – coefficient of thermal diffusivity, V – medium velocity (matches surfacing speed and wire feed speed), ρ – density, P – total power input to the product [W/m^3].

The contribution of convective transport in the melt is taken into account by the formulaλ_e $\lambda\ (2 - T_L/T)$, where$\lambda_{eff}$is the thermal conductivity of a liquid, T_L – liquidus temperature.

The latent heat of fusion and crystallization was taken into account by introducing the effective heat capacity:

$$\frac{p[((T \quad t)/(T \quad))]}{\sqrt{\ }(\quad s)} \tag{2}$$

where C_0 is heat capacity depending on temperature, H_f – latent heat of fusion, T_{melt} – melting point, which is assumed to be average in the range from solidus temperature T_S to liquidus temperature T_L [12].

In view of symmetry, it is sufficient to include only half of the entire object in the computational domain for description.

Mixed type boundary conditions were used. Thermodynamic boundary conditions on the front surface (surface of the deposited metal):

$$— \quad — \quad (T \quad)$$

in the symmetry surface

$$—$$

and on all other surfaces

$$— \quad (T \quad),$$

where $r = \sqrt{(x - x_0)^2 + (y - y_0)^2}$ is the distance from the axis of the heat source to the considered point (x_0 and y_0 are the coordinates of the center of the source), α_0 – effective heat transfer coefficient taking into account heat loss due to convection and radiation. Table 1 and 2 show the parameters of the surfacing process and the physical characteristics of the deposited material.

Table 1. Parameters of the surfacing process.

Welding current, I (A)	Welding voltage, U (B)	Surfacing speed, V_{surf} (m/s)	Wirefeedspeed, V_{wf} (m /s)	Deposition time of one layer (s)	Pause time between layers (s)
115	17	$1\cdot10^{-2}$	$8\cdot V_{surf}$	16	60

The mass flux density of the i-th alloying element in the wire melting zone is determined by the flux density due to the evaporation of the J_{ev} i-th element, is directly proportional to the impurity concentration

$$\left({}^{S}/_{S}\right), \tag{3}$$

where J_0 is the evaporation flux density of the pure i-th element; S_0, S – respectively, the initial and current concentrations of the i-th element in the alloy.

In turn, the density of the diffusion vaporization stream of the i-th element is determined by the formula [13]:

$$\cdot \left(P(T)/\sqrt{\mu \cdot T}\right), \tag{4}$$

where μ is the mass in moles of the i-th alloy element; J_v – molar flow of the i-th alloy element; A – dimensional coefficient; $P(T)$ – partial pressure of the i-th component of the melt, which can be determined by the technique described in [14].

Thus, the task is to reduce to a joint solution of equations (1-4) and to determine the mass flow in the volume of the melt carried away from the surface of the melt:

$$\sum_{i=1}^{N}(J_0) \qquad — \tag{5}$$

where $\partial m/\partial t$ is the mass flow of metal vapor carried away from the surface of the melt.

The initial mass of alloying easily evaporated components in the volume of the molten bath was calculated by the formula

where $V_{melt}{\cdot}\rho$ is themass of alloy in liquid state,C_i^0 - initial concentration of alloying elements in the weld wire.
The change in the mass of alloying elements during surfacing over time is determined

Thus, we determine the final concentration of alloying elements in the weld bead

The shape of the weld bead was approximated by a semi-ellipse. The volume of deposited per unit time of the bead coincides with the volume of supplied filler material$\pi r_{wire}^2 V_{wf} = \pi a_{в} b_{в} V_{pl}/2$, where $a_{в}$ and $b_{в}$ are the width and the height of the weld bead, respectively.
Numerical implementation was carried out using the application software package Comsol 4.3 (heat transfer module). The process simulation area was covered with a three-dimensional grid inscribed in the calculation area. The grid had an uneven pitch (Fig. 2). In the zone of arc influence and filler wire feed, the maximum cell size was 0.5 mm, in the rest of the region - 2.5 mm.

During the deposition of aluminum wire, a liquid melt forms in the zone of action of the heating source, as shown in Fig. 3 (b), in this zone the phase transition from the liquid to the solid state is taken into account, as shown in Fig. 3 (a).

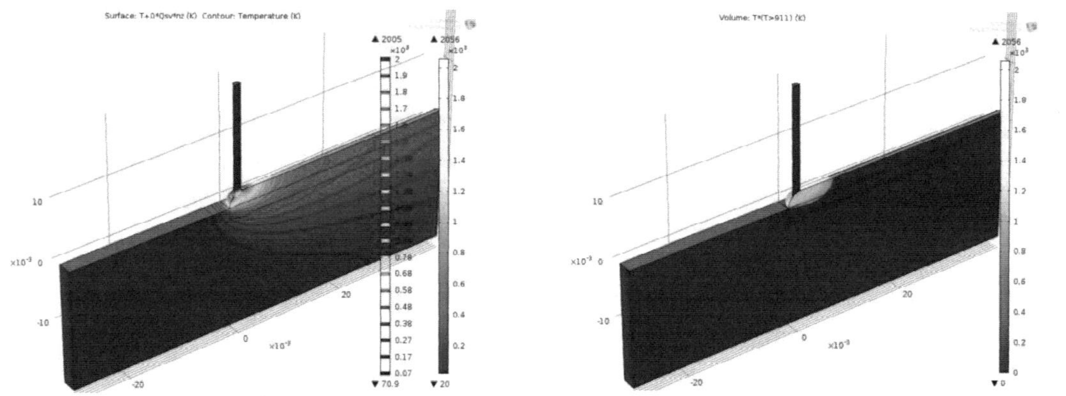

a) b)

Fig. 2. The results of numerical modeling:
(a) zone of metal in liquid and solid states, the maximum temperature of the solid state is 911 K,
(b) action of the heating source, maximum design temperature 2056 K.

With an increase in the power of the heating source during CMT surfacing, we observe a decrease in the concentration of magnesium in the deposited layer.

3. Conclusions

The most probable cause of such a difference in the concentration of magnesium in the deposited layers is the influence of thermal action from each subsequent layer and the high diffusion activity of magnesium. Magnesium in aluminum alloy acts as a hardener. To avoid obtaining deposited products

with reduced strength properties, it is necessary to mechanically remove the first layer deposited on the substrate.
A mathematical model has been developed that describes the processes of evaporation of easily evaporated components of an aluminum alloy, which allows predicting the chemical composition of the beads obtained by CMT surfacing. The mathematical model was verified by conducting a full-scale experiment and comparing the results of chemical analysis of the weld metal zones with the calculated values.

References

[1] Cong B. et al. Influence of cold metal transfer process and its heat input on weld bead geometry and porosity of aluminum-copper alloy welds 2016 *Rare Metal Materials and Engineering* **45** 3 pp 606-611
[2] Shchitsyn Y, Terentyev S, Neubybin S, Artemov A, Belinin D Formation of Structure and Properties of Steel 04CR18NI9 at Additive Production of Trainings 2018 *Vestnik of PNRPU. Engineering, materials science* **20** 3 pp 55-62
[3] Zhang H T, Feng J C, He P, Zhang B, Chen J M, and Wang L The Arc Characteristics and Metal Transfer Behavior of Cold Metal Transfer and its Use in Joining Aluminum to Zinc-Coated Steel 2004 *Mater. Sci. Eng.* **499** pp 111–113
[4] Feng J, Zhang H, and He P The CMT Short-Circuiting Metal Transfer Process and its Use in Thin Aluminum Sheets Welding 2009 *Mater. Des.* **30**
[5] Pickin C G, and Young K, Evaluation of Cold Metal Transfer (CMT) Process for Welding Aluminum Alloy 2006 *Sci. Technol. Weld. Joining* **11** pp 583–585
[6] Pires I, Quintino L, Amaral V, and Rosado T Reduction of Fume and Gas Emissions Using Innovative Gas Metal Arc Welding Variants 2010 *Int. J Adv. Manuf. Technol.* **50**
[7] Rao Z H, Zhou J, Liao S M, and Tsai H L Three-Dimensional Modeling of Transport Phenomena and Their Effect on the Formation of Ripples in Gas Metal Arc Welding 2010 *J. Appl. Phys.* **107** 054905
[8] Rao Z H, Liao S M and Tsai H L Effects of Shielding Gas Compositions on Arc Plasma and Metal Transfer in Gas Metal Arc Welding 2010 *J. Appl. Phys.***107** 044902
[9] Haidar J Prediction of Metal Droplet Formation in Gas Metal Arc Welding II 1998 *J. Appl. Phys.* **84**, pp 3530–3540
[10] Fan H G and Kovacevic R A Unified Model of Transport Phenomena in Gas Metal Arc Welding Including Electrode, Arc Plasma and Molten Pool 2004 *J. Phys. D: Appl. Phys.* **37** pp 2531–2544
[11] Hu J, and Tsai H L Heat and Mass Transfer in Gas Metal Arc Welding, Part I: The Arc 2007 *Int. J. Heat and Mass Transfer* **50** pp 833–846
[12] Rao Z H, Hu J, Liao S M and Tsai H L Modeling of the Transport Phenomena in GMAW Using Argon–Helium Mixtures. Part I. The Arc 2010 *Int. J. Heat Mass Transfer* **53** pp 5707–5721
[13] Xu G, Hu J and Tsai H L Modeling Three-Dimensional Plasma Arc in Gas Tungsten Arc Welding 2012 *ASME J. Manuf. Sci. Eng.* **134**(3) p 031001
[14] Xu G, Hu J and Tsai H L Three-Dimensional Modeling of Arc Plasma and Metal Transfer in Gas Metal Arc Welding 2009 *Int. J. Heat Mass Transfer* **52** pp 1709–1724

Acknowledgments

This work was financially supported by the Government of the Perm Krai under the agreement S-26/787 of 12/21/2017 and the Russian Foundation for Basic Research in the framework of project No. 18-08-01016 A.

IOP Conf. Series: Materials Science and Engineering **758** (2020) 012065 doi:10.1088/1757-899X/758/1/012065

Cracking Control in *Mezzanine Floor Slabs* using Rice Husk Ash and Polypropylene Fibers

B Cano[1], J Galarza[1], J Rodríguez[2] and F García[2]

1 Bachelor, Civil Engineering Program, Universidad Peruana de Ciencias Aplicadas, Lima, Perú
2 Full Professor, Civil Engineering Program, Universidad Peruana de Ciencias Aplicadas, Lima, Perú
Email: u201322444@upc.edu.pe
Abstract. The continuous population increase in recent years requires a greater number of households to be built quickly, with good materials and produced under quality standards that guarantee their manufacturing process. The prefabricated concrete, produced and supplied by concrete plants, is poured into the different structural elements, the mezzanine slabs being the most careful surfaces in the appearance of fissures; because being horizontal and having larger dimensions, the dimensional changes in the concrete appear more frequently due to the rapid loss of water from the surface of the concrete before setting; which generates superior stresses to the resistant capacity of the concrete at early ages, which affect the durability and reduce the resistance of the structures, causing greater economic expenses in maintenance and repairs. In the present investigation, 5%, 10% and 15% of rice husk ash was used as a replacement for cement and 900g/m^3 of polypropylene fiber; The results indicate that as the percentage of rice husk ash increases, there is a reduction in the slump and the crack fissures, and that the resistance to compression and flexion decreases, with respect to the concrete pattern.

1. Introduction

House constructions are constantly growing and the most used material that provides flexibility in its design is concrete [1], which after being placed on large surfaces, such as solid slabs, tends to crack due to plastic shrinkage [2]; the magnitude of the cracks depends on the ambient temperature, relative humidity and wind speed [3], producing in the structures the reduction of their serviceability and aesthetics [4]; The appearance of cracks in the concrete reduces its useful life and increases the permeability and the risk of penetration of environmental agents [5]. Rice husk residues are discarded in landfills and cause pollution [6], their ashes are used as partial cement replacement because they improve the quality and durability of concrete [7,8,9]; however, high amounts of replacement can negatively affect its resistance and permeability [10]. On the other hand, polypropylene fibers are used in the reduction of early cracking [5]; in the control of cracks, in the increase of durability and serviceability of the structure [12]. The particle size of rice husk ashes (RHA) is important because the smaller it is, the plastic shrinkage increases [13]; Compressive and flexural strengths are reduced by adding RHA, 10% being an optimal percentage [14,15]. On the other hand, synthetic fibers are used to prevent shrinkage cracking, polypropylene fibers stand out for their efficiency and can be applied in quantities by volume of 0.10% [16]. Fiber length is important in reducing shrinkage, with 19mm fibers better than 13mm fibers for volume replacements from 0.05% to 0.20% [17]. With regard to the compressive and flexural strengths, with the amount of 0.9% of total volume, resistances higher than conventional concrete are achieved [17]. The purpose of this work is to study the effect of RHA on the settlement, plastic shrinkage cracking, compressive strength and flexural strength in order to make a concrete to control the fissures of the mezzanine floor slabs.

2. Materials and Method

2.1 Materials

Andino Ordinary Portland Cement Type I was used; Natural fine aggregate from the "Trapiche" quarry; 1" natural coarse aggregate from the" Jicamarca "quarry; polypropylene (FPP) fibers of L = 12 mm, Ø = 0.012 mm, density 910 kg / m^3 and tensile strength of 480 MPa; rice husk ash (RHA) obtained from uncontrolled combustion in the open air (temperature of 500 - 600 oC) from the city of Chiclayo and sieved by the mesh N°30 (600 μm). The water was drinkable.

2.2 Method
The mix design was made for a concrete f'c = 210 Kg/cm^2 and according to [18]. Table 1 shows the designs of mixtures made for the mixing of standard concrete without inclusion of RHA or FPP. The elaborated mixtures are shown in Table 1 and in Table 2 the tests, standards, number of specimens and curing ages. The test specimens for the compressive strength test were cylindrical of 15 cm x 30 cm, for the flexural strength test with loads to the thirds of the light of 15 cm x 15 cm x 50 cm, which were demoulded at 24 hours and then rehearsed.

Table 1. Mix design for concrete f'c = 210 Kg/cm^2

Mix	OPC (Kg/m^3)	FA (Kg/m^3)	CA (Kg/m^3)	Water (L/m^3)	RHA (Kg/m^3)	FPP (Kg/m^3)
Pattern	366	844	944	224	0	0
RHA-5_FPP	347.7	844	944	224	18.3	0.9
RHA-10_FPP	329.4	844	944	224	36.6	0.9
RHA-15_FPP	311.1	844	944	224	54.9	0.9

Table 2. Tests Performed

Concrete Tests	Methods	N^o Specimens	Cure Age (days)
Slump	[19]	8	---
Plastic shrinkage cracking	[20]	2	1
Compressive strength	[21]	9	7, 14 y 28
Flexural strength	[22]	9	7, 14 y 28

3 Results and Analysis

3.1 Slump
Table 3 shows the effect of the percentage of RHA and fixed amount of FPP in the slump. It can be seen that as the percentage of RHA replacement increases, the slump decreases, by 0.50", 0.75" and 1" in relation to the pattern mixture. [23], for a ratio of w/c = 0.60, the slump decrease values of 0.4" and 0.7" are obtained for 10% and 15% with respect to traditional concrete. Similarly [24], for the same w/c ratio and a replacement of 10% of RHA, the slump decreases by 0.50". These trends are similar to those of the study. [25] explains that this behavior is due to the fact that RHAs have a high specific surface that attracts more water molecules, requiring increasing the water in the mixture to maintain an optimum level of workability.

3.2 Plastic Shrinkage
Table 3 shows the effect of the percentages of RHA and fixed amount of FPP on plastic shrinkage. It is observed that the crack reduction ratio (CRR) increases with the percentage of RHA, reaching the values of 26.67%, 40% and 78.32% for mixtures RHA-5_FPP, RHA-10_FPP and RHA-15_FPP with respect to the pattern sample. [26] perform concrete mixtures with 1% FPP in total volume, finding that after 24 hours the mixtures with FPP increase the CRR by 65% compared to traditional concrete. [27] performs concrete mixtures with a/c = 0.50 and cement replacements of 15% and 25% of fly ash with 0.05% of basalt fibers in total volume, obtaining that after 24 hours the CRR increases 50% in average. These trends are similar to those of the present study. According to [28], this behavior can be

IOP Conf. Series: Materials Science and Engineering **758** (2020) 012065 doi:10.1088/1757-899X/758/1/012065

attributed to the lower cement content compared to control mix, as well as to the pore size and grain size refinement processes which strengthen the mechanical interlocking in the transition zone.

3.3 Compressive Strength

Figure 2 shows the effect of cure time on compressive strength for different percentages of RHA and a fixed amount of FPP. It can be seen that as the curing time increases, the resistance increases for all the mixtures, the pattern mixture being the one that obtains the greatest resistance; reaching for the 28 days the value of 251.33 Kg / cm^2 which represents 9.32%, 18.12% and 28.53% more than the RHA-5_FPP, RHA-10_FPP and RHA-15_FPP mixtures. Similar results after 28 days of curing. [29], that partially replacing the cement with 5%, 10% and 15% of RHA obtains lower resistance values with respect to traditional concrete being these of the order of 25.13%, 30.57% and 31.05%. In the same way, [30] demonstrates for replacements of 10%, 20% and 30% the resistance with RHA, at the age of 28 days, is lower than traditional concrete in 8.88%, 9.96% and 27.20%. These trends are similar to those of the study. [23,30] indicates that this behavior is due to the increase in RHA content, the volume of capillary pores increases and Ca $(OH)_2$ accumulates in the interface, resulting in a less compact structure that causes the resistance to be less; that is, by adding RHA to the concrete, the porosity increases and the density decreases causing a lower resistance to compression.

3.4 Flexural Strength

Figure 3 shows the effect of curing time on flexural strength for different percentages of RHA and a fixed amount of FPP. It is seen that as the curing time increases, the resistance increases for all the mixtures, the pattern mixture being the one that obtains the greatest resistance; reaching for the 28 days the value of 27.40 Kg / cm^2 that represents 17.01%, 18.14% and 28.47% more than the RHA-5_FPP, RHA-10_FPP and RHA-15_FPP mixtures. These results are consistent with [23], which with 10% and 15% RHA for ages 7, 14 and 28 days finds that the flexural strength decreases with respect to the standard concrete by 1.85%, 2.96%, 16.57% and 2.78%, 3.94%, 27.68% for replacements of 10% and 15% respectively. In turn, [31] investigates the addition of RHA in flexural strength using 3 calcination methods (outdoors, in stoves and in ovens) for 5%, 10% and 15% replacement at 28, 90 and 150 days; obtaining that the resistance at 28 days for the 3 calcination methods with replacements of 5%, 10% and 15% RHA, is lower with respect to the control sample in average values of 35%, 48% and 52% . These trends are similar to those of the present study. [32,33] indicate that this behavior is due to the low pozzolanic activity of the RHA.

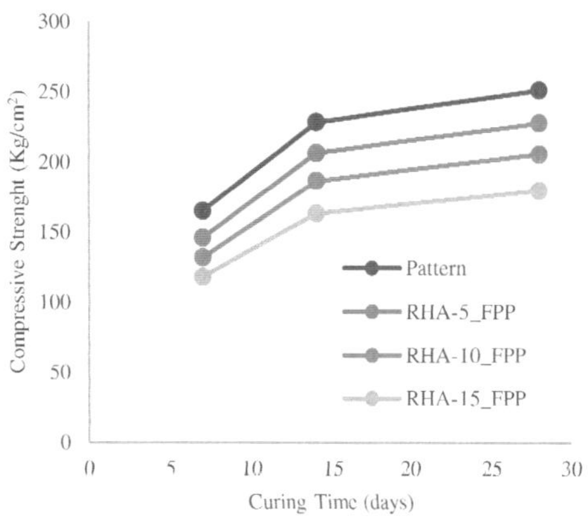

Figure 1. Effect of cure time on compressive strength

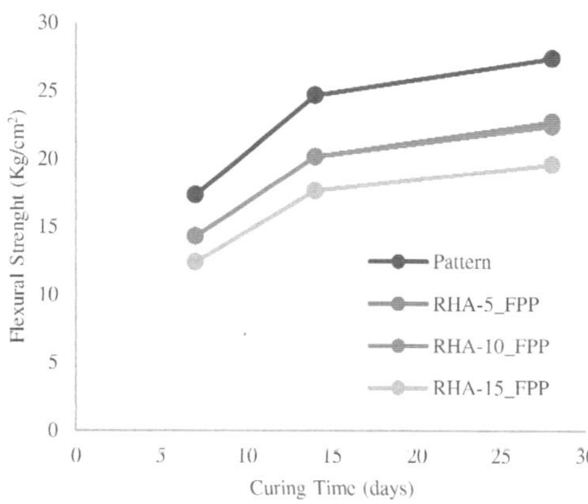

Figure 2. Effect of cure time on flexural strength

4. Conclusions

- The partial replacement of cement by 10% of RHA and 900 g/m^3 of FPP demonstrates that plastic shrinkage cracks decrease in size and width, preventing the entry of external agents that can damage the durability of the structure.
- The workability of the mixture decreases with the additions of RHA and FPP, because the hydrophilic nature of the RHA stimulate the absorption of water from the mixture, and the FPP produce higher friction in the mixture.
- Compressive strength and flexural strength decrease, with RHA particles favorable for fissure reduction and FPP reinforce concrete elements.
- Plastic shrinkage cracking decreases, indicating that the combined incorporation of RHA and FPP helps in reducing the crack fissures.

Table 3. Effect of the percentage of RHA and FPP on slump and CRR

Code	Slump (in.)	# Cracks	Crack Width (mm)		Average Crack Width (mm)	CRR (%)
			Mín.	Max.		
Pattern	1,5"	10	0,1	0,6	0,375	---
RHA-5_FPP	1"	10	0,05	0,5	0,275	26,67
RHA-10_FPP	0,8"	10	0,05	0,3	0,225	40,00
RHA-15_FPP	0,5"	10	0,05	0,1	0,0813	78,32

5. References

[1] Ceballos A 2016 El concreto, material fundamental para la infraestructura Construcción y tecnología en concreto *Revista cyt* 24-25

[2] Samman T, Mirza W and Waffa F 1996 Plastic shrinkage cracking of normal and high-strength concrete: a comparative study *ACI Mater J* **93** 36-40

[3] Shetty M 2006 Concrete technology theory and practice *New Delhi: Chand and Company Ltd*

[4] Wang K, Shah S and Puaksuk P 2001 Plastic shrinkage cracking in concrete materials influence of fly ash and fibres *ACI Mater J* **93** 36-40

[5] Soroushian P, Mirza F and Alhozaimy A 1995 Plastic shrinkage cracking of polypropylene fiber reinforced concrete *ACI Materials Journal* **92** 553–560

[6] Chindaprasirt P and Rukzon S 2008 Strength, porosity and corrosion resistance of ternary blend Portland cement, rice husk ash and fly ash mortar *Construction and Building Materials* **22** (8) 1601-1606

[7] Ganesan K, Rajagopal K and Thangavel K 2008 Rice husk ash blended cement: assessment of optimal level of replacement for strength and permeability properties of concrete *Construction and Building Materials* **22** (8) 1675 - 1683.

[8] Jauberthie R, Rendell F, Tamba S and Cisse I 2000 Origin of the pozzolanic effect of rice husks *Construction and Building Materials* **14** (8) 419-423

[9] Jauberthie R, Rendell F, Tamba S and Cisse I., 2003. Properties of cementerice husk mixture *Construction and Building Materials* **17** (4) 239-243

[10] Chatveera B and Lertwattanaruk P 2009 Evaluation of sulfate resistance of cement mortars containing black rice husk ash *Journal of Env Manag* **90** (3) 1435-1441

[11] Grzybowski M and Shah S 1990 Shrinkage Cracking of Fiber Reinforced Concrete *ACI Materials Journal* **87** (2) 138-148

[12] Naaman A, Wongtanakitcharoen T and Hauser G 2005 Influence of different fibers on plastic shrinkage cracking of concrete *ACI Mater J* **102** (1) 49–58

[13] Habeeb G and Fayyadh M 2009 Rice husk ash concrete: the effect of RHA average particle size on mechanical properties and drying shrinkage *Australian Journal of Basic and Applied Sciences* **3** (3) 1616-1622

[14] Mahmud H, Majuar E, Zain M and Hamid N 2004 Mechanical Properties and Durability of High Strength Concrete Containing Rice Husk Ash Proceedings of the Eighth CANMET *ACI International Conference on Fly Ash, Silica Fume, Slag and Natural Pozzolans in Concrete* 751-765

[15] Isberto C, Labra K, Landicho J and De Jesus R 2019 Optimized Preparation of Rice Husk Ash (RHA) as a Supplementary Cementitious Material *International Journal of GEOMATE* **16** (57) 56–61.

[16] Pelisser F, Neto A, Rovere H and Pinto R 2010 Effect of the addition of synthetic fibers to concrete thin slabs on plastic shrinkage cracking *Construction & Building Materials* **24** (11) 2171–2176

[17] Al R 2017 Influence of polypropylene fibers on the flexural behavior of reinforced concrete slabs with different opening shapes and sizes *Structural Concrete* **18** (6) 986–999

[18] Fernández M 2007 Hormigón *Construcción y tecnología en concreto* Servicio de Publicaciones – Colección Escuelas Madrid Octava Edición

[19] ASTM C143/C 143M-08. Standard Test Method for Slump of Hydraulic-Cement Concrete

[20] ASTM C 1579, Standard Test Method for Evaluating Plastic Shrinkage Cracking of Restrained Fiber Reinforced Concrete (Using a Steel Form Insert)

[21] ASTM C39 / C39M - 18 Standard Test Method for Compressive Strength of Cylindrical Concrete Specimens

[22] ASTM C78 / C78M-15b, Standard Test Method for Flexural Strength of Concrete (Using Simple Beam with Third-Point Loading)

[23] Siddika A, Al Mamun A and Ali H 2018 Study on concrete with rice husk ash *Inn. Infr. Solutions **3**(10) 1-9*

[24] Bheel N D, Meghwar S L, Abbasi S A, Marwari L C, Mugeri J A and Abbasi R A 2018 Effect of rice husk ash and water-cement ratio on strength of concrete *Int. Civil Eng. Journal* **4** 2373-2382

[25] Antiohos S K, Papadakis V G and Tsimas S 2014 Rice husk ash (RHA) effectiveness in cement and concrete as a function of reactive silica and fineness *Cem. Concr. Res. 61–62* 20-27

[26] Li X, Zhao J, Wang W and Jiang A 2006 Effects of Polypropylene Fiber on Plastic Shrinkage Crack and Mechanical Properties of Concrete *Key Eng. Mat.* ***324-325*** 487-490 10.4028/www.scientific.net/KEM.324-325.487

[27] Hemalatha, T and Ramesh G 2019 Mitigation of Plastic Shrinkage in Fly ash Concrete Using Basalt Fibres *Canadian Journal of Civil Eng.* 46. 10.1139/cjce-2018-0075

[28] Fapohunda C, Akinbile B and Shittu A 2017 Structure and properties of mortar and concrete containing rice husk ash as partial replacement of ordinary Portland cement - A review *Int. Journal of Sust. Built Env. (In Press)* 10.1016/j.ijsbe.2017.07.004.

[29] Kachwala A, Pammani A and Raval A 2015 Effect of rice husk ash as a partial replacement of ordinary Portland cement in concrete. *Int. Res. J. Eng. Technol. **2**(5) 175-177*

[30] Khassaf S I, Jasim A T and Mahdi F K 2014 Investigation The Properties Of Concrete Containing Rice Husk Ash To Reduction The Seepage In Canals. *Int. Journal of Sci & Tec. Res.* **3**(4)

[31] Ettu L, Amadip S, Onyezewep E, Awodiji C and Iwup U 2016 Flexural Strengths of Concretes Containing Rice Husk Ash from Different Calcination Methods *Int. Journal of Inn. Sci., Eng.g & Techn. **3**(12) 1-8*

[32] Hossain K M A 2003 Blended Cement Using Volcanic Ash and Pumice. *Cem. and Concr. Res. **33**(10)* 1601-1605

[33] Adesanya D A and Raheem A A 2009 A study of the Workability and Compressive strength characteristics of corn cob ash blended cement concrete. *Constr. and Build. Mat. **23** 311-317*

A regenerative chatter observer analysis for micro-milling

Di Liu[1], Xiaoli Liu[1], Xuming Pei[1], Caidong Wang[1] and Lumin Chen[1]

1 School of Mechanical and Electrical Engineering, Zhengzhou University of Light Industry, Zhengzhou 450002, China
E-mail: liuxiaolizzuli2018@163.com

Abstract. The regenerative chatter in micro-milling that is employed to produce small products with complex 3D geometries and high precision is gradually getting attention from researchers in the field of manufacturing, mechatronics and control for its severe damage to the cutting tool and micro-milling stability. To facilitate developing observer-based control strategy for suppressing the regenerative chatter, the integrating an adopted observer proposed for nonlinear systems with time delay and a practical regenerative chatter model for micro-milling is presented in this paper. And the observer analysis is shown. Simulation results are given to validate the integration scheme.

1. Introduction

It is well-established in literature that the regenerative chatter known as self-excited vibrations in micro-milling can lead to unexpected surface finish, rapid tool wear, and severely unstable cutting process. For increasing chatter-free material removal rate to improve cutting precision, cutting efficiency and cutting quality, various modelling methods and control strategies have been developed for suppressing or controlling regenerative chatter in micro-milling.

There are three important control strategies for suppressing regenerative chatter which include passive chatter suppression techniques that are related to choosing preferred cutting conditions (e.g., spindle speed,depths of cut, and tooth path) and applying dynamic vibration absorbers in cutting process, active chatter control methods that are based on using active actuators and structural optimization for designing optimized machine tool structure forms and cutter geometry (e.g., helix and pitch).

In the pioneering works on chatter suppression in micro-milling, Shakeri and Samani [1] designed two linear and two nonlinear vibration absorbers composed of mass, spring and dashpot elements in two intersecting directions by combing the second-order differential equations for illustrating the dynamics of micro-milling, and the impactions of tool run-out, cutting velocity and micro cutter edge radius. As a result, the effectiveness and the optimized parameters of the absorbers were shown. And it implied that the nonlinear absorbers were inclined to provide more chatter-free zone than the linear ones. Liu et al. [2] completed a study for controlling regenerative chatter in micro-milling by employing a novel simultaneous time-frequency control strategy. And it indicated from the simulations and experimental results that the tool chatter could be negated.To obtain the critical depths of cut and asymptotic spindle speed for regenerative chatter stability lobes in micro-milling with process damping effect, an improved mathematical model was formulated by Wang et al. [3]. According to the proposed mathematical model validated by comparison with prior works and experimental results, unstable chatter can be avoided by specifying the depths of cut and spindle speed for micro-milling. The active vibration control system integrating the novel design of spindle with piezoelectric stack actuators and the dynamic control strategy based on H2 optimal control scheme has been investigated by Aggogeri et al. [4], ensuring an appreciable decrement of the un-stability of chatter in micro-milling. The structural optimization of a 3-

axis miniaturized machine tool for micro-milling was presented by [5], which offered a promising methodology for chatter isolation. The dynamic surface controller with prescribed performance has been derived for controlling regenerative chatter in micro-milling by adopting the active control scheme with piezoelectric actuators as control elements in [6]. And the accomplished proof and simulation results indicated the boundness of all signals of the controlled system.

During the past several years, there have been several formulas developed for demonstrating the dynamics of regenerative chatter in micro-milling. An analytical regenerative chatter model based on velocity and chip load dependent cutting coefficients was established for depicting the regenerative chatter system with one degree of freedom in high-speed micromilling of Ti6Al4V in [7], wherein the experiments were carried out to validate the effectiveness of the established chatter model. Another work [8] on modeling of regenerative chatter in high-speed micromilling of Ti6Al4V incorporated the regenerative chatter system with two degrees-of-freedom and the cutting forces with segmented velocity–chip load dependent cutting coefficients for describing the dynamics of chatter, and employed the experimental results to illustrate the effectiveness of the created model. A characteristic equation for predicating the stability of dynamic chatter system in high speed micromilling of Ti6Al4V was evolved by including the effects of instantaneous chip thickness with dynamic run-out in [9], time delay of regenerative chatter, frequency response functions at tool tip, and immersion dependent directional cutting coefficient matrix. The results pointed out that the proposed model yield reasonably good estimates of the regenerative chatter stability, suggesting it can be extended for the dynamic modeling of chatter. Jin and Altintas [10] presented a practical equation for depicting the dynamic regenerative chatter system in micro-milling with process damping effect depended on the velocity of chatter vibration.

The innovative works introduced above can be utilized to develop dynamic models and control schemes for the regenerative chatter in micro-milling. There are also a number of outstanding research works on the prediction and analysis of chatter stability in micro-milling, facilitating the chatter suppression. While the significant contributions of the works mentioned above focused on the modeling, suppressing, predicting and analyzing of the chatter, developing observers that can be applied to develop controllers for chatter suppression should be payed attention to.

This paper will adopt the observer for nonlinear systems with time delay effect from the work by Ghanes et al. [11] and integrate it with the dynamic regenerative chatter system constructed by Singh et al. [7] for micro-milling, expecting to devote to the perspective of exploiting observer-based control approach for the chatter suppression.

This paper is organized as follows for presenting the integration scheme. The regenerative chatter system demonstration, the observer derivation and the observation error analysis are presented in Section 2. And the simulation results that illustrate the performances of the integration scheme are described in Section 3. The paper is concluded with a brief summary for the presented integration scheme and a perspective for possible observer-based controller developments in Section 4.

2. Observer analysis for regenerative chatter system in micro-milling

2.1 Dynamics depiction of regenerative chatter system

The following mass-spring-damper system in [7] is borrowed to describe the dynamics of regenerative chatter vibration which is assumed to be the micro-tool vibration with the single degree of freedom in Y-direction as shown in Figure 1.

$$\left.\begin{array}{l} \ddot{y}(t)+2\zeta\omega_n\dot{y}(t)+\omega_n^2 y(t)=\frac{\omega_n^2}{k_y}\sum_{j=1}^{N} F_{yj} \\ F_{yj}=F_{tj}\sin\phi_j-F_{rj}\cos\phi_j \\ F_{tj}=K_t a h(\phi_j) \\ F_{rj}=K_r F_{tj} \\ h(\phi_j)=[y(t)-y(t-\tau)]\cos(\phi_j) \\ \phi_j(t)=\Omega t+(j-1)\phi_p \end{array}\right\} \quad (1)$$

where y is the micro-tool vibration in Y-direction with $\dot{y}$ and $\ddot{y}$ as velocity and acceleration, N is the tooth number of the micro-mill, ζ, ω_n and k_y are the damping ratio, natural frequency and stiffness of the chatter system, F_{yj} is the cutting force caused by the jth tooth in Y-direction, F_{tj} and F_{rj} are the tangential force and radial force caused by the jth tooth, K_t and K_r are the tangential and radial cutting force coefficients, ϕ_j is the immersion angle, $h(\phi_j)$ is the dynamic chip load at ϕ_j, a is the axial depth of cut, Ω is the spindle speed, $\phi_p = 2\pi/N$ is the pitch angle of the micro-tool, t is the time parameter, τ is the time delay between two successive cuttings.

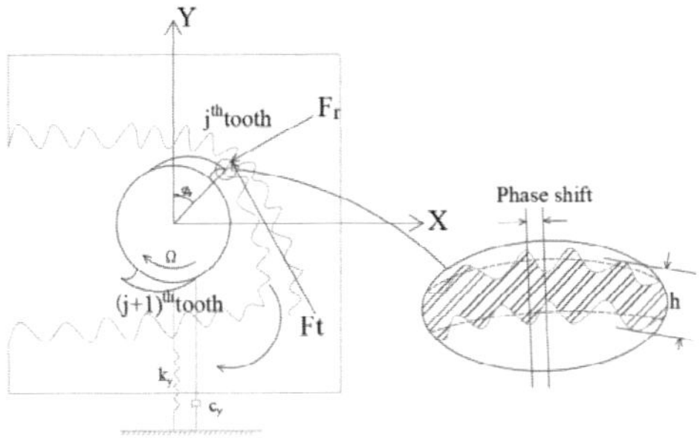

Figure 1. Lumped mass-spring-damper depiction of micro-milling [7]

The system (1) can be rewritten as

$$\Sigma: \begin{cases} \dot{y}(t) = Ay(t) + \Psi(y(t), y_\tau) \\ Y(t) = Cy(t) \end{cases} \tag{2}$$

with $y = \begin{bmatrix} y_1 \\ y_2 \end{bmatrix}$, $y_1 = y$, $y_2 = \dot{y}$, $y_\tau = \begin{bmatrix} y_{1,\tau} \\ y_{2,\tau} \end{bmatrix}$, $y_{1,\tau} = y_1(t-\tau)$, $y_{2,\tau} = y_2(t-\tau)$, $A = \begin{bmatrix} 0 & 1 \\ 0 & 0 \end{bmatrix}$, $C = [1 \quad 0]$, $\Psi(y(t), y_\tau) = \begin{bmatrix} 0 \\ \Psi_2(y_1, y_{1,\tau}, y_2) \end{bmatrix}$, $\Psi_2(y_1, y_{1,\tau}, y_2) = -2\xi w_n y_2 - w_n^2 y_1 + \frac{w_n^2}{k_y}\sum_{j=1}^{N} F_{yj}$.

2.2 Observer design

The observer which is adopted from [11] to estimate the chatter vibration for the system (2) when it is local or globally Lipschitz on $y \in R^n$ with regard to y and y_τ is expressed as

$$\Sigma_A: \begin{cases} \dot{\hat{y}}(t) = A\hat{y}(t) + \Psi(\hat{y}(t), \hat{y}_\tau) - \theta\Delta_\theta^{-1}S^{-1}C^T\{\hat{Y}(t) - Y(t)\} \\ \hat{Y}(t) = C\hat{y}(t) \end{cases} \tag{3}$$

where $\Delta_\theta = \begin{bmatrix} 1 & 0 \\ 0 & \frac{1}{\theta} \end{bmatrix}$ with the design parameter $\theta > 1$, and S is the unique symmetric positive definite solution of the matrix equation

$$S + A^T S + SA - C^T C = 0 \tag{4}$$

The observation error is defined as

$$\varepsilon = \hat{y} - y \tag{5}$$

with

$$\dot{\varepsilon} = \{A - \theta\Delta_\theta^{-1}S^{-1}C^T C\}\varepsilon + \Psi(\hat{y}, \hat{y}_\tau) - \Psi(y, y_\tau) \tag{6}$$

(5) can be transformed into the following equation

$$e = \Delta_\theta \varepsilon \tag{7}$$

Considering the forms of A ,C and Δ_θ ,

$$\left.\begin{array}{l} \Delta_\theta A \Delta_\theta^{-1} = \theta A \\ C = C\Delta_\theta^{-1} \\ \Delta_\theta \theta \Delta_\theta^{-1} = \theta \end{array}\right\} \tag{8}$$

are obtained. Combing (6), (8) and

$$\left.\begin{array}{l} \dot{e} = \Delta_\theta \dot{\varepsilon} \\ \Delta_\theta^{-1} e = \varepsilon \end{array}\right\} \tag{9}$$

yields the following expression

$$\dot{e} = \theta\{A - S^{-1}C^T C\}e + \Delta_\theta \Psi(\hat{y}, \hat{y}_\tau) - \Delta_\theta \Psi(y, y_\tau) \tag{10}$$

The Lyapunov–Krasovskii candidate functional is chosen as

$$V(e) = e^T Se + \int_{t-\tau}^{t} exp^{-\frac{\alpha}{2\tau}(t-\sigma)} e^T(\sigma) e(\sigma) d\sigma \tag{11}$$

where $\alpha > 0$ is a design parameter. Considering the time derivative of (11), the definition of S, (4), and (10), the following equation holds

$$\dot{V}(e) + \frac{\alpha}{2\tau}V(e) = -(\theta - \frac{\alpha}{2\tau})e^T Se - \theta e^T C^T Ce + e^T e - exp^{-\frac{\alpha}{2}} e_\tau^T e_\tau + 2e^T S\Delta_\theta\{\Psi(\hat{y}, \hat{y}_\tau) - \Psi(y, y_\tau)\} \quad (12)$$

with $e(t-\tau) = e_\tau$. According to the following formulas

$$\left.\begin{array}{l} \|\Delta_\theta\{\Psi(\hat{y}, \hat{y}_\tau) - \Psi(y, y_\tau)\}\| \le \kappa\|\Delta_\theta(\hat{y} - y)\| + \kappa\|\Delta_\theta(\hat{y}_\tau - y_\tau)\| \le \kappa\|e\| + \kappa\|e_\tau\| \\ \|(Se)^T\| \le \|e\|\|S\|_2 \le \lambda_2\|e\| \\ 2e^T S\Delta_\theta\{\Psi(\hat{y}, \hat{y}_\tau) - \Psi(y, y_\tau)\} \le 2\lambda_2\kappa\|e\|^2 + 2\lambda_2\kappa\|e\|\|e_\tau\| \\ 2\lambda_2\kappa\|e\|\|e_\tau\| \le \lambda_2\kappa(\|e\|^2 + \|e_\tau\|^2) \\ \lambda_1 e^T e \le e^T Se \le \lambda_2 e^T e \end{array}\right\} \quad (13)$$

where λ_1, λ_2 and $\|S\|_2$ are the minimum, maximum eigenvalues and 2-norm matrix of S, κ is a Lipschitz constant. (12) can be written as

$$\dot{V}(e) + \frac{\alpha}{2\tau}V(e) \le -[\lambda_1\left(\theta - \frac{\alpha}{2\tau}\right) - 1 - 3\lambda_2\kappa]\|e\|^2 - \left[exp^{-\frac{\alpha}{2}} - \lambda_2\kappa\right]\|e_\tau\|^2 \quad (14)$$

with $\left(\theta - \frac{\alpha}{2\tau}\right) > 0$.

Selecting θ and α satisfying $[\lambda_1\left(\theta - \frac{\alpha}{2\tau}\right) - 1 - 3\lambda_2\kappa] > 0$, then

$$\left.\begin{array}{l} \dot{V}(e) + \frac{\alpha}{2\tau}V(e) \le 0 \\ V(e) \le exp^{-\frac{\alpha}{2\tau}t}V(e_0) \end{array}\right\} \quad (15)$$

can be achieved.

2.3 Observation error analysis

Based on Lemma1 in [7], the following inequality can be proved

$$\left.\begin{array}{l} \int_{t-\tau}^{t} exp^{-\frac{\alpha}{2\tau}(t-\sigma)} e^T(\sigma)e(\sigma)d\sigma < \delta_M(\alpha,\tau) \max_{s\in[t-\tau,t]}\|e(s)\|^2 \\ \delta_M(\alpha,\tau) = \frac{2\tau}{\alpha}\left(1 - exp^{-\frac{\alpha}{2}}\right) \end{array}\right\} \quad (16)$$

Then

$$\lambda_1\|e(t)\|^2 \le V(e(t)) \le \delta_M^*(\alpha,\tau) \max_{s\in[t-\tau,t]}\|e(s)\|^2 \quad (17)$$

$$\delta_M^*(\alpha,\tau) = \lambda_2 + \delta_M(\alpha,\tau) \quad (18)$$

are obtained. Consequently, the above inequality and (15) yield

$$\left.\begin{array}{l} \|e(t)\| \le K(\alpha,\tau)exp^{-\frac{\alpha}{4\tau}t} \max_{s\in[-\tau,0]}\|e(s)\| \\ K(\alpha,\tau) = (\frac{\delta_M^*(\alpha,\tau)}{\lambda_1})^{1/2} \end{array}\right\} \quad (19)$$

which can be reformed as

$$\left.\begin{array}{l} \|e(t)\| \le K(\alpha,\tau)\, exp^{-\frac{\alpha}{4\tau}T_0}\,\delta_1 \le \delta_2, \forall t \ge T_0 \\ \max_{s\in[-\tau,0]}\|e(s)\| \le \delta_1 \end{array}\right\} \quad (20)$$

with δ_1 and δ_2 are any possible positive constant. And the inequality

$$\|\varepsilon(t)\| \le \theta\, K(\alpha,\tau)\, exp^{-\frac{\alpha}{4\tau}t} \max_{s\in[-\tau,0]}\|\varepsilon(s)\| \quad (21)$$

can be derived. Then, the observation error is bounded can be guaranteed for the condition when the system (2) is local or globally Lipschitz on $y \in R^n$ with regard to y and y_τ.

3. Numerical simulations

The parameters for depicting the dynamic regenerative chatter are chosen as $N = 2$, tool diameter $T_d = 600\mu m$, $K_t = 40$, $K_r = 28$, $\zeta = 0.07$, $w_n = 240HZ$, $k_y = 0.06N/m$. The condition 1 for the chatter system and the observer system are $\Omega = 5\times10^5 rad/min$, $a = 20\mu m$, $\mathrm{y}(0) = 2\times10^{-6}\mathrm{m}$, $\dot{\mathrm{y}}(0) = 1.5\times10^{-5}\mathrm{m/s}$, $\hat{\mathrm{y}}(0) = 1.9\times10^{-6}\mathrm{m}$, $\dot{\hat{\mathrm{y}}}(0) = 1.5\times10^{-5}\mathrm{m/s}$ and $\theta = 7\times10^6$, for which the simulation results are shown in Figure 2 and Figure 3. The simulations were realized by MATLAB/Simulink.

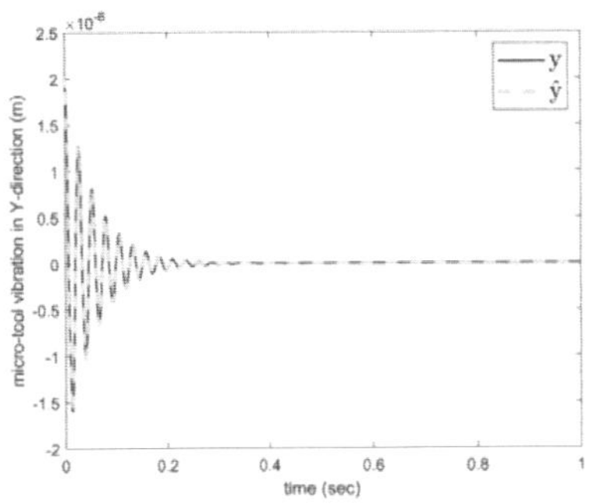

Figure 2. y and ŷ for initial condition 1

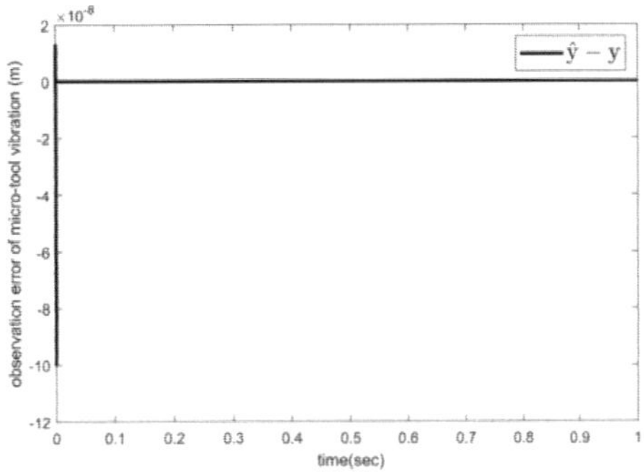

Figure 3. Observation error of micro-tool vibration for initial condition 1

The simulation results are displayed in Figure 4 and Figure 5 for the condition 2, where $a = 150\mu m$, $\Omega = 5 \times 10^3 rad/min$, $y(0) = 2.5 \times 10^{-6}$m and the other parameters are the same as that of the condition 1, and it is assumed that if y is larger than $1 \times 10^{-3} m$ which is the maximum tool vibration allowed during cutting, the cutting procedure will be terminated. For the condition 3 with the same parameters of the condition 2 except for $\hat{y}(0) = 2.5 \times 10^{-6}$m, the simulation result for observation error of micro-tool vibration is presented in Figure 6.

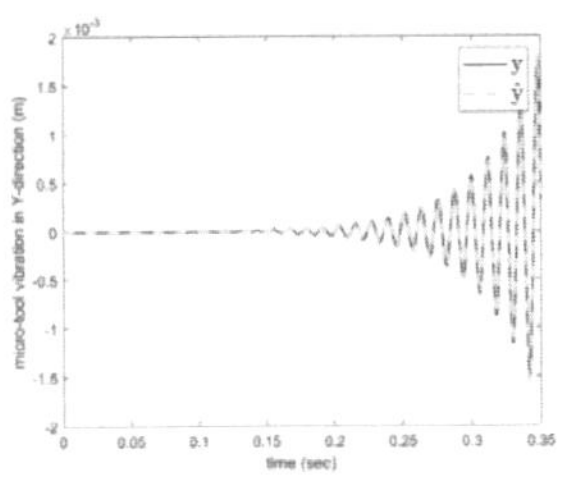

Figure 4. y and ŷ for condition 2

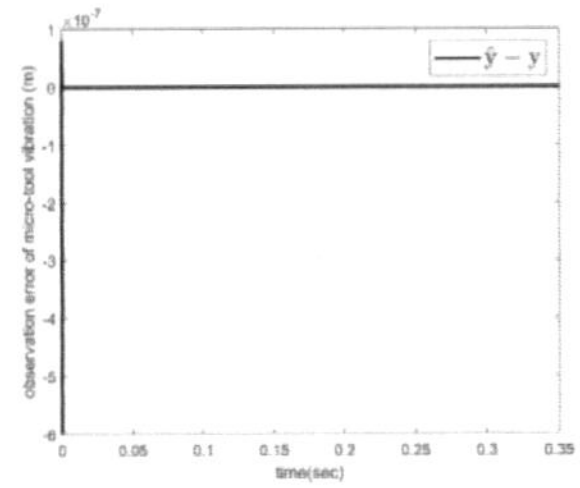

Figure 5. Observation error of micro-tool vibration for condition 2

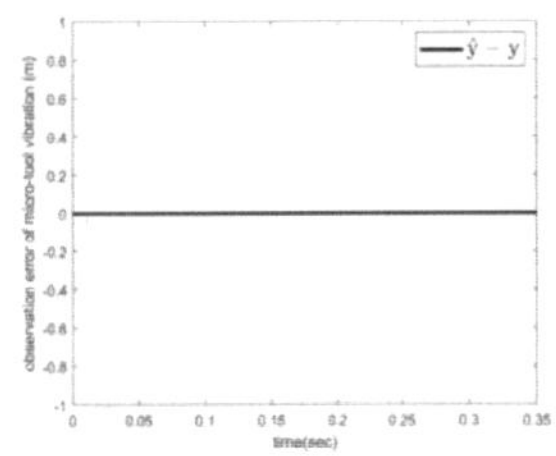

Figure 6. Observation error of micro-tool vibration for condition 3

It can be found from Figure 2- Figure 6 that the observer can lead to good estimate of chatter vibrations for different cutting conditions. Figure 5 and Figure 6 indicate the observation error of micro-tool vibration is affected by the initial value of approximated tool vibration.

4. Conclusions

In this paper, an integration scheme based on an observer for nonlinear systems with time delay and a regenerative chatter model for micro-milling was presented. And the observer analysis was provided. The simulation results showed the designed parameters can be used to estimate the chatter vibration when it satisfies the local or global Lipschitz conditions, showing possible applications in observer-based control strategy for suppressing regenerative chatter in micro-milling.

References

[1] Shakeri S and Samani F S 2017 Application of linear and nonlinear vibration absorbers in micro-milling process in order to suppress regenerative chatter *Nonlinear Dynam.* **89** pp 851–862

[2] Liu M K, Halfmann E B, and Suh C S 2014 Simultaneous time-frequency control of multi-dimensional micro-milling instability *J. Vib. Control* **20** pp 643-660

[3] Wang J J , Uhlmann E , Oberschmidt D, Sung C F and Perfilov I 2016 Critical depth of cut and asymptotic spindle speed for chatter in micro milling with process damping *CIRP Ann. -Manuf. Techn.* **65** pp 113–116

[4] Aggogeri F, Al-Bender F, Brunner B, et al. 2013 Design of piezo-based AVC system for machine tool applications *Mech. Syst. Signal Pr.* **36** pp 53-65

IOP Conf. Series: Materials Science and Engineering **758** (2020) 012066 doi:10.1088/1757-899X/758/1/012066

[5] Chen W, Bai Q, Yang K, Zhang Q, and Liang Y 2013 Dynamics design optimization and experimental validation of a miniaturized machine tool for micro-milling *Prod. Eng. Res. Devel.* **7(5)** pp 477-482
[6] Liu X, Su C Y, and Yang F 2017 FNN approximation-based active dynamic surface control for suppressing chatter in micro-milling with piezo-actuators *IEEE T SYST MAN CY-S* **47(8)** pp 1-14
[7] Singh K K, Kartik V, and Singh R 2015 Modeling dynamic stability in high-speed micromilling of TI-6Al-4 V via velocity and chip load dependent cutting coefficients *Int. J. Mach. Tool. Manu.* **96** pp 56-66
[8] Singh K K , Kartik V , Singh R 2017 Modeling of dynamic instability via segmented cutting coefficients and chatter onset detection in high-speed micromilling of Ti6Al4V *J. Manuf. Sci. E. -T. ASME* **139** pp 051005-1-13
[9] Singh K K, Kartik V, and Singh R 2019 Stability modeling with dynamic run-out in high speed micromilling of Ti6Al4V *Int. J. Mech. Sci.* **150** pp 677-690
[10] Jin X and Altintas Y 2013 Chatter stability model of micro-milling with process damping *J. Manuf. Sci. E. -T. ASME* **135(3)** pp 031011-9
[11] Ghanes M, Leon J D , Barbot J P 2013 Observer design for nonlinear systems under unknown time-varying delays *IEEE T AUTOMAT CONTR* **58(6)** pp1529-1534

Experimental and numerical investigation of cold roll forming process

Mohanraj Murugesana,1, Muhammad Sajjadb,1 and Dong Won Jung*,1
1Department of Mechanical Engineering, Jeju National University, Jeju-Island.
E-mail: jdwcheju@jejunu.ac.kr
Keywords: cold roll forming; tensile test; aluminum alloy; bow effect; adaptive meshing
Abstract. This research work is focused on the cold roll forming process to produce U-channel from the flat sheet. The mechanical properties of aluminum alloy was obtained from the tensile test and subsequently, the properties were incorporated into LS-DYNA to define the material plastic region. The forming process involves the progressive sheet deformation by passing it through a roll pass at constant roll speed. Each sets are designed at different bending angle to increase the deformation until the final geometry is achieved and here, the forming machine with four roll stands were used to manufacture the parts. As a sheet is going into different passes at the same time, the sheet plastic deformation is so complex and highly non-linear because of various stress and strain developments. From experimental results, at low roll speed (RS) the longitudinal bow was observed in the formed parts and to investigate, numerical simulation was modeled at various roll speed and found that as roll speed increasing the longitudinal bow noticed to be decreasing. To confirm the finite element results, one more experiment at high speed was performed and the obtained results showed better agreement with the simulation results. Before confirming the accuracy, many issues were encountered and modified such as mesh size at roller and blank, contact mechanism, choosing element formulation and friction coefficient. Furthermore, the adaptive meshing procedure was adopted to efficiently remesh the sheet at contact locations. The research procedures presented can be utilized to perform the forming process for any existed or new material to form a desired shape for a specific applications in the automobile and aerospace industries.

1. Introduction

The performance of automobile and aerospace parts is becoming the main consideration these days based on the safety regulations. Al6061 aluminum alloy is one of the most common and versatile material for the structural applications in both automobile and aerospace industries owing to its mechanical properties. In specific, its precipitation hardening alloy, containing magnesium and silicon alloy elements in major quantity which exhibits high strength, fatigue resistance, high level weldability, corrosion resistance and oxidation effect, etc. [1]. Al6061 aluminum alloy material is widely used in various industrial structural parts such as manufacturing trucks, railway vehicles and air frames as well as different aircraft parts like wings, tail sections, bulk-head, etc. These day to day applications are evident that long constant cross sectional structural shapes can be produced through the metal forming process like hot forging, stamping and roll forming process. Out of these metal forming process, roll forming process has gained more attention from both academic and industrial side considering its immense production capability on the long constant and variable cross sectional parts. Roll forming process involves a wide range of process that manufacture parts with various cross sections based on its applications. [2–4,5]. Normally, the desired parts are manufactured by passing the sheet metal into multiple roll stages. To perform the mass production, the passes for the suitable design are modeled in prior to control both time and cost consumption economically. However, the roll stage design with proper roll positions is a main problem, such as in the automobile and aerospace industries, because the parts are unique and the product life should be consistent throughout the cycle of usage. So considering the product shapes and quality production leads to put more effort on the flower design in order to control the defects and the shape error in the production unit. Due to modern technology, with the help of numerical tools, the new production methods can be developed to reduce the aforementioned issues and the mass production in terms of a quality production is possible nowadays. In particular, recently, several researchers has been published a great number of articles considering the roll forming process for a specific applications. [6,7–10,11,12,13,14].

Many studies have been carried out on bowing in conventional roll forming and flexible roll forming. Li et al. [14] investigated the performance of advanced high strength steels (AHSS) in chain-die forming process comparing with the conventional roll forming procedure by experiment and simulation. They reported that from the systematic comparison, apart from the bow effects on web and flange, the spring back effect after chain forming is found to be much larger than that of roll forming and requires more effort to achieve a desired part. In a roll formed profiles with variable cross section, park et al. [15] examined the longitudinal strain distribution in the concave and straight zones with the help of forming parameters of the incremental counter forming (ICF) process (an idle roll) during the roll forming. As a results, the shape error was corrected by reducing the longitudinal strain at flange and web sections using the ICF process. The twist effect in the asymmetrical channel sections made of the mild steel (st12) material with different flange lengths was studied by Tajik et al. [16] and found that the reason for the twist is due to the larger forming force on the long flange than the short flange. In addition, they concluded that the twist effect can be reduced by modeling asymmetrical forming rolls to form both flanges more symmetrically. In the flexible roll forming process, the effects of process variables on the longitudinal strain and longitudinal bow were examined both experimentally and numerically for U-sections made up of three different sheet materials and blank shapes such as trapezoid, convex and concave by woo et al. [17] After the detailed investigation, they introduced leveling roll into the forming process and as an outcome, the longitudinal bow was reduced with the use of three blank shapes and the leveling roll in the forming process. Badr et al. [18] investigated the effect of forming method on part shape quality in high strength Ti-6Al-4V sheet and reported that both experimental and numerical results suggest that spring back is lower in the constant radius forming method than that of the constant arc length procedure. In addition, they stated that the major shape defects in roll forming such as bow and spring back are not only influenced by material properties, the forming sequence and inter-stand distances and also due to the forming strategy.

The aim of the present work is to investigate the cold roll forming process to produce U-channel from the flat sheet by both experiment and simulation. The room temperature tensile test was performed to achieve the mechanical properties of aluminum alloy material. A roll forming process to obtain the U-channel was established and implemented into the finite element model using LS-DYNA tool. The longitudinal bow was measured from the experimental formed parts and compared with the numerical results. After the model verification, the evolution of longitudinal strain at lower and upper surfaces are measured at the bending section and discussed here. In addition, numerous simulations were performed to discuss the bow effect at different speeds and are experimentally compared using the verified numerical model.

2. Cold roll forming procedures

The chemical composition of Al6061 aluminum alloy, used in this research, is summarized in Table 1. In order to obtain the mechanical material properties, uniaxial tensile tests were conducted on Al6061 aluminum alloy sheet samples with the thickness of 2mm. Before modeling the experiment, the specimen samples were prepared with respect to the rolling direction based on the Korean Industrial Standard (KSB0802). The experiments were carried out using UH-FX Series Hydraulic Universal Tensile Testing (HUTT) machine equipped by an extensometer with the gauge length of 50 mm. The load–displacement curves were obtained from this test, and subsequently, the standard tensile test equations were employed to compute the stress–strain curves from the measurements and the samples used in the experiment are displayed in Fig.1(a).

Table 1 . Chemical composition of Al6061 alloy (in wt. %)

Al	Si	Fe	Cu	Mn	Mg	Cr	Zn	Ti
97.8	0.4-0.8	Max 0.7	0.15-0.40	Max 0.15	0.8-1.2	0.04-0.35	Max 0.25	Max 0.15

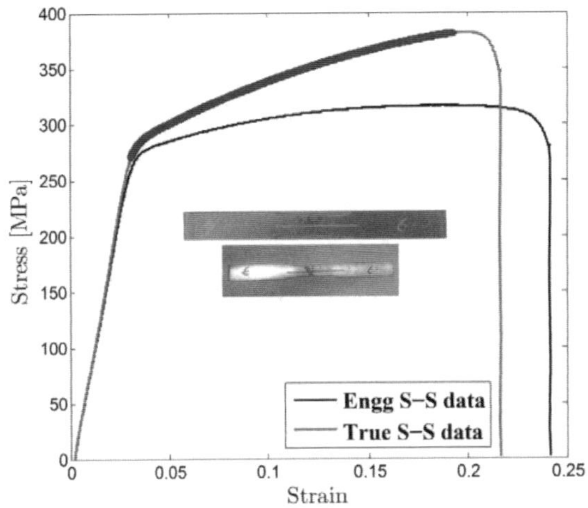

(a) Stress-strain curve

(b) Roll forming test setup

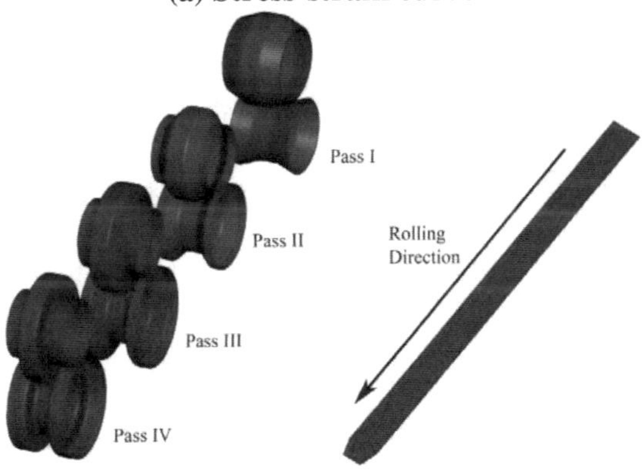

(c) Finite element model

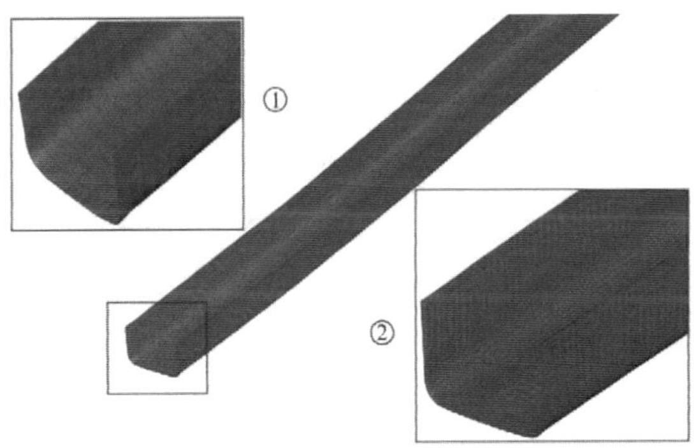

(d) Mesh methods (1) contact mesh (2) adaptive mesh

Figure 1 Cold roll forming procedures.

Table 2 Mechanical properties of Al6061 aluminum alloy

Density(kg m-3)	σ_u(MPa)	σ_y(MPa)	E(GPa)	υ	Elongation (%)
2710	340	271	68.6	0.33	19

The cold roll forming experimental set-up utilized in this research work is illustrated in Fig.1(b). The roll forming machine is consists of four stages and one guide in order to achieve the desired U-channel section in the final stage output. In detail, the roll stages were designed to produce different bending

IOP Conf. Series: Materials Science and Engineering **758** (2020) 012067 doi:10.1088/1757-899X/758/1/012067

angle at each stages and the bending angles were 22.5°, 45°, 67.5° and 90°, respectively. In this study, Al6061 aluminum alloy was adopted as a testing material and the material properties are summarized in Table.2. The Al6061 aluminum alloy specimen dimensions are: 700 mm length, 60 mm width and 2mm thickness and the each forming stage horizontal distance is 220 mm. During the forming process, the aluminum sheet was manually guided into the initial roll with the help of guide placed in front of the first roll stage as shown and highlighted in Fig.1(b). At first, the sheets were tested at low forming speed and from the formed product, the shape error was identified due to longitudinal bow effect and flange camber effect as depicted in Fig.2. To confirm this results, multiple specimens were tested under the same forming conditions and an outcome on the shape quality was same as aforementioned. In order to check the reason of this issues, the detailed study was conducted by examining the formed product, the defect positions and the roll positions such as horizontal and vertical roll distance, center-line alignment and roll gap. From the study, it was found that during the low speed forming, the sheet material takes so much time stay in the each roll stand before it move into other stages. Because of the slow forming time, the forming sheet was experiencing too much vertical roll forming force in the longitudinal direction and it causes the longitudinal bow phenomenon on the formed product. So, in the next step, the roll speed was increased and the same procedures were carried out to produce the U-channel sections. From the formed product, it was found that the longitudinal bow got reduced as depicted in Fig.2(c).

(a) Tested specimens

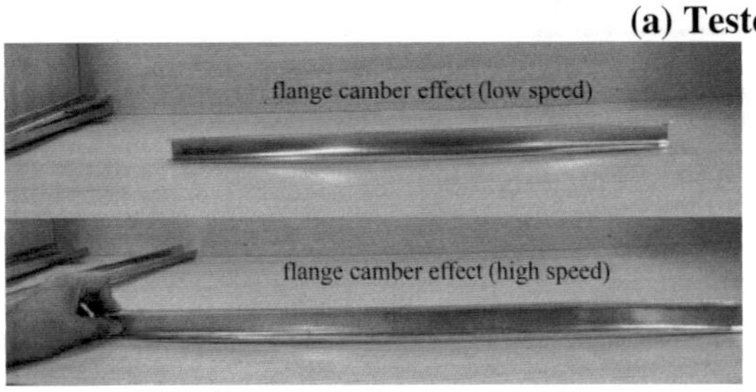

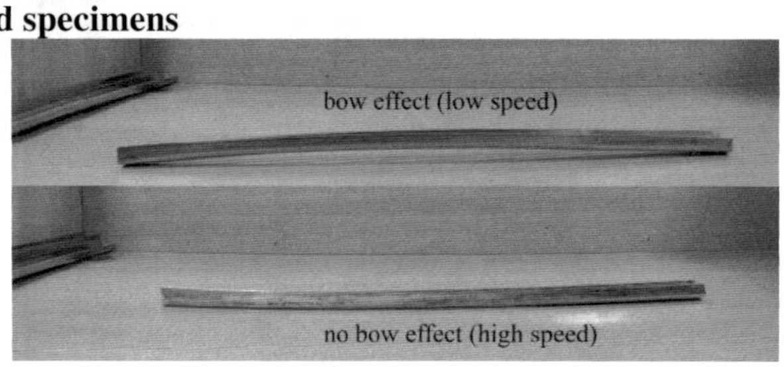

(b) Flange camber comparison **(c) Longitudinal bow comparison**

Figure 2 Manufactured U-channel parts.

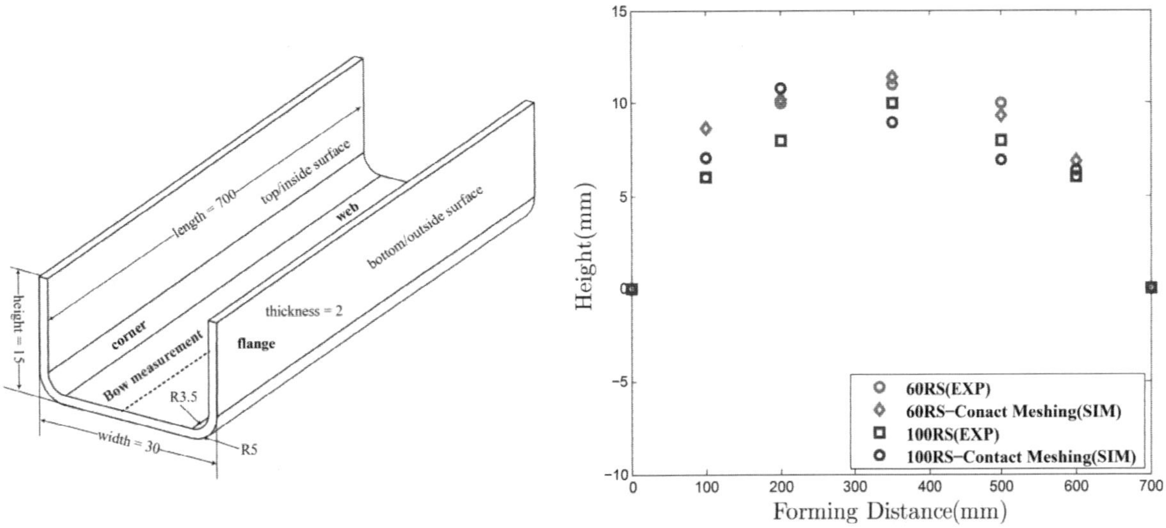

(a) CAD geometry with specifications **(b) Comparison of bow measurement**

Figure 3 Verification results between experiment and simulation

The numerical simulation of the cold roll forming process was modeled in LS-DYNA commercial tool as shown in Fig.1(c). Due to rigid boundary conditions on the rollers, each and every roll in the stand was defined by using a rigid material properties and the rigid boundary conditions (constrained displacements in all three directions, rotations constrained in *y* and *z* directions and no constraint in *x* direction) are applied in the global directions coordinated system. The same sheet specimen dimensions used in the experiment were adopted to develop the finite element (FE) blank model with Al6061 aluminum alloy material. In order to include the plastic deformation, the material properties and the power law equation parameters computed from the tensile test were incorporated into the numerical simulation through the material card (MAT018 MAT POWER LAW PLASTICITY). From literature survey, it was identified that roll forming is a pure bending method, so shell elements are well suited for pure bending method to effectively solve the cold roll forming problem. In element formulation, elform type 16 with 5 integration points was chosen as it can be useful when we consider both forming and spring back problem together. Also stiffness-based Hourglass control card with type 4 was used to control the unwanted deformation as it is generally more effective for structural parts when we consider shell elements. In addition, two kind of meshing methods such as contact area meshing and adaptive meshing procedure were adopted to check the accuracy of finite element solution compared with the experimental measurements as shown in Fig.1(d).

The numerical simulation results are useful to evaluate the deformation process and to find the high strain and stress regions to discuss about the problem in the forming process and further it can be modified by adjusting forming parameters such as bending sequence, roll speed, roll positions and roll passes distance, etc. After running the simulation at roll speed 60mm/sec, the numerical model was checked by comparing with the experimental ones. As seen in Fig.3(b), it was observed that the results found to be adequate with experimental measurements in terms of longitudinal bow measurements and the bow measurement place is highlighted in Fig.3(a). This results comparison proves that the numerical model developed can be further utilized to run for other roll speed in order to investigate the bow effect and also the model can be used to discuss the presence of stress and strain in the formed part. Numerous FE simulation were modeled considering different roll speed and the results from 100mm/sec roll speed was consider for the discussion eventually. Numerically it was found that the bow tends to be reduced and to confirm this, one more experiment was conducted. From observations, the results showed better agreement with the experimental data as illustrated in Fig.3(b).

IOP Conf. Series: Materials Science and Engineering **758** (2020) 012067 doi:10.1088/1757-899X/758/1/012067

The effective strain distributions were measured at front bending section along the rolling direction and the maximum effective plastic strains were found to be occurred at the bending section as shown in Fig.4(a). Furthermore, the effective plastic strain at each stand exit, Fig.4(b), is increasing whenever the material is actively reaches every roll stand and also noticed to be constant when it stays at roll contact of particular stand and this behavior is same at each and every stand. When the forming speed is high the strain also tends to be increasing but showed the same behavior as low forming speed as illustrated in Fig.4(b). In addition, plastic strains at lower and upper sides of the formed part is measured and depicted in Fig.4(c). Fig.4(c) explains that as lower surface experiences tension force, the strains at that region found to be positive and steady till the second roll stand and showed exponential growth at the last two roll stand. In a similar way, upper surface experiences compressive force and results in compressive strains in terms of negative sign and the behavior found to be symmetric.

Overall the parts can be manufactured to the desired shape with superior quality by means of proper FE modeling techniques, on the contrary it needs more computational time effort.

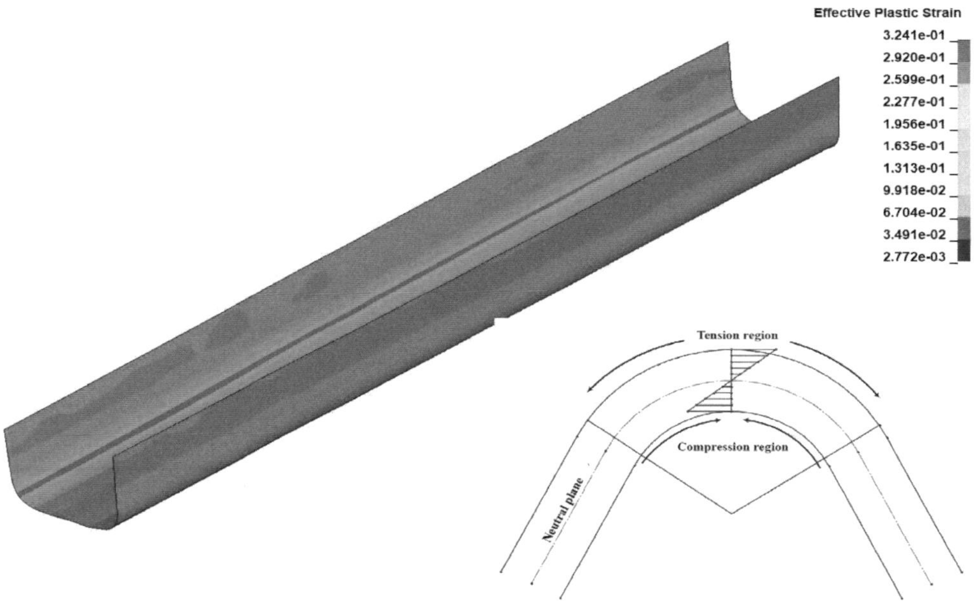

(a) Effective plastic strain at 100RS in U-channel formed part (post cut)

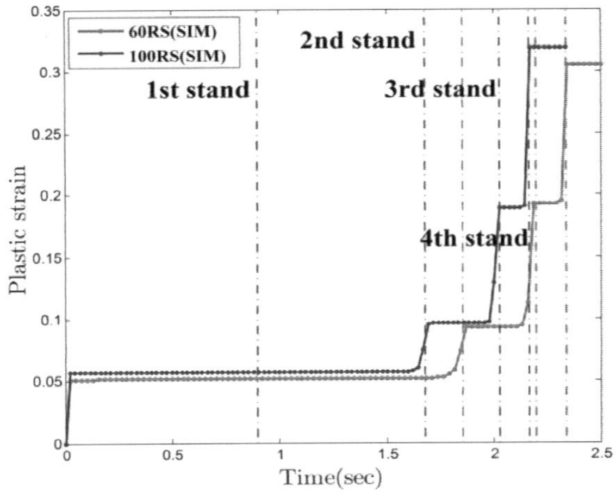

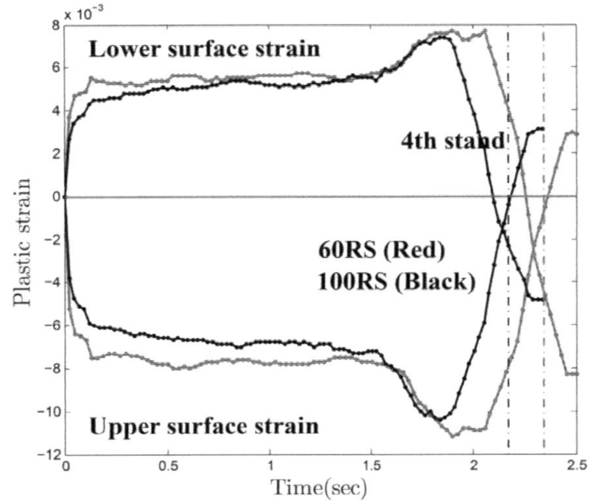

(b) Longitudinal strain at formed part **(c) Strain at upper and lower surface**

Figure 4 Numerical results about plastic strains in formed part at different speeds

3. Conclusions

The cold roll forming process using both experiment and numerical simulations were carried out to evaluate the forming process.

- Flower pattern for four stages of roll forming was developed considering U-channel section and further the dimensions were utilized to design the rollers.
- More than two samples were tested in the roll forming machine until get the high quality product and on the other hand, error in the roll positions and the roll gaps were rectified during the forming process.
- The tensile test at room temperature was conducted to obtain the mechanical properties and later incorporated into FE tool to define the material plastic region.
- The geometry of the U-channel roll formed product was predicted successfully using FEM simulation.
- Good results correlation in terms of longitudinal bow measurements was achieved between numerical simulation and experimental observations.
- Moreover, it was found that the longitudinal bow can be minimized at high roll speed and it was proved using the experimental measurements eventually.
- The presence of longitudinal strains in terms of each stand, lower and upper surfaces were discussed here.
- Overall, for numerical simulation, mesh should be finer at bend zones and can be coarse in the straight parts, where there will be no deformation, and shell elements are recommend.
- Further studies will be concentrating on investigating the formability in terms of the modified bending sequence to check the model improvement, springback and failure analysis in order to decrease the formability effects such as bow, edge waves and flange camber to increase the parts quality for the industrial applications.

References

[1] Marnette, J., Weiss, M., Hodgson, P. D. Roll-formability of cryo-rolled ultrafine aluminium sheet. Materials and Design 2014, 63, 471–478.
[2] Kim, D., Cha, M. H., Kang, Y. S. Development of the Bus Frame by Flexible Roll Forming. Procedia Engineering
2017, 183, 11–16.
[3] Sweeney K., Grunewald, U. The application of roll forming for automotive structural parts. Journal of Materials Processing Technology 2003, 132, 9–15.
[4] Park, J. C., Yang, D. Y., Cha, M. H., Kim, D. G., Namb, J. B. Investigation of a new incremental counter forming in flexible roll forming to manufacture accurate profiles with variable cross-sections. International Journal of Machine Tools and Manufacture 2014, 86, 68–80.
[5] Crutzen, Y., Boman, R., Boman,L., Ponthot, J. P. Continuous roll forming including in-line welding and post-cut within an ALE formalism. Finite Elements in Analysis and Design 2018, 143, 11–31.
[6] Espositoa, L., Bertoccoa, A., Sepea, R., Armentania, E. 3D strip model for continuous roll-forming process simulation. Procedia Structural Integrity 2018, 12, 370–379.
[7] YongSun, Y. L., Xiao, H., Bong, H. J., Shi, L., Li, S., Li, D., Ding, S., Wagoner, R. H. A numerical study on chain-die forming of the AHSS U-channel and contrast with roll formin. International Journal of Mechanical Sciences 2018, 135, 279–293.
[8] Kim, D., Cha, M., Kang, Y. S. Development of the Bus Frame by Flexible Roll Forming. Procedia Engineering
2017, 183, 11–16.
[9] Sun, Y., Luzin, V., Daniel, W. J. T., Meehan, P. A., Zhang, M., Ding, S. Development of the slope cutting method for determining the residual stresses in roll formed products. Measurement 2017, 100, 26–35.
[10] Badra, O. M., Rolfeb, B., Weissa, M. Effect of the forming method on part shape quality in cold rollforming high strength Ti-6Al-4V sheet. Journal of Manufacturing Processes 2018, 32, 513–521.
[11] Woo, Y., Kang, P., Oh, I., Moon, Y. Flexible roll forming of double layered blank. Procedia Manufacturing 2018,
15, 775–781.
[12] Chen, W., Jiang, J., Li, D., Zou, T., Peng, Y. Flower pattern and roll positioning design for the cage roll forming process of ERW pipes. Journal of Materials Processing Tech 2019, 264, 295–312.
[13] Ma, Z., Luo, Y., Wang. Y., Mao, J. Geometric design of the rolling tool for gear roll-forming process with axialinfeed. Journal of Materials Processing Tech 2018, 258, 67–79.
[14] Liu, X., Cao, J., Chai, X., Liu, J., Zhao, R., Kong, N. Investigation of forming parameters on springback for ultra-high strength steel considering Young's modulus variation in cold roll forming. Journal of Manufacturing Processes 2017, 29, 289–297.
[15] Park, J. C., Yang, D. Y., Cha, M. H., Kim, D. G., Namb, J. B. Investigation of a new incremental counter forming in flexible roll forming to manufacture accurate profiles with variable cross-sections. International Journal of Machine Tools and Manufacture 2014, 86, 68–80.
[16] Tajik, Y. T., Naeini, H. M., Tafti, R. A., Bidabadi, B. S. Constitutive Relationship Modeling and Characterization of Flow Behavior under Hot Working for Fe–Cr–Ni–W–Cu–Co Super-Austenitic Stainless Steel. Thin-Walled Structures 2018, 130, 395–404.
[17] Woo, Y. Y., Han, S. W., Hwang, T. W., Park, J. Y., Moon, Y. H. Characterization of the longitudinal bow during flexible roll forming of steel sheets. Journal of Materials Processing Technology 2018, 252, 782–794.
[18] Badra, O. M., Rolfeb, B., Weissa, M. Effect of the forming method on part shape quality in cold rollforming high strength Ti-6Al-4V sheet. Journal of Manufacturing Processes 2018, 32, 513–521.

FM analysis of aging properties of UV531/SBS modified bitumen

FENG Zhengang [1,2], CAI Fengjie[2], YAO Dongdong[3], LI Xinjun[2]

[1] Guangxi Key Lab of Road Structure and Materials, Nan'ning, 530007
[2] School of Highway, Chang'an University, Xi'an, 710064
[3] Jilin Provincial Transport Scientific Research Institute, Changchun, 130012
Corresponding author: FENG Zhengang
E-mail: zgfeng@chd.edu.cn

Abstract. Ultraviolet absorber UV531 and linear SBS modified bitumen were selected to prepare SBS modified bitumen via melt blending. Thermal-oxidative and ultraviolet (UV) aging of SBS modified bitumen were simulated by the film oven test (TFOT) and accelerated UV aging. The influence of UV531 on the micromorphology of SBS modified bitumen before and after aging was analyzed by fluorescent microscopy (FM). Combined with the three indicators and aging properties of modified bitumen, the relationship between microstructure and properties of SBS modified bitumen and UV531/SBS modified bitumen was discussed. The FM images show that the UV531 improves the thermal-oxidative aging property of SBS modified bitumen, but it has little effect on UV aging.

1. Introduction

SBS modified bitumen is more and more widely used in the construction of high-grade highway in the world. Some road agencies and petrochemical departments have invested a lot of manpower in the development, performance evaluation and application of SBS modified bitumen [1,2]. As for the problem that the service life of SBS modified bitumen is easily affected by aging, researchers have carried out a lot of anti-aging studies for SBS modified bitumen [3,4]. Adding anti-aging agents, such as antioxidant, layered silicate, nanometer material and ultraviolet absorber to SBS modified bitumen is the most widely used way to improve its aging resistance [5-7]. The results show that SBS modified bitumen with appropriate dosage of anti-aging agent possesses better aging resistance to heat, oxygen or ultraviolet (UV) light.

As a highly effective aging resistant agent, UV531 has been used in the modification of SBS modified bitumen. It is reported that UV531 can effectively improve the thermal-oxidative and UV resistant aging properties of SBS modified bitumen through conventional test and rheological property test method [8]

The physical properties and aging resistance of UV531/SBS modified bitumen must relate to its structural characteristics, so the microstructure of UV531/SBS modified bitumen needs to be analyzed to reveal action mechanism of UV531 on SBS modified bitumen. So far, the methods to analyze the microstructure of SBS modified bitumen include scanning electron microscopy (SEM), atomic force microscope (AFM), fluorescence microscope (FM) and other digital image processing technology [9,10]. Among these methods, the fluorescence microscope is the quickest, most acceptable and convenient method to analyze morphology of SBS modified bitumen, because it can observe the homogeneity and structure of SBS modified bitumen in the raw state, and investigate the morphology of the SBS modified bitumen by determining the dispersion state of bitumen and SBS modifier, and characterize the nature of the continuous and discontinuous phase [11]. Even though, the compatibility and the action mechanism of UV531 on SBS modified bitumen is still not very clear at present due to the extremely complicated chemical composition and structure of SBS modified bitumen.

In view of these, this paper prepared UV531/SBS modified bitumen via melt blending with base bitumen, SBS, and UV531. The thermal-oxidative and UV aging processes of UV531/SBS modified bitumen were simulated by the thin film oven test TFOT) and accelerated UV aging test. FM was used to evaluate the morphology and distribution of UV531/SBS modified bitumen before and after aging by assessing the state of dispersion of SBS modifier and bitumen phases. Combining with the fundamental properties (penetration, ductility, softening point) of UV531/SBS modified bitumen by using conventional test methods, the effect of UV531 on the micromorphology of SBS modified bitumen before and after aging was confirmed.

2. Materials and Methods

2.1. Materials

The bitumen, AH-70, was provided by Panjin Northern Bitumen Co., Ltd., China. Physical properties of the bitumen are shown in Table 1.

Table 1. Physical properties of the AH-70

Bitumen	Penetration (25°C，0.1mm)	Ductility (10°C，cm)	Softening point (°C)
AH-70	64	60.1	47.4

SBS modifier, linear SBS 1401, was supported by Hunan Yueyang Baling Petrochemical Co., Ltd. The physical properties are shown in Table 2.

Table 2. Physical properties of the SBS

SBS	300% constant elongation stress (MPa)	Tensile strength (MPa)	Elongation at break (%)	Hardness (HA)	Permanent set at break (%)	Melt mass flow rate (g/10 min)
1401	⩾3.0	⩾20.0	⩾680	⩾85	⩽60	0.10~5.00

The UV absorber UV531 was supplied by Nanjing Milan Chemical Co., Ltd., China. The physical properties of the UV531 are shown in Table 3.

Table 3. Physical properties of the UV531

UV absorber	Appearance	Melting point（°C）	Light transmittance（%）	
			450nm	500nm
UV531	Light yellow needle-like powder	48	90	≥95

2.2. Preparation of UV531/SBS modified bitumen

The UV531/SBS modified bitumen was prepared via melt blending using a high shear mixer. To begin with, the AH-70 matrix bitumen was heated to 180°C. Then the 4wt% SBS modifier and 0.1wt% sulfur were added into matrix bitumen, and the mixture was blended at 4000rpm for 1h. Then, high shear mixer was used to shear for 1h at 4000rpm. It was further stirred by mechanical mixer for 1.5h to complete the preparation of SBS modified bitumen. After that UV531 with a dose of 0.6% by mass of SBS modified bitumen was added into SBS modified bitumen slowly. The temperature was maintained at 180°C, and the high-speed shearing machine was kept at 2500rpm for 0.5h to obtain the UV531/SBS modified bitumen. The SBS modified bitumen without UV531 was treated by the same preparation method to obtain the control sample.

2.3. Aging experiment

According to ASTM D1754, the thin film oven test (TFOT) was carried out in type 82 bitumen film oven. The well-mixed bitumen weighed (50±0.5) g was placed in iron pan with inner diameter of 140mm and depth of 9.5mm, and the thickness of bitumen film was about 3.2mm. Then, the samples were heated in the oven for 5h with temperature at (163±1) °C.

IOP Conf. Series: Materials Science and Engineering **758** (2020) 012068 doi:10.1088/1757-899X/758/1/012068

The UV aging test was performed in a chamber (50×60×50 cm) with an UV lamp of 300W, and the wavelength of UV radiation was 280～400nm. The UV intensity of the UV lamp was (25±0.4) W/m2. The oven was equipped with digital automatic temperature-control system. The UV aging test process was as follows: the TFOT aged bitumen samples were placed in the ultraviolet aging oven to continue the test. The aging temperature was kept at 60°C and the aging time was 6 days.

2.4 Fluorescent microscopy (FM)

The change of shape, size, quantity and distribution of SBS modifier in the blending and aging processes of SBS modified bitumen was tested by fluorescence microscopy. SBS modified bitumen samples were examined at room temperature under the microscope with fluorescent light (generated from a high pressure Xenon lamp) at magnification levels of 400×. A drop of hot bitumen was placed on the glass microscope slide to form a transparent film. Then a coverslip was covered on the film to observe the phase morphology under fluorescent microscopy.

2.5 Physical properties test

The penetration (25°C), ductilty (5°C) and softening point of the SBS modified bitumen before and after aging were tested according to the standards ASTM D5, ASTM D113 and ASTM D36, respectively.

3. Results and Discussion

3.1. Effect of UV531 on morphology of SBS modified bitumen before aging

The bitumen material will not appear fluorescence under the ultraviolet light emitted by the fluorescence microscope, while SBS modifier can emit fluorescence under the ultraviolet light. Therefore, the appearance and phase state of SBS modifier in the bitumen can be visually observed by using the fluorescence microscopy.

The microstructure of SBS and UV531/LSMB before aging is shown in figure 1 and figure 2. The small particles are evenly distributed, the bitumen is continuous phase and presented as black, the modifier is dispersed phase and presented as fluorescent yellow, and the SBS modifier is evenly dispersed in the modified bitumen. The figure shows that the addition of UV531 makes the SBS modifier particles in LSMB smaller and the number increased, and the contact area between the modifier and bitumen increased, indicating that UV531 changes the dispersion state of SBS polymer in bitumen.

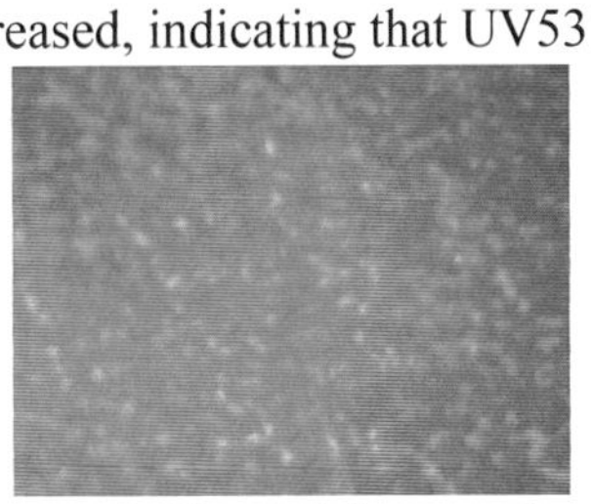

Figure 1. The FM image of unaged SBS modified bitumen

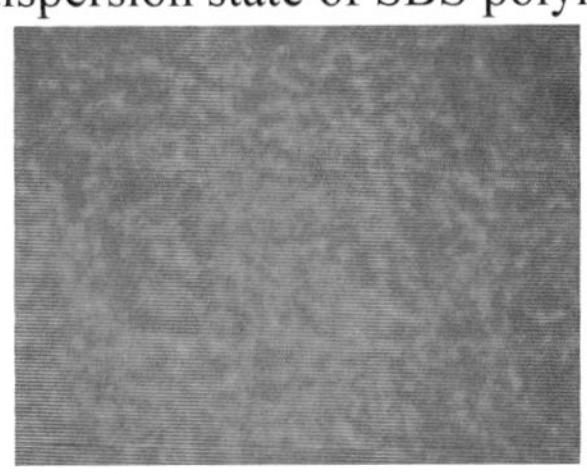

Figure 2. The FM image of unaged UV531/ SBS modified bitumen

As shown in Table 4, the addition of UV531 increased the penetration and elongation of SBS modified asphalt, while the softening point decreased. The increase of penetration degree of SBS modified asphalt indicated that the addition of UV531 reduced the viscosity of SBS modified asphalt, and the increase of its elongation indicated that UV531 could improve the low temperature ductility of SBS modified asphalt and improve the low temperature performance of SBS modified asphalt. The addition of UV531 had an adverse effect on the high-temperature performance of SBS modified asphalt, but the effect was small. In general, the addition of UV531 improved the physical properties of SBS modified asphalt, which was consistent with the conclusion obtained by fluorescence microscope.

Table 4. Physical properties of the SBS modified bitumen

Content of UV531(%)	Penetration (25°C，0.1mm)	Ductility (5°C，cm)	Softening point (°C)
0	46	19.0	53.1

0.6	56	19.5	50.0

3.2. Effect of UV531 on morphology of SBS modified bitumen after TFOT aging

The microstructure of TFOT aged samples is shown in figure 1 and figure 2. After TFOT aging, SBS modifiers in SBS modified bitumen and UV531/ SBS modified bitumen agglomerate, crosslink and degrade, the number of SBS modifiers decreases and the size of particles becomes larger. In the two states, the bitumens are both continuous phases, and the SBS modifiers are both dispersed phases. However, the particles of SBS modifier increase after UV531 is added to the modified bitumen, which indicates that UV531 prevents the degradation of SBS modifier and improves the performance of SBS modified bitumen. It implies that UV531 has a promoting effect on the heat and oxygen aging resistance of SBS modified bitumen.

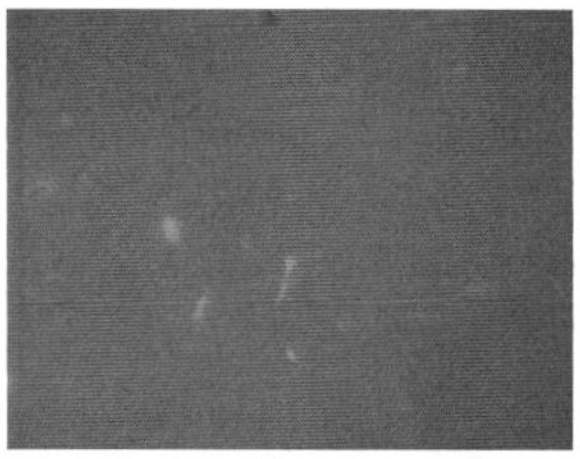

Figure 3. The FM image of TFOT aged SBS modified bitumen

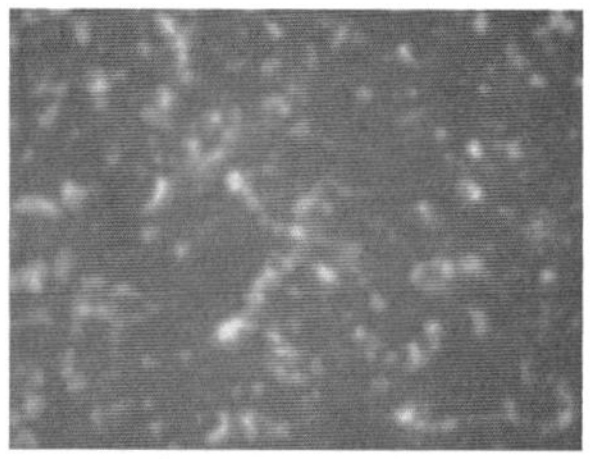

Figure 4. The FM image of TFOT aged UV531/ SBS modified bitumen

Combined with residual penetration ratio (PRR) and softening point increment (△S), the influence of UV531 on aging performance of SBS modified bitumen is verified and calculated. PRR and △S can reflect the susceptive degree of aging. The formulas are as (1)~(2)

$$\text{PRR} = \frac{\textit{Aged penetration value}}{\textit{Unaged penetration value}}\, x\ 100 \quad (1)$$

$$\triangle S = \textit{Aged softening point value} - \textit{Unaged softening point value} \quad (2)$$

As illustrated in table 5, after TFOT aging, the PRR of modified bitumen with UV531 increases and the SPI decreases compared with the modified bitumen without UV531. It demonstrates that the addition of UV531 improves the heat aging performance of SBS modified bitumen, which is consistent with the results obtained by FM.

Table 5. The influence of UV531 on TFOT aging performance of SBS modified bitumen

Content of UV531（%）	PRR	△S（℃）
0	85	7.8
0.6	90	4.8

3.3. Effect of UV531 on morphology of SBS modified bitumen after UV aging

In figure 5 and figure 6, there is almost no SBS modifier in the fluorescence micrograph after UV aging. The reason lies in that the aging of modified bitumen dissolves the insoluble polybutadiene segment in SBS modifier into matrix bitumen, thus showing the disappearance of polymer particles.

Figure 5. The FM image of UV aged SBS modified bitumen

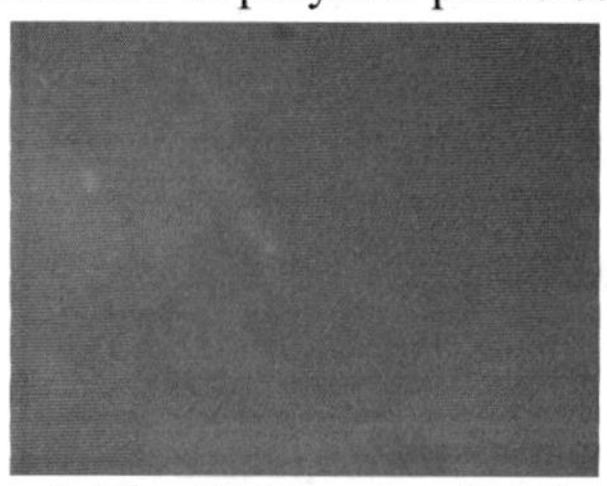

Figure 6. The FM image of UV aged UV531/ SBS modified bitumen

As can be seen from Table 6, after UV aging, UV531/SBS modified bitumen has only slight changes in PRR and SPI compared with SBS modified bitumen without UV531, which is also consistent with the FM images of SBS modified bitumen obtained after UV aging.

Table 6. The influence of UV531 on UV aging performance of SBS modified bitumen

Content of UV531（%）	PRR	ΔS（℃）
0	59	10.6
0.6	61	9.7

4. Conclusions

(1) The addition of UV531 changed the dispersive state of SBS modified bitumen and improved the physical properties of SBS modified bitumen to some extent. Before aging, SBS modifier particles in the UV531/modified bitumen become smaller and increase. The bitumen and modifier form a two-phase continuous network structure, which improves the physical properties of SBS modified bitumen. After TFOT aging, the addition of UV531 increases the number of SBS modifier particles, indicating that UV531 has degradation effect on the thermal-oxidative aging resistance of SBS modified bitumen. After UV aging, the polymer particles basically disappear, indicating that UV531 has little effect on the modified bitumen.

(2) The conclusions obtained from the conventional three indexes are basically consistent with the conclusions obtained from the FM images, which indicates that the microstructure of SBS modified bitumen has a good correspondence with its macroscopic properties and can qualitatively reflect the physical properties of modified bitumen.

Acknowledgements
This work is supported by the Transportation Science & Technology Project of Jilin Province (2018-1-4, 2018-1-7), the Fundamental Research Funds for the Central Universities (300102219216), and the Open Fund of Guangxi Key Lab of Road Structure and Materials (2015gxjgclkf-001). The authors gratefully acknowledge their financial support.

References

[1] Airey G D 2003 *Rheological properties of styrene butadiene styrene polymer modified road bitumens* Vol 82 (Amsterdam: Elsevie) pp 1709-1719

[2] Lee S, Amirkhanian S N and Kim K W 2009 *Laboratory evaluation of the effects of short-term oven aging on asphalt binders in asphalt mixtures using HP-GPC* Vol 23 (Amsterdam: Elsevie) pp 3087-3093

[3] Zhang D, Zhang H and Shi C 2017 *Investigation of aging performance of SBS modified asphalt with various aging methods* Vol 145 (Amsterdam: Elsevie) pp 445-451

[4] F Durrieu, Farcas F and Mouillet V 2007 *The influence of UV aging of a styrene/butadiene/styrene modified bitumen: comparison between laboratory and on site aging* Vol 86 (Amsterdam: Elsevie) pp 1446–1451

[5] Ouyang C 2006 *Improving the aging resistance of styrene-butadiene-styrene tri-block copolymer modified asphalt by addition of antioxidants* Vol 91 (Amsterdam: Elsevie) pp 795-804

[6] Yu JY et al 2009 *Effect of organo-montmorillonite on aging properties of asphalt. Construction and Building Materials* Vol 23 (Amsterdam: Elsevie) pp 2636–2640

[7] Feng Z G et al 2013 *Effect of ultraviolet aging on rheology, chemistry and morphology of ultraviolet absorber modified bitumen* Vol 46 (France: Rilem) pp1123–1132

[8] Feng Z G et al 2015 *Effect of Ultraviolet Absorber UV531 on Properties of SBS Modified Bitumen* Vol 37 (Wuhan: Journal of Wuhan university of technology) pp53-58

[9] Yu X, Burnham N A and Tao M 2015 *Surface microstructure of bitumen characterized by atomic force microscopy* Vol 218 (Amsterdam: Elsevie) pp17-33

[10] Sengoz B and Isikyakar G 2018 *Analysis of styrene-butadiene-styrene polymer modified bitumen using fluorescent microscopy and conventional test methods* Vol 150 (Amsterdam: Elsevie) pp424-432

[11] Dong F Q et al 2016 *Rheological behaviors and microstructure of SBS/CR composite modified hard asphalt* Vol 115 (Amsterdam: Elsevie) pp 285-293

IOP Conf. Series: Materials Science and Engineering **758** (2020) 012069 doi:10.1088/1757-899X/758/1/012069

Investigation of bow defect in roll forming

Muhammad Sajjada,1, Mohanraj Murugesanb,1 and Dong Won Jung*,1
1Department of Mechanical Engineering, Jeju National University, Jeju, South Korea
Email: jdwcheju@jejunu.ac.kr
Abstract Even though New materials are coming into the market Aluminum Alloy is still widely used in Aerospace, Automobile and defense industries. The main reason is these alloys are light weight, high strength and easy to recycle. Many traditional methods like punching, shearing and blanking has been widely used previously to deform these sheets in different industries. Meanwhile, roll forming which is die less processing method, has been used for manufacturing of parts for automobiles and aircrafts structure parts. In this paper, combination of numerical simulation and experimentation of Aluminum alloy 5052 for bow reduction has been studied for U-shape sections. From results it was concluded that the bow defect is significantly reduced by increasing the forming speed of forming rollers.

1. Introduction

Due to the increasing demand of improvement in automobile industries make new ways for the manufacturing processes which is having safety of labors, energy efficient and low threads to environment. With the emergence of new alloys makes these process more interesting for all area of automotive and structural application. New aluminum alloy can achieve the same strength as well as effectively reduce the material thickness while having perfect corrosion resistance. Demand of new materials forming technique has been widely increased in the market, so the insiders pay close attention to new technology. Roll forming is the method for these alloys to deform into its final shape with smooth surface and good tolerance. Roll forming is a series of mechanical process, in which a sheet of different thickness and different materials are undergoes with plastic deformation to achieve its desired finals shape. This process consists of two main types cold and hot roll forming. In cold roll forming the sheet is deform at room temperature while in hot roll forming the sheet is preheated to some temperature and then undergoes the deformation process. It is widely used in automobile and construction industries where sheet is to be deform in specific shapes.

Roll forming production line is more effective and the process is continuous which also make it has a higher economy. The length of the sheet can be adjusted without to having to replace the rolling tool. Compared with the stamping and bending process, roll forming is more flexible, especially for the small-scale production. Traditional molding and other produce any cross-sectional profiles, while punching and bending section to produce only relatively simple [1-6]. The most important advantage of cold-formed profiles is relatively uniform hardening properties and high surface quality, plated metal and enamel processed substantially without leaving scratches. Cross-section cold-formed steel is widely used in various industrial fields, especially automobile manufacturing and construction, the traditional cold-formed profiles or closed beam-like openings in high demand.

G.Nefuss and so on [7] have apply the elastic-plastic method to the roll forming simulation in 1996.The sheet is considered to be thin and the neutral layer is coons curve. In 1998, neixue [8] and so on used the computer to simulate the pipe and assume that there is only plane strain exist. From which they get the force and moment across the profile without considering the interaction of the rolls. In 2004, A.Alsamhan ,I.Pillinger and P.Hartely [9] applied remeshing techniques to the roll forming process with the sheet friction calculation. Zhang Dongjuan and Cui Zhenshan [10] had developed an analytical model for predicting the sheet spring back after U-bending based on the Hill's 48 yield criterion and plane strain condition. Also, three materials hardening –kinetic, isotropic and combined hardening has been used to consider the effect in the bending process. R.Safdarian and H.Moslemi Naeini [11] studied the effect of some roll forming parameters of channel section are investigated on the edge longitudinal strain and bow defect of products. The relationship between the parameters and defect have been recovered with simulation and experiment method. In previous study many researchers try to modify many parameters in order to minimize the bow defect in final profile of roll forming. However, there other method through which we can also minimize the bow defect. In our research study we are investigating the effect of roll speed on final profile of aluminum alloys. Results shows that the roll speed has significant effect on the final profile of forming sheet.

2. Experiments

2.1 Roll forming machine specifications

Experimental tests were performed on a cold roll forming machine with four stands. The roll stands are located at equal distance from each other while distance between two successive roll stands is 220mm. In detail, the roll stages were designed to produce different bending angle at each stage and the bending angles were 22.5^0 (1st stage), 45^0 (2nd stage), 67.5^0(3rd stage) and 90^0 (4th stage), respectively.

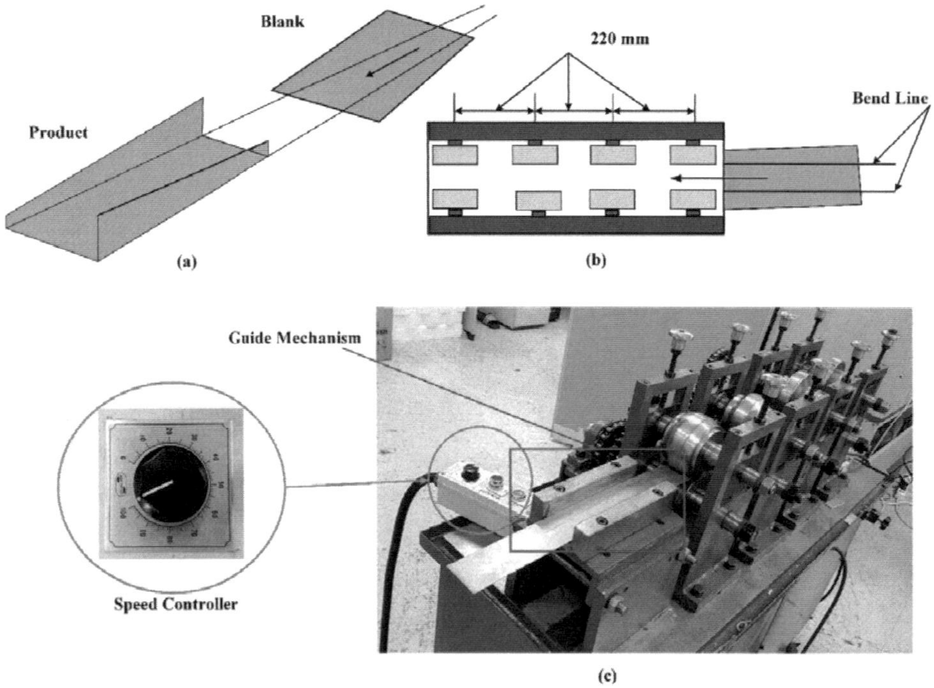

Fig. 1 Roll forming machine specification and guiding mechanism

Using this apparatus, we can deform sheet of different metals into U-channel sections. In some cases, in order to feed the initial strip aligned to the production line, a guide mechanism was used as shown in Fig. 1. Roll forming machine can be operate for different speed up to 100 mm/s according to our need of experiment as shown in above figure.

2.2 Measure of the roll forming defect and material specification

Roll forming is a complex deformation process, which involves large displacement, finite strain and the problems of contact and friction between strip and rolls. This process exhibits obvious geometry, physical and boundary nonliterary. The complex processes contain many defects like twist, spring back while most common among them is bow defect. Which is the longitudinal twist of deformed sheet after roll forming process as shown in Figure .2.

Table. 1 Material chemical and Mechanical properties

Material: AA5052								
Composition (wt. in %)	Si	Fe	Cu	Mg	Zn	Cr	Al	Mn
	0.05	0.2	0.03	3.2	0.12	0.23	balance	0.08
Mechanical Properties	Density (g/cc)	Poisson's ratio	Young's modulus (GPa)	Tensile strength (MPa)	Yield strength (MPa)	Strain hardening coefficient	Ductility (%)	-
	2.66	0.33	70.3	356	155	0.16	12	-

IOP Conf. Series: Materials Science and Engineering **758** (2020) 012069 doi:10.1088/1757-899X/758/1/012069

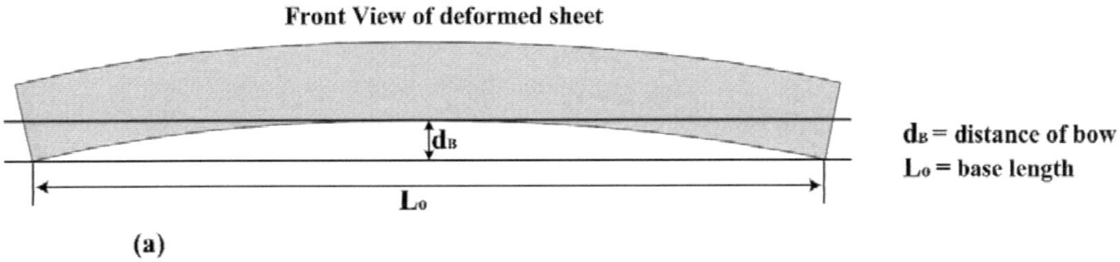

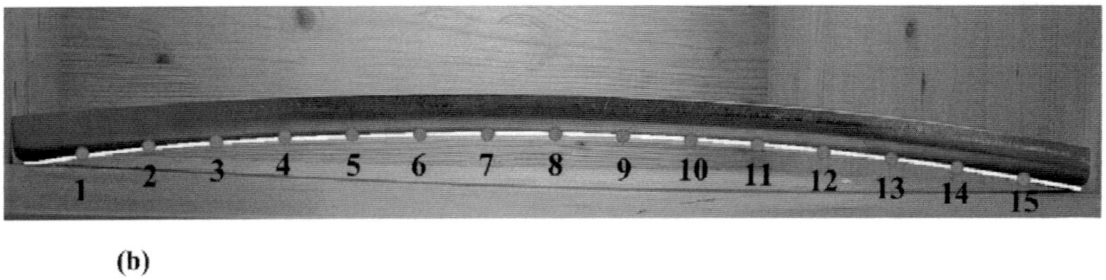

Fig. 2 Bow defect in aluminum roll forming

We are using a cost-effective method in which the bow defect is measure with the help of 15 different points on deformed sheet with equal distance from each other as shown. In the first step the sheet is place on the uniform surface, that all the vertices of the deform sheet touch the flat surface as shown in Figure. 2(a). In the next step two major quantities has been calculated, L_o which is the horizontal length of uniform surface (base line) and d_B is the vertical distance of bow from base line. The values of bow has been measured for 15 points on each sheet for three different roll speed.

In this study, Al6061 aluminum alloy was adopted as a testing material and the chemical composition and the material properties are summarized in Table 1. The Al6061 aluminum alloy specimen dimensions are: 700 mm length, 60 mm width and 1mm thick.

2.3 Finite element analysis of roll forming process

During the roll forming process, plastic strains are produced in different areas of the strip. Through these deformations, the strip is formed to make the desired shape. By investigating the effective stress distributions that occur on the strips, it is possible to determine how the bow is happened to strip deformations. Therefore, three models were simulated, according to experimental tests, using the ANSYS/LS_DYNA finite element software. Figure. 3 (a) shows the forming process model of a strip in a four-stand roll forming line.

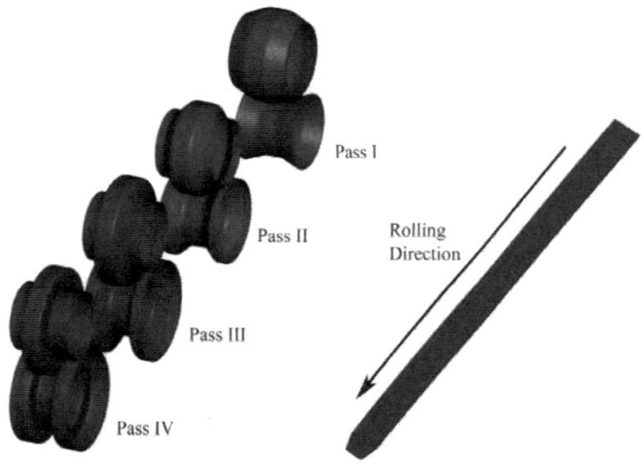

(a) FE model description

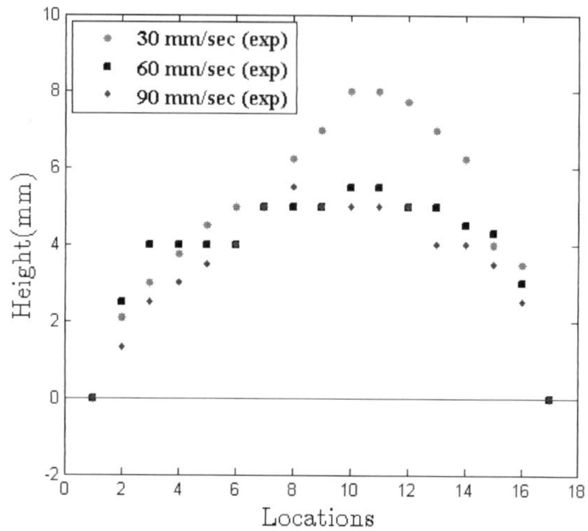

(b) bow measurement (mm) at different speeds

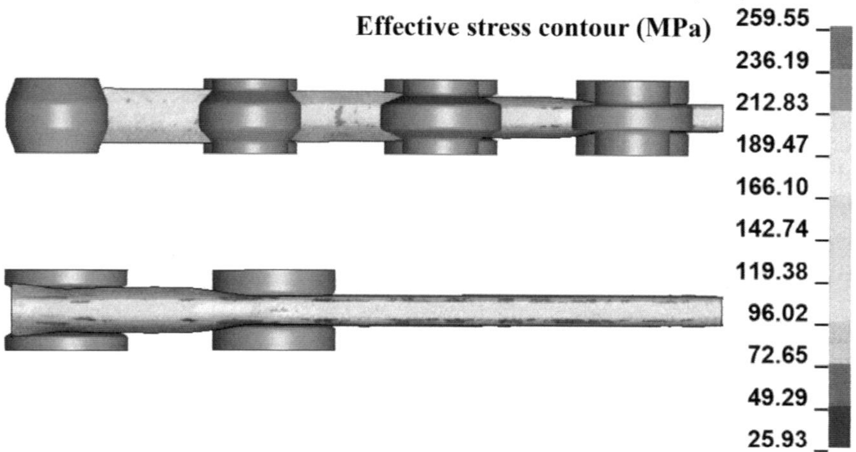

(c) Stress contour in formed profile

Figure 3. Finite element model (FEM) results of cold roll forming process for AA5052 material.

As shown in Figure 3 (c), all effective stress results were obtained on the strip length and its values is maximum at those points where the sheet passes the rolls. The bow defect values for 30, 60 and 90mm/sec roll speed are shown in Figure .3(b) which shows that the bow defect is symmetric in 90 mm/sec roll speed. In addition, at high roll speed, the bow height tends to be reduced than that of other two low roll speeds.

Conclusion

In this investigation, the effects of different roll speed on the amount of bow to produce a symmetrical U-shape section using the cold roll forming process were studied experimentally. Also, through the finite element method, the effect of roll speed for 30, 60 and 90mm/sec and effective stress distribution within the strips was studied by using FE model. Then, using the results from both simulation and experimentation, the effect of speed in bow was investigated. Which led to demonstrate the relationships between the bow at different roll speed for three AA5052 specimens. Results show that the bow value is inversely proportional to the roll speed during the roll forming process. By increasing the roll speed the bow defect is decreased by considerable amount, also the bow is more uniform and symmetric in high roll speed than compared to low speeds.

References

[1] A.Abvabi, B.Rolfe , P.D.Hodgson , M.Weiss, The influence of residual stress on a roll forming process, International Journal of Mechanical Sciences, Vol.10, pp.124–136, 2015

[2] Mahmoud Yaakoubi, Mounir Kchaou, Fakhreddine Dammak, Simulation of the thermo mechanical and metallurgical behavior of steels by using ABAQUS software, Computational Materials Science,Vol. 68, pp.297–306, 2006

[3] Behnam Davoodi, Behrooz Zareh-Desari, Assessment of forming parameters influencing spring-back in multi-point forming process: A comprehensive experimental and numerical study, Materials and Design, Vol.59, pp.103–114, 2014

[4] Xuechun Li, Yuying Yang, Yongzhi Wang, Effect of the material-hardening mode on the spring back simulation accuracy of V-free bending, Journal of Materials Processing Technology, Vol.123. pp.209–211, 2002

[5] ZhengyangCai, KeshanDiao , XiangdongWu, Constitutive modeling of evolving plasticity in high strength steel sheets. International Journal of Mechanical Sciences, Vol. 107, pp.43–57, 2006

[6] M. Lindgren, Cold roll forming of a U-channel made of high strength steel, Journal of Materials Processing Technology, Vol. 186, pp.77–81,2007

[7] K. B. Kim, W. S. Lee, Y. Y. Kim, "Investigation of Electric Vehicle Performance Affected by Cabin Heating", Journal of the Korea Academia-Industrial, Vol. 14, No. 10, pp. 4679-4684, 2013

[8] D. W. Lee, D. H. Oh, and Y. J. Jee, "Investigation of R134a heat pump system for zero emission vehicle", Proceedings of KSAE 2011 Autumn Annual conference, Vol. 11, pp. 595-600, 2011

[9] A. A lsamhan,I. Polliinger, P. Hartely, The development of real time re-meshing technique for simulating cold-roll-forming using FE methods, Journal of Materials Processing Technology, Vol. 147, pp. 1-9, 2004

[10] Zhang Dongjuan, Cui Zhenshan, Ruan Xueyu and Li Yuqiang, An analytical model for predicting spring back and side wall curl of sheet after U-bending, Computational Materials Science, Vol. 38, pp. 707–715, 2007

[11] R. Safdarian, H.MoslemiNaeini, The effects of forming parameters on the cold roll forming of channel section, Thin-Walled Structures, Vol. 92, pp. 130–136, 2015

IOP Conf. Series: Materials Science and Engineering **758** (2020) 012070 doi:10.1088/1757-899X/758/1/012070

Effect of superplasticizer in anti-washout admixture on paste

Ming Zhang[1,2] and Wuxiang Wang[1,2]

[1]State Key Laboratory of Green Building Materials, Beijing, China
[2]College of Civil Science and Engineering, Yangzhou University, Yangzhou, China
E-mail: zhangmingjob@sina.com

Abstract. Anti-washout admixture used for anti-washout underwater concrete is composed of flocculant (e.g., paste stabilizer), superplasticizer, setting retarder, strengthening agent, etc. Good compatibility among those constitutes is necessary to release the vast material potential of anti-washout admixture. In the study, effects of polycarboxylate superplasticizer on fresh cement paste mixed paste stabilizer of various ionic type were experimentally investigated. Based on the experimental results, the optimizations on the ionic type and molecular weight of paste stabilizer, regarding the compatibility with polycarboxylate superplasticizer, were hence obtained. Additionally, suitable content ranges of polycarboxylate superplasticizer, molecular weight and content of paste stabilizer have been determined through systematic tests, such as saturation point test, fluidity test, and anti-washout test.

1. Introduction

With the gradually scarce underground natural resources, aquatic environment resources such as river and ocean resources are being founded ceaselessly. Nowadays, humans are even more in-depth exploiting river and ocean resources, which also promotes the development of hydraulic engineering and marine engineering technologies. Technologies of anti-washout underwater concrete turn it into reality to pour concrete underwater, which avoids cofferdam, foundation seepage control, foundation pit drainage and other related engineering measures and consequently saves a large amount of manpower and material resources [1, 2], These technologies are frequently applied to marine environment structures than to river structures [3]. Due to the limited strength of anti-washout underwater concrete affected by washout loss and water infiltration, steel bar corrosion embedded in the concrete may occur during casting and setting period [4, 5], so how to make sure that the residual stresses meet the engineering requirements is critical technology to the concrete [6, 7].

2. Materials and methods

2.1 Materials and mix design

The cement type is chosen as P· II 52.5 grade Portland cement; polycarboxylate superplasticizer, solid-containing content is 25%, and water reducing rate is 30 % when the content is 1.0% of cement by weight; the setting retarder is sodium tripolyphosphate (Na5P3O10), regent; stabilizers include anionic stabilizers named SSA40, SSA60, SSA 80 consequently, which average molecular weight are 4, 6, and 8 million; Cationic stabilizers named SSC40, SSC60, which average molecular are 10 million, and nonionic stabilizer named SSN100, which average molecular weight is 10 million.

2.2 Methods

Fluidity and fluidity loss test, determination of anti-washout of paste and shear stress and viscosity test were carried out according to reference [8].

3. Results and discussion

3.1 Saturation Point Test

Prepare cement paste with the W/B ratio of 0.45 and the setting retarder of 4.0 wt.‰. Different mass combinations of superplasticizer are mixed into mixing water before poured into mixer. The fluidity as show in figure 1., the superplasticizer reaches to saturation point when the content achieves 0.4%. As shown in figure 2., the saturation points of superplasticizer is between 1.0% and 1.5% for different contents of paste stabilizer SSN100, which suggests that the paste stabilizer content does not have obvious influence on the saturation point.

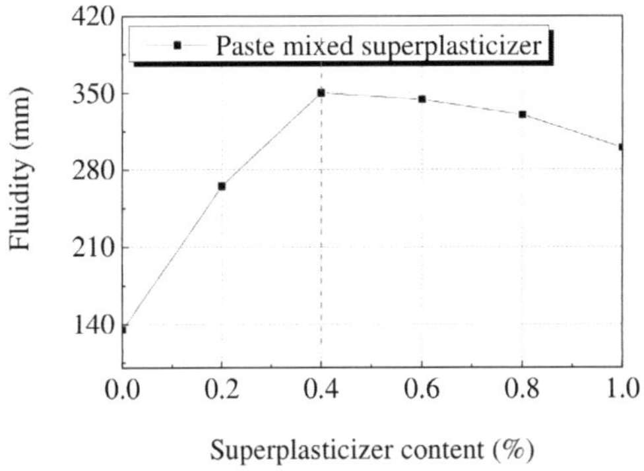

Figure 1. Initial fluidity of control paste mixed different content of superplasticizer

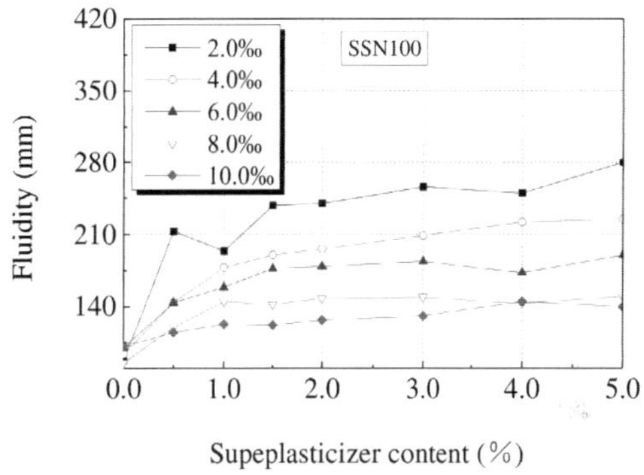

Figure 2. Initial fluidity of paste mixed paste stabilizer SSN100

The change regulations of fluidity are similar to that of different ionic type of paste stabilizer (see figure 3, 4). The increase of molecular weight has little effect on saturation point of superplasticizer. This indicates that the saturation point is within the value of 1.0% and 1.5%, although different molecular weight, lager molecular weights lead to lager tackifying effect.

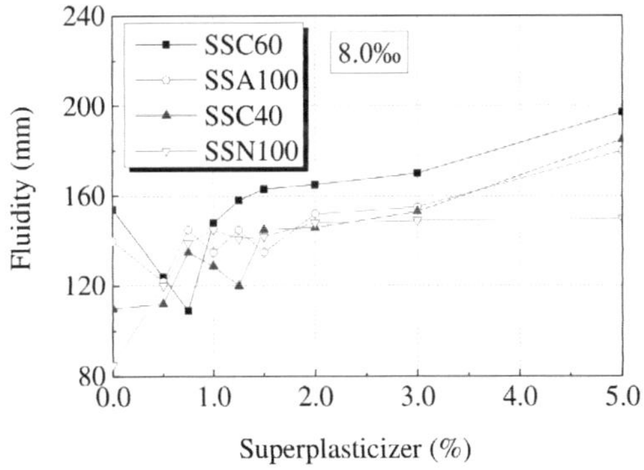

Figure 3. Saturation point of superplasticizer in paste mixed different ionic type of paste stabilizer with ten million molecular weight

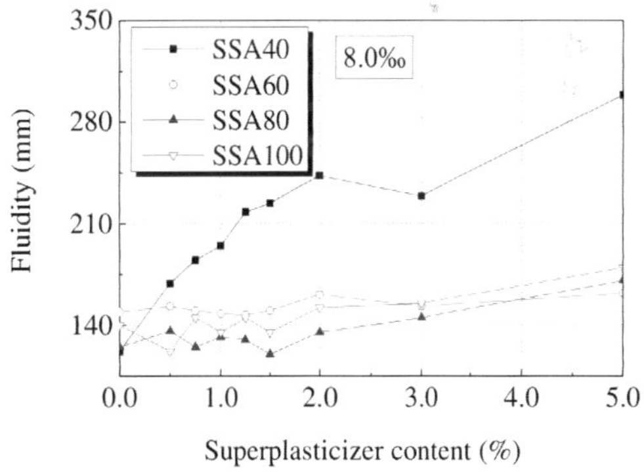

Figure 4. Saturation point of superplasticizer in paste mixed different molecular weight of paste stabilizer SSA

3.2 Fluidity Loss Test

Figure (5). expresses the fluidity loss of paste mixed 0.0‰~10.0% of paste stabilizer SSN100. The higher content of paste stabilizer SSN100, the smoother development of fluidity, and consequently the lower of fluidity loss.

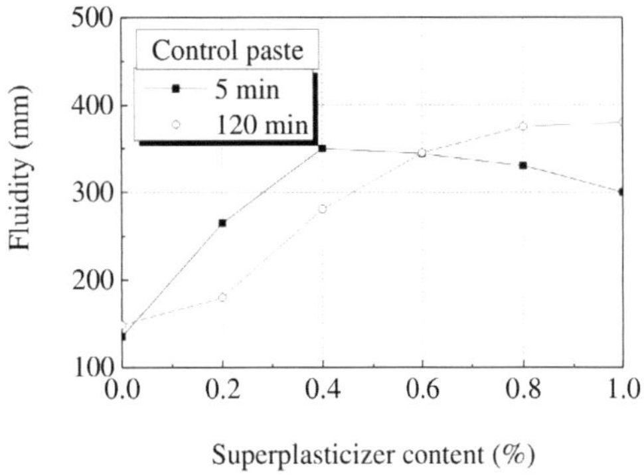

(a) 0‰

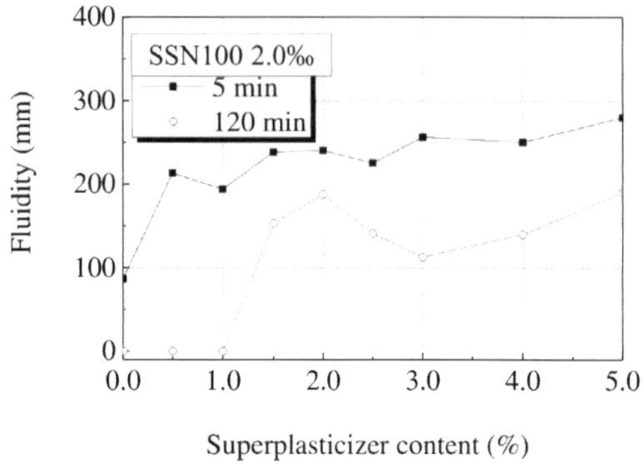

(b) 2.0‰

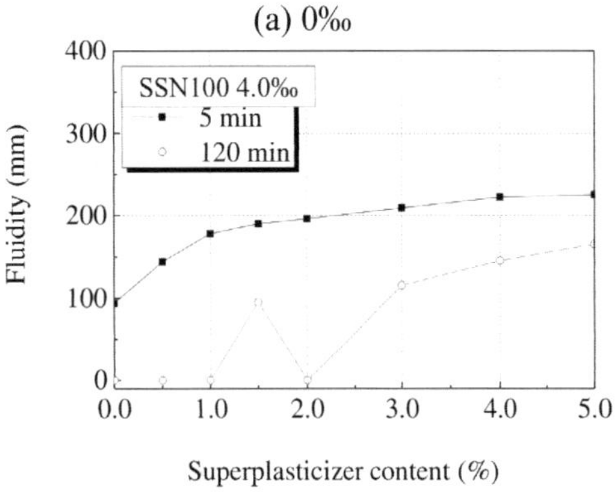

(c) 4.0‰

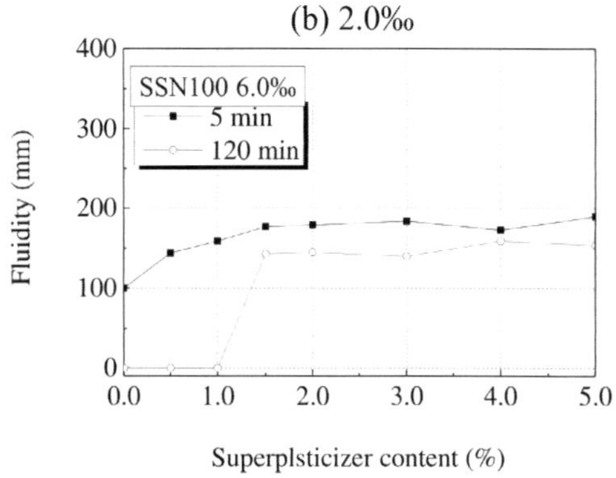

(d) 6.0‰

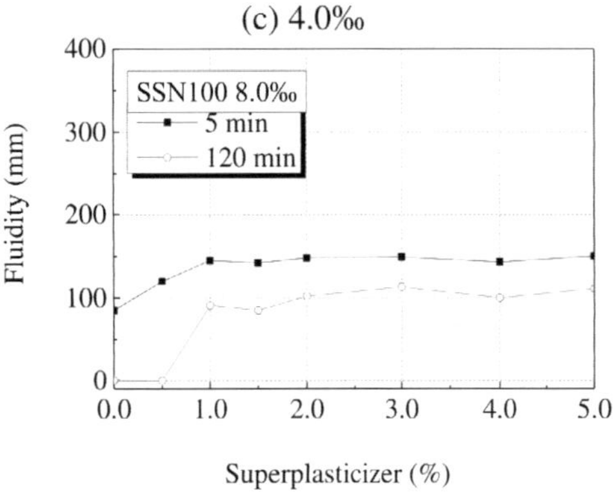

(e) 8.0‰

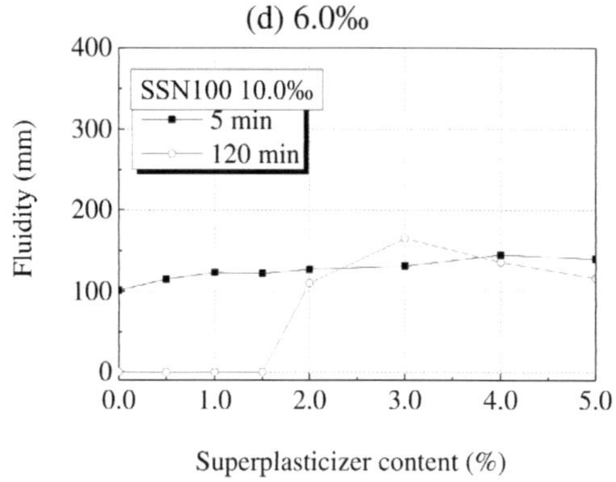

(f) 10.0‰

Figure 5. Fluidity of paste mix different content of paste stabilizer SSN100

3.3 Anti-Washout Test of Paste

Figure 6. is the water pH value that immerses the paste mixed 0.0%～5.0% content of superplasticizer and keeps the content of paste stabilizer SSN100 constant. Comparing to figure (6), the water pH value increases with the rise of superplasticizer content when it is lower than the saturation point (i.e., 1.0%～1.5%). The water pH value does not increase obviously when the content of superplasticizer exceeds the saturation point. To the same content of superplasticizer, the higher content of SSN100, the lower pH value of water. To sum up, SSN100 has good anti-washout property. Changing rules of water pH value are like that of fluidity.

As shown in figure 7, the turbidity does not increase obviously when the content of superplasticizer is less than 3.0%, but does increase obviously when the content of superplasticizer exceeds 3.0%. As shown in figure 8, in the same manner, the turbidity increases with the rise of superplasticizer content varying between 1.0% and 5.0%.

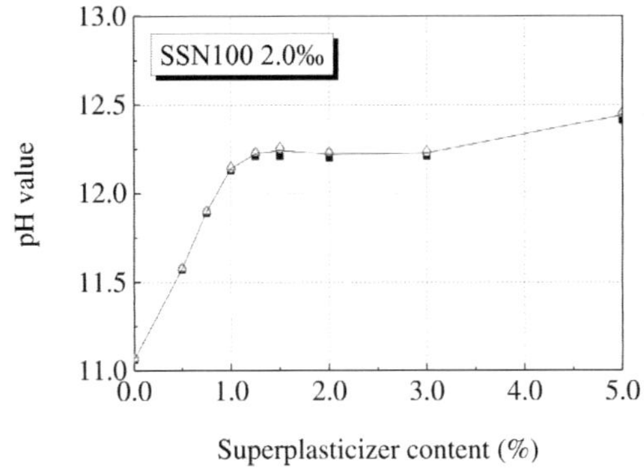

(a) SSN100 2.0‰

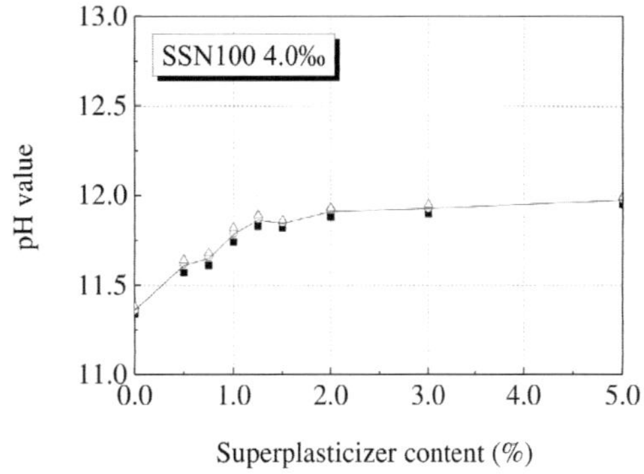

(b) SSN100 4.0‰

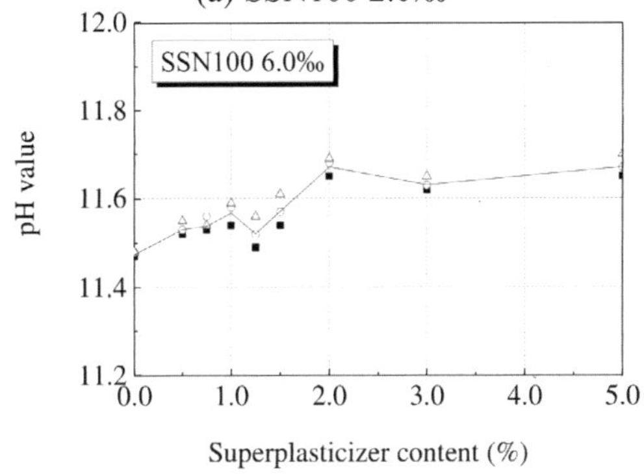

(c) SSN100 6.0‰

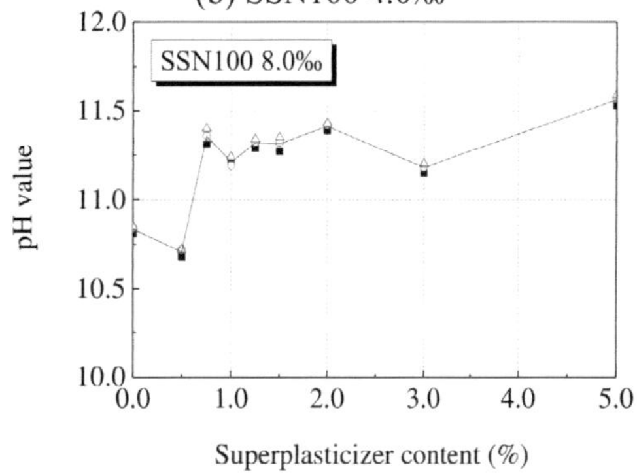

(d) SSN100 8.0‰

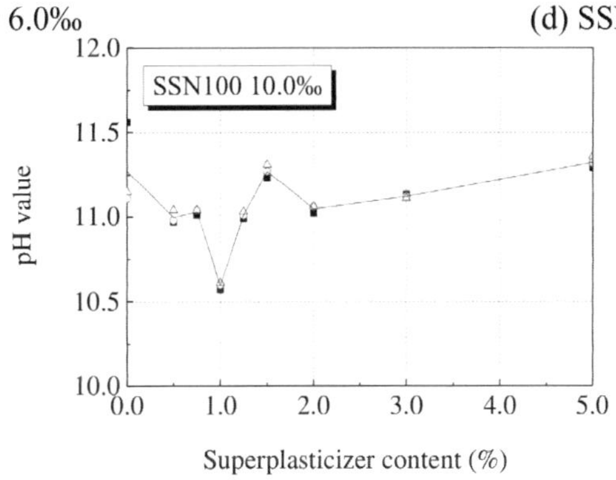

(e) SSN100 10.0‰

Figure 6. pH value of water immersed paste mixed different content of superplasticizer

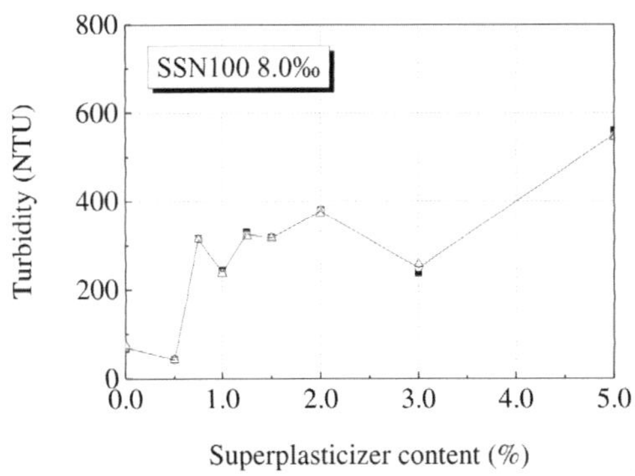

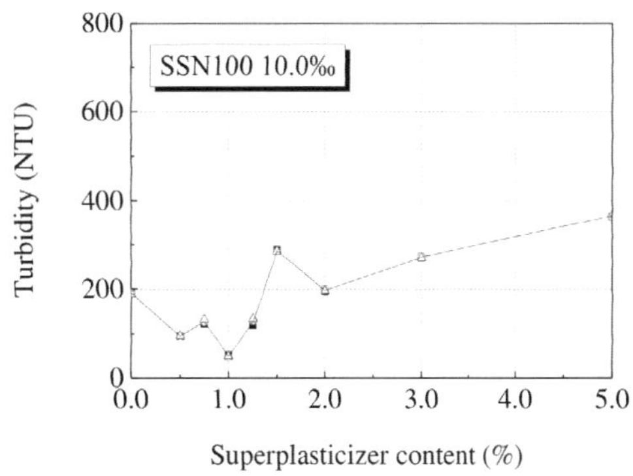

Figure 7. Turbidity of water after immersed paste mixed 8.0‰ of paste stabilizer SSN100

Figure 8. Turbidity of water after immersed paste mixed 10.0‰ of paste stabilizer SSN100

Figure (9). is the suspension particle content of water after immersed paste mixed different content of superplasticizer. The suspension particle content linear increase with the rise of superplasticizer content when less than 6.0‰, which means that the paste stabilizer has not played the act of anti-washout effect within the content, and that suspension particle content does not increase obviously at 8.0‰ and 10.0‰.

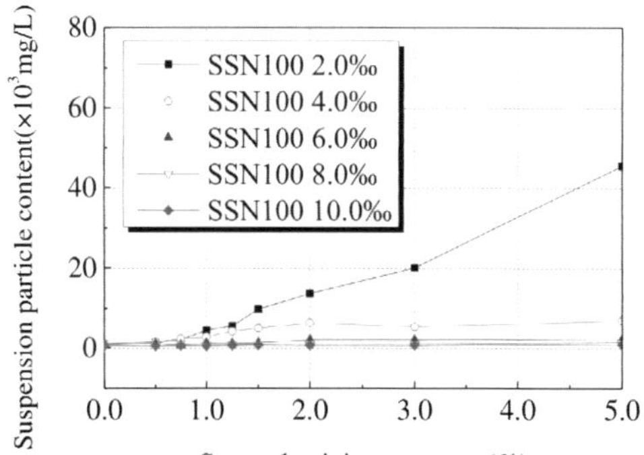

Figure (9). Suspension particle content of water after immersed paste mixed superplasticizer

Conclusion

Regarding the paste mixed anionic, cationic, and nonionic paste stabilizer, the saturation point of polycarboxylate superplasticizer is between 1.0% and 1.5%. The ionic type has little effect on polycarboxylate superplasticizer.

To the paste mixed different molecular weight and content of anionic, cationic of paste stabilizer, the fluidity of paste completely loses within 30 min. But to the paste mixed nonionic paste stabilizer, the paste of the same kind shows good fluidity within the test content scope of superplasticizer. This indicates bad compatibility between anionic, cationic stabilizer with polycarboxylate superplasticizer, but good compatibility between nonionic stabilizer with polycarboxylate superplasticizer.

For the content of polycarboxylate superplasticizer lower than saturation point, anti-washout properties of paste mixed nonionic paste stabilizer decrease with the rise of superplasticizer. To the content of polycarboxylate superplasticizer over the saturation point, anti-washout properties no longer decrease obviously, but the shear stress and viscosity decrease slightly with the rising content of polycarboxylate superplasticizer.

Acknowledgement

The study of this paper is Supported by the Opening Project of State Key Laboratory of Green Building Materials, Science Foundation of Ministry of Housing and Urban-Rural Development of the People's Republic of China (No. 2016-K4-016), China postdoctoral science foundation (No. 2015M581702), and Yangzhou Science and Technology Planning Project (No. YZ2017290).

Reference

[1] C.S. Jiang, L.N. Lu, S.B. Guan 2004 Journal of Wuhan University, vol 19 p 66
[2] G. X. Chen, Z. Q. Du, R Yang 2012 Concrete p 117.
[3] H. Y. Moon, K. J Shin 2007 Construction and Building Materials vol 21 p 98
[4] J. J. Assaad, C. A. Issa 2013 Material and structure vol. 46 p 1613
[5] H. Y. Moon, K. J. Shin 2006 Cement and Concrete Research vol 36 p 521
[6] Hughes BP 1961 Civil Eng Public Rev vol 56 p 633
[7] Sonebi M, Khayat KH 1999 ACI Mater J vol 96 p 519
[8] Juan Liu 2015 Hunan University p 16

IOP Conf. Series: Materials Science and Engineering **758** (2020) 012071 doi:10.1088/1757-899X/758/1/012071

Effect of early strength agents on specified density concrete

Shuaishuai Yan[1,2] and Ming Zhang[1,2]

[1] State Key Laboratory of Green Building Materials, Beijing, China
[2]College of Civil Science and Engineering, Yangzhou University, Yangzhou, China
E-mail: zhangmingjob@sina.com

Abstract. Early strength agents are beneficial to improving early strength of concrete, thereby shorting curing time as well as production efficiency of concrete products, and consequently saving production costs. This study investigated the influence of triethanolamine, potassium carbonate, Calcium formate and Potassium aluminum sulfate on properties of specified density concrete, based on orthogonal experiment, composite early strength agent is determined.

1. Introduction

Specified density concrete is the concrete that apparent density is 1950 kg/m^3 ~ 2300 kg/m^3 by in addition of moderate amount of lightweight aggregate into ordinary aggregate. Specified density concrete has many good qualities such as high strength, light weight and long durability, which results in the concrete has broad prospect of application on high rise building, bridge and marine engineering. Like other concrete materials, mechanical properties is one of typical properties. Main effect factors of mechanical properties include strength and water absorption of ceramisite, water to binder (W/B), and sand ratio, et al.

When specified density concrete is damaged by compressing, generally the damage begins from ceramsite, then the crack gradually runs through the whole concrete matrix [1], and the higher cylinder compressive strength of ceramsite, the higher strength of specified density concrete. Replacement rate of ceramsite has important effect on strength of concrete too, with the rise of replacement rate of ceramsite the mechanical properties of concrete decreases [2].

More Higher water absorption of ceramsite, more beneficial to late mechanical properties, which because high water absorption means high water storage, and consequently more beneficial to improving interfacial transition zone as well as mechanical properties.

There is inverse relationship between W/B and mechanical properties of specified density concrete, with the increase of W/B, the mechanical properties decrease, and the ratio of compressive strength to flexural is between 10 and 19 [3].

Within certain scope, increase of sand ratio is good to improve the compatibility of specified density concrete as well as mechanical properties [4], however, if sand ratio too high, the density difference between mortar and ceramsite increases, and the homogeneity of concrete becomes worse similarly to mechanical properties.

Within certain scope of content, the common mineral admixtures such as fly ash and slag can decrease the density of cement mortar, thereby relieve the layering trend of between ceramsite and mortar, which lead to the improving of mechanical properties.

2. Materials and methods

2.1. Materials and mix design

Cement was P·O 42.5 ordinary Portland cement. river sand was medium sand. crushed strong size was 5 mm~10 mm. grade I fly ash was obtained from thermal power plant. spherical calcined clay Ceramsite was 5 mm ~ 10 mm diameter, mixing water was tap water. the early strength agents were Triethanolamine, potassium carbonate, Calcium formate and Potassium aluminum sulfate, all of them were chemical regents. The mix proportion of specified density concrete as shown in Table 1.

Table 1. Mix proportion of specified density concrete (kg)

Cement	Fly ash	sand	Crushed stone	ceramsite	water
320	80	730	834	104	112

2.2. Methods

According to specified density concrete mix proportion in Table 1, and GB/T 50082-2009 [5], concrete cubes with the dimensions of 100 × 100 × 100 mm^3 were prepared and demoulded right after 1 d, the samples were cured in concrete standard curing room (more than 95%RH and 20 ± 2 ℃) until the compressive strength measurements were carried out at 3 d and 7 d.

3. Results and discussion

3.1. Triethanolamine

As shown in figure 1, mechanical properties of concrete increase with the dosage of triethanolamine within the scope of 0.00% ~ 0.03%, but inverses within the scope of 0.03% ~ 0.09%. during hydration process of cement, triethanolamine forms complex ions with Fe^{3+} and Al^{3+}, the complex ions can accelerate the dissolution of C_3A and C_4AF and generate calcium sulphoaluminate, consequently reduce the concentration of Fe^{3+} and Al^{3+} in solution, which is conducive to the hydration of C_3S and early strength development of cement [6].

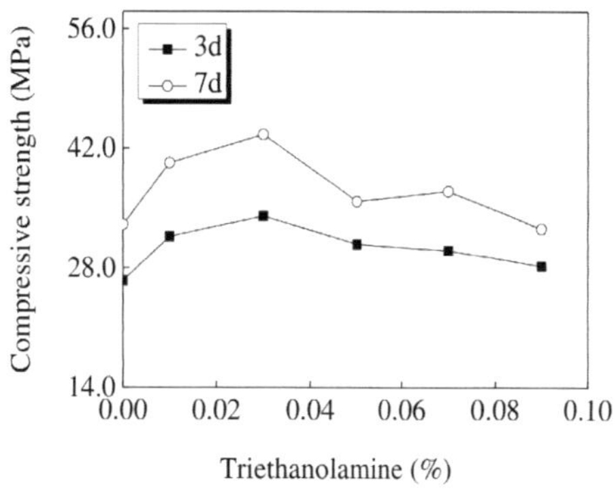

(a) Compressive strength

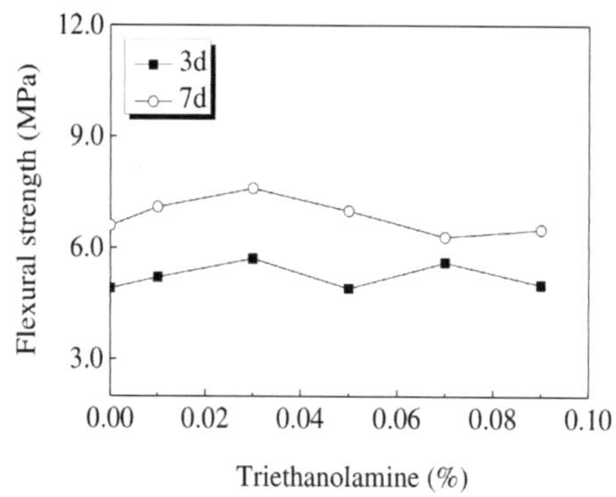

(b) Flexural strength

Figure 1. Mechanical properties of specified density concrete mixed triethanolamine.

3.2. Potassium carbonate

Potassium carbonate reacts with calcium hydroxide in cement paste and generate calcium carbonate, which reduces the concentration of calcium hydroxide, and increases the solid ratio in the solution accordingly improves early strength [7]. However, with the increase dosage of potassium carbonate, mechanical properties of concrete decrease. as shown in figure 2, when the dosage of potassium is between 0.0% and 1.0%, both the compressive strength and flexural strength are improved, and reach to the peak value at the dosage of 1.0%, which means the optimal dosage of potassium carbonate is 1.0%. excessive addition will lead to negative effect on neither compressive or flexural strength of specified density concrete.

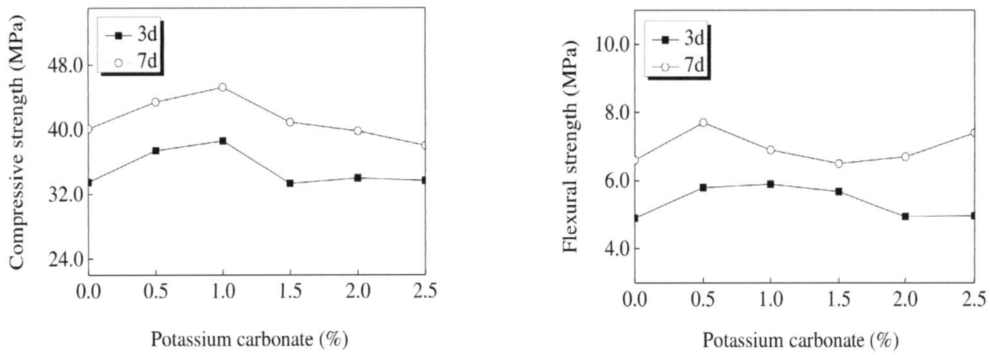

(a) Compressive strength (b) Flexural strength

Figure 2. Mechanical properties of specified density concrete mixed potassium carbonate

3.3. Calcium formate

According to figure 3, when the dosage of calcium formate is from 0.0% to 1.5%, the mechanical properties of specified density concrete show an upward trend, which is mainly due to the role of formate ions, it generates products limilar to AFt and AFm and penetrate into hydration layer of C_2S and C_3S, resulting in the hydration of C_2S and calcium hydroxide is accelerated, so as to improve the early strength. When dosage of calcium formate between 1.5% and 2.5%, mechanical properties of specified density concrete decreases.

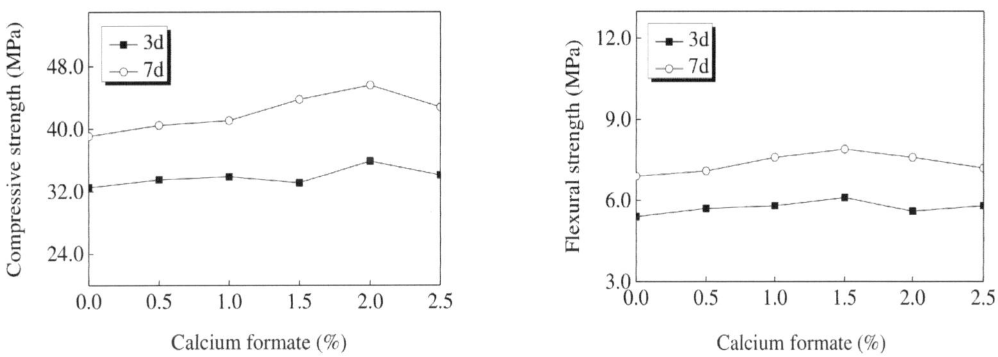

(a) Compressive strength (b) Flexural strength

Figure 3. Mechanical properties of specified density concrete mixed calcium formate

3.4. Potassium aluminum sulfate

Figure 4 is mechanical properties of specified density concrete mixed different dosage of potassium

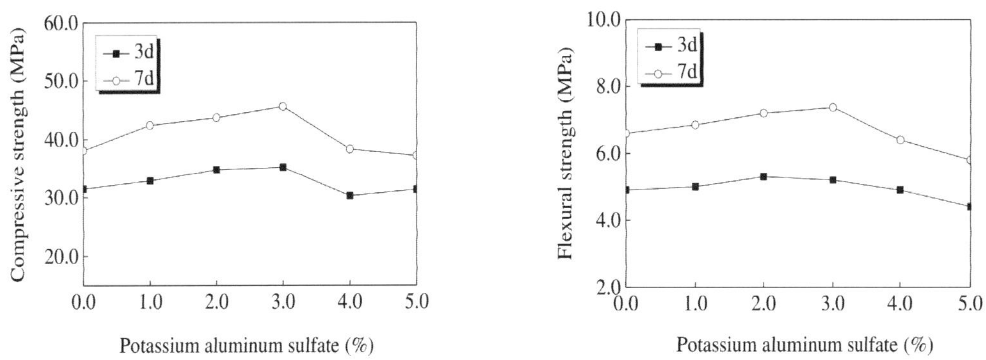

IOP Conf. Series: Materials Science and Engineering **758** (2020) 012071 doi:10.1088/1757-899X/758/1/012071

(a) Compressive strength (b) Flexural strength

Figure 4. Mechanical properties of specified density concrete mixed potassium aluminium sulfate

aluminum sulfate, as shown in the figure, under dosage of 3.0%, the mechanical properties of specified density concrete increase with the dosage of potassium aluminum sulfate, however the mechanical properties decrease when the dosage of potassium aluminum sulfate beyond 3.0%, which means the optimal dosage of potassium aluminum sulfate is 3.0%.

3.5. Orthogonal experiment

Table 2 is orthogonal experiment and results of early strength, according to the range analysis (R value), the order of influence factors on 3 d compressive strength are $R_b > R_d > R_c > R_a$, namely potassium carbonate > potassium aluminum sulfate > calcium formate > triethanolamine, and the best composite is $A_3B_1C_1D_2$ based on R value. Similarly the order of influence factors on 7 d compressive strength are $R_b > R_d > R_c > R_a$, namely potassium carbonate > potassium aluminum sulfate > calcium formate > triethanolamine, $A_2B_1C_1D_2$ is the best composite based on R value.

Table 2. Orthogonal experiment and results

No.	(A) Triethanolamine (%)	(B) Potassium carbonate (%)	(C) Calcium formate (%)	(D) Potassium aluminum sulfate (%)	Compressive strength (MPa) 3d	7d
1	1(0.01)	1(0.50)	1(1.50)	1(1.00)	40.91	52.62
2	1	2(1.00)	2(2.00)	2(2.00)	41.43	42.42
3	1	3(1.50)	3(2.50)	3(3.00)	30.20	43.16
4	2(0.03)	1	2	3	40.95	49.19
5	2	2	3	1	36.55	47.06
6	2	3	1	2	37.61	45.45
7	3(0.05)	1	3	2	42.22	55.03
8	3	2	1	3	38.55	44.82
9	3	3	2	1	34.44	38.98
$\bar{K}_1$	37.51	41.34	39.01	37.28		
$\bar{K}_2$	38.37	38.84	38.91	40.42		
$\bar{K}_3$	38.74	34.08	36.32	36.55		
R	1.23	7.26	2.69	3.87		

Table 3 and table 4 are gest group of composite agent and optimal dosage of single early strength agent and the compressive strength of specified density concrete mixed those agents.

Table 3. Best group of composite agent and optimal dosage of single early strength agent

Best group	(A) Triethanolamine (%)	(B) potassium carbonate (%)	(C) Calcium formate (%)	(D) Potassium aluminum sulfate (%)
$A_2B_1C_1D_2$	0.01	0.50	1.50	2.00

As shown in table 4, the specified density concrete mixed composite early strength agent ($A_2B_1C_1D_2$) than that of whichever single early strength agent no matter at 3 d or 7 d.

Table 4. Compressive strength of specified density concrete mixed composite and single agent (MPa)

Curing age	$A_2B_1C_1D_2$	Single early strength agent Triethanolamine	potassium carbonate	Calcium formate	Potassium aluminum sulfate
3d	39.42	32.21	36.63	35.42	34.16
7d	48.33	39.65	43.27	42.41	41.65

4. Conclusions

(1) to the addition single early strength agent, mechanical properties of specified density concrete increase with the dosage of early strength agent within certain scope, however the rule inverse if the dosage of early strength agent over certain value.
(2) the composite early strength agent is more effective than whichever single agent, and the best composite is Triethanolamine : potassium carbonate : Calcium formate : Potassium aluminum sulfate =0.01%：0.50%：1.50%：2.00%.

Acknowledgments
The study of this paper is Supported by the Opening Project of State Key Laboratory of Green Building Materials, Science Foundation of Ministry of Housing and Urban-Rural Development of the People's Republic of China (No. 2016-K4-016), China postdoctoral science foundation (No. 2015M581702), and Yangzhou Science and Technology Planning Project (No. YZ2017290).

References

[1] Bing Han, Tian Yu Xiang 2016 Construction and Building Materials vol 23 p 338
[2] Yankun Zhang, Yuchen Wang, Xiaolong Wu 2016 Journal of North China University of Technology vol 28 p79
[3] Durga Chaitanya Kumar Jagarapu 2019 Materials Today: Proceedings, p 742
[4] Xin Yu, Hongying Li 2017 Jiangsu construction material p 21
[5] GB/T 50082-2009 2009 China Architecture & Building Press p 6
[6] Wei Wang, Yanan Li 2019 Development Guide to Building Materials p 93
[7] Yafei Wu, Deren Liu 2016 bulletin of the Chinese Ceramic Society p3351

IOP Conf. Series: Materials Science and Engineering **758** (2020) 012072 doi:10.1088/1757-899X/758/1/012072

Study on anti-permeability of specified density concrete

Shuaishuai Yan[1,2] and Ming Zhang[1,2]

[1]State Key Laboratory of Green Building Materials, Beijing, China
[2]College of Civil Science and Engineering, Yangzhou University, Yangzhou, China
E-mail: zhangmingjob@sina.com

Abstract. Anti-permeability is one of critical properties of concrete, as a porous material, ceramsite in specified density concrete leads to many performance changes compare with ordinary concrete, such as mechanical properties, volume stability, durability, et al. combined with micro-test technology, this study investigated the effect of replacement rate and pre-wetting degree of ceramsite, water to binder (W/B) ratio, the content of fly ash, and the composite of aggregate on specified density concrete, and concluded appropriate range of above narrated factors.

1. Introduction

Concrete is the largest amount of building materials, with the advantages of a wide range of raw material sources, low production energy consumption as well as good durability, concrete is widely used in civil engineering, traffic engineering, hydraulic engineering, nuclear power plant engineering and other fields [1]. Specified density concrete is the concrete with a range of apparent density between 1950 kg/m^3 ~ 2300 kg/m^3 by replacing an amount of ordinary aggregate with light weight aggregate [2]. Due to high weight, difficult in transportation and hoisting, Ordinary concrete members are more and more not satisfied the requirement of prefabrication industry. Apparent density of light weight aggregate concrete is not higher than 1900 kg/m^3, light weight makes its prefabricated members easy to transportation and construction [3]. But due to low strength of light weight aggregate concrete, its member prone to damaged during the transportation, storage as well as hoisting. Compare with ordinary concrete, the addition of light weigh aggregate makes specified density concrete lower weight, and compare with light weight aggregate concrete, specified density concrete improves mechanical properties, the light weight and high strength of specified density concrete match the requirement of prefabrication members, which leads to the specified density concrete has broad application prospect.

Permeability refers to how difficulty that air and liquid permeate, diffuse, or migrate in concrete under influence of external environment. Anti-permeability of concrete is related to pore structure, some factors of specified density of concrete, such as aggregate, mineral admixtures, W/B ratio, et al, effect the size of and distribution of internal pore, and then effect the anti-permeability of concrete.

2. Materials and methods

2.1. Materials and mix design

The raw materials were P·O 42.5 ordinary Portland cement, grade Ⅰ fly ash, medium particle size river sand, clay ceramsite (5~10 mm and 5~15 mm particle size) and crushed stone (5~10 mm and 5~20 mm particle size). The initial mix proportion of ordinary concrete is shown in Table 1, and specified density concrete obtained by replacing certain amount of crushed stone with ceramsite.

Table 1. Mix proportion of specified density concrete (kg)

Cement	Fly ash	sand	Crushed stone	ceramsite	water
320	80	730	1191	-	118

2.2. Methods
Preparing and anti-permeability test of specified density concrete samples were carried out according to GB/T50082-2009 [4].

3. Results and discussion

3.1. Effect of volume replacement rate of ceramsite on anti-permeability of concrete
figure 1 is water permeability height of specified density concrete mixed with 0% ~ 40% of ceramsite, as shown in Figure 1, when the volume replacement rate is between 0% and 20%, the water permeability height decreases, which means the anti-permeability is improved. Ceramsite has the function of backwater effect, and beneficial to the interfacial transaction zone (ITZ), consequently concrete become more compact and anti-permeability is improved. When the volume replacement rate of ceramsite is between 20% and 40%, with the volume replacement rate of ceramsite from 20% up to 40%, the water permeability height is from 50 mm rises up to 96 mm, which means anti-permeability of specified density concrete is reduced.

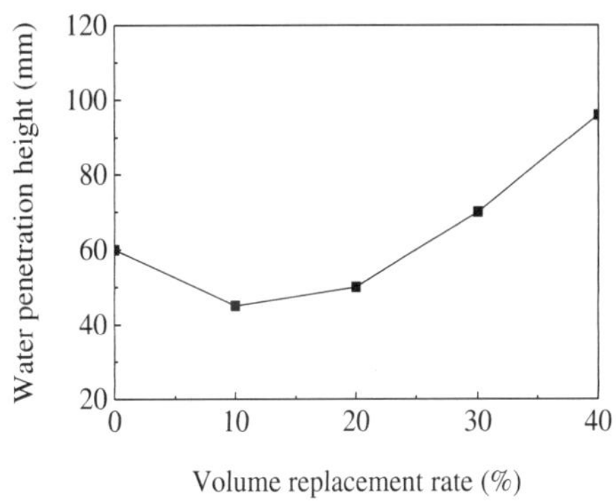

Figure 1. Water permeability height of specified density concrete with different volume replacement rate of ceramsite

The figure 2 is ITZ between aggregate and hardened cement paste, as shown in the Figure 2, the ITZ

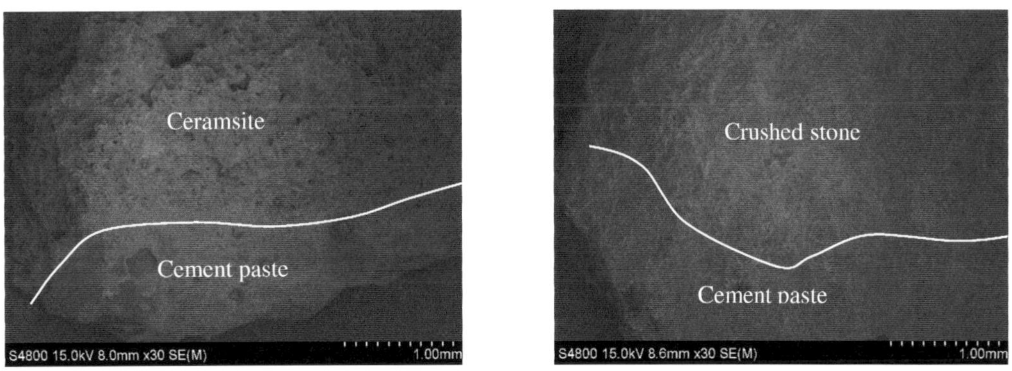

(a) ITZ between ceramsite and cement paste (×30) (b) ITZ between crushed stone and cement paste (×30)

(c) Micro structure of ITZ between ceramsite and cement paste (×5000)

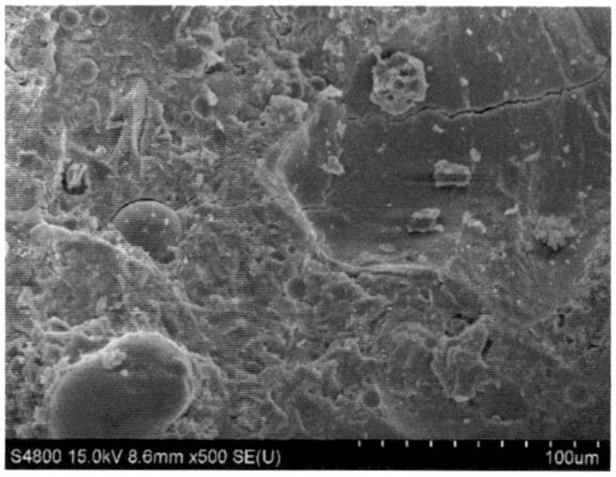

(d) Micro structure of ITZ between crushed stone and cement paste (×5000)

Figure 2. ITZ of between aggregate and cement paste

between crushed stone and hardened cement paste is obvious, and the bonding is loose. But the ITZ between ceramsite and hardened cement paste exists a large amount of hydration products such as C-S-H and a small amount of AFm, the hydration products form the compact three-dimensional structure, and thereby improve anti-permeability of specified density concrete.

3.2. Effect of pre-wetting degree of ceramsite on anti-permeability of concrete

Figure 3 is water penetration height of specified density concrete mixed ceramsite of pre-wetting time, as shown in the figure, the effect degree of pre-wetting degree on water penetration height of concrete are different with curing age. To the specified density concrete cured 7 d, the concrete mixed ceramsite of pre-wetting 0 h has lowest water penetration height. To the specified density concrete crued 28 d, the water penetration height of the concrete mixed ceramsite of pre-wetting 0 h do not change obviously, but that of concrete mixed ceramsite of pre-wetting 24 h obtains lowest water penetration height, and the anti-permeability of concrete reach to best.

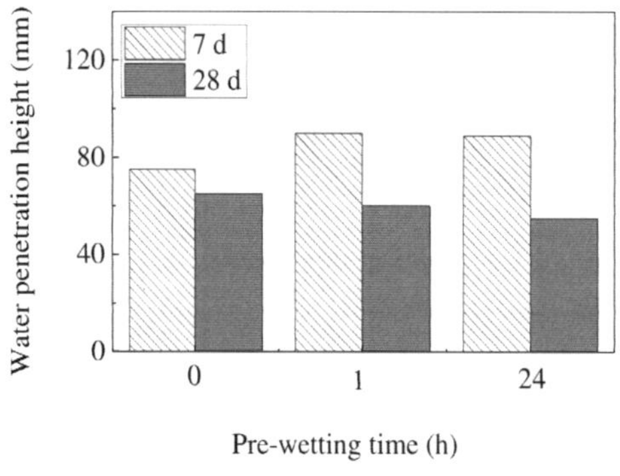

Figure 3. Water penetration height of specified density concrete mixed ceramsite of different pre-wetting time

Figure 4 is SEM images of specified density concrete mixed ceramsite of different pre-wetting time, according to figure 4(a), to the concrete mixed ceramsite of pre-wetted 0 h, the quantity of hydration products at ITZ is few so that the microstructure is loose. As shown in figure 4(b), with the pre-wetting time prolong, the quantity of hydration products at ITZ increase, and C-S-H begins fill into ITZ, results in microstructure of ITZ become compact. As shown in figure 4(c), ceramsite of pre-wetting 24 h make the quantity of hydration products increase obviously, those hydration products fill into ITZ, and form compact spatial network, and which indicates the pathway of water penetration decrease and anti-permeability increase. The change of the microstructure of ITZ can explain the internal reason that anti-permeability of specified density concrete increases with the increase of pre-wetting degree.

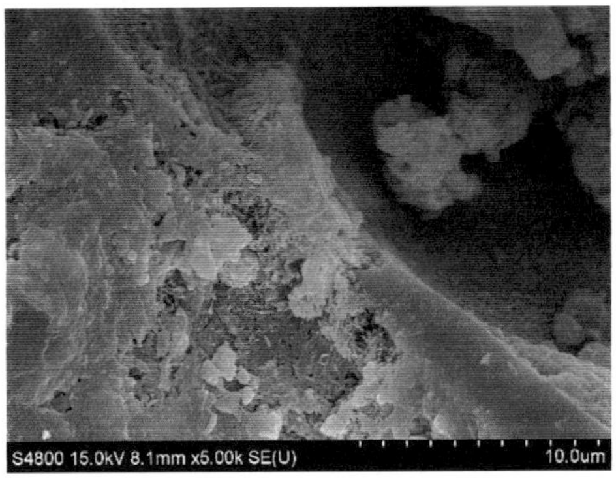

(a) 0 h (×5000)

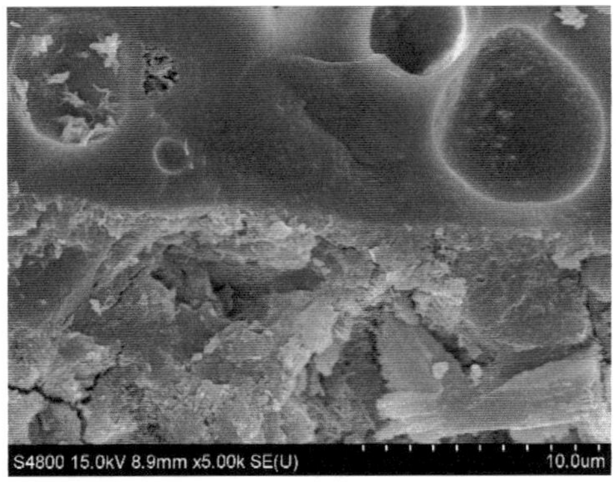

(b) 1 h (×5000)

(c) 24 h (×5000)

Figure 4. SEM images of specified density concrete mixed ceramsite of different pre-wetting time

3.3. Effect of W/B on anti-permeability of concrete

Figure 5 is water penetration height of specified density concrete with different W/B, as shown in the figure, when W/B is 0.30, water penetration height is lowest, which means the anti-permeability is best. With the increase of W/B, the anti-permeability decrease, the low W/B of concrete has less free water inside, and the connected pores caused by free water migration and bleeding are also reduced, and the interior is become dense and anti-permeability is improved [5]. With the increase of W/B, excessive moisture exists in ceramsite as free water, and after evaporating, the free water generates

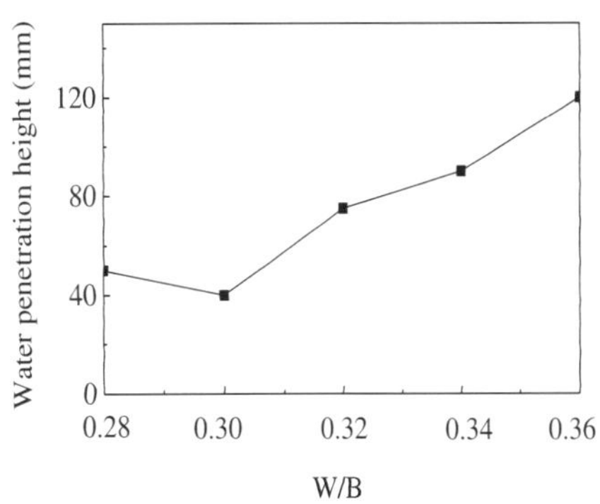

Figure 5. Water penetration of specified density concrete with different W/B

pores, and the penetration paths increase, which is disadvantage to anti-permeability. The higher W/B, the lower anti-permeability.

3.4. Effect of fly ash on anti-permeability of concrete

Figure 6 is water penetration height of specified density concrete mixed with different content of fly ash, as shown in the figure, with the content of fly ash between 0% and 30%, water penetration height decreases and anti-permeability improves, however, the rule is inversed when the content of fly ash over 30%.

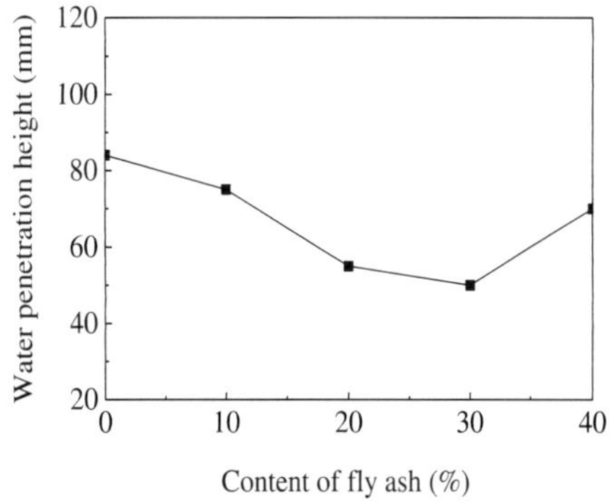

Figure 6. Water penetration height of specified density concrete mixed different content of fly ash

4. Conclusions

With the replacement rate of ceramsite between 0% and 20%, the anti-permeability of specified density concrete improves with the replacement rate of ceramsite, and the rule inverses when the replacement rate over 20%.

To the concrete cured 7 d, the anti-permeability of specified density concrete mixed ceramsite of pre-wetted 0 h is highest, and to the concrete cured 28 d, that of the concrete mixed ceramsite of pre-wetted 24 h is highest.

Anti-permeability of specified density concrete in inverse proportional with W/B.

Acknowledgments

The study of this paper is Supported by the Opening Project of State Key Laboratory of Green Building Materials, Science Foundation of Ministry of Housing and Urban-Rural Development of the People's Republic of China (No. 2016-K4-016), China postdoctoral science foundation (No. 2015M581702), and Yangzhou Science and Technology Planning Project (No. YZ2017290).

References

[1] Yuanchun Liu, Zhenfeng Huang 2019 Northern Architecture vol 4 p47
[2] JGJ51-2002 2003 China Architecture & Media Co. Ltd p2.
[3] Saraber.A, Overhof. R, Green T 2012 Waste Management vol 32 p144
[4] GB/T 50082-2009 2010 China Architecture & Media Co. Ltd p 6
[5] Yifan Gu, Jiaguo Sun 2019 Low Temperature Architecture Technology vol 41 p1

IOP Conf. Series: Materials Science and Engineering **758** (2020) 012073 doi:10.1088/1757-899X/758/1/012073

Wavy water thin-film characteristics on horizontal plate using PLIF Experimental

Mohamed Moustafa[1,2,3], Tian Ruifeng[1] and Islam S. Emam[4,5]

1 Harbin engineering university, college of nuclear science and technology, fundamental science nuclear safety, Nantong St., Harbin 150001, Heilongjiang, P.R.C

2 Mechanical Engineering Department, Faculty of Engineering, Alexandria University, Egypt.

3 Ministry of Electricity and Energy, Arab Republic of Egypt.

4 College of Materials Science and Chemical Engineering, Harbin Engineering University, Harbin 150001, China

5 Materials Engineering Department, Zagazig University, Zagazig, 44519, Egypt

E-mail: moustafa@hrbeu.edu.cn

Abstract. Nonlinear Thin liquid film is commonly applied in numerous engineering fields, including the Passive Containment Cooling System of the nuclear power plant and steam dryer in the steam generator of nuclear power plants. The counter gas flow effect through a rectangular channel was inspected, the flow is motivated by the pressure gradient and shear gas stress while the gravity force does not influence the flow. Different flow rates of gas were set to different Reynolds numbers from 6950 to 34800 with influence on a different range of input water Reynolds number from 170 to 875 respectively. The data was collected at different sections from inlet flow until the separation section. The thickness of liquid film investigated using (PLIF) experiments. The thickness has been calculated via digital image processing and MATLAB code. Film thickness was obtained as a function of time and frequency. Results were analyzed using statistic tools such as property density function (PDF), fast Fourier transform (FFT), and power spectral density (PSD).

1. Introduction

Investigation of moving film thickness on the surface represents an important part of the numerous engineering fields. For example, the Passive Containment Cooling System (PCCS) is using water as a reactor coolant. In the aspect of safety, economical and NPP design, flow characteristics in PCCS are of immense importance. This type of research aims to get full experimental results on flow performance of the small fluid film on the surface. In the latest years, different types of research have been conducted, numerical and experimental. Different experimental designs are used to investigate and examine the thickness of water film with different properties such as the experiment's geometry, different fluid parameters, etc. The measurement system used in this research is Planar Laser-Induced Fluoresce (PLIF) method. The examination model is attached to a CCD camera and laser source. The obtained results have been treated with MATLAB software.

2. Background

On a large scale in the different industrial applications, including all sorts of equipment where the thin liquid sheet is occurring, like distillation, lubrication, heating, and cooling processes, the different thickness of liquid film is outflowing down an inclined plate, because of its comparatively minor and high mass flow. Blocken et al. [1] studied the flows water of the rain on the fronts of the buildings. One of the usages of thin liquid film is washing the dust from fixed plates of steam turbines and in the PCCS. Observation of the thin liquid film is possible over large number of the phenomena where hydrodynamic film is passing over some solid shape and we can determine contact angle between liquid and solid. Paras et al. [2] studied formation and development of the rivulets liquid film with observation of creation of cyclic waves. For many years, scholars have conducted extensive research on wall liquid membranes. Hartley et al. [3] proposed a force balance standard when investigating stable dry spots which formed via isothermal liquid film rupture. Since then, the researchers have always concluded multi aspects such as thermal effect, airflow shearing, and special-shaped slab via the force balance standard. The force factor proposes various forms of force balance models. Zuber et al. [4] considered into calculation the thermal effects as liquid film vapor pressure, thermal capillary forces and examination the liquid film thermal cracking via the force balance standard. Murgatroyd et al. [5] proposed a force balance model driven by airflow shear force considering the imbalance between liquid-gas two-phase interface shear force and liquid-solid interface wall viscous force. Joo et al. [6]investigated the minor disturbance of the flow film and established a surface wave disturbance standard, based on the long-wave model of free flow film. Ye Xuemin et al.[7] investigated the wave instability of the falling film on the inclined plate and used the equation of Orr-Sommerfeld to obtain the long-wave transmission speed. The investigation displays that the gas-liquid interface shear force and the longitudinal gravity component defined the liquid film flow state; the liquid film flow state, the direction, and gas flow size determined the liquid film variation instability. Lory et al. [8] determined a problematic situation of film split at the corner of the plate-based the surface wave disturbance standard. Usually, the flow direction of gas and liquid has three kinds of relationships: same, vertical and opposite. The most significant modification between the three statuses, the liquid film flow driving force is different, and the cracking mechanism is various. At current, wall liquid film flow research mostly focuses on the first two methods, and there is less investigation on the third case. The liquid film of the steam separator wall at the steam generator of the power station boiler and the nuclear power plant flows downward by the action of gravity and is subjected to the airflow.

3. Experiment facility

This experiment has been performed using two-phase, air-water and system as shown in Figure 1. The system is a rectangle channel with air-inlet cross-section 10 cm x 2 cm and 25 cm long with a 30-degree angle at the corner and water-inlet cross-section 0.2 cm x 10 cm. The observed segment is composed of the organic glass and steel support.

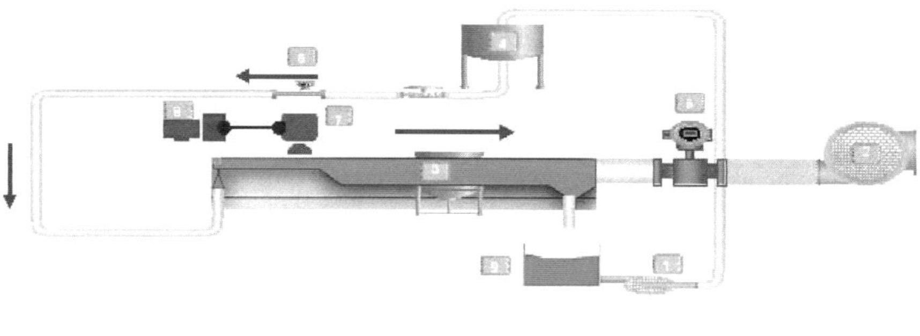

Figure 1 Sketch of the experimental system

1. Pump
2. Air-vacuum
3. Rectangular Channel
4. Main tank
5. Air-flow meter
6. Water-flow meter
7. High-speed camera
8. Computer
9. Drain tank

Three points were selected at identified distances from the corner to study generated wave, as shown in Figure 2.

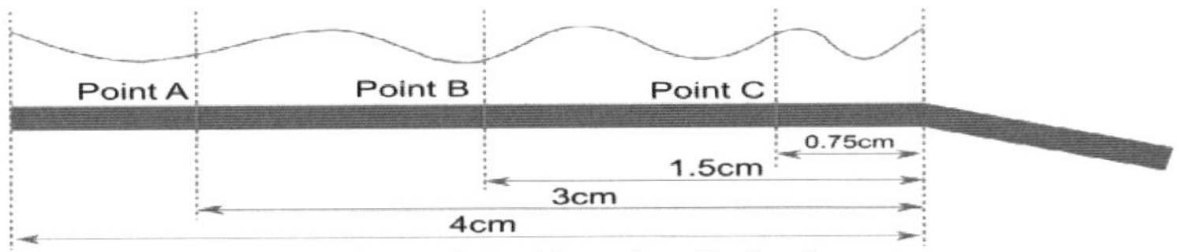

Figure 2 Position of studied points

As shown in Figure 3, the laser emits light of specific wavelength and turns the point source into a line source through a plan-convex cylindrical lens. The line source laser is irradiated onto the film in which the fluorescent dye is dissolved through the corrugated plate made of plexiglass and a flat, bright region is formed on the liquid film as shown in Figure 4.

Figure 3 CCD, laser, and model

Figure 4 Film image

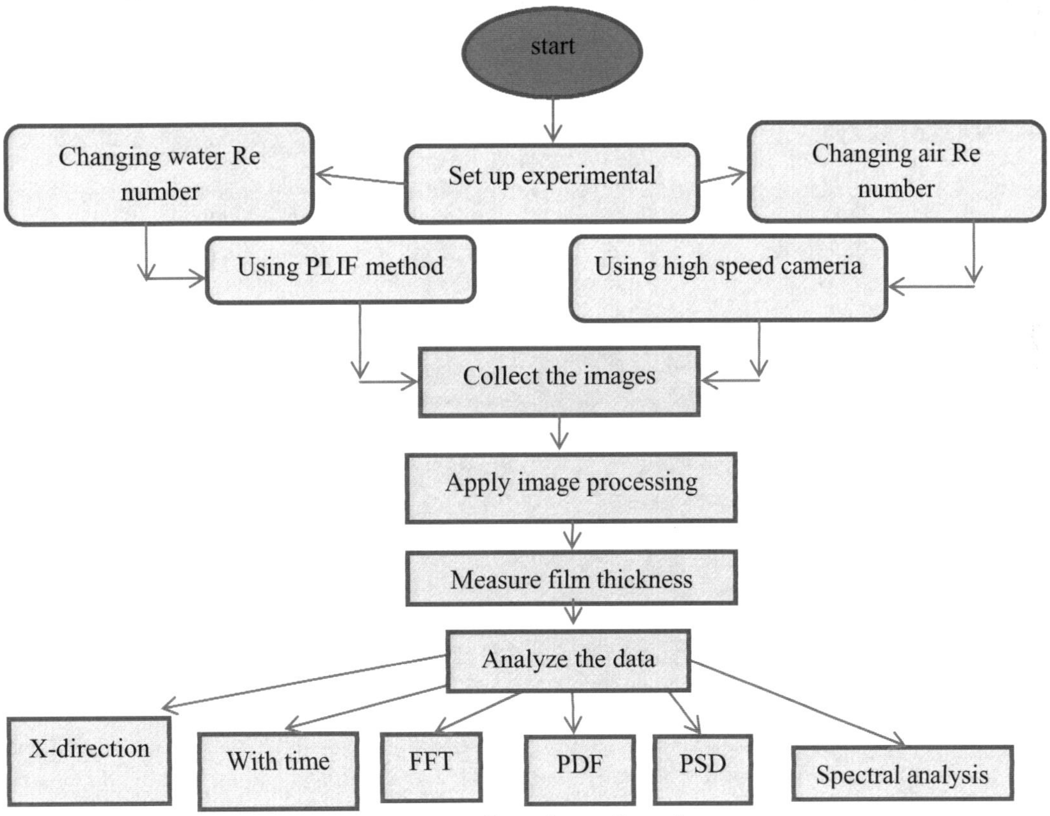

Figure 5 Experiment flow chart

4. Results and Discussion

IOP Conf. Series: Materials Science and Engineering **758** (2020) 012073 doi:10.1088/1757-899X/758/1/012073

To investigate film characteristics, we apply a different inlet velocity for water and gas.

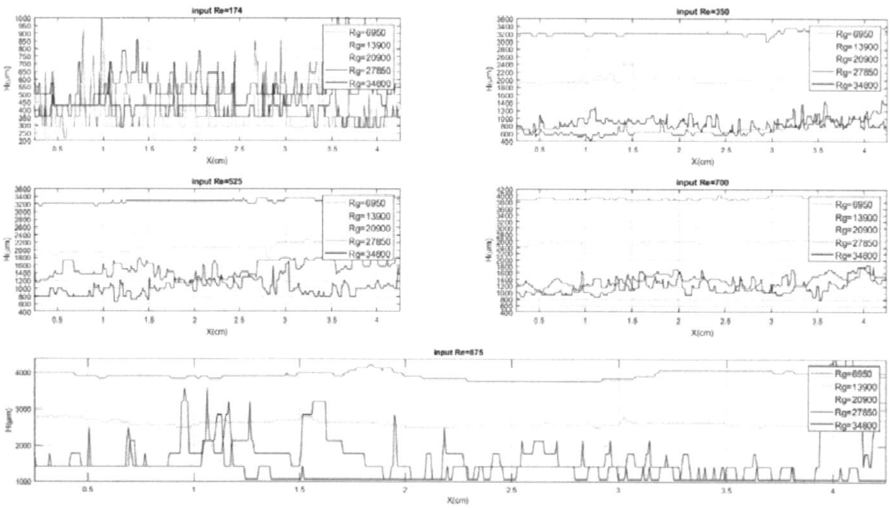

Figure 6 Water film thickness with respect to the x-direction

4.1 The behavior of the wave with the X direction

Figure 6 represents the wave shape of the film. Due to the problem of data recording, the slow-motion video has been taken in total of two seconds. After one second of the movie duration, a picture was taken. Results are recorded for five different inflow gas Reynolds (Re) number (6950, 13900, 20900, 27850 and 34800). At 174 water Re number and with a range of flow gases, solitary wave was detected for all five inflow gases; For water Re number within range of 350-700 and for 13900 and 20900 air Re number solitary wave and the capillary waves were detected and for high gas Reynolds number (27850 and 34800), capillary wave and airily waves appeared, while at 6950 air Re number, airily wave is observed. At 875 water Re number and for all five inflow air Re number airily wave is detected.

4.2 The behavior of the wave with respect to time

The evaluating points are curtained as Figure 7 for 2 seconds, where the sampling frequency is 3.2 kHz. The behavior of the wave is almost the same at all three positions (A, B and C). Wave behavior is not simple harmonic motion as presented in Figure 6. At 20900 air Re number, the behavior of capillary wave shows compressing passing with increase water Re number between two peaks. With low water Re number, the passing of the wave is direct without comprising passing wave. The film thickness fluctuation increases with increasing Re number.

4.3 The behavior of the wave with FFT analysis

FFT is demarcated as a process used to estimate a scale from a time domain to a frequency domain. The FFT spectrum analyzer input samples calculate the signal magnitude values and show the phase angle and values of that wave. It offers a decomposition of frequency in harmonics, hence can be studied individually. Therefore, it becomes more significant than the statistical analysis of a wave elevation and then is relatively essential to calculate a probability distribution function (PDF).

The water film frequency distinguishing is not precise at low Re numbers, the fluctuation of FFT is very high and the magnitude is small comparing with higher Re number.

At higher Re number, as shown in Figure 8, with a range of frequencies from 0Hz – 250Hz the fluctuation of FFT and the magnitude are very high, while with the range of frequencies from 250Hz – 2800Hz, the fluctuation of FFT is very high while the magnitude is small. So to study that curves more clearly we will apply PDF analysis.

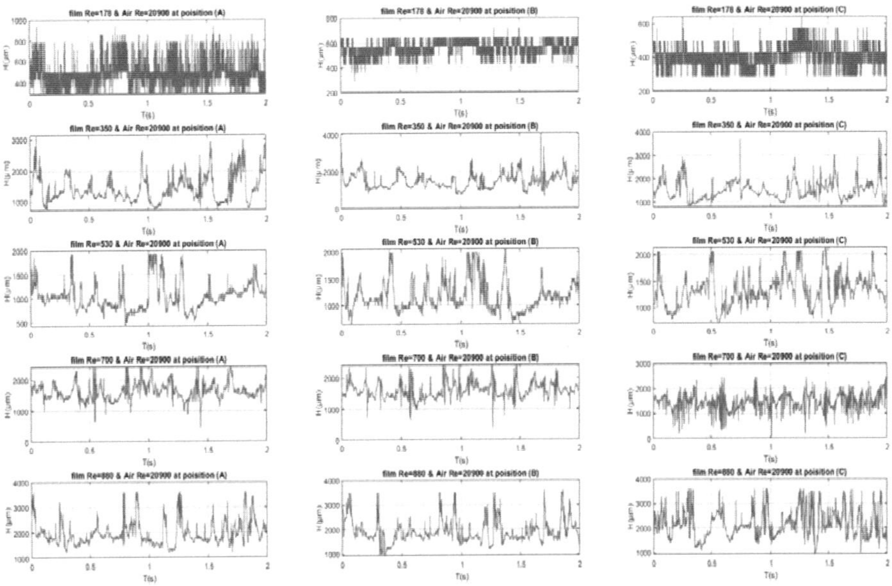

Figure 7 Film thickness with respect to time

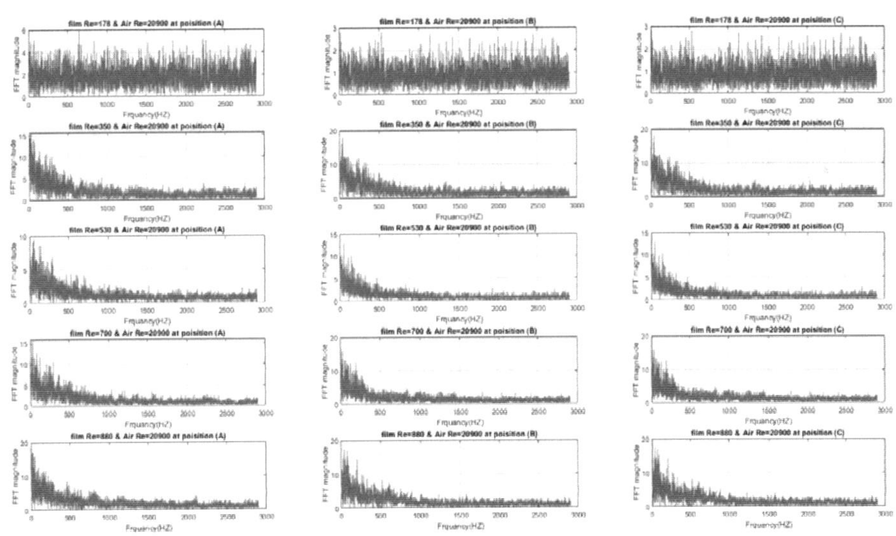

Figure 8 FFT with respect to frequency

4.4 The behavior of the wave with PDF analysis

The curves of PDF have been shown in Figure 9, with presented peak positions with the highest probability density. With low water Re number the PDF magnitude is very high, however, the density of peaks is small. At 174 water Re number we can notice peaks at two frequency values, in comparison with increasing water Re number to 350, we can see that the density of peaks of PDFs is increasing for different frequency values in sequence. We can interpret that the comprising passing wave at high water Re number is causing higher density of PDFs.

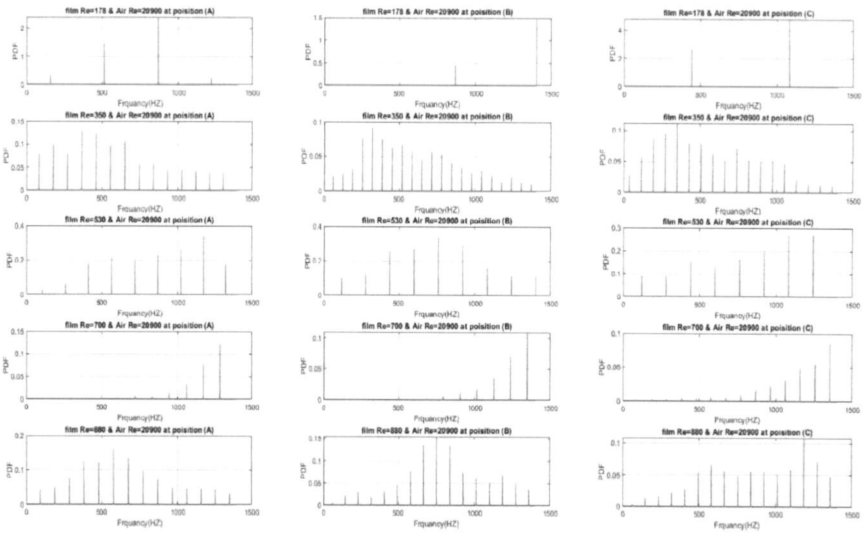

Figure 9 PDF with respect to frequency

4.5 The behavior of the wave with PSD examination

As shown in Figure 10, the power spectral density profile of the film, for different water Re numbers is approximately similar. The wave distribution all over the frequency set shows the huge-frequency features of film thickness variation. For each graph, between 0 kHz and 0.25 kHz, the curve falls rapidly and after the fall is more gradual. Moreover, in that range the height of the PSD stays considerably within a line of change. This Figure reveals that this signal has multi-frequency with energy variation.

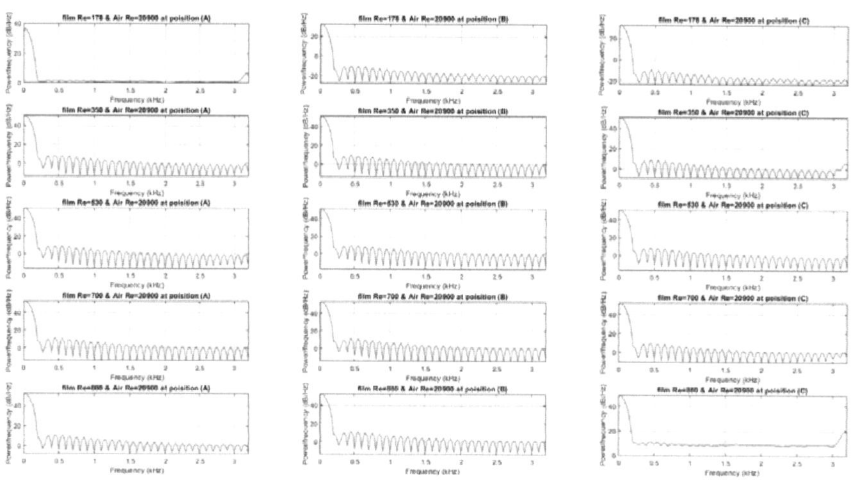

Figure 10 Power spectral analysis

4.6 The behavior of the wave with spectral analysis

Film frequency is spread all over the frequency set as shown in Figure 11, considering the multi-frequency film features variation, although frequency set is not clear. With increasing the water Re number the multi-frequency scattering is more evident in the spectral diagram. Continuous spreading happens through all diagram area and it is regular spreading. The specific frequency spreading of the film thickness at all Reynolds numbers, that is, the characteristic frequency has a particular time periodicity.

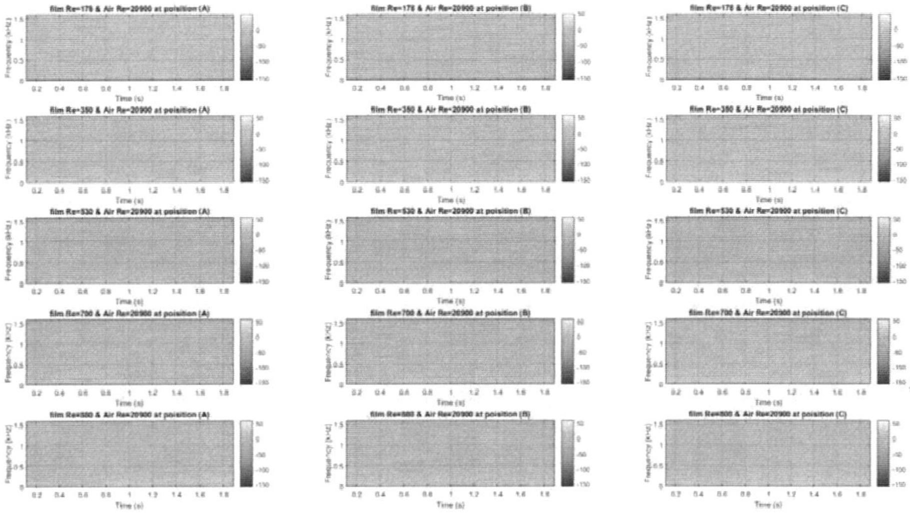

Figure 11 Spectrogram

5. Conclusion and summary

Research on water film features over the horizontal plate is investigated in this paper using (PLIF) method. The film thickness is estimated by the application of time domain, FFT analysis, probability density function, and spectrogram and power spectral density. Water film fluctuation characteristics are investigated and a brief illustration of the conclusion is pointed as:

1. The wave has changed in the x-direction and depending on the combination of air and water Re number, the different types of waves have been detected.
2. The wave of the film changed with time at three-point of study and the behavior of the wave at each point is the same for each case, but the waves that occur are not simple harmonic motion.
3. There is no clear periodical time for the studied wave. Some waves like solitary, capillary, and airily wave are caught on film of wave behavior.
4. With PDF analysis we can notice the compressing wave and PSD analysis reveals that the water film surface has multi-frequencies.

Reasons for different wave occurrences at different air and water Re numbers have not been part of this research and we propose that the next research on thin liquid film should be to investigate the roughness of the plate and relation between air shear stress and free water surface.

Acknowledgments

The authors want to thank Rawya Zamzam and Menna Allah Moustafa and Emir Hanic for their support and assistance. We would like to show appreciation the College of Nuclear Science and Technology, HEU and China Scholarship Council for supporting this research

References

[1] Blocken B, Derome D, Carmeliet J. Rainwater runoff from building facades : A review[J]. Building and Environment, Elsevier Ltd, 2013, 60: 339–361.

[2] Paras S V, Karabelas A J. Karapantsios,'~" karabelas~[J]. 1989, 15(1): 1–21.

[3] Hartley D E. CRITERIA FOR THE BREAK-UP OF THIN LIQUID LAYERS FLOWING ISOTHERMALLY OVER SOLID SURFACES TW = dX S P [WY) I ~ / ~[J]. 1964, 7: 1003–1015.

[4] Zuber N. FILMS OF DRY PATCHES FLOWING FORMING IN LIQUID OVER HEATED Ix[J]. 1966, 9: 897–905.

[5] Murgatroyd W, Road M E, London E. THE ROLE OF SHEAR AND FORM FORCES IN THE STABILITY[J]. 1965.

[6] Joo S W, Davis S H, Bankoff S G. Long-wave instabilities of heated falling films : two-dimensional theory of uniform layers[J]. Harbin Engineering University, 1991, 230: 117–146.

[7] YE Xuemin, WANG Songling, ZHANG Ying et al. Hydrodynamics and Instability of Shear Interfacial Waves of Liquid Films[J]tle[J]. Journal of Shanghai Jiaotong University, 2008, 42(8): 1243–1253.

[8] LLORY D, Le COZ J F H C. Liquid film atomization on wall edges-separation criterion and droplets formation model[J][J]. Journal of Fluids Engineering, 2002(124): 565–575.

[9] Calculation G F. A model for liquid film flow and separation in a wave-plate mist eliminator[J]. 2005(May).

Influence of Cr^{3+} Doping on Photocatalytic Hydrogen Production Performance of TiO_2 Film Growing in Situ on the Surface of Ti Plate

Xuefeng Bai[1,2], Wei Wu[2]

1 Institute of Petrochemistry, Heilongjiang Academy of Sciences, Harbin 150040, China
2 School of Chemistry and Material Sciences,Heilongjiang University,Harbin,150070, China
E-mail: tommybai@126.com

Abstract: The Cr^{3+} doped TiO_2 films (Cr@TiO_2) were prepared by in-situ growing on the pure Ti plate substrate in the electrolyte solution of Na_3PO_4 containing a certain amount of $K_2Cr_2O_7$ using micro-arc oxidation (MAO) technique. The Cr@TiO_2 were characterized by SEM, XPS, XRD and DRS. The influence of Cr^{3+} doping amount on the morphology, structure and photocatalytic hydrogen production performance of Cr@TiO_2 was studied. The characterization results showed that Cr^{3+} doping into TiO_2 film not only affects the crystallinity of TiO_2, but also affects the composition of crystal phase, and the maximum absorption edge of Cr@TiO_2 reflects an obvious redshift, reaching around 650 nm of visible light with the increase of Cr^{3+} content. The Cr^{3+} doping was good for the improvement of photocatalytic performance. The Cr@TiO_2 prepared under the condition of $K_2Cr_2O_7$ adding amount of 0.29 g/L exhibited the highest rate of 0.378 mol/ cm^2•h.

1. Introduction

Since the photocatalytic splitting of water on TiO_2 electrodes was found by Fujishima and Honda in 1972[1], a great many research have been carried out in theory and application of the photocatalytic reaction. At present, the environmental pollution has become an obstacle to sustainable development. Hydrogen as a clean fuel is the best way to solve environmental pollution [2].

The powdered TiO_2 exhibits excellent mass transfer performance, but low efficiency of light utilization and difficult cycle use of catalysts for suspension system has limited its application. The TiO_2 film can better solve the above problems and is beneficial to industrialization [3].

The production methods of TiO_2 film include the CVD technology [4], sol-gel method [5],sputtering and electrochemical method[6]. The micro-arc oxidation (MAO) technique allows the in-situ growing of ceramic-like film with excellent chemical durability and bonding strength on the surface of metals in electrolyte solutions[7~8]. Metal-doped TiO_2 film on Ti surface can be prepared by adding metallic and nonmetallic elements to the electrolyte solution [9].

In this research, MAO technique was used to prepare the Cr^{3+} doped TiO_2 film(Cr@TiO_2) by using the pure Ti plate as substrate. The effect of Cr^{3+}doping amount on the morphology, structure and photocatalytic hydrogen evolution performance of TiO_2 film was investigated.

2. Experimental

2.1. Preparation of Cr@TiO_2 film

All chemicals used are chemically pure. Commercial pure Ti plate as the substrate needs to be cleaned by a mixed solution of $HNO_3/HF/H_2O$. The MAO treatment of Ti plate was carried out in 0.11 mol/L Na_3PO_4 electrolyte solution containing a certain amount of $K_2Cr_2O_7$ using HG-I device as power source with current density of 4 A/cm^2 for oxidation time of 9 min. The final Cr^{3+} doped TiO_2 film(Cr@TiO_2) was obtained by washing and drying.

2.2. Characterization of Cr@TiO_2 film

The X-ray diffraction spectra (XRD) were obtained on a Bruker D8 Advance X-ray diffractometer using Cu Kα irradiation. The patterns were recorded from 10° to 70° (2θ) . The UV-Vis diffuse reflectance spectra (DRS) were completed on Shimadzu UV-2450 spectrophotometer in the scanning range from 250 to 700 nm. The experimental data were obtained by transferring from reflection to absorption by Kubelka-Munk method. The scanning electron microscopy(SEM) was characterized on a HITACHI S-4800 instrument operated at 20 kV. The X-ray photoelectron spectroscopy analysis (XPS) was completed on a Thermo Electron instrument.

2.3. Evaluation of photocatalytic hydrogen production performance

Photocatalytic experiments were carried out in an inner-irradiation quartz reactor. A 250 W high-pressure lamp was been used as light source. In order to get visible light, 1 M $NaNO_2$ solution was added into an inner circulating system to remove the UV part of the light with a wavelength less than 400 nm. The above-prepared Cr@TiO_2 films were added into aqueous mixed solution of 0.25 M Na_2SO_3 and 0.35 M Na_2S. Before opening light source, N_2 was purged through as-above prepared solution to remove oxygen. The photocatalytic reaction temperature was controlled at 30 ± 5℃.Photocatalytic hydrogen production performance was evaluated by collecting the amount of hydrogen evolution by drainage extraction method. Hydrogen purity analysis was conducted on GC with thermal conductivity detector .

3. Results and discussion

3.1. Structure

Figure 1 showed XRD patterns of Cr@TiO_2 with different Cr^{3+} doping amount. It was shown in Figure 1 that with the increase of the addition amount of $K_2Cr_2O_7$ in the electrolyte, the characteristic peak strength of anatase located in 2θ=25.28° gradually weakens and shifts towards a small angle. When the addition amount of $K_2Cr_2O_7$ exceeds 0.58g /L, the characteristic diffraction peaks of rutile phase (110) crystal plane located at 2θ=27.45° and rutile phase (101) crystal plane at 2θ=36.09° can be found in the XRD patterns, indicating that the addition of $K_2Cr_2O_7$ not only affects the crystallinity of TiO_2, but also affects the composition of crystal phase. When the addition amount of $K_2Cr_2O_7$ exceeded 0.87g /L, the characteristic peak of anatase disappeared, indicating that the addition of $K_2Cr_2O_7$ hindered the growth of anatase TiO_2 in the system and affected the crystallinity of TiO_2. When the addition of $K_2Cr_2O_7$ exceeded 1.16 g/L, no other peaks were found except the basal peak. The color of prepared TiO_2 film changed from grey to golden, but the color of the film did not change significantly with the addition amount of $K_2Cr_2O_7$. When the addition amount of $K_2Cr_2O_7$ exceeded 1.45g /L, micro-arc discharge could not be generated, and the generated golden yellow film was broken off on the surface of the substrate, exposing the substrate. TiO_2 thin film prepared under such conditions could not be used for photocatalytic evaluation.

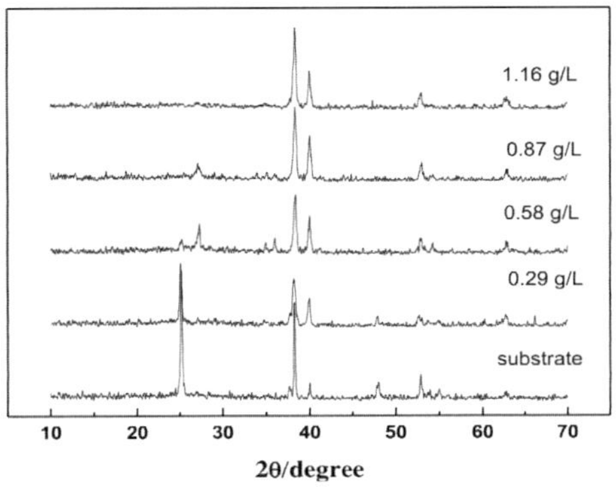

Figure1. XRD patterns of Cr@TiO_2 films prepared at different concentrations of $K_2Cr_2O_7$

3.2. Optical Properties

Figure2 showed the DRS spectra of Cr@TiO_2 with different Cr^{3+} doping amount. With the increasing of the addition amount of $K_2Cr_2O_7$ into the electrolyte, the maximum absorption edge of Cr@TiO_2 showed an obvious red shift, reaching around 650 nm of visible light. The reason for this phenomenon is due to the formation of impure energy levels by the Cr^{3+} doping in the forbidden band, reducing band gaps to enhance utilization of visible-light. However, when the addition amount of $K_2Cr_2O_7$ exceeds 0.87g /L, anatase phase TiO_2 no longer exists in the film layer, and rutile phase TiO_2 plays the role of light absorption in the film layer. Since no EDTA complexant was added to the electrolyte during Cr^{3+} doping, the color of the solution did not change after the reaction. It indicated that Cr^{3+} was not deposited on the surface of film, but was doped into the lattice of TiO_2.

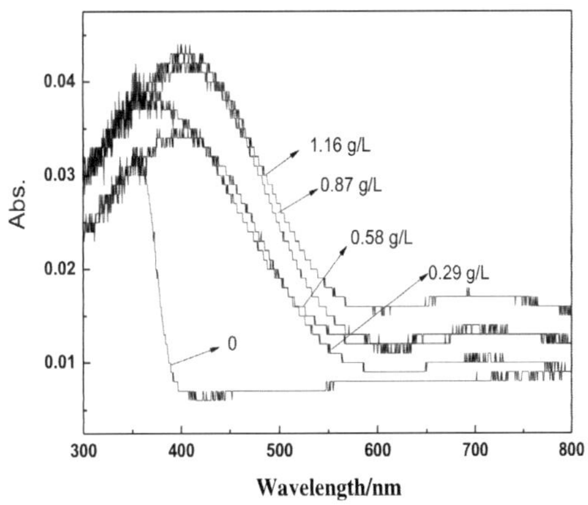

Figure 2.UV–Vis absorption spectra of Cr@TiO_2 films prepared at different concentrations of $K_2Cr_2O_7$

3.3. Morphology

Figure 3 showed that SEM images of Cr@TiO_2 with different Cr^{3+} doping amount. It was seen from Figure 3 that with the addition amount of $K_2Cr_2O_7$ into electrolyte, the corrosion degree of TiO_2 film surface increases, indicating that $K_2Cr_2O_7$ will cause corrosion of the film during the process of Cr^{3+} doping. When $K_2Cr_2O_7$ addition amount is less than 0.58 g/L, there are a lot of holes with the diameter of 2~4 μm on the film surface due to the electric breakdown. Damage on the film surface morphology caused by corrosion is seen at the adding amount of more than 0.58 g/L. When the amount of $K_2Cr_2O_7$ continues to increase more than 1.45 g/L, the voltage to 100 V or so will stop growing, micro-arc discharge cannot occur. No oxided film can be formed on the surface of the substrate, leaving only tiny discharge channels.

IOP Conf. Series: Materials Science and Engineering **758** (2020) 012074 doi:10.1088/1757-899X/758/1/012074

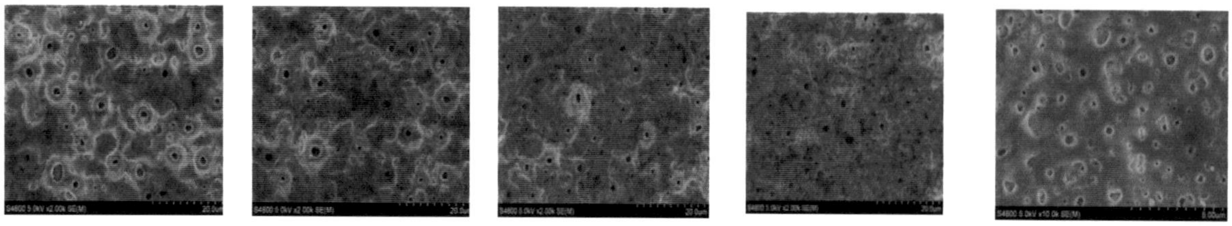

a b c d e

Figure 3. SEM images of Cr-TiO_2 films prepared at different concentrations of $K_2Cr_2O_7$ (a)0.29 g/L(b) 0.58 g/L (c)0.87 g/L (d)1.16 g/L (e) 1.45 g/L

3.4. Surface State Analysis

Figure 4 (a~d) showed that XPS analysis of Cr@TiO_2 films with different Cr^{3+} doping amount. The characteristic peaks of survey spectrum are seen in Figure 4 (a). The P2p diffraction peak at 134.7 ev was also found in the XPS spectra. As shown in the Cr2p spectra of Figure 4(b), there are two peaks at 576.4 eV and 586.4 eV corresponding to the $Cr2p_{1/2}$ and $Cr2p_{3/2}$, indicating Cr^{3+}doping into the film layer, and two peaks at 579.7 eV and 589.1 eV corresponding to the $Cr2p_{3/2}$ and $Cr2p_{1/2}$, indicating that Cr^{6+} is contained in the film layer. The diffraction peaks at 458.8 eV and 464.5 eV in Figure 4(c) correspond to $Ti2p_{3/2}$ and $Ti2p_{1/2}$ respectively, indicating that the valence of Ti is 4+.The binding energy of O in Figure 4(d) indicates that only lattice oxygen exists in the film, and no chemisorption oxygen exists.

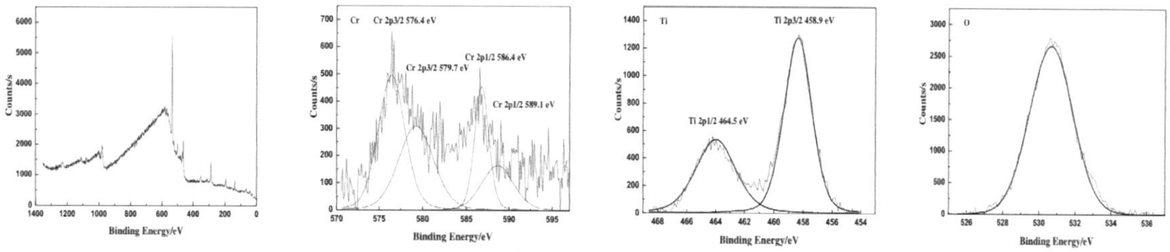

a b c d

Figure4. XPS spectra of Cr@TiO_2 films prepared at different concentrations of $K_2Cr_2O_7$ (a) Survey spectrum, (b) Cr, (c) Ti (d) O

3.5. Photocatalytic Hydrogen Production Performance

Figure 5 showed the photocatalytic hydrogen evolution rate of with different Cr^{3+} doping amount. H_2 production rate showed a trend of increasing first and then decreasing with the increasing of Cr^{3+} doping amount. The Cr@TiO_2 with the addition of $K_2Cr_2O_7$ of 0.29 g/L exhibited the highest rate of 0.378 mol/ cm^2•h. This was because the increase of the concentration of $K_2Cr_2O_7$ in the electrolyte hindered the growth of anatase phase TiO_2, and seriously corroded the film layer, causing the generated film to break away from the substrate, which directly affected the photocatalytic activity.

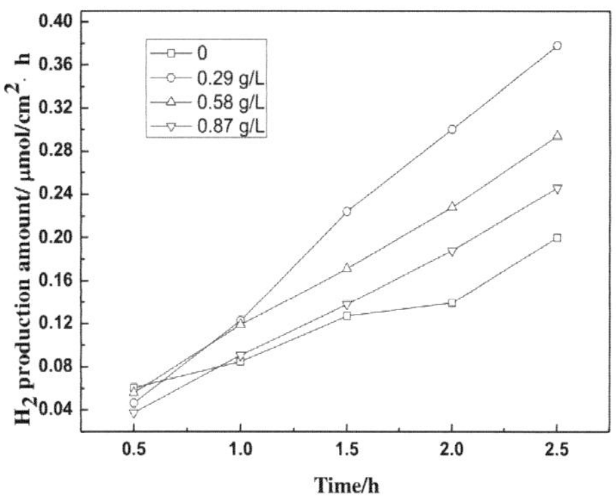

Figure5. H_2 production rate over the Cr@TiO_2 films prepared at different concentrations of $K_2Cr_2O_7$

4. Conclusions

The Cr^{3+} doped TiO_2 film on the pure Ti plate have been prepared by micro-arc oxidation process (MAO). The addition of $K_2Cr_2O_7$ not only affects the crystallinity of TiO_2, but also affects the composition of crystal phase. The Cr^{3+} doping into TiO_2 film results in the formation of impure energy levels in the forbidden band, reducing band gaps to enhance utilization of visible-light. The maximum absorption edge of Cr@TiO_2 reflects an obvious red shift, reaching around 650 nm of visible light with the increase of Cr^{3+} content. The Cr^{3+} doping was good for the improvement of photocatalytic performance. The Cr@TiO_2 with the addition amount of $K_2Cr_2O_7$ of 0.29 g/L exhibited the highest rate of 0.378 mol/ cm^2•h.

5. References

[1] Fujishima, A Honda K, 1972 *Nature* **238** 37

[2] Modisha P M, Ouma N M, Garidzirai R, Wasserscheid P, 2019 *Energy & Fuels* doi:10.1021/acs.energyfuels.9b00296

[3] Bensouici F, Bououdina M, Dakhel A A, Tala-Ighil R, 2016 A*pplied Surface Science* doi:10.1016/j.apsusc.2016.07.034

[4] Rasoulnezhad H, Kavei G, Ahmadi K, Rahimipour M R, 2017 *Applied Surface Science* 408 1

[5] Qiao L,Xie F,M Xie,Gong C, 2016 *Transactions of Nonferrous Metals Society of China* 26 2109

[6] Seeger S, Ellmer K, Weise M, Gogova D, 2016 *Thin Solid Films* 605 44

[7]Wang Y,Yu H, Che C n,Zhao Z, 2015 *Materials & Design* 85 640

[8] Liu S, Li B, Liang C,Wang H, Qiao Z,2016 *Applied Surface Science* 362 109

[9]L Zhang,Q Gao,Y Han, 2016 *Journal of Materials Science & Technology* **32** 919

IOP Conf. Series: Materials Science and Engineering **758** (2020) 012075 doi:10.1088/1757-899X/758/1/012075

Improved mechanical properties of a high plasticity clay soil by adding recycled PET

A Alvarez1, J Sosa2, G Duran[3] and L Pacheco[4]

1 Anthony Robert J Alvarez Cabrera, UPC, Bachelor degree, Peru.
2 Jose Antonio Sosa Gonzales, UPC, Bachelor degree, Peru.
3 Gary Duran Ramírez, UPC, Professor, Peru.
4 Lidia Miranda Pacheco, UPC, Professor, Perú
Email: u201211400@upc.edu.pe[1], u201416653@upc.edu.pe[2], gary.duran@upc.pe[3], ppcilpac@upc.edu.pe[4]

Abstract. To the problems posed by clay soils of high plasticity, such as vulnerability to variations in volume, producing differential settlements causing damage to structures, it is due to factors such as this it was decided to seek an alternative enhancement to parameters such as increased resistance, friction angle and cohesion of a high plasticity clay soil (CH). In the execution of civil works, these parameters generally do not conform to the requirements of the project, for that reason should seek a solution for the stability of the soil involved in any construction. The constant search for improving environmental conditions has led to the use of waste plastic to reinforce the soil and improve its mechanical and physical properties. That is why according to our proposal we try to increase data for the use of this alternative material as a soil improvement proposal, obtaining interesting and different results to that found in other investigations that used similar materials. We chose crushing polyethylene terephthalate (PET) obtaining a recycled PET 3-5 mm length as a reinforcing material to stabilize and improve the mechanical properties of the soil. According to standard proctor assays and direct shear, it was determined that the optimum percentage to be added is from 1% to study the soil because of better behaviors were observed study material with respect to the other mixtures.

1.Introduction

Techniques soil improvement introduced by geotechnical engineering have been developed to continuous with the passage of time, one of these inputs to soil improvement with the addition of natural fibers (twigs, grass, leaves). According to studies a major problem worldwide is the production of plastic materials such production of non-biodegradable has been increasing and thus waste the same as polyethylene terephthalate (PET). This residue, in the world of engineering, is commonly used in the manufacture of geosynthetic materials for which is multes, filtered and extruded into fibers polyester smooth appearance, good tensile strength and low absorption capacity water, with which geotextiles type and reinforcing geogrids are made. On the other hand, with the addition of this residue seeks to mitigate the environmental impact, so it is expected that the use of the same is an option as soil improvement technique to ensure stability and useful life of the structures. S Peddaiah [1] This paper presents a detailed study on the behavior and the use of plastic waste in improving the soil, experimental research on the ground, results reinforced plastic showed that plastic can be used as an effective stabilizer in order to find a problem of waste disposal, as well as an economical solution for stabilization of weak soils. M Mirzababaei [2] on the basis of the experiments carried out in this study, the optimal fiber to be added to increase the shear resistance of the clay soil with an initial vacuum ratio of e0 = 0.64 was found to be 0, 25% with 140%, 81% and 23% increase in the ratio of the tension on the ground without problems in the normal effective tensions of 50, 100 and 200 kPa, For all soil samples tested in this study, it found that the stress ratio at 200 kPa of normal effective stress is in the ranges of 0.45 and 0.60. Thus J Hoseein [3]It aimed to analyze the use of plastic materials waste for reinforcing clay soils, which is why to investigate the compressibility and behavior to the undrained shear clay soil mixed with plastic waste mixed study material with various amounts of plastic waste (0, 0.5, 1.0, 1.5 and 3.0% by dry weight). According to the results, the waste plastics do not affect the characteristics of clayey soil compaction considerably. Addition of waste plastics exceeding a specific value (1.0% in this investigation) changes the behavior of undrained samples from dilatative contraction. Also, B Mishra [4] was based on the effect of soil properties to performing the addition of polyethylene terephthalate (PET) in combination with fly ash applied to a base of flexible pavement, it is why the properties studied were the resistance to the cutting module cutting using clayey soil. The proportions ranging from 0% to1.6% by weight of the earth with an increase of 0.4% from PET fibers while 0% to 20% by weight of the soil with a 5% increase in if fly ash. The optimum amount to be found, recycled fibers 1.2% with 15% fly ash by weight of the floor, which showed improved strength parameters of the sub-base soil. Finally, Zukri [5] discloses that many countries are concerned about the problems caused by polyethylene terephthalate (PET) and polypropylene (PP), this economic and population growth development is due. For this reason the authors presented results of an investigation into the use of fiber products such as PET bottles and PP products in order to improve the engineering properties of clay soil Pekan. Finally concluded that the fibers of PET and PP can be used successfully as reinforcing materials for stabilizing clay soils, the use of these compounds waste as alternative materials for stabilizing clay soils is reasonable and cost effective because it they are constantly available.

2. MATERIALS

2.1 High plasticity clay soil (CH)

A clay soil high plasticity which has the physical characteristics required for continued research was obtained.

2.2 PET crushed

Furthermore, the PET factory Gexim SAC, which was ground to have dimensions between 3 -5 mm was obtained.

3. EXPERIMENTAL PROCEDURE

3.1 Elaboration of soil mixtures

For this investigation five mixtures were made, which is presented in Table 1:

Table 1. Names and mixtures nomenclatures

Name	Nomenclature
Natural soil	S100 + 0% PET
Natural soil + 0.5% PET	S100 + 0.5% PET
Natural soil + 1.0% PET	S100 + 1.0% PET
Natural soil+ 2.5% PET	S100 + 2.5% PET
Natural soil+ 3.5% PET	S100 + 3.5% PET

3.2 Physical characterization

3.2.1 Grain size by sedimentation

Because our soil is very fine sought to make a distribution of particle size distribution by testing by hydrometer.

3.2.2 Atterberg limits

To determine if the soil with which you are working is a high plasticity clay chose to find the LL, LP and IP.

IOP Conf. Series: Materials Science and Engineering **758** (2020) 012075 doi:10.1088/1757-899X/758/1/012075

3.2.3 *Specific gravity (Gs)*

This test is responsible for determining the specific gravity of the solids in a soil sample the particles are less than 4.75mm.

3.3 Mechanical characterization

3.3.1 Standard Proctor

Assay Proctor standard for the 5 mixtures because it seeks to make a computer data about the material, which is why the compaction test is not performed by testing modified Proctor, which is done trials have performed high demand loads.

3.3.2 Direct Shear

The direct shear test was performed for the five mixtures, in order to determine their friction and cohesion parameters of the above.

4 Results and Analysis

4.1 Granulometry

In Figure 1 the granulometric curve shown hidrometer, obtaining the percentages of materials found 200 mesh before and after it. The results were the following:

Table 2. Percentage composition of materials in the sample.

RESULTS	
Gravels (%)	0.00
Arenas (%)	13.52
Silts (%)	42.50
Clays and Colloids (%)	43.98

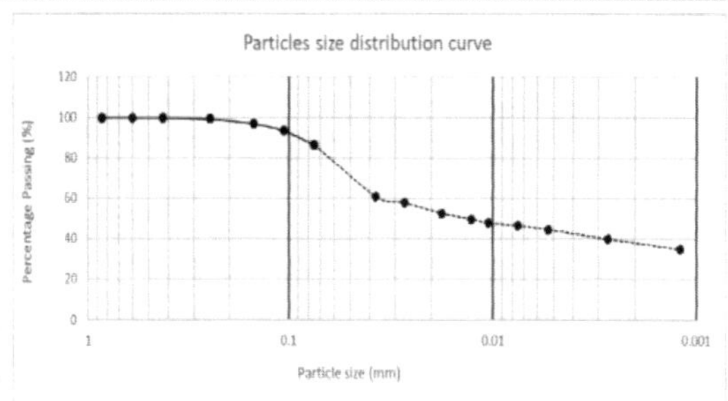

Figure 1. Particles size distribution curve

4.2 Atterberg limits

Table 3. Summary of Atterberg limits

Atterberg limits	Percent
Liquid limit	57.52
Plastic limit	18.41
Plasticity index	39.11

Note: the liquid, plastic and plasticity index of the study material was determined limit.

After obtaining the values obtained, we proceeded to enter the plasticity chart the liquid limit and plasticity index. Thus determining that it is a high plasticity clay (CH).

4.3 Rating Gs

Of assay specific gravity was obtained as result a value of 2.71, thus being this value is compared with a table of average values according to ASTMD 854 standard being verified for the second time is a clay of high plasticity as study material.

4.4 Standard Proctor Test

Table 4. Results Standard Proctor Test

Nomenclature	Max. Dry Density (g / cm3)	Optimal cont. hum. (%)
S100 + 0% PET	1,618	20.25
S100 + 0.5% PET	1,612	21.70
S100 + 1.0% PET	1,606	22.23
S100 + 2.5% PET	1,495	26.00
S100 + 3.5% PET	1,477	26.75

Note: We determined the maximum dry density and moisture content for each optimal mixture.

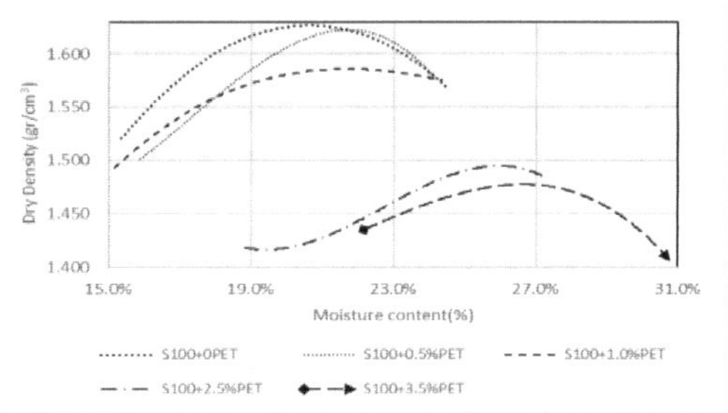

Figure 2. Compaction curves

A progressive decrease is evidence of the maximum dry density of the samples with additions of PET, this occurs because the moisture content increase as the percentage of PET reinforcement are added to the soil samples. This occurs because the particles of the material, which are waterproof, do not allow the passage of water is so retained. Also, the addition of progressively higher percentage of reinforcement there is a separation of the particles of the ground which causes the decrease maximum density of the samples dry.

4.5 Direct shear test

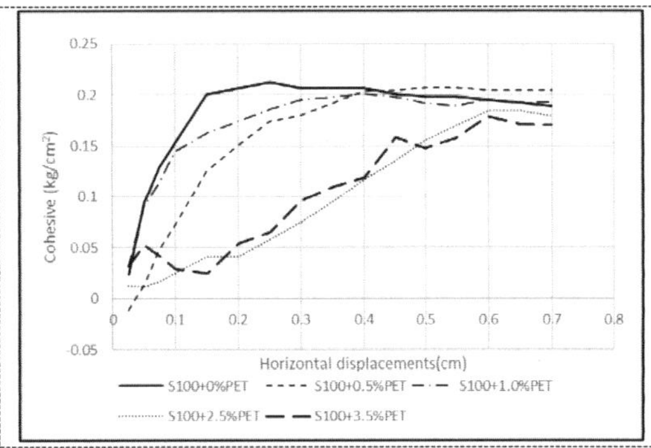

Figure 3. Cohesive Parameters

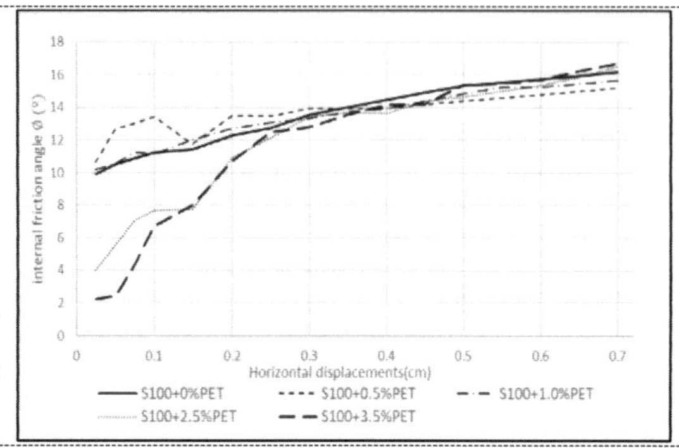

Figure 4. Parameters friction angle

It is observed in Figure 3 that for the S100 + 0% PET the cohesive is increasing at least to 0.25 cm horizontal displacement and then reduced to 0.18885 and seek constant trend C. To S100 + 0.5%PET a progressive increase is observed of the cohesive which gives signs of great improvement of the study material added to the PET, surpassing the cohesion of the S100 + 0%PET residual stress. Finally, the S100 + 3.5% PET is observed discontinuities in the graph and verifying that the particles of the mixture have failed the cohesive than the S100 + 0% PET, can say that it is not an optimal mix.

Also in Figure 4 to S100 + 0%PET, the angle of internal friction for displacing 0.25 cm is 12.73° indicating that this value is increasing until 16.18° angle is achieved through displacement of 0.80 cm. Furthermore, for other mixtures is noted that the graph does not change much because all friction angles have similar values in shift of 0.70 cm, just variations observed at the start of the assay these being smaller in the mixtures S100 + 2.5% PET and S100 + 3.5%PET.

4.6 Volumetric analysis

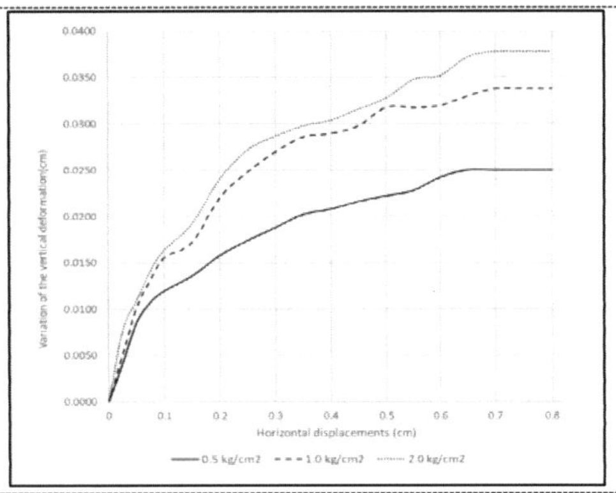

Figure 5. Volumetric analysis of the S100 + 0% PET

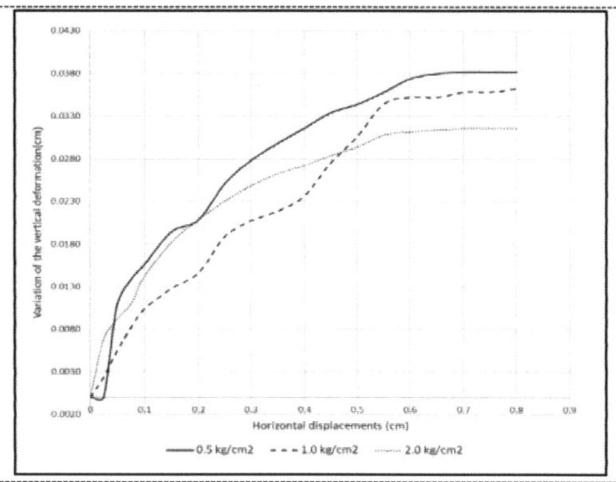

Figure 6. Volumetric analysis of S100 + 1.0% PET

To Figure 5 the mixture S100 + 0% PET shows that for 0.5 kg/cm^2 there is a constant volumetric change also to 1.0 kg / cm^2 are observed slight bumps that can be a rearrangement of the particles of the study material, culminating in 2.0 kg / cm^2 is also evidence constant volumetric change according to the horizontal displacement.

According to Figure 6 for mixing S100 + 1.0% PET in the effort of 0.5kg / cm2 at the beginning of the test the contraction of the material to then follow constant is observed until the end of the test, to the effort of 1.0 kg/cm^2 is evidenced a volumetric uniform behavior compared to other levels of confinement, however for the horizontal displacement of 0.4 centimeters expansion of the study material is slightly more pronounced and comes to exceed the expansion of the sample to a level of containment of 2.0 kg/cm^2. However, for this level of confinement a constant expansion is evident, but with less variation of vertical deformation, thus showing that this effort regarding S100 + 0%PET there is a significant improvement to the addition of recycled PET.

CONCLUSION

1. It was achieved characterize the soil as a materialinorganic red clay with high plasticity white (CH), medium high compressibility and expansion characteristics of permeable drainage. To themselvesand it found that polyethylene terephthalate was ground to about 3 to 5 mm. Also, 98.57% of the material passes No. 8 mesh which meet the requirements set for the completion of the investigation.
2. Obtaining curves compaction made with assay Proctor standard, according to ASTM D 698 standard, evidenced a progressive decrease of the maximum dry density of the samples with the additions of PET, this is logical because the content humidity (%) increased as the percentages are added to PET reinforcing soil samples.
3. The comparison of the mechanical parameters obtained from the direct shear test, it is concluded that there are improvements with regard to cohesion and decreased friction angle as claimed [3] these analyzed at each time instant in which they occur the horizontal displacements.
4. Of the direct shear test it determined that both parameters cohesion and angle of friction, their leakage resistances of the different mixtures approximate horizontal displacements larger.
5. It was determined that the volumetric analysis for levels of greater confinement as 2.0 kg/cm^2 a considerable improvement is observed in the S100 + 0% PET as this develops a more uniform behavior and less normal deformation compared to theS100 + 0% PET. Also, to levels lower confinement as 0.5kg/cm^2 a significant increase is observed in the normal strain of the S100 + 0%PET compared to the S100 + 1.0%PET.
6. It was determined that the optimal addition of reinforcement is 1.0% PET because this adopts a better performance, more uniform and less vertical variation.

BIBLIOGRAPHY

[1] Peddaiah S., Burman S. A. and Sreedeep 2018 Experimental Study on Effect of Waste Plastic Bottle Strips, *Geol Geotech Eng* **36** pp. 2907-20

[2] Mizababaei M., Arulrajah A., S. and Aldava Horpibulsuk M. 2017 Shear strength of a fiber-reinforced clay at large shear displacement When subjected to different stress histories *Geotexmem* pp. 1-8

[3] Hossein Soltani J. 2016 Compressibility and Shearing Behavior of Clayey Soil Reinforced by Plastic Waste Int J Civ Eng **14** (**7**) pp 479-89

[4] Brajesh M. and G. Kumar Mohit 2018 Use of randomly oriented polyethylene terephthalate (PET) fiber in combination with fly ash in subgrade of Flexible pavement *Construction and Building Materials* **190** pp 95-107

[5] Zukri A., R. Nazir and Nabilah An Experimental Study F. 2017. On Stabilization Of Pekan Clay Using Polyethylene And Polypropylene *American Institute of Physics* pp 1-8

IOP Conf. Series: Materials Science and Engineering **758** (2020) 012076 doi:10.1088/1757-899X/758/1/012076

Numerical analysis and Experimental Investigation of Lateral vibration on Drill String under Axial Load Constrained with Horizontal Pipe

Baojin Wang[1*] Zhongyang Wang[2] Anhe Zhu[3] Wang Hongda[4] Fushen Ren[5*]

1,2,5.Department Mechanical Science and Engineering, Northeast Petroleum University, China
3. Xuzhou Construction Machinery Group Co.,China
4. Research Institute of Oil Production Engineering,Daqing Oilfield Co Ltd.,China
E-mail: 1.bjwangbaojin@126.com;5.renfushen@126.com

Abstract. Horizontal well technology is an important means to improve drilling efficiency and oil and gas production, but it is easy to generate the lateral vibration of the drill string. This strong lateral vibration of the drill string will further lead to the whirl of the drill string, which will accelerate the wear of the drill string and further lead to the failure of the drill string. According to the rotor dynamics theory and fluid dynamics, the whirling equation of the drill string in horizontal wells is established. The effects of the weight of the drill string, wall friction, internal and external drilling fluid flow velocity and fluid resistance on the whirling of the drill string are studied. The mechanism of inducing the whirling of the horizontal drill string and the boundary conditions for judging the whirling are analyzed. Based on the existing dynamic simulation test equipment of the horizontal drill string in the laboratory, the experimental research on the whirling characteristics of the horizontal drill string is carried out, and the influence laws of the weight on bit and the rotation speed on the whirling of the drill string are studied. The results show that with the development of axial load, the intense of lateral vibration of drill string increases.

1. Introduction

Drill string is located in a very narrow circular space during drilling. Inside and outside of the drill string are filled with drilling fluid. The angle of inclination of horizontal well varies greatly. Due to the force acting on the drill string itself, the drill string in the horizontal section is almost flat against the lower wall. During the rotary drilling process, the friction of the drill string is serious, showing obvious non-linear contact characteristics. At the same time, the force and motion characteristics of the horizontal drill string are very complex due to the interference of ground equipment, formation and fluid viscous resistance. The drill string is subjected to transverse vibration, longitudinal vibration, torsional vibration, whirl and coupling vibration among them.

Jansen[1-3]presented the dimensionless whirling motion equation of drill string according to the nonlinear rotor dynamics theory and carried out numerical simulation. A.P.Christoforou[4] analyzed the whirling characteristics of bottom hole assembly by means of Rayleigh beam model. The research results well reflected the whirling characteristics of bottom drill string. Chen et al.[5]discussed the dynamic characteristics of drill bit during whirling of drill string. The research results show that the backward whirl of the drill string will hinder the transmission of the WOB of the bottom hole drill bit by the surface equipment, making the WOB obtained by the drill bit extremely small in actual operation. It is considered that the whirl of the drill string is a severe lateral vibration.

Y.A .Khulief[6]studied the coupling vibration of the lateral vibration and whirl and conducted finite element research. Hayat[7] analyzed and discussed the dynamic characteristics of the bottom hole assembly through combining test and simulation analysis. The influence of different friction coefficients on the whirling trajectory of drill string was discussed during the simulation analysis and simulation test. Gao Deli [8] et al. discussed the causes of different whirl and studied the effect of drill string whirl on borehole deviation prevention. Guan Zhichuan et al. [9,10] studied the whirl characteristics of drill string at the bottom of vertical well using simulation test equipment, analyzed the influence of different WOB and rotating speed on the whirl trajectory of drill string. Xiao Wensheng [10] analyzed the whirl mechanism of drill string under the action of internal and external drilling fluid and friction. The conditions for determining the whirl instability of the drill string are obtained. Di Qinfeng [11] et al. established the dynamic simulation model of the bottom hole assembly. The bottom hole assembly is simplified to a rotor model. And the whirl characteristics of the drill string under different WOB are studied. Yao Yonghan[12] et al. analyzed the whirling track, whirling speed, radial track and radial speed of drill collar by using the compiled simulation program. From the research status of drill string lateral vibration and whirl at home and abroad, it can be seen that the research on drill string lateral vibration is mostly focused on vertical well. The research results have been very rich and perfect. Although domestic and foreign scholars have also done a lot of research on horizontal well drill string lateral vibration, these researches are far from enough compared with vertical wells.

In this paper, non-linear factors such as weight of drill string, wall friction, internal and external drilling fluid velocity, density, structural nonlinearity and contact nonlinearity are considered. Based on Hamiltonian principle, the fluid-solid coupling vibration equation of drill string lateral vibration is established. The influence of weight on bit(WOB) and rotation speed on the lateral vibration characteristics of drill string is analyzed and discussed by the research method combining experiment and theoretical analysis.

2. Mathematical Model

Transverse vibration of drill string is a vibration perpendicular to the axis of drill string before deformation. Considering the difference in dynamic characteristics between the horizontal drill string and the vertical drill string, this paper focuses on the analysis of the lateral vibration characteristics of the drill string in the direction of gravity, and simplifies the drill string into an elastic beam with equal cross section hinged at both ends, as shown in figure 1.

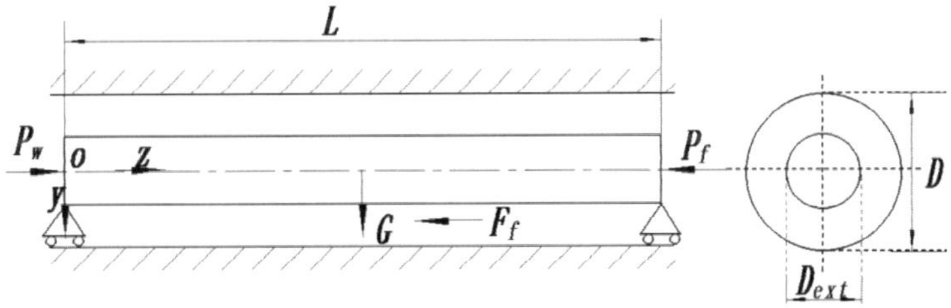

1.

Figure 1. Drill string model

The quality of drill string and drilling fluid in drill string per unit length is

$$m_1 = m_d + m_a = p_d \pi (D_{ext}^2 - D_{int}^2)/4 + p\pi D_{int}^2/4 \tag{1}$$

Where, m_d is the mass of drill string per unit length. m_a is the mass of drilling fluid per unit length in drill string. D_{int} is inner diameter of drill string. ρ is the density of drilling fluid.

Let drilling fluid flow along the axial direction of the drill string. And the speed of the transverse vibration of the drill string in the axial direction can be written as

$$\dot{y} = \frac{\partial y}{\partial t} \tag{2}$$

The velocity of drilling fluid in the axial direction is V_1. The velocity of drilling fluid in the axial direction is the combined velocity of drilling fluid velocity and drill string velocity in the axial direction.

$$V_{1y} = \dot{y} + V_1 y' = \frac{\partial y}{\partial t} + V_1 \frac{\partial y}{\partial z} \tag{3}$$

Where, $V_1 = \dfrac{4Q}{\pi D_{int}^2}$.

Analyzing the micro-element dx, the total kinetic energy of the micro-element can be written as

$$dT = \{\frac{1}{2} m_d \dot{y}^2 + \frac{1}{2} m_a [V_1^2 + (\frac{\partial y}{\partial t} + V_1 \frac{\partial y}{\partial z})^2]\}dz \tag{4}$$

Total potential energy is

$$dU = \left\{\frac{1}{2} EI\left(y''\right)^2 + p_d Agydz\right\}dz \tag{5}$$

Work done by non-conservative force is written as

$$\delta W = -\int_{z_1}^{z_2} \begin{pmatrix} P_w - P_f \cos(\omega t) \\ -F_f - p_{int} A_{int} \end{pmatrix} y'' \delta y dz \tag{6}$$

E is the function of $y(z,t)$. The equation can be given according to Hamilton principle.

$$\delta E = 0 \tag{7}$$

$$\begin{aligned} \delta E = \delta \int_{t_1}^{t_2} \int_{z_1}^{z_2} & \{\frac{1}{2} m_d y_t^{\,2} + \frac{1}{2} m_a [V_1^2 + (\frac{\partial y}{\partial t} + V_1 \frac{\partial y}{\partial z})^2] \\ & -\{\frac{1}{2} EI(y')^2 + \rho_d Agydz\}\}dzdt \\ & -\int_{t_1}^{t_2} \int_{z_1}^{z_2} \left(F - F_f - p_{int} A_{int}\right) y'' \delta y dzdt \\ & = 0 \end{aligned} \tag{8}$$

Where, $F = P_w - P_t \cos(\omega t) - F_f$.

The variational form of equation (8) can be written as

$$\int_{t_1}^{t_2} \int_{z_1}^{z_2} \{m_d \dot{y} \delta \dot{y} + m_a V_{1y} \delta V_{1y} - EIy'' \delta y'' + F_Z y' \delta y' + p_d Ag \delta y\} dzdt \tag{9}$$

Where, $F_Z = P_w - P_t \cos(\omega t) - F_f - p_{int} A_{int}$, $\delta y'' = \dfrac{\partial^2 \delta y}{\partial z^2}$, $\delta \dot{y} = \dfrac{\partial \delta y}{\delta t}$.

The equation can be obtained after integrating equation (9).

$$\int_{t_1}^{t_2} \int_{z_1}^{z_2} (m_1 \ddot{y} + 2m_a V_1 \dot{y}' + m_a V_1^2 y'' + EIy'''' + F_Z y'' + p_d Ag) \delta y dzdt = 0 \tag{10}$$

Due to the arbitrariness of δy, the differential equation of the coupled vibration of the drill string can be obtained considering the factors of drilling fluid, weight on bit, friction and weight of the drill string.

$$EI \frac{\partial y^4}{\partial z^4} + \left(m_a V_1^2 + F_z\right) \frac{\partial^2 y}{\partial z^2} + 2m_a V_1 \frac{\partial^2 y}{\partial z \partial t} + m_1 \frac{\partial^2 y}{\partial t^2} + p_d Ag = 0 \tag{11}$$

The micro-element section dz is taken for analysis. Tthe quality of drilling fluid inside and outside the drill string per unit length is given as follows

$$m_2 = m_d + m_a + m_b \tag{12}$$

The mass of drilling fluid in annulus space is given as

$$m_b = \rho \frac{\pi (D^2 - D_{ext}^2)}{4} \tag{13}$$

Flow speed of drilling fluid in annulus space can be written as

$$V_2 = \frac{Q}{A_2 - A_1} \tag{14}$$

Where, $V_2 = \frac{Q}{A_2 - A_1}$, $A_1 = \frac{D_{ext}^2}{4}$, $A_2 = \frac{D^2}{4}$.

Drilling fluid resistance is

$$f = C_v \dot{y} \tag{15}$$

Where, C_v is the viscosity coefficient of drilling fluid.

When considering the liquid pressure in the annulus, the additional force exerted by the annular drilling fluid on the drill string is

$$c_m m_b \left(\frac{\partial^2 y}{\partial t^2} + 2V_2 \frac{\partial^2 y}{\partial z \partial t} + V_2^{\,2} \frac{\partial^2 y}{\partial z^2}\right) + c_v \frac{\partial y}{\partial t} + p_{ext} A_{ext} \frac{\partial^2 y}{\partial z^2} \tag{16}$$

Finally, the differential equation can be obtained.

$$\begin{gathered} EI \frac{\partial^4 y}{\partial z^4} + 2\left(m_a V_1 + C_m m_b V_2\right) \frac{\partial^2 y}{\partial z \partial t} + \left(F_T + m_a V_1^{\,2} + C_m m_b V_2^{\,2}\right) \frac{\partial^2 y}{\partial z^2} \\ + \left(m_d + m_a + C_m m_b\right) \frac{\partial^2 y}{\partial t^2} + \rho_d A g = 0 \end{gathered} \tag{17}$$

Were, $F_T = P_w - P_t \cos(\omega t) - F_f + f + p_{ext} A_{ext} - p_{int} A_{int}$.

According to the method of separating variables, the frequency equation of drill string transverse vibration is obtained.

$$w_n = \frac{n\pi^2}{L^2} \sqrt{\frac{EI}{m_2}} \sqrt{n^2 - \frac{L^2\left(m_a V_1^{\,2} + C_m m_b V_2^{\,2} - \rho I w^{\,2} + F + \rho_d A g\right)}{\pi^2 EI}} \tag{18}$$

3. Experiment and Simulation Result

The test equipment is designed according to the similarity theory. Thus, there is a similarity ratio between the test parameters and the actual parameters.

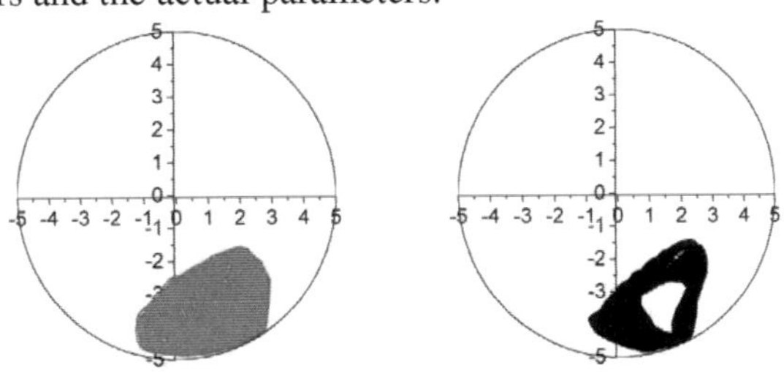

(a)experiment simulation

WOB=0.5kg

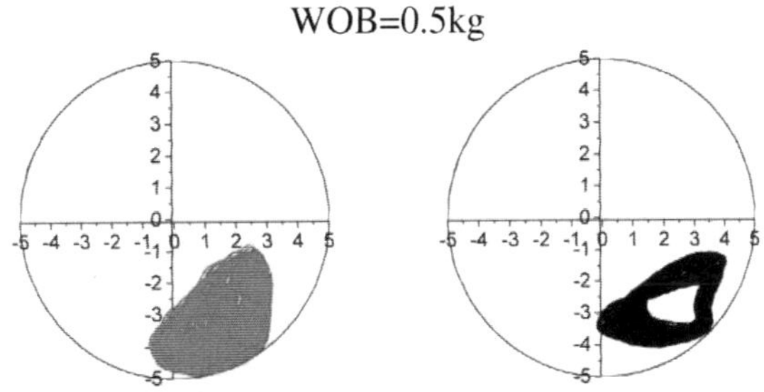

(b)experiment simulation

WOB=1.5kg

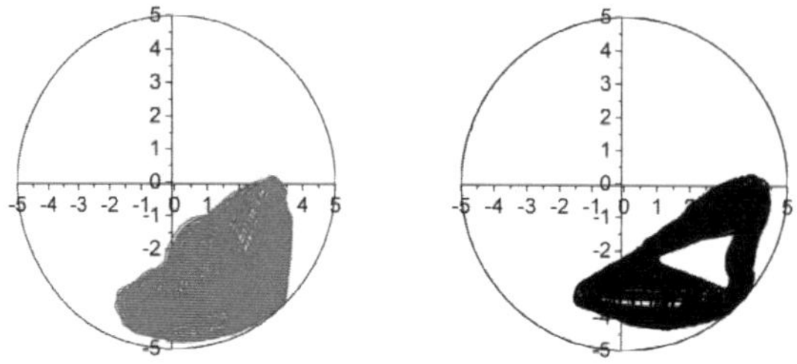

(c)experiment simulation

WOB=2.5kg

Figure 2. Axis trajectory of drill string

As shown in figure 2, red and black lines are experiment and simulation results, respectively. And the rotional speeds applied in experiment and simulation are 100r/min and 35.7r/min. The axial load applied in experiment are 0.5kg,1.5kg and 2.5kg. The axial load applied in simulation are 45kN,135kN and 225kN. The results show that the intense of whirling of drill string increases with the development of axial load. However, the increase is not significant.

4. Conclusions

In the actual drilling process, larger WOB can increase whirling. Thus, the smaller WOB should be applied. And then, the impact and friction between the drill string and the borehole wall reduce. Moreover, the service life of drill string is prolonged.

5. Acknowledgments

The authors would like to acknowledge financial support by Natural Science Foundation of China, Project 11372071. Natural science foundation of Heilongjiang province, Project LH2019A003. Fundamental Research Funds for Provincial Undergraduate Universities of Heilongjiang Province, Project 2019QNL-13.

6. References

[1] Jansen J D 1991 Non-linear rotor dynamics as applied to oil-well drill string vibration *Journal of sound and vibration.*

[2] Jansen J D 1991 Whirl and chaotic motion of stabilized drill collars *SPE.*

[3] Jansen J D 1992 Whirl and chaotic motion of stabilized drill collars *SPE Drilling Engineering.*

[4] Christoforou A P, Yit A S 1997 Dynamic modeling of rotating drillstrings with borehole interaction *Journal of Sound and Vibration.*

[5] Dykstra M W, Chen D C, Warren T M 1994 Experimental evaluations of drill bit and drill string dynamics *SPE.*

[6] Khulief Y A,A l-Su laiman F A 2007 Vibration analysis of drillstrings with self-excited stick-slip oscillations *Journal of Sound and Vibration.*

[7] Hayat Melakhessou, Alain Berlioz, Guy Ferraris 2003 A nonlinear well-drillstring interaction model *Journal Vibration and Acoustics.*

[8] Gao D L,Gao B K,Geng R P 1996 Analysis of Eddy Characteristics of Drill String *Oil Drilling & Production Technology.*

[9] Guan Zhichuan,Jin Yanxin,Wang Yifa 2003 Experimental research on motion behavior of bottom drill string in straight hole *ACTA Petroleum Sinica.*

[10] Xiao Wensheng,Zhang Yang,Zhong Yifang 2004 Annulus whirling motion analysis of the rotary drill string under the action of rubs between well-walls and drilling fluids *China Mechanical Engineering.*

[11] Di Qinfeng,Wang Wenchang,Yao Yonghan 2010 Dynamics model and whirl features simulation of bottom hole assembly *Journal of China University of Petroleum.*

[12] Yao Yonghan, Di Qinfeng,Zhu Weiping 2010 Analysis of whirling properties of bottom hole assembly *Petroleum Drilling Technology.*

IOP Conf. Series: Materials Science and Engineering **758** (2020) 012077 doi:10.1088/1757-899X/758/1/012077

Comprehensive Planning of New Plant Under Industry 4.0

Zhang Tiefeng[1], Tong Yifei[1*], Zhang Yuanhao[1] and Zhang Xu[2]

[1]School of Mechanical Engineering, Nanjing University of Science and Technology, Nanjing, China
[2]AECC SOUTH INDUSTRY CO., LTD, Zhuzhou, China
E-mail: 809758956@qq.com
*Corresponding author

Abstract. The article carries out layout design optimization and warehouse improvement for HF Company. First of all, taking the production management process of the new plant as the research object, set a comprehensive planning of the plant area from the aspects of product mix, production management and logistics management. According to the planning objectives, detailed design, data collection for production forecasting are proposed based on SLP method and mapping of the new plant layout, combined with lean thinking and Industry 4.0 concept. Finally, the expected goals and improve production efficiency are achieved.

1. Introduction

With the rapid development of manufacturing industry, how to deliver goods faster and reduce costs is a problem that all enterprises must pay attention to. Reasonable workshop layout is of great significance to give full play to production capacity and improve product quality [1]. As far as manufacturing mode is concerned, it has shifted from mass production to just-in-time and lean production, and further to computer integrated manufacturing and agile manufacturing. Ma proposed the workshop layout for agile manufacturing, elaborated the workshop layout of the new model, and proposed the idea of distributed layout based on agile manufacturing according to the philosophy of lean production [2]. In terms of computer assistance, MRP-II, ERP, CIMS and other software and facility planning complement each other. Balakrishnan and Cheng [3] solved the DPLP problem combined with mathematical algorithm [4]. Today, facility planning is no longer limited to enterprises, including social services and even supply chain rivals.

Facility planning plays an important role in enterprises. How to achieve the maximum work efficiency in the shortest time is the problem that every enterprise expects to solve [5]. Optimization and improvement of layout become the top priority. In the process of layout, many enterprises have already used SLP, and achieved very successful results. In this paper, SLP method is applied to design and optimize the overall layout of the new factory [6, 7].

2. Plant layout planning

2.1. Analysis of logistics relationship among operation units

Using the logistics relationship classification table and related tables, non-logistics relationships are analyzed. The calculation process is as follows: by consulting and counting the inventory list and historical data of each database in previous years, the materials of the raw material warehouse and the semi-finished product warehouse are transported to the production workshop in a tray manner, the raw material warehouse is transported to the production workshop for 5678 pallets, and the semi-finished product warehouse is 3210 pallets. And the packaging materials stored in the packaging material warehouse are shipped to the production workshop in one direction. The equivalent material flow rate is 1065 pallets, and the equivalent material flow rate in the production workshop is 5678+3210+1065=8533. Take this as an example and multiply the annual equivalent material flow of each region by the distance between each operating unit to get the logistics intensity.

Table1: Logistics intensity from - to (unit: tray·m)

	Serial number	1	2	3	4	5	6	7
	from \ to							
1	Raw materials library			448562				
2	Semi-finished products lib			224700				
3	production workshop				379718.5	60705		181760
4	Quarantine						849033.5	
5	Packaging material library							
6	Shipping department							
7	Carton recycling area							

According to the function interval flow and the operation batch, the logistics strength analysis is carried out by the logistics relationship classification table, which is represented by the symbol AEIOU as in Tab. 2.

Table2: Logistics intensity from - to (unit: tray·m)

The serial number	Logistics route	logistics intension	level
1	4-6	849033.5	A
2	1-3	448562	E
3	3-4	379718.5	E
4	2-3	224700	I
5	3-7	181760	I
6	3-5	60705	O

2.2. Analysis of non-logistics relationship among operating units

In the shop floor layout, there is also a non-logistics relationship between the work units. The non-logistics relationship of the operating unit is similar to the logistics relationship, and is also divided into five different levels which is represented by AEIOU.

2.3. Comprehensive relationship among operation units

In the production arrangement, the relationship between logistics and non-logistics operation units is commonly different. In order to outline the comprehensive relationship between different operational units, it is essential to combine logistics and logistics, in order to find out the comprehensive relationship between layouts.

The calculation of the comprehensive relationship is as follows:

$$TR_{ij} = m \times MR_{ij} + n \times NR_{ij} \tag{1}$$

In this paper, A=4, E=3, I=2, O=1, U=0 and X=-1 are used as the amount of the correlation degree. And through the suggestion of the study supervisor, the logistics relationship and non-logistics relationship m:n=2:1 are brought into the formula, and the calculation table of the relationship is achieved.

Table3: Comprehensive relational computation

Unit pair	Relation grade				Comprehensive relationship	
	Logistics weighted values：2		Non-logistics weighted values：1			
	grade	Points	grade	Points	grade	Points
1-2	U	0	O	1	O	1
1-3	E	3	A	4	A	10
…	…	…	…	…	…	…
7-9	U	0	U	0	U	0
8-9	U	0	I	2	O	2

According to the scores, the comprehensive relationship table of operation units is achieved.

Table4: Operational units

number	Name of operation unit
1	1 Raw materials library
2	2 Semi-finished products library
3	3 The production workshop
4	4 quarantine
5	5 Packaging material library
6	6 Shipping department
7	7 Carton recycling area
8	8 Conference office area
9	9 Staff rest area

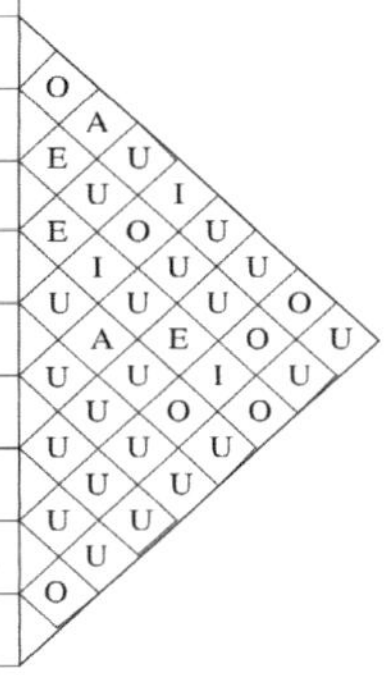

3. Layout design

In this paper, the unit of work unity relationship is transformed into a form of a triangular matrix to quantify the level.

Table5: Sorting table for comprehensive approximation of operating units

from \ to	1	2	3	4	5	6	7	8	9
1 Raw materials library		O 1	A 4	U 0	I 2	U 0	U 0	O 1	U 0
2 Semi-finished products library	O 1		E 3	U 0	O 1	U 0	U 0	O 1	U 0
3 The production workshop	A 4	E 3		E 3	I 2	U 0	E 3	I 2	O 1
4 quarantine	U 0	U 0	E 3		U 0	A 0	U 0	O 1	U 0
5 Packaging material library	I 2	O 1	I 2	U 0		U 0	U 0	U 0	U 0
6 Shipping department	U 0	U 0	U 0	A 0	U 0		U 0	U 0	U 0
7 Carton recycling area	U 0	U 0	E 3	U 0	U 0	U 0		U 0	U 0
8 Conference office area	O 1	O 1	I 2	O 1	U 0	U 0	U 0		O 1
9 Staff rest area	U 0	U 0	O 1	U 0	U 0	U 0	U 0	O 1	
Synthetic proximity	8	6	18	4	5	0	3	6	2
Sort	2	3	1	6	5	9	7	4	8

IOP Conf. Series: Materials Science and Engineering **758** (2020) 012077 doi:10.1088/1757-899X/758/1/012077

It can be seen that the unit with the highest comprehensive proximity is the serial number 3, which is placed in the middle position of the position map, and the work unit of the A grade is found to draw 1-3 and 4-6. The A-level work unit pairs are connected by four lines and separated by one-unit distance, then the E-level work unit pairs 2-7, 3-7, 3-4 are found. Connected by three lines, the working unit pairs are separated by two-unit distances. Finally, the working unit pairs of I and O are processed, respectively, with two lines and one line, and the working unit position correlation diagram is drawn.

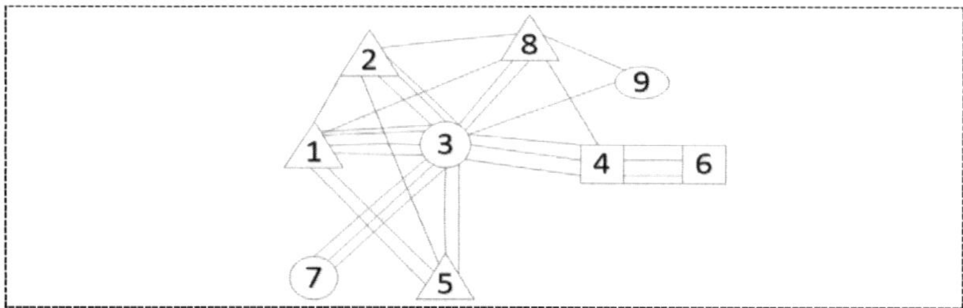

Figure 1. Location correlation diagram of operation unit.

According to (Fig.1), each work unit is designed according to the shape and proportion by area graphic method, and the general layout plan of the HF Company is obtained.

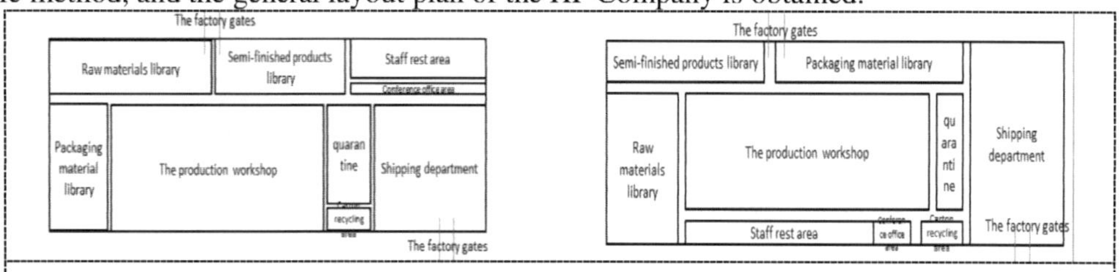

Figure 2. HF plant general layout plan (plan 1-left and plan 2-right)

The weighting factor method is used to prioritize the layout scheme and then optimize for the preferred scheme.

Table6: Weighting factor method to evaluate factor weights

Evaluation factors	A	B	C	D	E	F	G	sum	weight %	sort
facility utilization A		0	0.5	0.5	0.5	1	1	3.5	12.5	3
transport rationalization B	1		1	1	0.5	0.5	1	5	25	1
Investment cost C	0	0		0.5	0	0	0.5	1	5	6
Working environment D	0.5	0	0.5		0.5	0.5	1	3	15	4
Technological process E	0.5	0.5	1	0.5		1	1	4.5	22.5	2
Flexible F	0	0.5	0.5	0.5	0		1	2.5	12.5	5
The original layout mode G	0	0	0.5	0	0	0		0.5	2.5	7
Total								20	100	

Then, the two schemes are analyzed with 3 as the highest score and 1 as the lowest score.

Table7: Program evaluation preferred form

Evaluation factor	Weights	Point	product	Point	product
facility utilization A	12.5	2	25	2	25
transport rationalization B	25	3	75	2	50
Investment cost C	5	2	10	2	10
Working environment D	15	3	45	1	15
Technological process E	22.5	2	45	2	45
Flexible F	12.5	2	25	2	25
The original layout mode G	2.5	2	5	2	5
Total	100		230		175
Priority order		1		2	

IOP Conf. Series: Materials Science and Engineering **758** (2020) 012077 doi:10.1088/1757-899X/758/1/012077

Mainly comparing the differences between the two schemes, we draw the following conclusions.

The employee rest area and conference office area of Option 1 are located in the upper right corner to reduce the impact of noise generated by the production workshop on the rest staff, the waste area and the conference office are far apart, which can reduce the impact of pollution on the health of employees, It is placed adjacent to the package library to facilitate the management of warehouse personnel. In Scheme 2, there is still a long distance between the pairs of operating units with strong logistics correlation. Based on the above analysis, the first solution is the optimal layout scheme. And make improvements. According to the principle of smooth logistics, carry out u-shaped layout in the production workshop to save space.

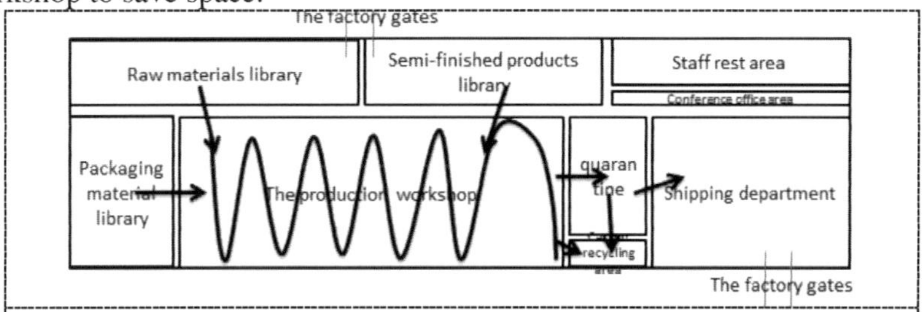

Figure 3. Schematic map of logistics

By optimizing the layout of the production area, the production cycle of the whole product is shortened, the logistics cost is reduced, the logistics is more economical and smoother, and the cross-flow of people is avoided, and the production efficiency of the factory is improved. Expected goals: (1) the production efficiency of the production line increased by 15%, (2) the annual logistics cost decreased by 10% after re-planning.

4. Conclusion

Producing products faster and more efficiently and improve space utilization are quite crucial to manufacturing companies worldwide. This paper introduces the SLP method used in facility planning, and comprehensively plans the entire new plant from the aspects of product, production process, logistics, etc. The comprehensive planning method of plant layout proposed in this paper shortens the cycle of product manufacturing, reduces logistics costs, avoids the intersection of people flow, and effectively improves the production efficiency of the factory.

It is hoped that this study can bring new changes to the enterprise, and through its comprehensive workshop planning, it can improve the chaotic production model in the old factory area, make the workshop logistics smoother, and finally get a future factory that is more in line with lean thinking.

Acknowledgments

This work was financially supported by the Fundamental Research Funds for the Central Universities (No. 30919011205).

References

[1] Yi Shuping Guo Fu 2014 Basic Industrial Engineering *China Machine Press* pp 20-22
[2] Yao Wei 2009 Research on Layout and Optimization of Digital Production Workshop *Xidian University*.
[3] Tompkins J A 2010 Facilities Planning *Wiley* pp 344-347
[4] Stock G N Greis N P Kasarda J D 2000 Enterprise Logistics and Supply Chain Structure: the role of fit *Journal of Operations Management* (vol 18) pp 531-547.
[5] Dong Hai 2004 Application of Computer Simulation Technology in Facility Planning and Design *Journal of Shenyang University Natural Science* (vol 16) pp 35-37.
[6] Chen Bing 2008 Based on Green Manufacturing Workshop Facility Planning Research. *Wuhan: Wuhan University of Science and Technology*
[7] Yuan Yujiao Ye Chunming Sun junhua 2016 Research on Mobile Phone Shopping Web Design Based on Moving Line Analysis and System Layout Design *Ergonomics* (vol 22) pp 32-38

IOP Conf. Series: Materials Science and Engineering **758** (2020) 012078 doi:10.1088/1757-899X/758/1/012078

Research of elevator real time monitoring method based on spark

Huang Jianpeng[1], Jiang Xiyang[2*], Tong Yifei[1] and Gu Feng[3]
[1] Special Equipment Safety Supervision Inspection Institute of Jiangsu Province , Nanjing, China
[2]School of Mechanical Engineering, Nanjing University of Science and Technology, Nanjing, China
[3]Nantong Vocational University, Nantong226007, P. R, China
E-mail: 2176609317@qq.com
*Corresponding author

Abstract. With the development of economy, elevator has become one of the essential means of transportation in people's life. However, due to the complexity of the elevator internal system, the sensors collect a large number of elevator operation data so that the reliability and real-time of data processing is very important for the safety monitoring of elevator operation. To solve this problem, this paper proposes an elevator real-time monitoring method based on spark. By using the stream data processing method provided in spark streaming, a large number of real-time data are divided into time windows, the data is divided into small batches of data below the second level and submitted to spark engine for analysis and processing, which greatly improves the speed and throughput of data processing.

1. Introduction

Elevator plays an increasingly important role in people's daily life. However, due to the large number of elevators, the maintenance quality of maintenance units and elevator manufacturers is uneven, the safety of elevators will be difficult to guarantee if the maintenance of elevators is supervised in the traditional way. In addition, due to the high frequency use of elevators, the number of old elevators is increasing, and there is vicious competition in the current elevator maintenance market, resulting in frequent accidents. Therefore, it is very important to carry out real-time and effective safety monitoring of elevator operation. At the same time, due to the elevator real-time flow data has the characteristics of high-speed production and large date volume, it is necessary to filter the invalid data in the massive real-time flow data when processing these elevator data, and detect the abnormal state in the elevator operation process in time, so as to facilitate the subsequent fault diagnosis and maintenance.

In recent years, many scholars have studied the processing technology of condition monitoring data in many aspects. Zhuang regards monitoring data as flow data, and proposes a real-time processing framework based on flow [1]. Zhang proposed a k-nearest neighbor algorithm based on MapReduce, which runs in a distributed environment, thus greatly improving the efficiency of the algorithm [2]. Papadimitriou designed a distributed clustering framework, which can be directly applied to large-scale data [3]. Zheng proposed a big data-based urban air quality inference system, which can infer urban air quality in real time by analyzing historical air quality data, urban traffic flow data, and personnel mobility data [4]. Kim built data mining algorithm in Hadoop, a distributed parallel processing framework, and used MapReduce parallel collaborative filtering algorithm to process massive twitter big data [5]. Hadoop is a parallel batch processing technology based on distributed file system, which is suitable for large-scale calculation of massive archived data. However, due to the large amount of disk operations required in the calculation process, which affects the calculation efficiency, it is usually used for offline analysis. In order to solve the problem of low performance of parallel computing caused by disk I / O, spark, a memory based parallel computing framework, has been widely used [6]. The variability and complex data relationship of elevator operation monitoring data are significantly different from the characteristics of Internet industry and manufacturing industry. Under the existing distributed data access framework, according to the characteristics of elevator monitoring data, a massive real-time data access scheme based on memory or efficient disk access needs to be studied, this paper discusses the real-time processing technology of elevator monitoring data. Under the framework of Hadoop technology, Kafka message queue is used to realize the real-time access of massive data, spark streaming is used to complete the analysis of process data, and spark is used to complete the data batch processing [7,8]. It improves the real-time property of data access and processing, meets the needs of real-time processing of large quantities of data in the monitoring system, and explores the application and promotion of real-time analysis technology based on big data in the field of elevator operation monitoring.

2. Analysis of real-time monitoring method of elevator running state

2.1. Overall research route

The overall analysis architecture of this paper is shown in Figure 1, which is mainly composed of five functional modules: data acquisition layer, data transmission layer, data processing layer, data storage layer and data application layer.

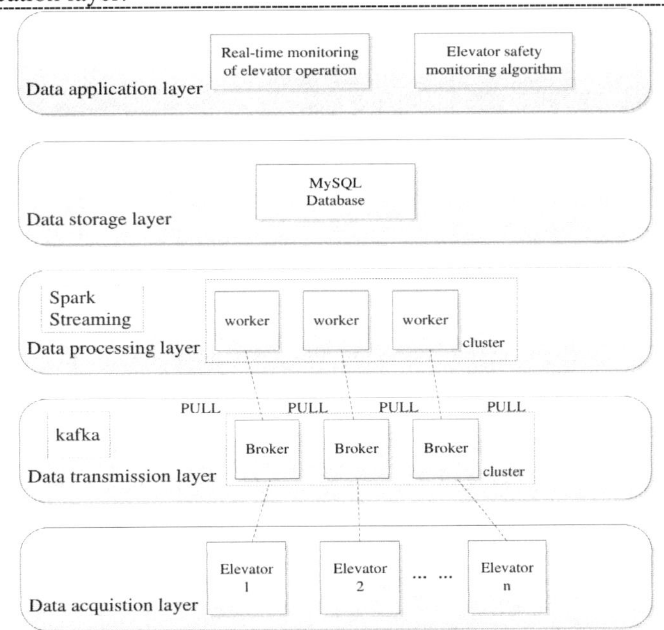

Figure 1. The overall analysis architecture

2.2. Data transmission layer: elevator monitoring data real-time transmission technology

The real-time access of elevator operation status monitoring data includes the collection, transmission and storage of status monitoring data, and provides data input for other real-time processing frameworks (such as spark streaming processing framework). Traditional elevator condition monitoring data transmission mainly uses the method of direct connection with distributed database or real-time database to store, but there are some problems, such as the speed of data collection and data writing does not match, excessive consumption of database space and so on. Kafka is a distributed and high-throughput open source message system supported by Apache foundation. By reducing the complexity of message queuing system and improving the performance, scalability and throughput of real-time access system, Kafka can solve the delay problem caused by massive state monitoring data written into HDFS in real-time storage, and ensure that all collected state monitoring data can be obtained to real-time processing.

Therefore, this paper uses Kafka as a bridge between data distribution layer and data processing layer, classifies messages in Kafka cluster through topic, and consumers and producers use topic to publish and subscribe to topics of messages, and uses partition to group topics, as shown in Figure 2. When a producer sends a message, it sends the message to a specific partition according to a specific distribution policy. Kafka can ensure the order of messages in the same packet. In the application of detection algorithm to detect elevator fault, it is necessary to detect the state transition process of elevator operation, so the order of messages is very important in the processing of elevator real-time data. In order to ensure the order of the same elevator data when being processed, this paper uses elevator ID as the partition basis, and divides the elevator real-time data by calculating the hash value of elevator ID To ensure that the data of the same elevator is divided into the same partition. In order to ensure fault tolerance, the partitioned data will be saved on multiple nodes in the form of copies, so that the data will not be lost when some nodes go down.

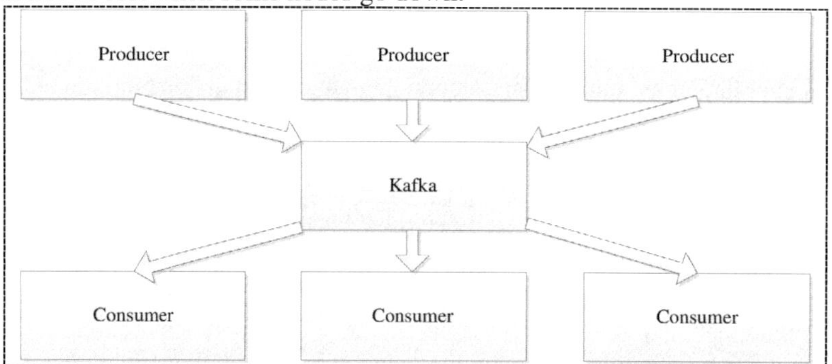

Figure 2. Distributed subscription architecture of Kafka

2.3. Data acquisition layer: building finite state machine for elevator operation

Through the monitoring terminal to collect the real-time state of the elevator running, because of the amount and complexity of the monitoring data, it brings a lot of challenges for the diagnosis of the elevator state. However, the elevator can be described with a certain state at every time in the whole operation cycle. When a specific event (such as elevator being called up and down, door opening at the arrival station, etc.) occurs, the state transition will occur. In order to solve the above problems, aiming at this characteristic of elevator operation, this paper intends to use the finite state machine (FSM) to model the elevator operation state and its state transition process. The finite state machine of elevator running state can be defined as a quintuple:

$$M = (S, \Sigma, \delta, S_0, F) \tag{1}$$

Where S represents the elevator system operation set, that is, the elevator stop state, elevator operation state, elevator arrival state and other elevator operation processes defined in this paper; Σ represents the system input set, and the input of the elevator system is composed of the signals of the elevator control system: up and down call signal, door opening and closing signal, floor button signal, elevator arrival signal and leveling completion signal; δRepresents the state transition function, that is, the elevator system changes the state into the next state through the input signal in one state s_0 Indicates the initial state of the elevator; F Indicates the elevator termination status. Combined with the finite state machine model of elevator operation, the process of elevator operation state transfer is shown in Table 1 below:

Table1: Table of elevator operation state transfer

Original state	Input	Target status

Original state	Input	Target status
Elevator stop	Up (down) signal	Elevator operation
	Up (down) signal and on the current floor	Elevator opening
Elevator operation	Arrival	Elevator arrival
Elevator arrival	Stop at the district level layer	Elevator opening
Elevator opening	Execution of the door opening command	Completely opening
Completely opening	Timeout or close button is pressed	Elevator closed
Elevator closed	Execution of the closing command	Completely closed
Completely closed	No other instructions	Elevator stop
	Up (down) signal	Elevator operation

2.4. Data processing layer: Spark streaming data processing technology

The elevator data collection system transmits the real-time data of the elevator running state to the analysis platform through the monitoring platform designed in this paper. Because each piece of data sent by the data distribution system to the message middleware only represents the state of the elevator running at a certain time, and does not directly distinguish the elevator in a specific running process, Therefore, in order to judge the running state of the elevator and carry out fault detection, it is necessary to collect the data for a period of time to determine the collection time interval, which has a great impact on the success rate and efficiency of detection. If the time is too small, it may not be able to effectively determine the running state change process of the elevator, and if the time is too long, it will affect the execution efficiency of the algorithm Therefore, this paper proposes to use the sliding window mechanism in the preprocessing of stream data. First, the micro batch data is divided according to the set sliding window size, and then each data is processed by spark engine in a similar batch processing way. As shown in Fig. 3, it is a schematic diagram of sliding window. The sliding window is 3, and the sliding step is 2. That is to say, the calculation is performed every 3 time units, and the length of each time interval is 2. By setting the window size to control the time interval of single detection, the fault detection algorithm detects the state change of elevator in this period according to the data in this time interval. The sliding step size can avoid the problem that the fault can not be detected due to the time span of elevator operation.

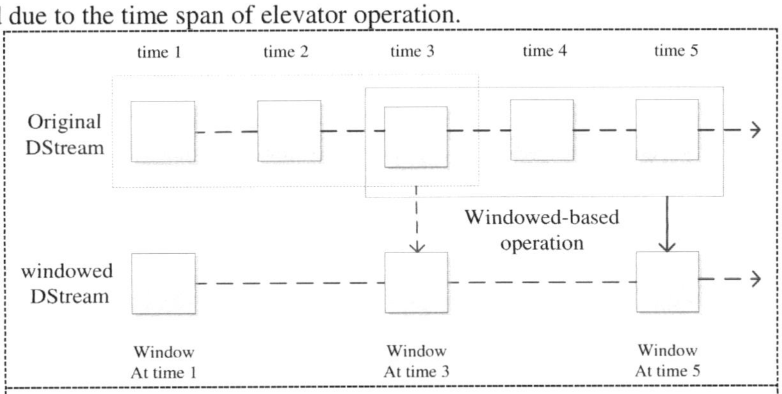

Figure 3. Sliding window mechanism

2.5. Data application layer: Elevator Fault Monitoring Algorithm Based on flow data

Step1. First, the number of spark streaming partition will be divided according to the partition of Kafka, and the micro batch data will be divided using the sliding window mechanism.

Step2. Take the micro batch data as the processing unit, use map to process each stream data, extract the elevator ID in the data as the key value, and filter the useful data as the value.

Step3. Use the groupByKey operator to gather the data according to the key value, that is, the elevator ID. since Kafka can ensure the order of messages in the same partition, it will still ensure the order after aggregation.

Step4. Use mapToPair operator to aggregate the data in the form of < key, value > into the form of < key, list < value > >. At this time, the 'list' corresponding to each 'key' is the data contained in the elevator operation represented by the key.
Step5.Then using map operator, according to the data in each list, analyze whether the running state of the current elevator is abnormal.
Step6.Finally, use foreachRdd and foreachPartition to read the results in each partition and save the results to the database.

3. Conclusion

This paper proposes to use the finite state machine (FSM) to model the elevator operating state, and establish the state change process of the entire elevator operating cycle. At the same time, according to the characteristics of a large number of real-time data generated during the elevator operation, this paper adopts the mixed mode of Kafka, Hadoop, spark streaming and spark to establish a real-time processing framework for elevator state monitoring. Kafka message queue is used to complete real-time data transmission to prevent data loss caused by data access rate mismatch. The sliding window mechanism in spark streaming is used to process the data in batches to realize the real-time processing of the flow data, so as to ensure that all monitoring data and alarm information can be processed in a timely and effective manner, and stored in the distributed file system or real-time database; Finally, through the memory based distributed processing framework spark, combined with mapToPair, reduceByKey and other operators to realize the in-depth analysis of the processed state monitoring data, which provides a new idea and method for the real-time safety monitoring of elevator operation state

Acknowledgments
This work was financially supported by the Fundamental Research Funds for the Central Universities (No. 30919011205).

References
[1] ZHUANG Xue-yin ZHANG Li WENG Xiao-qi 2013 Realtime stream data processing framework for complex equipment condition monitoring *Computer Integrated Manufacturing Systems* (vol 19) pp 2929-2939
[2] Zhang C Li F 2012 Efficient parallel kNN joins for large data in MapReduce *Proceedings of the 15th International Conference on Extending Database Technology ACM* pp 38-49.
[3] Papadimitriou S Sun J 2008 Disco: Distributed co-clustering with map-reduce: A case study towards petabyte-scale end-to-end mining *Eighth IEEE International Conference on IEEE* pp 512-521
[4] Zheng Y Liu F Hsieh H P 2013 U-Air:when urban air quality inference meets big data *ACM SIGKDD International Conference on Knowledge Discovery and Data Mining. ACM* pp 1436-1444
[5] Kim Y Shim K 2011 TWITOBI: A Recommendation System for Twitter Using Probabilistic Modeling pp 340-349
[6] Zaharia M Chowdhury M Franklin M J 2010 Spark: cluster computing with working sets *Usenix Conference on Hot Topics in Cloud Computing USENIX Association* pp 10-10
[7] Kafka A 2014 A high-throughput distributed messaging system *URL: kafka. apache. Org* (vol5)
[8] Zaharia M Xin R S Wendell P 2016 Apache spark: a unified engine for big data processing *Communications of the ACM* (vol59) pp 56-65

The research status of key problems of experimental equipment for tube hydroforming

Kunyan Zhang[1], Lianfa Yang[2*], Zhanbin Chen[3], Jinjie Huang[4], Jianping Ma[5], Pengju Zong[6], Jingyu Jiang[7], Jun Liu[8] and Xianzhang Li[9]

Mechanical and Electrical Engineering, Guilin University of Electronic Technology, Guilin, Guangxi, 541004, China
E-mail:[1]z_kunyan@163.com,[2]y-lianfa@163.com,[3]C_zhanbin@163.com, [4]JinjieHuang231@126.com,[5]majianping1229@163.com,[6]452667657@qq.com,[7]j_jingyu@163.com. [8]1015190927@qq.com, [9]229198641@qq.com

Abstract: Tube hydroforming technology is an advanced manufacturing technology adapted to lightweight structure. The technology has the advantages of low production cost, high material utilization rate and light part quality. Forming equipment is the basis for implementing the technology, the performance of the equipment greatly affects the product quality. However, the key problems affecting the performance of the equipment include: loading path, axial feeding and sealing method, etc. The article is for the purpose of key question carrying on the system combing with the discussion. Among them, several hydraulic pressure control methods are summarized, the influence of different loading path on tube hydroforming is shown, A loading method that axial stress - circumferential stress is always proportional in the forming process is proposed. Meanwhile, the control method of axial feed is summarized, it is expressed that axial feeding can improve tube forming ability. Besides, five kinds of common pipe sealing methods were compared, the reliable and effective sealing structure is summarized.

1. Introduction

Tube hydroforming technology is a kind of plastic forming technology for producing thin-walled metal components with complex sections. This technology has the characteristics of low production cost, high material utilization, light weight of parts and high rigidity. The technology can realize component structure integration and lightweight manufacturing. It has been widely used in aviation, automobile, home appliances and other fields [1]. It has been widely used in aviation, automobile, home appliances and other fields. The process of this technology is: The tube fitting of a certain length is put into the die and high-pressure liquid is injected into the tube, at the same time, axial force is provided at both ends of the tube to make the tube deform and fill the die cavity. According to the

IOP Conf. Series: Materials Science and Engineering **758** (2020) 012079 doi:10.1088/1757-899X/758/1/012079

above forming principle, it is necessary to solve the problems of high-pressure fluid, axial feeding, sealing technique and so on. Scholars at home and abroad have found that different loading paths have different effects on tube hydroforming performance. Axial feeding will improve the forming effect. Leakage is the "common fault" of the hydraulic system, and good sealing is the guarantee of the system [2]. This article summarizes the recent development of tube hydroforming technology in loading path, axial feeding and sealing method is reviewed. The development and application of forming equipment at home and abroad are discussed, and the future development trend of tube hydroforming is prospected.

2. Loading paths

During tube hydroforming, the liquid should be flushed into the inner cavity of the tube while the die is closed, the production and control of liquid pressure will directly affect the forming effect of tube. The loading path refers to the relationship between the hydraulic pressure and the axial feed during forming. Scholars at home and abroad have put forward such methods as broken line loading, pulsating loading and impact loading. In 2005, Imaninejad et al. [3] conducted the bulging experiment on A6063 circular tube with diameter of 38mm, wall thickness of 3.81mm and length of 200mm by using a press with polyline loading method. The results show that the effect of this loading method on the uniformity of tube wall thickness is not obvious, but it can effectively improve the sticking of tube. In 2001, Rikimaru et al. [4] adopted the pulsating loading mode, which can improve the hydroforming performance of tubes, the research shows that the pulsating loading mode can make better use of wrinkles to improve the tube forming. In 2017, Ngaile et al. [5] provided high-pressure liquid through supercharger, the liquid pressure transformation is realized by the proportional valve, the pressure sensor detects the pressure in real time, converts the simulation signal into digital signal and transmits it to PLC for logic operation and outputs the controllable liquid pressure curve. In 2011, Lin [6] was adjusted by PID controller and then acted on the pilot proportional solenoid relief valve. The relief valve regulated the pressure to realize the unloading of the low-pressure chamber of the booster cylinder. In this way, the output pressure gradually increases according to the pulsation of the set curve, this system use PID control system, but most of the debugging use the empirical method, and the ratio, integral, differential three parameters set too dependent on the field debugging. In 2007, K. Mori et al. [7] et al. Used servo punch to generate pulsating hydraulic pressure for pulsating hydroforming, Compared with the non-pulsating hydroforming, it is found that under the condition of pulsating hydroforming, there will be micro wrinkles in the process of tube forming, and the wrinkles have the characteristics of repeated generation and disappearance, which makes the wall thickness of tube deformation area change more evenly, Therefore, it is believed that pulsating hydraulic pressure is helpful to improve the forming performance of tube. In 2014, Huang cm et al. [8] applied the impact force to the liquid with a punch, and carried out the research on the impact hydroforming, this technology has been applied in the field of automobile and aviation industry because of its fast forming speed, high forming quality and low cost. It is proved that the folding line loading can improve the film sticking property of tubes, wrinkle can be used to improve formability under pulsating loading, impact loading can improve production efficiency.

3. Axial feeding

According to the theory of plastic forming, plastic flow occurs when materials are deformed, this theory has been widely used in THF, by applying axial force to both ends of the tube, the material flows to the bulging area. It can avoid the fracture of the bulging area due to the thin deformation wall thickness, relieve the tensile stress of the tube in the bulging area, and effectively improve the forming performance. If the material is not replenished in time, the tube wall will become thinner and fracture. Excessive feeding can cause wrinkling of the tube. Therefore, the precise control of axial feeding is very important. In 2006, Li et al. [9] analysis the relationship between the internal pressure and the axial load based on the plasticity theory, put forward the method of controlling the axial feed and optimized the load mode. It is proved by simulation and experiment that axial feeding can improve tube forming effect. In 2017, Ken-ichi et al. [10] performed bulging analysis on C1220-H and SUS304 tubes with a diameter of 0.5mm and a wall thickness of 0.1mm. It is concluded that SUS304 has better formability than C1220-H. However, the use of gear propulsion to provide axial movement, feed accuracy is lower, and the comparability of the final experimental results will be reduced. The structure is shown in Figure 1. In 2007, Hartl et al [11]. Used the spindle gear driven by gear to provide axial force to carry out the hydroforming of small tubes. In 2017, Dai [12] conducted experiments on 304 tube with diameter of 2mm and wall thickness of 0.3mm, it is found that axial feeding can also improve the formability. The power source is provided by the stepping motor, and the feeding quantity is precisely controlled, which effectively improves the feeding accuracy. However, due to the length of the tube is 50mm, the span of the bulging area is 5mm, the guide area is 22.5MM long, and the friction force is too large. The tube in the guide area will have a large compression strain, which will reduce the actual material entering the bulging area, and the wall thickness will be excessively thinned and cracked. Therefore, the length of the guide area should be reduced appropriately. It is proved that precise control of axial feeding can improve the formability of tube.

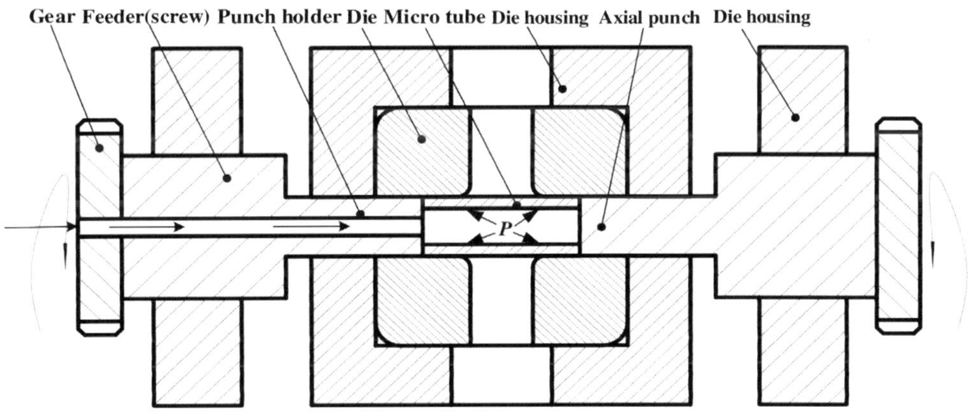

Figure 1. Structure diagram of axial feed provided by gear.

4. Sealing methods

During the tube hydroforming process, leakage will seriously affect the forming quality. Serious external leakage will affect the working performance of the hydraulic pump and the stability of the actuator. Internal leakage will cause the system not to work properly. In 2010, Qin et al. [13] summarized the sealing methods commonly used in THF. The structure is shown in Figure 2. Figure 2 (c) and (d) have shown poor wear resistance in actual production, as shown in Figure 2 (b), the tapered

IOP Conf. Series: Materials Science and Engineering **758** (2020) 012079 doi:10.1088/1757-899X/758/1/012079

punch shows good sealing and wear resistance, but the material can't be put into the bulging zone, in the actual mass production, it is proved that the appropriate punch design in Figure 2 (a). The wear here is mainly due to the sharp angle around the contact area between the punch and the pipe end, When the punch contacts the end of the tube, this angle is pressed into the front of the tube to form a seal. In 2012, Lungershausen et al. [14] avoided the difficulty of machining the sealing groove, and ensure the sealing effect, the structure is shown in Figure 3. At the same time, the problems of difficult processing, demoulding and sealing are solved. It can be seen that the traditional conical structure can be used in the hydroforming of micro tubes. In 2013, Ngaile et al. [15] successfully avoided the difficult problem of micro pipe sealing, they developed a floating die set, the structure is shown in Figure 4. The punch, die and feed system are submerged in the high-pressure liquid chamber, the outer surface of the tube is coated with Teflon, successful completion of T-shaped Y-shaped hydroforming. It provides a new scheme for micro tube hydroforming. Reasonable sealing method can solve the problems of system leakage and demoulding.

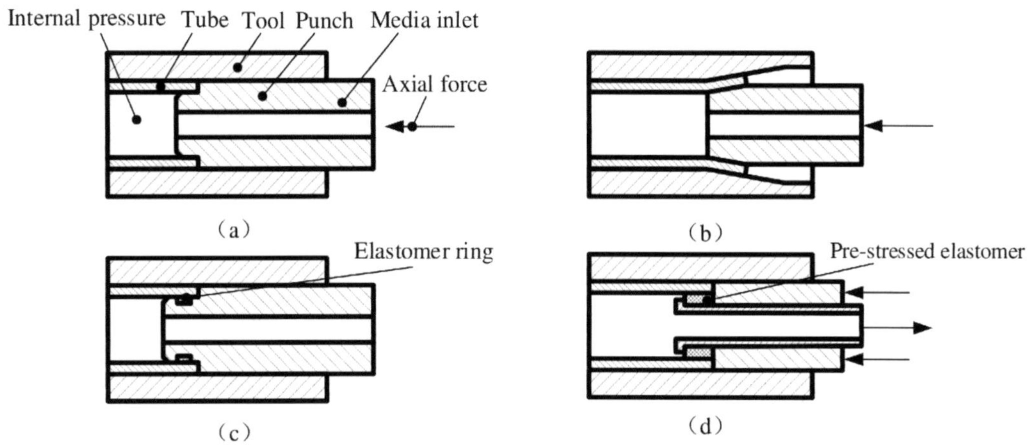

Figure 2. Common sealing principle.

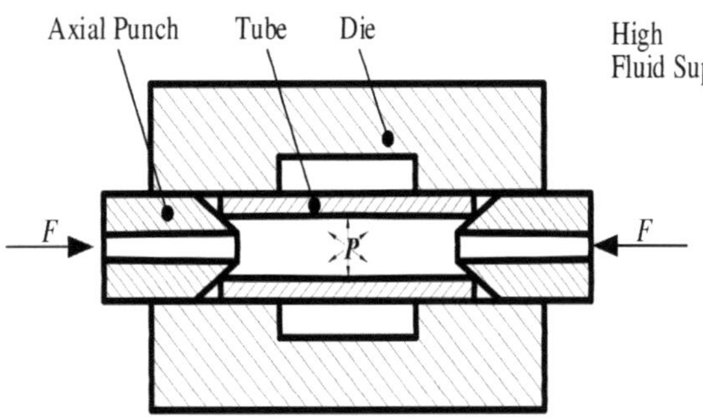

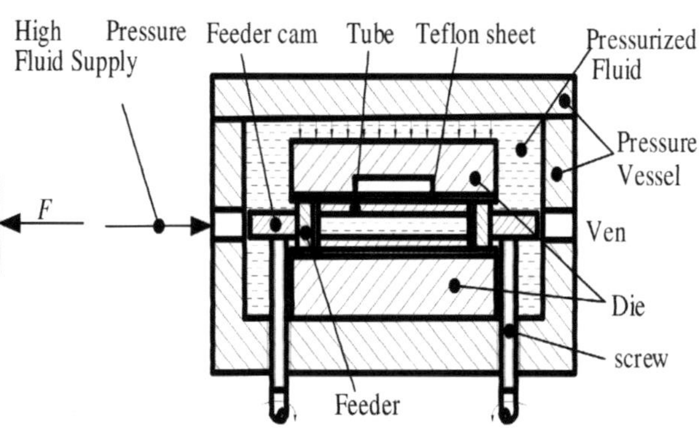

Figure 3. Sealing structure of conical punch.

Figure 4. Structure of the floating die.

5. Conclusion

This article mainly introduces the key problems of the tube hydroforming experimental equipment in three aspects: loading path, axial replenishment and sealing method. This article be aimed at broken line loading, pulsating loading and impact loading respectively. Pulsating loading can make use of wrinkling to improve the forming performance of tube. Impact loading can greatly improve production efficiency. Based on the above loading method, a proportional loading method is proposed: The axial stress is always proportional to the circumferential stress at the highest point of expansion, this loading path can be used to predict the deformation characteristics, it has a good guiding effect on actual production. It has been proved by scholars that the controllable feed rate can significantly improve the axial feeding, use servo motor can improve the accuracy of control. The structure of punch seriously affects the sealing of hydraulic system, it has been proved by scholars' practice: Conical structure can achieve good sealing, and easy to process, convenient to take off the die, At the same time, this kind of structure can be applied to micro pipe.

References

[1] Rong H S, Yang L F and Mao X C 2011 Development of Experimental Devices for Tube Hydro-Forming *J. Adv. Mater. Res*, 1351-54.

[2] Volder M D and Reynaerts D 2009 Development of a hybrid ferrofluid seal technology for miniature pneumatic and hydraulic actuators *J. Sens and Actuators, A: Physical,* **152**,234-40.

[3] Imaninejad M, Subhash G and Loukus A 2005 Loading path optimization of tube hydroforming process *J. International Journal of Machine Manufacture,* **45 (2)**,1504-14.

[4] Rikimaru T, Ito M. 2001 Hammering hydroforming of tubes *J. Press Working,***7**,39.

[5] Ngaile G, Lowrie J 2017 Punch Design for Floating Based Micro-Tube Hydroforming Die Assembly *J. Journal of Materials Processing Technology.***253,** 168-177

[6] Lin X B, Wang B D and Liu Y 2011 Critical Problems in Tube Hydroforming System with Function of Pulsating Loading *J. Mechanical Engineering & Automation,*.**6**,90-91

[7] Mori K, Maeno T and Maki S et al 2007 Mechanism of improvement of formability in pulsating hydroforming of tubes *J. International Journal of Machine Tools & Manufacture*, **47(6)**,978-84.

[8] Huang C M, Liu J W and Zhong Y Z et al. 2014 Exploring liquid impact forming technology of the thin-walled tubes *J. Applied Mechanics & Materials*, 841-844.

[9] Li X J, Zhou X B and Lang L H 2006 Process parameters and limits in tube hydro-forming with axial feeding *J. Journal of Beijing University of Aeronautics & Astronautics,* **32**,475-80.

[10] Ken-ichi M, Hideki S and Kenta I et al. 2017 Factors Influencing the Forming Characteristics in Micro Tube Hydroforming by Ultra High-Forming Pressure *J. Procedia Engineering,* **207**,2334-39.

[11] Hartl C, Lungershausen J, Eguia J, et al. 2007 Micro hydroforming process and machine system for miniature/micro products[C]/Proc. of Int. Conf 7th euspen, Bremen. 2,69-72.

[12] Dai Z Y, Yang C.2017 Influence of axial feeding on the hydroforming of micro-tube J. Journal of Plasticity Engineering, **24(4)**,47－53 .

[13] Qin Y 2010 *Overview on Micro Manufacturing* M. Elsevier Science, Oxford, UK, 146–161.

[14] Lungershausen J Fundamental 2012 Studies on Tube Hydroforming and Laser Assistance in the Manufacture of Micro-Parts *D. University of Strathclyde*, Glas-gow.

[15] Ngaile G, Lowrie J 2014 Micro-Tube Hydroforming System Based on Floating Die Assembly *J. Springer Vieweg,427-32*

Acknowledgments

This work was financially supported by the National Natural Science Foundation of China (51564007), Natural Science Foundation of Guangxi Province (2017GXNSFAA198133).

IOP Conf. Series: Materials Science and Engineering **758** (2020) 012080 doi:10.1088/1757-899X/758/1/012080

Synthesis of SAPO-41 Molecular Sieves with Different SiO_2/Al_2O_3 Ratios and their Catalytic Performance of n-Decane Hydroisomerization

Han Wang, Chunmu Guo, Jinying Meng,Ting Li, Le Luo, Quan Wang, Wei Wu*
International Joint Research Centre for Catalytic Technology; School of Chemistry and Material Sciences, Heilongjiang University, Harbin 150080, China
E-mail: wuwei@hlju.edu.cn

Abstract: Using di-n-propylamine (DPA) as a template, SAPO-41 molecular sieves with SiO_2/Al_2O_3 ratios (n) of 0.3, 0.5, and 0.7 were synthesized. Their structure, morphology and acidity were characterized by XRD, N_2 physical adsorption, SEM, and Py-IR. All the synthesized samples have characteristic diffraction peaks corresponding to the AFO topological structure, which means that samples are pure phase SAPO-41 molecular sieve. The samples exhibited a highly ordered prismatic morphology with regular arrangement of individual grains, wrapped in flakes, and possessed complete channel structure. With the increase of n, the specific surface area and the amount of B acid showed a downward trend after increasing. Hydroisomerization reaction results of n-decane over SAPO-41 loaded with 0.5wt% Pd (Pd/SAPO-41(n)) showed that the higher the amount of B acid, the higher the conversion of n-decane,and high specific surface area and moderate B acid are favorable for yield of isomerization products. When the conversion of n-decane catalyzed by Pd/S41(0.3)-P was 88.9%, the yield of hydroisomerization products reached 83.3%.

1. Introduction

Iso-alkanes can be obtained by hydroisomerization of *n*-alkanes. With increasing of *iso*-alkanes content, the freezing point or pour point of diesel oil or lubricating oil is significantly reduced, reduce and the fluidity is greatly improved, while maintaining a high cetane number.[1~2]. For low-carbon alkanes, the octane number of gasoline increases significantly,which meets automotive requirements for high quality fuel [3~4].

In the process of hydroisomerization of long-chain *n*-alkanes, cracking reactions are often accompanied, which reduces the selectivity of branched alkanes and yield of fuel oil [5~6]. It is important to choose a molecular sieve carrier with a suitable pore structure and acidity.

In this paper, a series of SAPO-41 molecular sieves with different silicon contents by changing SiO_2/Al_2O_3 in the initial gel used di-n-propylamine (DPA) as a template. The influences of silicon content on structure, morphology and acidity of SAPO-41 molecular sieves and the catalytic performance of n-decane hydroisomerization were discussed.

2. Experimental

Using di-n-propylamine (DPA) as a template, SAPO-41 molecular sieves with different silicon contents (SiO_2/Al_2O_3 ratio (n) are 0.3, 0.5, 0.7, named S41(n)-P) and the bifunctional catalysts of

SAPO-41 loaded 0.5wt%Pd (0.5Pd/S41(n)-P) were systhezed, referring to [7]. The process of hydroisomerization of n-decane catalyzed by 0.5Pd/S41(n)-P was carried out, referring to [8].

3. Results and discussion

3.1. Characterization of S41(n)-P

The structure, morphology, and acidity of the synthesized S41(n)-P were characterized by means of XRD, SEM, N_2 physical adsorption, and Py-IR,the results were shown in Fig.1~4 and Table 1~2.

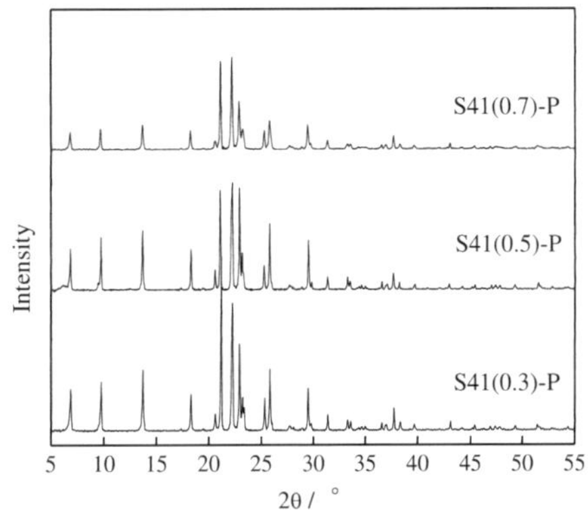

Fig. 1 XRD patterns of S41(n)-P samples

It can be seen from Fig.1 that compared with the standard spectrum of AlPO4-41 molecular sieve, all the synthesized S41(n)-P samples have characteristic diffraction peaks corresponding to the AFO topology without the finding of diffraction peaks heterocrystals, indicating that the synthesized samples are pure SAPO-41 molecular sieve. With the increase of silicon content, the characteristic diffraction peak intensity of the sample decreases, indicating that the crystallinity of the sample decreases with the increase of silicon content. This may be due to the fact that when the silicon content in the initial gel is higher, silicon is not conducive to entering the molecular sieve framework and forming non-skeletal species.

Fig. 2 SEM images of S41(n)-P samples.(a) S41(0.3)-P; (b) S41(0.5)-P; (c) S41(0.7)-P.

The SEM images of S41(n)-P shown in Figure 2 represent the typical morphology of SAPO-41 molecular sieves, which all show a regular arrangement of single crystal, a sheet-like wrapping, and a highly ordered prismatic structure. With the increase of silicon contents in the initial gel, the morphology of the S41(n)-P sample did not change significantly, but irregular small crystal aggregates appeared on the sample surface, which may be part of the non-skeletal species mentioned above.

It can be seen from the N_2 adsorption-desorption isotherms of S41(n)-P samples shown in Fig.3 that the isotherms of all samples have the characteristics of type I isotherms presented by microporous molecular sieves.

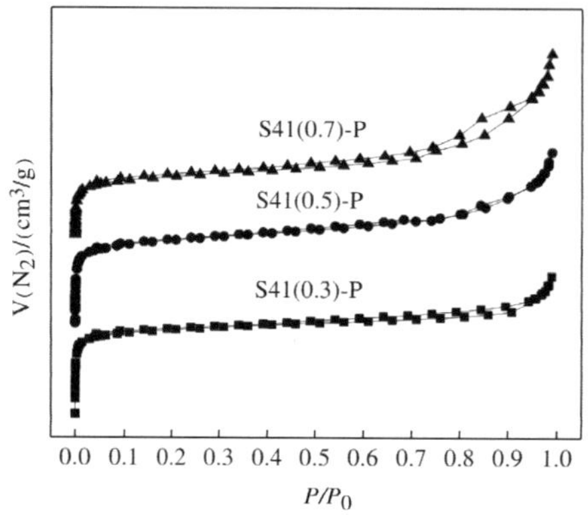

Fig. 3 N_2 adsorption-desorption isotherms of S41(n)-P samples

The S41(0.7)-P sample showed obvious hysteresis loops at the relative partial pressure $P/P_0 = 0.5\sim0.9$, which may be due to the existence of structural defects in the sample or the formation of intergranular pores with small crystals.

It can be known from Table 1 that as the silicon content increases, the microporous specific surface area and micropore volume of the sample increase, indicating that the channel structure of S41(n)-P is more complete. When the silicon content further increases, the microporous specific surface area and microporous volume of the sample decrease,which may be due to the clogging of the pores due to the formation of some non-skeletal species.

Table 1 N_2 adsorption data of S41(n)-P samples

Samples	Surface area(m^2/g)			Pore volume (cm^3/g)		
	BET[a]	Micropore[b]	External	Total[c]	Micropore[b]	Mesopore
S41(0.3)-P	215	190	25	0.129	0.074	0.055
S41(0.5)-P	236	196	40	0.165	0.077	0.088
S41(0.7)-P	167	131	36	0.166	0.052	0.114

[a]BET method, [b]t-plot method, [c]Volume adsorbed at $P/P_0 = 0.99$.

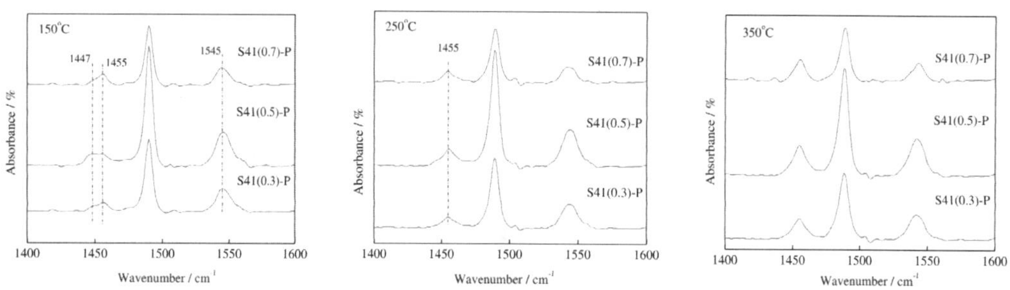

Fig. 4 Py-IR spectra of S41(n)-P samples

It can be seen from Fig.4 that when the pyridine desorption temperature is 150℃, two absorption peaks appear in the vibration region of 1440~1462 cm^{-1}.The peak at 1447 cm^{-1} is due to the interaction between pyridine and Si-OH, and the peak 1455 cm^{-1} is attributed to the variable-angle vibration of the CH bond of the pyridine molecule adsorbed on the L acid site, that is, the interaction between pyridine and the L acid site. As the desorption temperature of pyridine increases, the absorption peak at 1447 cm^{-1} decreases. This is due to the decrease in the adsorption of pyridine molecules and Si-OH at high temperature, which results in the desorption from the surface of Si-OH. The acid sites for desorbing pyridine at relatively low temperatures belong to weaker acid sites, while the acid sites for desorbing pyridine at higher temperatures belong to stronger acid sites. In S41(n)-P, the L acidic site is derived from a framework defect site or a non-framework species in the molecular sieve, and the B acidic site is derived from Si substituted P or Al and P to generate a bridge hydroxyl group in Si-OH-Al. The acid amount of the sample was calculated from the peak area of the pyridine infrared spectrum, and the results are shown in Table 2.

Table 2 Acid distribution of S41(n)-P samples

Samples	Brönsted acid (μmol/g)		
	150℃	250℃	350℃
S41(0.3)-P	30.23	29.03	21.30
S41(0.5)-P	44.62	41.34	28.94
S41(0.7)-P	20.71	17.01	11.64

It can be seen from Table 2 that as the silicon content increases, the amount of B acid on the sample increases. This is because more silicon enters the molecular sieve framework and generates more negative charges, and more hydrogen protons are needed to balance the framework charge,which leads to more B acid sites generated on the molecular sieve. With the increase of SiO_2/ Al_2O_3 ratio in the initial gel, Si follows the SM3 mechanism, that is, the proportion of two Si replacing one P and one Al into the molecular sieve framework increases, forming more border area, more strong B acid sites are generated. However, when the SiO_2/ Al_2O_3 ratio in the initial gel was increased to 0.7, the amount of B acid in the sample decreased. This may be due to the excessive amount of silicon in the initial gel, which prevents it from entering the molecular sieve framework, forming some non-framework silicon species in the molecular sieve product.

3.2. 0.5Pd/S41(n)-P catalytic hydroisomerization

Fig. 5 shows the results of n-decane conversion and isomerization selectivity at different reaction temperatures in the n-decane hydroisomerization reaction under the action of the bifunctional catalyst 0.5Pd/S41(n)-P.

It can be seen from Fig.5 that as the reaction temperature increases, the conversion of n-decane on each catalyst increases. In the high temperature region, the increase in the conversion of n-decane tends to be gentle, which may be due to the influence of the diffusion rate at high conversion. The order of the reactivity of each catalyst is 0.5Pd/S41(0.5)-P>0.5Pd/S41(0.3)-P>0.5Pd/S41(0.7)-P,which is consistent with the order of its B acid amount. When the conversion of n-decane is greater than 55%, the selectivity order of isomerization on the catalyst is 0.5Pd/S41(0.3)-P> 0.5Pd/S41(0.5)-P> 0.5Pd/S41(0.7)-P. 0.5Pd/S41(0.5)-P has a large amount of strong B acid, which leads to an intensified cracking reaction. Due to the presence of some non-skeletal species on 0.5Pd/S41(0.7)-P, the pore channels are blocked, and the microporous channels are smaller,which limits the diffusion of reaction intermediates and products, and increases the probability of cracking reactions.

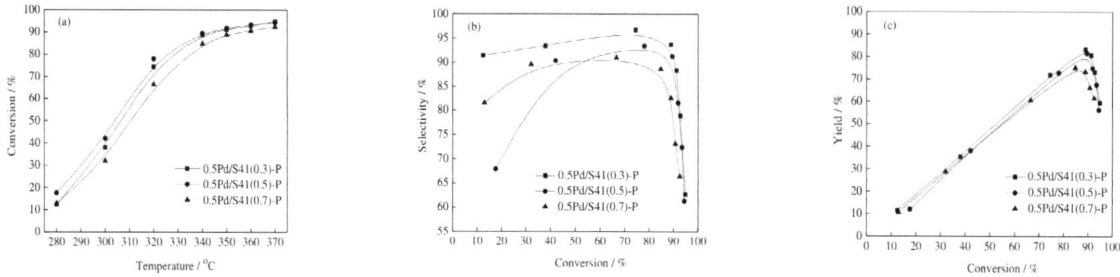

Fig. 5 Catalytic performance of 0.5Pd/S41(n)-P in n-decane hydroisomerization

(a) Conversion of n-decane; (b) Selectivity for i-decane; (c) Yield of iso-decane.

0.5Pd/S41(0.3)-P with suitable acid strength and pore volume can effectively inhibit the cracking reaction under certain active conditions, and therefore shows the highest selectivity of hydroisomerization products.It can be seen from Fig.5(c) that the yield of the isomerization product on 0.5Pd/SAPO-41 first increases and then decreases with the increase of the n-decane conversion, which is consistent with The conversion mechanism of n-alkanes on bifunctional catalysts. The carbocations obtained from the dehydrogenation of n-decane at the metal site undergo isomerization or cracking reaction at the acid site. When the reaction temperature is higher, the conversion of n-decane is increased, but the cracking reaction is intensified, resulting in a decrease in the yield of the isomerization product.

Among three catalysts, the highest hydroisomerization yield on 0.5Pd/S41(0.3)-P reached 83.3% at the n-decane conversion of 88.9%.

4. Conclusions

The pure phase SAPO-41 molecular sieves with difference SiO_2/Al_2O_3 ratios were synthesized using DPA as a template. With the increase of SiO_2/Al_2O_3 ratios , the specific surface area and the amount of B acid showed a downward trend after increasing. The conversion of n-decane rise with the increase of B acid amount,but high specific surface area and moderate B acid are favorable for yield of hydroisomerization products. Pd/S41(0.3)-P is the best catalyst and the yield of isomerization products reached 83.3% at conversion of n-decane of88.9%.

5. References

[1] Meller E, Sasson Y, Z. Aizenshtat.2016 *RSC Adv.* **6** 103149

[2] Parmar K, Pant M, John K.2015 *J. Mol. Catal.* A **47** 404

[3] Kim J, Han,J Kim 2018 *ACS Catal.* **8** 10545

[4] Yang Z,LiuY,LiY 2018 *J Ener Chem* **28** 23

[5] Yang L,Wang W,Song X,2019 *Fuel Proc Tech* **190** 13

[6] Gao L, Shi Z,Ubong Jerome Etim2019 Fuel **252** 653

[7] Wei X， Kikhtyanin O V， Parmon V N， et al. 2018 *J Porous Mater* **25** 235

[8] Wu H, Zhang J, Zhao A,et al. 2017 *J Porous Mater* **24** 437

6. Acknowledgements

This work was financially supported by National Natural Science Foundation of China (No.21676074), National Ministry of Education "Silk Road 1+1" Research Cooperation Project and Innovation and entrepreneurship training program for college students in Heilongjiang province of China （201910212105）

IOP Conf. Series: Materials Science and Engineering **758** (2020) 012081 doi:10.1088/1757-899X/758/1/012081

Post-cracking Poisson Ratio of Concrete in Steel-Concrete-Steel Panels Subjected to Biaxial Tension Compression

Manish Prasad[1] and Song XiaoBing[2,3]

1 Master's student, Shanghai Jiao Tong University, China
2 Asst. Professor, Shanghai Jiao Tong University, China
e-mail: xbsong@sjtu.edu.cn

Abstract. Consideration of post-cracking Poisson ratio of concrete is found significant in predicting the behavior of Steel-Concrete-Steel (SCS) panels under biaxial tension compression. Concrete reinforced with face steel plates was found to be under confinement which deviates the established nature of the post-cracking Poisson ratio. In order to account for such confinement, an extension to previously proposed expression for the Poisson effect in reinforced concrete (RC) is presented in this paper to be used for the analysis of SCS in which the reinforcing elements, steel plates, provide confinement in the principle tensile direction. A tri-linear model is proposed for the Poisson ratio and then compared with the experimental results.

1. Introduction

An SCS panel is a composite in which core concrete is reinforced with two face steel plates connected to each other with ties and to the core concrete with shear studs at regular intervals. SCS is being used for different purposes such as nuclear power plants, off-shore structures, high rise buildings and others because of its improved structural capacity and modular usability in construction.

An accurate and efficient analysis of SCS panels demands for a precise modeling of the steel plates and the core concrete. Steel plate models have been almost perfected in past researches but concrete, however, due to its discontinuity after cracking, lacks appropriate 2D models to be used for panels. Different orthotropic models have been presented in past for cracked concrete but most of them were formulate for RC panels which involve rebar reinforcement. Vecchio's [1] model for concrete considered the strength degradation after cracking, strength enhancement due to biaxial or triaxial compression, and concrete lateral expansion. Soften Membrane Model (SMM) [2] by Zhu and Hsu incorporates softening and lateral expansion of concrete.

Even though these models incorporate concrete expansion and confinement effects, an explicit expression for position ratio has not been put forward. Average Poisson ratio for cracked concrete, as evidenced by Zhu and Hsu [3], can exceed the maximum Poisson ratio of 0.5 for continuous material and can be as large as 1.9. They gave an expression for Zhu/Hsu ratio (Eq.(2)), for average Poisson ratio in cracked concrete as a function of principle tensile strain ε_1 and its behavior is dependent on the yielding of the steel bar. Poisson ratio being a property of concrete should not be dependent of any other material but itself. Also, Zhu/Hsu ratio starts increasing from 0.2 even before cracking, which might be acceptable in RC but in SCS a constant value of Poisson ration 0.2 before cracking should be considered.

[3] Author to whom any correspondence should be addressed.

Vecchio's [1] model does account for both Poisson ratio and the confinement but the value of Poisson ratio is limited to 0.5 as given by Eq.(3).

$$\nu_{12} = \begin{cases} 0.2 + 850\varepsilon_1, \ \varepsilon_1 \le \varepsilon_y \\ 1.9, \ \varepsilon_1 > \varepsilon_y \end{cases} \tag{1}$$

$$\nu_{12} = \begin{cases} 0.2 + 850\varepsilon_1, \ \varepsilon_1 \le \varepsilon_y \\ 1.9, \ \varepsilon_1 > \varepsilon_y \end{cases} \tag{2}$$

$$\nu_{12} = \begin{cases} \nu_o, & 0 > \varepsilon_{c2} > \dfrac{\varepsilon_p}{2} \\ \nu_o\left[1 + 1.5\left(\dfrac{2\varepsilon_{c2}}{\varepsilon_p} - 1\right)^2\right] \not> 0.5, & \dfrac{\varepsilon_p}{2} > \varepsilon_{c2} \end{cases} \tag{3}$$

Here, ν_{12} is the average Poisson ratio caused by the expansion of concrete in direction-1 due to the stress in direction-2, ν_o is the initial value of the Poisson ratio, ε_1 is the strain in principal tensile direction, ε_y is the yielding strain of the rebar reinforcement, ε_{c2} is the strain in principal compressive direction and ε_p is the peak compressive strain.

In this paper, the expression for Zhu/Hsu ratio will be modified to incorporate the observed confinement in concrete from the Huang's et al. [4] experimental program of SCS panels. Thus, modified expression for the average Poisson ratio will be used to analyze Huang's et al. [4] panels and compared with the experimental results.

2. Experimental Work

Table 1. Test Specimen Properties and Principle Variables [4].

Specimen	Concrete	Steel plate		Steel plate ratio (ρt)	Bs/ts	Load increment ratio (ΔC/ΔT)	Load modes
	fcu (MPa)	ts (mm)	fy (MPa)				
S3-10	42	2.95	310	1.14%	25.4	-1/0	U.C.
S6-10	41	6	275	2.31%	12.5	-1/0	
S3-51	42	2.95	310	1.14%	25.4	-5.2/1	P.S.
S5-41	41	4.75	300	1.83%	15.8	-3.7/1	
S6-41	41	6	275	2.31%	12.5	-3.5/1	
S6-81	41	6	275	2.31%	12.5	-8/1	
S5-11	41	4.75	300	1.83%	15.8	-1/1	S- P.S.-C.
S6-21	41	6	275	2.31%	12.5	-1.8/1	
S6-01	41	6	275	2.31%	12.5	0/1	U.T.

U.C.—Uniaxial compression S-P.
S.-C.—Sequential loading.
P.S.—Proportional loading in stepped increment.
U.T. —Uniaxial Tension.

Nine SCS panels with properties as shown in Table 1 were tested with the testing apparatus as shown in Figure 3. To obtain the behavior of panels under different loading conditions, a proportional loading with wide range of compression to tension ratio (C/T) were adopted. Different loading modes are shown in Figure 2 For uniform transfer of load from the loading device to the specimen, each specimen was designed to have compression and tension load transfer zones as shown in Figure 1.

3. Equilibrium and Compatibility Equations

For SCS panels [4], tensile load was applied in direction-1 and compressive load was applied in direction-2 as shown in the Figure 3 Applied load directions were the principle directions, so the equilibrium equations can be written as Eq.(4) and Eq.(5), where σ_1 , σ_1^c and $\rho\sigma_1^s$ are equivalent

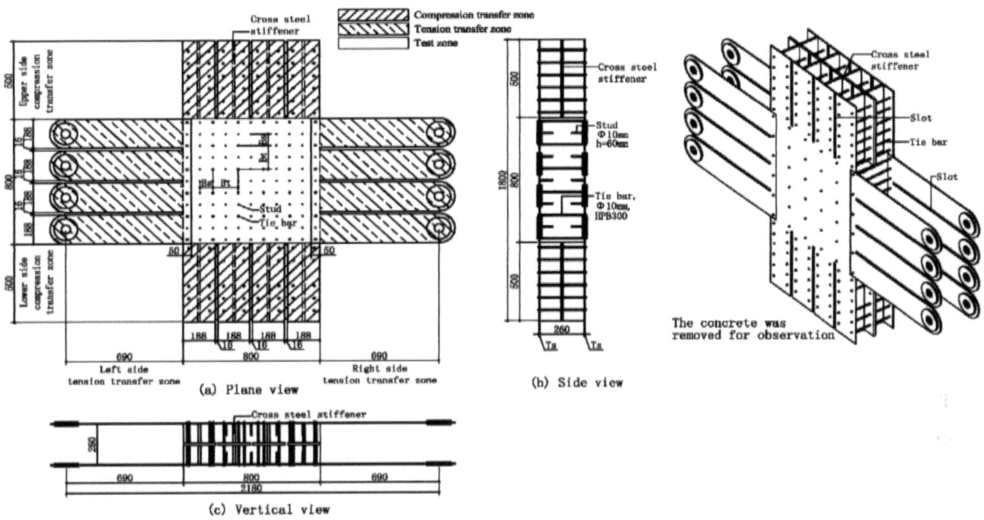

Figure 1. Test Specimen Properties [4].

stresses in direction-1 on the panel, concrete and steel plate respectively. Similarly, σ_2, σ_2^c and $\rho\sigma_2^s$ are equivalent stresses in direction-2 on the panel, concrete and steel plate respectively. Equivalent stress is the stress on the element equivalent to the stress on concrete element of same thickness. ρ is the reinforcement ratio which is the ratio of steel plates thickness $2t_s$ to the concrete thickness t_c. Since the applied load directions coincides with the principal directions, compatibility equations can be written as Eq.(6) and Eq.(7), where ε_1, ε_1^c and ε_1^s are the average strains in direction-1 of the panel, concrete and steel plate respectively. ε_2, ε_2^c and ε_2^s are the average strains in direction-2 of the panel, concrete and steel plate respectively. Average strain is the total strain summed over several cracks.

$$\sigma_1 = \sigma_1^c + \rho\sigma_1^s \tag{4}$$

$$\sigma_2 = \sigma_2^c + \rho\sigma_2^s \tag{5}$$

$$\varepsilon_1 = \varepsilon_1^c = \varepsilon_1^s \tag{6}$$

$$\varepsilon_2 = \varepsilon_2^c = \varepsilon_2^s \tag{7}$$

4. Stress-strain Relationship of Concrete

Constitutive law of concrete in compression as proposed by Belarbi and Hsu [5] is adopted and modified based on the Huang's et al. [4] experimental data. From the analysis of experimental data, it was observed that a confining phenomenon exists along with the softening phenomenon in concrete due to the difference in Poisson ratio of steel plate and that of concrete. A modified stress strain relationship of

IOP Conf. Series: Materials Science and Engineering **758** (2020) 012081 doi:10.1088/1757-899X/758/1/012081

concrete is proposed as Eq(8), where ζ is the combined stress softening and confining parameter and ξ is the combined strain softening and confining parameter.

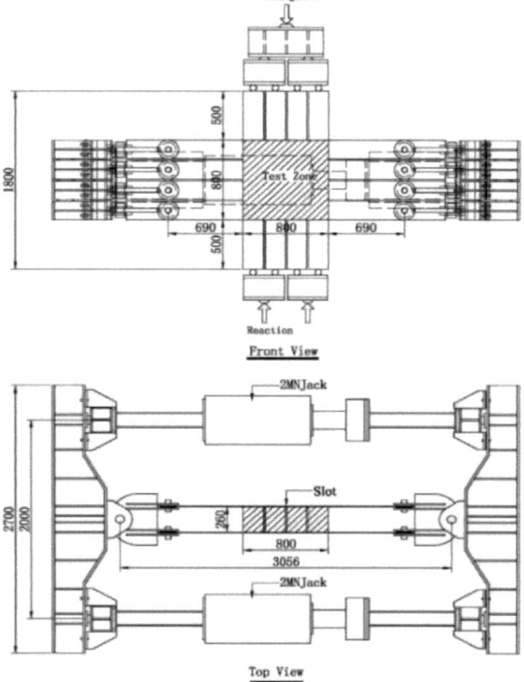

Figure 3. Load Setup and Testing Apparatus [4].

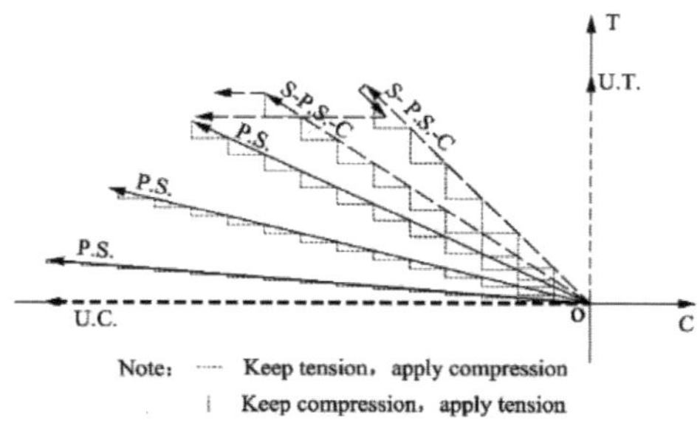

Figure 2. Loading Modes [4].

$$\sigma_2^c = \zeta f_c' \left[2\frac{\bar{\varepsilon}_2^c}{\xi \varepsilon_o} - \left(\frac{\bar{\varepsilon}_2^c}{\xi \varepsilon_o} \right)^2 \right] \tag{8}$$

$$\zeta = \left(1 - 12\frac{\sigma_C^1}{f_C'} \right) \left(\frac{1}{\sqrt{1 + 200\varepsilon_1}} \right) \tag{9}$$

$$\xi = 2\zeta \tag{10}$$

Here, σ_2^c is compressive stress in principal direction-2, f_c' is compressive strength of the concrete, $\bar{\varepsilon}_2^c$ is the uniaxial compressive strain in principal direction-2, and ε_1 is the tensile strain in principle direction-1. Stress strain relationship of concrete in tension as proposed by Belarbi and Hsu [6] and Pang and Hsu [7] is adopted and modified to incorporate the observed confinement. A modified relationship is presented in Eq.(11).

$$d\sigma_1^c = E_c' d\bar{\varepsilon}_1^c \tag{11}$$

$$\text{where } E_c' = 4 f_{cr} \frac{\varepsilon_{cr}^{0.4}}{\varepsilon_1^{1.4}}$$

Here, σ_1^c is tensile stress in principal direction-1, f_{cr} is cracking stress the concrete, $\bar{\varepsilon}_1^c$ is the uniaxial tensile strain in principal direction-1, and ε_{cr} is cracking tensile strain.

IOP Conf. Series: Materials Science and Engineering **758** (2020) 012081 doi:10.1088/1757-899X/758/1/012081

5. Stress-strain Relationship of Steel Plate

Before yielding a linear relationship (Eq.(12)) and after yielding Prandtl-Reuss model [8] for plane stress case (Eq.(13)) is adopted.

$$\begin{Bmatrix}\sigma_1\\ \sigma_2\end{Bmatrix}=\left\{\frac{E}{1-\nu^2}\begin{bmatrix}1 & \nu\\ \nu & 1\end{bmatrix}\right\}\begin{Bmatrix}\varepsilon_1\\ \varepsilon_2\end{Bmatrix} \tag{12}$$

$$\begin{Bmatrix}d\sigma_1\\ d\sigma_2\end{Bmatrix}=\left\{\frac{E}{1-\nu^2}\begin{bmatrix}1 & \nu\\ \nu & 1\end{bmatrix}-\frac{1}{s}\begin{bmatrix}t_1^2 & t_1t_2\\ t_1t_2 & t_2^2\end{bmatrix}\right\}\begin{Bmatrix}d\varepsilon_1\\ d\varepsilon_2\end{Bmatrix} \tag{13}$$

Where, $t_1=\frac{E}{1-\nu^2}\left(s_1+\nu s_2\right)$, $t_2=\frac{E}{1-\nu^2}\left(s_2+\nu s_1\right)$, $s=s_1t_1+s_2t_2$, ν is the Poisson ratio of the steel plate, and s_1, s_2 are the deviatoric stresses.

6. Modification of the Zhu/Hsu ratio

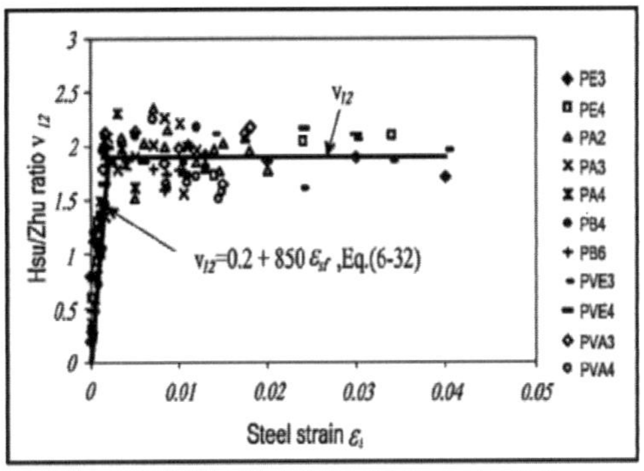

Figure 4. Zhu/Hsu ratio [3].

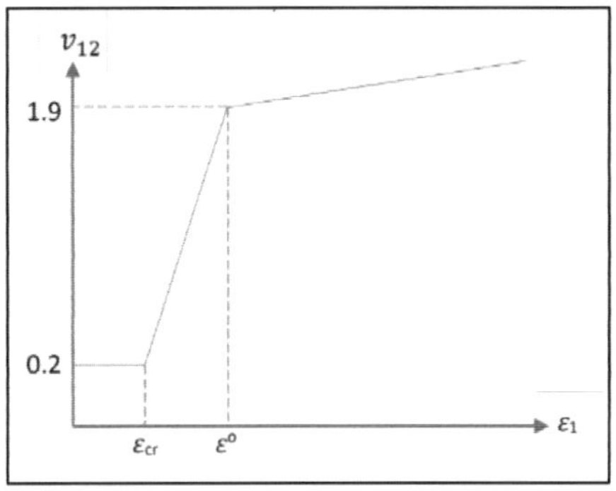

Figure 5. Modified Zhu/Hsu ratio.

Poisson ratio of concrete after cracking can be measured as Zhu/Hsu ratio as shown in Figure 4 Eq(2) suggested for Hsu/Hsu ratio was formulated from an extensive test program of 12 full-scale reinforced concrete panels. The expression for Zhu/Hsu ratio was in fact obtained for RC panel but was used for concrete in SMM model because of the fact that measured Zhu/Hsu ratios were not influenced by the reinforcement ratio. It is, however, apparent that if the Zhu/Hsu ratio is measured for plain concrete, it would be influenced by the presence of reinforcement. Moreover, the suggested expression for Zhu/Hsu ratio is dependent on the yielding property of the reinforcement. This conclusion makes it difficult for the suggested expression for Zhu/Hsu ratio to be implemented for concrete when it is reinforced with any other material whose properties are not similar with steel rebar. Therefore, the expression might need some modification when implemented for panels with different kind of reinforcement.

In modeling of the concrete for SCS panel, Poisson effect has a significant influence. Reinforcing steel plate has a Poisson ratio of 0.3 before yielding and about 0.5 after yielding. Concrete has a Poisson ratio of 0.2 before cracking and the average Poisson ratio or the Zhu/Hsu ratio after cracking keeps increasing even beyond 0.5. Due to this difference in the Poisson ratios, the expansion in the principal direction-2 due to the applied stress in principle direction-1 is different for concrete and steel plate. Thus, with compatibility condition imposed, concrete is under tension when Poisson ratio of concrete is smaller than that of steel plate, and is under compression when the Poisson ratio of concrete is larger than that of steel plate.

Behavior of concrete is very sensitive to the Poisson ratio in SCS panel, both before and after cracking. Before cracking, concrete is isotropic and the Poisson ratio should be constant, however, Eq.(2) shows that it begins increasing even before cracking. Even though the change in Poisson ratio is small before cracking, as given by Eq.(2), if it exceeds the value 0.3(Poisson ratio of Steel plate) before yielding, the stress condition of concrete might change as explained in the earlier section. Therefore, it is essential for the concrete model for SCS to have a Poisson ratio of 0.2 before cracking. Thus, the first line of the trilinear model, as shown in Figure 5, proposed for the modified Zhu/Hsu ratio represents a constant value of 0.2 until the concrete cracks.

After cracking, the Zhu/Hsu ratio increases rapidly from a value of 0.2 to a relatively large value of 1.9. This rapid increase to a value of 1.9 might have caused because of the brittle nature of the concrete and the width of continuously forming cracks in the concrete. Zhu and Hsu [3] proposed a steep slope of 850 for v_{12} versus ε_1 curve. This slope was defined by and the value of v_{12} near yielding strain of the reinforcement as observed from the experimental work [3]. The slope for plain concrete, however, could have been underestimated because of the bond strength provided by the reinforcement which tends to restrict the cracking. So, if it were for plain concrete, a steeper slope could have been more plausible. Based on this assumption and analysis of the Huang's et al. experimental data [4] , a slope of 1889 for v_{12} versus ε_1 curve is suggested as shown in Figure 5 by the second line. This line was assumed to continue till it reaches a value of 1.9 similar as in Zhu/Hsu ratio.

A constant value of 1.9 for the average Poisson ratio was proposed followed by the yielding of the reinforcement. Not only relating the property of the concrete to the property of reinforcement but also a constant value of the average Poisson ratio, when not implemented for RC panels, seems a bit inappropriate. The behavior of the average Poisson ratio reaching a value of 1.9, and after that can be explained without relating it to the yielding of reinforcement. The increasing nature of v_{12} can be related with the formation of new cracks after each loading step. Formation of the cracks and its severity can be measured as a damage parameter ε_1 and when concrete reaches a certain damage level ε^o, the formation of new cracks in SCS panels can be assumed to slow down. Hence, it can be represented with a gentle slope as shown in Figure 5 by the third line instead of a horizontal line as in Zhu/Hsu ratio.

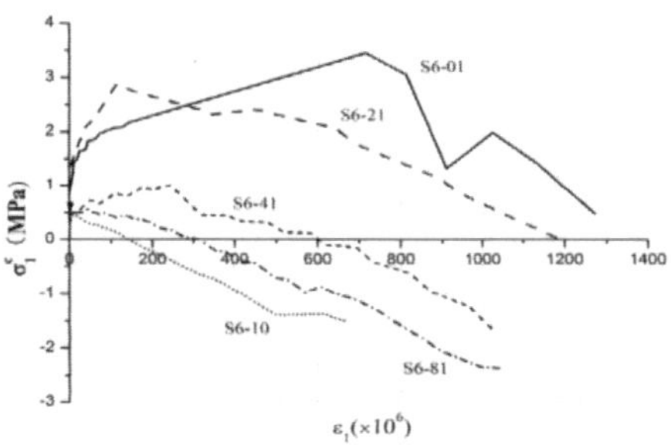

Figure 6. Averaged concrete tensile stress-strain relationship in SCS (before the yielding of steel plate) [4].

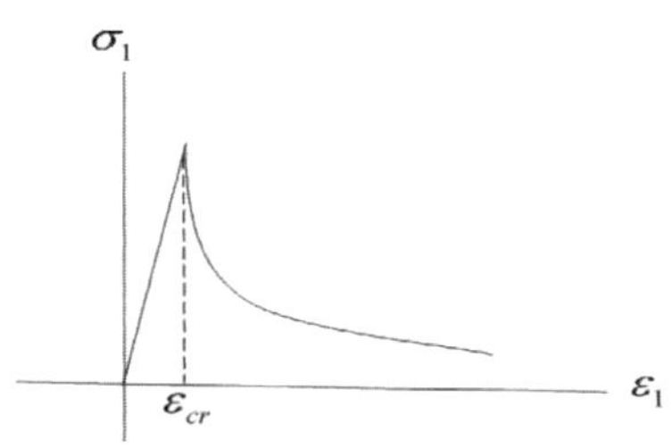

Figure 7. Tensile Stress-Strain Relationship of Concrete [7] [2].

$$v_{12} = \frac{\Delta\varepsilon_1}{\Delta\varepsilon_2} = \frac{\Delta\sigma_1}{E_1\Delta\varepsilon_2} \tag{14}$$

An increasing v_{12} after it reaches a value of 1.9 for concrete in SCS, unlike what suggested by Zhu and Hsu, can be further supported by the Eq(14). A relatively constant v_{12} in later stage of the Zhu and Hsu's experiment might have been observed because of the tensile nature of concrete in absence of confinement as shown in the Figure 7, in which $\Delta\sigma_1$ gets smaller with increasing applied load. However, in Huang's et al. experiment, $\Delta\sigma_1$ seems to be relatively constant even near the later stages of loading as shown in the Figure 6. Average Poisson ratio of panels S3 5.2-1 and S4.5 1-1from Huang's et al.

experiment is show in Figure 8. It can be seen from the figures that the average Poisson ratio of the panels kept increasing throughout the loading history. As the Poisson ratio of steel plate does not increase beyond 0.5, the resulting average Poisson ratio of the panel should be because of the concrete property. Therefore, based on these observations and the data of Huang's et al. experiment, a modified equation for Zhu/Hsu ratio is suggest for confined concrete in SCS panels as Eq.(15). Conventional values for v_{21} as in Eq.(16) is used. With v_{12} and v_{21} , uniaxial strains can be calculated using Eq.(17) and Eq.(18).

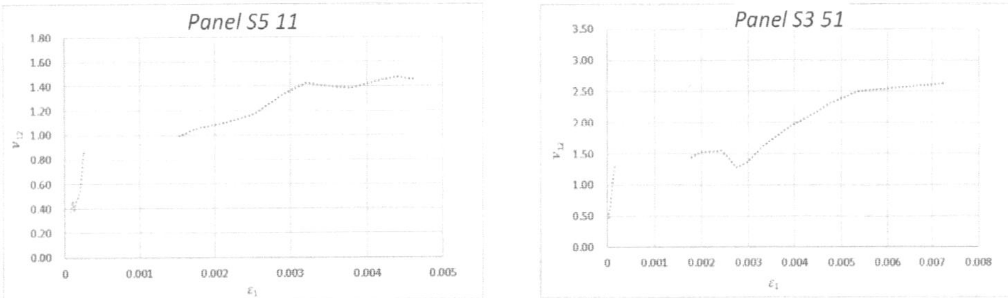

Figure 8. Average Poisson Ratio of the Panels [4].

$$v_{12} = \begin{cases} 0.2 & \varepsilon_1 \le \varepsilon_{cr} \\ 0.2 + 1889(\varepsilon_1 - \varepsilon_{cr}) & \varepsilon_{cr} < \varepsilon_1 \le \varepsilon^o \\ 1.9 + 80(\varepsilon_1 - \varepsilon^o) & \varepsilon_1 > \varepsilon^o \end{cases} \quad (15)$$

$$v_{21} = \begin{cases} 0.2, & \varepsilon_1 \le \varepsilon_{cr} \\ 0, & \varepsilon_1 > \varepsilon_{cr} \end{cases} \quad (16)$$

$$\bar{\varepsilon}_1^c = \frac{\varepsilon_1}{1 - v_{12}v_{21}} + \frac{v_{12}\varepsilon_2}{1 - v_{12}v_{21}} \quad (17)$$

$$\bar{\varepsilon}_2^c = \frac{\varepsilon_2}{1 - v_{12}v_{21}} + \frac{v_{21}\varepsilon_1}{1 - v_{12}v_{21}} \quad (18)$$

7. Application of the Modified Zhu/Hsu ratio and Results

The modified Eq.(15) is implemented along with the equilibrium, compatibility and constitutive equations of concrete and steel plate to calculate the response of SCS panels of Huang's et al. [4] experiment. An iterative procedure is adopted to solve for the unknowns. Load deformation curve obtain was plotted against the experimental results in Figure 9 which shows a very good agreement between two data.

8. Conclusion

Post cracking Poisson effect in concrete suggested as Zhu/Hsu ratio by Zhu and Hsu [3] was modified based on the experimental data of nine SCS panels to be used for the concrete model. The use of thus modified Zhu/Hsu ratio in the analysis of SCS panels not only increased the result accuracy but also increased the calculation efficiency. Incorporation of modified Zhu/Hsu ratio also led to a better understanding of the confinement developed in the concrete in SCS panels. Moreover, this study helped us to understand that the relationship developed for the Poisson effect from the test program of RC other than steel rebars.

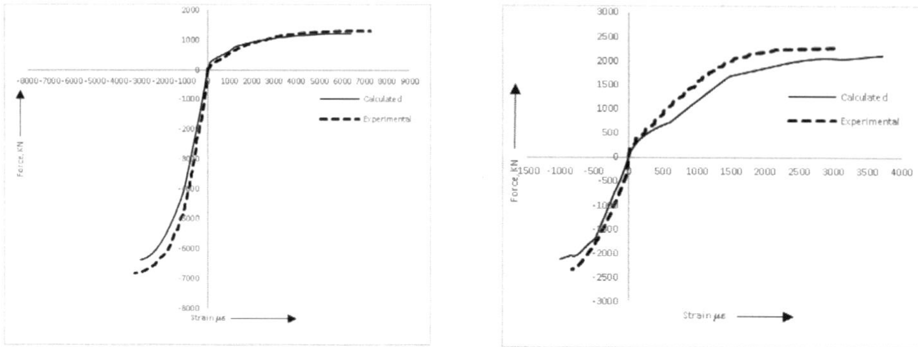

Panel S3 51 *Panel 5 11*

Figure 9. Calculated Vs. Experimental results with modified Zhu/Hsu ratio.

Acknowledgements

The experimental research presented in this paper was funded by the Shanghai Nuclear Engineering Research and Design Institute (SNERDI).

References

[1] F. Vecchio, "Finite Element Modeling of Concrete Expansion and Confinement," *Journal of Structural Engineering,* pp. 2390-2406, 1992.

[2] T. Zhu and R. Hsu, "Softened Membrane Model for Reinforced Concrete Elements in Shear," *ACI Structural JournalACI Structural Journal,* vol. 99, no. 4, pp. 460-496, 2002.

[3] R. Zhu and T. Hsu, "Poisson Effect in Reinforced Concrete Membrane Elements," *ACI Structural Journal,* pp. 631-640, 2002.

[4] C. Huang, S. J. Chen, Y. Leng, C. Qian and X. B. Song, "Experimental research on steel-concrete-steel sandwich panels," *Journal of Constructional Steel Research,* p. In progress, 2019.

[5] T. Hsu and A. Belarbi, "Constitutive Laws of Softened Concrete in Biaxial Tension-Compression," *American Concrete Institute- Structural Journal,* pp. 562-572, 1995.

[6] A. Belarbi and T. Hsu, "Conctitutive laws of Concrete in Tension and Reinforcing Bars Stiffened by Concrete," *ACI Srtuctural Journal,* pp. 465-474, 1994.

[7] X. Pang and T. Hsu, "Behaviour of Reinforce Concrete Menbrane Elements in Shear," *ACI Structural Journal,* pp. 665-679, 1995.

[8] W. Chen, Constitutive Equations for Engineering Materials, vol. 1, New York: John Wiley & Sons, Inc, 1994.

Roughened bed stilling basin and its hydraulic jump characteristics

Yakun Liu[1], Di Zhang[1*], Jian Wu[1], Dong Zhang[1] and Miaomiao Yang[1]

1 School of Hydraulic Engineering, Dalian University of Technology, Dalian, China
E-mail: di.zhang@dlut.edu.cn; liuyakun@dlut.edu.cn

Abstract: The roughened-bed stilling basin, serving as one new type of energy dissipator, has been analyzed both theoretically and experimentally in the literature. This study deduces the calculation formula of the length of the curve C_0 between the flow-depth contraction section (*i.e.* h_c) and the section just before the hydraulic jump (*i.e.* h_1) when the same elevation is employed for both the bottom of roughened-bed stilling basin and the river-bed surface in the downstream. Further, by approximately assuming the equation of the resistance force exerted on the flow and caused by the bed roughness, this study obtains the expression of the ratio between the flow depth before the hydraulic jump and that behind the hydraulic jump, namely h_1/h_2. With the aid of the experimental results in the existing literature, it can be concluded that, when compared with the traditional smooth-bed hydraulic jump, the sequent depth and the length of the rolling hydraulic jump can be reduced by about 20~30% and 20~50%, respectively, in the condition of the chopping roughness riverbed. Additionally, by combining with the hydraulic model tests, this study indicates that adding the bed-roughness can eliminate the secondary hydraulic jump, improve the flow regime and (under the circumstance of the sill-roughness stilling basin) decrease the length of the stilling basin up to about 40%~50%.
Keywords: Stilling Basin; Hydraulic Jump; Hydraulic Characteristics; Roughened Bed; Sequent Depth; Energy Dissipation

1. Introduction

As for medium or low head weir flow and orifice outflow, the hydraulic jump energy dissipation is frequently used with the aid of the stilling basin [1-4]. Usually, the conjugate depth h_c'' of the flow depth in the contraction section (*i.e.* h_c) is greater than the flow depth of the channel in the downstream (*i.e.* h_t, corresponding to the discharge considered). Therefore, the bench-type or sill-type stilling basin is often constructed for the purpose of increasing the water depth up to approximate $1.05\,h_c''$ within the stilling basin and further avoiding the occurrence of the repelled downstream hydraulic jump [5-8].

As shown in *Figure* 1, the total length of the repelled downstream hydraulic jump includes two parts: 1).The length of the curve C_0 (*i.e.* l_{C0}) stretching between the contraction section and the cross-section just in front of the hydraulic jump. 2). The length of the rolling hydraulic jump (*i.e.* l_{rj}) spanning between the cross-section just before the hydraulic jump (whose water depth is denoted by h_1) and that just behind the hydraulic jump (whose water depth is expressed by h_2). Obviously, if the value of l_{C0} is very large, it is not economical to construct such a long stilling basin for the repelled downstream hydraulic jump. Thus, in practical engineering it is common that the channel bottom or the river bed is intentionally roughened (*viz.* increasing the friction drag) in order to reduce the values of l_{C0} and l_{rj} in order to keep the length of the stilling basin within the acceptable range [1, 5, 9, 10].

The influence of different roughness degrees on the ratio h_2/h_1 and the roller length l_{rj} for the hydraulic jumps on corrugated beds has been studied experimentally in *Ref.* [1], whose results indicate that the increase of the roughness can bring down the aforementioned two values. Besides, from the water-surface equation deduced in this paper, it is clear that the increase of the roughness can also lead to the reduction of the length of the curve C_0.

Combining with the model test of a practical engineering, this study qualitatively compares the hydraulic characteristics of the roughened bed stilling basin with that of the traditional stilling basin. It can be concluded that the addition of the bed roughness can result in a shorter length of the stilling basin, a smooth flow regime and a satisfactory energy dissipation effect.

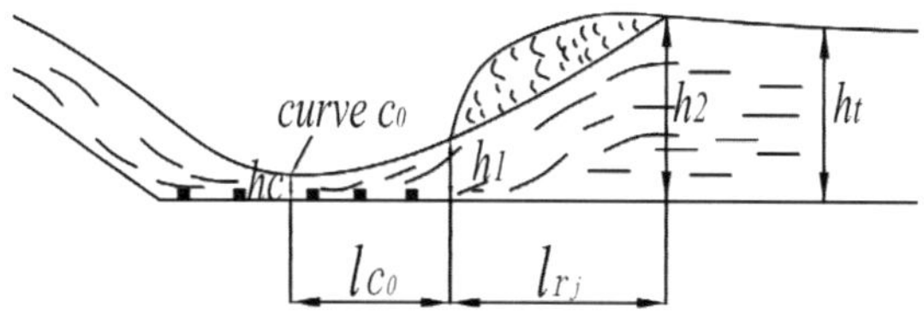

Figure 1. The schematic diagram of the repelled downstream hydraulic jump

2. Hydraulic design of the roughened bed stilling basin

2.1. The designing thought of the stilling basin

The designing thought of the stilling basin is described as following: After obtaining the contraction water depth h_c in the downstream of outlet structures, to compare the downstream channel flow depth h_t and the conjugate water depth h_c''. If $h_t \le h_c''$, either the repelled downstream hydraulic jump or the critical hydraulic jump happens, and it is necessary to construct a stilling basin for the energy dissipation. However, when $h_t > h_c''$, the submerged hydraulic jump takes place, and the downstream water will be pushed towards the upstream under the effect of the hydrodynamic pressure. In practical engineering, under the circumstance of $h_t \le h_c''$, the stilling basin is usually constructed to guarantee a smooth connection between the sequent depth h_2 and the downstream flow depth h_t [9-10].

As displayed in *Figure* 1, the downstream channel flow depth h_t and the Froude number F_r are required in the course of designing the roughened bed stilling basin. The sequent depth h_2 corresponding to h_1 can be calculated by:

$$h_2 = \frac{h_1}{2}\left(\sqrt{1+8(F_{r1})^2} - 1\right) \tag{1}$$

2.2 The length of the curve C_0

Assuming that the breadth of channel (*i.e. b*) is much larger than the depth (*i.e. h*), the describing equation of the water surface curve is:

$$\frac{dh}{dl} = \frac{i - i_f}{1 - F_r^2} \tag{2}$$

[9]

where l is the length along the flow direction, i is the base slope and i_f is the slope of friction.

$$i_f = \frac{n^2 q^2}{h^{10/3}} \tag{3}$$

$$F_r^2 = \frac{h_{cr}^3}{h^3} \tag{4}$$

$$q^2 = h_{cr}^3 \cdot g \tag{5}$$

where n is the coefficient of roughness, q is the discharge per unit width and h_{cr} is the critical depth. For a smooth-bed stilling basin with *i*=0, substituting *equations* (3)~(5) into *equation* (2) , and further defining:

$$Y = \frac{h}{h_{cr}} \quad , \quad L = \frac{l}{h_{cr}} \tag{6}$$

Equation (2) switches to:

$$dL = -\frac{1}{K_n}(Y^{\frac{10}{3}} - Y^{\frac{1}{3}})dY \tag{7}$$

$$K_n = \frac{n^2 g}{h_{cr}^{\frac{1}{3}}} \tag{8}$$

Integrating *equation* (2) along the l_{C0} line, we obtain:

$$K_n L_{C_0} = \left(\frac{3}{4}Y^{\frac{4}{3}} - \frac{3}{13}Y^{\frac{13}{3}}\right)\Big|_{Y_c}^{Y_1} \tag{9}$$

$$l_{C_0} = L_{C_0} \cdot h_{cr} = \frac{h_{cr}}{K_n}\left(\frac{3}{4}Y^{\frac{4}{3}} - \frac{3}{13}Y^{\frac{13}{3}}\right)\Big|_{Y_c}^{Y_1} \tag{10}$$

Along the whole curve C_0, the value of Y is smaller than 1. Consequently,

$$l_{C_0} < \frac{0.5192 h_{cr}}{K_n} \tag{11}$$

Considering that K_n is directly proportional to n^2, l_{C_0} is inversely proportional to n^2, which means that increasing n can decrease the length of the curve C_0 dramatically. In order to obtain the value of Y_c, it is necessary to compute the value of h_c, which can be expressed by the following empirical equation [11]:

$$h_c = E\left(\frac{1}{3} - \frac{2}{3}\sin\left(\frac{\pi}{6} - \frac{1}{3}\cos^{-1}(1 - 13.5\beta)\right)\right) \tag{12}$$

where

$$\beta = \frac{q^2}{2gE^3} \tag{13}$$

and E is the head above the bottom of the pool.

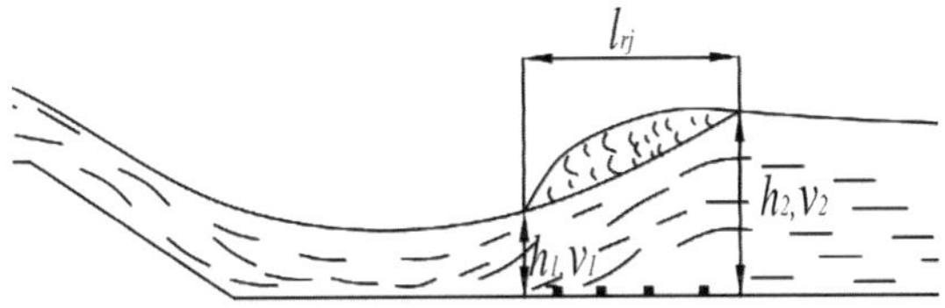

Figure 2. The schematic diagram of the rolling hydraulic jump on roughened bed

2.3 The sequent depth ratio and the length of the rolling hydraulic jump on roughened bed

The rolling hydraulic jump on roughened bed is demonstrated in *Figure* 2, and the corresponding momentum governing equation is:

$$\frac{\rho g b h_1^2}{2} - D - \frac{\rho g b h_2^2}{2} = \rho q b (v_2 - v_1) \tag{14}$$

where D is the roughness resistance exerted on the water flow within the rolling jump region, which can be approximately defined as (from *Ref.* [1]):

$$D = \frac{C_D \rho g b h_1^2}{2} \tag{15}$$

where C_D is the corresponding drag coefficient and b is the breadth of the river bed.
Letting:

$$\eta = \frac{h_2}{h_1} \tag{16}$$

Substituting *equation* (15) into *equation* (14) and using *equation* (16), we can obtain:

$$\eta^3 - A\eta + B = 0 \tag{17}$$

$$A = 1 + 2F_{r1}^2 - C_D \tag{18}$$

$$B = 2F_{r1}^2 \tag{19}$$

$$F_{r1} = \frac{v_1}{\sqrt{gh_1}} \tag{20}$$

$$\eta = \frac{2\sqrt{A}}{\sqrt{3}} \cos\left(\frac{1}{3}\cos^{-1}(-1.5\sqrt{3}M)\right) \tag{21}$$

$$M = \frac{B}{A^{3/2}} \leq 0.385 \tag{22}$$

Only a small number of studies have been carried out for the hydraulic jump on roughened bed in the literature. The diagrammatic sketch of the roughened-bed configuration employed in *Ref.* [1] is illustrated in *Figure* 3 and the experimental results are listed in *Table* 1.

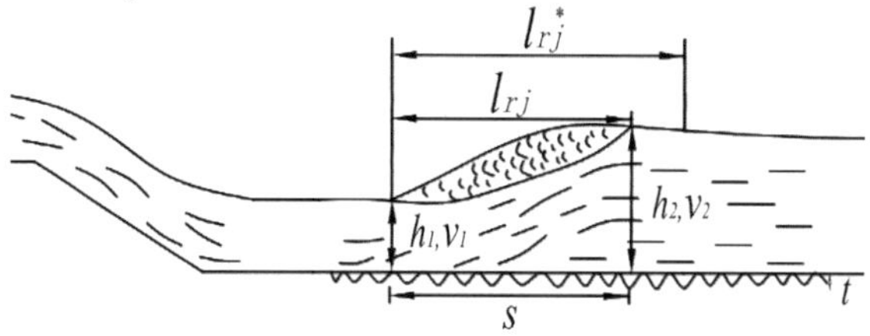

Figure 3. The diagrammatic sketch of the roughened bed adopted in *Ref.* [1]

Table 1. Characteristics of the hydraulic jump on the roughened-bed (b= 0.446 *m*)

$t(mm)$	$s(mm)$	$h_1(mm)$	$v_1(m/s)$	F_{r1}	$h_2(m)$	η	$L_{rj}(m)$	h_2^*	$L_{rj}^*(m)$	h_2/h_2^*	L_{rj}/L_{rj}^*	C_D
13	68	25.4	2	4	0.104	4.09	0.31	0.132	0.5	0.79	0.62	8.45
13	68	25.4	2.5	5	0.128	5.04	0.41	0.167	0.69	0.77	0.59	15.68
13	68	25.4	3	6	0.145	5.71	0.48	0.203	0.88	0.71	0.55	27.79
13	68	25.4	3.49	7	0.188	7.4	0.61	0.239	1.06	0.79	0.58	31
13	68	50.8	2.82	4	0.21	4.13	0.75	0.263	0.96	0.8	0.78	8.19
13	68	50.8	4.07	5.8	0.31	6.1	1.22	0.39	1.59	0.79	0.77	20.04
22	68	50.8	2.82	4	0.21	4.13	0.61	0.263	0.96	0.8	0.64	8.19
22	68	50.8	4.07	5.8	0.31	6.1	1.02	0.39	1.59	0.79	0.64	20.04

h_2^* and L_{rj}^* in *Table* 1 are the sequent depth of the traditional smooth-bed hydraulic jump and the length of its rolling jump region, respectively. Their values can be calculated by *equation* (23) and *equation* (24) [12].

$$h_2^* = \frac{h_1}{2}\left(\sqrt{1+8F_{r1}^2} - 1\right) \tag{23}$$

$$L_{rj}^{*\,[12]} = \begin{cases} 8(F_{r1}-1.5)h_1\,, & 2.5 < F_{r1} < 8 \\ (160\tanh(\dfrac{F_{r1}}{20})-12)h_1, & F_{r1} < 15, \dfrac{h_1}{b} < 0.1 \\ (100\tanh(\dfrac{F_{r1}}{12.5})-12)h_1, & F_{r1} < 15, 0.1 < \dfrac{h_1}{b} < 0.7 \end{cases} \tag{24}$$

It can be concluded from *Table* 1 that the value of h_2/h_2^* (on the chopping roughness bed shown in *Figure* 3) lies in the range of 0.71~0.80, which means h_2 has been shortened by approximately 20~30% relative to h_2^*. Besides, L_{rj}/L_{rj}^* is equal to about 0.52~0.78, which indicates the length of the rolling jump region can be reduced by approximately 20~50% for the roughened bed. Technical indexes make it clear that it is potential to put the roughened-bed stilling basin into engineering use. The drag coefficient in *Table* 1 is calculated with the aid of the experimental η, F_{r1} and *equations*. (17)~(19). In *Ref.* [1], the C_D value for the chopping roughness bed is summarized as:

$$C_D = (F_{r1}-1)^2 \tag{25}$$

Table 1 also indicates that, when the height of chopping roughness t increases from 13*mm* to 22*mm*, C_D and h_2/h_2^* stay almost the same while L_{rj}/L_{rj}^* gets much smaller. Data in *Table* 1 have provided the influence degree of the roughened bed on the characteristics of the hydraulic jump.

Figure 4. The secondary hydraulic jump generated after the end of the traditional smooth-bed stilling basin.

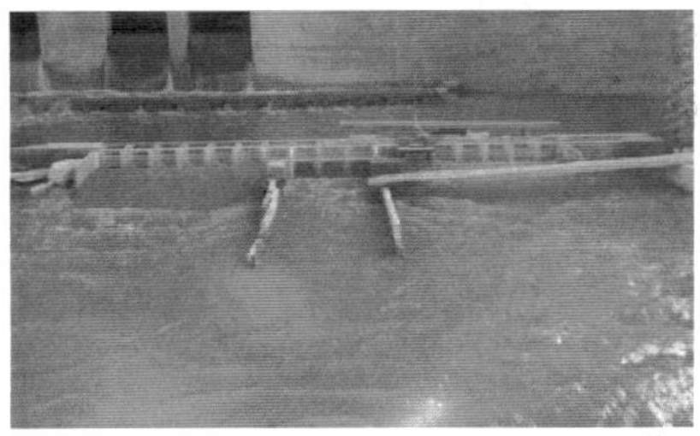

Figure 5. Comparison of the hydraulic jump characteristics between the roughened-bed stilling basin and the traditional smooth-bed stilling basin.

3. Application in practical engineering projects

Combining with a practical engineering, the stilling basins with and without artificial roughness are tested by the hydraulic model to examine the effect of the artificial roughness on the flow characteristics of the hydraulic jump. The scale of length is 1/70. By setting five different roughness degrees within the stilling basin, the length of the hydraulic jump can be significantly reduced.

By comparing *Figure* 4 and *Figure* 5, it can be concluded that the secondary hydraulic jump may be produced after the end of the traditional smooth-bed stilling basin, which is not safe in practical engineering. Extending the length of the stilling basin is an alternative but expensive way. Therefore, adding the roughness degree of the bed may be a better approach. From the *Figure* 4 and *Figure* 5, it is obvious that the addition of the bed roughness can reduce the flow depth at the end of the still basin, keep the original length of stilling basin unchanged and at the same time keep the flow regime smooth.

In view of the fact that the chopping-roughness stilling basin base slab is difficult to construct in practical engineering projects, this study will further investigate the hydraulic-jump characteristics on the sill-roughness stilling basin base slab, which is relatively more feasible to construct in reality.

3.1 The drag coefficient C_D *and the roughness coefficient* n *on sill-roughened bed*

Figure 6 illustrates the schematic of the test case adopted to measure the roughness coefficient of the sill-roughness bed. The height of the sill-roughness structure is either 50/3*mm* or 62/3*mm*, and the streamwise length and the interval distance of this structure is 10*mm* and 83.33*mm* respectively. The width of the water channel is equal to 0.567*m*, and the flow discharge per unit width, denoted by q, can be calculated by u_1, defined as the flow velocity at the upstream section of the riverbed part considered (*i.e.* l=0.426*m*). The experimental results, such as the water depth at different sections (*i.e.* h_1 and h_2), the equivalent roughness coefficient n and the equivalent drag coefficient C_D are given by *Table* 2.

In *Table* 2, the following parameters are defined: $Y_1 = \frac{h_1}{h_{cr}}$, $Y_2 = \frac{h_2}{h_{cr}}$, $L = \frac{l}{h_{cr}}$, and

$$n = \frac{h_{cr}^{\frac{1}{3}}\left(\left(\frac{3}{4}Y_2^{\frac{4}{3}} - \frac{3}{13}Y_2^{\frac{13}{3}}\right) - \left(\frac{3}{4}Y_1^{\frac{4}{3}} - \frac{3}{13}Y_1^{\frac{13}{3}}\right)\right)^{\frac{1}{2}}}{gL} \tag{27}$$

$$C_D = \frac{(1+2F_{r1}^2)Y - Y^3 - 2F_{r1}^2}{Y}, \quad Y = \frac{Y_2}{Y_1} \tag{28}$$

It can be concluded that, for all the test cases considered in this study, the values of n and C_D don't experience significant variation when the height of the sill-roughness structure is changed from 50/3*mm* to 62/3*mm*. A similar formulation (to *equation* (25)) can be deduced from *Table* 2:

$$C_D = 1.18(F_{r1} - 1)^2 \tag{29}$$

More experimental tests are necessary in order to further calibrate the above C_D equation.

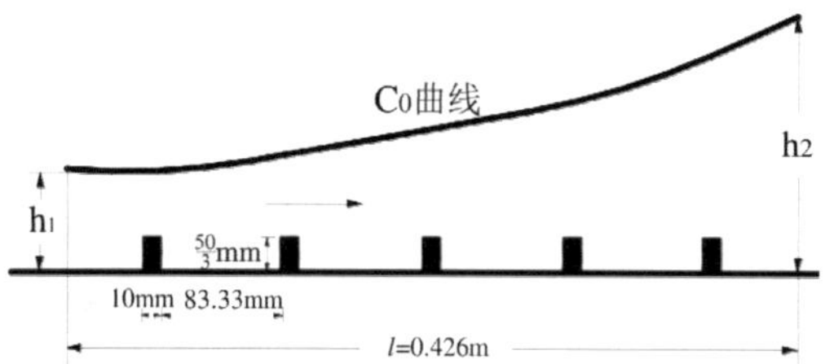

Figure 6. Schematic of the test case adopted to measure the roughness coefficient of the sill-roughness bed.

Table 2. The experimental results of the sill- roughness bed test case

t(mm)	*s(mm)*	$h_1(m)$	$h_2(m)$	$v_1(m/s)$	Fr_1	$h_{cr}(m)$	Y_1	Y_2	L	Y	C_D	n
16.67	83.33	0.118	0.150	1.58	1.47	0.153	0.774	0.978	2.785	1.265	0.31	0.061
16.67	83.33	0.156	0.178	1.80	1.46	0.200	0.778	0.887	2.128	1.141	0.22	0.061
16.67	83.33	0.146	0.165	1.92	1.61	0.200	0.729	0.822	2.128	1.128	0.31	0.063
20.67	83.33	0.114	0.142	1.58	1.49	0.149	0.766	0.952	2.855	1.242	0.32	0.061
20.67	83.33	0.115	0.143	1.60	1.50	0.151	0.763	0.946	2.826	1.241	0.34	0.062
20.67	83.33	0.160	0.183	1.79	1.43	0.203	0.787	0.899	2.095	1.142	0.21	0.061
20.67	83.33	0.154	0.178	1.86	1.51	0.203	0.759	0.874	2.095	1.151	0.28	0.066

3.2 Comparison of the energy dissipation characteristics between two types of stilling basin

With the purpose of making the description more convenient and more distinct, the following engineering project is analyzed: the elevation of the weir crest of the overflow dam is 99.8*m*, the check reservoir water level is 115.0*m*, the corresponding flow discharge per unit width in the stilling basin is q=106m^3/s, the water level in the downstream river channel is 34*m*, the downstream river-bed elevation is 10*m*, as displayed by *Figure* 7.

3.2.1 Design of the traditional stilling basin

In practical engineering projects, the trial and iterative method is utilized, in other words, a depth of the stilling basin (denoted by *d*) is firstly assumed, and then we examine that whether this depth can result in a slightly submerged hydraulic jump within the stilling pool. If the aforementioned requirement can be met, the assumed *d* is the actual depth needed. Otherwise, a iteration procedure is adopted to obtain the correct depth of the stilling basin.

Supposing *d*=7.6*m*, it can be deduced that the total head above the stilling pool base slab is equal to *E*=115-2.4=112.6*m*. With the aid of *equatoin* (4), *equation* (12) and *equation* (23), it can be acquired that h_c=2.275*m*, F_{rc}=9.863, $h_c^{'}$=30.615*m*. Considering that

$$\Delta z = \frac{q^2}{2g}(\frac{1}{(\varphi \times h_t)^2} - \frac{1}{(\sigma_j \times h_c'')^2}) \tag{30}$$

where $\varphi = 0.95$ is the velocity coefficient of the broad crested weir, and $\sigma_j = 1.05$ is the submer-gence factor of the hydraulic jump occurring within the stilling pool. Consequently, it can be obtained from *equation* (30) that $\Delta z = 0.548m$. Seeing that:

$$d = \sigma_j h_c'' - h_t - \Delta z \tag{31}$$

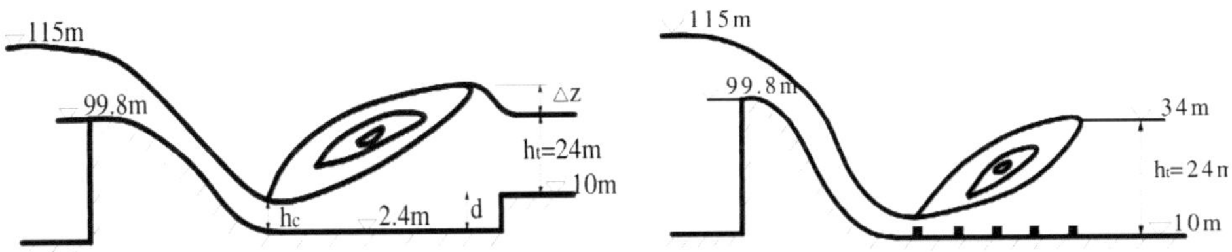

(*a*) the traditional stilling basin (*b*) the sill-roughness stilling basin

Figure 7 Schematic of the design of the traditional stilling basin and the sill-roughness stilling basin

Substituting $h_c^{'}$, h_t and Δz, we can get the computed depth of the stilling basin $d^{'}$ =7.598*m*, which means that the initial assumption of *d*=7.6*m* can meet the design criterion (*i.e.* $d \approx d^{'}$). Then, the length of the traditional stilling basin can be calculated according to the following equation [13, 14].

$$l_s = 6.0(h_c'' - h_c) \approx 170.04m \tag{32}$$

3.2.2 Design of the sill-roughness stilling basin

As illustrated by *Figure* 7(*b*), for the sill-roughness stilling basin, *d*=0 and h_c=2.362*m* (ignoring the loss of water head and calculated from *equation* (12)). At this time, the water depth in the contraction section is actually the water depth just in front of the hydraulic jump (*i.e.* h_1=h_c), and the water depth in the downstream river channel is in fact the sequent depth of the hydraulic jump (h_2=h_t). Therefore, the corresponding sequent depth ratio is:

$$\eta = \frac{h_t}{h_c} = 10.161, \quad F_{r1} = F_{rc} = 9.323 \tag{33}$$

Substituting *equation* (33) into *equation* (17), we can get the roughness drag coefficient C_D =54.487.

Referring to *equation* (25) and *equation* (29), for this project it can be presumed that:

$$C_D = \alpha(F_{r1} - 1)^2 \tag{34}$$

Therefore, $\alpha = \frac{C_D}{(F_{r1}-1)^2}$ =0.787. Taking into account that α =1.0 for the chopping roughness bed In *Ref.* [1] (*i.e. Table* 1 in this study) and α =1.18 for the sill-roughness river bed (*i.e. Table* 2 and *Figure* 6 in this study), it can be inferred that it is not difficult to construct the roughened river-bed with α =0.787 in practical engineering projects.

When the critical hydraulic jump happens (*i.e.* α =0.787), in view of that the shortening coefficient (defined as $\xi = L_{rj}/L_{rj}^{*}$ here) of the roughed-bed stilling basin is equal to about 0.52~0.78 (known from *Table* 1), for the sake of conservation, taking ξ =0.70 in this study. Consequently, the length of the roughened stilling basin is:

$$l_s' = 6.0(h_c'' - h_c)\xi = 90.88m \approx 53\% \cdot l_s \tag{35}$$

Obviously, by roughening the river bed, the length of the stilling basin can be reduced as much as about 50% (*i.e.* $l_s'/l_s = \frac{90.88}{170.04} \approx 53\%$) when compared with the traditional smooth-bed stilling pool.

Being different from the design of the traditional stilling basin, in this situation the sequent depth h_2

and the corresponding Froude number Fr_2 are known in advance, but the water depth and the Froude number just in front of the hydraulic jump (*i.e.* h_1 and Fr_1) need to be determined. Hence, when solving the sequent depth ratio η, *equation* (21) is not able to be directly applied because the unknown quantity Fr_1 is included in the equation of C_D, as shown by *equation* 25, *equation* 28, *equation* 29 and *equation* 34. Considering that:

$$\eta^3 = \frac{F_{r1}^2}{F_{r2}^2} \tag{36}$$

The hydraulic jump equation can be rewritten as:

$$\frac{F_{r1}^2}{F_{r2}^2} - (1 + 2F_{r1}^2 - \alpha(F_{r1} - 1)^2)(\frac{F_{r1}}{F_{r2}})^{\frac{2}{3}} + 2F_{r1}^2 = 0 \tag{37}$$

When the repelled downstream hydraulic jump occurs (*i.e.* α <0.787, taking α =0.6 as an example), due to Fr_2=0.288, it can be deduced that Fr_1=7.78. Thus, the corresponding $h_1 = (\frac{q^2}{gF_{r1}^2})^{\frac{1}{3}}$ =2.665*m*, and the critical water depth h_{cr} =10.460*m*. In addition, $Y_c = \frac{2.362}{10.46} = 0.2258$, $Y_1 = \frac{2.665}{10.46}$ =0.2548, and consequently the length of the curve stretching between the contraction section and the cross-section just in front of the hydraulic jump (*i.e.* l_{C_0}) can be obtained by *equation* (10):

$$l_{C_0} = L_{C_0} \cdot h_{cr} = \frac{h_{cr}}{K_n}(\frac{3}{4}Y^{\frac{4}{3}} - \frac{3}{13}Y^{\frac{13}{3}})\Big|_{Y_c}^{Y_1} = \frac{0.0414}{n^2} \tag{38}$$

It can be presumed that *n*=0.06 from *Table* 2, Thus:

$$l_{C_0} = \frac{0.0414}{n^2} = 11.49m,$$

$$l'_{rj} = 6.0(h''_c - h_c)\xi = 89.61m,$$

$$l''_s = l_{C_0} + l'_{rj} = 101.10m \approx 59.5\% \cdot l_s \tag{39}$$

It can be concluded from the above discussion that, when the roughness scale of the downstream river-bed is large enough, the critical hydraulic jump can be formed, and the length of the roughened stilling basin is about 53% of that of the traditional stilling pool. However, when the roughness scale of the downstream river-bed is not large enough, the repelled downstream hydraulic jump develops, and the length of the roughened stilling basin is about 60% of that of the traditional stilling basin.

3.2.3 Comparison of the energy dissipation ways

The two stilling basins (demonstrated by *Figure* 7(*a*) and *Figure* 7(*b*)) share the same working condition, namely the equivalent reservoir water level, the same flow discharge per unit width, the identical downstream river-bed elevation and the equivalent water level of the downstream river channel. Therefore, their energy dissipation ratios are identical, although their energy dissipation ways are not exactly same. *Figure* 8 illustrates the specific energy at different typical sections of the stilling basin shown by *Figure* 7. In *Figure* 8, the point *A* stands for the contraction section (namely the section just in front of the hydraulic jump) in *Figure* 7(*a*), and its water depth and specific energy are expressed by $h_{c,A}$ and E_{SA}, respectively. The section just after the hydraulic jump lies in the point A'', the conjugate water depth is equal to $h_{c,A}'' = h_t + d + \Delta z$, and the corresponding specific energy is expressed by $E_{SA''}$. Therefore, the energy dissipation ratio is equal to $E_{SA} - E_{SA''}$ for the traditional stilling basin displayed by *Figure* 7(*a*). Due to the fact that the elevation of the stilling pool base slab in *Figure* 7(*b*) is *d*(*m*) higher than that in *Figure* 7(*a*), the specific energy at the contraction section of the flow in *Figure* 7(*b*) (*i.e.* the point *B*) is $E_{SB} = E_{SA} - d$, and the corresponding contracted water depth is represented by $h_{c,B}$. If the critical hydraulic jump occurs in *Figure* 7(*b*), the contraction section *B* is corresponding to the section just before the hydraulic jump, and the section just behind the hydraulic jump is denoted by the point B'' (the relevant sequent depth is $h_{c,B}''=h_t$). Thus, the energy dissipation ratio is equal to $E_{SB}-E_{SB''}$ in *Figure* 7(*b*) when the critical hydraulic jump happens. In view of $E_{SB}=E_{SA}-d$ and $E_{SB''}=E_{SA''}-d$, it can be concluded that $E_{SA}-E_{SA''}=E_{SB}-E_{SB''}$, which means that the energy dissipation ratios are equivalent for the aforementioned two stilling basins.

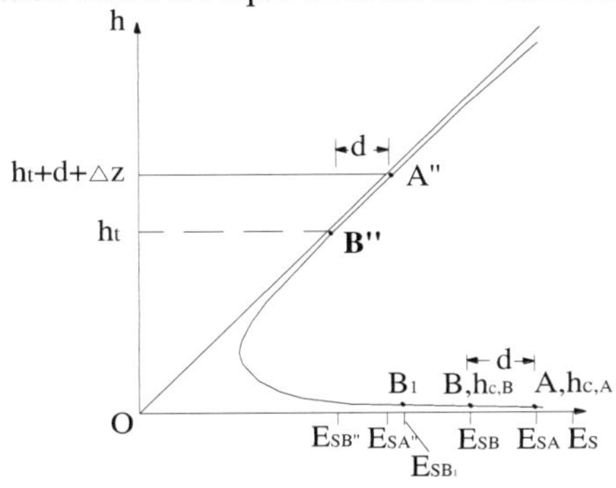

Figure 8. The specific energy at different typical sections of the stilling basin shown by *Figure* 7

If the repelled downstream hydraulic jump occurs in *Figure* 7(*b*) and the point B_1 (in *Figure* 8) indicates the section just in front of the hydraulic jump, BB_1 is corresponding to the curve C_0 in *Figure* 1 and the section just behind the hydraulic jump is still located at B'' (thus $h_t=h_{B1}''$). At this time, the energy dissipation consists of two parts. The first part is the energy dissipation along the curve C_0, being equal to $E_{SB}-E_{SB_1}$. The second part is the energy dissipation through the hydraulic jump, denoted by $E_{SB_1}-E_{SB''}$. At this moment, the water-depth difference caused by the hydraulic jump has been reduced to h_t-h_{B1}.

When the water outflow from the stilling basin in *Figure* 7(*a*), its specific energy will be transformed from the point A'' to the point B'' in *Figure* 8. During this process, a portion of potential energy will be transformed into kinetic energy once again, giving rise to additional disturbance of the water flow. As a consequence, although the nominal energy dissipation ratios are identical for *Figure* 7(a) and *Figure* 7(b), the former's real energy dissipation ratio will be smaller that that of the latter, especially for the low water-head outlet structure the aforementioned discrepancy will be even more significant [15, 16].

4. Conclusion

Compared to the traditional smooth-bed stilling basin, there is a higher ratio of damping on the roughened bed stilling basin. With the aid of the theoretical analysis and the hydraulic model experiments, it has been proved that the addition of the chopping roughness on the riverbed can decrease the sequent depth (*i.e.* h_2/h_2^* <0.71~0.80) and reduce the length of the rolling hydraulic jump region (*i.e.* L_{rj}/L_{rj}^* <0.52~0.78) and that of the curve C_0 (*i.e.* l_{C_0} is inversely proportional to n^2). Besides, it has been shown that the roughened bed stilling basin can result in a better flow regime between the sequent depth h_2 and the downstream water depth h_t (*e.g.* eliminating the secondary hydraulic jump). In addition, this study indicates that adding the sill-roughness structure on the downstream river-bed can decrease the length of the stilling basin up to about 40%~50% when compared with the smooth-bed stilling basin. These conclusions are of important reference value for the design of the flood-discharge energy-dissipation stilling basin in practical engineering projects.

Acknowledgments
The authors gratefully acknowledge the financial support provided by the National Natural Science Foundation of China (51909024), the Natural Science Foundation of Liaoning Province (20180540132、

2019- ZD-0013), the Fundamental Research Funds for the Central Universities (DUT17RC(3)100), and the Special Research Fund for Key Laboratories of DUT (DUT19LAB35).

References

[1]. Ead, S.A., Rajaratnam,N., "Hydraulic jumps on corrugated beds.", *Journal of Hydraulic Engineering*, 2002, **128(7)**,656-663.
[2]. Chen, J.G., Zhang, J.M., Xu, W.L., Wang, Y.R., "Numerical simulation of the energy dissipation characteristics in stilling basin of multi- horizontal submerged jets." *Journal of Hydro-Dynamics, Ser. B*, 2010, **22**(5),732-741.
[3]. Huang, Q.J., Wu, J.H., "Research on contraction energy dissipators." *Journal of Hohai University (Natural Sciences)*, 2008, **2**, 015. (In Chinese)
[4]. Wu, J.H., LiCheng, C.Z., Ma F., Li, T.C., Wu, W.W., "Hydraulics of crest spillway with large unit discharge and low Froude number.", *Journal of Hydrodynamics, Ser. B*, 2015, **27**(2), 242-247..
[5]. Nikmehr, S., Tabebordbar, A., "Hydraulic jumps on adverse slope in two cases of rough and smooth bed." *Research Journal of Applied Sciences, Engineering and Technology*, 2010, 2(1), 19-22.
[6]. Aboulatta, N., Ezizah, G., Yousif, N., Fathy, S., "Design of stilling basins using artificial roughness". *International Journal of Civil and Environmental Engineering*, 2011, **3**, 65-71.
[7]. Qu L., Li R., Li J., Li K., Deng Y. "Field observation of total dissolved gas supersaturation of high-dams.", *Science China Technological Sciences*, 2011, **54**(1), 156-162.
[8]. Li R., Li J., Li K., Deng Y., Feng J., "Prediction for supersaturated total dissolved gas in high-dam hydropower projects.", *Science in China Series E: Technological Sciences*, 2009, **52**(12), 3661-3667.
[9]. Liu Y., "Hydraulics", *China Water &Power Press*, 2008, 3, 296-302. (In Chinese)
[10]. Zeng Y., Huai W., "Application of artificial neural network to predict the friction factor of open channel flow.", *Communications in Non- linear Science and Numerical Simulation*, 2009, **14**(5), 2373-2378.
[11]. Design specification for concrete gravity dams, Water conservancy industry standard of the People's Republic of China, SL319-2005, *China Water & Power Press*, 2005.10. (In Chinese)
[12]. Ni H., Liu Y., "A generalized explicit solution of the sequent depth ratio for the hydraulic jump." *Journal of Hydrodynamics, Ser. B*, 2005, **17**(5): 596-600.
[13]. Hager W.H., Bremen R., Kawagoshi N. "Classical hydraulic jump: length of roller." *Journal of Hydraulic Research*, 1990 28(5), 591-608.
[14]. Pillai N.N., Goel, A., Dubey A.K. "Hydraulic jump type stilling basin for low Froude numbers." *Journal of Hydraulic Engineering*, 1989, 115(7), 989-994.
[15]. Dong J.R., Li Y.X. "Research on the hydraulic characteristics of the stilling basin with serrated tail-weir." *Journal of Hydraulic Engineering*, 1985 8, 51-57. (In Chinese)
[16]. Bejestan, M.S., Neisi, K., "A new roughened bed hydraulic jump stilling basin." *Asian Journal of Applied Sciences*, 2009, **2**(5), 436- 445.

IOP Conf. Series: Materials Science and Engineering **758** (2020) 012083 doi:10.1088/1757-899X/758/1/012083

Evaluation of the geotechnical behaviour of a volcanic soil wall with additions of lime and cement against landslides

C Davila[1], R Vera[2], L Pacheco[3] and G Duran[4]

1,2 Bachelor, Universidad Peruana de Ciencias Aplicadas, Lima, Perú.
3 Part time profesor, Universidad Peruana de Ciencias Aplicadas, Lima, Perú.
4 Full time profesor, Universidad Peruana de Ciencias Aplicadas, Lima, Perú.
E-mail:[1]u201322301@upc.edu.pe,[2]u201414317@upc.edu.pe,[3]pccilpac@upc.edu.pe, [4]gary.duran@upc.edu.pe

Abstract. The construction of earth walls can be a significant response to prevent the next landslides from reaching the road and avoid accidents. Therefore, a material of the same slope was used and reinforced with mixtures of lime and cement, with this same reinforced material a mechanically stabilized hypothetical earth wall (MSE) was developed. An analysis of the original slope was developed to check if there was a possible failure through its safety factor. Then, a hypothetical wall was developed with a floor reinforced with mixtures, in order to assess its overall safety factor and its maximum landslides. According to the results, in principle it was determined that the dosage M-3 / C-4-4 improves in a range of 30% to 37% the friction angle. In addition, it was found that a reinforced wall, that is to say with Lime and cement additions, presents a better behaviour. In its effect, its displacements are about 8 mm and have a global factor of 1.23.

1. Introduction

Surface erosions and landslides are the products of the various gravitational forces that occur in situ, especially in steep slopes and / or on hillsides forming granular soil. Consequently, the soil with a low degree of compactness has a minimum resistance to any destabilizing force causing surface detachment (landslides), the behavior is mainly due to the gravitational predominance of granular soils[1]. In Peru there are several areas with problems of instability of slopes and / or slopes, and we refer to the Moquegua region that presents land roads on rugged terrain, high rainfall and this adds to the lack of infrastructure to anticipate damage caused for a type of landslide, to this is added, the predominant soil in the southern part of Peru, mostly of volcanic origin. In reality the origin, of most of these types of soil, belong to outcrops of sedimentary rocks (sandstones and siltstones), so that in rainy seasons they are easily removed as detritus flows. In sum, these volcanic soils within their characteristics are found to have low density and large grain porosity [2]. Geotechnical engineering offers solutions for various problems that occur in slopes. As mentioned, there are structural and non-structural methods, in this first solution are presented that leads to the development of steel frame systems, gabions, concrete blocks, improved soil walls, etc. Mechanically Stabilized Walls (MSW) is geotechnical structures that highlight the use of armed land to form a supporting wall. These types of structures have an advantage in terms of structural strength, durability of the material, mitigates environmental damage and the security it provides to the road network [1]. According to mentions that despite having certain advantages, these structures have failures if they are not evaluated correctly. In an analysis of a sample that was Mechanically Stabilized Walls (MSW), about 65% of walls were identified with heights between 4-12 m that failed. For the stabilization of the bases of a soil, different types of binders are used, tales such as lime or cement; these aggregates when finished mixing, devices such as filling empty spaces and particle binding; these particularities improve the strength parameters of the material; to be more exact the angle of friction and cohesion. This document will evaluate the geotechnical conditions of the proposal for a structure of a Mechanically Stabilized Earth Wall (MSW), in addition, that this material will be composed of a mixture of lime and cement.

2. Experimental Program

2.1. Materials

2.1.1. Granular Soil

The material of the area that will make up the earth wall is located in the Moquegua region, Peru; of the departmental highway MO-154 its sexagesimal coordinates are as follows 16°44'23.2"S 71°00'18.3"W. The physical properties of the granular soil were identified and investigated in accordance with the technical standards. ASTM. The next table (Table 1) will show the characteristics (figure 1).

Table 1 Basic characteristics

Mixture	LL (%)	LP (%)	IP (%)	SUCS	AASHTO	PUS. (kg/cm^3)	compacted unit weight (kg/cm^3)	% Retained 200
M-1	N.P.	N.P.	N.P.	SP	A1-b	0.887	1.06	0.11

2.1.2. Cement

The cement used is the Portland Type I of the Pacasmayo brand; it was acquired in Promart department stores (figure 3).

2.1.3. Lime

Lime in use is a compound made up of traces of calcium carbonate, calcium hydroxide and particles of disintegrated rocks. The lime was acquired from the department store of Promart brand "Martell"(figure 2). According to NTP 339.176 the potential Hydrogen (pH) It was performed to find optimal lime content, since this pH should not exceed 12.4 so that it has so that the mixture does not have high acidity content [3].

Figure 1. Volcanic Soil

Figure 2. Lime

Figure 3. Cement

2.2. Mix Preparation

First mixtures were prepared according to the following table (Table 2), in order to perform the modified Proctor tests and to find the right combination to continue with the tests. Then, it developed shear strength of soil by direct shear test, for the same dosage and the best combination was sought to use its parameters in the plaxis models.

Table 2 Dosage of mixture and Symbology

Mixture	Soil (%)	Lime (%)	Cement (%)	Simbology
Volcanic Soil	100	0	0	M-1
Mixture 1	100	4	2	M-2/C-4-2
Mixture 2	100	4	4	M-3/C-4-4
Mixture 3	100	4	6	M-4/C-4-6
Mixture 4	100	4	8	M-5/C-4-8

3. Methodology

3.1. Laboratory Test

To know the physical properties of Sandy soils and the mixtures were performed the following tests:

Table 3 Tests made to found the geotechnical properties

Test Name	Norm
Standard Test Method for Particle-size Analysis of soils	NTP 339.128 (ASTM D422)
Standard Test Methods for Laboratory Compaction Characteristics of Soil Using Modified Effort	NTP 339.141 (ASTM D1557)
Standard Test Method for Direct Shear Test of Soils Under consolidated Drained Conditions	NTP 339.1171 (ASTM D3080)

3.2. Slope

The geological, hydrogeological and mechanical data of the slope were necessary for the elaboration of this investigation. This data will be used for numerical modeling using Slide V 6.0 for the processing of a safety factor [4]. The slope has 2 strata, the smaller one which belongs to the road (MO-154) and the larger one dimension belonging to the land (T-1), the parameters of shear resistance to cutting (friction angle) are 33 and 38 degrees respectively, and the cohesion for both cases is zero. Due to the presence seismic of our study region it should have been added PGA information our design, this is understood as the Horizontal Maximum Acceleration and vertical within the, this parameter was got of the Peruvian Technical Norm of resistance seismic E0.30. With all the data, in addition of the different natural soils characteristics indicated, it will proceed to the modeling of the slope.

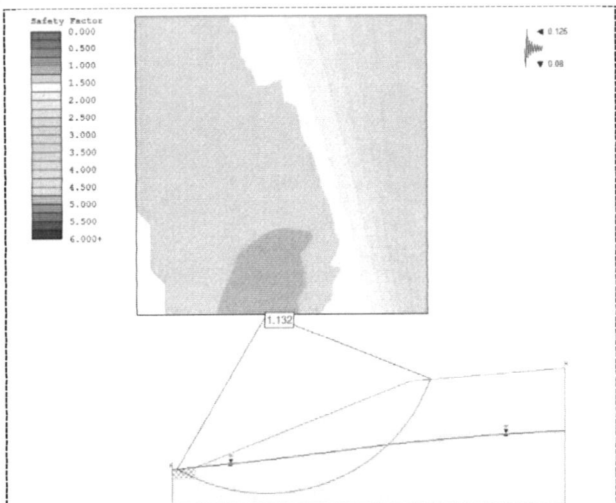

Figure 4. The security factor was obtained for a Morgesten-Price method in the Slide program, it is 1.132, also slope analyzed indicated have a slip potential.

3.3. Reinforced Soil Wall

Armed Earth walls also known as Mechanically Stabilized Walls (MSE) is a construction system that has many advantages such as improved strength and stability of the structure, durability of the materials used, better resilience, aesthetics, limited environmental footprint and profitability. Numerical modeling is a useful tool that allows the analysis of complex systems, such as the interaction of structural elements and soil. In this way, with the "PLAXIS" program, the analysis of the mechanically stabilized wall can be performed in order to obtain the deformation and stress behavior generated [5]. For the pre-dimensioning, the percentage of the slope is 38%, the height of the wall is 3.00 m and a lower base of 2.00 m, for better construction process, geotextiles were placed for each 20 cm high block and bracing the main. On the other hand, for the selection of the floor material that will comprise the wall, the mixture M-3 /C-4-4 was chosen, since it is the dosage that best responded to the tests and presents a better behavior.

4. Analysis and results

It was determined in the granulometry tests that varied with respect to the M-1 sample; this will be shown in the following table and the granulometric curves in Figure 2. This is due to the fact that by introducing lime into the sample it works for the rearrangement of particles and cement as a binder of these, this increases its coefficient of uniformity and reciprocally the coefficient of curvature.

Table 4 SUCS classification of natural soil and mixture

classification	SUCS
M-1	SP
M-2/C-4-2	SP-SM
M-3/C-4-4	SW-SM
M-4/C-4-6	SP-SM
M-5/C-4-8	SM

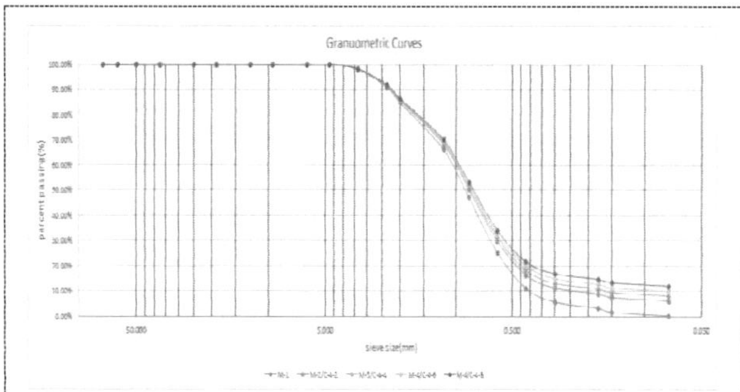

Figure 5. Granulometric curves of volcanic soil and mixtures

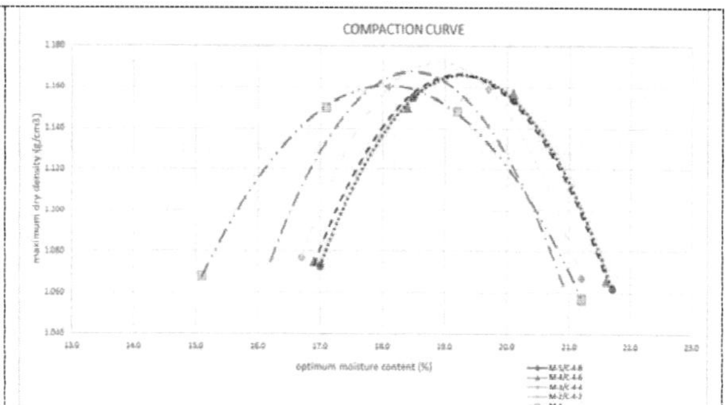

Figure 6. Compaction curves of volcanic soil and mixtures

In the modified Proctor test, improvements of maximum dry density with respect to the natural soil were obtained; this because the binders interact with the soil as grain binders in the M-1 sample, also the addition of lime allows the rearrangement of particles and covers the empty spaces of said sample.

Table 5 Maximize dry density and optimal moisture content of natural soil and mixtures

	M-1	M-3/C-4-4	M-3/C-4-4	M-4/C-4-6	M-5/C-4-8
MDS(g/cm^3)	1.161	1.167	1.172	1.168	1.165
OCH (%)	18	18.4	18.8	19.1	19.2

IOP Conf. Series: Materials Science and Engineering **758** (2020) 012083 doi:10.1088/1757-899X/758/1/012083

With the data obtained from the shear tests, the fault envelopes for dosing were performed (Figure. 5) where the friction angle and cohesion parameters for each dosage will be obtained from the direct shear tests (table 6.)

Table 6 Friction angle for each dosage

	M-1	M-2/C-4-2	M-3/C-4-4	M-4/C-4-6	M-5/C4-8
Ø	38.1	44.5	49.6	49.0	46.7
Cohesion	0	0.14	0.24	0.25	0.27

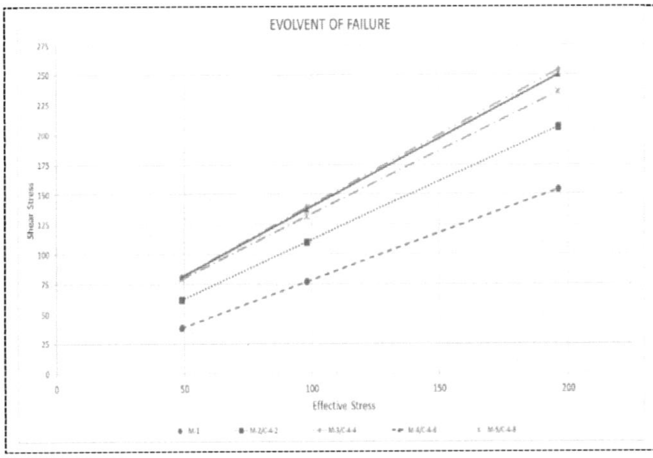

Figure 7. it was obtained that the percentages of 4% lime and 4% cement obtain a better performance in combination with the soil. The resistance parameters, in this case the increase in the friction angle for the M-2 / C-4-2 mixture is improved by 17%, with the mixture of M-3 / C-4-4 and M-4 / C-4-6 of cement its friction parameter improves by 30% compared to the natural one. Finally, with a mixture of M-5 / C4-8 the friction increases by 23% in relation to the natural soil.

The results of the modeling carried out in the Plaxis program for a ground wall with the combination M-3 / C-4-4 indicate that for the horizontal displacements (Figure. 8) of the vertical face on the wall it presents a maximum horizontal displacement 8.93 mm. Finally, the global security factor (Figure. 7) has a value of 1.2343, which according to "AASHTO LRFD Bridge Design and AASHTO LRFD Bridge Construction Specifications" establishes a global Safety Factor that is 0.75 for defined geotechnical wall parameters and slope that do not support structural elements.

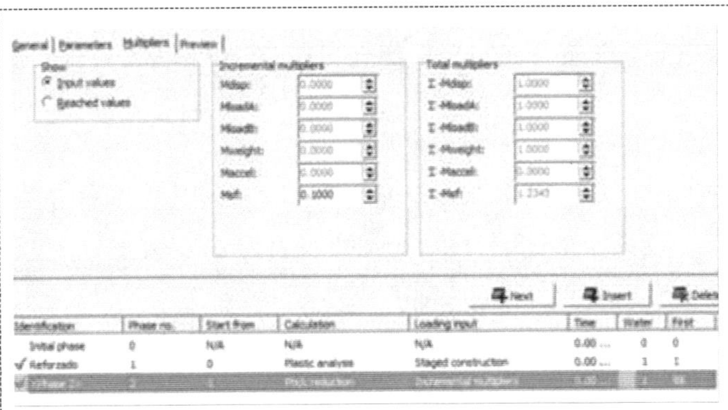

Figure 8. Global safety factor of the wall

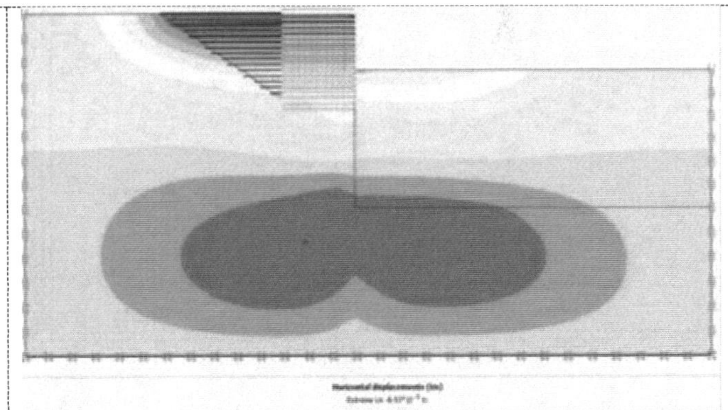

Figure 9. Horizontal Displacement in Reinforced Wall

5. Conclusions

- The maximum optimum density of the mixture M-3 / C-4-4 increases by 0.011 g / cm3 with respect to M-1, this because the binders interact with the soil as grain binders in this sand,

Additionally to adding them Fine to soil composition, these results are expected given the type of soil of the M-1.

- In the tests of direct shear, the maximum resistance to the shear stress was presented when the horizontal displacement is 2.25mm. When the normal stress applied is 49kPa, the shear strength of the M-3 / C-4-4 mixture compared to natural soil (M-1) increased from 38.5kPa to 79.3kPa, in the same way when the stress Normal applied is 98.1kPa increased from 76.3kPa to 138.9kPa and when normal effort is 196.1kPa increased from 153.9kPa to 253.9kPa

- The friction angle of the optimum mix M-3 / C-4-4 has an efficiency percentage of 30% with respect to M-1, this percentage was higher than the other doses.

- The safety F.S of the wall with the mixture M-3 / C-4-4 is greater than the required 0.75 being this of 1.23 reason why the wall will be in correct operation and will stop the landslides of the Slope.

- The construction of this wall contributes to the conservation of the environment, since the materials to be used do not come from the activity that most critically impacts the "Quarry exploitation" environment. However, quality control in a construction process is of vital importance in terms of the optimal dosage presented and the compaction that must reach the maximum dry density determined in the laboratory.

References

[1] R. M. Koerner and G. R. Koerner, "A data base, statistics and recommendations regarding 171 failed geosynthetic reinforced mechanically stabilized earth (MSE) walls," *Geotext. Geomembranes*, vol. 40, pp. 20–27, 2013.

[2] S. Yokohama, "Effects of compaction conditions on undrained strength and arrangements of soil particles for Shikotsu volcanic soil," *Volcan. Rocks Soils - Proc. Int. Work. Volcan. Rocks Soils, 2015*, no. September, pp. 103–104, 2015.

[3] G. Odar *et al.*, "Stabilization method with lime in subgrades for rigid pavements designed by AASHTO 93 in road projects with the presence of bofedales Método de estabilización con cal en subrasantes para pavimentos rígidos diseñados por AASHTO 93 en proyectos viales con presencia de bofedales," no. July 2019, pp. 24–26.

[4] H. Ghiassian, M. Jalili, and D. Kasemi, "Model study of sandy slopes under uniform seepage and reinforced with anchored geosynthetics," *Geosynth. Int.*, vol. 16, no. 6, pp. 452–467, 2009.

[5] T.-L. Chau and T.-H. Nguyen, "Study the influence of adherence edge to steel strip and soil interaction in Mechanically Stabilized Earth Wall with a self-made strip," *Lect. Notes Civ. Eng.*, vol. 54, pp. 757–762, 2020.

Seismic Effect Analysis of the Adjacent Longitudinal Box Culvert for the Utility Tunnel

Zenagebriel Gebremedhn [1], Guofu Qiao [1,2] and Lu Sun[3]

[1] PhD Candidate, School of Civil Engineering, Harbin Institute of Technology, Harbin,150090, China,

[2] Professor, School of Civil Engineering, Harbin Institute of Technology, Harbin, 150090, China,

[3]Associate Professor, Engineering School , Harbin University , Heilongjiang Key Laboratory of Underground Engineering Technology, Harbin, 150086, China,

Abstract
The utility tunnel has the vital significance of the underground public facilities construction for the modernized urban construction inevitable trend. The traditional utility tunnel was used as one of the approaches to achieve sustainable developments. The whole energy supply and distribution system in china will need to be restructured and reworked. Thus, there is still a limited research available on the seismic effect of the utility tunnel. So that, this paper aims to fulfil this research need by investigating the FEA of the adjacent longitudinal box culvert for the utility tunnel is deployed under the circumstance of seismic effect through ABAQUS software concerning to Amplitude A, B and C with time period of 53.47sec, 53.73sec and 53.79sec. The result indicates that, the seismic stress of the box culvert specifically the box culvert including certain soil parts with seismic time and load were found grey zone which is needed greatest warning zone before failure. Finally, the displacement time-history are then studied in the paper which meets the requirements duration from 10 to 20 times. The displacement change is caused by the subsidence of the soil foundation due to the displacement values on the corresponding corner points are much different.

Keywords: Box culvert, FEM model, Seismic effect analysis, Underground structure, Utility tunnel.

1. Introduction

Utility tunnel is a type of underground structure that accommodates various public facilities. It is an important component of urban lifeline infrastructure.[1-3] Whose security and stability not only for the normal running of the city has a vital role, and to ensure that urban disaster prevention capabilities, meaning even more important. Jorge from the perspective of human health and safety risks of the utility tunnel may make specific requests.[4] In addition to this, the utility tunnel is the most sustainable development of urban underground space facilities, will be incorporated into urban underground space planning. At the same time, utility tunnel has so many advantages there easy to maintain and will not cause traffic jams, noise and environmental pollution, can easily be added municipal pipe and so on.[5] However, there is still limited research to date on the seismic analysis of the utility tunnel. The shape of utility tunnels is circular and multi-circular, the box shape tunnels are used in underground cavities, subway stations, especially in shallow buried mine tunnels. These utility tunnels have been constructed by either the cut-and-cover method or shield tunneling method depending on the geological conditions. The cut-and-cover method is used to construct the precast box culvert utility tunnel. Meanwhile, this paper focusses on the box-shaped and shallow-buried concrete structures. Because it is preferable due to simple construction, reduces construction time and maintenance cost as well as low seismic risk.[6-8] At present, the seismic design of underground structures commonly used in China mainly adopts three methods: static method, quasi-static method and dynamic response analysis method. This paper mainly used the quasi-static method which is namely the displacement method. The reaction displacement method simulates the structure elements, and the soil is replaced with a spring of equivalent stiffness. The method of applying

IOP Conf. Series: Materials Science and Engineering **758** (2020) 012084 doi:10.1088/1757-899X/758/1/012084

the seismic load of the reaction displacement method is to first study the motion of the three-dimensional soil layer, and apply the maximum displacement of the upper and lower bottom positions of the soil layer to the end of the soil spring away from the structure. Finally, the utility tunnel has many advantages over traditional method of laying lifelines. This paper established a three dimensional finite element model through ABAQUS software concerning to Amplitude A, B and C, as well as the seismic effect of the utility tunnel which is made up of adjacent longitudinal precast box culvert with the flexible joint. Besides, the soil in a concerning seismic load has been simulated to obtain stress and displacement time-history. The finite element model is preferable to simulate the numerical solution of the problems under the condition of assembled box culvert using ABAQUS software, and the results provide to the analysis of the longitudinal precast box culvert and construction practices.[9-13]

2. Materials and Methods

2.1 Geometrical Parameters

In order to study the behavior of all the possible standard adjacent box culverts three dimensional finite element model were prepared for the standard size of longitudinal box culvert as per GB50010-2010.[14] This model included three dimensional solid elements soil (C3D8) and three dimensional solid elements for concrete floor reaction (C3D8) and Box culvert (C3D4) having geometric and material linearity. Moreover, the elements are standard meshed separately. So that, the mesh element generated high on the part soil and least on the part floor reaction. In addition to this, the total mesh generates 357,119 elements and 101,390 nodes. Thus, the stress and displacement was studied for all the load steps of double box tested using brick element including flexible connection concerning to soil beneath, side soil, adjacent box culvert and floor reaction. [15, 16]

2.2 Material Properties

The absence of some material property data, the properties parameter of the subsoil and the culvert concrete were calculated by FEM analysis. This method based on the observational property data (soil, box culvert, floor reaction) was used to gained displacement of the culvert section from the FEM analysis, and results was reasonable. [17] Methods comparing with each other, the finite element ABAQUS software was used to model and analyze concerning to Amplitude A, B and C. The displacement of two culverts including reaction floor with overlying soil on both side was calculated when the two culverts were flexible connected and the gap between them was tied using screw bolt. The corresponding stress and displacement diagram was obtained. Because the elastic modulus of box culvert and soil is four orders of magnitude different, the elastic displacement of soil is observed, and the adjacent box culvert is considered as a flexible connection, and its elastic displacement or deformation can be neglected. Moreover, the basic material parameters shown in Table 1 and the single detailed section size shown in Table 2 as well as it's cross section of utility tunnel including its corner numbering also shown in Figure 1. [14]

Table 1 Material property of the instance modeling

Instance	Size(m)	Unit type	Modulus of elasticity(Pa)	Poisson's Ratio	Density (Kg/m3)
Soil	30x30x30	C3D8	4E7	0.3	2000
Box culvert	2X2	C3D4	3.25E10	0.3	2400
Floor Reaction	30X5.22	C3D8	3.25E10	0.3	2400

IOP Conf. Series: Materials Science and Engineering **758** (2020) 012084 doi:10.1088/1757-899X/758/1/012084

Table 2 Single box culvert section size

Inner Size (m)	Width(m)	Height (m)	Length (m)	Thickness		
				Roof (m)	Floor(m)	Side wall (m)
2x2	2	2	2	0.2	0.2	0.18

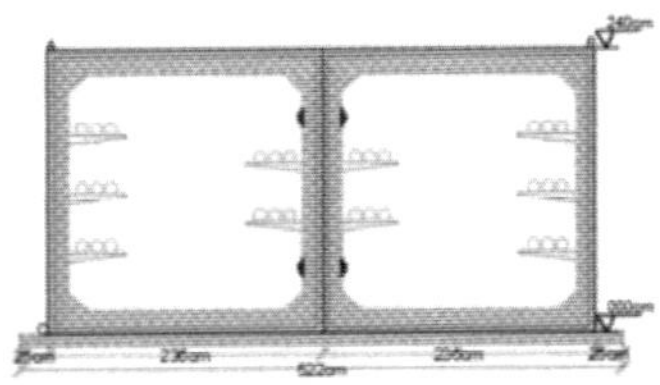

Figure 1 Cross section of utility tunnel and its corner numbering

2.3 Seismic Load

2.3.1 Damping Calculation

Damping source system is difficult to quantify. Damping forces in the structural vibration may come from many sources of the energy loss during hysteretic loading, viscoelastic material properties and external joint friction. Among these damping forces, the resistance form may be different from each other. Damping form called viscous damping is the most widely used, in which the damping force is velocity proportional.[18-20] For seismic actions, usually explicit method will be taken as the first choice in ABAQUS combined with Rayleigh damping. It is considered that the damp ratio of the adjacent longitudinal box culvert is between 2% to 5%. For the first-order vibration modes of the adjacent box culverts, the uniform value of material damping parameters is 5% of the ultimate damping 0.528. According to the orthogonality between the mass matrix and stiffness matrix, the fraction of critical damping of a given mode can be written as obtained to be used as input values. Solve the equations as follows;

$$\xi = \frac{\alpha}{2\omega_1} + \frac{\beta\omega_1}{2} = \frac{\alpha}{2\omega_2} + \frac{\beta\omega_2}{2}\zeta \tag{1}$$

$$\alpha = \frac{2\zeta\omega_1\omega_2}{\omega_1 + \omega_2}, \beta = \frac{2\zeta}{\omega_1 + \omega_2} \tag{2}$$

Where α is the parameter of mass proportional damping with 1.605 and β is the parameter of stiffness proportional damping with 0.174, $\omega_{1/2} = 3.04\pi$ which is a frequency measured in rad/s. Its value is the first and second frequency obtained by modal analysis. In general, the mass proportional damping will damp the lower frequencies and the stiffness proportional damping will damp those higher frequencies.[10, 16, 18]

2.3.2 Modal Analysis

The main function of the medium modal analysis is to determine the natural frequency and mode shape of the structure, because many dynamic analyses transient dynamics analysis, harmonic response analyses, etc. are based on the results of modal analysis. So that, the basic equations solved by typical undamped modal analysis are classical eigenvalue problems.[10]

$$K\varphi_i = \omega_i^2 M\varphi_i \quad (3)$$

Where K is the stiffness matrix; φ_i is the mode shape vector of the first order (eigenvector); $\omega_i = 2\pi f$ which is the natural frequency of the first order mode (ω_i^2); and M is the mass matrix.

3. Results and Discussion

3.1 Stress analysis of box culvert under earthquake

The output database files in ABAQUS were read by visualization module to create contour plots, animations, XY plots, and tabular output of the results. The displacement is not supported by visualization mode, so it is read in data file, which identifies the displacement elements and the level of stress at that point. [21, 22] According to the results of ABAQUS software calculation, the stress of the adjacent longitudinal jointed box culvert is obtained, as shown in Figure 2.

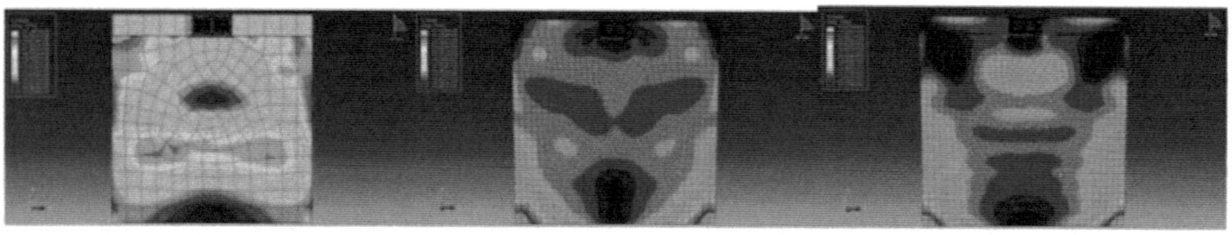

(a) Amplitude A with 53.47sec **(b)** Amplitude B with 53.73sec **(c)** Amplitude C with 53.79sec

Figure 2 The stress time-history for Amplitude A, B and C

According to Figure 2 (a), that the stress of the box culvert is symmetrically distributed as a whole, but contrary to the front stress distribution, the stress in the box culvert and certain soil parts which indicate the grey zone is large, and the black zone which is located at the foundation soil also small. Thus, the maximum stress indicates the largest absolute value appearing at the box culvert of amplitude A with 577KPa. The black zone has the smallest stress value which is located at the foundation soil. Based on the Figure 2 (b), the maximum stress indicates the largest absolute value appearing at the box culvert of amplitude B with 580KPa. The black zone has the smallest stress value which is located at the foundation soil. Based on the Figure 2 (c), the maximum stress indicates the largest absolute value appearing at the box culvert of amplitude C with 581KPa. The black zone has the smallest stress value which is located at the foundation soil of amplitude C. The following describes the stress of the box culvert under earthquake action of the amplitude A with time period of 53.47sec, amplitude B with time period of 53.73sec and amplitude C with time period of 53.79sec as well as its stress time-history graphs are as shown in Figure 3.

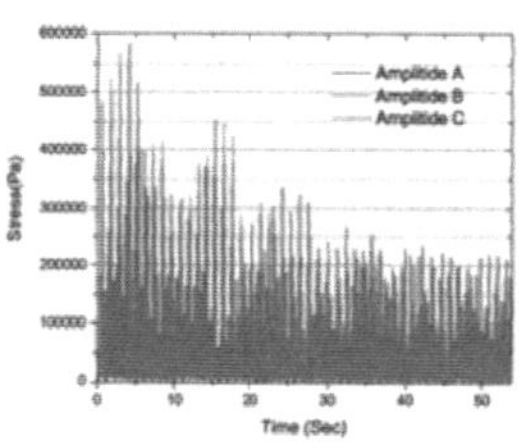

Figure 3 The stress time-history for Amplitude A,B and C

IOP Conf. Series: Materials Science and Engineering **758** (2020) 012084 doi:10.1088/1757-899X/758/1/012084

3.2 Deformation analysis of box culvert under earthquake

The adjacent box culvert analyzes the deformation or displacement of the corner point which is called upper and lower part of the displacement time-history. The deformation or displacement of each corner point is calculated by the ABAQUS software. According to the ABAQUS software visualization, the deformation or displacement is positive downward. The following describes the deformation of the box culvert under earthquake action, and its displacement time-history graphs are as shown in Figure 4.

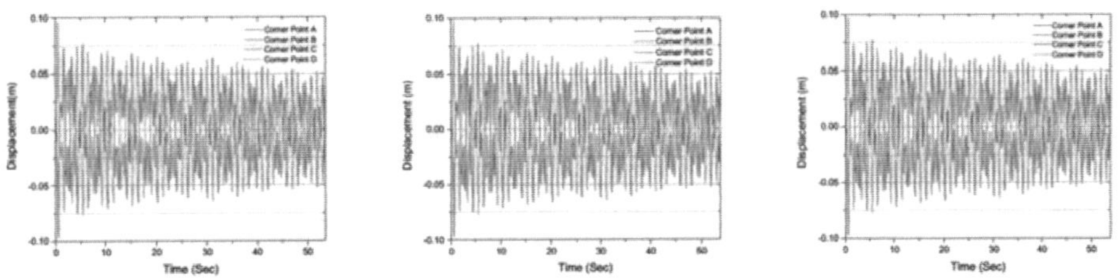

(a) Amplitude A with 53.47sec **(b)** Amplitude B with 53.73sec **(c)** Amplitude C with 53.79sec

Figure 4 The displacement time-history for each corner point and Amplitude A,B and C

According to Figure 4 (a), the displacement distribution of the box culvert parts are different. However, the displacement values on the corresponding corner points of A-C and B-D are much different. The displacement value between the left (Corner point A and C) and right (Corner point B and D) part direction of the amplitude A with time period of 53.47sec is 0.153m. Based on the Figure 4 (b), the displacement value between the left (Corner point A and C) and right (Corner point B and D) part direction of the amplitude B with time period of 53.73sec is 0.154m. Based on the Figure 4 (c), the displacement value between the left (Corner point A and C) and right (Corner point B and D) part direction of the amplitude C with time period of 53.79sec also 0.154m. It is obvious that the deformation of the box culvert is large, and the displacement change is caused by the subsidence of the soil foundation.

4. Conclusions

This study developed the seismic effect analysis of the adjacent longitudinal box culvert for the utility tunnel and the following conclusions are drawn;

A. The seismic stress in the box culvert and certain soil parts are indicated by grey zone. So that, it needs greatest warning zone before failure. This provides great convenience to study the seismic stress distribution of the adjacent longitudinal box culvert structures.

B. The deformation or displacement of the box culvert is large, and the displacement change is caused by the subsidence of the soil foundation. Because the displacement values on the corresponding corner points are much different.

Acknowledgment

This research was supported by the National Key Basic Research Program of China (973 Program CSC No.2016GXZ133).

References

[1] J. C. E. a. V. C. Julian C.P., "Analysing utility tunnels and highway networks coordination dilemma," *Tunnelling and Underground Space Technology*, pp. 185-189, 2009.

[2] Q. H. C. Qian, X.Q., "Situation, Problems and Countermeasures of Utility Tunnel Development in China and Abroad," *Chinese Journal of Underground and Engineering*, pp. 191-194, 2007.

[3] L. Jiang, J. Chen, and J. Li, "Seismic response of underground utility tunnels: shaking table testing and FEM analysis," *Earthquake Engineering and Engineering Vibration,* vol. 9, pp. 555-567, December 01 2010.
[4] J. C. E. J. C.P, " Indoor atmosphere hazard identification in person entry utility tunnels," *Tunnelling and Underground Space Technology,* pp. 426-434, 2005.
[5] O. B. a. F. B. B. Ludovic L., "Promoting the urban utilities tunnel technique using a decision-making approach," *Tunnelling and Underground Space Technology,* pp. 79-83, 2004.
[6] T. Alkhrdaji, and A. Nanni, "Design, Construction, and Field-Testing of an RC Box Culvert Bridge Reinforced with GFRP Bars " *Non-Metallic Reinforcement for Concrete Structrues-FRPRCS-5,* pp. 1055-1064 2001.
[7] X.-s. Cheng, C. H. Dowding, and R.-r. Tian, "New methods of safety evaluation for rock/soil mass surrounding tunnel under earthquake," *Journal of Central South University,* vol. 21, pp. 2935-2943, July 01 2014.
[8] S.-l. Chen and M.-w. Gui, "Seismic performance of tunnel lining of side-by-side and vertically stacked twin-tunnels," *Journal of Central South University of Technology,* vol. 18, pp. 1226-1234, August 01 2011.
[9] J. R. A. J. Wang Shuhong, Wang Pengyu, Liu Weihua, "Simulation of Force Performance of Precast Rectangular Box Culvert and Its Potential Failure Mode," *Journal of Northeastern University (Natural Science),* pp. 260-265, 2018.
[10] G. Yungao, "Mechanical analysis and reliability study of box culvert with low reinforcement ratio under earthquake " *Central South University (in chinese),* 2010
[11] G. Rui, "Research on Key Technology of Super-long Box Culvert Jacking in Expressway Shallow Covering Soil," *Chang'an University,* 2014.
[12] J. L. Qingxia YUE, "Seismic Analysis of Utility Tunnel Considering Wave Passage Effect," *4th International Conference on Earthquake Geotechnical Engineering* pp. 1-8, 2007.
[13] GB/T17742, "Chinese seismic intensity scale " *Published in Beijing,(chinese),* 2008.
[14] GB50010, "Design concrete structures," *Publishing House of building Industry in China, Beijing (in Chineses),* 2010.
[15] A. Abolmaali, & Garg, A., "Shear Behavior and Mode of Failure for ASTM C1433 Precast Box Culverts," *Journal of Bridge Engineering,* vol. 13, pp. 331-338, 2008.
[16] R. I. Pawtucket, "Abaqus/CAE Users Guide," 2014.
[17] X. L. G. C. X. W. a. Y. Liu, "Settlement Prediction and FEM Analysis of Culvert Section under High-Filled Embankment," *International Pipelines Conference 2008,* 2008.
[18] J. D. Xiao Ming Chen , Hu Qi, Yun Gui Li, "Rayleigh Damping in Abaqus/Explicit Dynamic Analysis," *Applied Mechanics and Materials,* vol. 627, pp. 288-294, 2014.
[19] J. P. R. Clough, "Dynamics of Structures,," *McGraw Hill Kogakusha,* 1975.
[20] E. Wilson, "Three-Dimensional Static and Dynamic Analysis of Structures: a Physical Approach with Emphasis on Earthquake Engineering," *Computers and Structures,* 1998.
[21] K. Garg Anil, & Abolmaali, A, "Finite-Element Modeling and Analysis of Reinforced Concrete Box Culverts," *Journal of Transportation Engineering,* vol. 3, pp. 121-128, 2009.
[22] S. Jie, "Large-section pipe shed Numerical simulation of box culvert structure with box culvert jacking," *Henan Science,* vol. 03, pp. 457-460, 2007.

IOP Conf. Series: Materials Science and Engineering **758** (2020) 012085 doi:10.1088/1757-899X/758/1/012085

Study and assessment of (FeCrAl) and (SiC) as candidate materials for accident tolerant fuel cladding in LWRs

Mamoun I A Sagiroun[1,2,*], Cao Xinrong[1] and Anuor A T Ayoub[1]

[1]College of Nuclear Science and Technology, Harbin Engineering University, Harbin, China, 150001

[2]Sudan Atomic Energy Commission, Khartoum, Sudan

Email: sagiroun005@hotmail.com

Abstract. Iron-chromium-aluminum (FeCrAl) alloys and Silicon-carbide (SiC) are considered the best choice proposed as alternative cladding materials to replace Zirconium (Zr) alloys to improve the accidents tolerance in light water reactor (LWR) cladding which have much highly oxidation resistance in high-temperature steam than Zr alloys, which lead to maintaining high performance under normal operations and enhancing safety margins during severe accidents. In this work great results were showed in the thermo-physical properties for the candidate materials compared with zirconium.

1. Introduction

After the Fukushima-Daiichi accident in 2011, the demand of increasing the safety margin in nuclear industry by proposed Accident Tolerant Fuels (ATF) for fuel and cladding materials, the main goal of ATF philosophy is increasing the controlling time during accidents in Light Water Reactors (LWRs) and improving the performance of reactors during normal operations and operational transients [1]. Figure. 1 Describes the core degradation processes under loss of coolant conditions during accidents condition which lead to elevated temperature in LWRs core, the blue curve show how the ATFs cladding mitigation the degradation processes inside the reactor core [2].

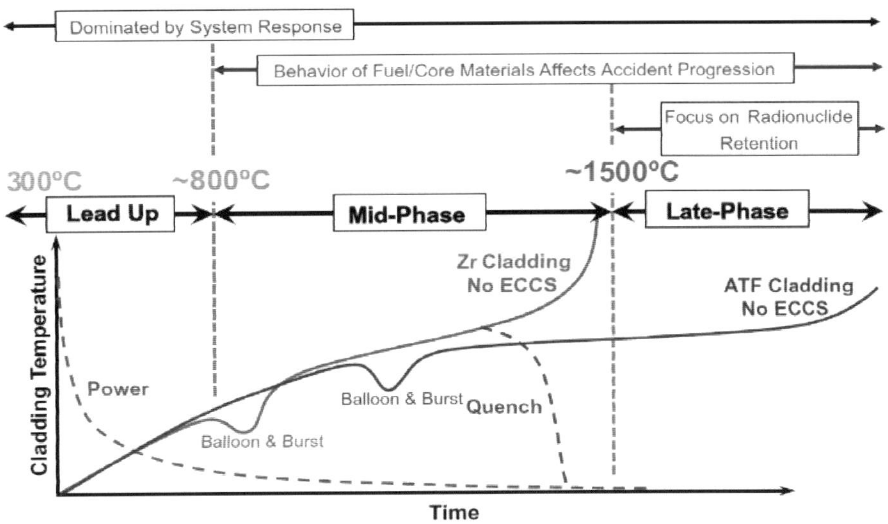

Figure 1. degradation processes inside LWRs core during coolant-limited

The efforts of development ATFs were focused to a main three issues: a) Eliminate the hydrogen production by reducing the oxidation rate of cladding with high-temperature steam; b) improve the thermal, mechanical and chemical properties of cladding materials to withstand the failure; c) modify the current fuel-cladding systems for lower operation temperatures [3]. Figure. 2 shows the different concepts of accident tolerant fuel under research now which include the fuel design and cladding materials [4]. For fuels, focused on advanced UO2 formulations, high density U fuels (nitride and silicide) compositions and fully ceramic microencapsulated (FCM) fuels [5]. for cladding research and development is focused on improved composition of current Zr alloys [6], coated Zr alloys like MAX phase coatings [7], FeCrAl alloys as replacement or coating on Zr alloys [8][9], Cr coatings [10], Mo coatings [11], Polycrystalline diamond coatings [12], and ceramic (SiC/SiC composite) cladding materials [13]. In this paper the thermo-physical behavior of two types of materials FeCrAl and SiC compared with zirconium were discussed including the advantages and the challenges.

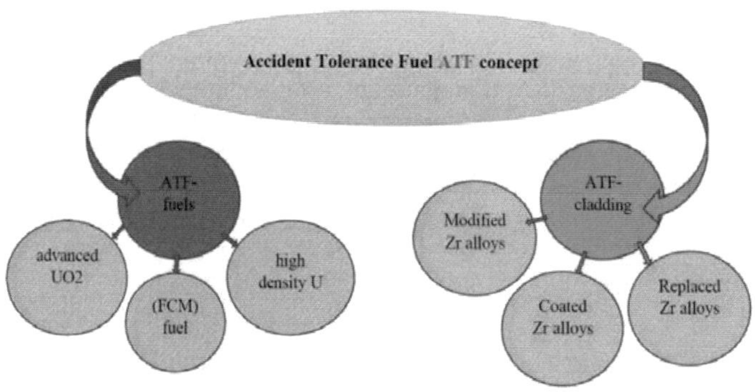

Figure 2. Clarified sketch of current Accident Tolerant Fuel concepts

2. Desirable Properties of fuel and Cladding:

As show in Table 1. The desirable properties for both fuel pellet and cladding always together for lower and higher under the normal or transient operations, however, sometime a property is desirable under normal condition but not desirable under other operation conditions and vice versa. High thermal conductivity is desired to limit the fuel centerline temperature, slow oxidation rate and high strength are desirable in cladding materials at all operation scenarios to limit the rapid degradation in the cladding, for LOCA and SBO, high heat capacity and low creep rate of cladding are desired but insensitive in the fuel pellet [14].

Table 1. Desirable properties of the fuel pellet and cladding materials under NO: normal operation, LOCA: loss-of-coolant accident and SBO: station blackout, (H & L) indicates for high, low respectively and (-) indicates insensitivity [14].

Fuel Cladding				Fuel Pellet		
	NO	LOCA	SBO	NO	LOCA	SBO
Thermal Conductivity	H	H	-	H	H	-
Strength	H	H	H	-	-	-
Oxidation Rate	L	L	L	-	L	L
Creep Rate	H	L	L	H	-	-
Heat Capacity	-	H	H	-	H	H

3. Properties of ATF candidate cladding materials

To create accident tolerant fuel in LWRs fuel cladding, many alternative materials to Zr alloys are being investigated which have >100X improvement in oxidation resistance at high temperature compared to current Zr-based cladding [2]. Figure 3. showed highly results in some thermo-physical properties of two types of candidate materials FeCrAl and SiC compared with zirconium which were amply discussed in references [15][16].

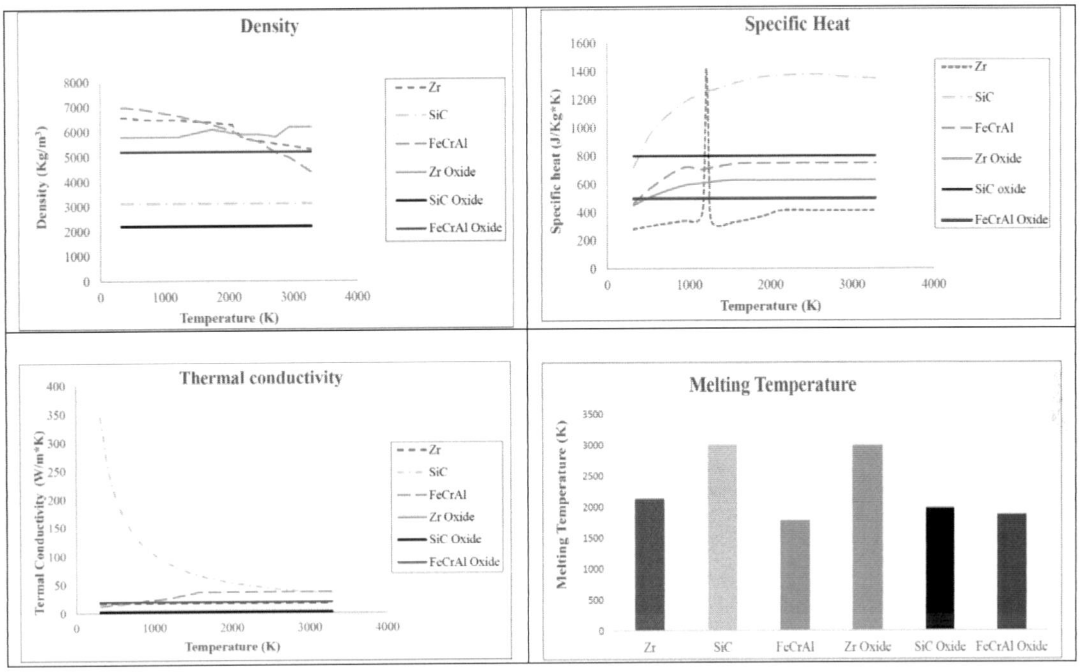

Figure 3. thermo-physical properties of candidate materials compared with zirconium

Table 2. Described some properties of candidate alternate cladding materials and cladding currently uses zircaloy, all candidate materials have significantly slower oxidation kinetic in high-temperature steam than zircaloy according to parabolic oxidation rate constant (POR) at 1200 °C. For SiC has the lower thermal absorption cross-section than Zircaloy, also SiC have chemically inaction, a high radiation resistance, reduced hydrogen generation under off-normal conditions, ability to withstand higher temperatures and considered to be generally stable in nuclear waste. For metallic candidate FeCrAl show macroscopic neutron thermal cross section (Σthermal abs) nearly ~22× greater than Zircaloy [17]. To withstand this neutronic penalty it will be necessary reduces the cladding walls thickness to nearly (~300μm) half the thickness of Zr alloys (~600μm) which current uses to have similar neutron absorption [18], also higher enrichment fuel with slightly larger pellets will be necessary [19].

Table 2. Summary of relevant data for cladding material options [17].

Clad Material	Density [g/cm^3]	Composition [wt.%]	$\Sigma^{thermal\ abs}$ [cm^{-1}]	POR constant in 1200 °C steam [mg]
Zircaloy	6.56	98.26 Zr, 1.49 Sn, 0.15 Fe, 0.1 Cr	0.0028	6.5x10^{-4}
FeCrAl	7.1	75 Fe, 20 Cr, 5 Al	0.0634	1.8x10^{-6}
SiC	2.58	70.08 Si, 29.92 C	0.0021	3.7x10^{-7}

Figure 4. described two types of candidate cladding materials when tested in flowing steam with elevated temperature and pressure at different experiments, in figure 4. a, both alumina formers and silica formers candidate claddings produced oxides outperformed zircaloy, which assumed to be suitable materials to withstand the degradation under abnormal operations condition in LWRs [20], also Pint et al. from ORNL their found the significantly reduced in the hydrogen production from candidate materials compared with zircaloy used as current cladding material, showed in figure 4. b [21].

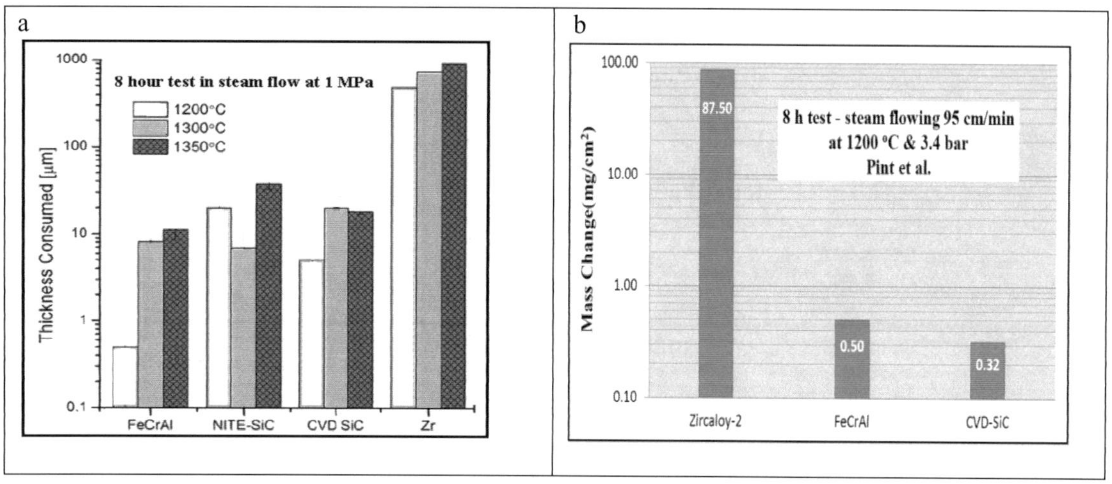

Figure 4. (a) Thickness loss and (b) Oxidation mass change of candidate materials

4. The challenges of candidate materials

As previously discussed, there are some challenges showed when compared the new materials with Zircaloy must solved before goes on to be used as a cladding material in commercial reactors, for FeCrAl material:

- FeCrAl alloys show a thermal neutron cross-section much greater than Zr alloys, this decreases the reactivity of the core, subsequently decreasing the potential cycle length for operation[22].
- FeCrAl cladding may release more tritium to the coolant than Zirconium alloys one solution for this problem might be coating inside part of the tube with alumina [23].
- Also the alpha prime precipitation in FeCrAl after neutron irradiation which is still under investigation [24].

For SiC there are two challenges requiring further examination before approved SiC as cladding materials:

- Water cooling caused aqueous corrosion.
- Microcracking during normal operating conditions.

5. Conclusion

One of approach of developing ATFs cladding LWRs is replace Zr alloys by new materials have highly oxidation resistance to delay and reduce the heat and hydrogen generated during accidents conditions. FeCrAl & SiC claddings, is most important materials will be uses to improving the performance of normal operation and withstand the rapid degradation at accident conditions in LWRs, which increase the safety margin during sever accidents which lead to high progress in nuclear industry. Some challenges meet FeCrAl such as neutronic penalty which overcoming by high enrichment and increased fuel, also decreasing the thickness of cladding might be effective to reduce this neutronic penalty.

Acknowledgments
This paper is funded by the International Exchange Program of Harbin Engineering University for Innovation-oriented Talents Cultivation.

References:

[1] N. OECD, "State-of-the-Art Report on Light Water Reactor Accident-Tolerant Fuels," 2018.

[2] S. J. Zinkle, K. A. Terrani, J. C. Gehin, L. J. Ott, and L. L. Snead, "Accident tolerant fuels for LWRs: A perspective," *J. Nucl. Mater.*, vol. 448, no. 1–3, pp. 374–379, 2014.

[3] Z. Liu *et al.*, "The development of cladding materials for the accident tolerant fuel system from the Materials Genome Initiative," *Scr. Mater.*, vol. 141, pp. 99–106, 2017.

[4] J. Ko *et al.*, "Development and testing of multicomponent fuel cladding with enhanced accidental performance," *Nucl. Eng. Technol.*, 2019.

[5] IAEA, "Accident Tolerant Fuel Concepts for Light Water Reactors," *URL, https//www.iaea.org/publications/10972/accident-tolerant-fuel-concepts-for-light-water-reactors.*

[6] Z. Duan *et al.*, "Current status of materials development of nuclear fuel cladding tubes for light water reactors," *Nucl. Eng. Des.*, vol. 316, pp. 131–150, 2017.

[7] B. R. Maier, B. L. Garcia-Diaz, B. Hauch, L. C. Olson, R. L. Sindelar, and K. Sridharan, "Cold spray deposition of Ti2AlC coatings for improved nuclear fuel cladding," *J. Nucl. Mater.*, vol. 466, pp. 712–717, 2015.

[8] D. Pan, R. Zhang, H. Wang, C. Lu, and Y. Liu, "Formation and stability of oxide layer in FeCrAl fuel cladding material under high-temperature steam," *J. Alloys Compd.*, vol. 684, pp. 549–555, 2016.

[9] W. Zhong, P. A. Mouche, X. Han, B. J. Heuser, K. K. Mandapaka, and G. S. Was, "Performance of iron-chromium-aluminum alloy surface coatings on Zircaloy 2 under high-temperature steam and normal BWR operating conditions," *J. Nucl. Mater.*, vol. 470, pp. 327–338, 2016.

[10] J. Bischoff *et al.*, "AREVA NP's enhanced accident-tolerant fuel developments: Focus on Cr-coated M5 cladding," *Nucl. Eng. Technol.*, vol. 50, no. 2, pp. 223–228, 2018.

[11] B. Cheng, Y. J. Kim, and P. Chou, "Improving Accident Tolerance of Nuclear Fuel with Coated

Mo-alloy Cladding," *Nucl. Eng. Technol.*, vol. 48, no. 1, pp. 16–25, 2016.

[12] I. Kratochvílová *et al.*, "Nanosized polycrystalline diamond cladding for surface protection of zirconium nuclear fuel tubes," *J. Mater. Process. Technol.*, vol. 214, no. 11, pp. 2600–2605, 2014.

[13] G. Singh, R. Sweet, N. R. Brown, B. D. Wirth, Y. Katoh, and K. Terrani, "Parametric Evaluation of SiC/SiC Composite Cladding with UO2 Fuel for LWR Applications: Fuel Rod Interactions and Impact of Nonuniform Power Profile in Fuel Rod," *J. Nucl. Mater.*, vol. 499, pp. 155–167, 2018.

[14] K. A. Terrani, "Accident tolerant fuel cladding development: Promise, status, and challenges," *J. Nucl. Mater.*, vol. 501, pp. 13–30, 2018.

[15] X. Wu *et al.*, "Preliminary safety analysis of the PWR with accident-tolerant fuels during severe accident conditions," *Ann. Nucl. Energy*, vol. 80, pp. 1–13, 2015.

[16] L. J. Ott, K. R. Robb, and D. Wang, "Preliminary assessment of accident-tolerant fuels on LWR performance during normal operation and under DB and BDB accident conditions q," *J. Nucl. Mater.*, vol. 448, no. 1–3, pp. 520–533, 2014.

[17] J.J. Powers, A. Worrall, K.R. Robb, N.M. George, and G.I. Maldonado, "ORNL Analysis of Operational and Safety Performance for Candidate Accident Tolerant Fuel and Cladding Concepts," *IAEA TECDOC Ser. 2016. 253.*

[18] H. Kim, H. Kim, J. Yang, and Y. Koo, "On the Minimum Thickness of FeCrAl Cladding for Accident-Tolerant Fuel," *Nucl. Technol.*, vol. 198, pp. 342–346, 2017.

[19] K. A. Gamble, J. D. Hales, K. A. Gamble, and J. D. Hales, "Preliminary Modeling of Accident Tolerant Fuel Concepts under Accident Conditions.," *United States: N. p.,* 2016. .

[20] Bragg-Sitton and Shannon, "Advanced LWR Nuclear Fuel Cladding System Development Technical Program Plan. Light Water Reactor Sustainability Program," *US Dep. Energy*, no. Idaho National Laboratory. External Report. INL/MIS-12-25696, 2012.

[21] J. Wang *et al.*, "Accident tolerant clad material modeling by MELCOR : Benchmark for SURRY Short Term Station Black Out," *Nucl. Eng. Des.*, vol. 313, pp. 458–469, 2017.

[22] K. A. Terrani, S. J. Zinkle, and L. L. Snead, "Advanced oxidation-resistant iron-based alloys for LWR fuel cladding," *J. Nucl. Mater.*, vol. 448, no. 1–3, pp. 420–435, 2014.

[23] R. B. Rebak, G. Electric, K. L. Ledford, and G. Electric, "Improving Nuclear Power Plant Safety with FeCrAl Alloy Fuel Cladding," no. January, 2017.

[24] C. P. Massey *et al.*, "Post irradiation examination of nanoprecipitate stability and α ′ precipitation in an oxide dispersion strengthened Fe-12Cr-5Al alloy," *Scr. Mater.*, vol. 162, pp. 94–98, 2019.

IOP Conf. Series: Materials Science and Engineering **758** (2020) 012086 doi:10.1088/1757-899X/758/1/012086

Experimental and Theoretical Investigation of Nonlinear Dynamic characteristics in an Oil-free Moving Coil Linear Compressor

Jian Sun[1,2*], Chengzhan Li[1,2], Jianguo Li[2] and Jinghui Cai[1,2]

1. University of Chinese Academy of Sciences, Beijing 100049, China
2. Key Laboratory of Space Energy Conversion Technologies, Technical Institute of Physics and Chemistry, CAS, Beijing 100190, China

Email: sunjian16@mails.ucas.ac.cn

Abstract. The gas pressure, piston displacement and current response characteristics of an oil-free moving coil linear compressor driven by sinusoidal voltages at different frequencies have been experimentally studied. The results show that the response curves of gas pressure, piston displacement and current have nonlinear distortions deviating from the sinusoidal waveform. Among them, the piston displacement waveform has relatively small deviation from sinusoid, and the total harmonic distortion rate (THD) of piston displacement has a maximum value of 6.2% at 50Hz; the current waveform has relatively large deviation from sinusoid, and the THD of current has a maximum value of 73.7% at 60Hz. At this time, the current has two peaks in one cycle. It was analyzed that the nonlinear distortion of displacement and current occurs only in the compression process in a working cycle. In addition, a nonlinear dynamic model based on MATLAB software was established to analyze current distortion. The simulated values of gas pressure, piston displacement and current curves have a good coincidence with measured values. Finally, it is present that the severe distortion of the current curve is mainly caused by the non-linearity of the gas pressure.

1. Introduction

Vapor compression refrigeration（VCR）is a very effective way for electronic cooling. As the core component of the VCR system, the compressor directly affects the energy efficiency ratio (EER) of the system. Liang et al. (2014) [1] pointed out that linear compressors have great potential for electronic cooling because of their unique structural advantages compared to conventional compressors. The linear compressor uses a linear oscillation motor to directly drive the piston, which simplifies the transmission mechanism, eliminates/reduces the lateral force between the cylinder and the piston and greatly improves the mechanical efficiency; resonance operation effectively amplifies the piston stroke; the use of large radial-axial stiffness ratio mechanical springs ensures that the piston does not touch the cylinder wall surface and avoids the use of lubricant. Oil-free operation improves the heat transfer performance of heat exchangers, widens the choice of refrigerants and simplifies the refrigeration system. In addition, the clearance and stroke of the linear compressor are adjustable, and the cooling capacity can be easily adjusted according to actual thermal loads (Kim and Kim (2014)) [2].

The gas pressure of linear compressors is non-linear (non-sinusoidal); its piston is free and the piston offset changes with operating conditions; these characteristics increase the difficulty of studying non-linear gas pressure. Thus, a number of research are conducted to focus on the nonlinear dynamics of linear compressors. Schmidt (1956) [3] first proposed the idea of simplifying non-linear gas force into a damping and a spring; Kim et al. (2009) [4] linearized the gas force using the average method; Choi and Kim (2000) [5] proposed the Fourier method to linearize the gas force; Liang (2018) [6] presents a method to calculate gas spring stiffness at high pressure ratios; Zou et al. (2018) [7] proposed an online method for measuring gas spring stiffness based on vector graphics. In experimental measurements, Tang (2010) [8] installed a force sensor between the motor bearing and the compressor piston to measure the gas force; Li et al. (2019) [9] installed a high-frequency pressure sensor in the cylinder wall to directly measure the gas pressure in the compression chamber

From the literature survey, it is clear that a lot of studies have investigated on non-linear gas force by experimental testing and theoretical modelling. However, there is a lack of research on the nonlinear response characteristics of linear compressors driven by sinusoidal voltage. In order to study the issue, the gas pressure in the compression chamber (the gas pressure), piston displacement and current response curves of a moving coil linear compressor driven by sinusoidal voltages were measured simultaneously. A nonlinear dynamic model was established and the gas pressure, piston displacement and current curves by simulation were compared with the measured curves.

2. Nonlinear Dynamic Modeling of Linear Compressor

Figure1.shows a schematic diagram of the simplified dynamic model. The electrical model is simplified as a resistance-inductance-motor structure, and the mechanical model is simplified as a mass-spring-damper structure.

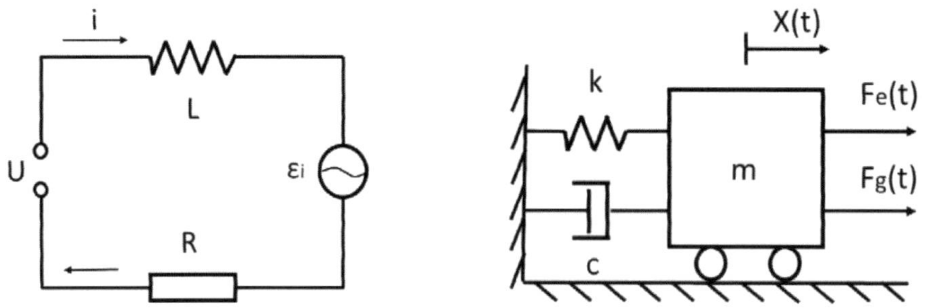

Figure 1. Dynamic model. (a) Electrical model. (b) Mechanical model.

The equations are listed as follows:

$$Ri + L\frac{\mathrm{d}i}{dt} + K_e i = U \tag{1}$$

IOP Conf. Series: Materials Science and Engineering **758** (2020) 012086 doi:10.1088/1757-899X/758/1/012086

$$m\frac{d^2x}{dt^2}+c\frac{dx}{dt}+k_s x=k_e i+F_g \quad (2)$$

Where i is the current, R is the resistance, L is the inductance, k_e is the electromagnetic force coefficient, U is the driving voltage; x is the piston displacement, m is the total moving mass, c is the damping coefficient, k_s is the mechanical spring coefficient, and F_g is the gas force.

The gas force can be expressed in the following:

$$F_g(t)=\left[P_c(t)-P_s\right]A_P \quad (3)$$

Where P_c is the gas pressure in the compression chamber, P_s is the suction pressure, and A_p is the cross-sectional area of the piston.

Figure2. reflects the working process of the compression chamber. The compression process and the expansion process was treated as polytropic processes, and suction and exhaust valves are treated as ideal valves without pressure loss. Thus the instantaneous gas pressure in the compression chamber can be obtained. The instantaneous pressure P_C of the compression chamber is:

$$P_c(t)=\begin{cases}\left[\dfrac{x_0+X}{x_0-x(t)}\right]^n P_s & \text{compression process}\\ P_d & \text{exhaust process}\\ \left[\dfrac{x_0-X}{x_0-x(t)}\right]^n P_d & \text{expansion process}\\ P_S & \text{suction process}\end{cases}$$

where x_0 is the initial clearance length , X is half the stroke and n is the polytropic index

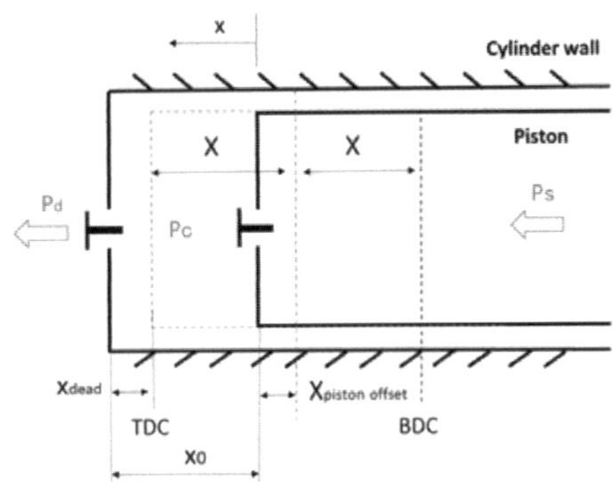

Figure 2. Compression chamber diagram

3 Experimental Setup

3.1 Test system

Figure3. is a schematic diagram of a VCR system driven by a linear compressor. The system consists of a linear compressor, a condenser, a drying filter, a throttle valve and an evaporator.

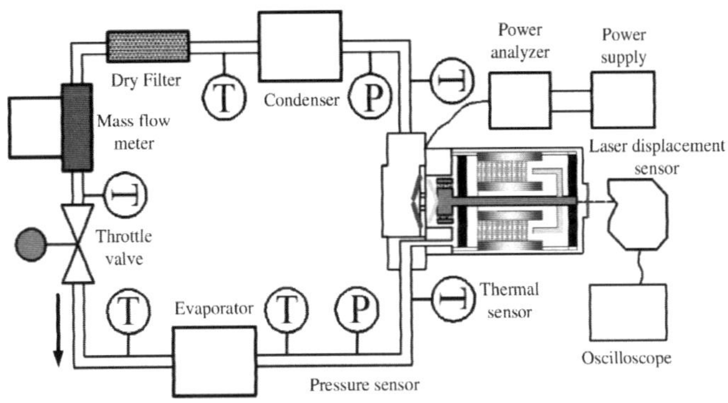

Figure 3. Test system chart of linear compressor

Figure 4. shows the power supply and online monitoring system of the linear compressor. The power supply system consists of a signal generator, a NF power supply and a power analyzer. The working process is that the signal generator generates a sinusoidal voltage signal, which is amplified by the NF power supply, and the electrical parameters of the amplified electrical signal are measured by a power analyzer.

The online monitoring system consists of a voltage sensor, a current sensor, a laser displacement sensor and a high-frequency dynamic pressure sensor. It can monitor the compressor's excitation signal (voltage) and response signal (piston displacement, current, the gas pressure) at the same time. A voltage sensor and a current sensor are installed between the power analyzer and the linear compressor in order to monitor the voltage and current signals input to the linear compressor; a set of laser displacement sensor is installed to obtain piston displacement online; A high-frequency dynamic pressure sensor is installed on the cylinder to monitor the gas pressure signal in compression chamber online， the installation position is shown in Figure 5.

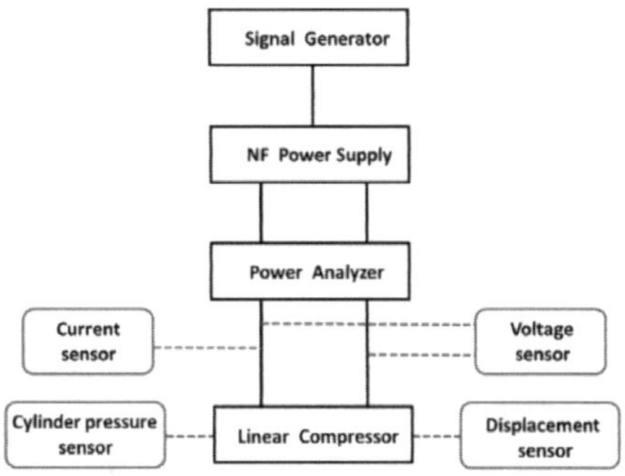

Figure 4. The power supply and online monitoring system

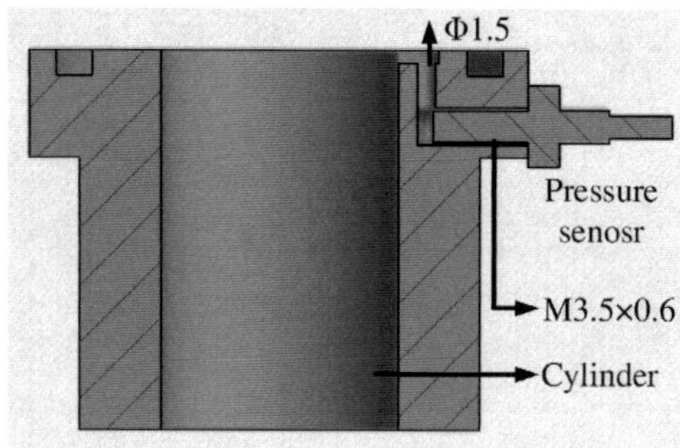

Figure 5. The installation position of the pressure sensor

3.2 Test conditions

Table 1 provides the test conditions of linear compressor. The work fluid is R134a.

Table 1 Test conditions

Items	Value(Unit)	Items	Value(Unit)
Evaporation temperature	5 (℃)	Refrigerant charge	110(g)
Condensation temperature	35(℃)	Stroke	8(mm)
Superheating temperature	5 (℃)	Operating frequency	50-70(Hz)
Supercooling temperature	5 (℃)	Ambient temperature	25(℃)

4. Results and Discussion

4.1 Experimental study on response characteristics

The linear compressor is driven with a sine voltage. The gas pressure, current, and piston displacement response curves can be obtained online at the same time.

Figure 6. shows the displacement response curves at different frequencies. It can be seen that the waveform of the displacement response curve at different frequencies is similar to sine. Figure 7. shows the current response curves of the compressor at different frequencies. It can be seen that the current response curves at different frequencies have non-linear distortions that deviate from the sinusoidal curve. The current distortion is particularly serious and two wave crests appear within one cycle at 60 Hz.

Figure 8.shows the gas pressure curve in different frequencies under the same working conditions, the four working stages (compression(a-b), exhaust (b-c), expansion(c-d) and suction(d-a1)) of the linear compressor can be clearly distinguished. The slope of the compression and expansion process increases with frequency. During the exhaust process, the cylinder pressure will bulge and then decrease. This is due to the sudden opening of the exhaust valve resulting in a sudden pressure change in the compression chamber. During the suction process, the suction valve fluctuates slightly. It is obvious that the gas pressure is non-linear.

Figure 9. shows the piston displacement response curve, current response curve and standard sine curve at 60Hz. The current curve and the displacement curve in Figure9. are divided into four processes according to the time points in Figure 8. In the compression process (a-b), the measured displacement curve does not coincide with the standard sine curve, and the displacement curve is higher than the sine curve. At this time, the current curve is seriously distorted. During the exhaust process (b-c), the nonlinear distortion of the measured displacement curve is small, and the current curve has no nonlinear distortion. During the expansion process (c-d) and suction process (d-a1), the measured displacement curve and current curve also have no nonlinear distortion.

To quantitatively illustrate the difference of the displacement and current curve compared with the standard sine curve, the Total Harmonics Distortion (THD) is introduced (Tang 2014) [8], and the expression of THD is:

$$THD_u = \frac{D_H}{D_1} = \frac{\sqrt{\sum_{h=2}^{\infty}(D_h)^2}}{D_1} \tag{4}$$

Where D_1 is root mean square (*RMS*) value of the fundamental wave, D_H is root mean square value of the high order harmonic component.

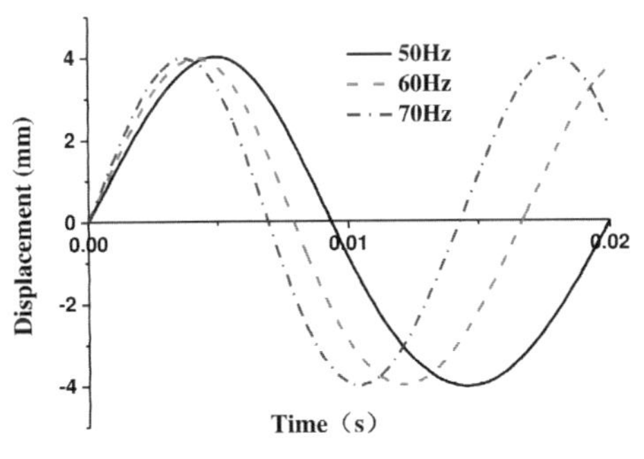

Figure6. The displacement curve in different frequencies

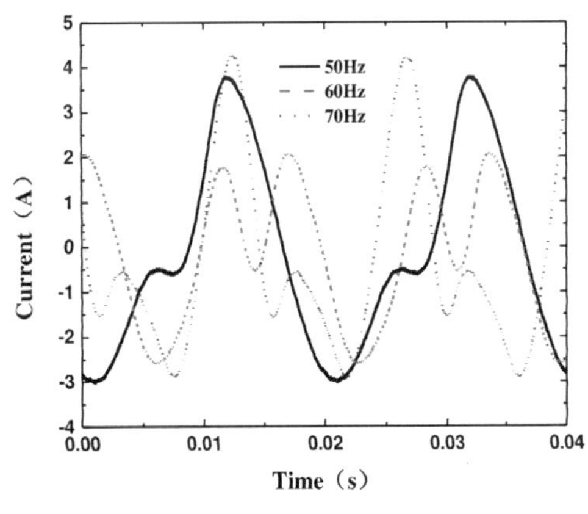

Figure7. The current curve in different frequencies

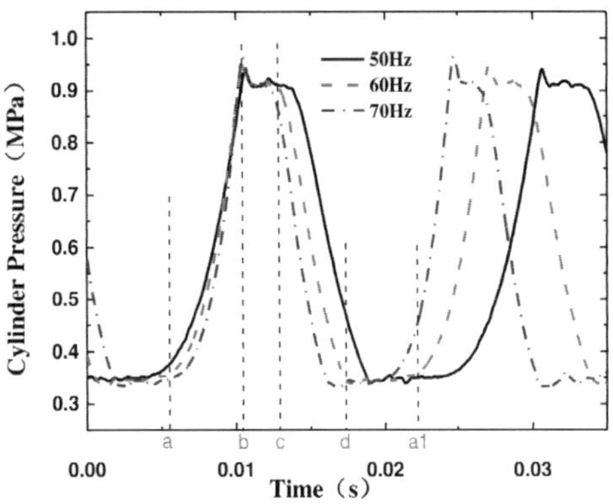

Figure8. The gas pressure curve in different frequencies

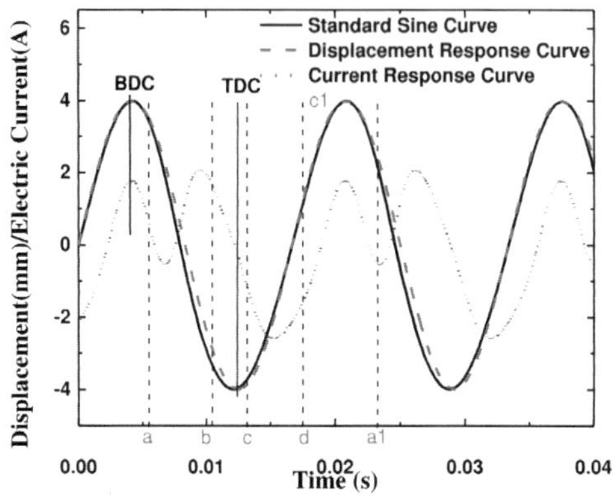

Figure 9. The displacement and current synchronous curve at 60Hz

Figure 10. shows the THDs of the displacement and current curves at different frequencies. It can be seen that the distortion rate of the displacement is small, and the distortion rate decreases with increasing frequency. In general, the THD of the displacement is within 7%, but the current has relatively large distortion at different frequencies, and the distortion rate is as high as 73.7% at 60Hz.

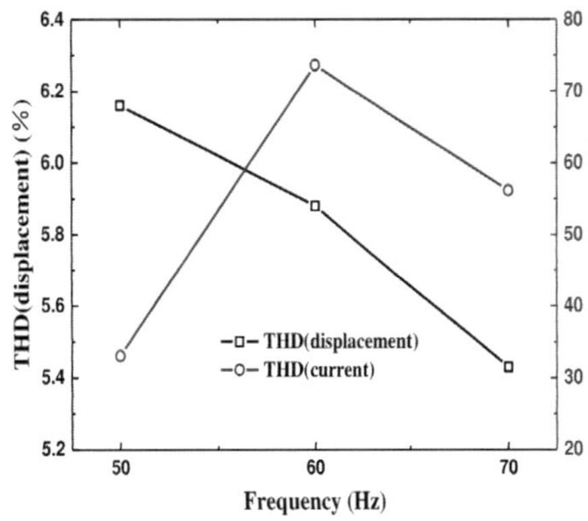

Figure10. The THD of the displacement and current curves at different frequencies.

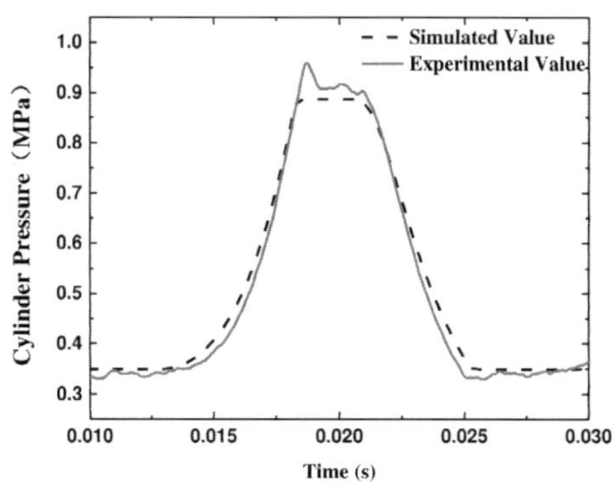

Figure11. The comparison between the calculation and measurement curve of gas pressure

4.2 Numerical Simulation Based on Nonlinear Dynamic Model

The nonlinear characteristics of the linear motor are ignored in the model. The parameters such as motor constant, equivalent resistance and equivalent inductance are treated as constants, and the damping coefficient in the mechanical system is treated as constant. Therefore, there only is non-linear gas loading in the model. This model takes 60Hz as an example to analyze the cause of current distortion. Figure11.shows the comparison between the calculated gas pressure and the actual measured gas pressure. Since the suction and exhaust valve is treated as an ideal valve, the difference between the two during the suction and exhaust process is large. Figure13.shows the calculated displacement curve, measured displacement curve and standard sine curve. It can be seen that both the calculated and measured displacement curves deviate from the standard sine curve at the ellipse marked in the Figure12. Figure 13. shows the comparison between the calculated and measured current curves, it can be seen that the calculated current curve also shows a significant double-peak distortion. This phenomenon shows that the cause of the large distortion of current is mainly the non-linear characteristics of the gas load, rather than the non-linear characteristics of the linear motor.

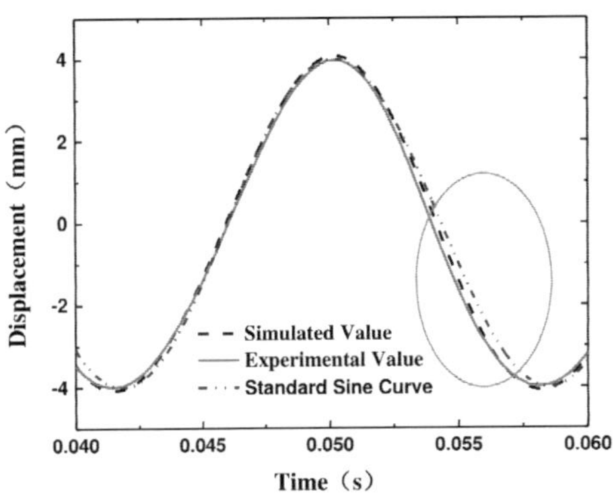

Figure12. The comparison between calculation and measurement curve of displacement

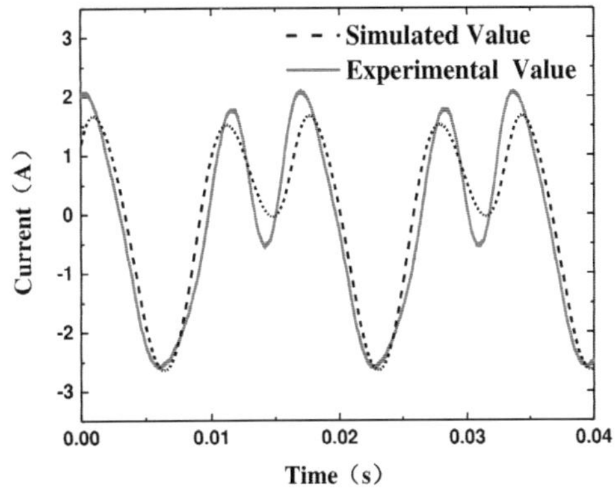

Figure13. The comparison between calculation and measurement curve of current

5. Conclusion

In this paper, the nonlinear response characteristics (the gas pressure, displacement and current response) of an oil-free moving coil linear compressor driven by sinusoidal at different frequencies are experimentally studied and theoretically analyzed based on the nonlinear dynamic model. The following conclusions can be drawn from this research:

1) The response characteristics of moving coil linear compressors (piston displacement curve, current curve and the gas pressure curve) have been experimentally measured simultaneously. The results show that the gas pressure curve, displacement curve and current curve are nonlinear (non-sinusoidal) when the moving coil linear compressor is driven by sinusoidal voltage. Among them, the current waveform has relatively large deviation from sinusoid.
2) By introducing the Total Harmonics Distortion (THD), the nonlinear distortion of waveforms is quantitatively evaluated. When the driving frequency is 60Hz, the THD of current has a maximum value of 73.7%, and the current has two peaks in one cycle. The THD of piston displacement has a maximum value of 6.2% at 50Hz.
3) In a working cycle, the nonlinear distortion of displacement and current occurs only in the compression process, and the displacement waveform and current waveform have almost no distortion in other processes.
4) The severe distortion of the current curve is mainly caused by the non-linearity of the gas pressure.

References

[1] Liang K, Stone R, Dad M and Bailey P 2014 A novel linear electromagnetic-drive oil-free refrigeration compressor using R134a. *International Journal of Refrigeration.* **40** 450-459

[2] Kim J K and Kim J B 2014 Modulation characteristics of a linear compressor for evaporating and condensing temperature variations for household refrigerators. *International Journal of Refrigeration.* **40** 370-379

[3] Schmidt T E 1956 Oscillating compressors in refrigerators. *Kaeltetechnik.* **8** 93-99

[4] Kim H, Roh C and G Kim J K 2009 An experimental and numerical study on dynamic characteristic of linear compressor in refrigeration system. *International Journal of Refrigeration.* **32** 1536-43

[5] Choi G and Kim K 2009 Analysis of nonlinear dynamics in a linear compressor. *International Journal of Japan Society of Mechanical Engineers.* **43** 545-52

[6] Liang K 2018 Analysis of oil-free linear compressor operated at high pressure ratio for household refrigeration. *Energy.***151** 324-31

[7] Zou H M, Li X, Tang M S, Wang M and Tian C.Q 2018 Online measuring method and dynamic characteristics of gas kinetic parameters of linear compressor. *Measurement.* **125** 545-53

[8] Tang M S 2014 *Research on the Nonlinear Characteristic and Control System of Linear Compressor* (China: University of Chinese Academy of Sciences)

[9] Li C Z, Li J G, Tang L F, Sun J, Zou H M and Cai J H 2019 Effects of the driving voltage waveform on the performance of vapor compression cycle system driven by the moving coil oil-free linear compressor. *International Journal of Refrigeration.* **108** 200-208

IOP Conf. Series: Materials Science and Engineering **758** (2020) 012087 doi:10.1088/1757-899X/758/1/012087

Characteristic enhancement of InGaN-based light emitting diodes grown on pattern sapphire substrates

Huanyou Wang[1,*], Gui Jin[1,2], Qiaolai Tan[1,3]

1 Academy of Electronic Information and Electrical Engineering, Xiangnan University, Chenzhou, 423000 ,China
2 School of Information and Optoelectronics Science and Technology, South China Normal University, Guangzhou, 510631, China
3 Institute of Physics and Information Science, Hunan Normal University, Changsha, 410081,China
E-mail: whycs@163.com

Abstract. Blue light-emitting diodes (LEDs) with an InGaN multi-quantum well (MQW) structure were fabricated on cone-shaped patterned sapphire substrate (PSS) using a single growth process of metal organic chemical vapor deposition (MOCVD). The PSS was proved to be an efficient method to decrease the threading dislocation (TD) density in GaN epifilm with the lateral growth mode on PSS. The LED designed on PSS increased the electroluminescene (EL) intensity. The internal quantum efficiency is increased by reducing the dislocation density, and light extraction efficiency is also enhanced owing to PSS.

1. Introduction

GaN and related compound semiconductors have been attracting great attention in optoelectronic device applications, for example light-emitting diodes (LEDs), laser diodes (LDs), and high-power and high-temperature electronics [1-3]. It is well known that high-density threading dislocations are inherent in epitaxial GaN films on sapphire substrates, due to the large difference in lattice constant and thermal expansion coefficient between the epitaxial layer and sapphire substrate. High dislocation density will affect the device properties, for instance device lifetime, electron mobility, and the quantum efficiency of radiative recombination. Thus, further more decreasing the dislocation density is an important problem for manufacturing high-performance InGaN LEDs. To enhance the crystalline quality for GaN-based epitaxial layers on sapphire substrate, different epitaxy methods have been recommended, for example extension lateral overgrowth (ELO) [4], cantalever extension (CE) [5], defection selected passivation [6], micro-scale SiN_x or SiO_x patterned mask [7], plastic stress relaxation via buried AlGaN crazing [8], and patterned sapphire substrate (PSS) [9]. The maskless, growth uninterruption and single-step overgrowth technique is commendatory, which can decrease the TDs and add entire internal reflection of light owing to geometrical action [10, 11]. The various PSS were designed [12-15] which could be effectively enhance the external efficiency in InGaN-based LEDs grown on PSS through the effect of the optical reflection from the side edge of the etched sapphire.

In this paper, the manufacture, optical and electrical properties of InGaN LEDs are discussed. The performance advance of the LED grown on PSS by contrasting LED on conventional planar sapphire substrates (CSS) is described here. Moreover, the microcosmic structures of LEDs on PSS were researched through a transmission electron microscopy (TEM).

2. Experiment

The samples were all grown over 2-in (0001) sapphire substrates with MOCVD system. Before the MOCVD growth, the PSS is prepared using dry etching to form a cone-shaped patterned array. The sapphire substrate was etched employing BCl_3/Cl_2 in an inductively coupled-plasma etcher. After that, the scanning electron microscopy (SEM) is used for inspecting the PSS, and inspecting micrograph is shown in figure 1. The diameter and height of each cone were about 2.5 μm and 1.5 μm, and interval is 0.5 μm. During the MOCVD growth, the Trimethylgallium, trimethylindium, and ammonia were used as precursors for Ga, In, and N, respectively. Biscyclopentadienyl magnesium and Si_2H_6 were used as p- and n-type dopants. H_2 and N_2 were used as carrier gases. The reactor pressure was maintained at 100 Torr. The LED structure is as following: a 25 nm thick GaN nucleation layer, a 2.5 μm thick un-doped GaN layer, a 2 μm thick highly doped n-type GaN layer, fifteen pairs of InGaN (2.5 nm)/GaN (12.5 nm) MQWs, a 25nm thick p-type AlGaN electron blocking layer, and a 100 nm thick Mg-doped GaN layer. In order to obtain better coalescence on PSS, the u-GaN growth temperature was rised up 1120 °C. LEDs with 300 × 300 μm^2 sizes were achieved via a normative photoetching technics until the n-GaN was exposed. Both p- and n-contacts were located on the epitaxial surface. The indium tin oxide (ITO) layer was deposited to form a p-side transparent contact layer and a current spreading layer, then the Cr/Au layer was deposited on the ITO layer to form the p-side and the n-side electrodes.

These samples were characterized by transmission electron microscopy (TEM) to reveal microstructure. TEM samples were processed with standard mechanical polishing and Ar ion-milling techniques to achieve electron transparency. The EL intensity of the LED lamp was observed utilizing an integrated sphere detector and the observed deviation was about 5%.

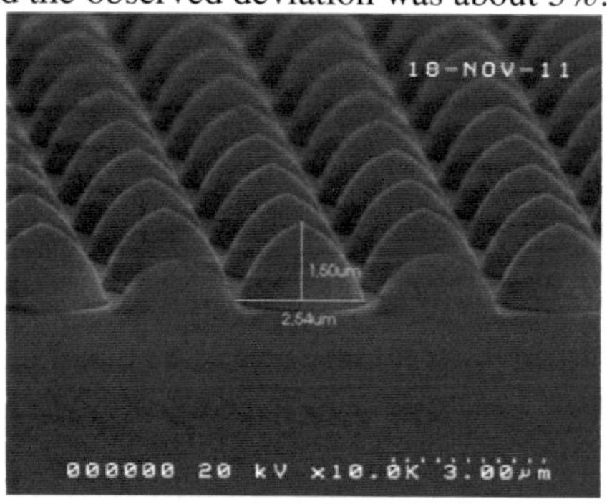

Figure 1. Plan-view SEM images of PSS.

3. Results and Discussion

To clarify the relation between the dislocation and microstructure, cross-sectional TEM observation for GaN on PSS was achieved. figure 2 shows cross-sectional TEM images of GaN on PSS. The GaN was assorted into three parts: (A) a GaN region (in circle 1 in figure 2(b)) with a high dislocation density where many dislocations begin from the interface owing to the lattice mismatch between GaN and sapphire, as usually observed when GaN is grown on a flat sapphire substrate, (B) a GaN region (in circle 2 in figure 2(b)) with few dislocation, where there is no vertical disseminate of dislocations because dislocations engendered anear the cone border bent as progressing GaN epitaxial growth, and (C) a GaN region (in circle 3 in figure 2(b)) with high dislocation density on the top of the cone area, this result is attributed to full coalescence by the lateral overgrowth of the GaN film on the trench region. It should take notice that the bent dislocations and presence of voids as indicated by arrows in figure 2(a) can effectively reduce the TDs by ELO. Thus, the ELO caused by the cone shaped pattern is responsible to the specific way of defect reduction and for the strain relaxation in GaN epifilms on PSS. In addition, a recrystallized GaN initiation layer (as indicated by arrows in figure 2(b)) is also observed as it grew into three-dimensional islands on the bulging patterns. Contrasting difference in the two-beam bright field images, we infered that this initiation islands had a different crystallographic orientation relationship with the later GaN layer.

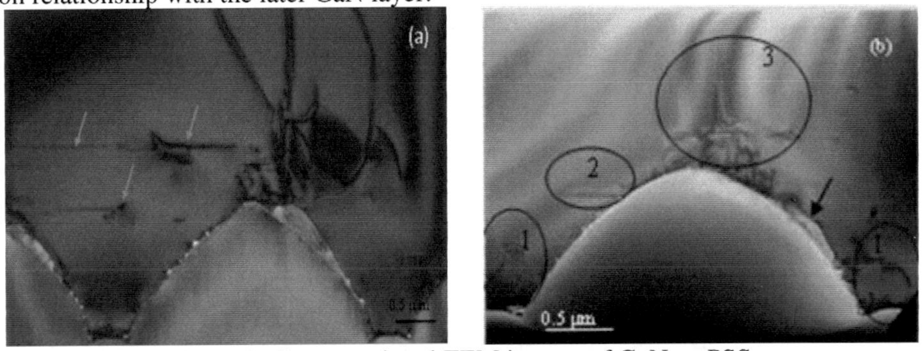

Figure 2. Cross-sectional TEM images of GaN on PSS.

Figure 3 shows the different luminescence ability for different inject currents. The luminous intensity of the LEDs on PSS and CSS at 40 mA injection current were 320.6 and 198.4 mcd. The luminescence ability of LED on PSS was 1.62 times higher than that of LED on CSS at the applied current of 40 mA. Through restaining the total internal reflection at cone-shaped interface and enhancing crystal quality owing to decreasement of TDs via PSS, the light extraction efficiency is increased, which results in larger luminous intensity for LED grown on PSS. The main cause of enhancement of luminous intensity is that the photons obtain a large number of opportunities to seek out the escape cone. The PSS can induce photons that were initially radiated out the escape return to the escape cone on LED surfaces. The chance of photons escaping from the LED on PSS is enhanced, thus light extraction is more effective. When reaching 78 mA for LED on PSS and 72 mA for LED on CSS, the luminous intensities begin to reduce although the injection current rises. The decline of luminous intensity relates to the thermal dissipation. Because when carriers recombinate in defects and/or dislocations, this recombination is non-radiative, yet non-radiative recombination will generate heat. The energized carriers that get the heat are more likely to escape from the InGaN/GaN MQW active layer. Due to the thermal action at high temperature, the effective radiation recombination rate will decrease. Because the PSS can effectively reduce the dislocation density, the heat dissipation effect for LED on PSS is superior to LED on CSS. Therefore, the maximum luminous intensity for LED on PSS is also above LED on CSS at a bigger injection current.

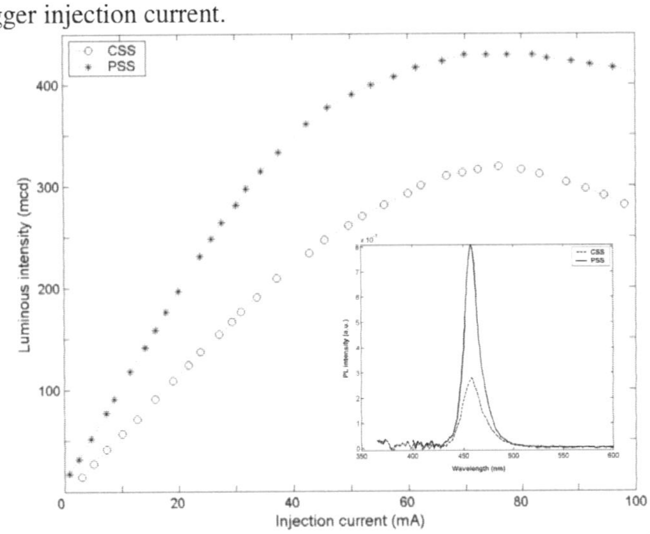

Figure 3. Luminous intensity of LEDs grown on PSS and CSS, the inset shows the room temperature EL spectra for LED on PSS and CSS.

The inset in figure 3 presents room temperature EL spectras of LED on PSS and on CSS at a forward current of 20 mA. A little blue shift of approximate 4 nm was viewed for the LED on PSS. This alteration can be relate to the band filling effect and screening effect in a piezoelectric field quantum well. The larger piezoelectric field is caused by the larger lattice mismatch in InGaN-based multiple quantum wells. The EL intensity of the LED on PSS is higher about 48% than that of LED on CSS as an injection current 20 mA. Moreover, the full width at half maximum for LED on PSS and on CSS is 22 nm and 34 nm respectively. It is well known that dislocations are non radiative recombination centers, while nonradiative recombination directly affects luminous efficiency. Therefore, the key improvement in EL intensity is mainly traceable to the reduction of TDs for LED on PSS. The another reason for the improvement of light extraction efficiency is effect of optical back scattering, because PSS can raise the chance of optical back scattering for LED on PSS.

4. Conclusion

We studied the strong association between the structure and optical characters of InGaN-based blue LEDs on PSS. The LED on PSS exhibit improved performance compared to the LED on CSS. The ELO caused via the cone shaped pattern is responsible for the specific way of defect reduction and for the strain relaxation in GaN epifilms on PSS. The enhancement of EL intensity of LED on PSS depends on upgrading of the quality of GaN epitaxial layer and high light extraction efficiency by PSS. Owing to the effect of facets of PSS, the light can be effectively induced to the top escape-cone of the LED surface.

Acknowledgments

This work was supported by Science Foundation of Education Department of Hunan Provincial (18A463), the Chenzhou Science and Technology Project (jsyf2017002).

References

[1] Nakamura S, Senoh M, Iwasa N, Nagahama S, Yamada T and Kukai T 1998 *Jpn. J. Appl. Phys.* **37** L479
[2] Tadatomo K, Okagawa H, Ohuchi Y, Tsunekawa T, Jyouichi T 2001 *Physica Status Solidi.* **188** 121
[3] Jeon H C, Rark C J, Cho H Y, Kang TW, Kim T W, Oh J E and Na J H 2005 *J. Korean Phys. Soc.* **47** S489
[4] Zheleva T S, Nam O H, Bremser M D and Davis R F 1997 *Appl. Phys. Lett.* **71** 2472
[5] Follstaedt D M, Provencio P P, Missert N A, Mitchell C C, Koleske D D, Allerma A A and Ashby C I H 2002 *Appl. Phys. Lett.* **81** 2758
[6] Lo M H, Tu P M, Wang C H, Cheng Y J, Hung C W, Hsu S C, Kuo H C, Zan H W, Wang S C, Chang C Y and Liu C M 2009 *Appl. Phys. Lett.* **95** 211103
[7] Sakai A, Sunakawa H, and Usui A 1997 *Appl. Phys. Lett.* **71** 2259
[8] Bethoux J M, Vennegues P, Natali F, Feltin E, Tottereau O, Nataf G, Mierry P De and Semond F 2003 *J. Appl. Phys.* **94** 6499
[9] Lee Y J, Hwang J M, Hsu T C, Hsieh M H, Lee B J, Lu T C, Kuo H C and Wang S C 2006 *IEEE Photon. Technol. Lett.* **18** 1152
[10] Yamada M, Mitani T, Narukawa Y, Shioji S, Niki I, Sonobe S, Deguchi K, Sano M and Mukai T 2002 *Jpn J Appl Phys.* **41** L1431
[11] Hsu Y P, Chang S J, Su Y K 2004 *J Cryst. Growth* **261** 466
[12] Lee J H, Oh J T, Park J S, Kim J W and KimY C, Lee J W and Cho H K 2006 *Phys. Status Solidi (c)* **3** 2169
[13] Lee J H, Oh J T, Choi I S and Kim Y C 2007 *J. Korean Phys. Soc.* **51** S249
[14] Song J C, Lee S H, Lee I H, Seol K W and Lee C R 2007 *Journal of Crystal Growth* **308** 321
[15] Jeonga S M, Kissinger S, Kim D W, Lee S J, Kim J S, Ahn H K and Lee C R 2010 *Journal of Crystal Growth* **312** 258

IOP Conf. Series: Materials Science and Engineering **758** (2020) 012088 doi:10.1088/1757-899X/758/1/012088

Testing research on friction and wear performance of microstructure surface of unfolding wheel for bearing steel ball detection

Pan Chengyi , Tong Yuanqi and Cao Guanqun

School of Mechanical and Power Engineering, Harbin University of Science and Technology, Harbin, 150080, China.
E-mail: panchengyi2010@163.com

Abstract. The unfolding wheel of AVIKO series steel ball surface quality automatic detector is expensive and easy to wear, which seriously restricts its application and development. In this paper, the wear reason and motion form of the unfolding wheel are analyzed, the unfolding contact model is simplified reasonably, and the corresponding experimental equipment and scheme are designed. The microstructures with different shape parameters are machined on the test pieces surface by laser machining technology, and the theoretical calculation formula for the weight of wear is obtained through friction and wear tests and data fitting. The test results show that compared with the smooth test piece, the test pieces with microstructure surface can increase the friction coefficient and reduce the degree of wear, and the diamond microstructure has the best effect on reducing the wear weight. The fitted formula provides theoretical and experimental basis for the follow-up research and the design of microstructure parameters of unfolding wheel.

1. Introduction

Steel ball is an important rotating part of rolling bearing, the quality of the steel ball will directly affect the bearing precision and life. In addition to the geometric accuracy, dimensional accuracy, material and heat treatment, another important index for evaluating the quality of steel balls is surface quality [1]. The steel ball and the unfolding wheel are both smooth surfaces, which are easy to slip in the unfolding process, and the contact collision and movement form cause the unfolding wheel to easily wear and failure. It has been shown that microstructure surface can improve the friction and wear properties of materials[2]. Tatsuya Sugihara et al. [3] machine microstructures on the carbide milling cutters and find that the wear of the blade surface is reduced and the cutting performance is improved. Miao Jiazhi et al. [4] find that the cylinder liner with the microstructure surface can reduce the friction coefficient by 41.83%, and increase by 33.68% compared to the smooth surface, it is determined by the microstructure parameters. Qi Ye et al. [5] conclude that the microstructure can alleviate the stress concentration in the contact area and reduce the wear within a certain range of areal density through the ring-block linear contact friction and wear test. Based on the above research, the microstructure is applied to the surface of the unfolding wheel, and its friction and wear properties are studied by tests.

2. Unfolding principle and simplified contact model

The unfolding mechanism of AVIKO series steel ball surface quality automatic detector is mainly composed of driving wheel, unfolding wheel and supporting wheel. When the steel ball is unfolded, driven by the motor, the driving wheel rotates around the fixed axis x_3, and at the same time drives the steel ball to rotate around the axis x_1. The rotating steel ball drives the unfolding wheel to rotate around the axis x_2. At the same time, under the function of the unfolding wheel with special geometry, the steel ball rotates around the axis z_1, realizing the full surface unfolded of the steel ball. In order to facilitate the test, after considering the friction and contact form between the unfolding wheel and the steel ball, the contact is simplified to the model shown in Fig. 1.

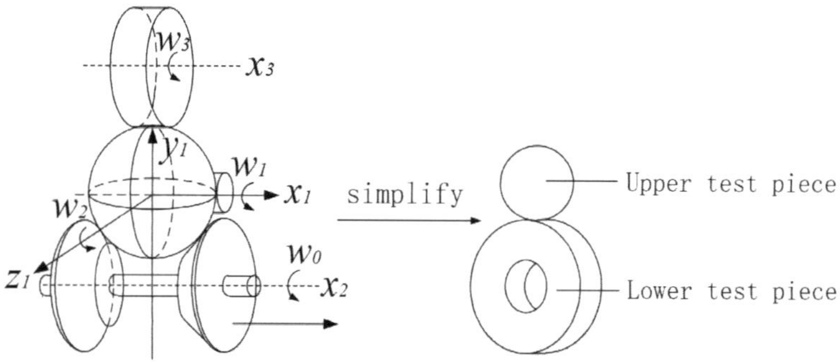

Figure 1. Unfolding principle and simplified contact model

3. Test

3.1. Test pieces preparation

The material of the upper test piece used in the test is GCr15, with a diameter of 12.7mm and a hardness of HRC62~65. The material of the lower test piece is T10A, and the hardness reaches HRC51~56. As shown in Fig. 2, the circular, square, and diamond pit microstructures are processed on the peripheral surface of the lower test pieces by the laser marking machine, and the pits depth is all 50μm. Table 1 shows the microstructure parameters of different test pieces.

Figure 2. Lower test piece with circular microstructures

Table 1. Microstructure parameters of different lower test pieces

Number	Shape	Single size(μm)	Area(μm²)
0	Smooth	none	none
1	Circular	100(diameter)	7850
2	Circular	150(diameter)	17663
3	Circular	200(diameter)	31415
4	Square	88.6(side length)	7850
5	Square	132.9(side length)	17663
6	Square	177.2(side length)	31415
7	Diamond	164.9/95.2(long/short axis)	7850
8	Diamond	246.9/148.5(long/short axis)	17663
9	Diamond	329.8/195.4(long/short axis)	31415

3.2. Test equipment and conditions

As shown in Fig. 3, this test uses a specially constructed point contact friction and wear tester on cylindrical surface microstructures, which is mainly composed of a friction and wear test device, a friction performance detection device and a screw adjusting device. The friction and wear test device designed according to the simplified contact model realizes contact and friction between the upper and lower test pieces; The friction performance detection device measures the friction force between the upper and lower test pieces by the S-shaped sensor and the force transmission rod. The screw adjusting device is used to adjust the steel ball to the proper point contact position on the surface of the lower test pieces. S-shaped sensor measures the friction force every 0.1 seconds, and transmits the measured data to the computer through the serial port in real time. Table 2 shows the test conditions.

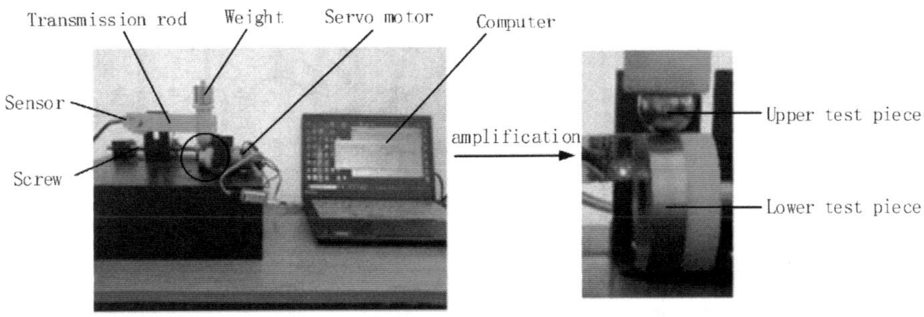

Figure 3. Cylindrical surface microstructure point contact friction and wear tester

Table 2. Conditions of friction and wear test

Temperature(°C)	Motor speed(r/min)	Time(min)	load(N)	Friction conditions
30	1400r/min	120	2/3	Dry friction

In the test, the weight loss of the lower test pieces is used to evaluate the wear resistance, the smaller the wear weight, the better the wear resistance. BSM220.4 electronic analytical balance is used to measure the weight of the lower test pieces before and after wear tests, and the measurement accuracy is 0.1mg. Before weighing, use the ultrasonic cleaner to clean for 20min and blow dry to avoid the interference of environmental humidity and debris on the weight.

3.3. Test results and analysis

3.3.1 Friction coefficient: In the process of test, the points that sudden increase of friction force due to the influence of human behavior should be removed. After sorting and converting a large number of measured friction force data under 2N load, the partial friction coefficient fluctuation curve as shown in Fig. 4 is obtained (The figure shows the friction coefficient of No. 3 test piece). Although the friction coefficient fluctuates, it generally tends to be a stable value. After calculating the average value, the friction coefficients of all test pieces are filled in table 3.

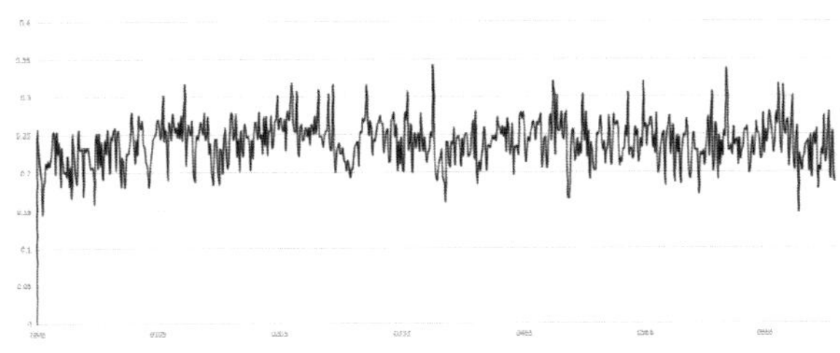

Figure 4. Partial friction coefficient fluctuation curve of No.3 test piece

Table 3. Friction coefficient of different test pieces

Area(μm²) \ Shape	Circular	Square	Diamond	Smooth
7850	0.22	0.27	0.27	
17663	0.23	0.22	0.25	0.18
31415	0.25	0.24	0.26	

The data in table 3 are classified and extracted according to the friction coefficients of the circular, square, and diamond microstructures under different areas, and connected into lines to obtain the chart shown in Fig. 5. It can be seen that compared with smooth test piece, all the test pieces with microstructures can increase the surface friction coefficient. This is because the microstructures change the surface morphology of the test pieces, increase the roughness of the surface of the friction pair, and reduce the contact area between the upper and lower test pieces. As a result, the normal pressure increases, so the friction force increases. The friction coefficient is related to the shape and area of the microstructures. Among the three shapes of microstructures, the diamond microstructure has the best friction enhancement effect, but the relationship between the microstructure area and the friction coefficient is not obvious.

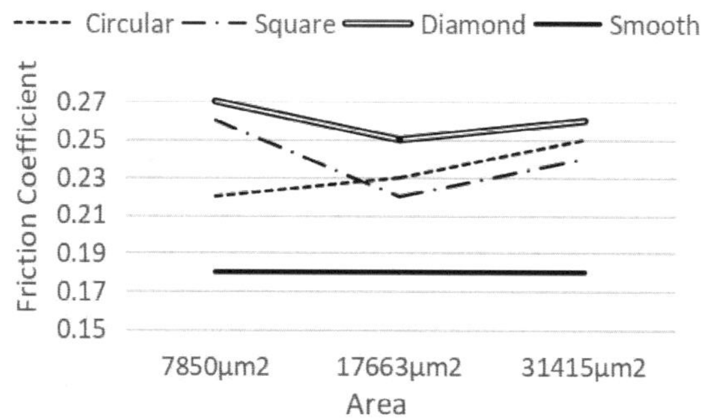

Figure 5. Friction coefficient of different microstructures shapes under different areas

3.3.2 Wear weight: In order to quantify the shapes of the microstructures, the ratio of S/d^2 is adopted to represent the shape parameters of the microstructures. S is the area of the microstructures, d is the distance between the longest two points on the pits surface, d is the diameter in the circle, the diagonal in the square, and the long axis in the diamond. According to the microstructure area classification, the wear weight of ten group of test pieces are listed in table 4.

Table 4. Wear weight of different microstructure test pieces

Area (μm^2)	Number	Shape	Shape parameter	Wear weight (g)	
				2N	3N
7850	1	Circular	0.785	0.0057	0.0061
	4	Square	0.502	0.0045	0.0055
	7	Diamond	0.289	0.0041	0.0044
17663	2	Circular	0.785	0.0052	0.0050
	5	Square	0.502	0.0036	0.0040
	8	Diamond	0.289	0.0025	0.0028
31415	3	Circular	0.785	0.0042	0.0045
	6	Square	0.502	0.0023	0.0027
	9	Diamond	0.289	0.0017	0.0022
none	0	none	/	0.0156	0.0162

By converting the wear weight data in table 4 into the Fig. 6, it can be seen that no matter what the load, the wear weight of microstructure test pieces is lower than that of smooth test piece, and the wear weight decreases with the increase of microstructure area. This is because the laser processing technology can increase the surface hardness, harden the edges of the microstructure pits, and improve the wear resistance. In addition, the wear debris formed in the friction process is filled into the pits, which effectively reduces the wear of the abrasive particles. The larger the area of the micro pits, the larger the size of the contained abrasive particles, and the more the abrasive particles. Among the three kinds of microstructure shapes, no matter which area, the wear weight of diamond microstructure test pieces are lower than that of other test pieces, so diamond microstructure has the best wear resistance.

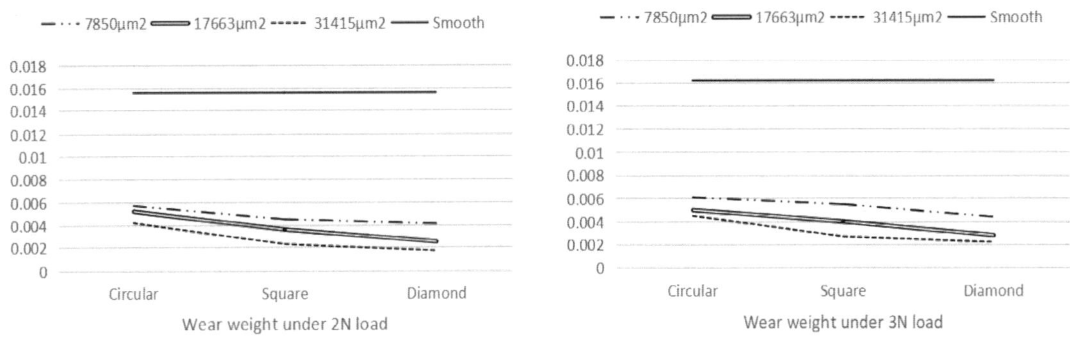

Figure 6. Wear weight of microstructure test pieces under different load

3.4. Fitting formula of wear weight

A total of 18 sets of data in table 4 are fitted to obtain the function formula of the wear weight W under the influence of three variable of the microstructure area S, the applied load N and the microstructure shapes X:

$$W = -8.599 * 10^{-8} S + 0.004 N + 0.005 X + 0.002 \tag{1}$$

The goodness index of fitting is R^2=0.9595, close to 1, which shows that there is a good correlation between the fitting results and the test data, so it can be used to calculate the wear weight of microstructures surface in the test.

4. Conclusions

1. The circular, square and diamond microstructures are machined by laser technology on the circumferential surface of the lower test pieces. Three kinds of microstructure areas are designed for each shape, and the friction and wear tests are carried out respectively. The results show that, compared with the smooth test piece, the microstructure test pieces can increase the friction coefficient and reduce the wear weight. Therefore, the application of microstructures to the surface of unfolding wheel can achieve the purpose of increasing friction and reducing wear.
2. With the increase of microstructure area, the weight of wear decreases. By comparison, diamond microstructure has the best effect on increasing friction and reducing wear, followed by square.
3. The fitted wear weight calculation formula can be used to calculate the wear weight under the influence of microstructure shapes, load and microstructure area, which can provide reference for the follow-up theoretical research.

References

[1] Zhao Gang, Wang Baoyi, Ma Songxuan. Theoretical analysis of meridian unfolding mechanismin in Aviko K steel ball appearance inspection machine. Journal of Sichuan University. 1997, 34(5): 635-639.
[2] Zeng Yawei, Chen Liyu, Yang Xiaming. State of surface texturing for improving friction performance. Tool technology. 2016, 50(5): 3-7.
[3] Tatsuya Sugihara, Toshiyuki Enomoto. Crater and flank wear resistance of cutting tools having micro textured surfaces. Precision Engineering. 2013, (37): 888-896.
[4] Miao Jiazhi, Guo Zhiwei, Yuan Chengqing. Effect of textured surface on the friction performance of cylinder liner-piston ring system in the internal combustion engine. Journal of tribology. 2017, (4): 465-471.
[5] Qi Ye, Chang Qiuying, Wang Bin, et al. Research on tribological behaviors of laser surface textures under dry sliding. Journal of China Ordnance. 2015, 36(2): 200-205.

Acknowledgments

This study is supported by the Natural Science Foundation of Heilongjiang Province of China. Roller Surface Microstructure and its Dry Friction and Wear Mechanism for Bearing Steel Ball Defect Detection (No.E2017052)

IOP Conf. Series: Materials Science and Engineering **758** (2020) 012089 doi:10.1088/1757-899X/758/1/012089

Influential Parameter analysis of spread GFRP stay-in-place form deformation

Zhaojie Tong[1,2] and Hailong Zhang[1]

1 Shenzhen Municipal Design & Research Institute Co., Ltd., Shenzhen, China
2 School of Civil Engineering, Chongqing University, Chongqing, China
E-mail: zongshi2006@163.com

Abstract. Recently spread GFRP stay-in place forms with T-ribs have been applied to bridge engineering in China. The deformation non-uniformity of spread GFRP plates in the deck transverse direction is more significant than that of narrow GFRP plates. In order to investigate the influential parameters of deformation, a 1.8 m simply supported GFRP plate was analysed using ANSYS. The analysis results show that (1) the maximum deflection of side GFRP plates is larger than that of middle plates; (2) the maximum deflection is affected by rib thickness, plate height, rib spacing and bottom plate thickness; (3) the non-uniformity of deflection is mainly affected by rib spacing and bottom plate thickness; (4) the rib thickness has little effect on the non-uniformity of deflection, and a higher GFRP plate results in a larger non-uniformity of deflection. Based on the analysis results, some measures on improving the flexural stiffness of GFRP plates in the transverse and longitudinal directions are proposed.

1. Introduction

In the past 10 years, considerable attention has been paid to GFRP-concrete composite structures. Among them, GFRP-concrete decks have attracted bridge engineers' interest because of excellent mechanical performance, good durability and convenient construction [1]. The GFRP-concrete deck includes GFRP plates and concrete. In the construction process, the GFRP plate could be used as the formwork to speed up the bridge construction. During the operation period the vehicle load is carried by GFRP plates and concrete. Spread GFRP plates are more convenient for workers than narrow plates, The deformation influential parameter of spread GFRP plates are more complex than that of narrow GFRP plates because of the non-uniformity of deformation. However, the study on the GFRP plate deformation is focused on the narrow plate. Recently, the GFRP-concrete decks with T-ribs has been applied in the steel-concrete composite girder bridge in China [2]. The study on the deformation influential parameter could provide reference for the design of spread GFRP plates with T-ribs.

According to the GFRP plate cross-section, the GFRP-concrete decks could be divided into composite decks with dual cavity system [3], corrugated GFRP-concrete decks [4], composite decks with T-ribs [5], and so on. At present, the mechanical performance of GFRP-concrete decks has been investigated in detail, which includes the static mechanical performance of simply supported decks [6], the static mechanical performance of continuous decks [7], and the dynamic performance of GFRP-concrete decks [8]. Researchers have also study the deformation of narrow GFRP plates [9]. However, there has been little about deformation influential parameter analysis of spread GFRP plates with T-ribs.

In this paper, the deformation influential parameters of spread GFRP plates with T-ribs have been studied by finite element simulation, which include thickness of T-ribs, width of GFRP plates, thickness of bottom plates, and height of GFRP plates. The relationships among above parameters, maximum deflection and non-uniformity of deformation have been investigated.

2. An example of GFRP plates and numerical model

According to Chinese standard JTG/T F50-2011, the deformation of GFRP plates cannot exceed 1/400 of the calculated span in the construction process of GFRP-concrete decks. When the adjacent rib spacing is large, the deformation non-uniformity of GFRP-concrete decks in the deck transverse direction is remarkable.

In this paper, the deformation of a 1.8 m simply supported GFRP decks with T-ribs was investigated based on finite element simulation. Fig. 1(a) shows the cross-section of the GFRP plate with T-ribs. In Fig. 1(a), *b* is the width of GFRP plate, t_{f1} is the thickness of GFRP bottom plates, t_{f2} is the thickness of flanges, b_2 is the width of flanges, b_3 is the width of T-ribs, *h* is the height of GFRP plates. Fig. 1(b) and (c) show the loading condition. The initial cross section parameters are as follows: *b* is 300 mm, t_{f1} is 6 mm, t_{f2} is 6 mm, b_2 is 60 mm, b_3 is 8 mm, and *h* is 80 mm. Then the effect of above parameters on deformation is investigated using ANSYS.

Fig. 2 shows the numerical model of GFRP plates. The GFRP plate was simulated by an eight node 3-D orthotropic element SOLID185. According to reference [2], the modulus of GFRP plates in the deck transverse direction is 1.08×10^4 MPa, and the modulus in the deck longitudinal direction is 3.26×10^4 MPa. A uniform load of 0.003952 MPa is applied to the top surface of the bottom plate, which is equal to the weight of concrete decks with the thickness of 15 cm. The numerical model is shown in Fig. 2.

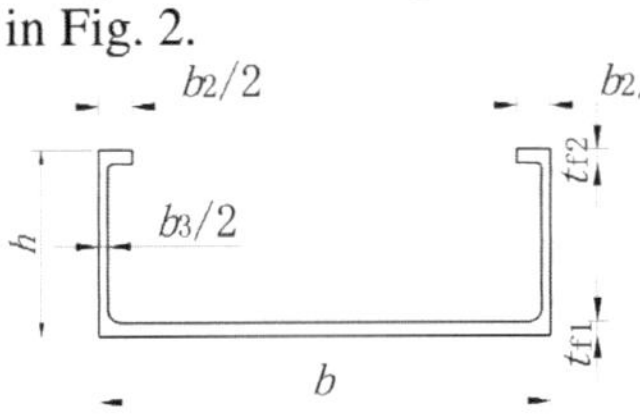

(a) GFRP plate cross-section

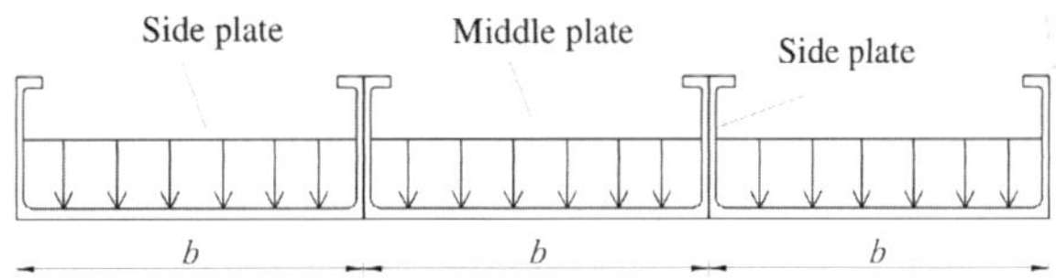

(b) Loading condition in the deck transverse direction

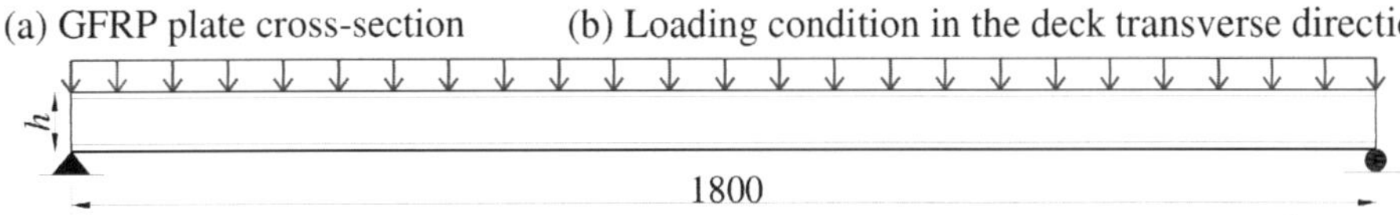

(c) Loading condition in the deck longitudinal direction

Figure 1. Loading condition of GFRP plates in detail/mm

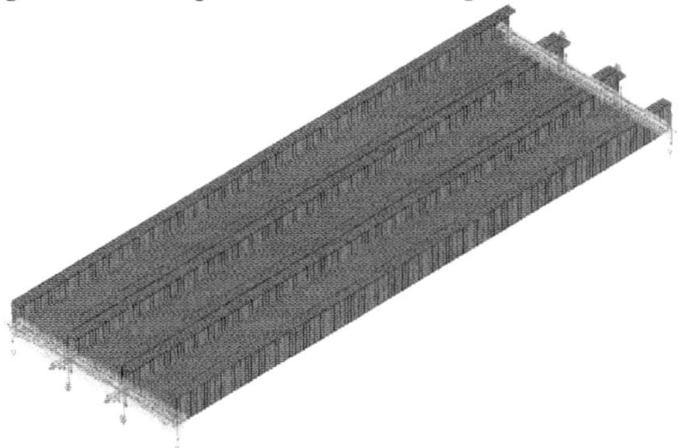

Figure 2. Finite element model

3. Parameter analysis

The deformation influential parameters were investigated, which included rib thickness, plate height, bottom plate thickness, and rib spacing. Fig. 3 shows the relationship between deflection and rib thickness. The non-uniformity of deflection under uniform load is impressive. The deflection of side plates is larger than that of middle plates because of the torque carried by side ribs. Moreover, a higher rib thickness resulted in a lower deflection. However, the non-uniformity of deflection in the deck transverse direction has not been improved.

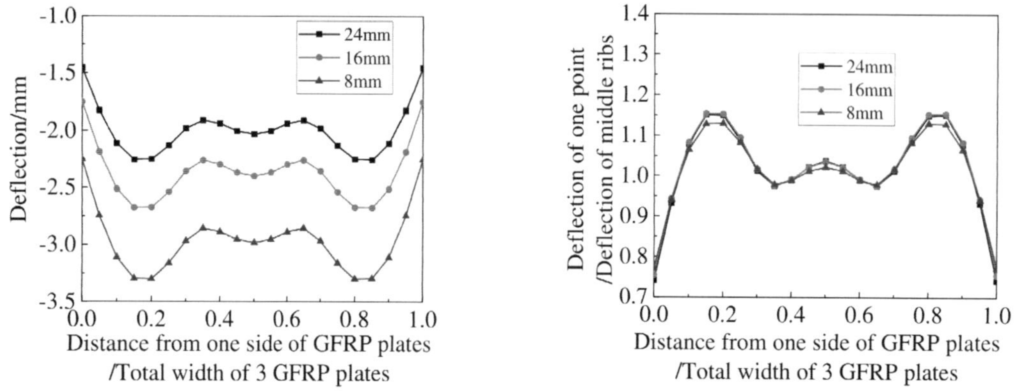

(a) Deflection distribution (b) Relationship between rib thickness and non-uniformity of deformation

Figure 3. Relationship between deflection and rib thickness

Fig. 4 shows the relationship between deflection and plate height. It could be seen from Fig. 4(a) that the maximum deflection of GFRP plates with the height of 90 mm is lower than that of GFRP plates with the height of 70 mm. A higher plate could reduce the deflection of GFRP plates. However, the deformation non-uniformity of GFRP plates increases with the increase of GFRP plate height. The increase of plate height is detrimental to the deformation uniformity.

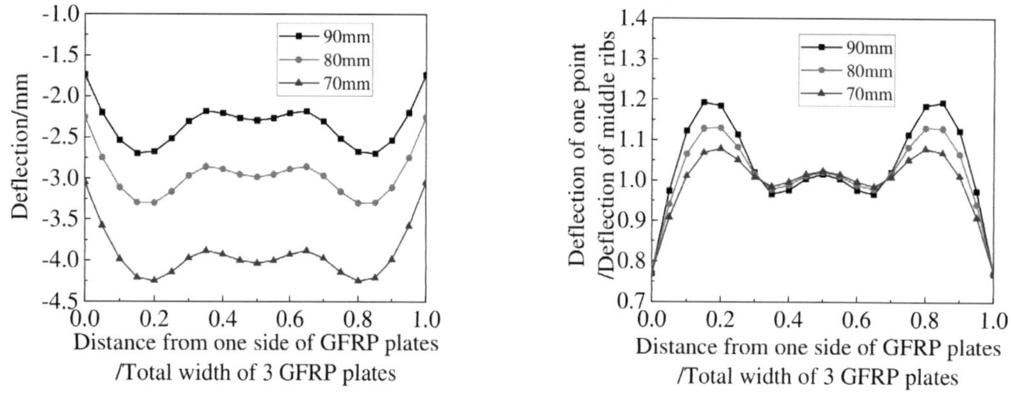

(a) Deflection distribution (b) Relationship between plate height and non-uniformity of deformation

Figure 4. Relationship between deflection and plate height

Fig. 5 shows the relationship between deflection and thickness of bottom plates. With the increase of bottom plate thickness, the maximum deflection and the deformation non-uniformity are effectively reduced. When the bottom plate thickness is 12 mm, the maximum deformation of middle plates is close to that of side plates. A thicker bottom plate enhances the flexural stiffness in the longitudinal and transverse direction. Therefore, the maximum deflection and the non-uniformity of deformation are reduced.

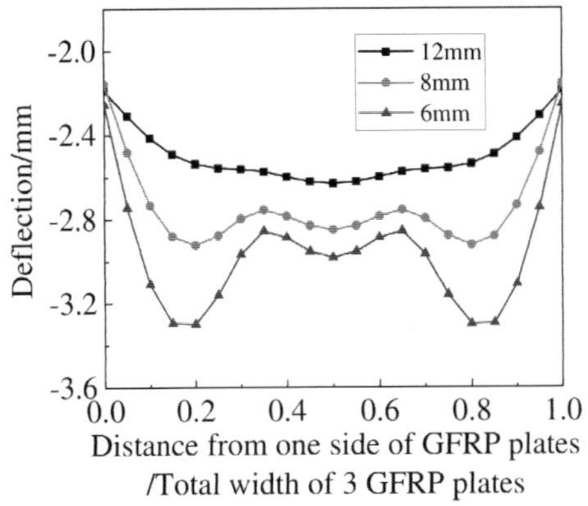

(a) Deflection distribution

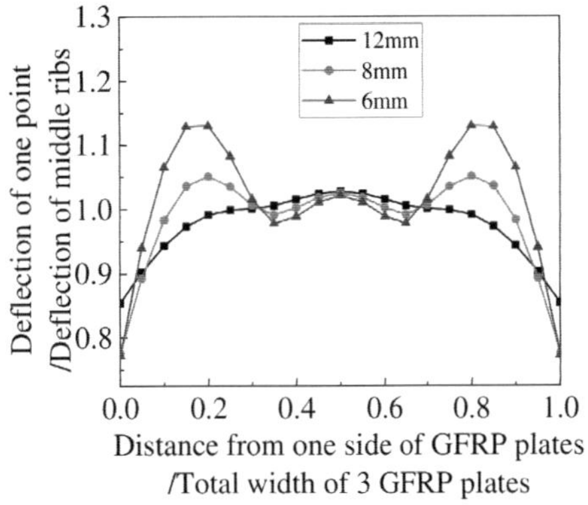

(b) Relationship between bottom plate thickness and non-uniformity of deformation

Figure 5. Relationship between deflection and thickness of bottom plates

Fig. 6 shows the relationship between deflection and rib spacing. The spacing between ribs has a significant effect on maximum deformation and deformation non-uniformity. As the average flexural stiffness of GFRP plates is increased with the decrease of rib spacing, the maximum deflection decreases. In addition, it can be seen from Fig. 6(b), the non-uniformity of spread GFRP plates is more significant than that of narrow GFRP plates, and the reduction of rib spacing could effectively improve the uniformity of deflection.

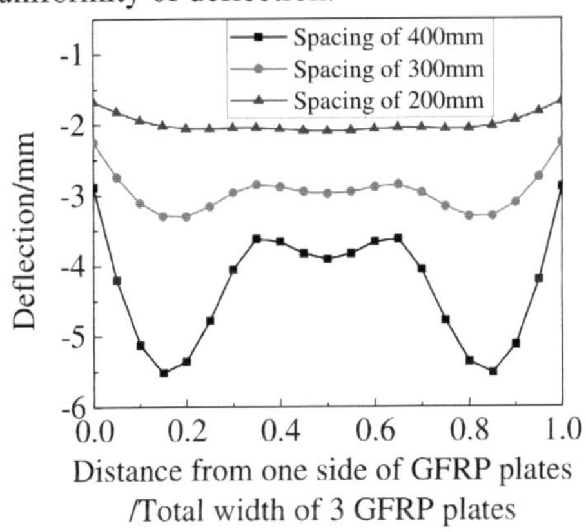

(a) Deflection distribution

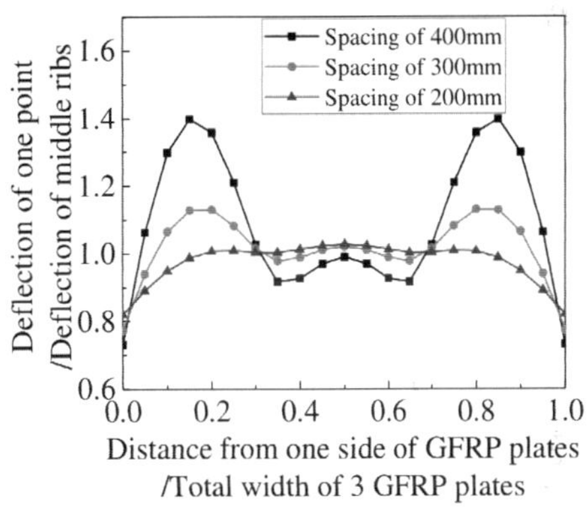

(b) Relationship between rib spacing and non-uniformity of deformation

Figure 6. Relationship between deflection and rib spacing

In general, the deflection non-uniformity of GFRP plates with a wider rib spacing is more remarkable than that of narrow GFRP plates. Although the increase of rib thickness, plate height, bottom plate thickness and the decrease of rib spacing could reduce the maximum deformation, different parameters have different effects on the deformation uniformity. The deformation uniformity is mainly affected by bottom plate thickness and rib spacing.

Based on the previous analysis, the following measures could be considered in the design and construction to improve the longitudinal and transverse stiffness of GFRP plates: 1) The side edge of GFRP plates could be placed on the top surface of cross beams. The side support could reduce the deformation of side plates. 2) The deformation uniformity in the deck transverse direction could be increased by increasing the thickness of GFRP bottom plates or reducing the spacing between GFRP ribs. 3) These methods could reduce the maximum deformation of GFRP plates, which include the increase of rib thickness, plate height, bottom plate thickness and the decrease of rib spacing.

4. Conclusions

1. The maximum deflection of side GFRP plates is larger than that of middle plates. A higher GFRP plate is beneficial to the total flexural stiffness, but it is detrimental to uniformity of deformation in the deck transverse direction.
2. The rib thickness, plate height, bottom plate thickness, and rib spacing have an significant influence on the maximum deflection of GFRP plates in the construction stage. The deflection could be reduced by increasing the rib thickness, plate height, and bottom plate thickness. In addition, a narrower GFRP plate results in a lower deflection.
3. The deflection non-uniformity of spread GFRP plates is more remarkable than that of narrow GFRP plates. The non-uniformity of deflection in the deck transverse direction is mainly affected by the bottom plate thickness and the spacing between T-ribs.

Acknowledgments
This research is funded by the Project funded by China Postdoctoral Science Foundation(Project No. 2019M653085) and financial support provided by China Communications Construction Company Ltd. (Program No. 271400140114).

References
[1] Tong Z J, Song X D and Huang Q 2018 *Steel Compos. Struct.* **26** 595-606.
[2] Tong Z J, Song X D and Huang Q 2019 *KSCE J. Civ. Eng.* **23** 3397-3408.
[3] Cho K, Park S Y, Kim S T, Cho J R and Kim B S 2013 *Compos. Part B-Eng.* **44** 679-685
[4] Fam A and Nelson M 2012 *J. Compos. Constr.* **16** 110-117
[5] Zuo Y, Liu Y and He J 2018 *Constr. Build. Mater.* **158** 728-741
[6] Bank L, Oliva M, Bae H, Barker J and Yoo S 2007 *Adv. Struct. Eng.* **10** 525-535
[7] Alagusundaramoorthy P, Harik I E and Choo C C 2006 *J. Bridge Eng.* **11** 384-393
[8] Cheng L J 2011 *Compos. Struct.* **93** 2895-2902
[9] He J, Liu Y Q, Chen A and Dai L 2012 *Constr. Build. Mater.* **26** 49-64

Experimental analysis of the addition of rice husk ash to the clayey subgrade of a road stabilized with lime

S Vizcarra[1], I Lujan[1], M Soto[1] y G Durán[1]

1 Faculty of Civil Engineering, UPC, Lima, Perú
E-mail: u201523064@upc.edu.pe/u201420769@upc.edu.pe/pccimsot@upc.edu.pe/
gary.duran@upc.pe

Abstract. There are many studies about how the addition of lime and rice husk ash (RHA) gives the soil a better mechanical behavior, particularly on clayey soils, where usually fine particles reach more than 75%. However, the soils with a small presence of fine particles (59-60%) do not have much research. This analysis evaluates the influence that RHA has on this kind of soil stabilized with 3% of lime. After the initial mix of soil-lime, CBR increased 11.2 times its initial value; within the addition of the ash, the CBR averaged between 45-50% up until 28% of RHA was added, where the results decreased considerably. Soil workability improved and the specimens with more ash resulted in a more granular material, with a group index value 0 following the AASHTO standards. The greatest CBR record was obtained with the specimen of 16% RHA, 3% lime and soil, reaching a 51.3% CBR, 1.58g/cm3 of MDD and 16.5% of OMC. Yet, it only showed a 1.55% more resistance than the lime-soil specimen. The CBR with more presence of RHA tends to decrease its value, therefore for silica-rich clayey soils, the addition of lime by itself should be enough for an adequate performance.

1. Introduction

There are currently a variety of problems facing civil engineering in terms of road infrastructure: cuts, landfills, earthworks, land stabilization, etc. As far as stabilization is concerned, this often turns out to be expensive because of the materials used to achieve adequate resistance of the ground. If stabilization is avoided, there is a risk that settlements will occur due to the mobile loads acting on a road [1].

Peru is a developing country where there is no adequate road infrastructure in many of its provinces. Focusing on the jungle region, it has more than 60% of unpaved roads. In the northeast of the country, 80% of unpaved roads are registered for the transit of vehicles. Studies conducted by the National Institute of Civil Defense reported the presence of soils with a high percentage of fines, to more granular clayey soils, CL according to SUCS classification [2].

There are procedures to stabilize a soil for road construction. Techniques such as physical and chemical stabilizations with additions of cement, asphalt, and natural pozzolanas like pond ash, coconut shell ash, and polymers are known [1, 3, 4, 7, 8]. However, the additional information that has been gaining strength in recent times is the rice husk ash (RHA) which use is becoming popular for roads construction.

The production of 80 tons of rice, generates 20 tons of rice husk, which can be reduced by up to 15% after incineration producing RHA [3]. It is well known, that the RHA is rich in silica, bordering 90% of its composition, which is a higher concentration of SiO_2 than fly ash [4, 9, 10]. This characteristic of the ash shows its benefits when mixed up with CaO, for the pozzolanic reaction that occurs when both silica and calcium are combined. However, it is important to mention that these pozzolanic reactions show more effectiveness in long terms. In fact, the immediate reaction is the cation exchange between soil and lime which leads to flocculation, lime carbonatation and a molecular restructuration of the soil [3, 5, 6].

Previous research reminds us of the enhancement in the workability of the soil and its resistance, produced by this type of combination (lime-RHA), also accompanied by cement and fibers [5]. Rice husk ash is found in abundance in the northern region of the Peruvian jungle, which is characterized as the largest rice producer in the country [14]. The combination of lime and ash in clayey soils of the area needs to have more research due to the increasing generation of this industrial waste product which is leading to health issues at the zone. In addition to its easy obtaining and abundance, soil stabilization using RHA reports an improvement in durability especially in extreme conditions such as those present in the Peruvian jungle [14].

This article shows the effects of RHA on lime stabilization of a clayey road with 49.1% of sand.

2. Experimental investigation

2.1. Materials used

Soil was obtained from a clayey road located at Tarapoto-Peru, this place is mainly known for its rainy days and high temperatures along the year. Late researches lead by the local National University, designated that the land is predominantly CL, also obtaining SC, ML, MH according to USCS [2, 14]. Two samples where extracted from the place and the respective classification was made, according to de ASTM and AASHTO parameters, the soil in study was sample 2 which characteristics are shown in table 1.

Table 1. Physical characteristics of soil.

Basic characteristics	Sample 1	Sample 2
Color	Dark grey	Dark brown
Depth (m)	2.0m	2.5m
Natural water content (%)	7	5
Passing 80μm sieve (%)	47.3	50.9
Liquid limit (%)	39	40
Plastic limit (%)	17	16
Plasticity index (%)	22	24
USCS Classification	SC	CL
AASHTO Classification	A-6 (4)	A-6 (5)

Table 2. Combination of samples

ID	Sample mixture (%)		
	Soil	RHA	Cal
R0C0	100	0	0
R0C3	97	0	3
R11C3	86	11	3
R16C3	81	16	3
R22C3	75	22	3
R28C3	69	28	3

The ash was obtained from a local rice miller, which incinerates the husk in standardized ovens. Finally, the lime used for stabilization was hydrated lime at 65% [11]. This product, accepted by the country's regulations for soil stabilization, can be obtained in the local market. Mineralogical characterization of the materials was performed as shown in tables 3 and 4. LOI indicates the loss due to the ignition generated in the X-ray fluorescence assay.

2.2. Stabilized soil combinations

For the analysis, the percentages of rice husk ash were 0, 11, 16, 22 and 28%. Lime percentage had a fixed value of 3% throughout the study, supported as an amount accepted by local road regulations [12, 13]. The details of the mixtures used are shown in table 2. A total of 6 samples were studied with the compaction and CBR test.

Table 3. Chemical properties of soil and RHA

Item	Formula	Sample 2 (%)	RHA (%)
1	SiO2	68.41	92.5
2	Al2O3	13.14	0.14
3	CaO	3.94	0.48
4	Fe2O3	3.32	0.39
5	K2O	1.26	1.9
6	MgO	1.14	0.6
7	TiO2	0.5	-
8	Na2O	0.26	-
9	SO3	0.12	0.07
10	P2O5	0.09	0.9
11	MnO	0.05	0.09
12	ZrO2	0.05	-
13	SrO	0.01	-
14	ZnO	0.01	0.01
15	Rb2O	0.01	-
16	Cl	0.01	0.02
17	LOI	7.7	2.9

Table 4. Hydrated lime properties

Item	Formula	%
1	Ca(OH)2	94.62
2	SiO2	2.50
3	MgO	1.50
4	CaCo3	1.00
5	Fe2O	0.20
6	Al2O	0.10
7	SO3	0.05
8	Cl	0.01
9	Other metals	0.03

2.3. Laboratory tests

For the classification of the soil and mixtures, the ASTM criteria were applied, using sieve granulometry tests and finding Atterberg's limits. Natural soil is characterized as a sandy clay (CL) according to the unified soil classification system (USCS) with 50.9% fine material, LL 40%, PL 16%, and PI 24%, with an A-6(5) AASHTO classification.

The compaction characteristics were obtained through the Modified Proctor test performed on all samples. Natural soil parameters were a Maximum Dry Density (MDD) of 1,826 g/cm^3 and an Optimum Moisture Content (OMC) of 14.80%.

The soil resistance characteristics were evaluated through the California Bearing Ratio test. Natural soil showed a 4.50% CBR. The procedure was performed on all samples to analyze the fluctuations in the CBR caused by the addition of RHA.

3. Results and discussion

3.1. Soil performance

Analyzing the compaction results shown in figure 1, we can see how the addition of ash causes the soil to require more water to reach its MDD. The OMC increases from 14.80% to 19.60%, between the natural soil and the R22C3 sample respectively. In the sample R16C3 -of maximum CBR- the values of MDD and OMC are 1.58g/cm3 and 16.50% respectively, in contrast to the mixture R0C3, the soil density decreases and more humidity is required to reach an approximate CBR of 50% as shown in figure 2.

Regarding the mixture R28C3 with the rest of the combinations used, the MDD continues to decrease in a controlled manner, however, the OMC increases drastically to 25.10%, generating a decrease in the CBR to 32.90%.

Flocculation between alumina and silica molecules with the calcium hydroxide creates an increase in the resistance due to the reactions between the lime and the soil's silica. With more silica added by the RHA, more pozzolanic reactions are available to react but this doesn't improve the performance significantly. The OMC of the R28C3 sample is due to the excess of water accumulated by the ash that hasn't reacted at all.

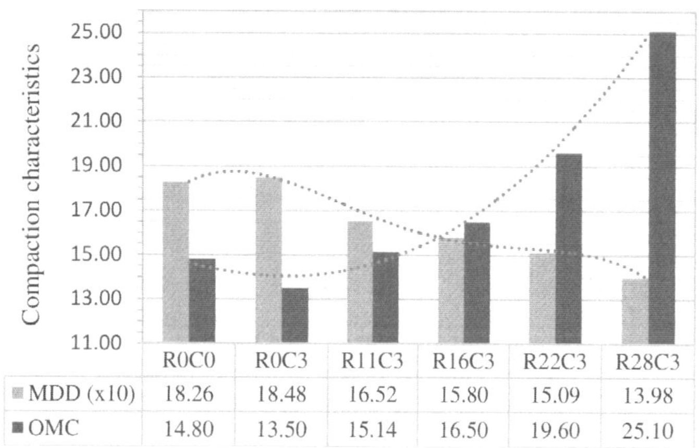

Figure 1. Effects of stabilizers on compaction parameters.

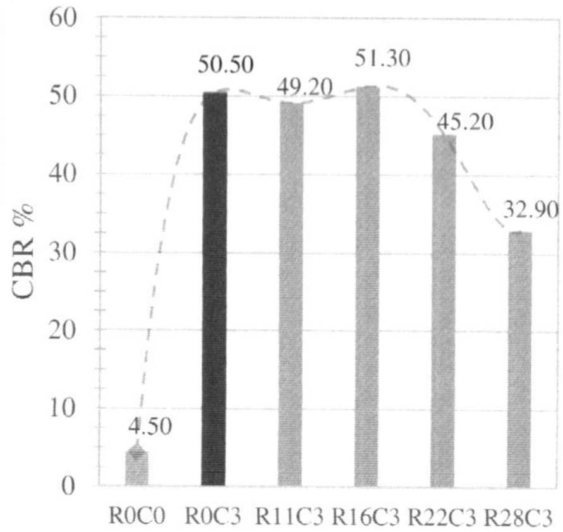

Figure 2. CBR of combinations.

3.2. Variation on Atterberg's limits

The results in figure 1 agree with the analysis of the limits shown in table 5. The R28C3 mixture shows an LL greater than 50%, which theoretically would be soil with high plasticity. However, it lacks on plasticity index and this is supported by the value of GI 0. The plasticity chart shown in figure 3 shows how the rest of the samples remain in the range of low plasticity soils.

The group index of the samples allows us to perform a more adequate analysis, showing that the soil tends to have thicker particles due to the ion exchange and the pozzolanic reactions that are caused by the additions.

Table 5. Atterberg's limits and soil classification

ID	LL	PL	PI	USCS	AASHTO
R0C0	40	16	24	CL	A-6 (5)
R0C3	40	25	15	CL	A-6 (3)
R11C3	43	31	12	ML	A-7-5 (2)
R16C3	46	33	13	ML	A-7-5 (2)
R22C3	46	NP*	NP*	ML	A-5 (0)
R28C3	67	NP*	NP*	MH	A-5 (0)

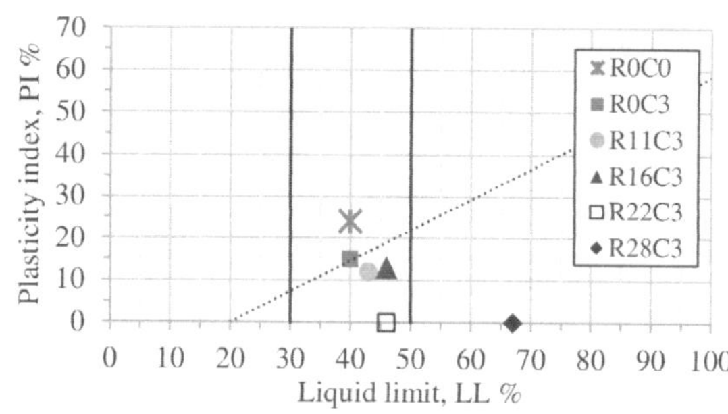

Figure 3. USCS plasticity chart.

4. Conclusions

The values of CBR decrease from the composition R16C3 onwards because the pozzolanic reactions between the CaO and the SiO_2 are stopped and the ash unreacted begins to remain in the internal structure of the compound. This goes side by side with the increase in OMC, since the ash absorbs more water, generating a gel that gives it resistance to the soil, but not as much as expected. The flocculation generated between the soil and lime generates an improvement of up to 11.2 times the value of the CBR at the beginning (R0C3) due to the internal restructuring that the soil and its silica-alumina compounds undergo. The addition of lime ash improves the durability of the soil and makes it easier to work by having more granular materials than at the beginning.

Consistency limits were 40% LL, 16% PL, 24% PI, resulting in a classification of sandy clay type A-6 (5) according to AASHTO. The GI decreases as more ash is added, indicating the low plasticity of the soil. The mineralogical analysis reflects that the clayey soil is rich in silica, so the addition of RHA does not imply a significant change in the SiO_2 percentage therefore, there was only a 1.55% improvement evidenced.

The MDD decreases from 1,826 g/cm^3 to 1,398 g/cm^3. On the other hand, the OMC increases from 14.80% to 25.10%, due to the flocculation generated by the addition of lime to the soil; The size of the nominal soil particles increases taking up more space and empty filling. At the same time, the pozzolanic reactions generated by the addition of rice husk ash cause the sample to absorb more water since this is essential for a pozzolanic reaction.

As well, the reduction in CBR from 22% of ash addition to the soil-lime combination demonstrates the absence of reactions and the excess ash remaining in the soil. However, pozzolanic reactions are progressive and improve over time, therefore the possibility is not ruled out that the CBR may improve in the future.

The results indicate that the ash has a low potential to be used as a stabilizer together with lime for silica-rich clay soils, since no improvement is demonstrated, on the contrary, the CBR decreases compared to the R0C3 combination. Even though, if resources are available, lime and ash can be used to decrease the permeability of the soil, giving it improved workability and generating its long duration for tropical climates where the soil is exposed to heavy rains.

In addition, the application of RHA in geotechnical projects adds value to this material, offering an environmentally appropriate destination for this type of waste. It is possible to carry out more tests with different dosages of lime especially evaluated in soils which percentage of material passing the mesh No. 200 is between 50%-40% since previous investigations have soil records with the presence of fines of up to 90%.

5. References

[1] Jagada J y Kulbir G 2006 *Journal of the Institution of Engineers (India): Civil Engineering Division* **87** 33–39
[2] INDECI 2004 *Mapa de peligros las ciudades de Tarapoto, Morales y La banda de Shilcayo* (San martín) p **1**
[3] Khelifa H, Mohamed G y Said K 2018 *Periodica Polytechnica Civil Engineering* **66** 1–10
[4] John S, Paul J y Chukwujama I 2018 *Journal of King Saud University – Engineering Sciences*
[5] Sujit K y Monowar H 2012 *J. Mater. Civ. Eng.* **24** 707–714
[6] Hali A, Adnan A y Chew K 1992 *Geotechnical and Geological Engineering* **10** 117 – 134
[7] Okagbue C y Yakubu J 2000 *Bulletin of Engineering Geology and Environment* **58** 107–113
[8] Al-Rawas A, Hago A y Al-Sarmi H 2005 *Building and Environment* **40** 681 – 687
[9] Manasseh J y Olufemi A 2008 *Electronic Journal of Geotechnical Engineering* **13** 1–12
[10] Hossain K, Lachemi M y Easa S 2007 *Resources Conservation and Recycling* **51** 711–731
[11] M y Calera Santa 2019 *Mineralogía de la composición de productos MyCal* p 1
[12] National Lime Association 2006 *Manual de estabilización de suelo tratado con cal-estabilización y modificación con cal* ed NLA pp 1-42
[13] Ministerio de Trasportes y Comunicaciones 2012 *Norma CE.020 Estabilización de suelos y taludes* (Lima) p 9-16
[14] Municipalidad distrital de El Eslabon 2016 *Mejoramiento del Servicio de Transitabilidad del camino vecinal: El Aguano, nuevo Huánuco, San José Bajo, Distrito de El Eslabon, Provincia de Huallaga – Región San Martin* pp 37-53

IOP Conf. Series: Materials Science and Engineering **758** (2020) 012091 doi:10.1088/1757-899X/758/1/012091

Proposal of Flowable Fill Designs for improvement of excavation and filling works of trenches in sanitation systems

J Cruz[1], J Ñiquin[1], I Bragagnini[2] and C Sotomayor[2]

1 Bachelor, Civil Engineering Program, Universidad Peruana de Ciencias Aplicadas, Lima, Perú
2 Full Professor, Civil Engineering Program, Universidad Peruana de Ciencias Aplicadas, Lima, Perú

Abstract. Population grow in recent years requires an extension of the current pipeline sanitary system. For this purpose, granular excavation and landfill works are associated with pedestrian traffic congestion. Therefore, it is necessary to develop an innovative and sustainable alternative to reduce the problems generated during the execution of the conventional process. This research proposes the use of flowable fill due to the multiple advantages offered by this material. On the one hand, it is economical for medium to large trench fill volumes, considering savings in labor (it is done with a small number of workers), in equipment (does not require the rental or purchase of compaction equipment) and in time (the pouring is done by directly pumping the mixture, from the mixing machines to the excavation). On the other hand, being self-compacting and self-leveling decreases the width of the trenches, reducing excavation and filling volumes; which, in turn, incur money savings. Also, this material guarantees work safety, since people are not required inside the excavation and fill in poorly accessible areas without any problem. Dosages were established for ten flowable fill mixtures with cement contents of 50, 60, 70, 80 and 90 kg of cement and a range of admixture from 1.75 to 2.00%; The results indicated that decreasing the fine aggregate - coarse aggregate ratio, the compressive strength of the mixtures increases and the slumps of the mixtures decreases, and the compressive strength increases directly proportional to the cement content.

1. Introduction

Conventional excavation and trench filling works for maintenance or installation of new pipes are a necessity to provide basic services for the population. Although conventional filling looks like a simple procedure, it demands a thorough control of the filling material, the condition of trench walls and the compaction process; to achieve the same soil conditions in their original state, and thus, avoid short and medium term slumps. Slumps that occur on soils under transit roads incur in cost overruns and traffic problems for vehicular and pedestrian traffic; due to excavation, filling and paving reworks [1].

Also, excavation works develop negative environmental impacts, since it is recurrent to leave the trenches open and expose them for a prolonged period. [1].

The flowable filling emerges as an alternative to avoid multiple problems which appears in excavation works and trench fillings, since it serves as a resistant and durable filling material for sanitation systems [2-5]. More importantly, it can be designed with an early strength able to support traffic loads within the first hours, after its pouring; and due to its self-compacting and self-leveling properties, ensures the alignment of the pipes over the time and avoids slumps possibilities [4,5,6].

2. Materials and Method

2.1 Materials

Lima Ordinary Portland Cement Type I was used; natural fine aggregate and 3/4” natural coarse aggregate from the “Jicamarca” quarry, with fineness modulus 2.93 and 6.73, respectively; superplasticizer (SP) and air entrainment (AE) admixture with a density of 1130 kg/m^3.

2.2 Method

The parameters considered for the realization of the mixing designs were cement content, superplasticizer and air entrainment admixture, the fine aggregate-coarse aggregate ratio, and the water-cement ratio as shown in Table 1.

Table 1. Parameters

Parameters	
Cement Content (kg)	50 - 90
SP - AE admixture (%)	1.75 - 2.00
F.A./C.A. ratio	2.33 - 4.00
W/C ratio	2.44 - 4.70

The properties of compressive strength, flowability, density and air content were evaluated according to technical specifications of the ACI 229R Committee as indicated in Table 2.

Table 2. Technical Specifications

Technical Specifications	
Flowability (in.)	8 - 12
Air Content (%)	8 - 15
Density (kg/m^3)	1800-2200
Compressive Strength (kg/cm^2)	5 - 14

Ten flowable fill mix designs were proposed show in Table 3. The water content varied in a range of 220 to 235 kg / m^3 and the superplasticizer and air entrainment admixture dosage in a range of 1.75 to 2.00% to obtain flow values ranging from 8 to 12 inches, a range necessary to ensure a fluid and self-leveling consistency of the mixture. From the point of view of practical construction, the most important property of the flowable fill is the capacity it has for self-compacting and self-leveling, and which can be conveniently used in trenches and narrow places where the use of labor and machines becomes complicated.

Table 3. Flowable fill Designs Chart

Summary Table of Flowable fill Designs to Prepare in Laboratory										
Mixture	M1	M2	M3	M4	M5	M6	M7	M8	M9	M10
Portland Cement Type I (kg)	50	50	60	60	70	70	80	80	90	90
Water (kg)	235	220	235	220	235	220	235	220	235	220
Fine Aggregate (%)	80	70	80	70	80	70	80	70	80	70
Coarse Aggregate (%)	20	30	20	30	20	30	20	30	20	30
SP - AE admixture (%)	2.00	2.00	2.00	2.00	2.00	2.00	2.00	2.00	1.75	1.75
Theorical Air (%)	10	10	10	10	10	10	10	10	10	10
F.A./C.A. ratio	4.00	2.33	4.00	2.33	4.00	2.33	4.00	2.33	4.00	2.33
W/C ratio	4.70	4.40	3.92	3.67	3.36	3.14	2.94	2.75	2.61	2.44

3. Results and Analysis

3.1. Flowability and Fine Aggregate – Coarse Aggregate Ratio

The flow values obtained at 0, 30, 60, 90 and 120 minutes for the mixtures made based on the fine aggregate - coarse aggregate ratios = 4.00 and 2.33, are shown in Figures 1 and 2, respectively.

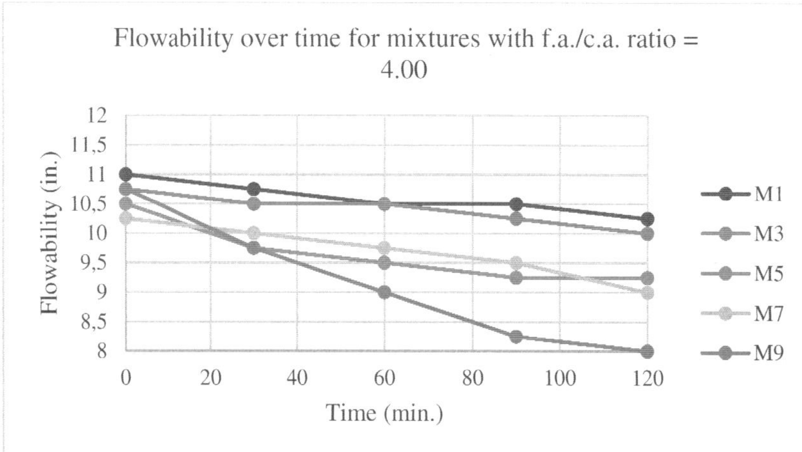

Figure 1. Flowability over time for mixtures in f.a./c.a. ratio=4.00

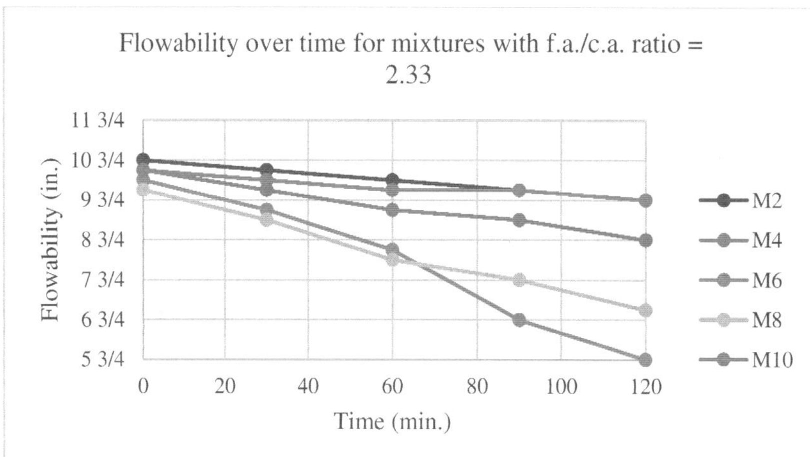

Figure 2. Flowability over time for mixtures with f.a./c.a. ratio = 2.33

The flow values obtained at 0, 30, 60, 90 and 120 minutes for the mixtures in relation to f.a./c.a.=4.00 varied from 11" to 8". and for mixtures in relation to f.a./c.a.=2.33 between 10 3/4" to 5 3/4".

All the mixtures with a f.a./c.a ratio = 4.00, met the flowability range established by ACI 229R committee (8" – 12"), ensuring the workability for the transport and pouring. However, in mixtures with a f.a./c.a. ratio = 2.33, reduced flowability values were observed. The lowest values were obtained for mixtures with low cement content and a f.a./c.a. ratio = 4.00, this prove that a high percentage of fine aggregate and a low cement content in the dosage increases the flowability of the flowable fill mixture.

3.2. Strength – Cement Content

The resistance results obtained at 2, 7, 14 and 28 days for the mixtures made based on the cement content (50, 60, 70, 80 and 90 kg) in the dosage is shown in Figure 3.

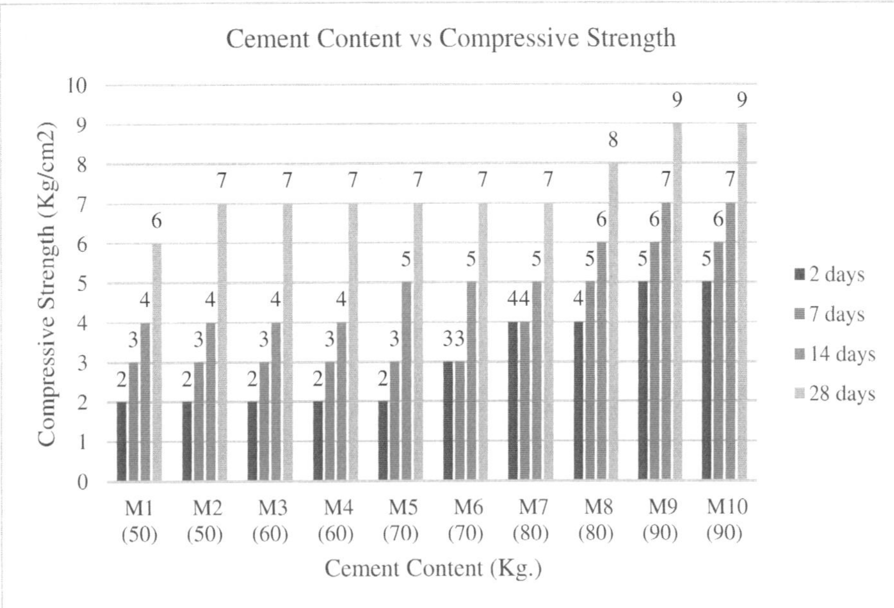

Figure 3. Cement Content vs Compressive Strength

The mixtures with 50 kg, 60 kg and 70 kg of cement content in their dosing developed a compressive strength of 2 to 7 kg / cm^2 in 28 days, mixtures with 80 kg of cement content in their dosing developed a resistance at compression of 4 to 8 kg / cm^2 in 28 days and mixtures with 90 kg of cement content in their dosage developed a compressive strength of 5 to 9 kg / cm^2 in 28 days.

The M6 and the mixtures with cement content in its dosage of 80 and 90 kg developed a resistance to compression at 2 days between 3 and 5 kg / cm^2, they reached the resistance of a natural soil that fluctuates between 3 and 7 kg / cm^2, which guarantees a rapid release of traffic after emptying the excavation.

All the mixtures developed a compressive strength at 28 days between 6 and 9 kg / cm^2, so it meets with the range established by the ACI 229R Committee (5 to 14 kg / cm^2), and ensures an easy removal of the material in case there is any change or failure in the sanitation system.

Figure 4. Compressive Strength Test

Figure 5. Slump Test

4. Conclusions

- The compressive strength not confined at 28 days of the mixtures varied from 6 - 9 kg / cm^2, meeting with the range established by the ACI 229R Committee (5 to 14 kg / cm^2). Higher 28-day strength values were achieved for mixtures with cement content in their 90 kg dosage.
- The flowability values of the mixtures with f.a./c.a. ratio = 4.00 varied from 8 - 11", meeting the range of flowability established by the ACI 229R committee (8 - 12").
- The lowest flowability values were achieved for mixtures with low cement content and f.a./c.a. ratio = 4.00.
- The mixtures with cement content in their dosage of 80 and 90 kg developed, at 2 days, a compressive strength between 3 - 5 kg / cm^2, they reached the resistance of a natural soil.

5. References

[1] ACI 229R-13: Report on controlled low-strength materials, ACI Committee 229, American Concrete Institute (ACI), Farmington Hills, MI, 2013.
[2] Naik T and Singh S 1997 Flowable slurry containing foundry sands *J. Mater. Civ. Eng.* **9**(2) 93–102.
[3] Naik T and Singh S 1997, Permeability of flowable slurry materials containing foundry sand and fly ash *J. of Geotech. And Geoenv. Eng.* **123**(5) 446-452.
[4] Dockter B 1998 Comparison of dry scrubber and class c fly ash in controlled low strength materials (CLSM) applications *The Design and Application of Controlled Low-Strength Materials (Flowable Fill),* ed. A. Howard and J. Hitch (West Conshohocken, PA: ASTM International) pp.13-26. https://doi.org/10.1520/STP13059S.
[5] Kaneshiro J, Navin S, Wendel L and Snowden H 2001 Controlled low strength material for pipeline backfill—specifications, case histories and lessons learned *Pipelines 2001: Advances in Pipelines Engineering and Construction, Pipeline Division Specialty Conference* pp. 1–13. 10.1061/40574(2001)53.
[6] Abelleira A, Barke N and Pickering D 1998 Corrosion activity of steel in cementitious controlled low-strength materials vs. that in soil, in: A.K. Howard, J.L. Hitch (Eds.), *The Design and Application of Controlled Low-Strength Materials (Flowable Fill),* ed. A. Howard and J. Hitch (West Conshohocken, PA: ASTM International) pp. 124-134.
[7] Trejo D, Folliard K and Du L 2004 Sustainable development using controlled low-strength material *Proc. of Intern. Workshop on Sust. Develop. and Concr. Techn.* 231–250.
[8] NRMCA. Guide specification for controlled low strength materials (CLSM). Report of national ready mixed concrete association, Specification Guide, 2006.

IOP Conf. Series: Materials Science and Engineering **758** (2020) 012092 doi:10.1088/1757-899X/758/1/012092

Effects of Steel Bracings in the Progressive Collapse Resistance of Reinforced Concrete Building

Bikram Shah[1] and Feng Xu[2]

1 Master's student, Shanghai Jiao Tong University, China
2 Asst. Professor, Shanghai Jiao Tong University, China
e-mail: bikram025@sjtu.edu.cn , f.xu@sjtu.edu.cn

Abstract. The bracing system used in the seismic retrofitting can also be used as a measure to resist the progressive collapse of the multi-storeyed structures. The nonlinear dynamic analysis is used to study the effects of floor wise and bay wise bracing systems in the progressive collapse resistance of seismic designed reinforced concrete buildings. The numerical model is created in SAP2000 and is analysed for three different column removal scenarios using UFC guidelines. The results showed the significant contribution of the bracing system to prevent the progressive collapse due to single removed column. The bay wise bracing system is more effective to reduce the deformations but there is no contribution of the bracings when the column is removed from the non-adjacent bay. The floor wise bracing is more reliable than bay wise bracings.

1. Introduction

The propagation of the failure from the critical members to the other members in a progressive manner leading to failure of the large proportion or complete collapse of the structure is called progressive collapse [4]. The collapse of the twin tower by the terrorist attack of world trade center in 2011 has created a great interest in the study of progressive collapse and in the design of the robust structures which can stand safe and stable even when some of the member fails.

The steel bracings are the most common retrofitting measure in lateral load resisting system. It provides additional stiffness by transferring tensile or compressive forces in the alternating cycles of load. Similarly, it can provide additional robustness by forming a truss like bridge across the missing elements and help in stress redistribution to the larger number of elements. The bracings decrease the stress of the horizontal members and prevent the initiation of the progressive collapse chain.

The General Services Administration [1] and Unified Facilities Criteria 4-023-03 [2] are the most commonly used guidelines for the progressive collapse analysis. Both guidelines use alternate load path method and the analysis could be done using linear and nonlinear static and dynamic method. Many researchers have studied the effectiveness of bracing in progressive collapse resistance of steel moment frames [5-7]. Similarly, the most of the studies of progressive collapse of reinforced concrete shows the resistance is provided mainly due to compressive arch action, tensile catenary action or tensile membrane action [9]. There are only few studies showing the contribution of steel bracings in reinforced concrete buildings. Qian et al. [9] experimentally studied the one quarter model of 2 bays, 3 story sub-frame of seismically designed eight storied building with different bracing configurations. The experiment showed that the eccentric cross bracing have higher dynamic load capacity while V braces and reversed V braces configurations have mild effect in improving the load resisting capacity. However, the study of the global behaviour of the braced reinforced concrete building in resisting progressive collapse is limited.

This paper aims to study the capability of the seismically designed and retrofitted structures in resisting progressive collapse. The nonlinear dynamic analysis of the eight-story seismically designed steel braced reinforced concrete frame is carried out to show the contribution of braces.

IOP Conf. Series: Materials Science and Engineering **758** (2020) 012092 doi:10.1088/1757-899X/758/1/012092

2. Analytical Model

Three dimensional reinforced concrete buildings are modelled and analysed by commercial finite element software SAP2000. The progressive collapse analyses of eight-story braced and unbraced buildings were carried out using nonlinear dynamic analysis prescribed by UFC guidelines [2]. The building was designed for the low seismic zone and the design parameters are shown in the table 1 and 2. Three types of column removal scenarios (a) corner column removal (b) Middle Column of shorter side and (c) Middle column of longer side are considered, one at a time as shown in the figure 1.

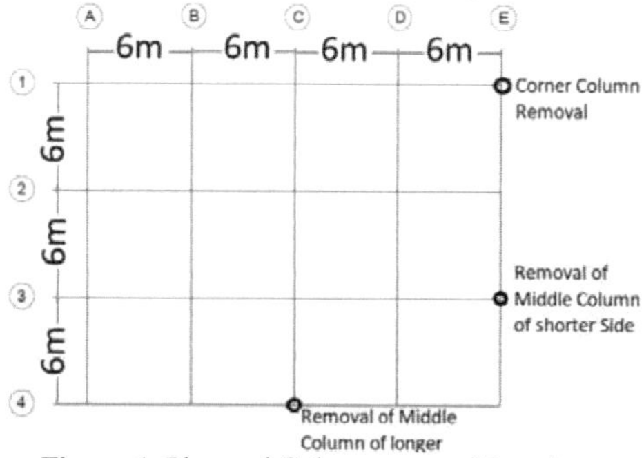

Figure 1. Plan and Column removal locations

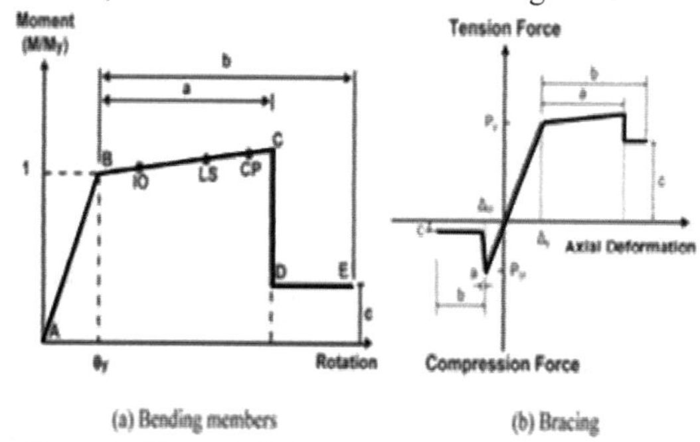

Figure 2. Hinge definitions for beams and bracings[11]

All the beams and columns are modelled by 2-noded beam elements with the beam column connection being moment resisting. The bracing members are added as pinned connection on both ends. The support condition at the foundation is modelled as fixed connection. The effect of membrane action of the reinforced concrete slab is not considered. The structure is designed as per the concept of strong column and weak beam. So, the column section remains elastic even after the load redistribution due to loss of column. Also, the plastic hinges are allowed to form in the beam body only. At large rotations, several cracks occur in the reinforced concrete beam section. This reduces the effective moment of inertia (I). So, the moment of inertia of the beams is taken as 0.3I as indicated in Table 10-5 of ASCE 41-13 [3].

Table 1 Design Parameters

Element	Size(mm)	Reinforcement
Beams	500 x 300	Top 3-16dia Bottom 3-12dia
Columns	500 x 500	8-16dia
Bracings (Tube)	HSS4-1/2x4-1/2x0.5 (inch)	Fe345 grade steel

Table 2 Structural Loadings

Live Load	3KN/m^2
Roof Live load	1.5KN/m^2
Wall load	10KN/m
Slab Load	3.125kN/m^2

Note: The dead loads of the modelled elements like beam, column and brace are taken automatically by SAP2000

The modulus of elasticity and compressive strength of concrete is taken as 25000 Mpa and 25 Mpa respectively. The yield strength of reinforcement is taken as 500 Mpa. The material non linearity is included in the model by defining the concentrated plastic hinges [3]. The nonlinear hinges of beam and braces are defined in the form of normalized force displacement curve (moment rotation curve for beam) as shown in the figure 2a and 2b. Points A, B, C, D and E represents the origin, yielding point, the ultimate capacity, residual strength and total failure (square shaped hinges shown in the figure) respectively. The points IO (Immediate Occupancy), LS (Life safety) and CP (Collapse Prevention) in line BC are the acceptance criteria and are considered acceptable by the UFC guidelines [2]. For the geometric nonlinearity, P-delta effects plus large displacements option available in SAP2000 is used [10]. In nonlinear dynamic collapse analysis, the simulation of sudden removal of column is done by replacing the column by its internal forces resulting from ASCE 7 load combination 1.2DL+0.5LL as

IOP Conf. Series: Materials Science and Engineering **758** (2020) 012092 doi:10.1088/1757-899X/758/1/012092

reaction and then this reaction force is suddenly dropped to zero in 0.01 second by using the time history function in SAP2000.

3. Results

The dynamic analysis of the unbraced frames showed that the hinges of the beams in the affected spans, reach the failure point E (square shaped hinge) as shown in the figure 4 in all the cases of column removal scenario. So, the building designed for low seismic zones are vulnerable to progressive collapse. The failure of one column leads to the collapse of all the members in the affected span. Hence, steel bracings are used as the remedy to provide resistance against the progressive collapse. Two types of bracing systems considered are (1) Floor wise bracing and (2) Bay wise bracing as shown in the figure 5 and figure 6 respectively.

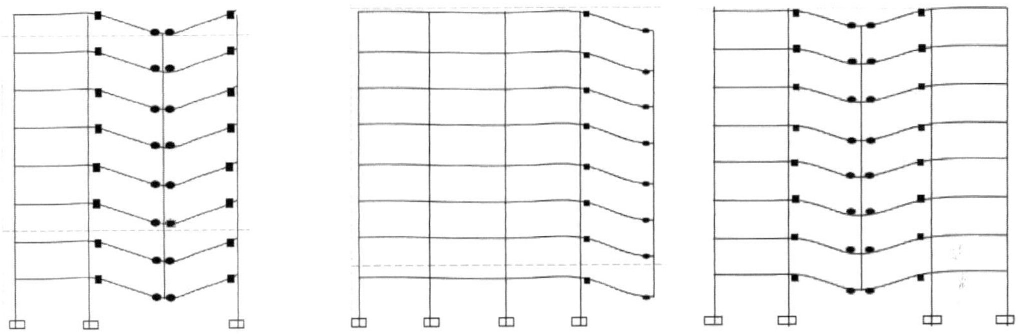

a) Middle Column of short side (b)Corner column removal c) Middle Column of long side

Figure 3. Deformed shape and Plastic Hinge states for Unbraced Frame

3.1. Floor wise bracing system

In all the cases of column removal scenario, the floor wise bracing system is capable to strengthen the building to resist the progressive collapse although few bracing members fail by buckling in compression. The plastic hinges formed in the beams are within the immediate occupancy (IO) range which meets the required acceptability limit defined by UFC guidelines.

The sudden removal of column creates the dynamic overload of the structure due to which the column removal point displaces downward to the peak displacements and then it oscillates and finally stabilizes to the fixed permanent deflection. The time histories of the displacements are shown by dotted line in figure 7(a), 7(b) and 7(c) for three cases of column removal. The peak displacements for corner column removal, middle column removal for shorter and longer side are -27.1mm, -32.2mm, and -31.1mm respectively.

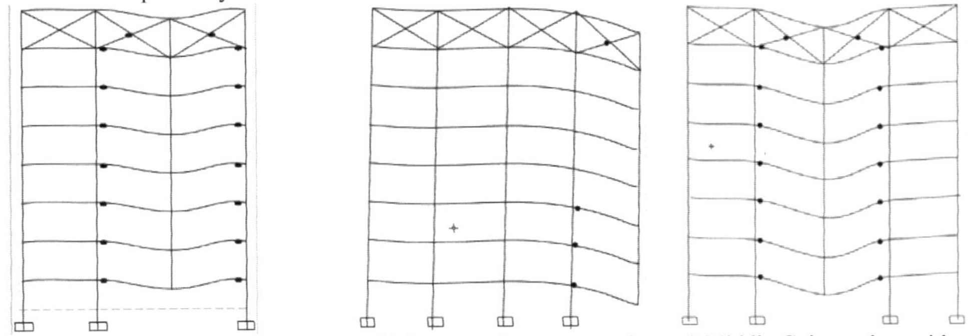

a) Middle Column of short side (b)Corner column removal c) Middle Column long side

Figure 4. Deformed shape and plastic hinge state for the floor wise bracing system

IOP Conf. Series: Materials Science and Engineering **758** (2020) 012092 doi:10.1088/1757-899X/758/1/012092

3.2 Bay wise bracing system

As shown in the figure 6(a) and 6(b), the bay wise bracings are capable of resisting the progressive collapse with the failure of few compression braces. The plastic hinges do not form in the beams and the displacements of the column removal point are relatively smaller than that of the floor wise bracing systems. The time histories of the displacements are shown by dashed line in figure 7(a), 7(b) and 7(c) for three cases of column removal. The peak displacements for corner column removal and middle column removal of shorter side are -11.8mm and -22.12mm respectively.

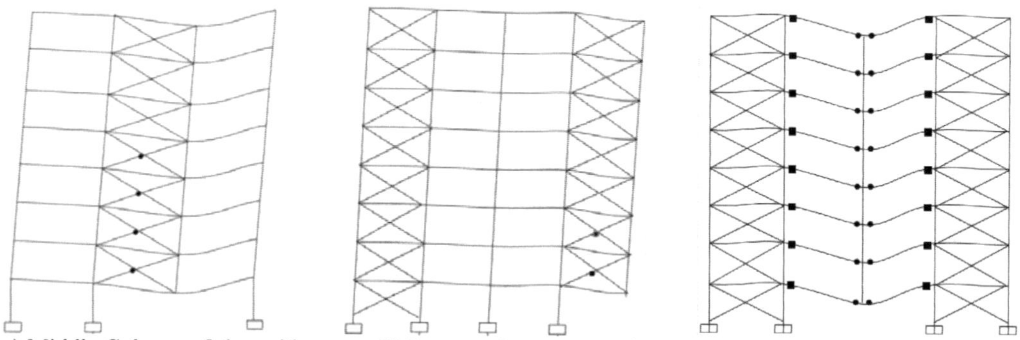

a) Middle Column of short side (b)Corner column removal c) Middle Column of long side

Figure 5. Deformed shape and plastic hinge state for the Bay wise bracing system

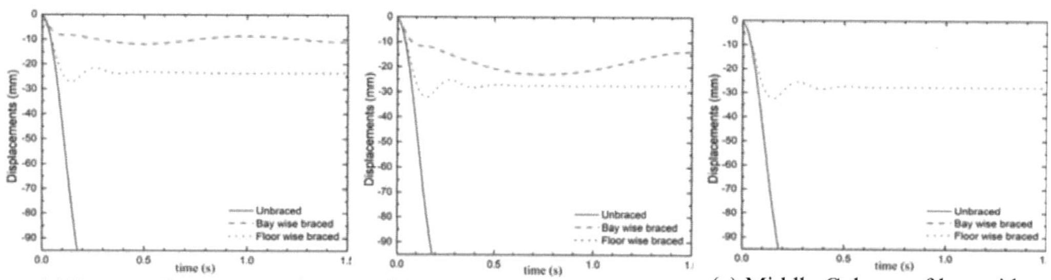

(a)Corner column removal (b) Middle Column of short side (c) Middle Column of long side

Figure 6. Comparison of time history for vertical displacement of column removal point

However, the removal of middle column of longer side shows the occurrence of progressive collapse as shown in figure 5(c)(square shaped hinges represent beam failure) because there is no bracing member near the removed column to provide the alternate path for the force redistribution.

4. Conclusion

a. The bracing systems are effective in reducing the deformations of the members significantly. So, the bracing system used for the seismic retrofitting can also help to prevent the progressive collapse.
b. The displacement is the least for the bay wise bracing configuration. But it is ineffective when the middle column of the longer side is removed.
c. The floor wise bracing reduce the deformations with lesser number bracing members. This type of bracing is more reliable because it has significant contribution in the progressive collapse resistance in all the cases of exterior and corner column removal. The deformations could be further reduced by adding such bracings to two or more floors (top and intermediate floors).

d. The number and severity of plastic hinges are greatly reduced when the bracings are included in the frame. So, the bracing system can be useful for the older building designs with insufficient seismic detailing.[9]

5. References

[1] General Service Administration. Alternate path analysis & design guidelines for progressive collapse resistance (GSA 2013). washington (DC).

[2] Department of Defence. Design of buildings to resist progressive collapse (UFC 4-023-03). washington (DC).

[3] American Society of Civil Engineers. Seismic rehabilitation of existing buildings (ASCE 41- 13). new york (NY).

[4] American Society of Civil Engineers. Minimum design loads for buildings and other structures (ASCE 7-10). new york (NY).

[5] Khandelwal K, El-Tawil S and Sadek F (2009) Progressive collapse analysis of seismically designed steel braced frames. Journal of Constructional Steel Research 65(3): 699–708.

[6] Chen J, Peng W, Ma R, et al. (2011) Strengthening of horizontal bracing on progressive collapse resistance of multistorey steel moment frame. Journal of Performance of Constructed Facilities 26: 720–724.

[7] Kim J, Lee Y and Choi H (2011) Progressive collapse resisting capacity of braced frames. The Structural Design of Tall and Special Buildings 20: 257–270.

[8] K. Marchand, A. McKay, and D. J. Stevens. Development and application of linear and non-linear static approaches in UFC 4-023-03. In Proceedings of the 2009 Structures Congress, pages 1–10.

[9] K. Qian, Y.H. Weng, B. Li, Improving Behavior of Reinforced Concrete Frames to Resist Progressive Collapse through Steel Bracings, J. Struct. Eng., 2019, 145(2): 04018248

[10] SAP2000 (Version 18) [Computer software]. Berkeley, CA: Computers and Structures, Inc.

[11] Kim, J., & Park, J. (2014). Design of special truss moment frames considering progressive collapse. International Journal of Steel Structures, 14(2), 331–343. doi:10.1007/s13296-014-2013-1

IOP Conf. Series: Materials Science and Engineering **758** (2020) 012093 doi:10.1088/1757-899X/758/1/012093

Low Permeability Concrete for *Buildings Located in Marine Atmosphere Zone* using Clay Brick Powder

M Castillo[1], K Hernández[1], J Rodríguez[2] and C Eyzaguirre[2]
1 Research Assistant, Civil Engineering Program, Universidad Peruana de Ciencias Aplicadas, Lima, Perú
2 Full Professor, Civil Engineering Program, Universidad Peruana de Ciencias Aplicadas, Lima, Perú
Email: u201410388@upc.edu.pe

Abstract. The concrete is not one hundred percent impermeable since the water that remains inside it causes its corrosion, in the case of reinforced concrete, exposed in an area of marine atmosphere, the sea salt mostly present in large particles of the marine spray, produce the reduction of the alkalinity of the concrete causing a rapid corrosion of the steel. There are buildings built in this marine area that have been designed without durability criteria, in which the use of pozzolanic materials is considered, for example, to fill the pores of the cement matrix and thus guarantee its impermeability. In the present study, the effect of clay brick powder (PLA) as a replacement for cement in concrete manufacturing is addressed, evaluating different characteristics of its components. The results indicate that pozzolanic activity and compressive strength increase, slump, voids content and the coefficient of permeability to water decreases.

1. Introduction

A durable concrete is one that has a dense structure with minimal total porosity and low permeability [1]; this being important in the concrete exposed to the water of the sea, because it delays the effect of sulfate attack and guarantees a minimum coating of steel [2]. The marine environment has different exposure areas of expure that influence the deterioration of concrete structures, being the marine atmosphere where the marine aerosol is formed and there is produces an increase in the concentration of salts that are transported inland [3]; these salts are less than 10 μm (floating and don't settle) and greater than 10 μm (precipitate), these being the ones that cause corrosion of the steel located a few hundred meters to the inland of the coast [4]. To minimize this corrosion, low permeability concrete made with calcined clay is used as a partial cement replacement [5]. Clay brick is used in massively in small and large-scale construction projects, with the worldwide annual production of 1391 billion [6]. In its, operations are carried out with high energy consumption, very polluting and harmful to the environment [7], recycling is done in small quantities [8]; some of the important environmental impacts tare the reduction of the natural soil level [9] and the generation of solid waste [10]. The use of cement has improved living conditions for humans, but it has also brought tremendous pressure on the energy supply and ecological environments [11], its production in 2018 was 4.1 billion metric tons [12] and during its production there are different air emissions that pollute the environment, such is the case for example of CO_2 that is responsible for global warming [13], [14]. Given this situation, an interesting alternative is to use PLA because to having a high porosity, high water absorption and is a used as an addition to cement [15]. The objective of this work is to study the influence of the percentage of PLA as a replacement for cement for the construction of buildings built in an area of marine environment addressing the study of pozzolanic activity, slump, compressive strength, voids content and coefficient of permeability to water.

2. Materials and Method

2.1 Materials

The cement is Andino Ultra HS; brick dust (PLA), its composition is shown in Table 1; the natural aggregates thick (19.5 mm) and fine are from the "Laura" quarry, their physical properties are shown in Table 2; the additive used was a Superplasticizer-Water Reducer; And the water was drinkable.

2.2 Method

For the pozzolanic activity index, 5 cm x 5 cm mortar cubes were prepared with 450 g of Sand and 112.5 ml of Water, using: for the MPLA-20 sample, 168.75 g (80% cement), 56.25 g (20% PLA); and for the MPLA-0 sample, 225 g (100% cement), for the calculation of the pozzolanic activity index. The mix design was made for a concrete f'c = 300 Kg / cm^2 according to [16]. 5 dosages were made with 0%, 5%, 15%, 25% and 30% of PLA, called CPLA-0 (standard mixture), CPLA-5, CPLA-15, CPLA-25 and CPLA-30 shown in the Table 1; the specimens after processing were demolded after 24 hours (Fig. 1) and then tested according to Table 2.

Table 1. Concrete mix design with f'_c= 300 kg/cm^2

Mix	Quantities in Kg for each cubic meter						
	W/C	Replacement PLA	Cement	Sand	Stone	Water	Super-plasticizer (ml)
CPLA-0	0.5	-	13.237	23.847	24.986	6.624	-
CPLA-5	0.512	0.962	12.575	23.588	24.715	6.443	151
CPLA-15	0.514	2.286	11.252	23.041	24.171	5.781	180
CPLA-25	0.516	3.609	9.928	22.523	23.628	5.119	200
CPLA-30	0.517	4.271	9.266	22.264	23.356	4.788	204

Table 2. Test performed

Test type	Days	Method	Measurement of samples	N° of specimens for test
Pozzolanic activity index	7, 28	[17]	5 cm x 5cm	3
Slump	-	[18]	-	-
Compressive strength	7, 14, 28	[19]	15 cm x 30 cm	3
Voids content	60	[20]	15 cm x 50 cm	1
Coefficient of permeability to water	60	[21]	15 cm x 20 cm	1

3. Results and Analysis

3.1. Pozzolanic Activity Index

Table 3 shows the effect of cure time on the pozzolanic activity index (IAP) for the MPLA-20 sample. It is seen that when the cure time varies, the IAP increases, reaching the value of 77.83% at 7 days and 171.08% at 28 days, giving an increase of 54.51% in relation to 7 days. As the IAP values for 7 and 28 days are greater than 75%, we conclude that MPLA-20 is a pozzolan. [22], evaluates the replacement of 20% of PLA in mortar, finding that the IAP for 7 and 28 days presents percentages of 100% and 101%, stating that it is a pozzolanic material (> 75%); The same author concludes that the content of PLA is sufficient for the hydration of the cement, that is, the formation of calcium hydroxide, silicate and aluminate improve the resistance properties. [23] also evaluates the replacement of 25% of PLA in the mortar finding that after 28 days the IAP is 100%, being considered a pozzolanic material (> 75%).

IOP Conf. Series: Materials Science and Engineering **758** (2020) 012093 doi:10.1088/1757-899X/758/1/012093

3.2. *Slump*

Table 4 shows the effect of the percentage of PLA addition on the settlement, it can be seen that the settlement decreases as the percentage increases, with decreases of 1.1" (16.92%) for the CPLA-30 and 0.5" sample (7.69%) for CPLA-25, CPLA-15 and CPLA-5, all these with respect to CPLA-0. [24], studies the settlement for different replacement percentages of ground PLA, finding that with 10% and 30% the settlement decreases by 2" and 5" with respect to the pattern; This behavior is due to the high absorption of the PLA which, by absorbing the water in the mixture, reduces its fluidity, so that its settlement is less and consequently the mixture is pasty and less workable. [25], evaluates the settlement for replacements of 20% and 100% of PLA in the form of coarse aggregate, obtaining a decrease of 0.01" and 0.7" for samples with 20% and 100%.

Table 3. Effect of curing time on pozzolanic activity index

	Compressive Strength (Pa)		Pozzolanic Activity Index (%)
Curing time	MPLA-0	MPLA-20	MPLA-20
7	19.888	39.729	77.83
28	15.479	67.968	171.08

3.3. *Compressive Strength*

Fig. 2 shows the effect of the test age on the compressive strength for different percentages of PLA. It´s observed that the compressive strength increases with the test age, obtaining a maximum resistance value at 28 days of 36.40 MPa for CPLA-25, representing a 28.39% increase with respect to CPLA-0; and for CPLA-30 the value of 14.53MPa is reached, which represents a decrease of 48.76% with respect to CPLA-0. [26] carries out tests with 10%, 20% and 30% of PLA for 7, 14 and 28 days, finding that resistance increases with age and that after 28 days with 20% of PLA resistance increases by 8.24% compared to the standard concrete, and with 30% PLA the resistance drops by 3.70%. [27], for a 25% replacement of PLA as a 5mm fine aggregate, it expresses a slight increase of 6% and 9% at 28 and 90 days respectively compared to the standard sample. However, for a 50% replacement, a decrease of 4.4% is observed for 28 days and 8% for 90 days. [28] says, that the increase in resistance with age is due, having more water storage, the particles take time to achieve adequate hydration, however, this result changes when of curing increase, reducing internal stress and contraction effects.

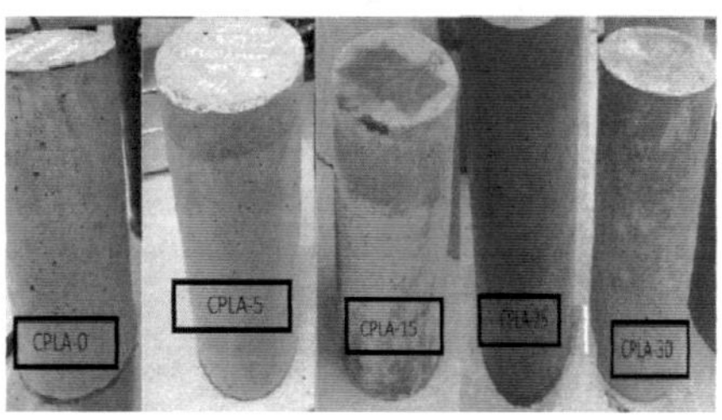

Figure 1. Specimens of concrete of 15 cm x 30 cm with PLA

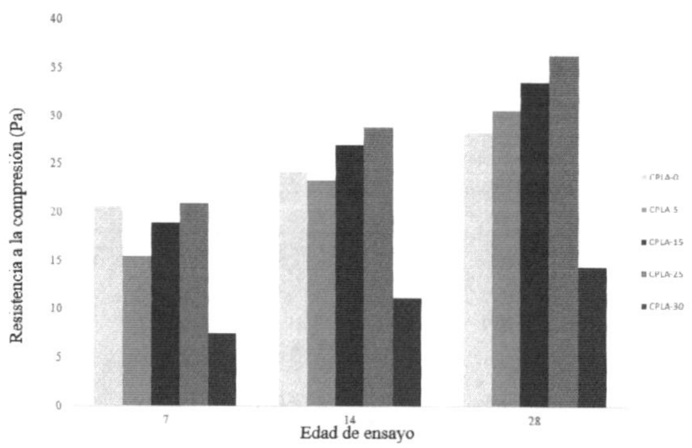

Figure 2. Effect of test age on compressive strength

3.4. Voids Content

Table 4 shows the effect of the percentage of PLA addition on the void content. Appreciating that as the PLA is increased to the mixture CPLA-5 and CPLA-25 the voids decrease and then increase for the mixture CPLA-30; reaching the highest decrease value of 13.40% and 20.80%, which represent - 8.84% and + 41.50% compared to CPLA-0. [29] evaluating the PLA for replacements of 10% and 20%, finds that the void content decreases 0.1% and 5% with respect to the pattern; this behavior is due to the finer particles of the brick filling the spaces left by the other materials reducing the volume of pores; also, the reaction of silica and alumina with calcium hydroxide generates silicon and aluminate that improve the structure. [30], studies the gaps for 25% and 50% replacements of clay brick, obtaining a decrease of -1% of gaps for 25% an increase of + 1% for 50% compared to the control; concludes that, the higher the brick content, the greater the increase in voids and density.

3.5. Coefficient of permeability to water

Table 4 shows the effect of the percentage of PLA on the coefficient of permeability to water (K) at 60 days. It is seen that as the PLA increases, K decreases by 1.07% and 1.78% for CPLA-5 and CPLA-25 with respect to CPL-0, reaching a low permeability; then increases the K to 5.07x10-11, representing an increase of 94.57% with respect to CPLA-0 and for an average permeability. [31] studies the K of a mixture with superplasticizer, coarse and fine brick aggregate, finding a reduction of 84% over the pattern; This behavior is because the superplasticizer affects the concrete, reducing the K and increasing its durability. [32] replaces 0, 40% and 100% to PLA at 28 days, finding that 40% and 100% of PLA increase by 109% and 179% over the pattern; This behavior is because PLA has a higher pozzolanic activity and that its smaller particle improves the density and properties of the mixture. [33] evaluates the replacement of 100% fine brick aggregate at 28 and 91 days, obtaining a decrease of 58% and 57% of K with the pattern; this is due to the pozzolanic reactions between the aggregate and the cement; that is, the chemical reaction of aluminum and brick silica with calcium from cement hydration, generates hydrated monosulfoaluminate, which together with the formation of calcium silicate hydrate gel decrease chloride penetration.

Table 4. Properties of concrete with influence of PLA

Mix	Settlement	Void Content	Coefficient of permeability to water (m/s)	Permeability
CPLA-0	6.5"	14.70%	2.81E-12	Low
CPLA-5	6"	14.62%	2.78E-12	Low
CPLA-15	6"	-	-	-
CPLA-25	6"	13.40%	2.76E-12	Low
CPLA-30	5.4"	20.80%	5.07E-11	Average

4. Conclusions

- The pozzolanic activity of the mortar increases what indicates that the behavior is uniform and positive of the cement with the PLA.
- The slump of the mixture decreases slightly due to the absorption of water presented by the PLA, making it a little less workable.
- The compressive strength increases for 25% of PLA which means that it can be used in buildings in an area of marine atmosphere.
- The voids content decreases due to the pozzolanicity of the PLA, which indicates that it´s suitable for making a waterproof concrete.
- The coefficient of permeability to water decreases which means that a concrete with good durability can be prepared.

5. References

[1] Kropp J, 2012, *Concrete*, Wiley-VCH Verlag GmbH & Co. KGA, p. 723-747
[2] ACI 357.1 R, 1997, State-of-the-art report on offshore concrete structures for the Artic, p. 1-11
[3] Meira G, Padaratz I, Alonso C, Andrade C, 2003, Efecto de la distancia al mar en la agresividad por cloruros en estructuras de hormigón en la costa brasileña, *Mat. de Const.*, 53, 271-272, p. 179-188
[4] Morcilo M, Chico B, Mariaca L, Otero E, 2000, Salinity in marine atmospheric corrosión: its dependence on the wind regime existing in the site, *Corrosion Science*, 42, January, p. 91-104
[5] Steven Kosmatka, Beatriz Kerkhoff, William Panarese, 2008, *Design and control of concrete mixtures*, fourth ed., Portland Cement Association, Illinois, p. 1-370.
[6] Rehman M, Ahmad M & Rashid K, 2020, Influence of fluxing oxides from waste on the production and physico-mechanical properties of fired clay brick: A review, *J of Buil Eng*, 27, 100965

[7] Hossain S, Mathur L, Majhi M, & Roy P, 2019, Manufacturing of green building brick: Recycling of waste for construction purpose, *J of Mat. Cycles and Waste Manag.*, 21, September, p. 281-292
[8] Alves A, Vieira T, De Brito J, & Correia J, 2014, Mechanical properties of structural concrete with fine recycled ceramic aggregates., *Const. and Build. Mat.*, 64, 103-113
[9] Riaz M, Khitab A, & Ahmed S, 2019, Evaluation of sustainable clay bricks incorporating brick kiln dust, *J of Build Eng*, 24, February, 100725
[10] Medeiros V, Pedroti L, Mendes B, Pitanga H & Silva T, 2019, Study of mixtures using simplex design for the addition of chamotte in clay bricks, *Int J of Appl Cer Techn*, Nov-Dec, p. 2349-2361
[11] Wang Y, Tan Y, Wang Y, & Liu, C, 2020, Mechanical properties and chloride permeability of green concrete mixed with fly ash and coal gangue, *Const. and Build. Mat.*, 233, February, 117166
[12] Filipak Vanin, 2020, Cement pastes modified by cellulose nanocrystals: A dynamic moduli evolution assessment by the impulse excitation technique, *Mat Chem & Phys,* 239, 122038
[13] Ghalehnovi M, Roshan N, Hakak E, Shamsabadi E & De Brito J, 2019, Effect of red mud (bauxite residue) as cement replacement on the properties of self-compacting concrete incorporating various fillers, *J of Cle Prod*, 240, December, 118213
[14] Kabir G & Madugu A, 2010, Assessment of environmental impact on air quality by cement industry and mitigating measures: A case study. *Envir Monit & Asses*, 160, 1-4, 91-99
[15] Bolouri J & Khayati M, 2012, Properties and performance of concrete made with recycled low-quality crushed brick. *J of Mat in Civil Eng,* 24, 4, 330-338.
[16] ACI, *Building Code Requements for Structural Concrete*, ACI.
[17] ASTM C 311, 2006, *Standard Test Methods for Sampling and Testing Fly Ash or Natural Pozzolans for Use in Portland-Cement Concrete*, p. 1-9
[18] ASTM C 143, 2010, *Standard Test Method for Slump of Hydraulic-Cement Concrete*, p. 1-6
[19] ASTM C 39, 2018, *Standard Test Method for Compressive Strength of Cyindrical Concrete Specimens*, p. 1-8
[20] ASTM C 642, 2006, *Standard Test Method for Density, Absorption, and Voids in Hardened Concrete*, American Society for Testing and Materials, p. 1-9
[21] UNE EN 12390-8, 2009, *Ensayos de hormigón endurecido. Parte 8: Profundidad de penetración de agua bajo presión*, Normalización Española, 2009.
[22] Bediako M, 2018, Pozzolanic potentials and hydration behavior of ground waste clay brick obtained from clamp-firing technology, *Case Studies in Cons Mat*, 8, June, p. 1-7
[23] Zheng L, 2011, *Mechanical Properties of Mortar with Recycled Clay-Brick-Powder*, ICCTP, 2011.
[24] Wang Ge, Sun Y, Wu X & Guan Y, 2015, Influence of ground waste clay brick on properties of fresh and hardened concrete, *Const and Build Mat*, 98, November, p. 128-136
[25] Alves A, Vieira V, De Brito T & Correia J, 2014, Mechanical properties of structural concrete with fine recycled ceramic aggregates, *Const and Build Mat*, 64, August, p. 103-113
[26] Letelier V, 2016, Mechanical properties of concretes with recycled aggregates and waste brick powder as cement replacement, *Proc Eng*, 171, p. 627-632
[27] Dang J & Zhao J, 2019, Influence of waste clay bricks as fine aggregate on the mechanical and microstructural properties of concrete, *Const and Build Mat*, 228, December, 116757
[28] Gonzalez-Corominas & Etxeberria M, 2014, Properties of high performance concrete made with recycled fine ceramic and coarse mixed aggregates, *Const and Build Mat*, 68, October, p. 618-626
[29] Schackow A, 2015, Influence of fired clay brick waste additions on the durability of mortars, *Cem & Conc Comp*, 62, September, p. 82-89
[30] Adamson M, 2015, Durability of concrete incorporating crushed brick as coarse aggregate, *Const and Build Mat*, 94, September, p. 426-432
[31] Bolouri J, Khayati M & Akrami N, 2006, Performance of concrete produced with crushed bricks as the coarse and fine aggregate, *The Geol Soc of London*, p. 1-10
[32] Zhu P, 2016, Investigation of using recycled powder from waste of clay bricks and cement solids in reactive powder concrete, *Const and Build Mat*, 113, June, p. 246-254
[33] Vieira T, Alves A, De Brito, Correia, 2016, Durability-related performance of concrete containing fine recycled aggregates from crushed bricks and sanitary ware, *Mat & Des*, 90, January, p. 767-776

IOP Conf. Series: Materials Science and Engineering **758** (2020) 012094 doi:10.1088/1757-899X/758/1/012094

Excited State Relaxation Dynamics of Silicon Nanosheets using Time-dependent Density Functional Tight Binding Theory

M Y Wang[1] and Q Wu[1,2*]

1 School of Science, Tibet University, Lhasa City, 850000, China
2 School of Environmental Science and Engineering, Tianjin University, Tianjin, 300072, China
Email: wuqi_zangda@163.com

Abstract. By combining time-dependent density functional tight binding (TD-DFTB) calculations for electrons with molecular dynamics simulations, time evolution of excited state relaxation and single electron self-trapping of silicon nanosheets (Si_nH_m) are investigated. We identify the atoms and bonds contributing to the excitations and observe the time evolution of charge density, indicating the migration of exciton from initial position to neighboring bonds. Different temperature 50K and 100K are attempted in the MD simulations, and the temperature is found to affect the electronic localization. Our simulations reveal the formation of exciton trapping and localization on certain bonds upon excitation within the timescale about 200-300fs. In summary, TD-DFTB excited state MD simulations can provide interesting insights into photoreactions and have crucial implications in modulating the exciton transport efficiency in strongly confined low-dimensional systems.

1. Introduction

Electronically excited states play a prominent role in many different areas of chemical and condensed matter physics and organic chemistry [1-2]. For example, luminescence energies and the resonance Raman spectra are associated with the displacement of the geometry in the excited state. The fluorescent properties of excited state systems are increasingly being employed as biomolecular probes [3-4]. Despite their obvious importance, a detailed investigation of the excited state remains a challenge to experimentalists and theoreticians alike. The interpretation of emission spectra is complicated by the fact that a detailed configurational representation of the experimental data requires knowledge of both the excited state and ground state energy surfaces [5].

Linear response time-dependent density functional theory (TD-DFT) based on Casida's equations [6] is probably the most widely used method for the simulation of processes involving excited states, such as photon absorption and emission [7-8]. Its success is due to the relatively modest numerical efficiency combined with the generally good level of accuracy. However, although the TD-DFT is much cheaper in terms of computational cost, simulations of nanosystems with hundreds of atoms are still out of reach. Hence, there is a clear need for computationally inexpensive electronic structure theory methods which can model the excited state potential energy surfaces. Based on the framework of the density functional based tight binding (DFTB) [9-10], a computationally very efficient alternative to TD-DFT has been developed in the form of time-dependent DFTB [11-12]. The time-dependent density functional tight binding (TD-DFTB) developed by T. A. Niehaus become widely used for the calculation of the vertical excitation energies and excited state properties of medium- to large-size molecular systems [13-21]. It has been shown that TD-DFTB yields reasonable agreement for optical spectra with experiments and first-principles calculations at a highly reduced cost in many cases.

However current TD-DFTB theory could only provide static excited state calculations, which is not efficient as dynamic information is totally lost. Photochemical reactions typically occur at femto- to picosecond timescales. As the relevant time and spatial resolutions are hard to access in experiment, molecular dynamics simulation is the best method of choice to study such ultrafast processes and resolve. Therefore, excited state molecular dynamic simulations are currently one of the frontiers of computational chemistry. These simulations demand a delicate combination of high-quality and high-speed methods to achieve a proper description of the excited-state time evolution of molecules. By combining time-dependent density functional tight binding (TD-DFTB) calculations for electrons with molecular dynamics simulations, here we investigate the capability of time-dependent density functional tight binding method to describe the time evolution of excited state potential energy surfaces. This yields the possibility of studying the trajectory following photo-excitation and of identifying the most important degree of freedoms involved in the excited state relaxation.

2. Methodology and model description

2.1. Time-dependent density functional tight binding theory

For a given structure, the total energy for an excited state is obtained in two steps. A self-consistent field (SCF) calculation is first conducted to obtain the single particle Kohn-Sham (KS) orbitals ψ_i and the corresponding KS energies ε_i. These single particle results are then corrected following TD-DFT linear response treatment of Casida6 错误!未定义书签。.

$$\sum_{kl\tau}[\omega_{ij}^2\delta_{jl}\delta_{\sigma\tau} + 2\sqrt{\omega_{ij}}K_{ij\sigma,kl\tau}\sqrt{\omega_{kl}}]F_{kl\tau}^I = \omega_I^2 F_{ji\sigma}^I \quad (1)$$

Where $\omega_{ij} = \varepsilon_j - \varepsilon_i$ and σ, τ are spin indices.

The coupling matrix is obtained by giving the SCF potential with respect to the electronic density change as follows:

$$K_{ij\sigma,kl\tau} = \iint \psi_i(\mathbf{r})\psi_j(\mathbf{r})(\frac{1}{|\mathbf{r}-\mathbf{r}'|} + \frac{\delta^2 E_{xc}}{\delta\rho_\sigma(\mathbf{r})\delta\rho_\tau(\mathbf{r}')})\psi_k(\mathbf{r}')\psi_l(\mathbf{r}')d\mathbf{r}d\mathbf{r}' \quad (2)$$

Equation (2) is further simplified in TD-DFTB approach using the Mulliken approximation [22-23].

From the results of Equation (1), oscillator strengths for singlet-singlet transitions are calculated according to Eq. (3):

$$f_I = \frac{2}{3}\omega_I \sum_{k=x,y,z}\left|\sum_{ij}\langle\psi_i|\mathbf{r}_k|\psi_j\rangle\sqrt{\frac{\omega_{ij}}{\omega_I}}(F_{ij\uparrow}^I + F_{ij\downarrow}^I)]\right|^2 \quad (3)$$

By solving the energies and oscillator strengths of ten lowest singlet excitations, we obtained the low energy part of the absorption spectrum for the ground state (S_0). Finally, the emission energies are evaluated by geometry optimization in the first excited singlet state (S_1), based on the assumption of rapid internal conversion, (i.e., the validity of Kasha's rule).

2.2 Molecular dynamics simulation

Our TD-DFTB molecular dynamics are purely adiabatic and proceed in the following way. The nuclear time evolution is treated classically by means of the Newton's equations, while the population of each electronic state is treated separately.

First, a simulation in S_0 is carried out in order to obtain an equilibrated system at the target temperature using a Berendsen thermostat. We then take random configurations (coordinates and velocities) from this simulation and vertically excite the system into a selected singlet state. The positions of the nuclei are updated according to the forces computed as the gradients of the excited state energy. In a second iteration, the excited state density and Kohn-Sham orbitals for the latest geometry are computed, and the TD-DFTB calculation provides the new excited state energy and the new forces on the nuclear. This approach allows us to adiabatically follow the instantaneous electronic excited state PES in the Born-Oppenheimer molecular dynamic scheme.

2.3 Application of TD-DFTB molecular dynamics simulations to silicon nanosheets

We reported the theoretical prediction of excited state relaxation and single electron self-trapping in Si nanosheets. In finite atomic cluster, we show that excitons are strongly localized around the Si-Si dimer with the bond distance enlarged. Analysis of molecular orbitals in the excited state exhibits a localized distribution along the distorted areas of the Si nanosheets. While a static exploration of the geometry provides only an incomplete study of relaxation on the excited state surface, molecular dynamics (MD) simulation is thus available for an understanding and confirmation of excited state relaxation at different temperature were thus carried out for S_1 state.

As an extension of our previous work on the effect of the edge width on exciton distribution in two-dimensional nanosheets, we herein explored the excited state relaxation of the most widely investigated Si (110) [24-27] oriented nanosheets using TD-DFTB molecular dynamic simulations. Detailed information about the computational model can be found in ref 20 . Due to restrictions in computational power, the excited state molecular dynamic simulations are only performed for Si sheets ($Si_{30}H_{48}$, $Si_{60}H_{82}$ and $Si_{90}H_{116}$). In order to examine the temperature effect on the excited state relaxation, MD simulations at 50K and 100K were attempted for each structure respectively.

The performance of TD-DFTB [28-31] against TD-DFT [12,32] has been characterized by many groups. For instance, the optical gap of Si_5H_{12} cluster was predicted as 6.4 eV [15], close to the experimental value of 6.5 eV [33]. For other larger clusters like $Si_{35}H_{36}$, the absorption energies (4.37eV) calculated using TD-DFTB coincides with the results (4.33 eV [34]) obtained using a high level *ab initio* multi-reference second order Møller–Plesset perturbation theory (MR-MP2). Meng *et al.* [35] also confirmed the reliability of TD-DFTB by comparing the calculated absorption energies with the results using PBE, which exhibits that the difference in absorption energies is only around 0.1eV. In general, these comparisons confirm that TD-DFTB offers acceptable accuracy to investigate the excited state properties of the hydrogenated silicon nanoparticles, and enables our study to be extended from several atoms to nanometer-size systems.

3. Results and discussion

3.1 MD simulation in S_1 state for $Si_{30}H_{48}$ at 50K and 100K

MD simulations in S_1 state of $Si_{30}H_{48}$ were performed at 50K and 100K, respectively. As Figure 1(a) shows, vital Si-Si bonds were labeled clearly to exhibit the excited state relaxation. The time evolution of bond length change of several Si-Si dimer are also shown in Figure 1(b). We observe that the Si-Si bond 2 extends to 2.79 Å at 82.5fs and exhibit an average bond length about 2.8 Å, which is consistent with our static calculation of the excited state (see Figure 5.3 in ref 20). While the other Si-Si bonds, for instance, bonds 2, 3, 4 and 5 in the vicinity remain almost unchanged. It is determined that the exciton is firstly localized on Si-Si bond 2. With the simulation time increases to 480fs, bond 2 exhibits an abrupt decrease in length and bond 3 becomes elongated, accordingly. The exciton is hence localized on bond 2 and bond 3.

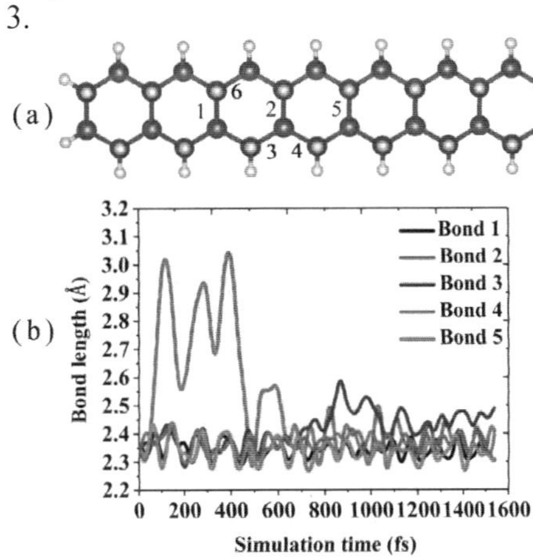

Figure 1. The structures of (a) SiNS $Si_{30}H_{48}$ and (b) time evolution of bond length change of labeled bonds at 50K. Si and H atoms are shown as blue and white balls, respectively. The numbers show the labeling of vital bonds of the SiNS.

IOP Conf. Series: Materials Science and Engineering **758** (2020) 012094 doi:10.1088/1757-899X/758/1/012094

To further interpret the optical properties of SiNSs, the LUMO (defined as the orbital which is occupied by one excited electron) in the S_1 state are plotted in Figure 2. Figure 2 (a) shows the non-localized distribution of LUMO in the vertical excite state of $Si_{30}H_{48}$. In the beginning of excitation, the LUMO in S_1 is distributed on several central Si bonds. In contrast, the molecular orbital of the structures at 115.5fs indicates a localized distribution and also a trapped exciton on the Si-Si bond. More interestingly, it is observed that the localization is not fixed on bond 2 anymore. With the elongation of bond 3, the trapped exciton has a shift in distribution when the simulation time is longer than 870fs, indicating the migration of electron from initial position to neighboring bonds. It is thus determined that the MD simulations in S_1 state yields reasonable forces and a reliable description of excited state relaxation compared with our static excited state calculations .

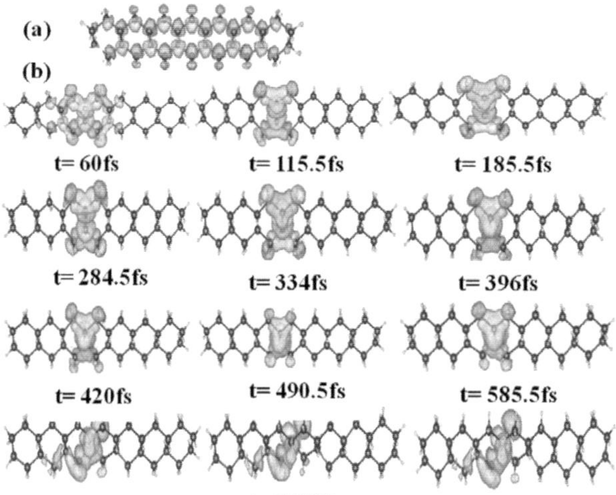

Figure 2. Frontier molecular orbitals of SiNS $Si_{30}H_{48}$ in the (a) ground state geometry and excited state geometries at different simulation time at 50K. Wave function isosurfaces are plotted at the same isovalue (0.013).

When the temperature is increased to 100K, it is bond 1 which has a large enlargement upon excitation (see Figure 3). Hence, the LUMO in S_1 exhibits a distribution mainly localized on bond 1 as shown in Figure 4. The resultant geometry is hence distorted along bond 2, indicating the instability of $Si_{30}H_{48}$ at 100K. Therefore we did not attempt a higher temperature in the MD simulations.

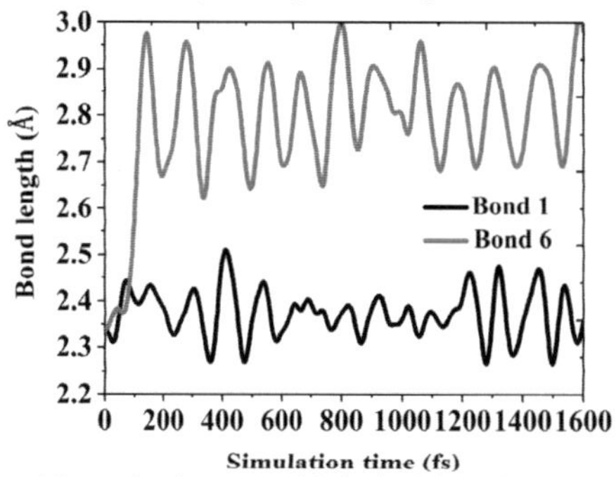

Figure 3. Time evolution of bond length change of labeled bonds at 100K for SiNS $Si_{30}H_{48}$.

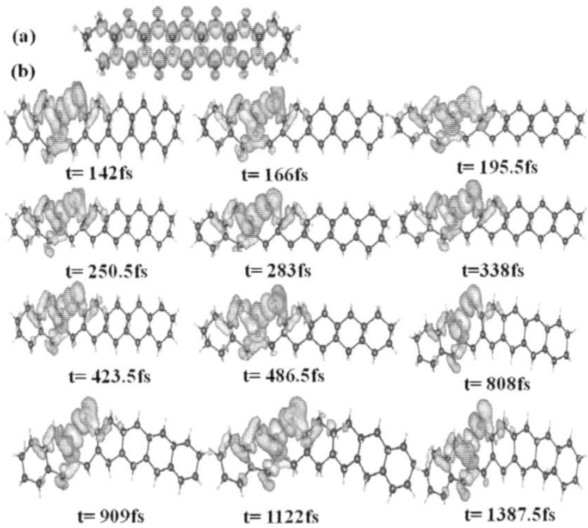

Figure 4. Frontier molecular orbitals of SiNS $Si_{30}H_{48}$ in the (a) ground state geometry and (b) excited state geometries at different simulation time at 100K. Wave function isosurfaces are plotted at the same isovalue (0.013).

3.2 MD simulation in S_1 for $Si_{60}H_{82}$ at 50K and 100K

By expanding the size of $Si_{30}H_{48}$, excited state MD simulations were also performed for larger structures $Si_{60}H_{82}$ and $Si_{90}H_{112}$. We also examine the time evolution of the critical bond length, and the distribution of molecular orbitals in the S_1 state.

For $Si_{60}H_{82}$, there is also one Si-Si bond becomes enlarged, which could be observed from the snapshots of the MD simulations. The time evolution of bond length change, as shown in Figure 5(b), indicates that bond 1 extends to a value around 2.8 Å. The simulated results have no discrepancies with our previous work [20]. LUMO of the first excited state S_1 still localizes on the Si-Si bond which has a large displacement. When the temperature is increased to 100K, it is bond 4 rather than bond 1 has an obvious enlargement, as shown in Figure 7. At about 780fs, bond 4 exhibits an abrupt decrease in length and bond 1, 2 and 5 become elongated accordingly to conserve the energy. A continuous increase in bond length of bond 4 ensued and hence the trapped exciton has a slight migration to the edge of NSs (see Figure 8).

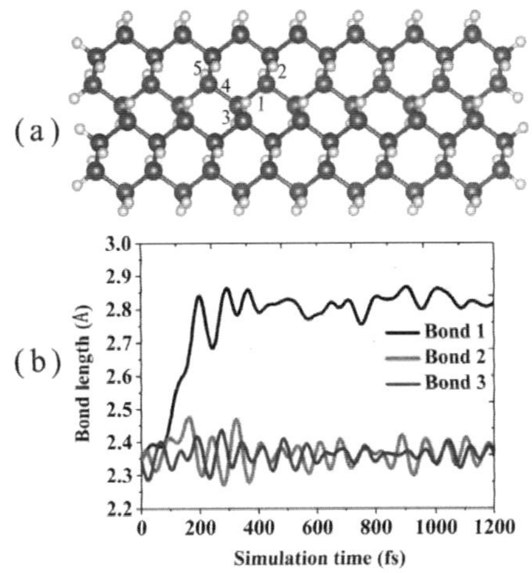

Figure 5. The structures of (a) SiNS $Si_{60}H_{82}$ and (b) time evolution of bond length change of labeled bonds at 50K.

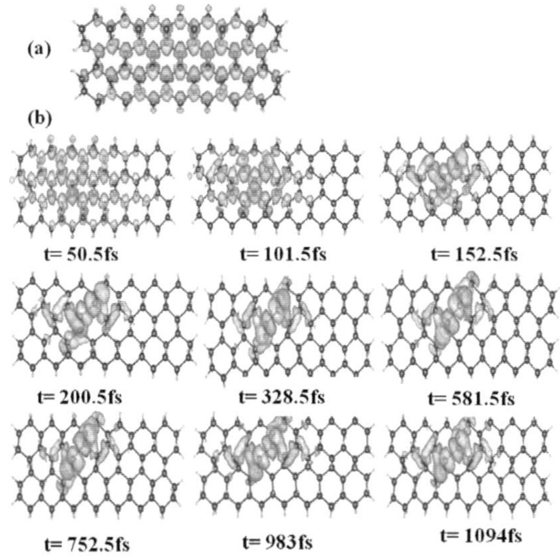

Figure 6. Frontier molecular orbitals of SiNS $Si_{60}H_{82}$ in the (a) ground state geometry and excited state geometries at different simulation time at 50K. Wave function isosurfaces are plotted at the same isovalue (0.013).

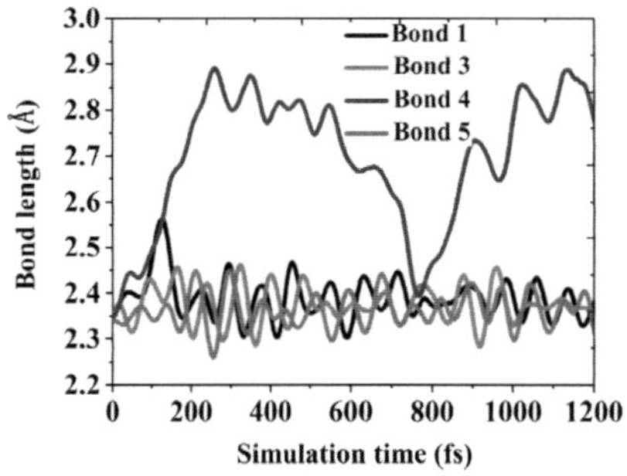

Figure 7. Time evolution of bond length change of labeled bonds at 50K for SiNS $Si_{60}H_{82}$.

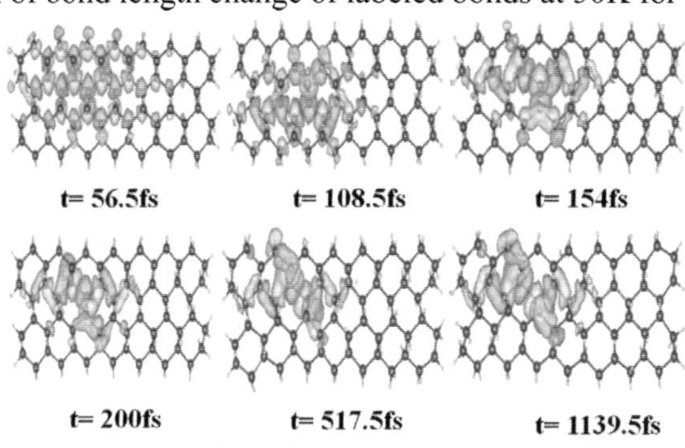

Figure 8. Frontier molecular orbitals of SiNS $Si_{60}H_{82}$ in the (a) ground state geometry and (b) excited state geometries at different simulation time at 100K. Wave function isosurfaces are plotted at the same isovalue (0.013).

3.3 MD simulation in S_1 state for $Si_{90}H_{116}$ at 50K and 100K

In the case of $Si_{90}H_{116}$, the structure distortions of $Si_{90}H_{116}$ distribute on the Si-Si bonds in the vicinity of center, thereby leading to a less deviation from the initial structure. In the initial 200fs in Figure 5.9, it is observed that how the exciton is trapped and localized on certain bonds after the system is excited. With the increase of simulation time, it exhibits a relatively stable distribution of LUMO in S_1.

In the case of $Si_{90}H_{116}$, Si-Si bond 4 as depicted in Figure 9(b) becomes enlarged, which could be observed from the snapshots of the MD simulations. The time evolution of bond length change, as shown in Figure 9(b), indicates that bond 1 extends to a value around 2.8 Å. While the central Si-Si bond has an average value around 2.35 Å. LUMO of the first excited state still localized on the Si-Si bond which has a large displacement (see Figure 10). In the initial 200fs, the process of exciton trapping on bond 1 is also observed and it finally stabilized on bond 1 and several neighboring bonds with the increase of simulation time. Similarly, the analysis of evolution of bond length change and the frontier molecular orbitals also indicate a localized distribution around bond 4 at 100K (see Figure 11 and 12).

In summary, our MD simulations reveal that the exciton is trapped as a function of the width and temperature. The process of trapping an exciton is observed within the timescale at around 200fs. We also reveal a possible migration of trapped exciton of $Si_{30}H_{48}$, which is helping in finding transport properties more conductive to technological application.

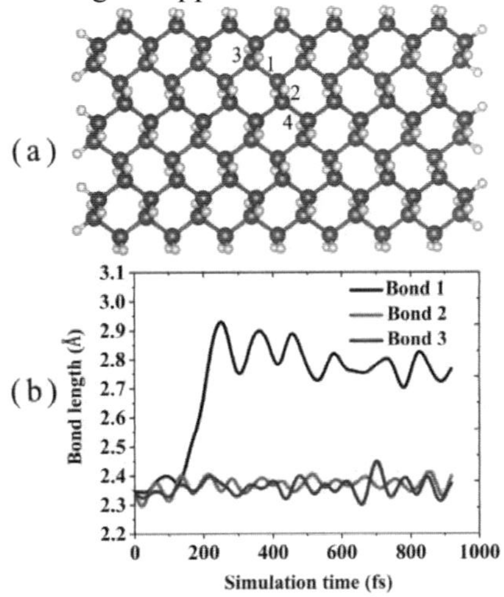

Figure 9. The structures of (a) SiNS $Si_{90}H_{116}$ and (b) time evolution of bond length change of labeled bonds at 50K.

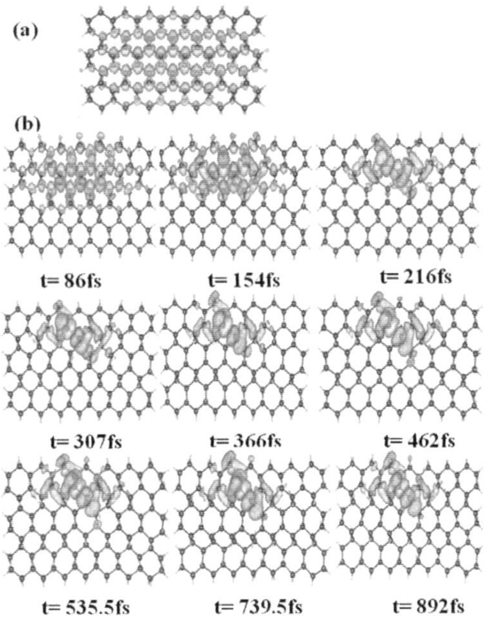

Figure 10. Frontier molecular orbitals of SiNS $Si_{90}H_{116}$ in the (a) ground state geometry and excited state geometries at different simulation time at 50K. Wave function isosurfaces are plotted at the same isovalue (0.013).

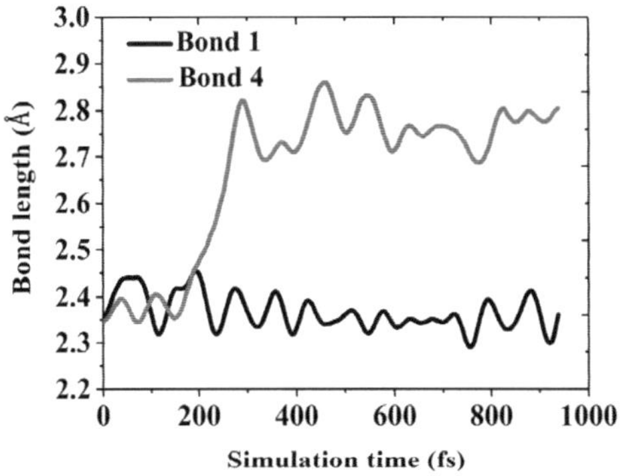

Figure 11. Time evolution of bond length change of labeled bonds at 100K for SiNS $Si_{90}H_{116}$.

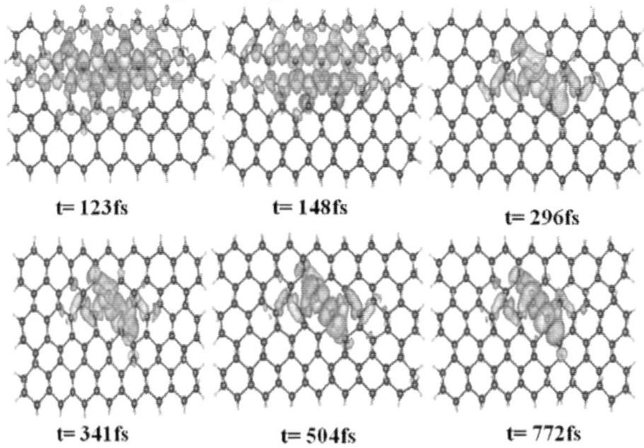

Figure 12. Frontier molecular orbitals of SiNS $Si_{90}H_{116}$ in the (a) ground state geometry and excited state geometries at different simulation time at 100K. Wave function isosurfaces are plotted at the same isovalue (0.013).

4. Conclusions

We have carried our excited state molecular dynamics simulations of silicon nanosheets with finite atomic models, in order to assess the performance of TD-DFTB for the description of excited state relaxation. Our excited state simulation show the relaxation of silicon nanosheets in S_1, which is dominated by a single Si-Si bond. It has a large enlargement in contrast to the initial structure, leading to localization of molecular orbitals on the elongated Si-Si bond. The localized distribution of frontier molecular orbitals and the sufficiently stretched central Si-Si bond signify formation of self-trapping of the exciton in the first excited state. The excited state MD simulations come to a coherent conclusion on the formation of the localized distribution of the frontier molecular orbitals and thus formation of the self-trapped excitons with our previous static investigation. We assert that the TD-DFTB approach proved as a reliable tool in predicting or reproducing the time evolution of excited state structural changes and potential energy surfaces. Since our current TDDFTB molecular dynamics are purely adiabatic, the effect of non-adiabatic coupling between two electronic states will be implemented in the future.

5. References

[1] Weiss U 1999 *Quantum Dissipative Systems* (Singapore: World Scientific)

[2] Klessinger M and Michl J 1995 *Excited States and Organic Molecules* (New York: Wiley-VCH)

[3] Dooley C T, Dore T M, Hanson G T, Jackson W C, Remington S J and Tsien R Y2004 *J. Biol. Chem.* **279** 22284.

[4] Hanson G T, Aggeler R, Oglesbee D, Cannon M, Apaldi R A, Tsien R Yand Remington S J 2004 *J. Biol. Chem.* **279** 13044

[5] Juan Manuel Ortiz-Sánchez, B. Denis, Levi C. T. Pierce, R. L. Phineus, Markwick and J. Andrew Mc Cammon 2012 *J Chem Theory Comput.* **8** 2752–2761

[6] Casida M E 1995 *Recent Advances in Density Functional Methods* (Singapore: World Scientific) Chapter 5 pp 155–192.

[7] Runge E and Gross E K U 1984 *Phys. Rev. Lett.* **52** 997
[8] M. E. Casida 1996 *Theoretical and Computational Chemistry* (Amsterdam: Elsevier) pp 391–439
[9] Porezag D, Frauenheim T, Köhler T, Seifert G and Kaschner R 1995 *Phys. Rev. B* **51** 12947
[10] Seifert G, Porezag D and Frauenheim T 1996 *Int. J. Quantum Chem.* **58** 185
[11] Niehaus T A, Suhai S, Della Sala F, Lugli P, Elstner M G and Frauenheim T 2001 *Phys. Rev. B,* ***63*** 085108
[12] T. A. Niehaus, *J. Mol. Struct.Theochem.* 2009, ***914***, 38-49.
[13] Joswig J O, Seifert G, Niehaus T A and Springborg M 2003 *J. Phys. Chem. B* **107** 2897
[14] Frenzel J, Joswig J O and Seifert G 2007 *J. Phys. Chem. C* **111** 10761
[15] Wang X, Zhang R Q, Niehaus T A, and Frauenheim T 2007 *J. Phys. Chem. C* **111** 2394
[16] Wang X, Zhang R Q, Lee S T, Niehaus T A and Frauenheim T 2007 *Appl. Phys. Lett.* **90** 123116
[17] Li Q S, Zhang R Q, Lee S T, Niehaus T A and Frauenheim T 2008 *J. Chem. Phys.* **128** 244714
[18] Zhang R Q, De Sarkar A, Niehaus T A and Frauenheim T 2012 *Phys. Status Solidi B* 249 401
[19] Fan G H , Li X, Liu J Y and He G Z 2012 *Comput. Theor. Chem.* **1030** 17
[20] Wu Q, Wang X H, Niehaus T A and Zhang R Q 2014 *J. Phys. Chem. C* **118** 20070-20076
[21 Wu Q, Huang B, Niehaus T, Yang X, Fan J and Zhang R Q 2015 *Phys. Chem. Chem. Phys.* **17** 10786
[22] Heringer D, Niehaus T A, Wanko M and Frauenheim T H 2007 *J. Comp. Chem.* ***28*** 2589-2601
[23] Fehér F 1997 *Molekülspektroskopische Untersuchungen auf dem Gebiet der Silane und der Heterocyclischen Sufane* pp1-55
[24] Yu D K, Zhang R Q and Lee S T 2002 *J. Appl. Phys.* **92** 7453-7458
[25] Lu G H, Huang M, Cuma M and Liu F 2005 *Surf. Sci.* **588** 61-70
[26] Visikovskiy A, Yoshimura M and Ueda K 2010 *Jpn. J. Appl. Phys.* **49** 08LB05
[27] Brázdová V and Bowler D R 2011 *Phys. Chem. Chem. Phys.* **13** 11367-11372
[28] Zheng G, Irle S and Morokuma K 2005 *Chem. Phys. Lett.* **412** 210-216
[29] Sattelmeyer K W, Tirado-Rives T and Jorgensen W L 2006 *J. Phys. Chem. A* **110** 13551–13559
[30] Otte N, Scholten M and Thiel W 2007 *J. Phys. Chem. A* **111** 5751–5755
[31] Goyal P, Elstner M and Cui Q 2011 *J. Phys. Chem B* 2011 **115**, 6790-6805
[32] Domínguez A, Aradi B, Frauenheim T, Lutsker V and Niehaus T A 2013 *J. Chem. Theory. Comput.* **9** 4901-4914
[33] Porezag D, Frauenheim T, Köhler T, Seifert G and Kaschner 1995 *Phys. Rev. B* **51** 12947-12957
[34] Zdetsis A D 2006 *Rev. Adv. Mater. Sci.* **11** 56-78
[35] Meng Y, Wu Q, Chen L, Wangmo S, Gao Y, Wang Z, Zhang R Q and Ding D 2013 *Nanoscale* **5** 12178-12184

Acknowledgments

The work was supported by the Natural Science Foundation of Tibet Autonomous Region (Grant NO. 2016ZR-TU-10), the Scientific Research Foundation for Returned Scholars of Tibet University (Study on the catalytic properties of transition metal oxide clusters) and the Doctoral Development Project of Zhu Feng Scholar of Tibet University.

IOP Conf. Series: Materials Science and Engineering **758** (2020) 012095 doi:10.1088/1757-899X/758/1/012095

Preparation of Eu(Avobenzone)$_3$TPPO by precipitation method and its luminescence performance

X Z Wei, S S Gao, P S Tang, L Q Zhang, R P Hou, and H F Chen

Department of Materials Engineering, Huzhou University, Huzhou 313000, P.R. China
E-mail: tangps@zjhu.edu.cn

Abstract. Using Eu_2O_3 as raw material, anhydrous ethanol as solvent system, 1-(4-Methoxyphenyl) -3-(4-tert-butylphenyl)-1,3-propanedione (Avobenzone) as first ligands and triphenylphosphane oxide (TPPO) as second ligands, the rare earth complex Eu(Avobenzone)$_3$TPPO has been successfully prepared by precipitation method. The Eu(Avobenzone)$_3$TPPO were characterized by fluorescence spectrum analysis (FS), Fourier transform infrared spectroscopy(FT-IR) and color coordinate calculation. The thermal stability of Eu(Avobenzone)$_3$TPPO was studied by thermogravimetry differential scanning calorimetry (TG-DSC). The results showed that Eu(Avobenzone)$_3$TPPO emitted strong red characteristic fluorescence, the optimal excitation wavelength was 397nm, the characteristic emission wavelength was 612 nm, and the color coordinates were (0.66, 0.34). Eu(Avobenzone)$_3$TPPO decomposes at about 200℃, it shows good thermal stability.

1. Introduction

Because the luminescence of rare earth ions has characteristics of long fluorescent life and high color purity[1], it has been widely used in the field of rare earth luminescent materials. In 1942, Weissman[2] first discovered that there was a stable conversion efficiency from covalent europium ions to ionic compounds. The fluorescence properties of rare earth ions could be enhanced by means of intra-molecular energy transfer, it also known as the 'antenna effect'. Since then, the studies on organic complexes of europium have developed rapidly. Trivalent europium ternary organic complexes showed high purity, stable chemical properties and good luminescence performance. Trivalent europium ternary organic complexes are widely used in fluorescent probes [3], biomedicine [4], optics and other fields. Therefore, the study of europium ternary complexes has important research value and significance.

2. Experimental details

2.1 Preparation of Eu(Avobenzone)$_3$TPPO

The following are the reagents used in the experiment. Eu_2O_3(purity99%), 1-(4-Methoxyphenyl) -3-(4-tert-butylphenyl)-1,3-propanedione, triphenylphosphane oxide, hydrochloric acid, anhydrous ethanol and sodium hydroxide were all analytically pure. The following are specific steps used in the experiment. 0.3519g Eu_2O_3 was dissolved in 20ml hydrochloric acid to obtain $EuCl_3$ solution, and $EuCl_3$ solution was evaporated to obtain $EuCl_3$. $EuCl_3$ was dissolved in 10ml anhydrous ethanol to form solution A. Avobenzone of 0.9312g was weighed and dissolved in 10ml anhydrous ethanol to obtain solution B. TPPO of 0.2783g was weighed and dissolved in 10ml anhydrous ethanol to obtain solution C. Mix solution B and C, then add the mixture into solution A drop by drop, and react at 60℃ for 30min at a constant temperature. Adjusting the pH to 6.5-7 with 1mol/L NaOH ethanol solution. The reaction solution was incubated at 60℃ for 2h and centrifuged to obtain pale yellow precipitation. The product was washed with anhydrous ethanol and dried in an oven at 60℃.

2.2 Experimental instruments and testing methods

The luminescence performance of Eu(Avobenzone)$_3$TPPO sample was characterized by sens9000/9003 model fluorescence spectrometer (Beijing Zhuoli Hanguang Instrument Co. Ltd.). The infrared spectra of Eu(Avobenzone)$_3$TPPO sample was determined by Nicolet 5700 Fourier of the US Thermo Corporation, it using the KBr pressing plate method. Eu(Avobenzone)$_3$TPPO sample was measured by the American TA-SDTQ600 thermal analyzer, the temperature range of the TG-DSC is from 20℃ to 600℃ and the heating rate is 10℃/min.

3. Results and discussion

3.1 Fluorescence spectrum analysis

Figure.1 is the excitation spectrum of Eu(Avobenzone)$_3$TPPO. It can be seen from Figure.1 that the excitation band of Eu(Avobenzone)$_3$TPPO is located at 300-600nm, and the optimal excitation wavelength is 397nm[5].

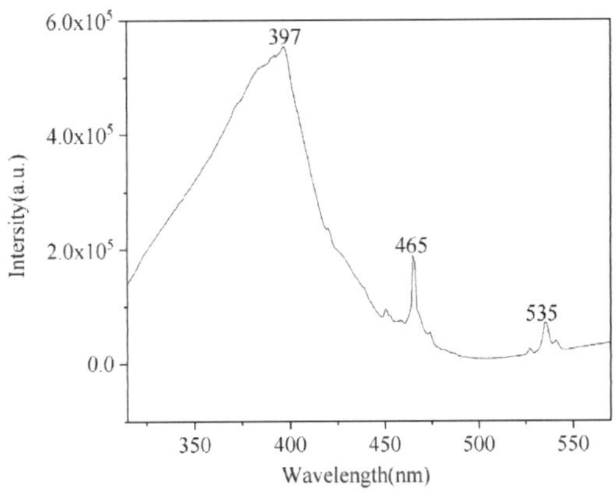

Figure.1 Excitation spectra of Eu(Avobenzone)$_3$TPPO

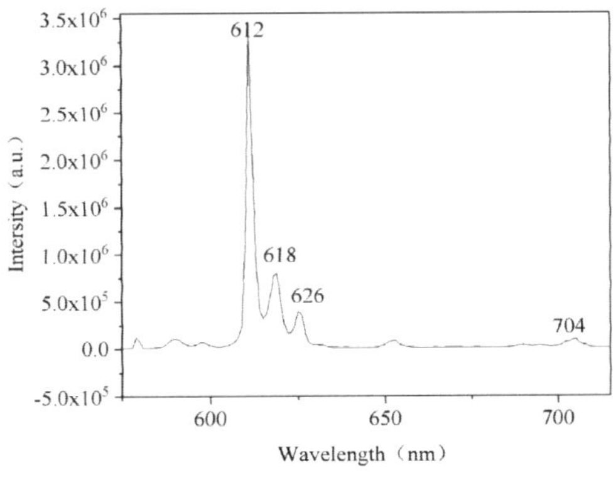

Figure.2 Emission spectra of Eu(Avobenzone)$_3$TPPO

Figure.2 is the emission spectrum of Eu(Avobenzone)$_3$TPPO. It can be seen from figure.2 that the excitation band of Eu(Avobenzone)$_3$TPPO is at 500~720nm, and 612nm is the characteristic emission wavelength. The characteristic emission peaks of Eu(III) ions are respectively located at 612, 618, 626 and 704nm, the $^5D_0$-$^7F_1$,$^5D_0$-$^7F_2$,$^5D_0$-$^7F_3$ and$^5D_0$-$^7F_4$ energy level transitions of Eu(III) ions are correspond to these characteristic emission peaks [6-7]. The strongest energy level transitions of $^5D_0$-$^7F_1$ are electric dipole moment transitions, which is the red strongest characteristic luminescence of central ion Eu(III).

According to the fluorescence emission data of complexes and ligands, CIE 1931 was used to calculate the CIE color coordinates of complexes and ligands. The color coordinate data were shown in table 1. According to the data in table 1, the color coordinate diagram drawn is shown in Figure.3. The ligand Avobenzone is in the green light zone, TPPO is in the yellow light zone, and Eu(Avobenzone)$_3$TPPO is in the red light zone. They have different color regions, so the ligand can be coordinated successfully on Eu^{3+}.

Table 1 color coordinates and color temperature data of complexes

category	Nomenclature	X coordinate	Y coordinate	Color temperature
First ligands	Avobenzone	0.36	0.59	5129K
Second ligands	TPPO	0.46	0.51	3313K
Complexes	Eu(Avobenzone)$_3$TPPO	0.66	0.34	12899K

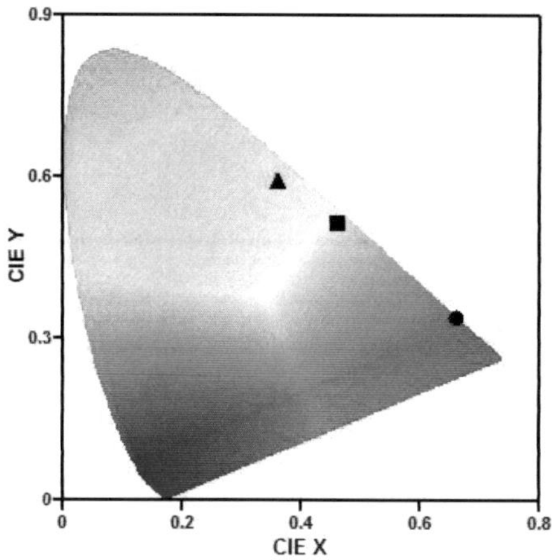

▲:Avobenzone ■:TPPO ●:Eu(Avobenzone)$_3$TPPO

Figure.3 Color coordinate complexes and ligand

3.2 Infrared spectral analysis

Figure.4 is the infrared spectrum of Eu(Avobenzone)$_3$TPPO. As can be seen from figure.4, the absorption peak of Eu(Avobenzone)$_3$TPPO at 3451cm^{-1} corresponds to the first ligand Avobenzone. Compared with the second ligand TPPO, the stretching vibration peak of the C=O bond in the activity moves from the original 1493cm^{-1} to 1431cm^{-1}, this situation is due to the Eu-O coordination bond formed by Eu(III) ions and ligand TPPO, which results in the decrease of P=O bond force constant and the absorption peak of Eu(Avobenzone)$_3$TPPO moves towards the short-wave number[8-9]. Thus, the coordination between TPPO and Eu(III) ion is successful.

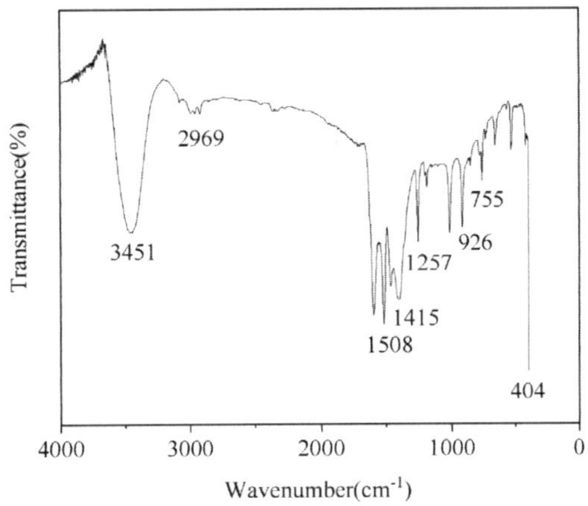

Figure.4 Infrared spectra of Eu(Avobenzone)$_3$TPPO

3.3 Thermogravimetric analysis

TG-DSC curve of Eu(Avobenzone)$_3$TPPO is shown in figure.5. As can be seen from figure.5, on the TG curve, the complexes has a loss of weight of about 8.7% in the range of 30℃~300℃, and on the DSC curve, there is a small endothermic peak at 241℃, which should be the removal and endothermic absorption of the complexes adsorbed water and other substances. As can be seen from the TG curve, the complexes began to decompose at 300℃, with a loss of weight of about 85.8% in the range of 300℃~600℃, while there was a small endothermic peak at 349℃ and a strong exothermic peak at 484℃ on the DSC curve, indicating that the decomposition process of the complexes was carried out step by step. At about 600℃, the complexes have been decomposed, and the rest should be Eu_2O_3, indicating the successful coordination between the partial distributor and europium. Moreover, there is basically no decomposition before 280℃, which indicates that Eu(Avobenzone)$_3$TPPO has a good thermal stability and meets temperature requirements for a lot of luminescent materials [10].

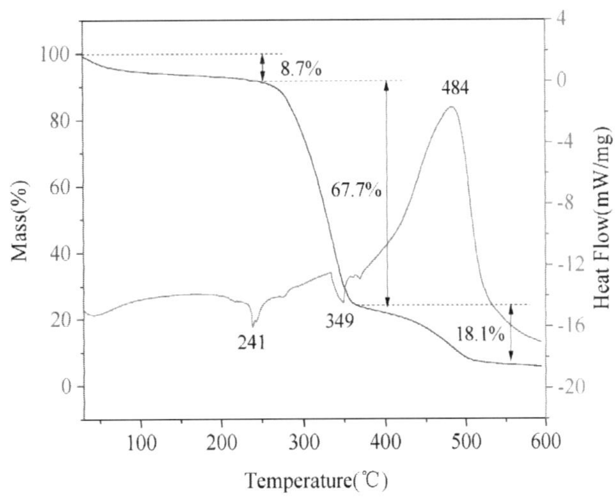

Figure.5 Eu(Avobenzone)$_3$TPPO thermogravimetric- ifferential scanning heat maps

4. Conclusions

The ternary complex Eu(Avobenzone)$_3$TPPO was prepared with Eu_2O_3 as europium source, 1-(4-Methoxyphenyl)-3-(4-tert-butylphenyl)-1,3-propanedione (Avobenzone) and triphenylphosphane oxide (TPPO) as preparation systems. Eu(Avobenzone)$_3$TPPO has good thermal stability and can emit characteristic red fluorescence. The strongest fluorescence emission peak is 612 nm, and the color coordinates are (0.66, 0.34). Eu(Avobenzone)$_3$TPPO is a kind of red fluorescent material with high luminous quantum efficiency of organic compounds and good stability of inorganic compounds.

Acknowledgements

The authors acknowledge education leading talent project of Huzhou South Taihu special plan and General Project of Zhejiang Provincial Department of Education (Y201737320).

References

[1] Guoqiang Wang, Xinghong Gong, Yujin Chen, et al. Novel red phosphors KBaEu$(XO_4)_3$ (X=Mo, W) show high color purity and high thermostability from a disorder chained structure. *Dalton Transactions*, 2017, **46(20)** 6776-6784.

[2] Giuliani A , Kamenetsky V S , Kendrick M A , et al. Oxide, sulphide and carbonate minerals in a mantle polymict breccia. Metasomatism by proto-kimberlite magmas, and relationship to the kimberlite megacrystic suite. *Chemical Geology*, 2013, **353** 4-18.

[3] Wang Z , Gao J , Zhang K , et al. Effective and efficient detection of pH fluctuations based on ratiometric metallic-ciprofloxacin architectures. *Optical Materials*, 2018, **81** 1-6.

[4] Tang Y , Mei R , Yang S , et al. Hollow $GdPO_4$,Eu^{3+} microspheres. Luminescent properties and applications as drug carrier. *Superlattices and Microstructures*, 2016, **92** 256-263.

[5] Dandekar M P , Kondawar S B , Itankar S G , et al. Luminescence Properties of Electrospun Nanofibers of Europium Complex Eu(TTA)$_3$Phen/Polymers. *Procedia Materials Science*,2015 , **10** 580-587.

[6] Zhang X , Tang J , Li H , et al. Red light emitting nano-PVP fibers that hybrid with Ag@SiO_2@Eu(TTA)$_3$phen-NPs by electrostatic spinning method. *Optical Materials*, 2018,**78** 220-225.

[7] Chigome S , Abiona A A , Ajao J A , et al. Synthesis and Characterization of Electrospun Poly(ethylene oxide)/Europium-Doped Yttrium Orthovanadate (PEO/YVO_4,Eu) Hybrid Nanofibers. *International Journal of Polymeric Materials*, 2010, **59(11)** 863-872.

[8] Wang B,Xiao L L,Xiang H X,et al.Enhanced Fluorescence Emission of Bonding Type Rare Earth Complexes Eu(MAA)$_3$Phen. *Materials Science Forum*,2017, **898** 1839-1843.

[9] X H Zhan, Z L Xiao, H H Zhan , et al . Syntheses and fluorescent properties of complexes of Eu(III) with HTTA, TPPO and benzoic acid. *Journal of Rare Earths*,2009, **27(3)** 369-372.

[10] Yu soff N H, Shualdi W, Azmi N A A N M, et al. Optical and Color Properties of Green and Red Phosphor Mixtures Doped into Hybrid Sol-Gel Matrix to Generate White Light. *Advanced Materials Research*,2016, **1133(12)** 419-423.

The performance of a novel inorganic organic composite drilling fluids loss reducer AM/AA/GO

Xiuping Chen [1, 2], Yunyue Zhang [3], Sheng Fan [1, 2], Weichao Du [4, *]

1 Research Institute of Petroleum Engineering, SINOPEC Northwest Oilfield Branch, Urumqi 830011, China;
2 Key Laboratory of EOR for Carbonate Fractured-Vuggy Reservoirs, SINOPEC, Urumqi 830011, China;
3 Shaanxi Coal Geological Oil and Gas Drilling and Exploitation Co. Ltd., Xian, 710054, China;
4 College of Chemistry and Chemical Engineering, Xi'an Shiyou University, Xi'an 710065, China
E-mail: duweichao@xsyu.edu.cn

Abstract. In this paper, an inorganic/organic water-based drilling fluids reducer AM/AA/GO was prepared by using acrylamide (AM), acrylic acid (AA) and graphene oxide (GO) as raw material. Performance evaluation showed that the AM/AA/GO has an excellent filtration loss reduce performance. It has a good salt resistance performance and its temperatures resistance ability was up to 140 ℃. The comprehensive results showed that AM/AA/GO is expected to be used in anti-collapse water-based drilling fluids.

Introduction

In the shale reservoir, the hydration expansion dispersion of wellbore is easily lead to the borehole collapse, block falling, sticking and other drilling projects in drilling engineering. Therefore, high requirements are put forward drilling fluids for loss reduce, shale inhibition and plugging ability [1]. The drilling fluids cake is the first drilling barrier to prevent the intrusion of water into oil and gas formation, therefor, the quality and the effective degree of drilling fluids cake will be directly affect the stability of wellbore, which is one of the most concerned indexe of drilling fluid performance to the drilling researcher [2]. As a fluids loss reducing agent, which participates in the cake formation and as an important part of mud cake, water-based drilling fluids loss reducer can greatly affects the filtration volume loss of drilling fluids [3, 4].

For a long time, there are many fluids loss reducers products, most of which are binary, ternary, quaternary, and even ternary co-polymers. The monomers are mostly acrylic acid (AA), acrylate (MA), acrylamide (AM) and functional monomer containing high-temperature-resistant function groups. These co-polymers undertake the task of reducing filtration loss within a certain period time and have promoted the progress of drilling engineering [5-7].

Nanomaterial have their own unique surface characteristics and thermal conductivity, and they have unparalleled material advantages in drilling fluid for rheology control, thermal conductivity and filtration loss performance, and have shown good development prospects in drilling fluids. Integrating the respective advantages of polymers and nanomaterial, nanoparticle/polymer composites have been initially explored in drilling fluids and have shown excellent performance [8]. But at the same time, it

IOP Conf. Series: Materials Science and Engineering **758** (2020) 012096 doi:10.1088/1757-899X/758/1/012096

is an indisputable fact that nanomaterial are easy to re-unite in the aqueous solution and are sensitive to salt, which has led to the limitation of their large-scale applications [9, 10].

Based on the above research background, in this paper, an novel water-based drilling fluids reducer AM/AA/GO was prepared by using acrylamide (AM), acrylic acid (AA) and graphene oxide (GO) as raw material. The performance of AM/AA/GO was evaluated by using filtration loss experiments.

1. Experimental

1.1. Materials

AM、AA、NaOH、(NH4)2S2O8、NaHSO3，Aladdin reagent，all the above reagents are analytically pure; graphite powder, Qingdao Chenyang Graphite Co., Ltd; Bentonite was purchased from Xinjiang Xiazi Street Bentonite Co., Ltd .; shale samples were taken from Chongqing, Drilling fluids materials were all purchased from Shandong Deshunyuan Co., Ltd.

1.2. Preparation of AM/AA/GO

1.7 g acrylic acid and 30 mL deionized water added into a round-bottomed flask, the solutions was adjusted to 7 by 25% NaOH, and 3.7 g acrylamide was add by following. Hereafter, the initiator K2S2O8 and NaHSO3 (n: n = 1: 1), which the initiator concentration is 0.8 wt% relative to the total monomer amount were added. The polymerization was carried out at 50 °C for 0.5 h while stirring, and then isopropanol was added to the solution, polymerization was preceded at 70 °C. The resulting product was obtained by repeatedly washing with ethanol to remove monomers, and the polymer AM/AA was obtained. Add 1g AM/AA and 20 mL deionized water in a round-bottomed flask, and then raised the temperature to 30 ° C, slowly add 0.7 g OP-10, and then add 0.75 g GO to the round-bottomed flask . Ultrasonic reaction was carried out for 2 h at a sealed condition and AM/AA/GO was obtained after washing, drying and shearing granulation.

1.3. Drilling fluids performance evaluation

1.3.1. Based muds preparation.

Add 10 L distilled water into the iron container and 400 g calcium bentonite was added in the mixing state. After that, 2.4g Na2CO3 was added and maintained the based muds 48 h.

1.3.2. Drilling fluids performance tests.

The performance tests of drilling fluids was according to the standard SY / T 5621-1993 of China.

2. Results and discussion

2.1. The effect of AM/AA/GO on rheological properties of based muds

Different amount of AM/AA/GO was added to the based muds which has been pre-hydrated for 48 h, and the effect result of AM/AA/GO on the rheology of based muds are shown in Table 1

Table 1 Effect of AM/AA/GO on rheology of based muds

system	AV/mPa·s	PV/mPa·s	YP/Pa	YP/PV	Gel/Pa/Pa
Based muds	5.5	5	0.5	0.1	1/1
Based muds +0.1% AM/AA/GO	7	5	2	0.4	1/1
Based muds +0.2% AM/AA/GO	10	7	3	0.4	1/2
Based muds +0.5% AM/AA/GO	12	8	4	0.5	2/3
Based muds +0.8% AM/AA/GO	15	10	5	0.5	5/8
Based muds +1.0% AM/AA/GO	23	14	9	0.6	7/13

It can be seen from Table 1 that with the increase addition amount of AM/AA/GO, the viscosity of the based muds system show a gradual rise trend. The rise of shear force is very conducive to the suspension of drilling fluid to drilling cuttings, which can bring cuttings out of the wellbore. When the dosage was more than 0.8%, the viscosity of based muds has been improved obviously. It can be seen that the most suitable dosage of AM/AA/GO in the based muds was 0.8%.

2.2. Effect of salt on rheological property of AM/AA/GO based muds

When drilling to the salt gypsum layer, the performance of drilling fluids will be changed due to the invasion of salts. In addition, AM/AA/GO contains graphene oxide which is sensitive to salts, so it is necessary to investigate its salt resistance ability. Figure.1 and Figure. 2 show the effect of NaCl and $CaCl_2$ on the rheological property of 0.8% AM/AA/GO based muds, respectively.

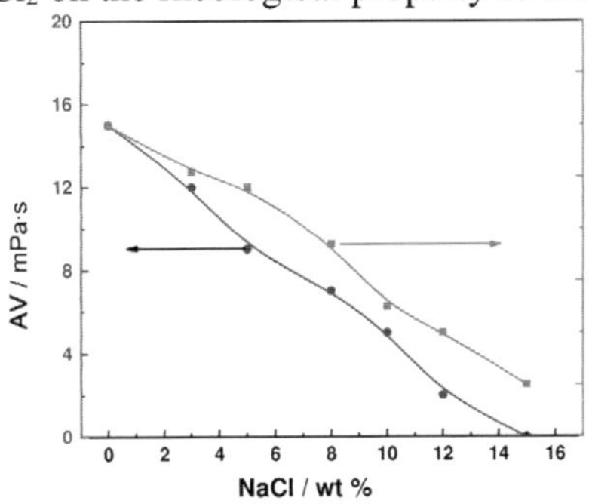

Figure 1. Effect of NaCl on rheological property of AM/AA/GO based muds

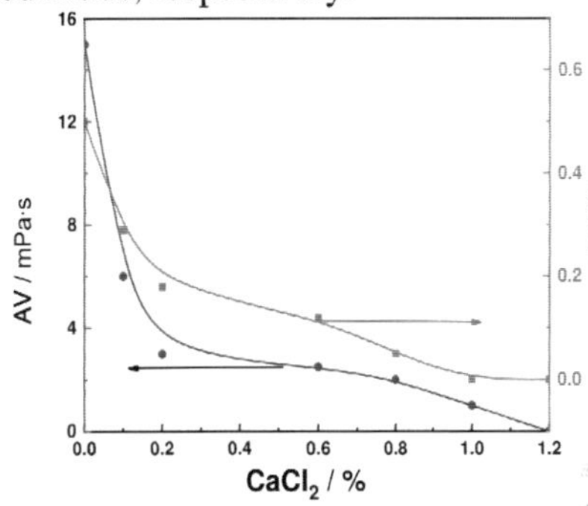

Figure 2. Effect of CaCl2 on rheological property of AM/AA/GO based muds

Inorganic salts can ionize cations in water, and cations have a great influence on the branched chain extension of polymers. The amide group of AM/AA/GO can hydrolyze carboxylic acid anion in aqueous solution. In addition, AM/AA/GO itself has carboxylic acid group. The salt in drilling fluid can make the double electric layer and hydration layer of carboxylic acid thin, weaken the repulsion of carboxylic acid group, and make the molecular chain of AM/AA/GO curl, thus losing the rheological property. It can be seen from Figure. 1 and Figure. 2 that with the addition increase of salts, the viscosity of AM/AA/GO based muds showed a trend of attenuation, but CaCl2 has a greater impact on it. When the dosage of NaCl is 12%, the apparent viscosity of AM/AA/GO based muds has been reduced to 2 m Pa•s, when the dosage of $CaCl_2$ is 1.0%, the suspension performance of the system has been lost.

2.3. Effect of salt on rheological property of AM/AA/GO based muds

On the one hand, temperature will affect the space expansion of polymer, on the other hand, it will promote the decomposition of polymer at high temperature. The effect of temperature on the rheological properties of 0.8 % AM/AA/GO based muds was investigated in the laboratory, as shown in Table 2.

Table 2 Effect of temperature on rheological property of AM/AA/GO based muds

condition	AV/mPa·s	PV/mPa·s	YP/Pa	YP/PV	Gel/Pa/Pa
Room temporature	15	10	5	0.5	5/8
100℃×16h	19	12	7	0.6	6/8
120℃×16h	15	9	6	0.7	4/5
130℃×16h	13	8	5	0.6	3/4
140℃×16h	9	6	3	0.5	1/2
150℃×16h	5	3	2	0.7	1/1

It can be seen from Table 2 that the rheological properties of 0.8% AM/AA/GO based muds show a downward trend with the increase of hot rolling temperature. However, the rheological property of drilling fluid increased slightly after 100℃×16h. The reason may be that under 100 ℃, the polymer chain of AM/AA/GO has not been fully extended, and AM/AA/GO has not fully developed its own viscosity increasing effect. With the further increase of hot rolling temperature, AM/AA/GO has begun to decompose, the spatial network state between polymers has been destroyed to a certain extent, and the viscosity of AM/AA/GO has begun to decline. When the hot rolling temperature reached to 140 ℃, the rheological property of the system has been completely destroyed. It can be seen that the temperature resistance of AM/AA/GO in the based nuds system can be as high as 140 ℃.

2.4. Effect of salt on rheological property of AM/AA/GO based muds

The filtration reduction performance of AM/AA/GO was studied in 4% based muds, and the result is shown in Figure 3

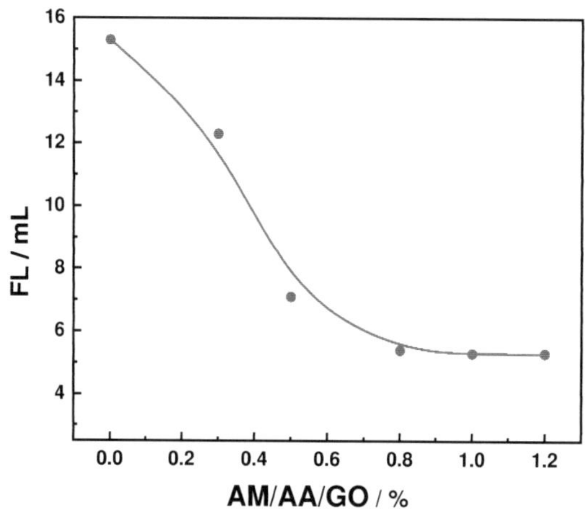

Figure 3. Fluids loss performance study of AM/AA/GO

AM/AA/GO was combination of inorganic and organic materials, its physical properties show the dual characteristics of the rigidity of inorganic particles and the deformability of polymer. It can reduce the porosity of mud cake by using the role of plugging and bridging of inorganic particles, and then play a role of filtration reduction. It can be seen from Fig. 3 that the filtration capacity of based muds was 15.3 mL. With the increase of the addition of filtrate reducer AM/AA/GO, the filtration amount of polymer based muds has been significantly reduced, when added mount of AM/AA/GO was 0.8%, the water loss was only 5.4 mL. AM/AA/GO / % FL / mL

3. Conclusion

In this paper, a novel water-based drilling fluids reducer AM/AA/GO was prepared with AM and AA and GO. The performance evaluation results showed that AM/AA/GO has a nice water loss reduction performance. AM/AA/GO can rely on inorganic / organic components to block the muds cake, reduced the water loss of drilling fluids and maintain the stability of the wellbore.

Acknowledgments

The authors would like to thank the Open Fund (PLC20190703) of State Key Laboratory of Oil and Gas Reservoir Geology and Exploitation (Chengdu University of Technology); Open Fund (PLN201814) of the State Key Laboratory of Oil and Gas Reservoir Geology and Exploitation (Southwest Petroleum University); Open Fund (KFKT2019-13) of the Key Laboratory of Auxiliary Chemistry and Technology for Chemical Industry, Ministry of Education Shaanxi University of Science and Technology; Xi'an Shiyou University 2019 Innovation and Entrepreneurship Training Program for College Students (201910705009) for their financial support.

References

[1] Gou S, Yin T, Xia Q and Guo Q 2015 *RSC Adv*. **41** 32064.
[2] Huang Y, Zhang D and Zheng W 2019 *J Appl Polym Sci*. **30** 47813.
[3] Salami O and Plank J 2012 *J Appl Polym Sci*. **4** 1449.
[4] Liu X, Liu K, Gou S, Liang L, Luo C and Guo Q 2014 *Ind Eng Chem Res*. **8** 2903.
[5] Rafati R, Smith S, Haddad A, Novara R and Hamidi H 2018 *J Petrol Sci Eng*.**161** 61-76.
[6] Hassani S, Amrollahi A, Rashidi A, Soleymani M and Rayatdoost, S 2016 *J Petrol Sci Eng*.**146** 183-190.
[7] Gou S, Yin T, Liu K and Guo Q 2015 *New J Chem*. **3** 2155.
[8] Aftab A, Ismail A, Ibupoto Z 2016 *Egy J Petrol*. **2** 291.
[9] Aftab A, Ismail A, Ibupoto Z, Akeiber H and Malghani M 2017 *Renew Sust Energ Rev*. **76** 1301.
[10] Abdo J and Haneef M 2013 *Appl Clay Sci*. **86** 76.

IOP Conf. Series: Materials Science and Engineering **758** (2020) 012097 doi:10.1088/1757-899X/758/1/012097

Low velocity impact analysis of high-order rectangular FG-CNTRC plates using the weak form QEM

Baojun Di[1], Qigao Hu[2], Zhiqiang Shen[2*], Wei Zhang[1] and Chongxu Wang[1]

1 National University of Defense Technology, Changsha, China
E-mail: zq_shen@outlook.com

Abstract. The functionally graded carbon nanotubes reinforced composites (FG-CNTRC) plates have excellent properties and broad application prospects. In the analysis of low velocity impact response of functionally graded carbon nanotubes reinforced composite plates, innovative numerical formulations are urgently needed. As a new high-order numerical method, the weak form quadrature element method (QEM) has its unique advantages and has been widely used in the engineering structural field. In this paper, the nonlinear Hertzian contact law is employed to describe the impact process between a rigid impactor and the rectangular FG-CNTRC plate, that was modelled with the Reddy's high-order shear deformation theory. By using the weak form QEM, the calculation model for the impact system is established, and the numerical solution is carried out with the help of Newmark-β step-by-step time domain integration method and *Newton-Raphson* iteration technology. The effectiveness and accuracy of the weak form QEM formulation were verified through the comparison with the existing literature and the relevant low velocity impact properties of FG-CNTRC plate are explored..

1. Introduction

Functional gradient carbon nanotube reinforced composite (FG-CNTRC) is a new type of advanced composite material using carbon nanotubes distributed in the matrix as reinforcements[1]. In order to give full play to the material advantages of carbon nanotubes and effectively improve the overall macro mechanical properties and designability of the composite, Shen Huishen first applied the concept of functional gradient to carbon nanotubes reinforced composite in 2009, and made a nonlinear bending analysis of the composite plate with gradient arrangement of carbon nanotubes reinforcement in the polyvinylidene chloride (PmPv) matrix[1]. Many scholars have done a lot of research on the bending, buckling and free vibration analyses of the composite beam, plate and shell structure, FG-CNTRC shows excellent thermal, mechanical and electrical properties, which indicates that it will have a broad application prospect in aerospace, energy engineering, bioengineering, military protection and other fields.

Generally speaking, when the duration of impact process is large enough to ignore the influence of strain rate and stress wave propagation of the impacted structural material, it can be called "low velocity impact". FG-CNTRC plate, as a structural component, is inevitably impacted by other small objects at low speed in the process of processing and utilizing. Especially when such components are directly used for shielding and protection purposes. Therefore, it is necessary to analyze the low-velocity impact dynamic response of this kind of plate.

Compared with a large number of research results in the field of low-velocity impact of traditional fiber-reinforced composite plates, the research work on low-velocity impact of FG-CNTRC plate structure is very limited. In the existing research, various kinds of analytical or numerical methods are used to obtained the solutions. Low-velocity impact is a time-dependent dynamic response process. To analyze this problem, we need to solve the original continuous problem discretely in both space and time domain. For the discrete solution of space domain, the displacement type control differential equation derived from Reddy's high-order shear deformation theory[2] needs to satisfy C^1 continuous condition on the element boundary, which increases the computational complexity of traditional finite element method. For example, the two-dimensional four node finite element rectangular element proposed by Ebrahimi et al.[3] requires 15 degrees of freedom for each node, which urges many researchers to adopt other numerical formulations in space domain, including the traditional Navier series method[4], the Rayleigh Ritz method[5,6], the new quadratic perturbation method[7], the differential quadrature method and some meshless method[8]. These methods have their advantages, but they also have shortcomings. As Navier series method, quadratic perturbation method is generally only suitable for plate structures with specific geometric features and boundary conditions; meshless method has strong adaptability to different geometric features and boundary conditions, but its preprocessing process is complex, and the boundary conditions are not easy to be directly applied[9]. The research on the low-velocity impact response of FG-CNTRC plate structure is not enough, and the demand for new and efficient numerical methods is still urgent.

Zhong Hongzhi[10] proposed a new numerical method in 2007, weak form quadrature element method (QEM), which has been widely used and developed in the field of engineering structure analyses. The weak form QEM is a new numerical method based on the weak form (control function) of the differential equation of the problem to be solved. It uses a unified element node to calculate the integral and differential in the control function of the problem. It has many advantages, such as accurate and efficient, direct implementation steps, simple and intuitive pre- and post- processing, etc. The application of this method shows its attractive advantages in dealing with complex geometry, inhomogeneous materials and nonlinear problems compared with the traditional finite element method. The calculation and analysis of linear elastic problems of structures by the method of quadrature element is mature, and the research focus is gradually turning to the field of quasi-static and dynamic nonlinear analysis.

In this paper, the nonlinear Hertzian contact law is employed to describe the impact process between a rigid impactor and the rectangular FG-CNTRC plate. Two different models for the impact system are considered. In model I, the impactor and FG-CNTRC rectangular plate are considered respectively, and the nonlinear impact contact force is considered as the external load; in model II, the impactor and FG-CNTRC rectangular plate are considered as the overall system, and the impact contact force is viewed as the internal force. By using the weak form quadrature element method, the calculation model of the system is established, and the solution is carried out by using the Newmark-β step-by-step time domain integration method and *Newton-Raphson* iteration technology. The effectiveness and accuracy of the weak form QEM formulation were verified through the comparison with the existing literature and the relevant low velocity impact properties of FG-CNTRC plate are explored.

2. Theoretical formulation

2.1. Material model

In this paper, FG-X rectangular plate adopts FG-X distribution and establishes coordinate system, as shown in the Figure1.

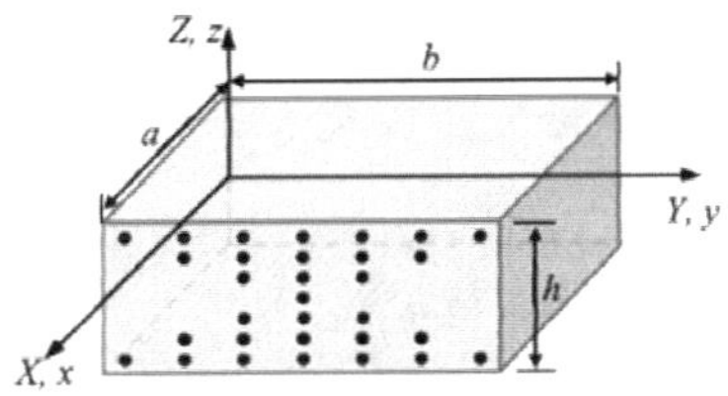

Figure 1. FG-X distribution of FG-CNTRC rectangular plate.

In this paper, we use the generalized mixed law model[1] to describe the FG-CNTRC plate. According to this model, Young's modulus and shear modulus of FG-CNTRC plate can be expressed as follows:

$$E_{11} = \eta_1 V_{CNT} E_{11}^{CNT} + V_m E^m,\ \eta_2 / E_{22} = V_{CNT} / E_{22}^{CNT} + V_m / E^m,$$
$$\eta_3 / G_{12} = V_{CNT} / G_{12}^{CNT} + V_m / G^m,\ \nu_{12} = V_{CNT}^* \nu_{12}^{CNT} + (1 - V_{CNT}^*) \nu^m. \tag{1}$$

In Eq. (1), E_{11}^{CNT}, E_{22}^{CNT} and G_{12}^{CNT} are the Young's moduli and shear modulus of the CNT respectively. The polymer matrix is assumed to be elastic and isotropic, with Young's modulus E^m and shear modulus G^m. Three efficiency parameters, η_1, η_2 and η_3, are introduced in order to consider the scale-dependent effect of the CNTs. The three efficiency parameters were determined by matching the Young's moduli and shear modulus of CNTRC observed from the molecular dynamics simulations with the numerical results from the extended rule of mixture[1]. Moreover, ν_{12}, ν_{12}^{CNT} and ν^m are the Poisson's ratio of the FG-CNTRC plates, CNT and matrix respectively..

2.2. Reddy's high-order plate theory

The Reddy's high-order shear deformation theory[2] is adopted to model the displacement field of the FG-CNTRC plate as follows:

$$d(x,y,z,t) = \begin{bmatrix} u + z\phi - (4/3h^2)z^3(\phi + w_{,x}) \\ v + z\varphi - (4/3h^2)z^3(\varphi + w_{,y}) \\ w \end{bmatrix} \tag{2}$$

where u, v and w are the displacement components of an arbitrary material point (x, y, z) on the neutral surface of the composite plate along the X, Y and Z coordinate directions at the time t; ϕ and -φ are the rotations of the normal to the neutral surface about the Y and X coordinate axis, respectively. In addition, $w_{,x}$ and $w_{,y}$ denote the first order derivatives of w with respect to the X, Y coordinate.

Thus, for small strains and moderate rotations problems as the paper concerned, the non-zero strains associated with the displacement field (Eq. (2)) can be expressed as

$$\varepsilon = \begin{bmatrix} \varepsilon_{xx} & \varepsilon_{yy} & \gamma_{xy} & \gamma_{yz} & \gamma_{xz} \end{bmatrix}^T = \bar{\varepsilon} + \tilde{\varepsilon} \tag{3}$$

where $\bar{\varepsilon}$ and $\tilde{\varepsilon}$ are the linear strain term and the nonlinear strain term respectively.

$$\tilde{\varepsilon} = \frac{1}{2}\begin{bmatrix} w_{,x}^2 & w_{,y}^2 & 2\,w_{,x}w_{,y} & 0 & 0 \end{bmatrix} \tag{4}$$

The strains in Eq. (4) are also known as von Kármán strains.

2.3. Impact contact law

The indentation $\alpha(t)$ is related to the impact load with the help of nonlinear Hertzian contact law [9],

$$F(t) = K_c \alpha(t)^p \tag{5}$$

$\alpha(t)$ can also be expressed as

$$\alpha(t) = s(t) - w(t) \tag{6}$$

Where, $s(t)$ is the displacement at the tip of the impactor, $w(t)$ is the displacement at the top of the plate.

2.4. The weak form quadrature element formulation

Considering the dynamic response of an FG-CNTRC plate, it is assumed that the solutions for the kinematic and stress variables at all time steps, from time t_0 to time t_n, have been solved, and that the solution for time t_{n+1} can be achieved employing the principle of virtual displacements,

$$\delta W_{\text{int}}^{t_{n+1}} = \delta W_{ext}^{t_{n+1}} \tag{7}$$

where $\delta W_{\text{int}}^{t_{n+1}}$ and $\delta W_{ext}^{t_{n+1}}$ are the internal and external virtual work expression at time t_{n+1} respectively. As for the Total Lagrange Formulation (T.L.), all variables in Eq. (7) are referred to the initial undeformed configuration, i.e. $t = t_0$, of the FG-CNTRC plate. Eq. (7) can be expressed as

$$\int_V \Delta\bar{\varepsilon}^T E\delta\Delta\bar{\varepsilon} dV + \int_V \delta\Delta\tilde{\varepsilon}^{T\ t_n}\sigma dV = \delta^{t_{n+1}} W_{ext} - \int_V \delta\Delta\bar{\varepsilon}^{T\ t_n}\sigma dV \tag{8}$$

First of all, similar to the finite element method, the weak form quadrature element method divides the whole physical domain of the problem to be solved into several continuous sub regions. In each sub region, the control equation is transformed from physical domain to unit square standard field by geometric mapping for the convenient of numerical integration and differential utilizing. After determine the number of integration points in the sub area, according to the Gauss-Lobatto integral rule[11], the integral term in Eq. (8) is expressed with the function value and their differential term at the numerical integral point. Take the one-dimension problem as example, the Gauss-Lobatto numerical integration formulation whose standard domain is the closed interval of [-1,1] is expressed as:

$$\int_{-1}^{1} f(\xi)d\xi = \sum_{i=1}^{N} W_i f(\xi_i) \tag{9}$$

where, ξ_i is the coordinate of the integral point i in the sub region, N is the total number of the integral points, and W_i is the integral coefficient at ξ_i.

Furthermore, by using the differential quadrature method (DQM) and generalized differential quadrature method (GDQM), the differential terms contained in the above equations are further expressed as function values at differential points which were the same with the numerical integral point. Similarly, take the one-dimension problem as example, The differential quadrature method is as follows:

$$\left.\frac{\partial^m f(\xi)}{\partial \xi^m}\right|_{\xi=\xi_i} = \sum_{i=1}^{N} h_i^{(m)} f(\xi_i) \quad (i = 1, 2, \cdots, N) \tag{10}$$

where $h_i^{(m)}$ is weighting coefficient and $f(\xi_i)$ is the function value at the point ξ_i.

After the above process, the Eq. (8) finally becomes the summation expression of the function values of the integral point. Next, using the transformation matrix, the expression in the standard domain is transformed into the expression in the physical domain, and the global incremental equation is obtained.

$$\boldsymbol{M}^{\ t_{n+1}}\ddot{\boldsymbol{d}} + \boldsymbol{K}\Delta\boldsymbol{d} = {}^{t_{n+1}}\boldsymbol{F}_{\text{ext}} - {}^{t_n}\boldsymbol{F}_{\text{int}} \tag{11}$$

Among them, $\boldsymbol{M}$ is the overall mass matrix, $\boldsymbol{K}$ is the overall stiffness matrix, $\ddot{\boldsymbol{d}}$ is the node acceleration vector, $\boldsymbol{F}_{\text{ext}}$ is the external force vector, $\boldsymbol{F}_{\text{int}}$ is the internal force vector.

Then, solve the above equation. Since the impact load is a variable vs time, it needs to be solved by the famous constant acceleration method (Newmark-β type time integration method). What must be done is to divide the whole process into multiple time steps. Because of the geometrical nonlinearity of the plate, the stiffness matrix $\boldsymbol{K}$ is a function of the displacement vector $\boldsymbol{d}$, so it is necessary to use *Newton-Raphson* method to iterate in each load step until the equation meets the required convergence criteria. Simultaneously, using Newmark-β time integration method, the acceleration and velocity are obtained immediately.

3. Numerical analysis and discussion

In this paper, a numerical example in the literatures[9,12] is selected and compared with the results in the literature to verify the quadrature element model of the low-speed impact response of the FG-CNTRC rectangular plate. In the calculation example selected in this paper, the relevant parameters are: the density of metal spherical impactor is 7960kg/m^3, the radius is 10mm, and the impact speed is 1.0 m/s. In the FG-CNTRC rectangular plate, the arrangement of carbon nanotubes is FG-X distribution. The length and width of the rectangular plate are both 0.5m, and the width thickness ratio is 30. The impact position of the impactor is the center point of the plate.

When the QEM is used for calculation, the FG-CNTRC rectangular plate is divided into one quadrature element. Four different schemes of integration points are adopted in the quadrature element, namely 7×7,9×9,11×11,13×13. As mentioned before, for the impact load in the research system, two different models are used to deal with it, and the response under the two models is obtained respectively, so as to further prove the effectiveness of the calculation model. The analysis results are as follows:

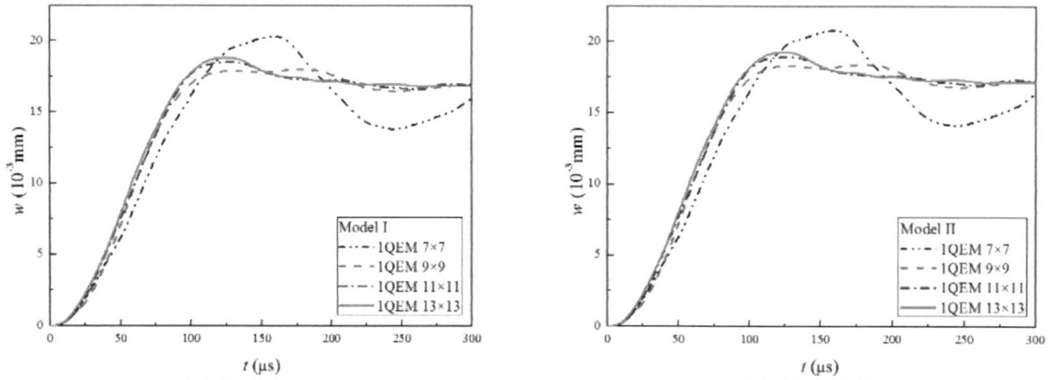

(a) Model I (b) Model II

Figure 2. The central deflection histories of the clamped FG-X rectangular plates.

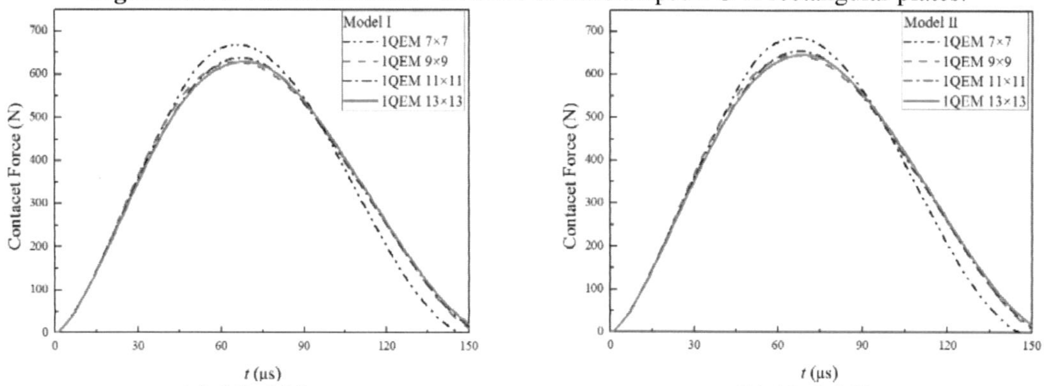

(a) Model I (b) Model II

Figure 3. The contact force histories of the clamped FG-X rectangular plates.

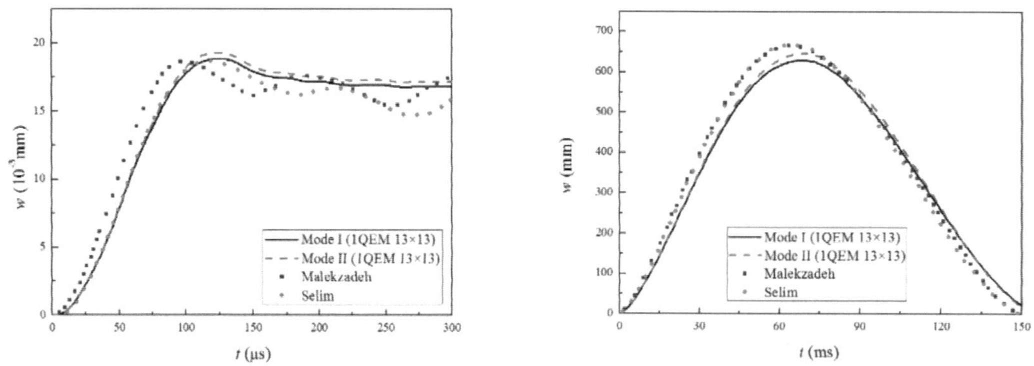

(a) The central deflection (b) The contact force

Figure 4. The comparison of the numerical results from the two models and related literatures.

The time history curves of deflection and impact contact force calculated by the two models are given in Figure 2-4 respectively. The results show that: (1) with the increase of the number of numerical integration points, the calculation values of deflection and impact contact force improve rapidly; in order to get the ideal calculation results, more than 11 × 11 numerical integration points are needed. (2) The calculation results in this paper are in good agreement with those in references[9,12]. The calculated values of deflection and impact contact force in model II are slightly higher than those in model I, and the relative difference between the two maximum values is not more than 3%.

4. Conclusion

In this paper, the problem of single low-velocity impact at the center of FG-CNTRC rectangular plate is studied by using the weak form quadrature element method. Through the research of this paper:

- The weak form quadrature element model of dynamic response of FG-CNTRC rectangular plate under single low-velocity impact load is established.
- The efficiency and applicability of the weak form quadrature element method in the geometric nonlinear problems of FG-CNTRC rectangular plates and the load nonlinear problems of impact are verified.
- It provides a new theoretical calculation method and reference data for the research and application of FG-CNTRC plate in impact dynamic response.
- It provides a solid theoretical basis for the study of FG-CNTRC plates with different shapes, continuous impact and eccentric impact.

5. References

[1] Shen H S. Nonlinear bending of functionally graded carbon nanotube-reinforced composite plates in thermal environments[J]. *Composite Structures*, 2009, 91(1): 9-19

[2] Reddy J N. Mechanics of laminated composite plates and shells: theory and analysis[M]. CRC press, 2004.

[3] Ebrahimi F, Habibi S. Low-velocity impact response of laminated FG-CNT reinforced composite plates in thermal environment[J]. *Advances in Nano Research*, 2017, 5(2):69-97.

[4] Feli S, Karami L, Jafari S S. Analytical modeling of low velocity impact on carbon nanotube-reinforced composite (CNTRC) plates[J]. *Mechanics of Advanced Materials & Structures*, 2017(6):1-13.

[5] Zarei H, Fallah M, Bisadi H, et al. Multiple impact response of temperature-dependent carbon nanotube-reinforced composite (CNTRC) plates with general boundary conditions[J]. *Composites Part B: Engineering*, 2017, 113: 206-217.

[6] Khalkhali A, Geran Malek N, Bozorgi Nejad M. Effects of the impactor geometrical shape on the non-linear low-velocity impact response of sandwich plate with CNTRC face sheets[J]. *Journal of Sandwich Structures & Materials*, 2018: 1099636218778998.

[7] Fan Y, Wang H. Nonlinear low-velocity impact analysis of matrix cracked hybrid laminated plates containing CNTRC layers resting on visco-Pasternak foundation[J]. *Composites Part B: Engineering*, 2017, 117: 9-19.

[8] Fallah M, Daneshmehr A R, Zarei H, et al. Low velocity impact modeling of functionally graded carbon nanotube reinforced composite (FG-CNTRC) plates with arbitrary geometry and general boundary conditions[J]. *Composite Structures*, 2018, 187:554-565.

[9] B.A. Selim, Zhang L W, Liew K M. Impact analysis of CNT-reinforced composite plates based on Reddy's higher-order shear deformation theory using an element-free approach[J]. *Composite Structures*, 2017, 170:228-242.

[10] Zhong H, Yu T. Flexural vibration analysis of an eccentric annular Mindlin plate[J]. *Archive of Applied Mechanics*, 2007, 77(4): 185-195.

[11] Davis PJ, Rabinowitz P. Methods of numerical integration Courier Corporation;2007.

[12] Malekzadeh P, Dehbozorgi M. Low velocity impact analysis of functionally graded carbon nanotubes reinforced composite skew plates[J]. *Composite Structures*, 2016, 140:728-748.

Synthesis and property of a novel clay anti-swelling agent

Weichao Du [1, 2, *], Jinwen Wei [1], Yunyue Zhang [3], Gang Chen [1], Jie Zhang [1]

1 Shaanxi Province Key Laboratory of Environmental Pollution Control and Reservoir Protection Technology of Oilfields, Xi'an Shiyou University, Xi'an 710065, China;
2 Key Laboratory of Auxiliary Chemistry and Technology for Chemical Industry, Ministry of Education, Shaanxi University of Science and Technology, Xi'an 710021, China;
3 Shaanxi Coal Geological Oil&Gas Drilling Co. Ltd., Xian 710054, China
E-mail: duweichao@xsyu.edu.cn

Abstract. In this work, a novel oil field clay anti-swelling agent triethylallyl ammonium bromide (TEAAB) was prepared. The effect of reaction condition on TEAAB yield was studied, and the optimal reaction conditions is reaction temperature 60 °C, reaction time 7 hours, and molar ratio of raw materials of (triethylamine: bromopropene) = 1.4. Anti-swelling results showed that TEAAB has an excellent anti-swelling performance and the anti-swelling rate is up to 98% when the dosage was 6%.

Introduction

Shale make up over 75% of the drilled formations, and over 70% of the borehole problems are related to shale instability clay [1]. Wellbore instability is one of the main problems that engineers meet during drilling, and it is an important factor to the reduction of reservoir permeability. When shale meets the working fluid outside, the water will be inevitable penetrates into the lattice structure of clay, which can cause the result that the volume of clay formation is increased, and leading to the decline of oil and gas reservoir permeability [2-4]. The degree costs for oil and gas wells drilling have continually increased over the past decades, exploration and development operations have expanded globally as the economics of exploring and producing for oil and gas have improved in drilling technology [5-6]. Thence, the reservoir protection is extraordinarily important in the exploration. Therefore, the development of novel and efficient clay anti-swelling agent become significant issue of oilfield workers [8-10].

The organic cationic and its polymer, due to the clay-stabilization ability is much more than inorganic salts and water-soluble polymer electrolyte, has served as the most developed and widest used anti-clay swelling reagent in the oil field. The clay surface will with a negative charge due to lattice substitute. A variety of clay anti-swelling agent have been implicated in the oilfield, however, the cation volume is too large to function well.

Therefore, in this work, we have designed a suitable small organic cation: triethylallyl ammonium bromide (TEAAB). The effect of reaction condition on the target yield was studied, and we have studied its anti-swelling performance.

1. Experimental

1.1. Materials

Triethylamine, allyl chloride, acetone, anhydrous ethanol, Chengdu Kelong Chemical Reagent Factory, the above reagent are all analytical grade, Sodium bentonite (Na-MMT) was purchased from Xiazi Street Bentonite Co., Ltd.

1.2. Preparation of TEAAB

Pre-heat the water bath to the required temperature, and accurate amount of triethylamine, acetone and bromopropene are added to the water bath, keep the condensation reflux at a certain temperature. Because the product is insoluble in acetone, a snow-white knot is obtained after the reaction is completed. Then, decant the acetone and dry the product in a vacuum oven to constant weight and calculate the yield.

1.3. Structure Characterization

TEAAB was firstly washed with acetone to remove the non-polarity impurities and then washed with ethanol to elute polarity impurities, after drying in the vacuum oven, FT-IR spectra were recorded via a WQF-520 Fourier transform infrared spectrometer in the wave number range of 4000-500 cm−1.

1.4. Anti-swelling performance evaluation of TEAAB

Weigh 0.50g bentonite and 10ml pure water to a 10 ml centrifuge tube and centrifugated for 15 min at 1500 r/min. Then read the volume V1 of the mixture. Repeat the previous step, added 10 ml kerosene and read the volume V2 of the mixture. Then added 10 ml TEAAB and read the volume V3 of the mixture. Anti-swelling rate was calculated by the following formula (1):

$$\sigma = \frac{V_1 - V_3}{V_1 - V_2} \tag{1}$$

2. Results and discussion

2.1. Effect of reaction conditions on TEAAB yield

2.1.1 Reaction temperature

The smooth progress of the reaction needs to be carried out at a certain temperature, but the quaternary ammonium salt will decompose at a certain temperature. Under the conditions of reaction time for 6 hours and molar ratio of raw materials was (triethylamine: bromopropene =1: 1.2), the effect of the reaction temperature on the product yield was studied.

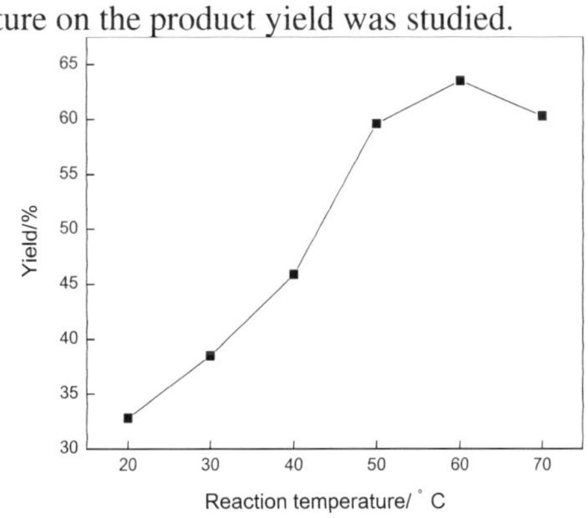

Figure.1 Effect of reaction temperature on TEAAB yield.

It can be seen from the results that as the reaction temperature increases, the yield of the product shows an upward trend. When the reaction temperature was 60 °C, the yield is the highest at 63.5%. As the reaction temperature was continue to increase, the yield was slightly reduced. The reason may be that the product has been decomposed at 70 °C. Therefore, the optimal reaction temperature was 60 °C.

IOP Conf. Series: Materials Science and Engineering **758** (2020) 012098 doi:10.1088/1757-899X/758/1/012098

2.1.2 Reaction time

The reaction time will greatly affect the product yield. When the reaction time is short, the reaction is not sufficient. When the reaction time is completed, it is not necessary to extend the reaction time. Under the conditions of reaction temperature at 60 °C and molar ratio of raw materials was (triethylamine: bromopropene =1: 1.2), the effect of the reaction time on the product yield was studied.

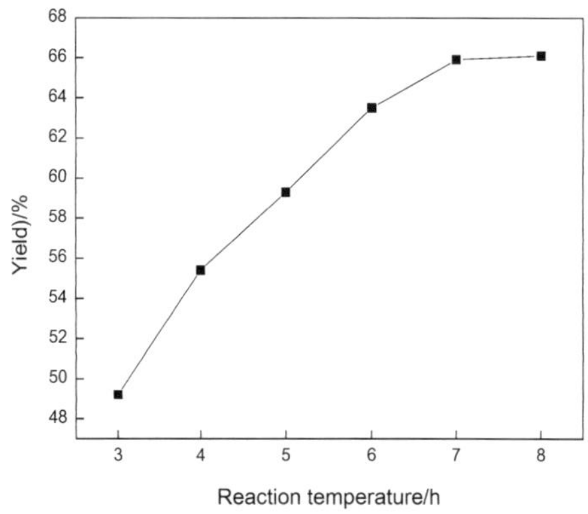

Figure.2 Effect of reaction temperature on TEAAB yield.

It can be seen from the Figure.2 that with the increase of reaction time, the yield of the product shows an upward trend, but when the reaction time exceeds 7 hours, the continued extension of the reaction time has little effect on the yield. Therefore, the optimal reaction time for this reaction is 7 h.

2.1.3 Reaction molar ratio

Both the reaction raw materials bromopropene and triethylamine are easy to volatilize. Therefore, the effect of the raw material ratio on the product needs to be examined. Under the conditions of reaction temperature at 60 °C and reaction time was 7 h, the effect of reaction molar ratio (triethylamine: bromopropene) on the product yield was studied, and the results are shown in Figure 3.

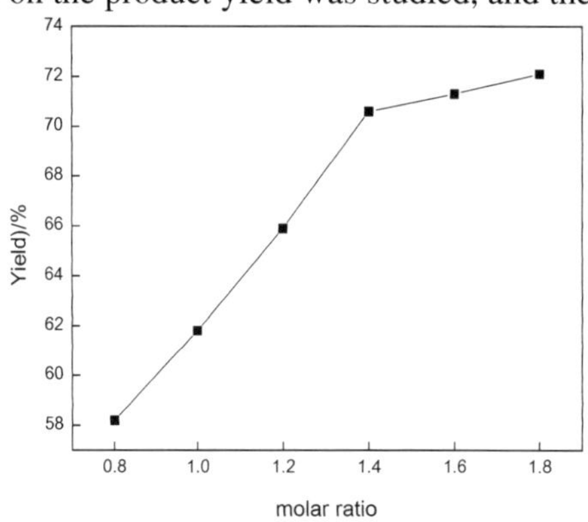

Figure.3 Effect of molar ratio on TEAAB yield.

From the results showed in Figure.3 we can conclude that as the molar ratio of the raw materials (triethylamine: bromopropene) increased, the yield of the product shows an upward trend. When the molar ratio was 1.4, the yield of the product was 71.3%. Therefore, the optimal reaction molar ratio (triethylamine: bromopropene) was 1.4.

It can be seen from the research results that the optimal reaction conditions for TEAAB are: reaction temperature is 60 ° C, reaction time is 7 hours, and molar ratio of raw materials of (triethylamine: bromopropene) is 1.4.

2.2. Anti-swelling performance evaluation

The anti-swelling performance of TEAAB with different concentrations on clay is shown in Figure 4. From left to right the concentration is 1%, 2%, 4%, 6%, 8%, and 10% and the corresponding anti-swelling rate is 94%, 95%, 95.6%, 97.2%, 98.0%, 98.0%. As the concentration increases, the anti-swelling effect has a clear upward trend. The anti-swelling performance of pure water TEAAB and kerosene is shown in Figure 5, from left to right is kerosene, 0.8 % TEAAB and pure water. It can be clearly seen that the product has a nice anti-swelling performance.

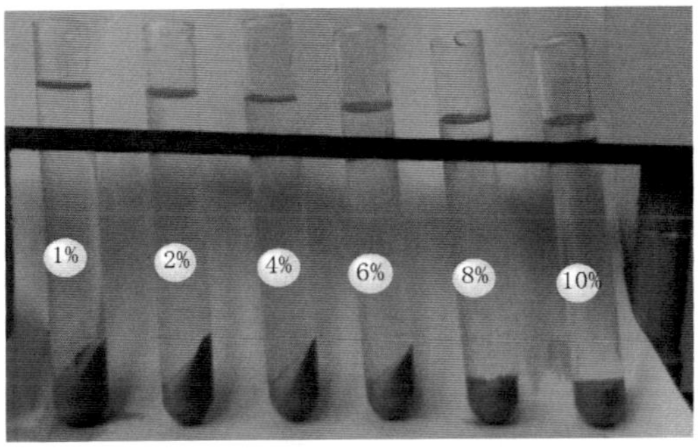

Figure.4 Effects of concentration on anti-swelling rate

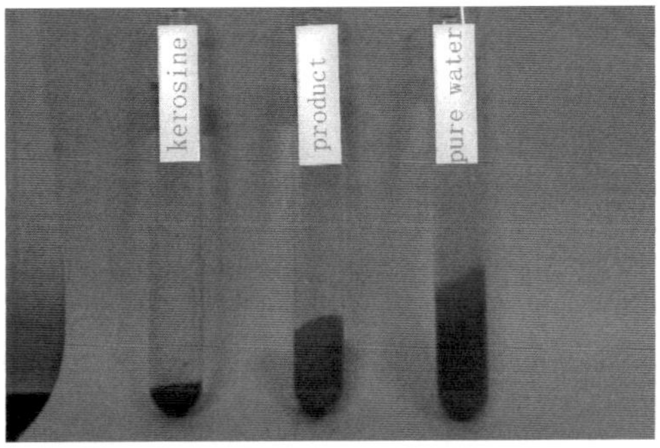

Figure.5 The anti-swelling capability of kerosene, product and pure water

3. Conclusion

In this work, a novel oil field clay anti-swelling agent triethylallyl ammonium bromide (TEAAB) was prepared from triethylamine and bromopropylene. The effect of reaction condition on TEAAB yield was studied. We have used centrifugal method to study its anti-swelling performance, and found that it has excellent anti-swelling performance. When the dosage is 6%, its anti-swelling rate can be as high as 98%. Results showed that TEAAB will be a nice anti-swelling agent and has a good application prospect in oil and gas field.

Acknowledgments

The authors would like to thank the Open Fund (KFKT2019-13) of the Key Laboratory of Auxiliary Chemistry and Technology for Chemical Industry, Ministry of Education Shaanxi University of Science and Technology; Xi'an Shiyou University 2019 Innovation and Entrepreneurship Training Program for College Students (201910705009) for their financial support.

References

[1] Du W, Pu X, Sun J, Luo X, Zhang Y and Li L 2018 *Adsorpt. Sci. Technol.* **36** 655.
[2] Zhang S, Du W and Liang F.S 2012 *Adv. Mater. Res.* **554-556** 783.
[3] Fu L, Liao K, Ge J, He Y, Wang X, Zhang S and Deng S 2019 *J Petrol Sci Eng.* **179** 958.
[4] Feng Q, Liu H, Peng Z and Zheng Y 2019 *Energ Fuel.* **33** 12202.
[5] Song Z, Zhang L, Huang L Liu X, Zhang J and Du W 2019 *J Macromol Sci A.* https://doi.org/10.1080/10601325.2019.1698962.
[6] Akhtarmanesh S, AmeriShahrabi M and Atashnezhad A 2013 *J Petrol Sci Eng.* **112** 290.
[7] Anderson R, Ratcliffe I and Greenwell H 2010 *Earth Sci. Rev.* **98** 201.
[8] Bera A, and Belhaj H 2016 *J. Nat. Gas Sci. Eng.* **34** 1284.
[9] Garea S, Ghebau A and Vasile E 2014 *Mater. Plast.* **51** 12.
[10] Abdo J and Haneef M 2013 *Appl Clay Sci.* **86** 76.

IOP Conf. Series: Materials Science and Engineering **758** (2020) 012099 doi:10.1088/1757-899X/758/1/012099

The Preparation and Characterization of Cu/AA6061 multi-layer Composte produced by Accumulated Roll Bonding

P Quang[1*] and P T Hao[1]

1 School of Material Science and Engineering, Hanoi University of Science and Technology, 1 DaiCoViet, Hanoi, Vietnam
*E-mail: quang.pham@hust.edu.vn

Abstract. The present study focuses on the preparation and characterization of Cu/AA6061 multi-layer composite materials produced by accumulated roll bonding (ARB). The bonding between metallic layers and the microstructure of samples was characterized by optical and scanning electron microscopy and tensile testing which were used to characterize the mechanical properties. After four ARB cycles, the interface only displays a little bit, showing that the good contact between layers were formed and the mechanical properties were improved. The both materials have a fine microstructure with grain sizes decreased with increasing cycles of ARB.

1. Introduction

Light alloys and light metals have low density and high strength-to-weight ratios. They are generally characterized by low toxicity in comparison to other heavy metals, but beryllium is an exception. Lightweight materials include aluminum, magnesium, titanium, and beryllium alloys. A composite material [1,2] is a combination of two or more materials that leads to better properties than an individual component. In this combination, each material retains its original chemical, physical and mechanical properties [3]. Two components of the composite material are reinforcements and matrices. The main advantage of lightweight composite materials is their rigidity, durability and high fatigue performance, combining with low density, which allows weight reduction during use compared to other bulk materials.

Knowing that large distortions at low temperatures can lead to significant changes in the microstructure of metallic materials [4-6] and metal matrix mixtures (MMC) [7-9]. Based on this particle transformation effect, bulk nanostructured materials are treated with some intense plastic deformation methods (SPD), such as channel angular pressing (ECAP), high-tension torsion, cumulative roll bonding, rolling equal angle, grooved rolling, equal pressing multi-channel angle, etc., was developed [10-14]. In different SPD methods, the accumulated roll bonding (ARB) is defined as a sheet metal forming process, in which a large plastic deformation is applied to a bulk process in order to make an ultra-fine grained materials. Accumulated rolling techniques are part of the development of conventional rolling methods to improve the mechanical properties of materials. It was developed by Saito and Tsuji [15-18], this technique involves the steps of: (1) cleaning the contact surfaces of two metal plates, (2) rolling two metal plates, (3) cutting adhesive metal sheets into two halves, (4) cleaning the surface and (5) continuing rolling them. This process is repeated according to the requirements of the product.

One cycle in the ARB technique and is repeated several times by the following procedure; cutting - cleaning the surface - stacking - rolling adhesive to achieve superfine structure and bond the desired layers shown in Figure 1. In the ARB technique, it is possible to use the same or different layers of materials. Recently, a number of studies on ARB techniques of different materials have been published [19-23].

In this paper, the study focuses on the preparation of Cu/AA6061 multi-layer composite materials as well as characterization of their microstructure and mechanical properties after having accumulated roll bonding.

2. Experimental work

Accumulated roll bonding was used to study the grain refinement potential. A detail study on microstructure, interfacial bonding and mechanical properties of Cu/AA6061 laminated composite deformed by this SPD technique was carried out. The experimental work was succeeded with processing a combination of two dissimilar materials: an aluminum alloy AA6061 (Al-Mg-Zn) and a commercial pure Cu (99.9 %). The initial thickness of each material was 2 mm. The ARB samples are cut perpendicularly to the original rolling direction. Two pieces of the original samples were erased by acetone. The samples were then stacked and heated in a furnace at 350 °C for 10 minutes and then rolled with a thickness of about 50 % reduced under non-lubricant conditions.

The microstructure evolution and interfacial bonding studies were carried out by using optical and scanning electron microscopy. Specimen preparation was done by mechanical grinding; polishing and etching with mixture of distilled water (100 ml), Hydrochloric acid (25 ml), Ferric chloride (5 grams) for copper and 0.5 % hydrofluoric acid for aluminum. Uniaxial tensile provide a precise estimation of mechanical properties. All the tensile tests have been carried out on the 809 Axial/Torsional Test System.

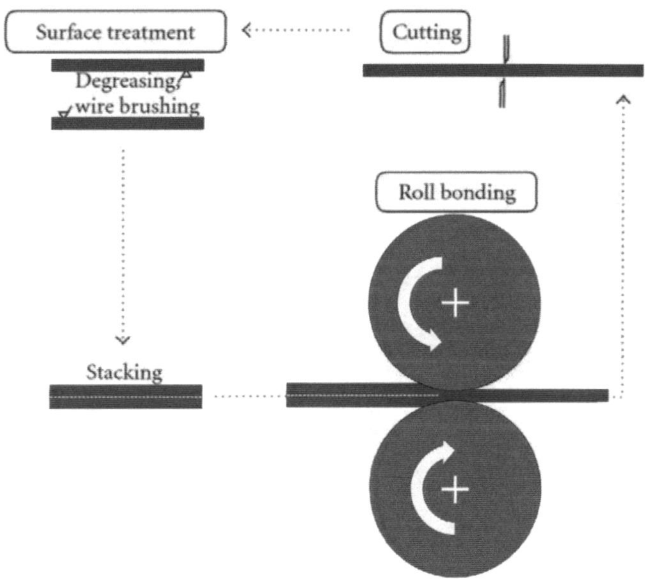

Figure 1. One cycle in the ARB technique. [13]

3. Results and Discussion

To process Cu/AA6061 multi-layer composite materials with stable, the ARB cross section processed strips is controlled as shown in Figure 1. To successfully process a large deformation, uniform and reduce residual stress in the rolling process, both Cu and AA6061 are heat treatment between ARB cycles.

The interface which is after the first cycle (Figure 2a) visible more the interface of after 3 ARB cycles (Figure 2c). After 4 ARB cycles (Figure 2d), the interface only displays a little bit, showing that the good contact between layers are formed. In this case, only a few unbound parts are seen in the sample.

IOP Conf. Series: Materials Science and Engineering **758** (2020) 012099 doi:10.1088/1757-899X/758/1/012099

Obviously, the next roll is enough to improve the alignment of the introduced interfaces in a previous cycle. The multi-layer of Cu/AA6061 combination are successfully processed according to the ARB process. By increasing the number of ARB cycles, the reinforcement layers in the aluminum matrix increase. The bonding between the reinforcement layers and the matrix layers are formed when the rolling force is applied on the sample. This causes the sample to undergo the severe plastic deformation process. In addition, the difference in mechanical properties, especially the hardness between the two materials, has created the appearance of the fracture of aluminum in the interlayer. The extrusion of the copper through the fractures into aluminum also occurs simultaneously. This creates the formation of bonding and coalescence in the interlayer.

When the samples are annealed between ARB cycles at 350 °C, residual stress has not been completely eliminated, but micro-cracks can significantly be reduced through the recovery process. Figure 3 and 4 shows that the grains are stretched in the rolling direction and reduced in size from the first ARB cycle to the fourth cycles. The microstructural analysis reveals that during further deformation small equiaxed grains appear. After four ARB cycles, the mean grain size reduced from (40 – 50) μm to (10 – 15) μm. The grain sizes in aluminum alloy are smaller than copper.

Base on the scanning electron microscopy (SEM) images of the grain structure of samples in the initial state and after 2, 4 ARB cycles, what can be easily seen is the change of the microstructure in the two materials. The grains are elongated along the rolling direction.

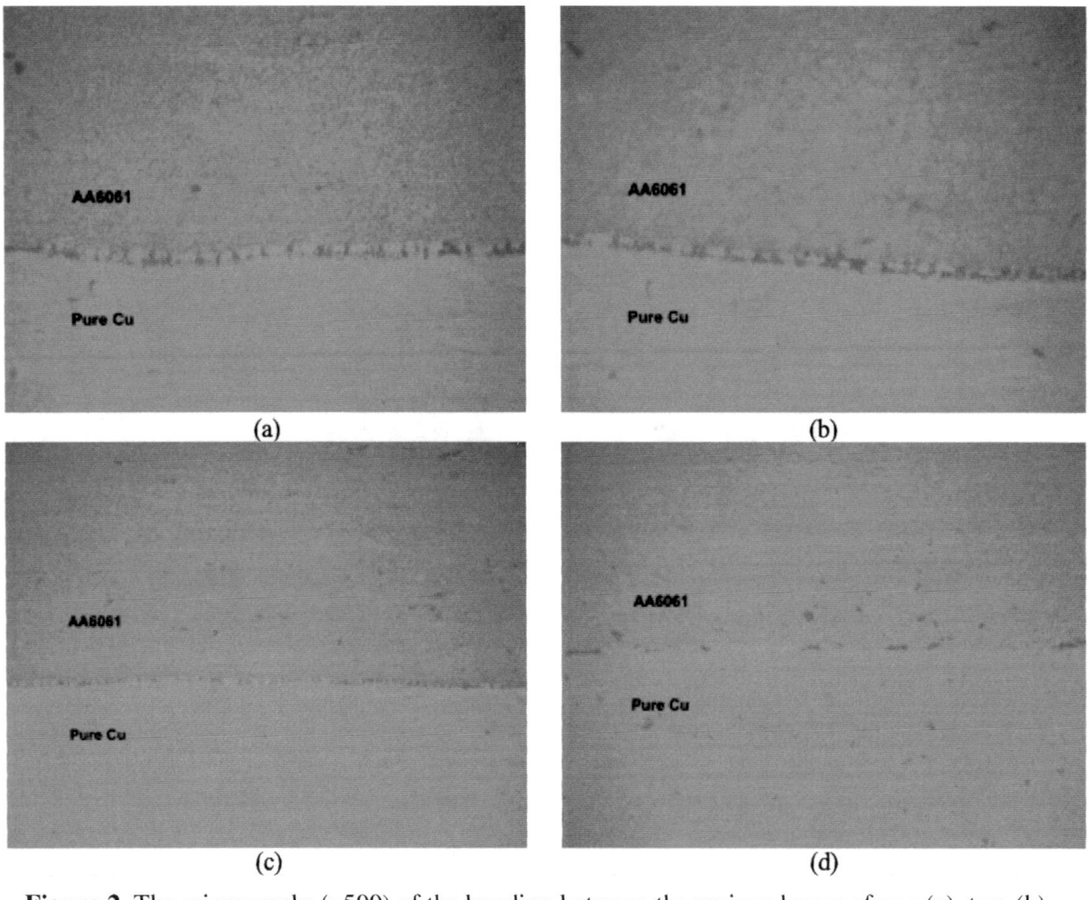

Figure 2. The micrographs (×500) of the bonding between the various layers of one (a), two (b), three (c) and four (d) ARB cycles

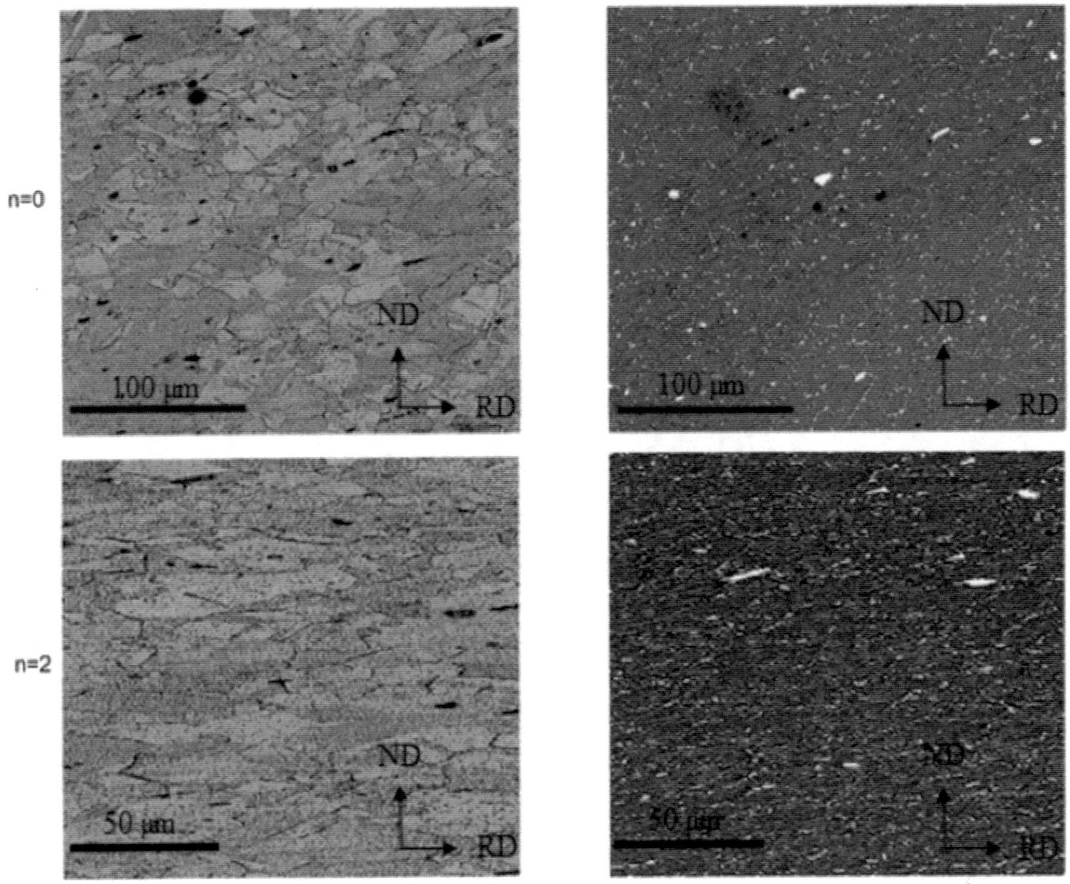

Figure 3. The scanning electron microscopy of pure Cu (left) and AA6061 (right) in befor and after 2 ARB cycles

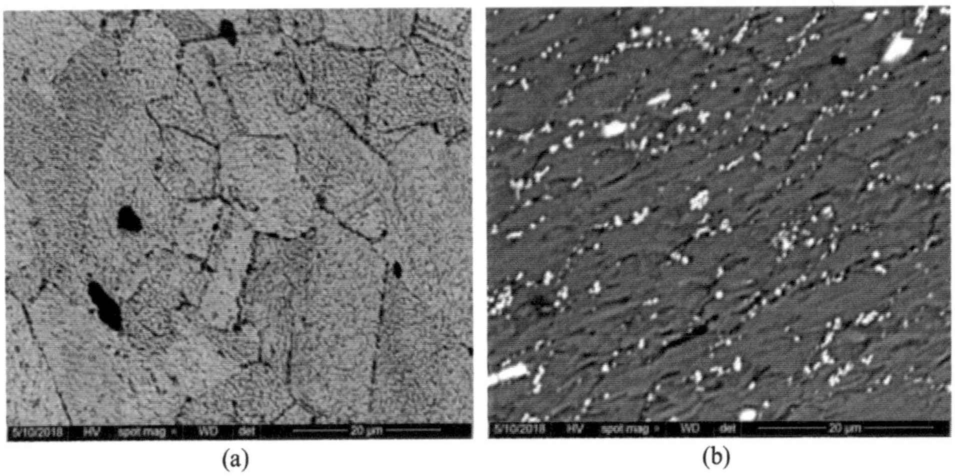

Figure 4. The scanning electron microscopy of pure Cu (a) and AA6061 (b) after 4 ARB cycles

The stress increased from 197 (Cu) and 233 (AA6061) to 247 MPa (4 cycles), so the stress of the after 4 ARB cycles is higher than initial single copper and aluminum alloy. On the other hand, the elongation value greatly increased from 33 % (AA6061) and 42 % (pure Cu) to more 70 % in the after 4 cycles of ARB (Figure 5). This can be explained as follows: with deformed aluminum alloys and copper, in addition to the movement of the deflection within each grains, the role of grain boundaries is not significant due to coarse grains. However, post-deformed aluminum and copper alloys with ARBs, have finer and finer grains with a higher angle of the coarse grains, hindering the movement of the stronger deflection, leading to a more durable material limit. The outstanding advantage of the durable increasing mechanism by making fine grained is to increase all indicators.

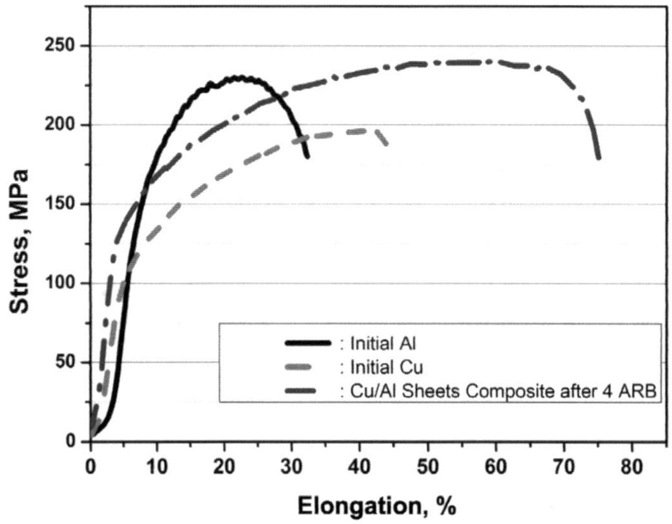

Figure 5. The stress – elongation of the initial copper and aluminum alloy and the Cu/AA6061 multi-layer composite materials after 4 ARB cycles.

In addition, the multi-layer composite materials can promote the advantages of each material separately. In this case, it is the combination of aluminum alloy strength and pure copper ductility. It can be seen that the uniform rolling of both aluminum alloy and copper in the rolling process is ensured by recovering them by heat treatment between ARB cycles. It is a typical phenomenon in SPD materials.

4. Conclusions

To maintain the stability of the layer in ARB composites of Cu/AA6061 multi-layer during processing. The weaknesses such as rolling speeds and intermediate annealing temperatures have been found to be effective because they prevent microscopic crack formation in Cu/AA6061 multi-layer composite materials.
By increasing the layer in the ARB process, the interface of the previous layer is improved. After four cycles, both materials have microscopic structure with reduced grain sizes.
Tensile strength increased moderately but the elongation increased significantly after ARB. This shows that when combining a variety of materials and increasing the number of layers by ARB technique, ductility has increased significantly.

Acknowledgments

This research was supported by the Vietnam National Foundation for Science and Technology Development **(NAFOSTED), No. 107.01-2018.25.**

References

[1] Ciach R 1998 *Advanced Light Alloys and Composites* (New York: © Springer Science + Business Media Dordrecht)
[2] Krishan K C 2012 *Composite Materials: Science and Engineering* 3rd Edition (New York: © Springer Science + Business Media Dordrecht)
[3] Gurdal Z et al. 1999 *Design and Optimization of Laminated Composite Materials*, (UK: John Wiley and Sons, Baffins Lane, Chichester, West Sussex P019 WD)
[4] Song H R , Kim Y S, Nam W J 2006 *Met. Mater. Int.* **12 (1)** 7–13

[5] Dang H T H, Thuy P T, Ngung D M, Quang P and Shchukin V Y 2019 *Acta Metallurgica Slovaca* **25** 48–54.
[6] Thuy P T, Hue D T H, Ngung D M and Quang P 2019 *IOP Conference Series: Materials Science and Engineering* **611** 012005 (Institute of Physics Publishing)
[7] Quang P, Jeong Y G, Hong S H and Kim H S 2006 *Key Eng. Materials* **326-328** 325-328.
[8] Quang P, Jeong Y G, Yoon S C, Hong S I, Hong S H and Kim H S 2007 *Materials Science Forum* **534-536** 245-248.
[9] Quang P, Jeong Y G, Yoon S C, Hong S H. and Kim H S 2007 *Journal of Materials Processing Technology* **187-188** 46-50
[10] Skrotzki W et all. 2014 *Adv. Eng. Mater.* **16** 1208-1216
[11] Quang P, Nghiep D M 2016 *Korean Journal of Metals and Materials* **54-3** 217-223.
[12] Segal V M 1995 *Mater. Sci. Eng.* **A197** 157-164
[13] Valiev R Z, R K Islamgaliev, Alexandrov I V 2000 *Progress in Materials Science* **45(2)** 103-189
[14] Valiev R Z 2001 *Metal Mater. Inter.* **7** 413-420
[15] Saito Y, Tsuji N, Utsunomiya H, Sakai T and Hong R G 1998 *Scripta Mater.* **39** 1221-1227
[16] Tsuji N, Saito Y, Utsunomiya H and Tanigawa S 1999 *Scripta Mater.* **40** 795-800
[17] Saito Y, Utsunomiya H, Tsuji N, Sakai T 1999 *Acta materialia* **47(2)** 579-583.
[18] Tsuji N, Saito Y, Lee S H, Minamino Y 2003 *Adv. Eng. Mater.* **5** 338-344
[19] Hsieh C C, Shi M S, Wu W 2012 *Metals and Materials Inter.* **18(1)** 1-6
[20] Wu K, Chang H, Maawad E, Gan W M Brokmeier H G, Zheng M Y 2010 *Materials Science and Engineering A* **527(13)** 3073-3078
[21] Mozaffari A, Danesh H, Janghorban K 2010 *J. Alloys and Compounds* **489** 103-109
[22] Ghalandari L, Mahdavian M M, Reihanian M 2014 *Materials Science & Engineering A* **593** 145-152.
[23] Pham Q 2019 *Materials Science Forum* **971** 9–14

AUTHOR INDEX